Stem Cell Anthology

Stem Cell Anthology

EDITED BY

Bruce M. Carlson, M.D., Ph.D.
University of Michigan

ELSEVIER

AMSTERDAM • BOSTON • HEIDELBERG • LONDON • NEW YORK • OXFORD
PARIS • SAN DIEGO • SAN FRANCISCO • SINGAPORE • SYDNEY • TOKYO

Academic Press is an imprint of Elsevier

Academic Press is an imprint of Elsevier
32 Jamestown Road, London NW1 7BY, UK
30 Corporate Drive, Suite 400, Burlington, MA 01803, USA
525 B Street, Suite 1900, San Diego, CA 92101-4495, USA

First edition 2010

Material in the work originally appeared in *Essentials of Stem Cell Biology* (Elsevier, Inc., 2006) *Essentials of Stem Cell Biology, Second Edition* (Elsevier, Inc., 2009), *Principles of Regenerative Medicine* (Elsevier, Inc., 2008), *Principles of Regenerative Biology* (Elsevier, Inc., 2007), *Principles of Tissue Engineering, Third Edition* (Elsevier, Inc. 2007, 2000, 1997), *Human Stem Cell Manual* (Jeanne F. Loring, 2007) and *Principles of Cloning* (Elsevier, Inc. 2002).

British Library Cataloguing-in-Publication Data
A catalogue record for this book is available from the British Library

Library of Congress Cataloging-in-Publication Data
A catalog record for this book is available from the Library of Congress

ISBN : 978-0-12-375682-4

For information on all Academic Press publications
visit our website at www.elsevierdirect.com

Typeset by Macmillan Publishing Solutions
www.macmillansolutions.com

Printed and bound in Canada

10 11 12 13 14 15   10 9 8 7 6 5 4 3 2 1

Contents

Contributors

Numbers in parentheses indicate the chapter number of the author's contribution

Russell C. Addis (8) John Hopkins University, School of Medicine, Baltimore, MD

Anthony Atala (17) Wake Forest Institute for Regenerative Medicine, Wake University School of Medicine, Winston-Salem, NC

Joyce Axelman (8) John Hopkins University, School of Medicine, Baltimore, MD

Anne G. Bang (12) Novocell, Inc., 3500 General Atomics Ct, San Diego, CA

Yann Barrandon (25) Laboratory of Stem Cell Dynamics, School of Life Sciences, Swiss Federal Institute of Technology Lausanne and Department of Experimental Surgery, Lausanne University Hospital 1015, Lausanne, Switzerland

Steven R. Bauer (32) Laboratory of Stem Cell Biology, Division of Cellular and Gene Therapies, Office of Cellular, Tissue and Gene Therapies, Center for Biologics Evaluation and Research, US Food and Drug Administration, Rockville, MD

Nissim Benvenisty (9) Department of Genetics, Silberman Institute of Life Sciences, The Hebrew University, 91904 Jerusalem, Israel

Paolo Bianco (2) Dipartimento di Medicina Sperimentale, Sapienza, Universita di Roma, Rome, Italy; Parco Scientifico Biomedico San Raffaele, Rome, Italy

Mairi Brittan (16) Centre for Gastroenterology, Institute of Cell and Molecular Sciences, Barts and the London School of Medicine and Dentistry, London, UK

Arnold I. Caplan (4) Skeletal Research Center, Case Western Reserve University, Cleveland, OH

Bruce M. Carlson (5) University of Michigan, Ann Arbor, MI 48109

Melissa K. Carpenter (12) Carpenter Group, 10330 Wateridge Circle #290, San Diego, CA

Fatima Cavaleri (11) Max Planck Institute for Molecular Biomedicine, Muenster, Germany

S.M. Chambers (15) Center for Cell & Gene Theraphy Baylor College of Medicine

Massimo Cimini (14) MaRS Center Toronto Medical Discover Tower

Gregory O. Clark (8) Division of Endocrinology, John Hopkins University, School of Medicine, Baltimore, MD

Michael F. Clarke (19) Stanford Institute for Stem Cell and Regenerative Medicine; Department of Medicine, Division of Oncology, Stanford University School of Medicine, Stanford, CA

Chad Cowen

Annelies Crabbe (13) Interdepartementeel Stamcelinstituut, Katholieke Universiteit Leuven, Belgium

Paolo De Coppi (17) Department of General Paediatric Surgery Great Ormond Street Hospital and Institute of Child Health

Natalie C. Direkze (16) Centre for Gastroenterology, Institute of Cell and Molecular Sciences, Barts and the London School of Medicine and Dentistry, London, UK; Histopathology Unit, London Research Institute, Cancer Research UK, London, UK

Yuval Dor (23) Department of Cellular Biochemistry and Human Genetics, The Hebrew University-Hadassah Medical School, Jerusalem, Israel

Margaret A. Farley (30) Yale University Divinity School, New Haven, CT

Loren J. Field (24) The Riley Heart Research Center, Herman B Wells Center for Pediatric Research; the Krannert Institute of Cardiology, Indiana University School of Medicine, Indianapolis, IN

Donald W. Fink (32) Cell Therapy Branch, Division of Cellular and Gene Therapies, Office of Cellular, Tissue and Gene Therapies, Center for Biologics Evaluation and Research, US Food and Drug Administration, Rockville, MD

Richard L. Gardner (1) University of Oxford, Dept of Zoology, Oxford, UK

John D. Gearhart (8) Institute for Cell Engineering, John Hopkins University, School of Medicine, Baltimore, MD

Pamela Gehron Robey (2) Craniofacial and Skeletal Diseases Branch, National Institute of Dental and Craniofacial Research, National Institutes of Health, Department of Health and Human Services, Bethesda, MD

Kiran Gollapudi (27)

Rodolfo Gonzalez (20) Program in Stem Cell Biology (Developmental & Regeneration Cell Biology), The Burnham Institue, La Jolla, CA

M.A Goodell (15) Center for Cell & Gene Therapy Baylor College of Medicine

Trevor A. Graham (16) Centre for Gastroenterology, Institute of Cell and Molecular Sciences, Barts and the London School of Medicine and Dentistry, London, UK

Ronald M. Green (29) Ethics Institute, Dartmouth College, Hanover, NH

George T.-J. Huang (27) Division of Endodontics Baltimore College of Dental Surgery University of Marlyand

Adam Humphries (16) Centre for Gastroenterology, Institute of Cell and Molecular Sciences, Barts and the London School of Medicine and Dentistry, London, UK

Pritinder Kaur (21) The University of Melbourne, Epithelial Stem Cell Biology Laboratory, Peter MacCallum Cancer Institute, East Melbourne, Victoria, Australia

Kathleen C. Kent (8) Johns Hopkins University, School of Medicine, Baltimore, MD

Candace L. Kerr (8) Department of Gynecology and Obstetrics, John Hopkins University, School of Medicine, Baltimore, MD

Irina Klimanskaya (7, 28) Advanced Cell Technology, 381 Plantation Street, Worcester, MA

Jennifer N. Kraszewski (8) Johns Hopkins University, School of Medicine, Baltimore, MD

Ren-Ke Li (14) MaRS Center Toronto Medical Discover Tower

Wan-Ju Li (27) Cartilage Biology and Orthopaedics Branch National Institute of Arthritis

William J. Lindblad (15) Department of Pharmaceutical Sciences Massachusetts College of Pharmacy & Health Sciences

John W. Littlefield (8) Johns Hopkins University, School of Medicine, Baltimore, MD

Ian Lyons (6) Stem Cell Innovations 11222 Richmond Ave Ste 180 Houston, TX77082, USA

Yoav Mayshar (9) Department of Genetics, Silberman Institute of Life Sciences, The Hebrew University, Jerusalem, Israel

Stuart A.C. McDonald (16) Centre for Gastroenterology, Institute of Cell and Molecular Sciences, Barts and the London School of Medicine and Dentistry, London, UK; Histopathology Unit, London Research Institute, Cancer Research UK, London, UK

Jill McMahon (7) Harvard University, 16 Divinity Ave, Cambridge, MA

Douglas A. Melton (23) Department of Molecular and Cellular Biology and Howard Hughes Medical Institute, Harvard University, Cambridge, MA

Patrea L. Pabst (31) Pabst Patent Group LLP

David P. Patterson (27) Cartilage Biology and Orthopaedics Branch National Institute of Arthritis

Ethan S. Patterson (8) Johns Hopkins University, School of Medicine, Baltimore, MD

Christopher S. Potten (3) EpiStem Limited, Incubator Building, Manchester, UK

Sean L. Preston (16) Centre for Gastroenterology, Institute of Cell and Molecular Sciences, Barts and the London School of Medicine and Dentistry, London, UK; Histopathology Unit, London Research Institute, Cancer Research UK, London, UK

Jean Pyo Lee (20) Program in Stem Cell Biology (Developmental & Regeneration Cell Biology) The Burnham Institue, La Jolla, CA

Mahendra Rao (6) Invitrogen 1600 Faraday Ave PO Box 6482 Carlsbad, CA

Ariane Rochat (25) Laboratory of Stem Cell Dynamics, School of Life Sciences, Swiss Federal Institute of Technology Lausanne and Department of Experimental Surgery, Lausanne University Hospital 1015, Lausanne, Switzerland

Michael Rothenberg (19) Stanford Institute for Stem Cell and Regenerative Medicine; Department of Medicine, Stanford University School of Medicine; Division of Gastroenterology and Hepatology, Stanford University School of Medicine, Stanford, CA

Michael Rubart (24) The Riley Heart Research Center, Herman B Wells Center for Pediatric Research, Indianapolis, IN

Hans Schöler (11) Max Planck Institute for Molecular Biomedicine, Muenster, Germany

Philip H. Schwartz (6) Center for Neuroscience Research, Children's Hospital of Orange County Research Institute, Orange, CA

Michael J. Shamblott (8) Institute for Cell Engineering, Johns Hopkins University, School of Medicine, Baltimore, MD

Daniel Skuk (26) Human Genetic Unit Centre de Recherche du CHUL

Evan Y. Snyder (20) Stem Cell Center Burnham Institute for Medical Research 10901 N. Torrey Pines Rd La Jolla, CA

Shay Soker (17) Wake Forest University School of Medicine Institute for Regenerative Medicine

David Tan (6) Singapore Biomedical Research Council 27 Jalan Layang Layang Singapore 598493

Doris A. Taylor (22) Center for Cardiovascular Repair

James A. Thomson (10, 46) Department of Biopharmaceutical Sciences, UCSF, San Francisco, CA

Jacques P. Tremblay (26) Human Genetic Unit Centre de Recherche du CHUL

Alan Trounson (18) Institute of Reproduction and Development, Monash University, Clayton, Victoria 3168, Australia

Rocky S. Tuan (27) Cartilage Biology and Orthopaedics Branch National Institute of Arthritis

Edward Upjohn (21) The University of Melbourne, Epithelial Stem Cell Biology Laboratory, Peter MacCallum Cancer Institute, East Melbourne, Victoria, Australia

George Varigos (21) The University of Melbourne, Epithelial Stem Cell Biology Laboratory, Peter MacCallum Cancer Institute, East Melbourne, Victoria, Australia

Catherine M. Verfaillie (13) Interdepartementeel Stamcelinstituut, Katholieke Universiteit Leuven, Belgium

Richard D. Weisel (14) MaRS Center Toronto Medical Discover Tower

James W. Wilson (3) EpiStem Ltd., M139XX, Manchester, UK

Nicholas A. Wright (16) Centre for Gastroenterology, Institute of Cell and Molecular Sciences, Barts and the London School of Medicine and Dentistry, London, UK; Histopathology Unit, London Research Institute, Cancer Research UK, London, UK

Munira Xaymardan (14) MaRS Center Toronto Medical Discover Tower

Junying Yu (10) The Genetics and Biotechnology Building University of Wisconsin-Madison Anthony Atala Department of Urology Wake Forest University

Andrey G. Zenovich (22) Center for Cardiovascular Repair

Introduction

Over the past decade, stem cell biology has been a dominant theme in biomedical research. Its prominence is due not only to the potential of employing stem cells in the treatment of many otherwise intractable diseases, but also to the public furor that has arisen because of the use of human embryos as the source of many lines of stem cells. The debate over the source and use of stem cells has been especially acrimonious in some political and religious circles, but even as the debate has continued, new recent technological developments have altered its character. Beneath the public aspect of the debate and discussion about stem cells, intense research activity has been taking place in laboratories throughout the world. The pace of progress in the field of stem cell biology has been breath-taking, in part because it is a new field, but also because of the relatively recent development of tools in the fields of molecular and cell biology that have direct application to stem cell research.

Despite the fact that in the field of hematology the existence of stem cells has been widely accepted for decades, the biomedical research community was for many years quite resistant to the notion. Since the major breakthroughs of the 1990s and early years of this century, however, the pendulum has swung the other way, and currently optimism about the therapeutic applications of stem cells, and even their existence in certain tissues and organs, has led to often unrealistic expectations in the eyes of both the public and the scientific community about the potential application of stem cell therapy for the treatment of human disease in the near future.

Within the past decade a new field, called regenerative medicine, has been created. It is now represented by research institutes or large interdisciplinary groups in almost all major medical centers throughout the world. A prime goal of regenerative medicine is to restore damaged tissues and organs through the applications of stem cell therapy, usually in combination with techniques of tissue engineering, especially in situations where natural biological regenerative processes are not sufficiently well developed to produce effective repair. The application of techniques of regenerative medicine typically involves the production of an artificial substrate or tissue matrix by tissue engineers and then the seeding of that matrix by stem cells before it is introduced into a pathological site within the body. In a number of cases, this is followed by the reconstruction within the body of a regenerated tissue that bears a remarkable similarity in both structure and function to the original tissue that had been damaged. In other cases, however, such faithful restitution does not occur, but yet, e.g. in the heart, some functional improvement may take place. Much yet remains to be learned about what happens after stem cells have been introduced into the body.

A major consideration in stem cell therapy is the source of stem cells. Although embryonic stem cells have the potential to differentiate into a large variety of specialized cell types, many biological issues, such as controlling the pathway of differentiation both before and after implantation into the body, face researchers. Naturally occurring stem cells have been found in an increasing number of adult tissues and organs, but other problems face those who wish to use these as sources of stem cell therapy. These include obtaining sufficient numbers of adult stem cells, as well as a relatively limited repertoire for differentiation in most types of adult stem cells.

Considerable research effort has gone into defining at both a cellular and molecular level what makes a cell behave as a stem cell. The results of this research have not only allowed better definitions of stem cell characteristics and better control over their behavior, but this information has provided the basis for the recent reports of conversion of certain types of adult cells into cells with many of the characteristics of embryonic stem cells.

This volume is a collection of chapters by leaders in the stem cell field. They place in perspective both the promise and the potential limitations of stem cell biology and regenerative medicine. It is designed to provide both an introduction to the field of stem cell biology for the non-specialist and an up-to-date summary of progress for those working in the field. An introductory section (I) introduces the reader to basic properties of stem cells and then describes in greater detail the major types of stem cells in both embryos and adults. This is followed by a practically-oriented section (II) that provides technical laboratory detail for those who are actually planning or carrying out stem cell research. Section III comprises a collection of chapters that, in addition to reviewing characteristics of a generic stem cell, discuss specific types of stem cells, their locations in the body and their biological properties. Section IV concentrates on stem cell biology as it applies to specific tissues and organs that seem at present to be most amenable to the application of regenerative medicine to specific types of pathology. These chapters are heavily oriented toward laboratory studies, because in very few cases are the techniques sufficiently well developed for application to humans, but they do give a good sense of perspective concerning the state of development of the field overall. The final section (V) is a collection of chapters that outline major ethical and religious concerns, as well as important regulatory considerations. Because of the rapidly changing legal environment as of this writing, chapters on the legal status of stem cells have not been included in this anthology.

Few aspects of biomedical science over the past half-century have captured the public imagination as much as stem cells. It is my hope that the chapters in this volume will provide the reader an accurate and accessible overview of the present status of the field, as well as a realistic depiction of the potential for the eventual clinical application of regenerative medicine.

Bruce M. Carlson

Foreword

Stem cells raise the prospect of regenerating failing body parts and treating a wide range of disorders that result from tissue loss or dysfunction. Over 16 million patients worldwide suffer from neurodegenerative disorders such as Parkinson's and Alzheimer's disease, over 200 million people suffer from diabetes, and millions more from cardiovascular disease, blindness, and other disorders that may one day be treatable using stem cell technology. With the advent of induced pluripotent stem (iPS) cells, we may even have a way to overcome the problem of immune rejection. In addition to being a promising—and potentially unlimited—source of cells for regenerative medicine and drug discovery, stem cells serve as an excellent model for studying both normal and abnormal vertebrate development.

The rapid progress and interest in stem cells has led to an avalanche of new knowledge and research tools that span multiple areas of scientific inquiry, ranging from cellular reprogramming to cloning to tissue engineering. Although books on various stem cell topics abound, *Stem Cell Anthology* combines in a single volume a collection of definitive chapters and essential subject matter from five leading stem-cell related volumes, including *Essentials of Stem Cell Biology, Principles of Tissue Engineering, Principles of Regenerative Medicine, Human Stem Cell Manual*, and *Principles of Cloning*.

Stem Cell Anthology successfully integrates this exciting area of biology, combining the prerequisites for a general understanding of adult, embryonic, and induced pluripotent stem cells. No topic in the field of stem cells is left uncovered, including methods for preparing, maintaining, and genetically manipulating stem cells and progenitor populations; the types, properties, and characteristics of stem cells; application of stem cells to specific organ systems and diseases, including neural and pancreatic stem cells; cancer; burns and ulcers, and cardiovascular and musculoskeletal repair; as well as a section on ethical, religious, and FDA product and pre-clinical regulatory considerations. The result is a comprehensive, easy to follow reference by the world's experts that will be useful to students, scientists, and clinicians alike.

Robert Lanza, M.D.

"Stemness": Definitions, Criteria, and Standards

INTRODUCTION

Stem cells have recently generated more public and professional interest than almost any other topic in biology. One reason stem cells capture the imagination of so many is the promise that understanding their unique properties may provide deep insights into the biology of cells as well as a path toward treatments for a variety of degenerative illnesses. And although the field of stem cell biology has grown rapidly, there exists considerable confusion and disagreement as to the nature of stem cells. This confusion can be partly attributed to the sometimes idiosyncratic terms and definitions used to describe stem cells. Although definitions can be restrictive, they are useful when they provide a basis for mutual understanding and experimental standardization. With this intention, I present explanations of definitions, criteria, and standards for stem cells. Moreover, I highlight a central question in stem cell biology, namely the origin of these cells. I also suggest criteria or standards for identifying, isolating, and characterizing stem cells. Finally, I summarize the notion of "stemness" and describe its possible application in understanding stem cells and their biology.

WHAT IS A STEM CELL?

Stem cells are defined functionally as cells that have the capacity to self-renew as well as the ability to generate differentiated cells (Weissman *et al.*, 2001; Smith, 2001). More explicitly, stem cells can generate daughter cells identical to their mother (self-renewal) as well as produce progeny with more restricted potential (differentiated cells). This simple and broad definition may be satisfactory for embryonic or fetal stem cells that do not perdure for the lifetime of an organism. But this definition breaks down in trying to discriminate between transient adult progenitor cells that have a reduced capacity for self-renewal and adult stem cells. It is therefore important when describing adult stem cells to further restrict this definition to cells that self-renew throughout the life span of the animal (van der Kooy and Weiss, 2000). Another parameter that should be considered is potency: Does the stem cell generate to multiple differentiated cell types (multipotent), or is it only capable of producing one type of differentiated cell (unipotent)? Thus, a more complete description of a stem cell includes a consideration of replication capacity, clonality, and potency. Some theoretical as well as practical considerations surrounding these concepts are considered in this chapter.

Self-renewal

Stem cell literature is replete with terms such as "immortal", "unlimited", "continuous", and "capable of extensive proliferation", all used to describe the cell's replicative capacity. These rather extreme and vague terms are not very helpful, as it can be noted that experiments designed to test the "immortality" of a stem cell would by necessity outlast authors and readers alike. Most somatic cells cultured *in vitro* display a finite number of (less than 80) population doublings prior to replicative arrest or senescence, and this can be contrasted with the seemingly unlimited proliferative capacity of stem cells in culture (Houck *et al.*, 1971; Hayflick, 1973; Hayflick, 1974; Sherr and DePinho, 2000; Shay and Wright, 2000). Therefore, it is reasonable to say that a cell that can undergo more than twice this number of population doublings (160) without oncogenic transformation can be termed "capable of extensive proliferation". In a few cases, this criteria has been met, most notably with embryonic stem (ES) cells derived from either humans or mice as well as with adult neural stem cells (NSCs) (Smith, 2001; Morrison *et al.*, 1997). An incomplete understanding of the factors required for self-renewal *ex vivo* for many adult stem cells precludes establishing similar proliferative limits *in vitro*. In some cases, a rigorous assessment of the capacity for self-renewal of certain adult stem cells can be obtained by single-cell or serial transfer into acceptable hosts, an excellent example of which is adult hematopoietic stem cells (HSCs) (Allsopp and Weissman, 2002; Iscove and Nawa, 1997). Adult stem cells are probably still best defined *in vivo*, where they must display sufficient proliferative capacity to last the lifetime of the animal. Terms such as "immortal" and "unlimited" are probably best used sparingly if at all.

Clonality

A second parameter, perhaps the most important, is the idea that stem cells are clonogenic entities: single cells with the capacity to create more stem cells. This issue has been exhaustively dealt with elsewhere and is essential for any definitive characterization of self-renewal, potential, and lineage. Methods for tracing the lineage of stem cells are described in subsequent chapters. Although the clonal "gold standard" is well understood, there remain several confusing practical issues. For instance, what constitutes a cell line? The lowest standard would include any population of cells that can be grown in culture, frozen, thawed, and subsequently repassaged *in vitro.* A higher standard would be a clonal or apparently homogenous population of cells with these characteristics, but it must be recognized that cellular preparations that do not derive from a single cell may be a mixed population containing stem cells and a separate population of "supportive" cells required for the propagation of the purported stem cells. Hence, any reference to a stem cell line should be made with an explanation of their derivation. For example, it can be misleading to report on stem cells or "stem cell lines" from a tissue if they are cellular preparations containing of a mixed population, possibly contaminated by stem cells from another tissue.

Potency

The issue of potency maybe the most contentious part of a widely accepted definition for stem cells. A multipotent stem cell sits atop a lineage hierarchy and can generate multiple types of differentiated cells, the latter being cells with distinct morphologies and gene expression patterns. At the same time, many would argue that a self-renewing cell that can only produce one type of differentiated descendant is none-the-less a stem cell (Slack, 2000). A case can be made, for clarity, that a unipotent cell is probably best described as a progenitor. Progenitors are typically the descendants of stem cells, only they are more constrained in their differentiation potential or capacity for self-renewal and are often more limited in both senses.

Definition

In conclusion, a working definition of a stem cell is a clonal, self-renewing entity that is multipotent and thus can generate several differentiated cell types. Admittedly, this definition is not applicable in all instances and is best used as a guide to help describe cellular attributes.

WHERE DO STEM CELLS COME FROM?

The origin or lineage of stem cells is well understood for ES cells; their origin in adults is less clear and in some cases controversial. It may be significant that ES cells originate before germ layer commitment, raising the intriguing possibility that this may be a mechanism for the development of multipotent stem cells, including some adult stem cells. The paucity of information on the developmental origins of adult stems cells leaves open the possibility that they too escape lineage restriction in the early embryo and subsequently colonize specialized niches, which function to both maintain their potency as well as restrict their lineage potential. Alternatively, the more widely believed, though still unsubstantiated, model for the origin of adult stem cells assumes that they are derived after somatic lineage specification, whereupon multipotent stem cells–progenitors arise and colonize their respective cellular niches. In this section, I briefly summarize the origin of stem cells from the early embryo and explain what is known about the ontogeny of adult stem cells focusing attention on HSCs and NSCs.

Stem Cells of the Early Embryo

Mouse and human ES cells are derived directly from the inner cell mass of preimplantation embryos after the formation of a cystic blastocyst (Papaioannou, 2001). This population of cells would normally produce the epiblast and eventually all adult tissues, which may help to explain the developmental plasticity exhibited by ES cells. In fact, ES cells appear to be the *in vitro* equivalent of the epiblast, as they have the capacity to contribute to all somatic lineages and in mice to produce germ line chimeras. By the time the zygote has reached the blastocyst stage, the developmental potential of certain cells has been restricted. The outer cells of the embryo have begun to differentiate to form trophectoderm, from which a population of embryonic trophoblast stem cells has also been derived in mice (Tanaka *et al.*, 1998). These specialized cells can generate all cell types of the trophectoderm lineage, including differentiated giant trophoblast cells. At the egg cylinder stage of embryonic development (embryonic day (E) 6.5 in mice), a population of cells near the epiblast can be identified as primordial germ cells (PGCs), which are subsequently excluded from somatic specification or restriction (Saitou *et al.*, 2002). PGCs migrate to and colonize the genital ridges, where they produce mature germ cells and generate functional adult gametes. PGCs can be isolated either prior or subsequent to their arrival in the genital ridges and, when cultured with appropriate factors *in vitro*, can generate embryonic germ (EG) cells (Matsui *et al.*, 1992; Resnick *et al.*, 1992). EG cells have many of the characteristics of ES cells with respect to their differentiation potential and their contribution to the germ line of chimeric mice (Labosky *et al.*, 1994; Stewart *et al.*, 1994). The most notable difference between ES and EG cells is that the latter may display (depending upon the developmental stage of their derivation) considerable imprinting of specific genes (Surani, 1998; Sorani, 2001; Howell *et al.*, 2001). Consequently, certain EG cell lines are incapable of producing normal chimeric mice.

Importantly, no totipotent stem cell has been isolated from the early embryo. ES and EG cells generate all somatic lineages as well as germ cells but rarely if ever contribute to the trophectoderm, extraembryonic endoderm, or extraembryonic mesoderm. Trophectoderm stem (TS) cells have been isolated, and these only generate cells of the trophectoderm lineage. It remains to be seen whether cells can be derived and maintained from totipotent

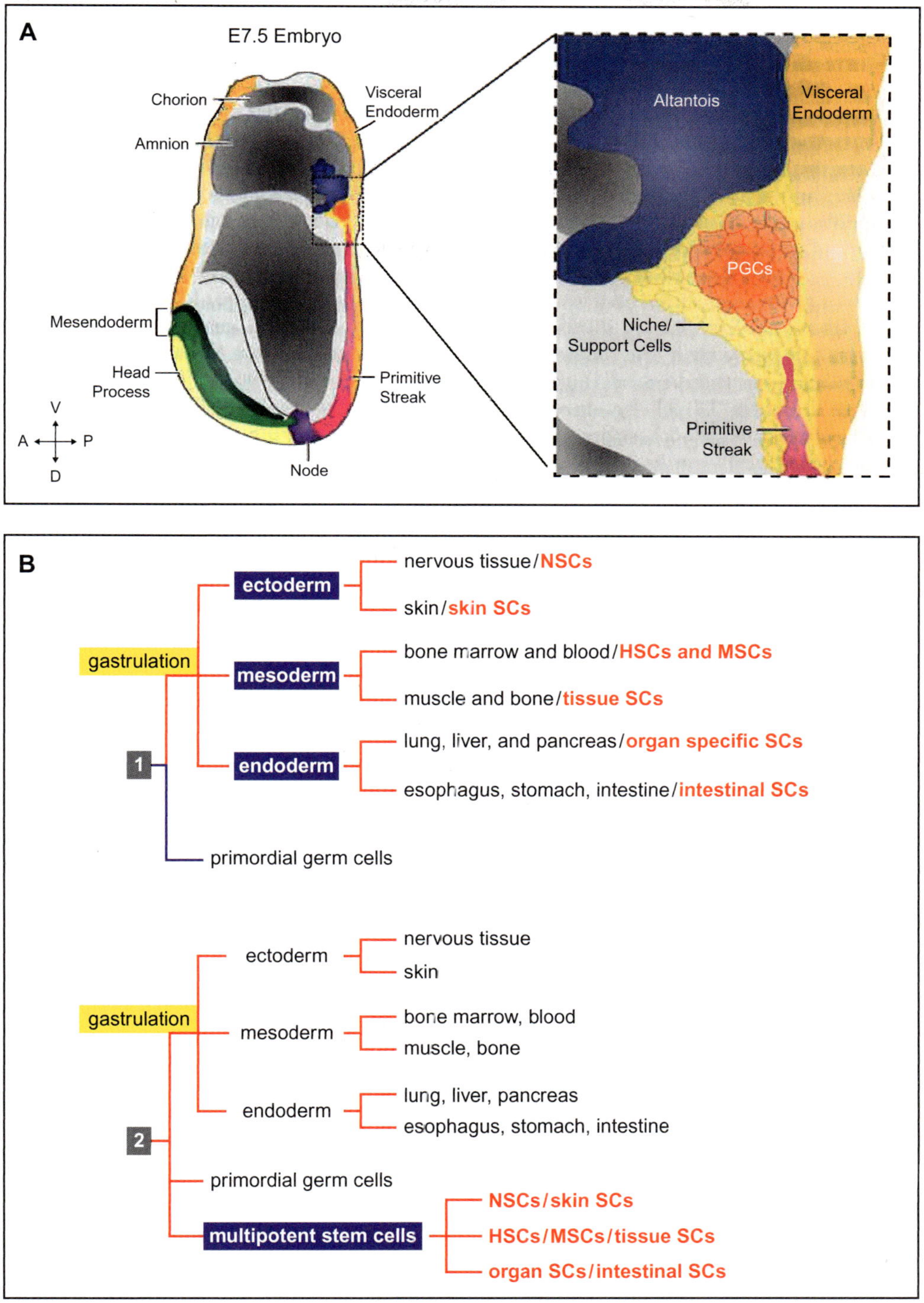

FIGURE 1 (A) Development of primordial germ cells. A schematic of an embryonic day 7.5 mouse embryo highlights the position of the developing primordial germ cells (PGCs) proximal to the epiblast. The expanded view on the right serves to illustrate the point that PGCs escape lineage commitment/restriction by avoiding the morphogenetic effects of migrating through the primitive streak during gastrulation. (B) Putative developmental ontogeny of stem cells. In lineage tree 1, the development of stem cells occurs after the formation of germ layers. These stem cells are thus restricted by germ layer commitment to their respective lineage (e.g., mesoderm is formed, giving rise to hematopoietic progenitors that become hematopoietic stem cells). Lineage tree 2 illustrates the idea that stem cells might develop similarly to PGCs, in that they avoid the lineage commitments during gastrulation and subsequently migrate to specific tissue and organ niches.

embryonic stages. Although our understanding of cell fates in the early embryo is incomplete, it appears that the only pluripotent stem cells found after gastrulation are PGCs (with the possible exceptions of multipotential adult progenitor cells (Jiang *et al.*, 2002) and teratocarcinomas). It may be that PGCs escape germ layer commitment during gastrulation by developing near the epiblast and subsequently migrate to positions inside the embryo proper. This developmental strategy may not be unique to PGCs, and it raises the interesting possibility that other stem cells might have similar developmental origins. Alternatively, it may be the case that adult stem cells are derived from PGCs. Although intriguing, it is important to stress that this idea lacks experimental evidence.

Ontogeny of Adult Stem Cells

The origin of most adult stem cells is poorly understood. With the issue of adult stem cell plasticity at the forefront, as described in this section, studies designed to elucidate the ontogeny of adult stem cells may help to reveal their specific lineage relationships and shed light on their plasticity and potential. Information on the origins of adult stem cells would also help to define the molecular programs involved in lineage determination, which may in turn provide insights into methods for manipulating their differentiation. To this end, I summarize what is known about the development of adult stem cells within the context of the hematopoietic and neural systems.

The development of hematopoietic cells in mice occurs soon after gastrulation (E7.5), although HSCs with the same activities as those in the adult have only been observed and isolated at midgestational stages (E10.5) (Orkin, 1996; Dzierzak, 2002; Weissman, 2000). These observations suggest that the embryo has a unique hematopoietic lineage hierarchy, which may not be founded by an adult-type HSC. Thus, hematopoiesis appears to occur at multiple times or in successive waves within the embryo, and the emergence of an HSC may not precede or be concomitant with the appearance of differentiated hematopoietic cells.

The first site of hematopoiesis in the mouse is the extraembryonic yolk sac, soon followed by the intraembryonic aorta–gonad–mesonephros (AGM) region. Which of these sites leads to the generation of the adult hematopoietic system and, importantly, HSCs is still unclear. Results from nonmammalian embryo-grafting experiments, with various findings in the mouse, suggest that the mammalian embryo, specifically the AGM, generates the adult hematopoietic system and HSCs (Kau and Turpen, 1983; Medvinsky *et al.*, 1993; Medvinsky and Dzierzak, 1996). Interestingly, the midgestational AGM is also the region that harbors migrating PGCs and is thought to produce populations of mesenchymal stem cells, vascular progenitors, and perhaps hemangioblasts (Molyneaux *et al.*, 2001; Minasi *et al.*, 2002; Alessandri *et al.*, 2001; Hara *et al.*, 1999; Munoz-Chapuli *et al.*, 1999). In the absence of studies designed to clonally evaluate the lineage potential of cells from the AGM, and without similarly accurate fate mapping of this region, it remains possible that all of the adult stem cell types thought to emerge within the AGM arise from a common unrestricted precursor. This hypothetical precursor could help to explain reports of nonfusion-based adult stem cell plasticity. The observed lineage specificity of most adult stem cells could likewise be attributed to the high-fidelity lineage restriction imposed on them by the specific niche they colonize or are derived from. Simple ideas such as these have not been ruled out by experimental evidence, underscoring both the opportunity and the necessity for further study of the developmental origins of adult stem cells.

A key lesson from studies of the developing hematopoietic system is that the appearance of differentiated cells does not tell us where or when the corresponding adult stem cells originate. Definitive lineage tracing, with assays of clonogenic potential, remains the method of choice for identifying the origin of stem cells. Another potential pitfall revealed by these studies is that the definition of the stem cell can make all the difference in its identification.

The development of NSCs begins with the formation of nervous tissue from embryonic ectoderm following gastrulation. Induction of the neural plate is thought to coincide with the appearance of NSCs as well as restricted progenitor types (Temple, 2001). The exact frequency and location of stem cells within the developing neuroepithelium remains unknown; specific markers must be discovered to fully unravel this question. An emerging view in the field is that embryonic neuroepithelia generate radial glial that subsequently develop into periventricular astrocytes and that these cells are the embryonic and adult NSCs within the central nervous system (Alvarez-Buylla *et al.*, 2001; Tramontin, 2003; Doetsch *et al.*, 1999; Gaiano and Fishell, 2002). Developing and adult NSCs also appear to acquire positional and temporal information. For example, stem cells isolated from different neural regions generate region-appropriate progeny (Kalyani *et al.*, 1998; He *et al.*, 2001; Anderson *et al.*, 1997). In addition, several studies suggest that temporal information is encoded within NSCs, that earlier stem cells give rise more frequently to neurons, and that more mature stem cells preferentially differentiate into glia (Temple, 2001; Qian *et al.*, 2000; White *et al.*, 2001). Moreover, more mature NSCs appear incapable of making cells appropriate for younger stages when transplanted into the early cerebral cortex (Desai and McConnell, 2000). Thus, the nervous system appears to follow a classical lineage hierarchy, with a common progenitor cell generating most if not all differentiated cell types in a regional- and temporal-specific manner. There may also be rare stem cells in the nervous system, perhaps not of neural origin, that have greater plasticity in terms of producing diverse somatic cell types and lacking temporal and spatial constraints (Weissman, 2000; Temple, 2001). There are several caveats that must be considered when describing the developmental origins of NSCs. First, disrupting the neuroepithelia to purify NSCs may have the undesirable effect of dysregulating spatial patterning acquired by these cells. Second, growth of purified NSCs in culture may reprogram the stem cells through exposure to nonphysiological *in vitro* culture conditions. Both of these problems can be addressed either by *in vivo* lineage tracing or by prospectively isolating NSCs and transplanting them into acceptable hosts without intervening culture.

Carefully designed experiments promise to answer questions important not only for stem cell biology but also for neuroembryology and development. These include which features of the developmental program are intrinsic to individual cells, which differentiation or patterning signals act exclusively to instruct specific cell fates, and how developmental changes in cell-intrinsic programs restrict the responses of progenitors to cell-extrinsic signals.

HOW ARE STEM CELLS IDENTIFIED, ISOLATED, AND CHARACTERIZED?

How stem cells are identified, isolated, and characterized are the key methodological questions in stem cell biology, so much so that subsequent chapters are devoted to addressing these problems in detail. Here, I briefly outline standards and criteria that may be employed when approaching the challenge of identifying, isolating, and characterizing a stem cell.

Embryonic Stem Cells

The basic characteristics of an ES cell include self-renewal, multilineage differentiation *in vitro* and *in vivo*, clonogenicity, a normal karyotype, extensive proliferation *in vitro* under well-defined culture conditions, and the ability to be frozen and thawed. In animal species, *in vivo* differentiation can be assessed rigorously by the ability of ES cells to contribute to all somatic lineages and produce germ line chimerism. These criteria are not appropriate for human ES cells; consequently, these cells must generate embryoid bodies and teratomas containing differentiated cells of all three germ layers. Moreover, as a stringent *in vivo* assessment of pluripotency is impossible, human ES cells must be shown to be positive for well-known molecular markers of pluripotent cells. These markers are defined as factors expressed consistently, and enriched, in human ES cells (Brivanlou *et al.*, 2003). As a substitute for whole-animal chimerism, human ES cells could be tested for their contributions to specific tissues when transplanted in discrete regions of nonhuman adults or embryos. A complementary analysis might include transplanting human ES cells into nonhuman blastocysts and evaluating their contribution to various organs and tissues, though this experiment has raised ethical concerns in some quarters. Finally, a practical consideration is the passage number of ES cells. Although it is important to establish the capacity of ES cells to proliferate extensively, it is equally important that low-passage cells are evaluated experimentally to guard against any artifacts introduced through *in vitro* manipulation.

Adult Stem Cells

The basic characteristics of an adult stem cell are a single cell (clonal) that self-renews and generates differentiated cells. The most rigorous assessment of these characteristics is to prospectively purify a population of cells (usually by cell surface markers), transplant a single cell into an acceptable host without any intervening *in vitro* culture, and observe self-renewal and tissue, organ, or lineage reconstitution. Admittedly, this type of *in vivo* reconstitution assay is not well-defined for many types of adult stem cells. Thus, it is important to arrive at an accurate functional definition for cells whose developmental potential is assessed *in vitro* only. Above all, clonal assays should be the standard by which fetal and adult stem cells are evaluated because this assay removes doubts about contamination with other cell types.

Two concepts about the fate or potential of stem cells have moved to the forefront of adult stem cell research. The first is plasticity, the idea that restrictions in cell fates are not permanent but are flexible and reversible. The most obvious and extreme example of reversing a committed cell fate comes from experiments in which a terminally differentiated somatic cell generates to another animal following nuclear transfer or cloning (Solter, 2000; Rideout *et al.*, 2001). Nuclear transfer experiments show that differentiated cells, given the appropriate conditions, can be returned to their most primal state. Thus, it may not be surprising if conditions are found for more committed or specified cells to dedifferentiate and gain a broader potential. A related concept is that of transdifferentiation. Transdifferentiation is the generation of functional cells of a tissue, organ, or lineage that is distinct from that of the founding stem cell (Liu and Rao, 2003; Blau *et al.*, 2001). Important issues here are whether the cells proposed to transdifferentiate are clonal and whether the mechanism by which they form the functional cell requires fusion (Medvinsky and Smith, 2003; Terada *et al.*, 2002; Wang *et al.*, 2003; Ying *et al.*, 2002). Experiments designed to carefully evaluate these possibilities will yield insight into the nature of stem cells.

STEMNESS: PROGRESS TOWARD A MOLECULAR DEFINITION OF STEM CELLS

Stemness refers to the common molecular processes underlying the core stem cell properties of self-renewal and the generation of differentiated progeny. Although stems cells in different cellular microenvironments or niches will by necessity have different physiological demands and therefore distinct molecular programs, there are likely certain genetic characteristics specific to and shared by all stem cells. Through transcriptional profiling, many of the genes enriched in ES cell, TS cell, HSC, and NSC populations have been identified (Ivanova *et al.*, 2002; Ramalho-Santos *et al.*, 2002; Tanaka *et al.*, 2002; Anisimov *et al.*, 2002; Luo *et al.*, 2002; Park *et al.*, 2002). By extending this approach to other stem cells and more organisms, it may be possible to develop a molecular fingerprint for stem cells. This fingerprint could be used as the basis for a molecular definition of stem cells that, when combined with their functional definition, would provide a more comprehensive set of criteria for understanding their unique biology. Perhaps more importantly, these types of studies could be used to help identify and isolate new stem cells. This goal is far from being accomplished, but the preliminary findings for specific stem cells have been described. The transcriptional profiling of stem cells has suggested that they share several distinct molecular characteristics. Stem cells appear to have the capacity to sense a broad range of growth factors and signaling molecules and to express many of the downstream signaling components involved in the transduction of these signals. Signal transduction pathways present and perhaps active in stem cells include TGF®, Notch, Wnt, and Jak/Stat family members. Stem cells also express many components involved in establishing their specialized cell cycles, either related to maintaining cell cycle arrest in G1 (for most quiescent adult stem cells) or connected to progression through cell cycle checkpoints promoting rapid cycling (as is the case for ES cells and mobilized adult stem cells) (Burdon *et al.*, 1999; Savatier *et al.*, 2002). Most stem cells also express molecules involved in telomere maintenance and display elevated levels of telomerase activity. There is also considerable evidence that stem cells have significantly remodeled chromatin acted upon by DNA methylases or transcriptional repressors of histone deacetylase and Groucho family members. Another common molecular feature is the expression of specialized posttranscriptional regulatory machinery regulated by RNA helicases of the Vasa type. Finally, a shared molecular and functional characteristic of stem cells appears to be their resistance to stress, mediated by multidrug resistance transporters, protein-folding machinery, ubiquitin, and detoxifier systems.

Although in its infancy, the search for a molecular signature to define stem cells continues. We have begun to understand in general terms what molecular components are most often associated with stem cells. In the future, it may be possible to precisely define stem cells as a whole and individually by their telltale molecular identities. Until that time, stemness remains a concept of limited utility with tremendous potential.

ACKNOWLEDGMENTS

I would like to thank Jayaraj Rajagopal and Kevin Eggan for helpful discussion and suggestions. I apologize to those authors whose work was inadvertently overlooked or omitted because of space limitations.

Douglas A. Melton, Ph.D
Chad Cowen, Ph.D

FURTHER READING

Alessandri, G., *et al.* (2001). Human vasculogenesis *ex vivo:* embryonal aorta as a tool for isolation of endothelial cell progenitors. *Lab. Invest.*, *81*, 875–885.

Allsopp, R. C., & Weissman, I. L. (2002). Replicative senescence of hematopoietic stem cells during serial transplantation: does telomere shortening play a role? *Oncogene*, *21*, 3270–3273.

Alvarez-Buylla, A., Garcia-Verdugo, J. M., & Tramontin, A. D. (2001). A unified hypothesis on the lineage of neural stem cells. *Nat. Rev. Neurosci.*, *2*, 287–293.

Anderson, D. J. *et al.* (1997). Cell lineage determination and the control of neuronal identity in the neural crest. *Cold Spring Harb. Symp. Quant. Biol.*, *62*, 493–504.

Anisimov, S. V. *et al.* (2002). SAGE identification of gene transcripts with profiles unique to pluripotent mouse R1 embryonic stem cells. *Genomics*, *79*, 169–176.

Blau, H. M., Brazelton, T. R., & Weimann, J. M. (2001). The evolving concept of a stem cell: entity or function? *Cell*, *105*, 829–841.

Brivanlou, A. H. *et al.* (2003). Stem cells: setting standards for human embryonic stem cells. *Science*, *300*, 913–916.

Burdon, T. *et al.* (1999). Signaling mechanisms regulating selfrenewal and differentiation of pluripotent embryonic stem cells. *Cells Tiss. Organs*, *165*, 131–143.

Desai, A. R., & McConnell, S. K. (2000). Progressive restriction in fate potential by neural progenitors during cerebral cortical development. *Development*, *127*, 2863–2872.

Doetsch, F. *et al.* (1999). Subventricular zone astrocytes are neural stem cells in the adult mammalian brain. *Cell*, *97*, 703–716.

Dzierzak, E. (2002). Hematopoietic stem cells and their precursors: Developmental diversity and lineage relationships. *Immunol. Rev.*, *187*, 126–138.

Gaiano, N., & Fishell, G. (2002). The role of notch in promoting glial and neural stem cell fates. *Annu. Rev. Neurosci.*, *25*, 471–490.

Hara, T., *et al.* (1999). Identification of podocalyxin-like protein 1 as a novel cell surface marker for hemangioblasts in the murine aorta–gonad–mesonephros region. *Immunity 11*, 567–578.

Hayflick, L. (1973). The biology of human aging. *Am. J. Med. Sci. 265*, 432–445.

Hayflick, L. (1974). The longevity of cultured human cells. *J. Am. Geriatr. Soc. 22*, 1–12.

He, W., *et al.* (2001). Multipotent stem cells from the mouse basal fore-brain contribute GABAergic neurons and oligodendrocytes to the cerebral cortex during embryogenesis. *J. Neurosci. 21*, 8854–8862.

Houck, J. C., Sharma, V. K., & Hayflick, L. (1971). Functional failures of cultured human diploid fibroblasts after continued population doublings. *Proc. Soc. Exp. Biol. Med. 137*, 331–333.

Howell, C. Y., *et al.* (2001). Genomic imprinting disrupted by a maternal effect mutation in the *Dnmt1* gene. *Cell 104*, 829–838.

Iscove, N. N., & Nawa, K. (1997). Hematopoietic stem cells expand during serial transplantation *in vivo* without apparent exhaustion. *Curr. Biol. 7*, 805–808.

Ivanova, N. B., *et al.* (2002). A stem cell molecular signature. *Science 298*, 601–604.

Jiang, Y., *et al.* (2002). Pluripotency of mesenchymal stem cells derived from adult marrow. *Nature 418*, 41–49.

Kalyani, A. J., *et al.* (1998). Spinal cord neuronal precursors generate multiple neuronal phenotypes in culture. *J. Neurosci. 18*, 7856–7868.

Kau, C. L., & Turpen, J. B. (1983). Dual contribution of embryonic ventral blood island and dorsal lateral plate mesoderm during ontogeny of hemopoietic cells in *Xenopus laevis*. *J. Immunol. 131*, 2262–2266.

Labosky, P. A., Barlow, D. P., & Hogan, B. L. (1994). Mouse embryonic germ (EG) cell lines: transmission through the germ line, and differences in the methylation imprint of insulin-like growth factor 2 receptor *(Igf2r)* gene compared with embryonic stem (ES) cell lines. *Development 120*, 3197–3204.

Liu, Y., & Rao, M. S. (2003). Transdifferentiation: Fact or artifact. *J. Cell Biochem. 88*, 29–40.

Luo, Y., *et al.* (2002). Microarray analysis of selected genes in neural stem and progenitor cells. *J. Neurochem. 83*, 1481–1497.

Matsui, Y. , Zsebo, K., & Hogan, B. L. (1992). Derivation of pluripotential embryonic stem cells from murine primordial germ cells in culture. *Cell 70*, 841–847.

Medvinsky, A. L., *et al.* (1993). An early preliver intraembryonic source of CFU-S in the developing mouse. *Nature 364*, 64–67.

Medvinsky, A., & Dzierzak, E. (1996). Definitive hematopoiesis is autonomously initiated by the AGM region. *Cell 86*, 897–906.

Medvinsky, A., & Smith, A. (2003). Stem cells: fusion brings down barriers. *Nature 422*, 823–835.

Minasi, M. G., *et al.* (2002). The mesoangioblast: A multipotent, selfrenewing cell that originates from the dorsal aorta and differentiates into most mesodermal tissues. *Development 129*, 2773–2783.

Molyneaux, K. A., *et al.* (2001). Time-lapse analysis of living mouse germ cell migration. *Dev. Biol. 240*, 488–498.

Morrison, S. J., Shah, N. M., & Anderson, D. J. (1997). Regulatory mechanisms in stem cell biology. *Cell 88*, 287–298.

Munoz-Chapuli, R., *et al.* (1999). Differentiation of hemangioblasts from embryonic mesothelial cells? A model on the origin of the vertebrate cardiovascular system. *Differentiation 64*, 133–141.

Orkin, S. H. (1996). Development of the hematopoietic system. *Curr. Opin. Genet. Dev. 6*, 597–602.

Papaioannou, V. (2001). Stem cells and differentiation. *Differentiation 68*, 153–154.

Park, I. K., *et al.* (2002). Differential gene expression profiling of adult murine hematopoietic stem cells. *Blood 99*, 488–498.

Qian, X., *et al.* (2000). Timing of CNS cell generation: a programmed sequence of neuron and glial cell production from isolated murine cortical stem cells. *Neuron 28*, 69–80.

Ramalho-Santos, M., *et al.* (2002). "Stemness": Transcriptional profiling of embryonic and adult stem cells. *Science 298*, 597–600.

Resnick, J. L., *et al.* (1992). Long-term proliferation of mouse primordial germ cells in culture. *Nature 359*, 550–551.

Rideout, W. M., 3rd, Eggan, K., & Jaenisch, R. (2001). Nuclear cloning and epigenetic reprogramming of the genome. *Science 293*, 1093–1098.

Saitou, M., Barton, S. C., & Surani, M. A. (2002). A molecular program for the specification of germ cell fate in mice. *Nature 418*, 293–300.

Savatier, P., *et al.* (2002). Analysis of the cell cycle in mouse embryonic stem cells. *Methods Mol. Biol. 185*, 27–33.

Shay, J. W., & Wright, W. E. (2000). Hayflick, his limit, and cellular ageing. *Nat. Rev. Mol. Cell Biol. 1*, 72–76.

Sherr, C. J., & DePinho, R. A. (2000). Cellular senescence: mitotic clock or culture shock? *Cell 102*, 407–410.

Slack, J. M. (2000). Stem cells in epithelial tissues. *Science 287*, 1431–1433.

Smith, A. G. (2001). Embryo-derived stem cells: of mice and men. *Annu. Rev. Cell Dev. Biol. 17*, 435–462.

Solter, D. (2000). Mammalian cloning: advances and limitations. *Nat. Rev. Genet. 1*, 199–207.

Stewart, C. L., Gadi, I., & Bhatt, H. (1994). Stem cells from primordial germ cells can reenter the germ line. *Dev. Biol. 161*, 626–628.

Surani, M. A. (1998). Imprinting and the initiation of gene silencing in the germ line. *Cell 93*, 309–312.

Surani, M. A. (2001). Reprogramming of genome function through epigenetic inheritance. *Nature 414*, 122–128.

Tanaka, S., *et al.* (1998). Promotion of trophoblast stem cell proliferation by FGF4. *Science 282*, 2072–2075.

Tanaka, T. S., *et al.* (2002). Gene expression profiling of embryoderived stem cells reveals candidate genes associated with pluripotency and lineage specificity. *Genome Res. 12*, 1921–1928.

Temple, S. (2001). The development of neural stem cells. *Nature 414*, 112–117.

Terada, N., *et al.* (2002). Bone marrow cells adopt the phenotype of other cells by spontaneous cell fusion. *Nature 416*, 542–545.

Tramontin, A. D., *et al.* (2003). Postnatal development of radial glia and the ventricular zone (VZ): a continuum of the neural stem cell compartment. *Cereb. Cortex 13*, 580–587.

van der Kooy, D., & Weiss, S. (2000). Why stem cells? *Science 287*, 1439–1441.

Wang, X., *et al.* (2003). Cell fusion is the principal source of bone marrow-derived hepatocytes. *Nature 422*, 897–901.

Weissman, I. L. (2000). Stem cells: units of development, units of regeneration, and units in evolution. *Cell 100*, 157–168.

Weissman, I. L., Anderson, D. J., & Gage, F. (2001). Stem and progenitor cells: origins, phenotypes, lineage commitments, and transdifferentiations. *Annu. Rev. Cell Dev. Biol. 17*, 387–403.

White, P. M., *et al.* (2001). Neural crest stem cells undergo cell-intrinsic developmental changes in sensitivity to instructive differentiation signals. *Neuron 29*, 57–71.

Ying, Q. L., *et al.* (2002). Changing potency by spontaneous fusion. *Nature 416*, 545–548.

Part I

An Introduction to Stem Cell Biology

Chapter 1

Pluripotential Stem Cells from Vertebrate Embryos: Present Perspective and Future Challenges

Richard L. Gardner

INTRODUCTION

Many have contributed to the various developments that brought recognition of the enormous potential of cells of early embryonic origin for genetic modification of organisms, regenerative medicine, and in enabling investigation of facets of development that are difficult to explore *in vivo*. However, historically, this field is firmly rooted in the pioneering work of Roy Stevens and Barry Pierce on mouse teratomas and teratocarcinomas, tumors which continued for some time after these workers had embarked on their studies to be regarded with disdain by many mainstream pathologists and oncologists. While Stevens developed and exploited mouse strains with high incidences of such tumors to determine their cellular origins, Pierce focused his attention on the nature of the cell that endowed teratocarcinomas with the potential for indefinite growth which the more common teratomas lacked. Conversion of solid teratocarcinomas to an ascites form proved a significant advance in enabling dramatic enrichment of the morphologically undifferentiated cells in such tumors which their stem cells were expected to be included among. Then, in an experiment of heroic proportions, Kleinsmith and Pierce showed unequivocally that, on transplantation to histocompatible adult hosts, individual morphologically undifferentiated cells could form self-sustaining teratocarcinomas that contained as rich a variety of differentiated tissues as the parent tumor. Hence, the embryonal carcinoma (EC) cell, as the stem cell of teratocarcinomas has come to be known, was the first self-perpetuating pluripotential cell to be characterized. Though teratocarcinomas were obtained initially as a result of genetically-determined aberrations in the differentiation of male or female germ cells, it was found later that they could also be established in certain genotypes of mice by grafting early embryos ectopically in adults. Adaptation of culture conditions to enable EC cells to be perpetuated in an undifferentiated state or induced to differentiate *in vitro* soon followed. Although the range of differentiation detected in these circumstances was more limited than *in vivo*, it could nevertheless be quite impressive. Research on murine EC cells, in turn, provided the impetus for obtaining and harnessing the human counterpart of these cells from testicular tumors for *in vitro* study.

One outstanding question regarding the use of murine EC cells as a model system for studying aspects of development remained, namely the basis of their malignancy. Was this a consequence of genetic change or simply because such "embryonic" cells failed to relate to the ectopic sites into which they were transplanted? The obvious way of addressing this was to ask how EC cells behave when placed in an embryonic rather than an adult environment. This was done in three different laboratories by injecting the cells into blastocyst stage embryos. The results from each laboratory led to the same rather striking conclusion. EC cells, which if injected into an adult would grow progressively and kill it, were able to participate in entirely normal development following their introduction into the blastocyst. Using genetic differences between donor and host as cell markers, EC cells were found to be able to contribute to most if not all organs and tissue of the resulting offspring. Most intriguingly, according to reports from one laboratory, this could very exceptionally include the germline. The potential significance of this finding was considerable in terms of its implications for possible controlled genetic manipulation of the mammalian genome. This is because it raised the prospect of being able to select for very rare events, and thus bring the scope for genetic manipulation in mammals closer to that in microorganisms.

There were problems, however. One was that the EC contribution in chimeric offspring was typically both more modest and more patchy than that of cells transplanted directly between blastocysts. The chimeras also not infrequently formed tumors, with those that proved to be teratocarcinomas often being evident already at birth. Therefore, it seems likely that regulation of growth of at least some of the transplanted EC cells failed altogether. Other chimeras developed more specific tumors such as rhabdomyosarcomas as they aged which were also clearly of donor origin, thereby revealing that the transplanted EC cells had progressed further along various lineages before their differentiation went awry. In extreme cases the transplanted EC cells disrupted development altogether so that fetuses did not survive to birth. Although the best EC lines could give very widespread contributions throughout the body of chimeras, they did so only very exceptionally. Finally, the frequency with which colonization of the germ line could be obtained with EC cells was too low to enable them to be harnessed for genetic modification. It seemed likely, therefore, that the protracted process of generating teratocarcinomas *in vivo* and then adapting them to culture militated against retention of a normal genetic constitution by their stem cells. If this was indeed the case, the obvious way forward was to see if such stem cells could be obtained in a less circuitous manner. This prompted investigation of what happens when murine blastocysts are explanted directly on growth-inactivated feeder cells in an enriched culture medium. The result was the derivation of lines of cells that were indistinguishable from EC cells in both morphology and expression of various antigenic and other markers, as well as in the appearance of the colonies they formed during growth. Moreover, like EC cells, these self-perpetuating blastocyst-derived stem cells could form aggressive teratocarcinomas in both syngeneic and immunologically compromised non-syngeneic adult hosts. They differed from EC cells principally in giving much more frequent and widespread somatic chimerism following reintroduction into the pre-implantation conceptus and, if tended carefully, also in routinely colonizing the germline. Moreover, when combined with host conceptuses whose development was compromised by tetraploidy, they could sometimes form offspring in which no host-derived cells were discernible. Thus, these cells, which exhibited all the desirable characteristics of EC cells and few of their shortcomings, came to be called embryonic stem (ES) cells. Once it had been shown that ES cells could retain their ability to colonize the germline after *in vitro* transfection and selection, their future was assured. Surprisingly, however, despite the wealth of studies demonstrating their capacity for differentiation *in vitro*, particularly in the mouse, the idea of harnessing ES cells for therapeutic purposes took a long time to take root. Thus, although Robert Edwards explicitly argued the case that human ES cells might be used thus more than 25 years ago, it is only within the past decade that this notion has gained momentum, encouraged particularly by derivation of the first cell lines from human blastocysts.

TERMINOLOGY

There is some confusion in the literature about terminology in discussing the range of different types of cells that ES cells are able to form, an attribute that, in embryological parlance, is termed their potency. Some refer to these cells as being totipotent in recognition of the fact that, at least in the mouse, they have been shown to be able to give rise to all types of fetal cells and, under certain conditions, entire offspring. However, this is inappropriate on two counts. First, totipotency is reserved by embryologists for cells that retain the capacity to form an entire conceptus, and thus give rise to a new individual, unaided. Apart from the fertilized egg, the only cells that have so far been shown to be able to do this are blastomeres from early cleavage stages. Second, murine ES cells seem unable to form all the different types of cell of which the conceptus is composed. Following injection into blastocysts, they normally give rise only to cell types that are products of the epiblast or fetal precursor cell lineage. While they can also form derivatives of the primitive endoderm lineage which, for some obscure reason they do much more readily *in vitro* than *in vivo*, they have never been convincingly shown to contribute to the trophectodermal lineage. Hence, a widely adopted convention is to describe ES cells as pluripotent stem cells, to distinguish them from stem cells like those of the hematopoietic system which have a narrower, but nevertheless impressive, range of differentiative potential. A further source of confusion is the surprisingly common practice of referring to cells, particularly putative ES cells from mammals other than the mouse, as totipotent because their nuclei have been shown to be able to support development to term when used for reproductive cloning.

Another facet of terminology relates to the definition of an ES cell, which again is not employed in a consistent manner. One view, to which the author subscribes, is that use of this term should be restricted to pluripotent cells derived from pre- or peri-implantation conceptuses that can form functional gametes, as well as the full range of somatic cells of offspring. While there are considerable differences between strains of mice in the facility with which morphologically undifferentiated cell lines can be obtained from their early conceptuses, competence to colonize the germline as well as somatic tissues seems nevertheless to be common to lines from all strains that have yielded them. This is true, for example, even for the non-obese diabetic (NOD) strain whose lines have so far been found to grow too poorly to enable their genetic modification.

It is, however, noteworthy that even in the mouse rates of malformation and perinatal mortality appear to be higher in EG than in ES cell chimeras. This may relate to erasure of imprinting in the germ line which seems to have begun by the time primordial germ cells have colonized the genital ridges or, for certain genes, even earlier. It is perhaps because of such concerns that the potential of EG cells for transgenesis in strains of mice that have failed to yield ES cells has not been fully explored. Interestingly, unlike in the mouse, EGL cell lines derived from genital ridges and the associated mesentary of 5–11 week human fetuses seem not to have embarked on erasure of imprinting. Obviously, it is important to confirm that this is the case before contemplating use of such cells as grafts for repairing tissue damage in humans.

FUTURE CHALLENGES

The value of ES and ESL cells as resources for both basic and applied research is now acknowledged almost universally. Present barriers to exploitation of their full potential in both areas are considered below, together with possible ways of addressing these. Fundamental to progress is gaining a better understanding of the nature and the basic biology of these cells.

BIOLOGY OF ES AND ES-LIKE CELLS

Germline Competence

Although murine ES cells have been used very extensively for modifying the genome, there are still a number of problems that limit their usefulness in this respect. Among these is loss of competence to colonize the germline, a common and frustrating problem whose basis remains elusive. It is not attributable simply to the occurrence of sufficient chromosomal change to disrupt gametogenesis, because it can occur in lines and clones that are found to be karyotypically normal. At present, it is not known whether it is due to failure of the cells to be included in the pool of primordial germ cells, or their inability to undergo appropriate differentiation thereafter, possibly as a consequence of perturbation of the establishment of genomic imprinting or its erasure. Even within cloned ES lines, cells have been found to be heterogeneous with respect to expression of imprinted genes. Given that many ES cell lines are likely to have originated polyclonally from several epiblast founder cells, there is the further possibility that they might, *ab initio*, consist of a mixture of germline-competent and non-competent sub-populations. Recent studies on involvement of BMP signaling in the induction of primordial germ cells have been interpreted as evidence against a specific germ cell lineage in mammals. Particular significance has been attached to experiments in which distal epiblast, which does not usually produce primordial germ cells, was found to do so when grafted to the proximal site from whence these cells normally originate. However, because of the extraordinary degree of cell mixing that occurs in the epiblast before gastrulation, descendants of all epiblast founder cells are likely to be present throughout the tissue by the time of primordial germ cell induction. Hence, yet to be excluded is the possibility that competence for induction is lineage-dependent and thereby segregates only to some epiblast founders cells. Because ES cell lines are typically produced by pooling all colonies derived from a single blastocyst, they might therefore originate from of a mixture of germline-competent and non-competent epiblast founder cells.

Male ES cell lines have almost invariably been used in gene targeting studies, even though this complicates work on X-linked genes whose inactivation may lead to cell-autonomous early lethality or compromised viability in the hemizygous state. Here, female (XX) lines would, in principle, offer a simpler alternative, except they are generally held to suffer partial deletion or complete loss of one X-chromosome after relatively few passages. It is, however, not clear how secure this conclusion is, because few references to their use have appeared in literature since the early reports in which consistent loss of all or part of one X was first documented. More recently, one of only two female lines tested was found to be germline-competent, but the entirely donor-derived litters were unusually small, raising the possibility not entertained by the authors that the line in question was XO. Interestingly, human ESL cell lines seem not to show a similar propensity for X-chromosome loss and, indeed, nor do murine EpiSc.

Origin and Properties of ES and ES-like Cells

It is evident from the earlier overview that there is considerable diversity even among eutherian mammals in the characteristics of cells from early conceptuses that can be perpetuated *in vitro* in a morphologically undifferentiated state. The reason for this is far from clear, particularly since the great majority of such cell lines have been derived at a corresponding stage, namely the pre-implantation blastocyst, often using inner cell mass tissue isolated from there. In the mouse, in contrast to their EC counterparts, ES cells have not been obtained from post-implantation stages, arguing that there is a rather narrow window during which their derivation is possible. What this relates to in developmental terms remains obscure, although the finding that ES cells can shift reversibly to a condition that shows altered colony morphology and gene expression, in conjunction with loss of ability to generate chimeras following blastocyst injection, offers a possible approach for addressing this problem. Whether the late blastocyst stage sets the limit for obtaining ESL cell lines has not yet been addressed critically in other mammals. However, the possibility that ESL cells in other mammals correspond to murine EpiSc clearly warrants further consideration.

Just as ES cell lines have been obtained from preblastocyst stages in the mouse, so have ESL cell lines from other mammals. However, in the case of the mouse, ES cell lines derived from early cleavage stages and morulae have not been found to differ in properties or developmental potential from those obtained from blastocysts, implying that all originate at the same very early stage in differentiation of the epiblast lineage. Thus claims that lines isolated from morulae have an advantage over those from blastocysts in being able to produce trophoblast have not been substantiated. However, species- as opposed to stage-related differences in the ability of cell lines to produce trophoblast tissue have been encountered. Early claims that mouse ES cells can form trophoblastic giant cells are almost certainly attributable to short-term persistence of contaminating polar trophectoderm tissue. Thus, production of such cells seems to be limited to the early passage of ES lines derived from entire blastocysts. It has never been observed with lines established from microsurgically isolated epiblasts. While the situation is not clear in many other species, in primates differentiation of trophoblast has been observed routinely in ESL cell lines established from immunologically-isolated inner cell masses. Moreover, differentiation of human cell lines to the stage of syncytiotrophoblast-formation has been induced efficiently by exposing them to BMP4.

Pluripotency

A seminal characteristic of ES or ESL cells is their pluripotency. The most critical test of this, which is not practicable in some species, particularly the human, is the ability to form the entire complement of cells of normal offspring. This assay, which was originally developed in the mouse, entails introducing clusters of ES cells into conceptuses whose development has been compromised by making them tetraploid, either by suppressing cytokinesis or through fusing sister blastomeres electrically at the two-cell stage. ES cells are then either aggregated with the tetraploid cleavage stages or injected into tetraploid blastocysts. Some of the resulting offspring contain no discernible host cells. It seems most likely that host epiblast cells are present initially and play an essential role in "entraining" the donor ES cells before being outcompeted, since groups of ES cells on their own cannot substitute for the epiblast or inner cell mass (author's unpublished observations). Selection against tetraploid cells is already evident by the late blastocyst stage in chimeras made between diploid and tetraploid morulae. Aggregating ESL cells between pairs of tetraploid morulae has been tried in cattle, but resulted in their contributing only very modestly to fetuses and neonates.

The second most critical test is whether the cells yield widespread, if not ubiquitous, chimerism in offspring following introduction into the early conceptus, either by injection into standard blastocysts or aggregation with morulae. The third is the formation of teratomas in ectopic grafts to histocompatible or immunosuppressed adult hosts, since it is clear from earlier experience with murine and human EC cells that a wider range of differentiation can be obtained in these circumstances than *in vitro*. For such an assay to be incisive it is necessary to use clonal cell lines and thus ensure that the diversity of differentiation obtained originates from one type of stem cell rather than a medley of cells with more limited developmental potential. While teratoma formation has been demonstrated with clonal ESL in the human, this is not true for corresponding cell lines in other species. A note of caution regarding the use of teratomas for assessing pluripotency comes from workers who found that hepatocyte differentiation depended not only on site of inoculation of mouse ES cells, but also the status of the host. Thus, positive results were obtained with spleen rather than hind-limb grafts, and only when using immmunologically-compromised nude rather than syngeneic mice as hosts.

Conditions of Culture

ES and ESL cells are usually propagated in complex culture conditions that are poorly defined, by virtue of including both growth-inactivated feeder cells and serum. This complicates the task of determining the growth factor and other requirements necessary for their maintenance, as well as for inducing them to form specific types of differentiated cells. While differentiation of murine ES cells in a chemically-defined medium has been achieved, their maintenance under such conditions has not. Murine ES cells can be both derived and maintained independently of feeder cells providing a cytokine that signals via the gp 130 receptor is present in the medium. However, whether the relatively high incidence of early aneuploidy recorded in the two studies in which LIF was used throughout in place of feeders is significant or coincidental is not clear. It is important to resolve this in order to learn whether feeder cells serve any other function than acting as a source of LIF or a related cytokine. Production of extracellular matrix is one possibility. However, species and cell-type variability is also a factor here since LIF is not required for maintaining human ESL lines, whose cloning efficiency is actually improved by omission of serum, although feeder cells are. The norm has been to use murine feeder cells both for obtaining and perpetuating ESL cell lines in other mammals, including the human. Recently, however, there has been a move to use feeders of human origin for human ESL cells. This is a notable development since it would obviously not be acceptable to employ xenogeneic cells for growing human ESL cell lines that were destined for therapeutic rather than laboratory use. The situation is somewhat confusing in the case of the pig since in one study, but not in others, porcine feeders were found to be necessary for deriving ESL cell lines which could then be perpetuated on murine STO cells thereafter. Moreover, among teleost fish, feeder-free conditions seem to be optimal for maintaining ESL cells in both the medaka and the sea bream, but possibly not in the zebra fish.

Susceptibility Versus Resistance to Derivation

A further area whose further investigation could be informative in facilitating the establishment of pluripotent stem cell lines in other species is the basis of susceptibility versus resistance to ES cell derivation in the mouse. Thus, while ES cell lines can be obtained very easily in 129 mice, and relatively so in C57BL/6 and a few additional strains, other genotypes have proved more resistant. Notable among the latter is the non-obese diabetic (NOD) strain from which, despite considerable effort, genetically-manipulatable lines have not yet been obtained. This is not simply related to susceptibility of this strain to insulin-dependent diabetes, because the ICR strain from which NOD was developed has proved to be equally refractory. However, refractoriness seems to be a recessive trait because excellent lines with high competence to colonize the germline have been obtained from [NOD × 129]F1 epiblasts. Moreover, this is not the only example where refractoriness has been overcome by inter-crossing. Interestingly, marked differences in permissiveness for ESL cell derivation have also been found among inbred strains of the medaka fish.

Human ES-like Cell

Mouse EC and ES cells have already been used extensively to study aspects of development which, for various reasons, are difficult to investigate in the intact conceptus. Exploiting corresponding cells for this purpose is even more pressing for gaining a better understanding of early development in our own species, given the relative scarcity of material, ethical concerns about experimenting on conceptuses, and statutory or technical limitations on the period for which they can be maintained *in vitro*. Obviously, in view of their provenance, human ESL cells are likely to provide a more apposite model system than human EC cells, which have mainly been used until recently. One concern here is that so-called "spare" conceptuses, i.e., those that are surplus to the needs of infertility treatment, are the sole source of material for producing human ESL cell lines. Since the conceptuses produced *in vitro* by IVF or related techniques that are judged to be of the highest quality are selected for treating infertility treatment, those used for deriving ESL lines tend to be of lower quality. Does this matter so far as the properties of the resulting cell lines are concerned, particularly if their use therapeutically is contemplated? Is the ability to form a blastocyst that looks satisfactory morphologically adequate, or will it prove acceptable to produce conceptuses specifically for generating ESL cell lines so that quality is less of a concern?

ES Cell Transgenesis

One important use of ES cell transgenesis is to obtain animal models of human genetic diseases. Since few would claim that the mouse is the ideal species for this purpose, the incentive for being able to undertake such studies in more appropriate or experimentally tractable mammals must remain a high priority. For example, given its widespread use for studying respiratory physiology, the sheep would obviously be a more relevant species than the mouse as a model system for studying cystic fibrosis. However, unless pluripotential cells that are able to colonize the germline can be obtained in other species, this particular approach to transgenesis will continue to be limited to the mouse. Although the feasibility of an alternative strategy, namely genetically modifying and selecting non-germline-competent cells such as fetal fibroblasts and then exploiting transfer of their nuclei to oocytes has been demonstrated, it is extremely demanding technically and entails considerable fetal attrition.

STEM CELL THERAPY

Potential Hurdles

One of the major present interests in ESL cells is the prospect of exploiting them therapeutically to repair damage to tissues or organs resulting from disease or injury. This poses a host of new challenges, not all of which have so far received the attention they deserve. Perhaps the most obvious one is whether it will be possible efficiently to obtain directed differentiation of stem cells to yield pure cultures of the desired type of more differentiated cells, as opposed to a mixed population. If the latter proves to be the case, the rigorous purging of cultures of residual undifferentiated or inappropriately differentiated cells will be necessary. How this is approached will depend on whether any contamination of grafts is acceptable and, if so, how much. One way in which this particular problem has been circumvented in murine model systems for *in vitro* differentiation of ES cells is to transfect them with the coding region of a gene for an antibiotic resistance or fluorescent protein coupled to the promoter of a gene that is expressed only in the desired type of differentiated cell. Recent advances have made it possible to carry out similar genetic modification of human ESL cells. While very effective selection of the desired type of differentiated cell may be achieved with this approach, it remains to be seen whether use of genetically modified cells will be acceptable in a clinical, as opposed to a laboratory, context.

A further important issue in contemplating stem cell therapy is the cycle status of the desired type of cell. In certain cases, including cells that are not post-mitotic in grafts may be highly undesirable or even hazardous. In others, the presence of such cells may be essential to meet the demands of tissue growth or turnover. The latter would depend on obtaining the differentiation of ESL cells to stem cells rather than fully differentiated cells of the desired type. Given the growing evidence that tissue stem cells require a specific niche for their maintenance, this could well prove difficult to achieve. Establishing and maintaining an appropriate niche *in vitro* so as to be able to enrich for tissue-specific stem cells is likely to pose a considerable challenge, and will unquestionably depend on better knowledge of the normal biology of individual tissues than we have at present.

Yet another important issue is whether engrafted cells will survive and function properly when placed in a damaged tissue or organ. In cases where the donor cells are to provide a hormone, neural transmitter, or soluble growth factor, it may be possible to place them at some distance from the site of damage. However, where this is not practicable, there remains the question of whether transplanted cells will fare any better than native ones in a tissue or organ that has been seriously damaged by disease or injury. If they do not, how can one circumvent this difficulty, bearing in mind that achieving complex organogenesis *in vitro* is still a rather remote prospect? Regarding neuro-degenerative disease, some progress has already been made in "cleaning up" sites of tissue damage. For example, antibody-mediated clearance of plaques from the brain in transgenic mice over-expressing amyloid precursor protein has been demonstrated. However, such intervention may not be necessary in all cases. Transplanting differentiated murine ES cells enriched for putative cardiomyocytes to a damaged region of the left ventricle in rats led concomitantly to a reduction in size of this region and to an improvement in the performance of the heart.

Significant progress with achieving limited organogenesis *in vitro* has been reported for the urinary bladder. Biopsies of this organ were obtained from patients in which it was defective, and the inner urothelial and outer muscle cells expanded separately in culture for seeding on the appropriate surfaces of a biodegradable hemi-bladder-shaped matrix. Following the suturing of such *in vitro*-produced hemi-bladders into the bladder, obvious improvement in its function was recorded. Obviously, orchestrating more complex organogenesis *in vitro* offers greater challenges for the future.

Therapeutic Cloning

Establishing ESL cell lines from blastocysts derived by nuclear replacement, so-called "therapeutic cloning," has been widely advocated as a way of tailoring grafts to individual patients, thereby circumventing the problem of graft rejection. While the feasibility of producing ES cells in this way has been demonstrated in the mouse, there is sharp division of opinion within the biomedical research community about whether such cells would be safe to use therapeutically. Particular concern centers on the normality of the donor genome regarding the epigenetic status of imprinted genes.

Embryonic Versus Adult Stem Cells

Concern about use of early human conceptuses as a source of stem cells had focused much attention on recent studies which suggest that so-called "adult stem cells" are much more versatile in their range of differentiation than has generally been supposed. There is continuing lively debate about interpretation of many of the findings, which certainly do not at present justify the common assertion that adult cells render the use of ESL cells for therapeutic purposes unnecessary. Of particular concern is a growing body of evidence that adult cells may not be changing their differentiated state as independent entities, but through fusing with cells of the type to which they are claimed to have converted.

There is a further, more general point that, with few exceptions, among which the hematopoietic system is the clearest example, evidence is lacking that cells from adult organs and tissues which can be propagated in culture actually functioned as stem cells *in situ*. Hence, adoption of the term "stem cell" for cells from many adult sources is questionable. It is possible, if not likely, that cells which are strictly post-mitotic in their normal environment can nevertheless be induced to resume cycling when removed from it and placed in an enriched culture medium which may contain growth factors to which they would not otherwise be exposed. Such cells might well lack features of true stem cells, such as accurate proofing of DNA replication, conservation of turnover through transit amplification of differentiating progeny, and maintenance of telomere length. They might therefore be severely compromised in their ability to function in grafts.

Recently, however, the distinction between adult and embryonic stem cells has become blurred as a result of the demonstration in both mouse and human that various types of adult somatic cells can be converted into pluripotent stem cells by transducing them with a small number of genes that are normally expressed in ES or ESL cells, respectively. Such induced pluripotential stem (iPS) cells offer a more realistic prospect than therapeutic cloning for producing specific patient-compatible tissue for grafting, providing the process of inducing them can be improved so as to avoid dependence on retroviral transduction of the requisite genes.

CONCLUSIONS

Since the pioneering studies of Steven and Pierce pointed the way in the 1950s and 1960s, impressive progress has been made in harnessing stem cells of embryonic, as opposed to fetal or adult, origin for both basic research and exploring new approaches to regenerative medicine. There is, however, still a great deal to be learnt about the origin and properties of such cells, as well as control of their self-renewal versus differentiation, if we are to take full advantage of what they have to offer. The effort of acquiring the necessary knowledge will undoubtedly provide us with the further reward of gaining deeper insight into the biology of stem cells in general.

KEY WORDS

Blastocyst The late pre-implantation stage of development when the embryo cavitates to form a hollow sphere bounded by a monolayer of trophectoderm, to part of the inner surface of which is attached a disk of cells called the inner mass cell. The former contributes exclusively to placental tissues and the latter both to the fetus and the placenta, plus additional extra-embryonic membranes.

Chimera An animal composed of cells originating from two or more embryos produced, for example, by injecting embryonic stem cells into a blastocyst of different genotype.

Embryonic stem cells (ES cells) Stem cells derived from the preimplantation embryo with an unrestricted capacity for self-renewal that has the potential to form all types of adult cells, including germ cells. Thus far, such cells have only be obtained from certain strains of mice.

Embryonic stem cell-like cells (ES-like cells) Stem cells that resemble murine embryonic stem cells, but whose potential to colonize the pre-implantation embryo and form all types of adults cells including germ cells has yet to be established.

Epiblast stem cells (EpiSc) Stem cells obtained from early postimplantation mouse embryos that resemble ESL cells from other species in terms of the conditions required for maintaining pluripotency and inducing differentiation.

Induced pluripotential stem cells (iPS cells) Differentiating or differentiated cells induced to return to a pluripotent state by manipulating them to express genes associated with pluripotency.

Pluripotency The potential to form most or all types of specialized cells: to be distinguished from totipotency, that is the ability to form a complete embryo capable of giving rise to offspring.

FURTHER READING

Amit, M., Carpenter, M. K., Inokuma, M. S., Chiu, C. P., Harris, C. P., Waknitz, M. A., Itskovitz-Eldor, J., & Thomson, J. A. (2000). Clonally derived human embryonic stem cell lines maintain pluripotency and proliferative potential for prolonged periods. *Dev. Biol.*, *227*, 271–278.

Andrews, P. W., Przyborski, S. A., & Thomson, J. A. (2001). Embryonal carcinoma cells as embryonic stem cells. In D.R Marshak, R. L. Gardner & D. Gottlieb (Eds), *Stem Cell Biology*, pp. 231–265. New York: Cold Spring Harbor Laboratory Press.

Bradley, A., Evans, M. J., Kaufman, M. H., & Robertson, E. J. (1984). Formation of germline chimaeras from embryo-derived teratocarcinoma cells. *Nature*, *309*, 255–256.

Brons, I. G., Smithers, L. E., Trotter, M. W., Rugg-Gunn, P., Sun, B., Chuva de Sousa Lopez, S. M., Howlett, S. K., Clarkson, A., Ahrlund-Richter, L., Pedersen, R. A., & Vallier, L. (2007). Derivation of pluripotential epiblast stem cells from mammalian embryos. *Nature*, *448*, 191–195.

Brook, F. A., & Gardner, R. L. (1997). The origin and efficient derivation of embryonic stem cells in the mouse. *Proc. Nat. Acad. Sci. USA*, *94*, 5709–5712.

Edwards, R. G. (1982). The case for studying human embryos and their constituent tissues *in vitro*. In R. G. Edwards & J. M. Purdy (Eds), *Human Conception* In Vitro, pp. 371–388. London: Academic Press.

Evans, M. J., & Kaufman, M. H. (1981). Establishment in culture of pluripotential cell line from mouse embryo cells from mouse embryos. *Nature*, *292*, 154–156.

Gardner, R. L. (2007). Stem cells and regenerative medicine: Principles, prospects and problems. *C. R. Biol.*, *330*, 465–473.

Kiessling, A. A., & Anderson, S. (2003). *Human Embryonic Stem Cells*. Sudbury, Massachusetts: Jones and Bartlett, Publishers.

Marshak, D. R., Gardner, R. L., & Gottlieb, D. (2001). *Stem Cell Biology*. Cold Spring Harbor, New York: Cold Spring Harbor Laboratory Press.

Pierce, G. B. (1975). Teratocarcinoma: introduction and perspectives. In M. I. Sherman & D. Solter (Eds), *Teratomas and Differentiation*, pp. 3–12. New York: Academic Press.

Robertson, E. J., Kaufman, M. H. Bradley, A., & Evans, M. J. (1983). Isolation, properties and karyotype analysis of pluripotential (EK) cell lines from normal and parthenogenetic embryos. In L. M. Silver, G. R. Martin & S. Strickland (Eds), *Teratocarcinoma Stem Cells*, pp. 647–663. Cold Spring Harbor, New York: Cold Spring Harbor Laboratory Press.

Teasr, P. J., Chenoweth, G. G., Brook, F. A., Davies, T. J., Evand, E. P., Mack, D. l., Gardner, R. L., & McKay, R. D. (2007). New cell lines from mouse epiblast share defining features with human embryonic stem cells. *Nature*, *448*, 196–199.

Yamanaka, S. (2008). Pluripotency and nuclear reprogramming. *Phil. Trans. R. Soc. Lond. B. Biol. Sci.*, *363*, 2079–2087.

Chapter 2

Postnatal Stem Cells

Pamela Gehron Robey and Paolo Bianco

INTRODUCTION

The years since the mid-1990s have witnessed the emergence of stem cell-based tissue engineering as a means to restore normal structure and function to tissues lost to trauma or disease. Successful application will rely on (1) further characterization of the biological properties of postnatal stem cells (what they can and cannot do); (2) optimization of their isolation and *ex vivo* expansion; (3) formulation of appropriate scaffolds and carriers, which may also include growth factors; (4) development of more advanced bioreactors to create tissues with appropriate biomechanical properties; and (5) determination of mechanisms by which to mobilize and activate endogenous stem cells. This chapter highlights these aspects of stem cell-based tissue engineering and presents examples of current approaches and applications.

At first inception, tissue engineering was based on the use of natural or synthetic scaffolds seeded with organ-specific cells *ex vivo*. This approach was somewhat distinct from guided tissue regeneration, which utilized scaffolds and/or bioactive factors to encourage local cells to repair a defect *in situ*. These two approaches are now merged in the current field of tissue engineering, which encompasses multiple and diverse disciplines to use cells, materials, and bioactive factors in various combinations to restore and even improve tissue structure and function. Stem cell-based tissue engineering represents a major turn in the conceptual approach to reconstruction of tissues. By expanding the repertoire of available cells, of potential targets, and of technological means of generating functional tissues *ex vivo* and above all by making it possible (or promising) to engineer tissues that otherwise would not be amenable to recreation, advances in stem cell biology are having a profound impact on tissue engineering in general. In many cases, tissue engineering seems to have more use for stem cells than does nature itself. This is especially apparent in the case of stem cells derived from tissues with low turnover or no apparent turnover at all. If neural stem cells were able to repair the loss of dopaminergic neurons in the intact brain *in vivo*, there would be no Parkinson's disease for which to envision stem cell-based therapies. Dental pulp stem cells do not regenerate primary dentin *in vivo*, but vast numbers of cells can be made from the pulp of a single extracted tooth to generate copious amounts of new dentin. Current definitions of stem cells in fact include a technological dimension in many cases (a cell that can be expanded *ex vivo* and bent to the generation of differentiating cells). Importantly, the two neighboring fields of cell therapy (reconstruction of functional tissues *in vivo* using cells) and tissue engineering proper (reconstruction of functional tissues using cells and something else) merge significantly once a stem cell angle is adopted for either, not only with respect to the ultimate goals, but also to several biotechnological aspects.

RESERVOIRS OF POSTNATAL STEM CELLS

Because of ethical constraints and for ease of harvest and control, current approaches to stem cell-based tissue engineering rely largely on the use of postnatal stem cells. Once restricted to a handful of constantly (and rapidly) self-renewing tissues, such as blood, skin, and the gastrointestinal tract, the repertoire of postnatal stem cells has expanded to include perhaps every single tissue in the body, regardless of the rate of tissue turnover or ability to regenerate (reviewed in Preston *et al.*, 2003; Wagers and Weissman, 2004). Not all of these tissue-specific stem cells, however, are equally accessible for safe harvest or available in sufficient quantity (or amenable for *ex vivo* expansion) to generate the number of cells needed for tissue regeneration. However, lessons on the dynamics of tissue homeostasis (growth and turnover) that can be learned from these cells have an obvious impact on the design of future tissue-engineering strategies none-the-less. In addition, mechanisms whereby postnatal cells maintain differentiated functions in tissues and organs are relevant to future embryonic stem cell-based approaches and can only be learned from postnatal cells (Wagers and Weissman, 2004).

Plasticity and Transdifferentiation–Implications for Tissue Engineering

The recent clamor about the potential *plasticity* (the ability of a cell to convert from one type to another) and *transdifferentiation* (conversion of a cell from one lineage to another) of certain classes of postnatal stem cells unquestionably adds further questions (Moore and Quesenberry, 2003). Rigorous proof of true plasticity and transdifferentiation have been lacking in many reports and can only be demonstrated by the ability of a single (clonogenic) cell to form cells of multiple different phenotypes and which can be shown to function as those different cell types. Some would also add that conversion must be more than an isolated phenomenon; it must occur at a high frequency and be persistent or stable (reviewed in Lakshmipathy and Verfaillie, 2005; Pauwelyn and Verfaillie, 2006). None-the-less, postnatal stem cell plasticity provides a new twist, once the perspective of usage for tissue reconstruction is directly addressed. It would be impractical, to say the least, to use neural stem cells for bone marrow transplantation; and a tissue engineer's enthusiasm for this unexpected finding may remain lukewarm. In contrast, deciding whether liver regeneration is successfully accomplished using previously unknown "hepatogenic" stem cells or transdifferentiation of hematopoietic stem cells (Lagasse *et al.*, 2000) or by reprogramming of a donor lymphocyte nucleus following cell fusion (Wang *et al.*, 2003) may not disrupt a tissue engineer's sleep, as long as the goal is met. Unquestionably, some examples of "unorthodox" differentiation of postnatal stem cells have been given (reviewed in Camargo *et al.*, 2004; Goodell, 2003; Wagers and Weissman, 2004). However, further investigation is needed to confirm not only the concept of postnatal stem cell plasticity, but specifically the technological aspects of its effective translation into clinical application, once proof of principle has been given. Even if true plasticity is a rare event, analyzing the way it occurs may provide clues as to how better to manipulate cell populations to increase its frequency.

Bone Marrow—A Source of Multiple Stem Cells

It has long been known that bone marrow is the home of at least two different types of stem cells, the hematopoietic stem cell (HSC, reviewed in Bryder *et al.*, 2006) and the skeletal stem cell (SSC, also known as the bone marrow stromal stem cell or "mesenchymal" stem cell; reviewed in Bianco and Robey, 2004), each able to reconstitute the hematopoietic and skeletal systems, respectively. Both systems are thought to be able to contribute differentiated cell types outside of their physiological progeny. Highly purified HSCs were reported to give rise to cardiomyocytes (Orlic *et al.*, 2001), as well as hepatocytes (Lagasse *et al.*, 2000) and a host of epithelial tissues (Krause *et al.*, 2001). Bone marrow stromal cells have been reported to generate functional cardiomyocytes *in vitro* (murine) (reviewed in Minguell and Erices, 2006) and to be capable of neural differentiation (reviewed in Phinney and Isakova, 2005). A rare subset of murine bone marrow cells (multipotential adult progenitor cells, MAPCs) have been reported to be almost as multipotent as ES cells (Jiang *et al.*, 2002). AC133 (CD133) positive endothelial progenitors are found in the marrow (Gehling *et al.*, 2000), and endothelial cells themselves may generate cardiomyocytes *in vitro* (Badorff *et al.*, 2003). Circulating, marrow-derived cells contribute to regeneration of skeletal muscle in response to injury (G. Ferrari *et al.*, 1998) and, in mouse models, of muscular dystrophy (Bittner *et al.*, 1999; Gussoni *et al.*, 1999). Donor-derived cells have also been detected in neuronal tissue, in newly formed vasculature, in the kidney, and even in the oral cavity following bone marrow or mobilized peripheral blood transplantation (reviewed in Poulsom *et al.*, 2002). Ideally, the identification of a single accessible site that would contain reservoirs of cells with pluri- and multipotentiality that are easily harvested in large quantities would mark a major advantage in tissue engineering. If substantiated, this wealth of observations would make bone marrow the central organ of tissue engineering and stem cell therapy (Figure 2-1). However, much more investigation is needed in order to substantiate these preliminary findings to determine the true efficacy of bone marrow for treatment of such a broad spectrum of abnormalities (reviewed in Pauwelyn and Verfaillie, 2006).

CURRENT APPROACHES TO TISSUE ENGINEERING

Tissue-engineering approaches to the regeneration of functional tissue using postnatal stem cells can be envisioned by three different scenarios: (1) expansion of a population *ex vivo* prior to transplantation into the host, (2) re-creation of a tissue or organ *ex vivo* for transplantation, and (3) design of substances and/or devices for *in vivo* activation of stem cells, either local or distant, to induce appropriate tissue repair (Figure 2-2). In all of these cases, considerable knowledge of the stem cell population's dynamics is required in order to predict and control their activity under a variety of different circumstances.

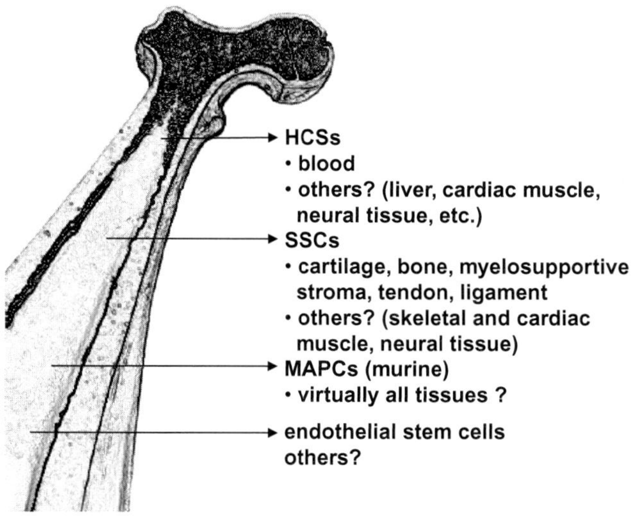

FIGURE 2-1 Bone marrow as a central source of postnatal stem cells. Bone marrow consists of at least two well-defined populations of postnatal stem cells, the hematopoietic stem cell (HSC) and the skeletal stem cell (SSC), both of which form numerous phenotypes within their cellular system but may also form cells outside of them. Multipotential adult progenitor cells (MAPCs) are a subset of bone marrow stromal cells and have been reported to form virtually all cell types in mouse. Endothelial precursors (AC133+) have recently been identified, and other types, such as a hepatocyte-like stem cell, may also exist. These remarkable findings, if verified, place bone marrow high on the list of tissues that are easily accessible in sufficient quantity for use in tissue engineering.

Ex Vivo Culture of Postnatal Stem Cells

Ex vivo expansion of tissue- or organ-specific cells, either used alone or added to carriers and scaffoldings or with growth factors at the time of transplantation, has been the primary approach in tissue engineering to date. However, *ex vivo* expansion in a fashion that maintains an appropriate proportion of stem cells within the population is a significant hurdle that must be overcome. For example, in spite of enormous effort, the culture conditions for maintaining HSCs (let alone expanding their number) are as yet undefined. It is perhaps for this very reason that currently there are only a handful of examples in which *ex vivo* expanded postnatal stem cells are used successfully to restore structure and function. The key to successful expansion will lie in understanding cell proliferation kinetics (asymmetric versus symmetric division) (Morrison and Kimble, 2006; Sherley, 2002). The efficacy achieved by the use of *ex vivo* expanded populations, whether stem or more committed in character, may also depend, at least in part, on the nature of the tissue under reconstruction. Within this context, the rate of tissue turnover most likely defines the rate of success. In tissues with a high rate of turnover, such as blood and skin, it is clear that long-term success depends on the persistence of a stem cell within the transplanted population. More committed progenitors may provide some short-term advantage; but without a self-renewing population, failure is ultimate. However, in tissues that turn over more slowly (e.g., bone), long-lasting benefit may be achieved with more committed populations of cells.

While optimizing culture conditions represents one hurdle, the issues of time and quantities represent others. In most cases, the amount of time required to generate the number of cells sufficient to repair defects induced by trauma or disease is in the order of weeks. In cases of trauma, this poses a large problem if the use of autologous stem cells is considered. For that reason, the use of allogeneic populations that could be used "off the shelf" from unrelated donors would be preferable, but this raises the issue of rejection, as in organ transplantation. Although it is suggested that some postnatal stem cells appear to escape from immune surveillance in allogeneic settings and are immunomodulatory (Le Blanc and Pittenger, 2005), definitive proof is lacking. Furthermore, differentiation of stem cells would imply the expression of a mature tissue-specific phenotype, including a complete histocompatibility profile (Fang *et al.*, 2004). Use of allogeneic cells would most likely require concomitant immunosuppressive therapy, which has its own list of side effects. Cotransplantation of allogeneic bone marrow to reconstitute the hematopoietic system has been proposed as a way of inducing tolerance (Weissman, 2005). Recent studies in organ transplantation in conjunction with bone marrow or mobilized blood transplantation after immune ablation indicate a substantial improvement in long-term survival. In this type of approach, allogeneic cells could be better envisioned, provided that a single donor is the source of cells for the tissue to be restored along with the immune system.

Delivery of Stem Cells

Delivery of bone marrow or mobilized blood by systemic infusion for restoration of the hematopoietic system by HSCs is clearly an example of the efficacy of this method of delivery, but it is perhaps the only example. The success of systemic infusion is based primarily on the fact that in terms of physical structure, blood is a simple, fluid entity. Although bone marrow transplantation has been attempted for the delivery of nonhematopoietic cells to sites of injury, it is not yet clear that there is long-term survival and engraftment of such cells (reviewed in Pauwelyn and Verfaillie, 2006). Reconstruction of two- or three-dimensional structures requires different approaches. Currently, there are a number of clinical applications, or ones soon to emerge, for the delivery of cell populations orthotopically into a site for tissue regeneration. In some cases, specific populations of cells are expanded *ex vivo* and delivered either alone or in conjunction with a natural or synthetic material. As important as it is to control the *ex vivo* behavior of the cell population, it is also important to

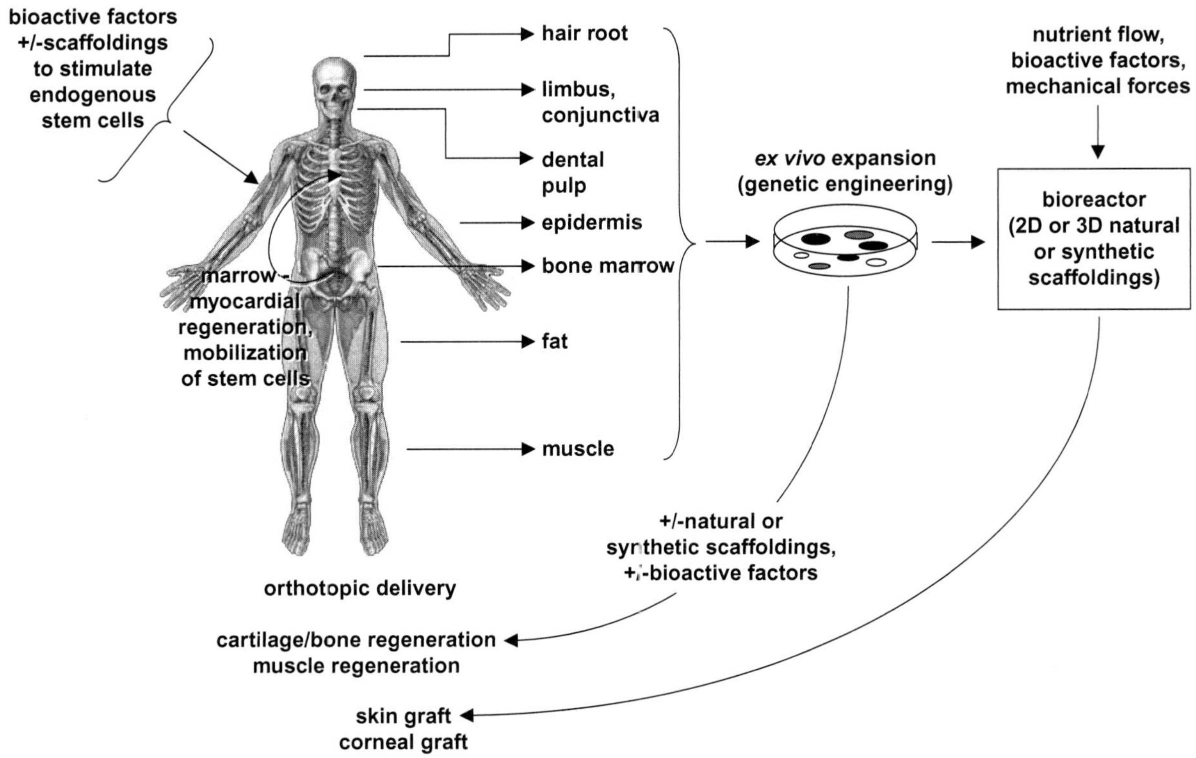

FIGURE 2-2 Current applications of postnatal stem cells in tissue engineering. While virtually all tissues in the body have cells with some regenerative capabilities, current postnatal stem cell–based strategies, or ones in the foreseeable future, rely on a relatively limited number of source tissues. Autologous bone marrow is currently in trial for myocardial regeneration. Most approaches utilize *ex vivo* expanded cell populations that are then delivered orthotopically in various combinations with bioactive factors and scaffolds, with skeletal and muscle regeneration. Generation of 2D and 3D structures *ex vivo* requires the use of bioreactors, in which cells are seeded onto scaffoldings and subjected to nutrient flow, bioactive factors, and mechanical forces to induce formation of functional tissue for transplantation.

take into consideration what the cells' response will be to the host environment. It is the host that must provide the signals that will dictate the differentiation if uncommitted populations are utilized. Furthermore, the recipient host tissue must also support the maintenance of the stem cell's niche, if not the creation of it *de novo*.

Reconstruction of the Skeletal Tissues—Bone, Cartilage, and Teeth

After numerous preclinical studies in a variety of animal models, it was demonstrated that bone marrow stromal cells, when used in conjunction with appropriate carriers, could regenerate new bone in critical-sized defects that would never heal without intervention (reviewed in Bianco and Robey, 2001; Cowan *et al.*, 2005). A wide variety of carriers were tested, ranging from collagen sponges and synthetic biodegradable polymers to synthetic hydroxyapatite derivatives, although, to date, none appear to be optimal (reviewed in El-Ghannam, 2005). A carrier that provides immediate stability, especially in the case of weight-bearing bones, and yet can be completely resorbed as new bone is formed and remodeled has not yet been fabricated. None-the-less, a small number of patients were treated with *ex vivo* expanded bone marrow stromal cells (BMSCs) in conjunction with hydroxyapatite/tricalcium phosphate ceramic particles, with good outcome (Quarto *et al.*, 2001), and other clinical studies are in progress. In addition to healing segmental defects, BMSCs were also used to construct vascularized bone flaps. In cases where morbidity of the recipient site is an issue, it can be envisioned that a new bone rudiment could be grown elsewhere in the body and then transferred, with vasculature intact (Mankani *et al.*, 2001). More recently, adipose tissue was identified as a source of stem cells that also have the potential to differentiate into bone and cartilage, at least, and may be equally as multipotential as BMSCs (reviewed in Gimble and Guilak, 2003). Naturally, if these studies are further substantiated, fat would become another prime source of easily accessible stem cells for tissue engineering (Parker and Katz, 2006).

Current cell-based therapy for the treatment of cartilage defects relies on growth of chondrocytes from a biopsy, followed by transplantation onto articular surfaces in conjunction with a periosteal flap (Brittberg *et al.*, 1994). While this procedure is practiced worldwide, long-term efficacy is questionable, and there is a clear need for better procedures. BMSCs and stem cells from adipose tissue are capable of forming cartilage in a test tube (Guilak *et al.*, 2004; Song *et al.*, 2004), but their use in the reconstruction of cartilage on articular surfaces will rely on inhibiting their further differentiation into hypertrophic chondrocytes. In addition, further development of suitable carriers that provide the appropriate microenvironment to maintain the cartilage phenotype and prevent hypertrophy is also required. Natural polymeric gels, such as hyaluronic acid, collagen, alginate, and chitosan, provide an adequate three-dimensional structure to maintain the chondrocyte phenotype, but these are not well modeled into specific shapes and have very poor biomechanical properties. For these reasons, synthetic biodegradable polymers, such as polylactic acid and polyglycolic acid, and mixtures of both that can be cross-linked and molded to form porous three-dimensional structures, are thought to provide more adequate scaffolding and have been used to construct cartilage with predefined shapes (reviewed in Frenkel and Di Cesare, 2004; Kuo *et al.*, 2006). In designing constructs of stem cells and scaffolds, the differences between the elastic cartilages of the ear and the nose and articular cartilage must also be considered to ensure that tissues form with the appropriate biomechanical characteristics.

Using techniques developed for the isolation and characterization of bone marrow stromal cells, dental pulp from the permanent dentition was found to contain a population of stem cells (dental pulp stem cells, DPSCs) that has the ability to form copious amounts of primary dentin and a pulplike complex on *in vivo* transplantation with hydroxyapatite/tricalcium phosphate ceramic particles (Gronthos *et al.*, 2000). Stem cells are also present in the pulp of deciduous teeth (SHED, stem cells from human, exfoliated, deciduous teeth) that have a high proliferative capacity, and, in addition to forming dentin, they may be able to form other cell types (Miura *et al.*, 2003). Further characterization of their differentiation potential is under way. Cells isolated from the periodontal ligament re-create a PDL-like structure and form cementum when transplanted in conjunction with hydroxyapatite/tricalcium phosphate (Seo *et al.*, 2004). Based on these findings, reconstruction of structures in the craniofacial complex can be envisioned (reviewed in Robey, 2005; Robey and Bianco, 2006).

Regeneration of Skeletal and Cardiac Muscle

Muscle contains a population of muscle-specific stem cells, the satellite cell (reviewed in Zammit *et al.*, 2006). Early attempts to use *ex vivo* expanded populations of allogeneic myoblastic cells derived from satellite cells to correct muscular dystrophies were all uniformly unsuccessful, due to immunological responses of the host to allogeneic cells (and the expression of a foreign protein by the donor cells) and the inability of donor cells to repopulate extensively (Cossu and Mavilio, 2000). In other studies, bone marrow transplantation provided a cell that could participate in the formation of new myotubes in a muscle injury model (G. Ferrari *et al.*, 1998) and in a model of muscular dystrophy (Gussoni *et al.*, 1999); however, the level of incorporation of donor cells was very low. Recently, "mesenchymal" stem cells isolated from the synovial membrane and sharing resemblances with bone marrow-derived stromal cells have been shown to repair dystrophic muscle in the mdx mouse (De Bari *et al.*, 2003). Mesoangioblasts, which are blood vessel-associated stem cells, may also be useful in skeletal muscle regeneration (Sampaolesi *et al.*, 2003).

Although skeletal muscle cells are functionally quite distinct from cardiomyocytes, autologous myoblastic cells have been viewed as a potential therapy for the treatment of myocardial infarct. Initial results suggest that there is functional improvement (Menasche *et al.*, 2003), even though the mechanisms are unclear, due to recent evidence that engrafted cells are not electromechanically linked with recipient cells (Leobon *et al.*, 2003). The utility of autologous bone marrow for myocardial regeneration is also currently receiving a great deal of attention, based on studies that attribute the ability of HSCs and BMSCs to form cardiomyocytes and the ability of marrow-derived endothelial precursors to participate in the development of new blood vessels in a number of animal models (reviewed in Itescu *et al.*, 2003). Based on these encouraging findings, transendocardial injection for the treatment of ischemic heart failure using marrow-derived cells is being tested in a number of small clinical trials. Patients receiving an autologous bone marrow transplant in conjunction with cardiac bypass surgery (Perin *et al.*, 2003) and those receiving AC133+ endothelial precursor cells alone (Stamm *et al.*, 2003) are reported to have increased cardiac perfusion and enhanced ventricular function. Interestingly, it appears that the use of many cell types results in a substantial improvement of cardiac function (Murry *et al.*, 2006). Whether they engraft at a substantial level and in fact differentiate into cardiomyocytes remains to be proven (Orlic, 2005; Stamm *et al.*, 2006). It should be considered that within this context, clinical benefit may be conveyed by a "helper" effect on putative local progenitors or by improvement of perfusion following angiogenesis induced by transplanted endothelial progenitors. Considering the high prevalence of ischemic heart disease, further testing of this application is warranted.

Ex Vivo *Reconstructions—Cells, Scaffolds, and Bioreactors*

The creation of tissues and even organs prior to transplantation is a rapidly expanding area of research. This approach relies on expanding cell populations on or in either natural or synthetic scaffolds in a bioreactor. Whether natural or synthetic, scaffolds must be biocompatable, bio-resorbable, and nonimmunogenic. Furthermore, they must also be instructive and dictate appropriate cell growth and differentiation, and they must support the recapitulation of a stem cell niche, which is essential for tissue renewal on transplantation. The most commonly used natural scaffolds include either individual purified extracellular matrix (ECM) proteins, such as collagen, fibronectin, and laminin (or peptides derived from them) or devitalized ECM from skin, sub-mucosa of the small intestine, urinary bladder, and others, and they can be autologous, allogeneic, or even xenogeneic. Such devitalized ECMs contain not only the structural proteins (which also have other important biological activities) that define the three-dimensional organization of the tissue, but also the local repertoire of growth factors that are stored within them (reviewed in Gilbert *et al.*, 2006). Various chemical and nonchemical treatments have been applied to ECMs and their components in attempts to modify their biomechanical and immunological properties on transplantation, but these result in less-than-desirable effects due to inactivation of their conducive and inductive properties and inhibition of their resorption and replacement. If left unmodified, ECMs generally promote infiltration and proliferation of cells and differentiation *in vitro* and *in vivo* and are ultimately turned over on transplantation (reviewed in Badylak, 2002). Synthetic scaffolds are designed not only to mimic the biological properties of ECMs, but also to have enhanced material properties appropriate for a particular tissue. Polyglycolic acids, polyhyroxyalkonates, and hydrogels are the most common examples, and all can be manipulated to form a broad range of structures with varying degrees of porosity and rigidity (Hollister, 2005). Scaffolds can also be formulated to include morphogenetic and growth factors or even naked plasmid DNA to transduce ingrowing cells to produce these important factors, such that they can become as instructive as ECMs (reviewed in Hench and Polak, 2002).

In addition to having appropriate scaffolds, the ability to construct tissues and organs *ex vivo* depends on the design of appropriate bioreactors. The initial designs provided primarily nutrient flow through the developing tissue. However, many tissues grown under these circumstances fail to achieve the biomechanical properties required for tissue function *in vivo*. For example, the construction of blood vessels using a variety of biomaterials and cell populations using culture perfusion yields structures that histologically resemble native blood vessels. Yet when transplanted, these constructs failed, due to their inability to withstand changes in pressure. A substantial improvement in such constructs was achieved by subjecting the developing tissues to pulsatile flow conditions (reviewed in Levenberg, 2005). Perfusion-type bioreactors support the growth of tissue with only a nominal number of cell layers and are amenable to construction of relatively simple structures. Construction of larger, more complex three-dimensional structures requires the development of even better bioreactors that will support cell survival to achieve a significant cell mass (reviewed in Martin *et al.*, 2004). Another critical aspect of using large constructs is the development of supporting vasculature. While constructs can be placed in vascular beds in some cases, it will be essential to design constructs that rapidly induce vascularization, perhaps by including the angiogenic factor, VEGF, or cells that produce it or even by including endothelial cells and their precursors, which tend to assemble themselves into primitive tubular structures that may allow for more rapid establishment of organ perfusion (reviewed in Griffith and Naughton, 2002).

Next to bone marrow transplantation, skin grafts perhaps provide the best example of the impact of postnatal stem cell biology on the success of cell-based tissue reconstruction. Although the use of *ex vivo* expanded keratinocytes for the generation of skin grafts was first brought to a clinical setting by the pioneering work of Green and coworkers (Green, 1989), persistent engraftment has not been routine. It is now recognized that both the expansion culture conditions in which the cells are grown and the scaffolding on which they are placed must support self-renewal of epidermal stem cells (cells that give rise to holoclones, as opposed to transiently amplifying meroclones or more differentiated paraclones) (Barrandon and Green, 1987). Expansion of keratinocyte populations on fibrin substrates was found to maintain the stem cell population and to improve long-term maintenance of skin grafts (Pellegrini *et al.*, 1999). A number of studies are now under way to molecularly engineer autologous epidermal stem cells to correct a gene defect causing severe blistering prior to generation of skin grafts (reviewed in S. Ferrari *et al.*, 2006). Currently, a number of commercial products are available (both autologous and allogeneic), some of which incorporate dermal fibroblasts in a collagen gel, which is then layered with *ex vivo* expanded epidermal cells (reviewed in Supp and Boyce, 2005). They provide coverage of severe wounds and are maintained while the recipient generates new skin. The standard starting material for establishing epidermal cultures is skin. However, recent evidence indicates that a stem cell located in the bulge region of the root sheet of hair follicles has the ability to form not only new follicles and sebaceous glands, but also interfollicular epidermis under certain circumstances (reviewed in Blanpain and Fuchs, 2005). These cells, which can be obtained noninvasively, are highly proliferative and have recently been used successfully to treat leg ulcers (Limat and Hunziker, 2002). Based on lessons learned from skin regeneration, stem cells in the limbus of the sclera and conjunctiva have been expanded in culture on fibrin substrates to generate sheets of epithelial cells able to reconstitute a damaged corneal surface, with dramatic beneficial clinical results (reviewed in Vascotto and Griffith, 2006).

Activation of Local and Distant Endogenous Stem Cells

While the *ex vivo* reconstruction of entire organs that are functional is the goal of the bioengineering world, what perhaps is even more challenging is the goal of inducing endogenous stem cells to become activated to reconstruct a tissue. Most tissues in the body display at least some sort of regenerative capability (Preston *et al.*, 2003; Tsonis, 2002) but in many cases remain insufficient to mount a spontaneous and successful repair response *in vivo*. The application or induction of growth factors and morphogens, perhaps in combination with appropriate scaffolds, might be envisioned to "encourage" local or distant progenitors to regenerate a functional tissue. This *in situ* process could occur through several different pathways, including activation of local or distant stem cells, transdifferentiation, or reprogramming, to generate adequate numbers of committed precursors. In all cases, a regulated morphogenetic process is needed to establish normal structure and function. Definition of the intrinsic properties of regenerative cells and extrinsic signals that may trigger a recapitulation of developmental processes is critical.

Local Cells

Based on reasoning that, following an injury, signals that normally activate a local cell to initiate repair would be either obliterated or not of a high enough magnitude, numerous studies attempted to provide appropriate morphogenetic and growth factors. However, this approach has been uniformly disappointing, most likely due to the short half-life of such soluble factors and the lack of an appropriate scaffolding. Not only does scaffolding provide a template for the organized outgrowth of local cells, but it also provides a substrate on which to stabilize and/or orient factors (or parts thereof) for appropriate presentation to a cell (reviewed in Lutolf and Hubbell, 2005; Rosso *et al.*, 2005). However, in using "smart" constructs, tissue regeneration of large defects may not be complete, due to exhaustion of the local stem cell population; hence the need to maintain a stem cell niche. Furthermore, some clues as to the negative impact of local constraints on the native, albeit incomplete, repair capacity of cells and tissues may be found in studies of spinal cord regeneration. Following injury, there is a phase characterized by neurite outgrowth from the severed end. However, it is short-lived due to the production of inhibitory factors by cells within the myelin sheath, and gliosis impedes neurite outgrowth (reviewed in Di Giovanni, 2006; Schwab, 2002). Thus, regeneration by endogenous cells must take into consideration not only the bioactive factors and scaffolding that are necessary to maintain an ingrowing population with the appropriate balance of stem cells, transiently amplifying cells, and differentiating cells, but also the inactivation of local inhibitory factors that work against the process.

Mobilization and Recruitment of Distant Stem Cells

Numerous studies have suggested that once introduced into the circulation, many types of cells have the propensity to end up at a diseased or injured site and to participate in local regeneration (Stocum, 2001), and it has been argued that HSCs might generate a broad variety of nonhematopoietic cell types. Ongoing tissue damage and repair, at the time when systemically infused stem cells reach a target organ, may be the critical determinant for triggering their homing, engraftment, and differentiation to local cell types.

Cytokine administration has become a well-established procedure for mobilization of HSCs into peripheral blood (reviewed in Winkler and Levesque, 2006), but it is not clear that other stem cell populations present in marrow would be equally as amenable to liberation by current procedures or whether different ones would have to be developed. Consequently, mobilization of marrow stem cells might provide a mechanism to enhance a local population and improve tissue regeneration, a hypothesis that might be tested in the context of extant local, tissue-specific damage and repair.

CONCLUSIONS

The recent advances in embryonic and postnatal stem cell biology have most certainly captured the public's attention and have raised expectations for miraculous cures in the near future. While certain applications that utilize postnatal stem cells are in practice or will be shortly, the majority are in their infancy and will take much more effort. However, the current sense of urgency in translating recent findings into clinical applications should not lead this field to repeat the mistakes experienced in the field of gene therapy. The development of tissue engineering as a viable medical practice must proceed in an evidence-based fashion in all of the associated disciplines that are involved, and several hurdles are yet to be overcome. First, our current understanding of postnatal stem cell biology is rudimentary, at best. The manner in which we isolate stem cells and manipulate them and their progeny appropriately relies heavily on a clear understanding of cell population kinetics. This also requires a complete understanding of their response not only to bio-active factors, scaffoldings, and delivery systems, but also to the host microenvironment in which they must survive and function. Second, more efficient bioreactors that adequately model the microenvironment and also allow for scale-up of the tissue-engineering process must be developed. Third, in order to translate what we learn into clinical application, development of appropriate preclinical models to prove the principle that stem cells do indeed have a positive biological impact is absolutely essential. In analyzing these models, stringent criteria must be defined to determine efficacy, and we must remain principled in assessing them. Exciting times, yes, but also a time for due diligence to bring what started off as a scientific curiosity into medical reality.

ACKNOWLEDGMENTS

The support of Telethon Fondazione Onlus Grant E1029 (to P.B.) and of the DIR, NIDCR of the IRP, NIH (P.G.R.) is gratefully acknowledged.

REFERENCES

Badorff, C., Brandes, R. P., Popp, R., Rupp, S., Urbich, C., Aicher, A., Fleming, I., Busse, R., Zeiher, A. M., and Dimmeler, S. (2003). Transdifferentiation of blood-derived human adult endothelial progenitor cells into functionally active cardiomyocytes. *Circulation 107*, 1024–1032.

Badylak, S. F. (2002). The extracellular matrix as a scaffold for tissue reconstruction. *Semin. Cell Dev. Biol. 13*, 377–383.

Barrandon, Y., and Green, H. (1987). Three clonal types of keratinocyte with different capacities for multiplication. *Proc. Natl. Acad. Sci. U.S.A. 84*, 2302–2306.

Bianco, P., and Robey, P. G. (2001). Stem cells in tissue engineering. *Nature 414*, 118–121.

Bianco, P., and Robey, P. G. (2004). Skeletal stem cells. *In* "Handbook of Adult and Fetal Stem Cells" (R. P. Lanza, ed.), pp. 415–424. Academic Press, San Diego.

Bittner, R. E., Schofer, C., Weipoltshammer. K., Ivanova, S., Streubel, B., Hauser, E., Freilinger, M., Hoger, H., Elbe-Burger, A., and Wachtler, F. (1999). Recruitment of bone-marrow-derived cells by skeletal and cardiac muscle in adult dystrophic mdx mice. *Anat. Embryol. (Berl.) 199*, 391–396.

Blanpain, C., and Fuchs, E. (2006). Epidermal stem cells of the skin. *Annu. Rev. Cell Dev. Biol. 22*, 339–373.

Brittberg, M., Lindahl, A., Nilsson, A., Ohlsson, C., Isaksson, O., and Peterson, L. (1994). Treatment of deep cartilage defects in the knee with autologous chondrocyte transplantation. *N. Engl. J. Med. 331*, 889–895.

Bryder, D., Rossi, D. J., and Weissman, I. L. (2006). Hematopoietic stem cells: the paradigmatic tissue-specific stem cell. *Am. J. Pathol. 169*, 338–346.

Camargo, F. D., Chambers, S. M., and Goodell, M. A. (2004). Stem cell plasticity: from transdifferentiation to macrophage fusion. *Cell Prolif. 37*, 55–65.

Cossu, G., and Mavilio, F. (2000). Myogenic stem cells for the therapy of primary myopathies: wishful thinking or therapeutic perspective? *J. Clin. Invest. 105*, 1669–1674.

Cowan, C. M., Soo, C., Ting, K., and Wu, B. (2005). Evolving concepts in bone tissue engineering. *Curr. Top. Dev. Biol. 66*, 239–285.

De Bari, C., Dell'Accio, F., Vandenabeele, F., Vermeesch, J. R., Raymackers, J. M., and Luyten, F. P. (2003). Skeletal muscle repair by adult human mesenchymal stem cells from synovial membrane. *J. Cell Biol. 160*, 909–918.

Di Giovanni, S. (2006). Regeneration following spinal cord injury, from experimental models to humans: where are we? *Expert Opin. Ther. Targets 10*, 363–376.

El-Ghannam, A. (2005). Bone reconstruction: from bioceramics to tissue engineering. *Expert Rev. Med. Devices 2*, 87–101.

Fang, T. C., Alison, M. R., Wright, N. A., and Poulsom, R. (2004). Adult stem cell plasticity: will engineered tissues be rejected? *Int. J. Exp. Pathol. 85*, 115–124.

Ferrari, G., Cusella-De Angelis, G., Coletta, M., Paolucci, E., Stornaiuolo, A., Cossu, G., and Mavilio, F. (1998). Muscle regeneration by bone marrow-derived myogenic progenitors. *Science 279*, 1528–1530.

Ferrari, S., Pellegrini, G., Matsui, T., Mavilio, F., and De Luca, M. (2006). Gene therapy in combination with tissue engineering to treat epidermolysis bullosa. *Expert Opin. Biol. Ther. 6*, 367–378.

Frenkel, S. R., and Di Cesare, P. E. (2004). Scaffolds for articular cartilage repair. *Ann. Biomed. Eng. 32*, 26–34.

Gehling, U. M., Ergun, S., Schumacher, U., Wagener, C., Pantel, K., Otte, M., Schuch, G., Schafhausen, P., Mende, T., Kilic, N., *et al.* (2000). *In vitro* differentiation of endothelial cells from AC133-positive progenitor cells. *Blood 95*, 3106–3112.

Gilbert, T. W., Sellaro, T. L., and Badylak, S. F. (2006). Decellularization of tissues and organs. *Biomaterials 27*, 3675–3683.

Gimble, J. M., and Guilak, F. (2003). Differentiation potential of adipose-derived adult stem (ADAS) cells. *Curr. Top. Dev. Biol. 58*, 137–160.

Goodell, M. A. (2003). Stem-cell "plasticity": befuddled by the muddle. *Curr. Opin. Hematol. 10*, 208–213.

Green, H. (1989). Regeneration of the skin after grafting of epidermal cultures. *Lab. Invest. 60*, 583–584.

Griffith, L. G., and Naughton, G. (2002). Tissue engineering—current challenges and expanding opportunities. *Science 295*, 1009–1014.

Gronthos, S., Mankani, M., Brahim, J., Robey, P. G., and Shi, S. (2000). Postnatal human dental pulp stem cells (DPSCs) *in vitro* and *in vivo*. *Proc. Natl. Acad. Sci. U.S.A. 97*, 13625–13630.

Guilak, F., Awad, H. A., Fermor, B., Leddy, H. A., and Gimble, J. M. (2004). Adipose-derived adult stem cells for cartilage tissue engineering. *Biorheology 41*, 389–399.

Gussoni, E., Soneoka, Y., Strickland, C. D., Buzney, E. A., Khan, M. K., Flint, A. F., Kunkel, L. M., and Mulligan, R. C. (1999). Dystrophin expression in the mdx mouse restored by stem cell transplantation. *Nature 401*, 390–394.

Hench, L. L., and Polak, J. M. (2002). Third-generation biomedical materials. *Science 295*, 1014–1017.

Hollister, S. J. (2005). Porous scaffold design for tissue engineering. *Nat. Mater. 4*, 518–524.

Itescu, S., Schuster, M. D., and Kocher, A. A. (2003). New directions in strategies using cell therapy for heart disease. *J. Mol. Med. 81*, 288–296.

Jiang, Q. J., Izakovic, J., Zenker, M., Fartasch, M., Meneguzzi, G., Rascher, W., and Schneider, H. (2002). Treatment of two patients with Herlitz junctional epidermolysis bullosa with artificial skin bioequivalents. *J. Pediatr. 141*, 553–559.

Krause, D. S., Theise, N. D., Collector, M. I., Henegariu, O., Hwang, S., Gardner, R., Neutzel, S., and Sharkis, S. J. (2001). Multi-organ, multi-lineage engraftment by a single bone marrow-derived stem cell. *Cell 105*, 369–377.

Kuo, C. K., Li, W. J., Mauck, R. L., and Tuan, R. S. (2006). Cartilage tissue engineering: its potential and uses. *Curr. Opin. Rheumatol. 18*, 64–73.

Lagasse, E., Connors, H., Al-Dhalimy, M., Reitsma, M., Dohse, M., Osborne, L., Wang, X., Finegold, M., Weissman, I. L., and Grompe, M. (2000). Purified hematopoietic stem cells can differentiate into hepatocytes *in vivo*. *Nat. Med. 6*, 1229–1234.

Lakshmipathy, U., and Verfaillie, C. (2005). Stem cell plasticity. *Blood Rev. 19*, 29–38.

Le Blanc, K., and Pittenger, M. (2005). Mesenchymal stem cells: progress toward promise. *Cytotherapy 7*, 36–45.

Leobon, B., Garcin, I., Menasche, P., Vilquin, J. T., Audinat, E., and Charpak, S. (2003). Myoblasts transplanted into rat infarcted myocardium are functionally isolated from their host. *Proc. Natl. Acad. Sci. U.S.A. 100*, 7808–7811.

Levenberg, S. (2005). Engineering blood vessels from stem cells: recent advances and applications. *Curr. Opin. Biotechnol. 16*, 516–523.

Limat, A., and Hunziker, T. (2002). Use of epidermal equivalents generated from follicular outer root sheath cells *in vitro* and for autologous grafting of chronic wounds. *Cells Tissues Organs 172*, 79–85.

Lutolf, M. P., and Hubbell, J. A. (2005). Synthetic biomaterials as instructive extracellular microenvironments for morphogenesis in tissue engineering. *Nat. Biotechnol. 23*, 47–55.

Mankani, M. H., Krebsbach, P. H., Satomura, K., Kuznetsov, S. A., Hoyt, R., and Robey, P. G. (2001). Pedicled bone flap formation using transplanted bone marrow stromal cells. *Arch. Surg. 136*, 263–270.

Martin, I., Wendt, D., and Heberer, M. (2004). The role of bioreactors in tissue engineering. *Trends Biotechnol. 22*, 80–86.

Menasche, P., Hagege, A. A., Vilquin, J. T., Desnos, M., Abergel, E., Pouzet, B., Bel, A., Sarateanu, S., Scorsin, M., Schwartz, K., *et al.* (2003). Autologous skeletal myoblast transplantation for severe postinfarction left ventricular dysfunction. *J. Am. Coll. Cardiol. 41*, 1078–1083.

Minguell, J. J., and Erices, A. (2006). Mesenchymal stem cells and the treatment of cardiac disease. *Exp. Biol. Med. (Maywood) 231*, 39–49.

Miura, M., Gronthos, S., Zhao, M., Lu, B., Fisher, L. W., Robey, P. G., and Shi, S. (2003). SHED: stem cells from human exfoliated deciduous teeth. *Proc. Natl. Acad. Sci. U.S.A. 100*, 5807–5812.

Moore, B. E., and Quesenberry, P. J. (2003). The adult hemopoietic stem cell plasticity debate: idols vs. new paradigms. *Leukemia 17*, 1205–1210.

Morrison, S. J., and Kimble, J. (2006). Asymmetric and symmetric stem-cell divisions in development and cancer. *Nature 441*, 1068–1074.

Murry, C. E., Reinecke, H., and Pabon, L. M. (2006). Regeneration gaps: observations on stem cells and cardiac repair. *J. Am. Coll. Cardiol. 47*, 1777–1785.

Orlic, D. (2005). BM stem cells and cardiac repair: where do we stand in 2004? *Cytotherapy 7*, 3–15.

Orlic, D., Kajstura, J., Chimenti, S., Limana, F., Jakoniuk, I., Quaini, F., Nadal-Ginard, B., Bodine, D. M., Leri, A., and Anversa, P. (2001). Mobilized bone marrow cells repair the infarcted heart, improving function and survival. *Proc. Natl. Acad. Sci. U.S.A. 98*, 10344–10349.

Parker, A. M., and Katz, A. J. (2006). Adipose-derived stem cells for the regeneration of damaged tissues. *Expert Opin. Biol. Ther. 6*, 567–578.

Pauwelyn, K. A., and Verfaillie, C. M. (2006). 7. Transplantation of undifferentiated, bone marrow-derived stem cells. *Curr. Top. Dev. Biol. 74*, 201–251.

Pellegrini, G., Ranno, R., Stracuzzi, G., Bondanza, S., Guerra, L., Zambruno, G., Micali, G., and De Luca, M. (1999). The control of epidermal stem cells (holoclones) in the treatment of massive full-thickness burns with autologous keratinocytes cultured on fibrin. *Transplantation 68*, 868–879.

Perin, E. C., Dohmann, H. F., Borojevic, R., Silva, S. A., Sousa, A. L., Mesquita, C. T., Rossi, M. I., Carvalho, A. C., Dutra, H. S., Dohmann, H. J., *et al.* (2003). Transendocardial, autologous bone marrow cell transplantation for severe, chronic ischemic heart failure. *Circulation 107*, 2294–2302.

Phinney, D. G., and Isakova, I. (2005). Plasticity and therapeutic potential of mesenchymal stem cells in the nervous system. *Curr. Pharm. Des. 11*, 1255–1265.

Poulsom, R., Alison, M. R., Forbes, S. J., and Wright, N. A. (2002). Adult stem cell plasticity. *J. Pathol. 197*, 441–456.

Preston, S. L., Alison, M. R., Forbes, S. J., Direkze, N. C., Poulsom, R., and Wright, N. A. (2003). The new stem cell biology: something for everyone. *Mol. Pathol. 56*, 86–96.

Quarto, R., Mastrogiacomo, M., Cancedda, R., Kutepov, S. M., Mukhachev, V., Lavroukov, A., Kon, E., and Marcacci, M. (2001). Repair of large bone defects with the use of autologous bone marrow stromal cells. *N. Engl. J. Med. 344*, 385–386.

Robey, P. G. (2005). Postnatal stem cells for dental and craniofacial repair. *Oral Biosci. Med. 2*, 83–90.

Robey, P. G., and Bianco, P. (2006). The use of adult stem cells in rebuilding the human face. *J. Am. Dent. Assoc. 137*, 961–972.

Rosso, F., Marino, G., Giordano, A., Barbarisi, M., Parmeggiani, D., and Barbarisi, A. (2005). Smart materials as scaffolds for tissue engineering. *J. Cell Physiol. 203*, 465–470.

Sampaolesi, M., Torrente, Y., Innocenzi, A., Tonlorenzi, R., D'Antona, G., Pellegrino, M. A., Barresi, R., Bresolin, N., De Angelis, M. G., Campbell, K. P., *et al.* (2003). Cell therapy of alpha-sarcoglycan null dystrophic mice through intra-arterial delivery of mesoangioblasts. *Science 301*, 487–492.

Schwab, M. E. (2002). Repairing the injured spinal cord. *Science 295*, 1029–1031.

Seo, B. M., Miura, M., Gronthos, S., Bartold, P. M., Batouli, S., Brahim, J., Young, M., Robey, P. G., Wang, C. Y., and Shi, S. (2004). Investigation of multipotent postnatal stem cells from human periodontal ligament. *Lancet 364*, 149–155.

Sherley, J. L. (2002). Asymmetric cell kinetics genes: the key to expansion of adult stem cells in culture. *Stem Cells 20*, 561–572.

Song, L., Baksh, D., and Tuan, R. S. (2004). Mesenchymal stem cell-based cartilage tissue engineering: cells, scaffold and biology. *Cytotherapy 6*, 596–601.

Stamm, C., Westphal, B., Kleine, H. D., Petzsch, M., Kittner, C., Klinge, H., Schumichen, C., Nienaber, C. A., Freund, M., and Steinhoff, G. (2003). Autologous bone-marrow stem-cell transplantation for myocardial regeneration. *Lancet 361*, 45–46.

Stamm, C., Liebold, A., Steinhoff, G., and Strunk, D. (2006). Stem cell therapy for ischemic heart disease: beginning or end of the road? *Cell Transplant 15*(Suppl. 1), S47–S56.

Stocum, D. L. (2001). Stem cells in regenerative biology and medicine. *Wound Repair Regen. 9*, 429–442.

Supp, D. M., and Boyce, S. T. (2005). Engineered skin substitutes: practices and potentials. *Clin. Dermatol. 23*, 403–412.

Tsonis, P. A. (2002). Regenerative biology: the emerging field of tissue repair and restoration. *Differentiation 70*, 397–409.

Vascotto, S. G., and Griffith, M. (2006). Localization of candidate stem and progenitor cell markers within the human cornea, limbus, and bulbar conjunctiva *in vivo* and in cell culture. *Anat. Rec. A Discov. Mol. Cell. Evol. Biol. 288*, 921–931.

Wagers, A. J., and Weissman, I. L. (2004). Plasticity of adult stem cells. *Cell 116*, 639–648.

Wang, X., Willenbring, H., Akkari, Y., Torimaru, Y., Foster, M., Al-Dhalimy, M., Lagasse, E., Finegold, M., Olson, S., and Grompe, M. (2003). Cell fusion is the principal source of bone-marrow-derived hepatocytes. *Nature 422*, 897–901.

Weissman, I. (2005). Stem cell research: paths to cancer therapies and regenerative medicine. *JAMA 294*, 1359–1366.

Winkler, I. G., and Levesque, J. P. (2006). Mechanisms of hematopoietic stem cell mobilization: when innate immunity assails the cells that make blood and bone. *Exp. Hematol. 34*, 996–1009.

Zammit, P. S., Partridge, T. A., and Yablonka-Reuveni, Z. (2006). The skeletal muscle satellite cell: the stem cell that came in from the cold. *J. Histochem. Cytochem. 54*, 1177–1191.

Chapter 3

Adult Epithelial Tissue Stem Cells

Christopher S. Potten and James W. Wilson

INTRODUCTION

Stem cell concepts have evolved dramatically over the last few years from the simple ideas in the literature in the mid-20th century. This has culminated in a rapid expansion of interest in both embryonic and adult tissue stem cells in the last five years with the development of interest in gene therapy, tissue engineering, and stem cell therapies. This chapter explores the evolution of stem cell concepts as applied to adult epithelial tissues. These tissues are characterized by a high degree of polarization and very distinct cell maturation and migration pathways, which permit the identification of specific locations in the tissues that represent the origins of all this cell movement. Cells at the origin of the migratory pathways must represent the cells on which the tissue is ultimately dependent and the cells that have a long-term (permanent) residence in the tissue, i.e., the stem cells. A variety of cell kinetic studies, together with lineage-tracking experiments, have indicated that in the intestine, the dorsal surface of the tongue, and interfollicular epidermis, the proliferative compartment of the tissue is divided into discrete units of proliferation. Each unit has its own stem cell compartment and adjacent family of stem cell progeny, including dividing transit cells and differentiated functional cells. In the skin the evolving stem cell studies suggest at least three distinct stem cell populations, providing a source of cells for the epidermis, for the growing hair follicle, and a reserve regenerative, highly potent population in the upper follicle region. In the small intestine there are indications that the stem cell compartment itself is hierarchical, with a commitment to differentiation occurring two to three generations down the lineage, resulting in a population of actual stem cells that perform their function in steady state and a population of potential stem cells that can be called into action if the actual stem cells are killed. Until recently there have been no reliable markers for adult intestinal stem cells. However, recent developments have indicated ways in which these cells may be identified. Cancer is rare in the small intestinal epithelium, which is surprising, since this tissue represents a large mass, with many stem cells dividing many times. This suggests that effective genome protective mechanisms have evolved, and some aspects of these have now been identified.

In the 1950s and '60s, all proliferating cells in the renewing tissues of the body were regarded as a having an equal potential to self-maintain, with one daughter cell (on average) from each division of a proliferative cell being retained within the proliferative compartment. Thus all proliferating cells were regarded as stem cells. It proved somewhat difficult to displace this concept. However, a groundbreaking paper by Till and McCulloch (1961) provided the first clear evidence for one of the replacing tissues of the body, the bone marrow, that not all proliferative cells are identical. The approach they employed was to study the cells that were capable of repopulating hemopoietic tissues, following cellular depletion of the tissue by exposure to a cytotoxic agent, i.e., radiation. Specifically, mice were irradiated to deplete their bone marrow of endogenous, functional hematopoietic precursors; then they were injected with bone marrow-derived precursors obtained from another animal. The exogenous cells were subject to a variety of treatments prior to transplant, including irradiation. It was found that the hemopoietic precursors circulated in the host and seeded cells into various hemopoietic tissues, including the spleen. Those cells that seeded into the spleen and possessed extensive regenerative and differentiative potential grew by a process of clonal expansion to form macroscopically visible nodules of haemopoietic tissue 10–14 days after transplant. By appropriate genetic or chromosome tracking (marking), it could be shown that these nodules were derived from single cells, i.e., were clones, and that further clonogenic cells were produced within the clones. The colonies were referred to as *spleen colonies*, and the cells that formed the colonies were called *colony-forming units (spleen)* (CFU-s). These experiments provided the theoretical basis for subsequent human bone marrow transplant studies. By a variety of preirradiation manipulations and pre- and posttransplantation variables, this technique led to our current understanding of the bone marrow hierarchies, or cell lineages, and their stem cells. These studies showed that

Stem Cell Anthology

this tissue contained undifferentiated self-maintaining precursor cells that generated dependent lineages able to differentiate down a range of different pathways, generating a variety of cell types. Recent studies have suggested that these CFU-s are not the ultimate hemopoietic stem cells but are part of a stem cell hierarchy in the bone marrow.

Such clonal regeneration approaches have subsequently been developed for a variety of other tissues, notably by the imaginative approaches adopted by Rod Withers for epidermis, intestine, kidney, and testis. These clonal regeneration approaches were summarized and collected in a book produced in 1985 (Potten and Hendry, 1985), but readers are also specifically referred to Withers and Elkind (1970), which deals with the gut, and Withers (1967), for studies on the epidermis. These approaches implicated hierarchical organizations within the proliferative compartments of many tissues (Potten and Hendry, 1985). The stringency of the criteria defining a clone varied enormously, depending as it did on the number of cell divisions required to produce the detectable clones. For epidermis and intestine, the stringency was high, since the clones could be large and macroscopic, containing many cells resulting from many cell divisions, and in fact were very similar in appearance to the spleen colony nodules.

One difficulty with the interpretation of clonal regeneration studies and the generality of their application to stem cell populations is that in order to see the regenerating clones, the tissue has to be disturbed, generally by exposure to a dose of radiation, and this disturbance may alter the cellular hierarchies that one wishes to study and will certainly alter the nature (e.g., cell cycle status, responsiveness to signals, susceptibility to subsequent treatment) of the stem cell compartment. This has been referred to as the biological equivalent of the Heisenberg uncertainty principle in quantum physics. However, these clonal regeneration assays do still provide a valuable and, in some places, unique opportunity to study some aspects of stem cell biology *in vivo*, i.e., by using this approach to look at stem cell survival and functional competence under a variety of conditions.

A DEFINITION OF STEM CELLS

There have been relatively few attempts to define what is meant by the term *stem cells*, which has resulted in some confusion in the literature and the use of a variety of terms, the relationship between which sometimes remains obscure and confusing. These include *precursors*, *progenitors*, and *founder cells*. The concept is further complicated by the use of terms such as *committed precursors* and *progenitors* and the sometimes confusing use, or implication, of the term *differentiation*. One of the difficulties in defining stem cells is that the definitions are often very context dependent; hence, different criteria are brought into the definition by embryologists, haematologists, dermatologists, gastroen-terologists, etc. In 1990, in a paper in *Development* (Potten and Loeffler, 1990), we attempted to define a stem cell. This definition was, admittedly, formulated within the context of the gastrointestinal epithelium, but we felt it had a broader application. The definition still largely holds and can be summarized as follows. Within adult replacing tissues of the body, the stem cells can be defined as a small subpopulation of the proliferating compartment, consisting of relatively undifferentiated proliferative cells that maintain their population size when they divide while, at the same time, producing progeny that enter a dividing transit population, within which further rounds of cell division occur together with differentiation events, resulting in the production of the various differentiated functional cells required of the tissue. The stem cells persist in the tissue throughout the animal's lifetime, dividing a large number of times, and, as a probable consequence of this large division potential, these cells are the most efficient repopulators of the tissue following injury. If this repopulation requires a reestablishment of the full stem cell compartment, the *self-maintenance probability* of the stem cells at division will be raised from the steady-state value of 0.5 to a value between 0.5 and 1, which enables the stem cell population to be reestablished while at the same time maintaining the production of differentiated cells to ensure the functional integrity of the tissue.

The consequences of this definition are obvious, namely that stem cells are:

- Rare cells in the tissue, vastly outnumbered by the dividing transit population, and are the cells on which the entire lineage and ultimately the tissue are dependent.
- The only permanent long-term residents of the tissue.
- Cells at the origin of any cell lineages or migratory pathways that can be identified in the tissue, producing progeny that differentiate.
- Cells that in steady state have a self-maintenance probability of 0.5, but this can be changed according to requirements.

The concept of *differentiation* enters into the definition of stem cells, and this also often leads to confusion. In our view, differentiation is a qualitative and relative phenomenon. Cells tend to be differentiated relative to other cells, hence, adult tissue stem cells may, or may not, be differentiated relative to embryonic stem cells (a point of current debate, bearing in mind the controversy in the literature concerning bone marrow stem cell *plasticity*). Stem cells produce progeny that may differentiate down a variety of pathways, leading to the concept of *totipotency* and *pluripotency* or *multipotency* of stem cells, in terms of their differentiation. This is actually a strange concept to apply to a stem cell since it is their progeny that differentiate and not the stem cells themselves. The fact is that the progeny can differentiate down more than one differentiated lineage, as is very obviously the case in the bone marrow, resulting in bone marrow stem cells being referred to as pluripotent; the initial dividing transit cells that initiate a lineage, which ultimately leads to specific differentiated cells, can be thought of as committed precursors for that lineage.

Some of the instructive signals for differentiation in the hemopoietic cell lineage are now well understood, but such signals for other tissues organized on a cell lineage basis have yet to be determined. There is much debate in the literature at present concerning the extent to which stem cells may be instructed to produce progeny of specific differentiated types and whether this is limited or unlimited. This topic is referred to as the *degree of plasticity* for stem cells. There are two very distinct issues here:

- The first is whether a stem cell, like a bone marrow stem cell, is ever instructed by its environment or niche naturally, or in laboratory or clinical situations, to make an apparently unrelated tissue cell type, such as a liver, intestinal, or skin cell, and whether it can regenerate these tissues if they are injured. A subsidiary question is not whether this ever happens normally in nature, but whether we, as experimentalists or clinicians, can provide the necessary instructions or environment for this to happen in a controlled situation.
- The second issue relates not only to the stem cells but also to the early progeny of stem cells from, for example, the bone marrow and whether these cells that circulate round the body and may end up in a distant tissue can end up expressing differentiation markers unrelated to the bone marrow cell lineages but specific to the tissue in which the cell then resides.

The former is an issue of plasticity of the bone marrow stem cells, and the latter may be more an issue of the plasticity of the bone marrow-derived cell lineages. If a bone marrow stem cell can ever be instructed to be a gastrointestinal stem cell, it should be capable of undertaking all the functional duties of a gastrointestinal stem cell, including the regeneration of the gastrointestinal epithelium if it is subsequently injured. The cloning of animals by nuclear transfer technology into egg cytoplasm clearly demonstrates that all nuclei of the body contain a full complement of DNA and that, under the right environmental conditions, this can be reprogrammed (or unmasked) by environmental signals to make all the tissues of the body. It must be remembered that such cloning experiments, e.g., Dolly the sheep, are rare and inefficiently produced events. They do, however, clearly indicate the enormous potential that can be achieved if we provide the necessary instructive reprogramming signals. This should enable us in the future to reproducibly instruct any adult tissue stem cell to make any tissue of the body. If and when this becomes the case, the distinction between embryonic stem cells and adult tissue stem cells may disappear.

HIERARCHICALLY ORGANIZED STEM CELL POPULATIONS

The issue here is what determines the difference between a dividing transit cell and a stem cell and whether that transition is an abrupt one or a gradual one. One can think of this transition as being a differentiation event that distinguishes a dividing transit cell from a stem cell. This is an old argument. Do differentiation signals act on preexisting stem cells, removing on average half the cells produced by previous symmetric divisions, or do the stem cells divide asymmetrically to produce a differentiated progeny at division and a stem cell? One possibility is that this distinction is made at the time that a stem cell divides. Indeed, does it need to divide to differentiate? In this case, such divisions must be regarded as asymmetric, with the dividing stem cell producing one stem cell (i.e., for self-maintenance) and one dividing transit cell. This type of asymmetric division may occur in some tissues, such as the epidermis. However, if this is the case, the stem cell must also retain the potential to alter its self-maintenance probability, which for an asymmetric division is 0.5 in steady state, and adopt a value somewhat higher than this if stem cells are killed and need to be repopulated. The current view regarding the bone marrow stem cells is that the transition between a stem cell and a dividing transit cell is a gradual one that occurs over a series of divisions within a cell lineage, which inevitably implies that one has a population of stem cells with a varying degree of stemness or, conversely, a varying degree of differentiation. For the bone marrow, one issue is whether experimentalists have ever identified the presence of the truly ancestral ultimate bone marrow stem cell. The difficulty here may be one of identifying and extracting such cells, the location of which is probably in the bone where they will be present in increasingly diminishing numbers, as one looks for the increasingly primitive cells.

Our current model for the gastrointestinal cellular organization, which is based on an attempt to accommodate as much experimental data as possible, is that the commitment to differentiation producing dividing transit cells does not occur at the level of the ultimate stem cell in the lineage but at a position two or three generations along the cell lineage. If such a concept is drawn as a cell lineage diagram, the proliferative units in the intestine, the crypts, each contain four to six cell lineages and, hence, four to six lineage ancestor stem cells but up to 30 second- and third-tier stem cells, which under steady-state circumstances are inevitably displaced and moved toward the dividing transit compartment. But if damage occurs in one or more of the ultimate stem cells, they can assume the mantle of the ultimate stem cell and repopulate the lineage (Potten *et al.*, 1997; Potten, 1998; Marshman *et al.*, 2002). This gives rise to the concept of *actual* vs. *potential stem cells* (see Figure 3-1), which is discussed later in this chapter.

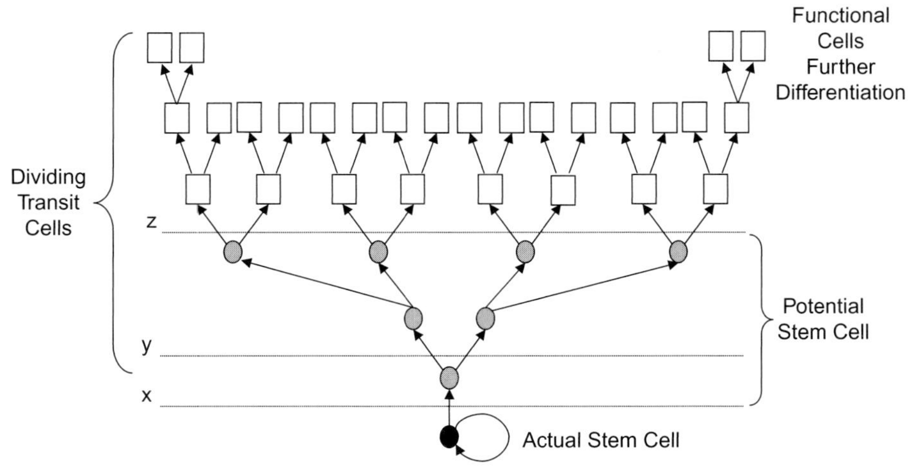

FIGURE 3-1 A typical stem cell-derived cell lineage, which may be applicable to most epithelial tissues of the body. The lineage is characterized by a self-maintaining lineage ancestor actual stem cell (black), which divides and produces a progeny that enters a dividing transit population. The number of cell generations in the dividing transit population varies from tissue to tissue. The commitment to differentiation that separates the stem cell from the dividing transit population can occur at the point of the actual stem cell division (x), in which case the stem cells are dividing asymmetrically on average. However, this commitment may be delayed to point y or z, generating a population of potential stem cells that can replace the actual stem cell if it is killed. Under normal steady-state circumstances, the potential stem cells form part of the dividing transit population and are gradually displaced down the lineage, undergoing further differentiation events if required to produce the functional mature cells of the tissue.

An analogy can be drawn here with the hierarchical structure within an organization such as the army, a concept that was discussed at the time that we were formulating the text for the *Development* paper in which we defined stem cells (Potten and Loeffler, 1990). In a military battlefield environment, the hierarchically organized army is under the control of and ultimately dependent on the highly trained (or so one hopes) general. In the event that the general is killed in the battlefield, there may be a reasonably well-trained colonel or major who can take over command and assume the insignia and uniform as well as the function of the general. In the event the colonel also should be killed, there may be lesser trained officers who will attempt to assume the mantle of command (e.g., a captain). Ultimately, the vast majority of the troops, the privates, would be insufficiently trained or experienced to be able to adopt the functional role of the commander. However, the Dolly the sheep scenario suggests that occasionally a private given a crash course in military strategy might function as the officer in command. The analogy could be taken even further to relate to the apoptosis sensitivity that is seen in the gastrointestinal ultimate stem cells. These cells appear to adopt a strategy with complete intolerance to any genetic damage and a reluctance to undertake repair, since this may be associated with inherent genetic risk they commit an altruistic suicide: the general who undergoes a nervous breakdown or suffers serious injury and has to be removed from command.

In the small intestinal crypts there have been in the past no useful markers that permit the stem cells to be identified and, hence, studied. However, such markers are now being identified. In the absence of markers, the small intestine proved to be an invaluable biological model system to study stem cells because the cells of the intestinal cell lineage are arranged spatially along the long axis of the crypt. This can be demonstrated by cell migration tracking and mutational marker studies. As a consequence, the stem cells are known to be located at very specific positions in the tissue (crypts): the fourth to fifth cell position from the crypt base in the small intestine and at the very base of the crypt in the mid-colon of the large intestine (see Figure 3-2) (Potten *et al.*, 1997; Potten, 1998; Marshman *et al.*, 2002).

SKIN STEM CELLS

The first suggestion that the proliferative compartment of the epidermis, the basal layer, was heterogeneous and contained only a small subpopulation of stem cells came with the development of the skin macrocolony (and subsequently microcolony) clonal regeneration assays developed by Withers (Withers, 1967; Withers and Elkind, 1970). This was soon combined with other cell kinetic and tissue organization data to formulate the epidermal proliferative unit (EPU) concept (Potten, 1974) (see Figure 3-3). This suggested that the basal layer consisted of a series of small, functionally and cell lineage-related cells, with a spatial organization that related directly to the superficial functional cells of the epidermis, the stratum corneum. The concept indicated that the epidermis should be regarded as being made up of a series of functional proliferative units, each unit having a centrally placed self-maintaining stem cell and a short stem cell-derived cell lineage (with three generations). The differentiated cells, produced at the end of the lineage, migrated out of the basal layer into the suprabasal layers in an ordered fashion, where

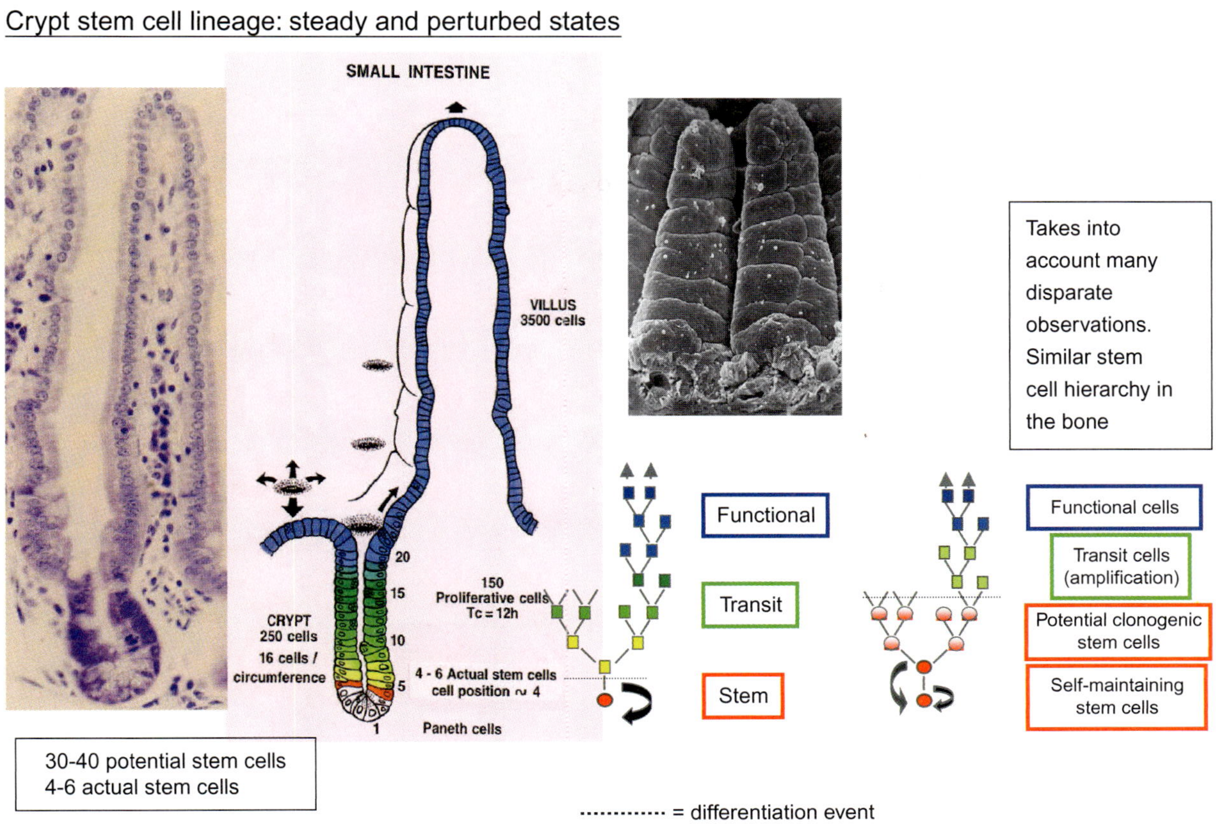

FIGURE 3-2 The cell lineage for the small intestinal crypts. It is postulated that each crypt contains four to six such lineages and, hence, four to six lineage ancestor actual stem cells, and there are about six cell generations in each lineage with at least four distinct differentiated cell types being produced. The attractive feature of this as a cell biological model system is that the position of a cell in a lineage can be related to its topographical position in a longitudinal section through the crypt.

further maturation events occurred, eventually producing the thin, flattened, cornified cells at the skin surface, which were stacked into columns (like a pile of plates), with cell loss occurring at a constant rate from the surface of the column (Figure 3-3). Such an organization is clearly evident in the body skin epidermis of the mouse, its ears, and a modified version of the proliferative unit can clearly be identified in the dorsal surface of the tongue (Hume and Potten, 1976). There has been, and continues to be, some debate as to whether this concept applies to human epidermis. It is clear that in many sites of the body in man a similar columnar organization can be seen in the superficial corneal layers of the epidermis. What is more difficult in humans is to relate this superficial structure to a spatial organization in the basal layer. However, the spatial organization seen in the superficial layers must have an organizing system at a level lower in the epidermis, and it does not seem unreasonable to assume that this is in the basal layer, as is the case of the mouse epidermis. The presence of units of proliferation in human epidermis has been demonstrated using β-galacosidase/GFP staining of skin in cultures and xenografts (MacKenzie, 1997; Ghazizadeh and Taichman, 2005).

A macroscopic, clonal regeneration assay for mouse epidermis was developed by Withers (1967), which generates nodules very similar in appearance to spleen colonies. Subsequently, Al-Barwari developed a microscopic clonal assay that required a shorter time interval between irradiation and tissue sampling (Al-Barwari and Potten, 1976). Together, these clonal regeneration assays were interpreted to indicate that only about 10% (or less) of the basal cells have a regenerative capacity, i.e., are stem cells.

The EPU stem cells must have an asymmetrical division mode under steady-state cell kinetics, since there is only one such cell per EPU. The epidermal microcolony assay developed by Al-Barwari suggests that, following injury such as irradiation, surviving EPU stem cells can change their division mode from asymmetric to symmetric for a period of time to repopulate the epidermis (i.e., change their self-maintenance probability from 0.5 to a value higher than 0.5). Al-Barwari's observations also indicated that a significant contribution to reepithelialization could come from the upper regions of the hair follicles. This has subsequently been expanded into a major area of interest: The hair follicle lineage stem cells have been shown to possess a wide range of regenerative activities, including production of epidermis. It was also clear from studies on the structural organization of the epidermis following injury that in order to reestablish the spatial distribution of stem cells, the epidermis undergoes a reorganization involving hyperplasia, during which

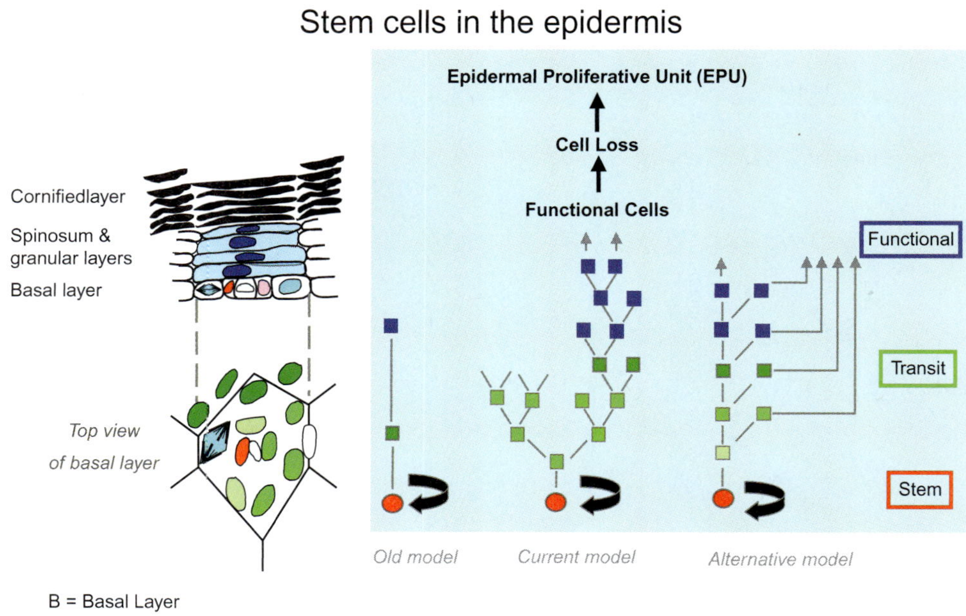

FIGURE 3-3 Diagrammatic representation of the cell lineage seen in the interfollicular epidermis and the relationship between the cell lineage and the spatial organization, characterized as the epidermal proliferative unit (EPU), as seen in section view (upper portion of the figure on the left) and in surface view in epidermal sheets (lower portion of the figure on the left). The old historical lineage model is shown on the left and an alternative model to the current one on the right.

stem cells are redistributed and eventually establish their EPU spatial configurations.

The skin clearly contains another important stem cell population, namely that associated with the growing hair follicles. Hair is produced over a protracted period of time by rapid divisions in the germinal region of the growing hair follicle (termed an *anagen follicle*). This hair growth may be maintained for long periods of time, three weeks in a mouse (where the average cell cycle time may be 12 h), months to years in man, and for more indefinite periods for some animal species, such as Angora rabbits, merino sheep, and some strains of dog. This high level of cell division in the germinal matrix of the follicle, which has a considerable spatial polarity like the intestinal crypt, must have a fixed stem cell population residing in the lowest regions of the germinal matrix that can maintain the cell production for the required period of time. Very little is known about these stem cells. The complication with hair follicles is that in mouse and man, the growing follicles eventually contain a mature hair and cell proliferation activity ceases. The follicle shrinks and becomes quiescent (a *telogen follicle*). The simplest explanation here is that the telogen follicle, which consists of far fewer cells in total than in a growing follicle, contains a few quiescent hair follicle stem cells that can be triggered back into proliferation at the onset of a new hair growth cycle. However, as discussed later, there is some controversy concerning this concept.

It is now very clear that the skin contains a third stem cell compartment, which is located in the upper outer sheath of the hair follicle, below the sebaceous glands. This is sometimes identifiable by virtue of a small bulge in the outer root sheath, and so this population of cells has been referred to as the *bulge* cells. What is very evident from a whole series of extremely elegant but complicated experiments is that these bulge cells possess the ability, under specialized conditions, to reform the hair follicle if it is damaged and also to contribute to the reepithelialization of the epidermis. It is cells from this region of the follicle that were probably responsible for the epidermal reepithelialization from follicles seen by Al-Barwari. Cells from the bulge can make follicles during development of the skin and also reestablish the follicles if they are injured. The controversy concerns the issue of whether bulge stem cells, which are predominantly quiescent cells, ever contribute to the reestablishment of an anagen follicle under normal undamaged situations. The simplest interpretation is that these cells are not required for this process, since in order for this to happen some very complex cell division and cell migratory and homing pathways have to be inferred. This goes somewhat against the concept of stem cells being fixed or anchored and also against the concept of keratinizing epithelia being a tightly bound strong and impervious barrier. What seems likely for the skin is that the EPU stem cell and the hair follicle stem cell have a common origin during the development of the skin to the bulge stem cells, which then become quiescent and are present as a versatile "reserve" stem cell population that can be called into action if the skin is injured and requires reepithelialization (see Figures 3-4 and 3-5) (Potten and Booth, 2002).

Another issue relating to the stem cells of the bulge region of the follicle is that they are defined, usually by immunohistochemistry, as being present as a "cluster" of

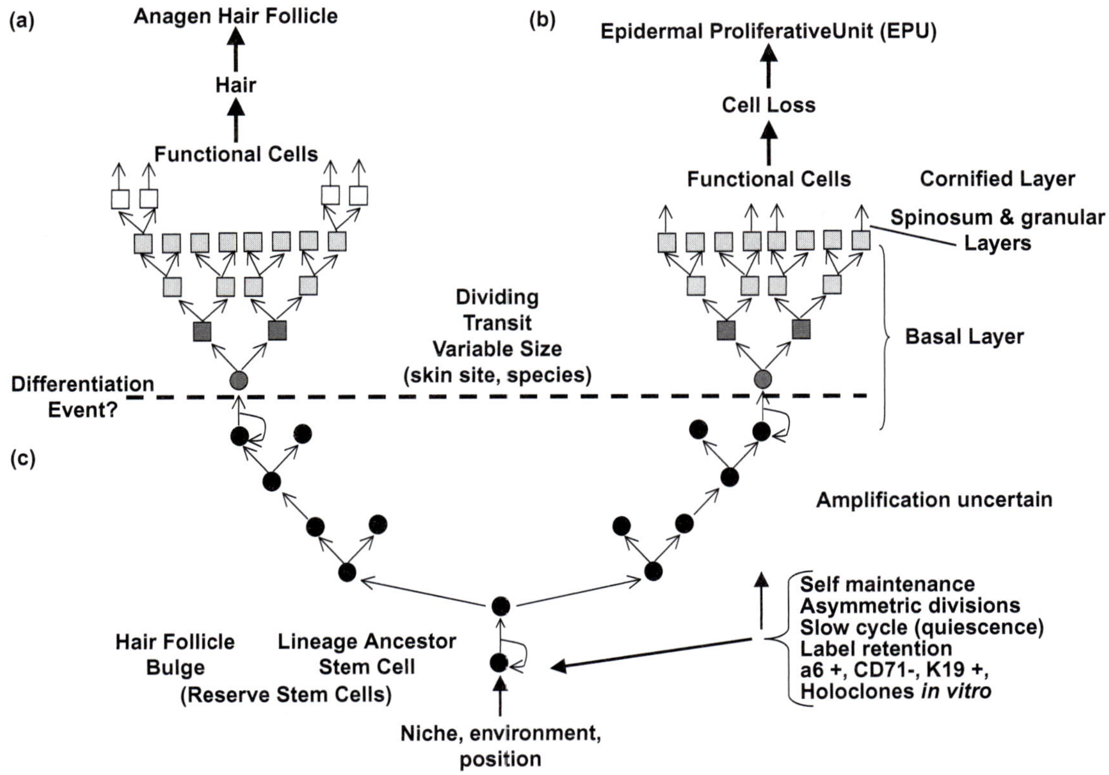

FIGURE 3-4 The complexity of the stem cell populations in mammalian skin as characterized in the mouse. (a) A distinct cell lineage is proposed for the interfollicular epidermis (EPU), (b) another is proposed for the matrix region of the growing hair follicle (anagen follicle), and (c) a potent reserve regenerative stem cell compartment, which resides in the upper/outer root sheath or bulge region of the hair follicle, is proposed. The stem cells in the bulge region can clearly regenerate the epidermis, the hair follicle, and probably other structures, such as the sebaceous glands.

largely quiescent cells. If these "reserve" cells are called into action, which cells divide? If the cells toward the center of the cluster divide, this will inevitably push the peripheral cells out of the cluster—so they were not stem cells in the first place. If the cells at the periphery divide, there is no need for the cells toward the center—so they do not satisfy stem cell criteria. The most likely explanation is that the cluster contains only a few (one or two?) *real* stem cells at the center and that the rest are quiescent early lineage cells.

Quiescence is a property often stated as applicable to stem cells. It is certainly true that in many, but not all, replacing tissues the stem cells are cycling more slowly than the transit cells (longer G1 phase). The *Oxford English Dictionary* defines *quiescence* as "a state of motionless, inertness, silence, dormancy, or inactivity." Stem cells are usually defined by virtue of what they can do. If they are doing nothing, can they be stem cells?

INTESTINAL STEM CELL SYSTEM

The intestinal epithelium, like all epithelia, is highly polarized and divided into discrete units of proliferation and differentiation. In the small intestine the differentiated units are the fingerlike villi protruding into the lumen of the intestine. These structures are covered by a simple columnar epithelium consisting of several thousand cells, which perform their specific function, become worn out, and are shed, predominantly from the tip of the villus. There is no proliferation anywhere on the villus. The cell loss from the villus tip is precisely balanced in steady state by cell proliferation at the base of the villi in units of proliferation called *crypts*. Each villus is served by about six crypts, and each crypt can produce cells that migrate onto more than one villus. The crypts in the mouse contain about 250 cells in total, 150 of which are proliferating rapidly and have an average cell cycle time of 12 hours. The cells move from the mouth of the crypt at a velocity of about one cell diameter per hour, and all this movement can be traced, in the small intestine, back to a cell position about four cell diameters from the base of the crypt. The very base of the crypt, in mice and humans, is occupied by a small population of functional differentiated cells called *Paneth* cells. Cell migration tracking and innumerable cell kinetic experiments all suggest that the stem cells that represent the origin of all this cell movement are located at the fourth position from the base of the crypt in the small intestine and right at the base of the crypt in some regions of the large bowel. The crypt is a flask-shaped structure,

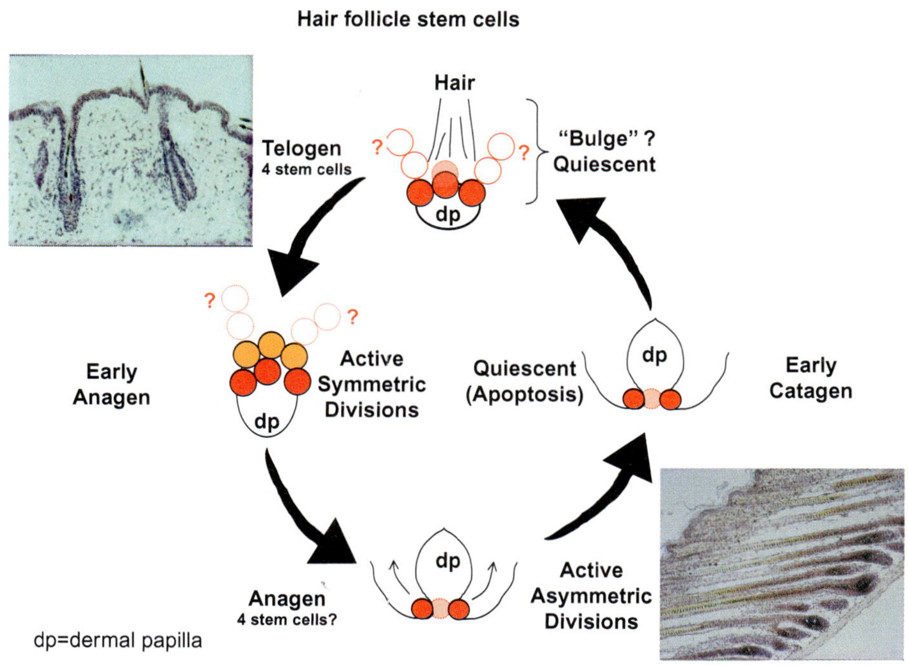

FIGURE 3-5 Diagrammatic representation of a growing anagen hair follicle and a resting or quiescent telogen follicle. Photomicrographs of representative anagen and telogen follicles are inset.

about 16 cells in circumference. Mathematical modeling suggests that each crypt contains about five cell lineages and, hence, five cell lineage ancestor stem cells. Under steady-state kinetics these cells are responsible for all the cell production, producing daughters that enter a dividing transit lineage of between six and eight generations in the small and large bowel, respectively (see Figures 3-1 and 3-2). The stem cells in the small intestine divide with a cycle time of approximately 24 hours; hence, in the lifetime of a laboratory mouse they divide about 1000 times. It is assumed that these cells are anchored or fixed in a micro-environmental niche that helps determine their function and behavior. The uniquely attractive feature of this model system, from a cell biological point of view, is that in the absence of stem cell-specific markers, the behavior and characteristics and response to treatment of these crucial lineage ancestor cells can be assessed by studying the behavior of cells at the fourth position from the bottom of the crypt in the small intestine. When this is done, one of the features that seems to characterize a small population of cells at this position (about five cells) is that they express an exquisite sensitivity to genotoxic damage, such as delivered by small doses of radiation (Potten, 1977; Hendry *et al.*, 1982; Potten and Grant, 1998). They appear to tolerate no DNA damage and activate a p53-dependent altruistic suicide (apoptosis). It is believed that this is part of the genome protection mechanisms that operate in the small intestine and account for the very low incidence of cancer in this large mass of rapidly proliferating tissue.

The clonal regeneration techniques for the intestine developed by Withers (Withers and Elkind, 1970) have been used extensively. These techniques suggest the presence of a second compartment of clonogenic or potential stem cells (about 30 per crypt) that possess a higher radioresistance and a good ability to repair DNA damage. These observations, together with others, suggest a stem cell hierarchy of the sort illustrated in Figures 3-1 and 3-2, with the commitment to differentiation that distinguishes dividing transit cells from stem cells, occurring about three generations along the lineage. Virtually identical lineage structures can be inferred for the colonic crypts (Cai *et al.*, 1997).

There has been an absence of stem cell-specific markers in the past. However, current work suggests that some may now be available. Antibodies to Musashi-1, an RNA-binding protein identified as playing a role in asymmetric division control in neural stem cells, appears to be expressed in very early lineage cells in the small intestine (see Figure 3-6) (Potten *et al.*, 2003).

Recent studies have indicated that the ultimate stem cells in the crypt possess the ability to selectively segregate old and new strands of DNA at division and to retain the old template strands in the daughter cell destined to remain a stem cell while passing the newly synthesized strands, which may contain any replication-induced errors, to the daughter cell destined to enter the dividing transit population and to be shed from the tip of the villus five to seven days after birth from division (Potten *et al.*, 2002). This selective DNA segregation process provides a second level of genome protection for the stem cells in the small intestine (Cairns, 1975), protecting them totally from the risk of replication-induced errors, thus providing further protection against carcinogenic risk and an explanation for the very low cancer incidence in this tissue (see Table 3-1). This mechanism of selective DNA segregation allows the template strands to be labeled with DNA synthesis

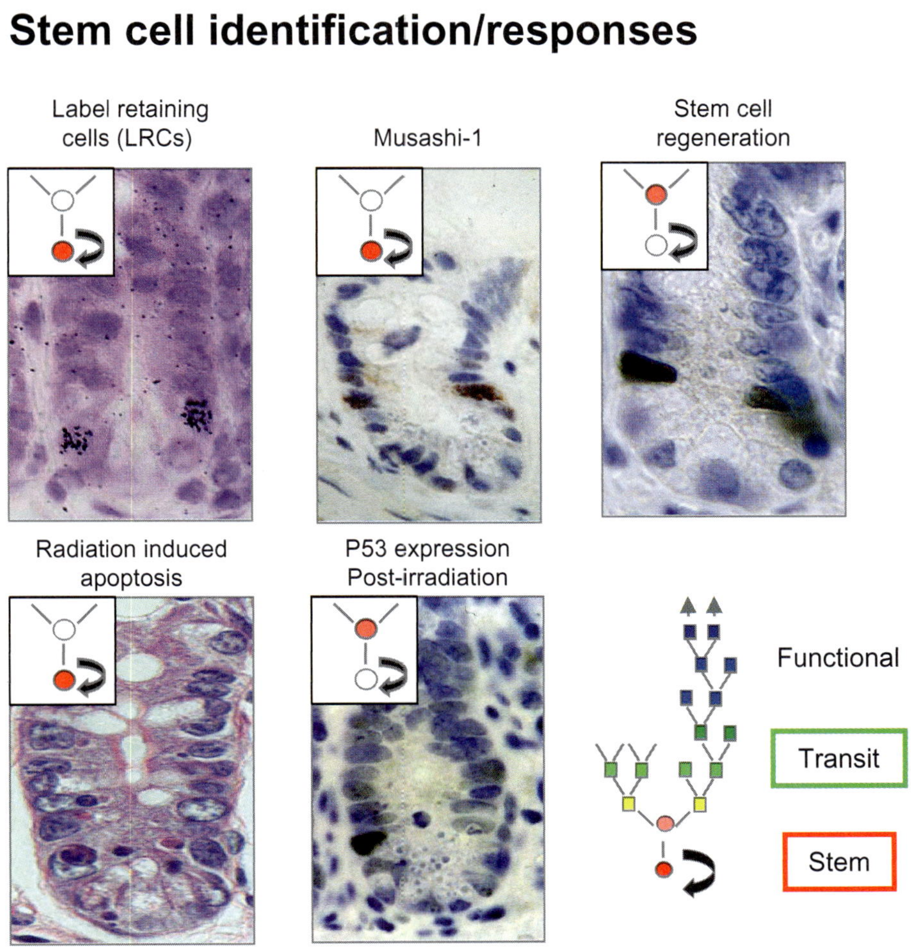

FIGURE 3-6 Photomicrographs of longitudinal sections of the small intestinal crypts from the mouse, illustrating a range of possible ways of identifying the stem cell compartment. Making use of the selective strand segregation hypothesis, template strands of DNA can be labeled, generating label-retaining cells at the fourth position from the bottom of crypts. Musashi-1, an RNA-binding protein, is expressed in early lineage cells and under some labeling conditions can show specificity for individual cells at around cell position 4. Part of the regenerative or potential stem cell compartment can be seen by S-phase labeling (Bromodeoyuridine labeling) at critical phases following cytotoxic injury, when these cells are called into regenerative mode. The example shown here is a labeling pattern at 24 hours after two doses of 5-fluorouracil, when the only cells in S phase are a few cells scattered around the fourth position from the base of the crypt. As part of the genome protective mechanism it is postulated that the ultimate lineage ancestor stem cells have an exquisite sensitivity to radiation and the induction of genome damage. When this happens, the cells commit suicide via apoptosis, which can easily be recognized and occurs at about the fourth position from the base of the crypt. These cells do not express p53 protein, at least at the times studied and as detectable by immunohisto-chemistry. However, some cells do express p53 protein at high levels following radiation exposure, and it is postulated that these are the surviving potential stem cells in cell cycle arrest, to allow for repair prior to entering rapid regenerative cell cycles. Under appropriate immunohistochemical preparative procedures, individual wild-type P53 protein-expressing cells can be seen at around cell position 4. The cartoons represent the cell within the lineage, which is thought to be labeled by each marker (stem cell = red; stem cell daughter/early progenitor = pink).

markers at times of stem cell expansion, i.e., during late tissue development and during tissue regeneration after injury. The incorporation of label into the template strands persists (label-retaining cells), thus providing a truly specific marker for the lineage ancestor cells (see Figure 3-6). This figure also illustrates some other ways in which intestinal stem cells may be distinguished from their rapidly dividing progeny.

The selective segregation of "old" and "new" DNA strands at mitosis is appearing to be a universal feature of the rare ultimate adult tissue stem cell. In addition to the small intestinal epithelium, this process has been detected

TABLE 3-1 Why Do Small Intestinal Stem Cells Not Develop More Cancers?

When one considers that:
• the tissue is 3–4 times greater in mass (length) • the cells are proliferating 1.5 times more rapidly • there are 2–3 times more total stem cells • there are 3–4 times more stem cell divisions in a lifetime
Compared with the large intestine, the small intestine has 70 times fewer cancers.

in breast epithelial stem cells, both *in vivo* and *in vitro*, in neural stem cells *in vitro*, and in some muscle satellite cells (Merok *et al.*, 2002; Clarke *et al.*, 2005; Karpowicz *et al.*, 2005; Pare and Shirley *et al.*, 2006; Shinin *et al.*, 2006). It can also be inferred from work on the dorsal surface of the tongue and epidermis.

STEM CELL ORGANIZATION IN FILIFORM PAPILLAE ON THE DORSAL SURFACE OF THE TONGUE

Oral mucosae are keratinizing, stratified epithelia, similar to epidermis in their structural organization. The dorsal surface of the tongue is composed of many small, filiform papillae that have a very uniform shape and size. Detailed histological investigations, together with cell kinetic studies performed by Hume (Hume and Potten, 1976), showed that each papilla is composed of four columns of cells: two dominant and two buttressing columns. The dominant anterior and posterior columns represent modified versions of the epidermal proliferative units and were called *tongue proliferative units*. The cell migratory pathways were mapped (like the studies in the intestinal crypts), which enabled the position in the tissue from which all migration originated to be identified, this being the presumed location of the stem cell compartment. The lineage characterizing this epithelium is similar to that seen in the dorsal epidermis of the mouse, i.e., self-replacing asymmetrically dividing stem cells, occurring at a specific position in the tissue, producing a cell lineage that has approximately three generations (Figure 3-7). The stem cells here have a particularly pronounced circadian rhythm (Potten *et al.*, 1977a).

GENERALIZED SCHEME

It thus appears that for the major replacing tissues of the body, hierarchical or cell lineage schemes appear to explain the cell replacement processes. These schemes may involve isolated single stem cells that, under steady-state circumstances, must be presumed to divide asymmetrically, producing a dividing transit population. The size of the dividing transit population differs dramatically from tissue to tissue. The number of generations defining the degree of amplification that the transit population provides for each stem cell division is related inversely to the frequency that stem cells will be found within the proliferating compartment (see Figure 3-8). For some systems, such as the bone marrow and the intestine, the commitment to differentiation that separates the dividing transit compartment from the stem cell compartment appears to be delayed until a few generations along the lineage. This generates a stem cell hierarchy, with cells of changing (decreasing) stemness or, conversely, increasing commitment, leading to the concept of committed precursor cells. In the small intestine this delay in the commitment to differentiation to a dividing transit population provides the tissue with a reserve population of potential stem cells that can repopulate the tissue if the lineage ancestor cells are destroyed, an added level of tissue protection in this extremely well-protected tissue. With regard

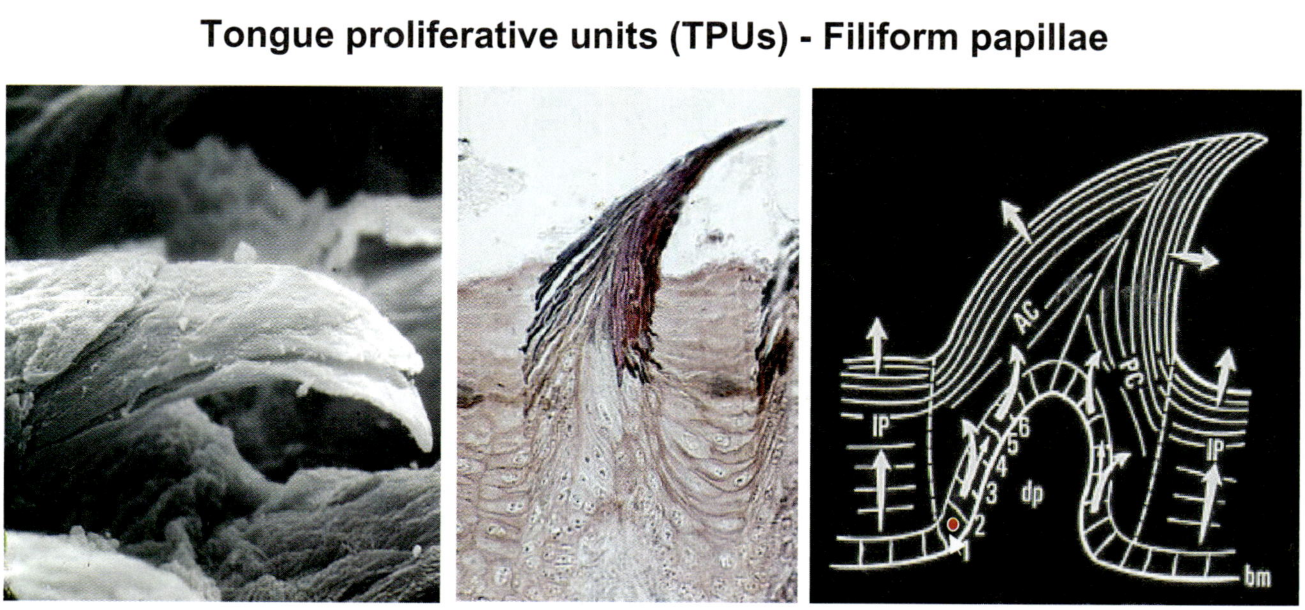

FIGURE 3-7 A scanning electron micrograph of (left) and histological section through the dorsal surface of the tongue (center), and a diagrammatic representation of this tissue (right), showing the tongue proliferative units (the dominant anterior column, AC, and posterior column, PC). Cell migratory pathways have been identified based on cell positional analyses and cell marking and on the location of the stem cells identified in the basal layer (red diamond—cell position 1). The stem cells in this tissue express one of the strongest circadian rhythms in proliferation seen anywhere in the body.

to the bone marrow, committed precursors (or even earlier cells) appear to circulate in the blood and may lodge in various tissues. Given appropriate microenvironments and local signals, some of these lodged cells may be instructed to differentiate down unusual pathways. This has prompted research into using such cells to correct genetic deficiencies. Although the transdifferention theory is attractive, recent research indicates that the apparent plasticity of stem cells may be less clear-cut. Transplantation experiments in mice with specific gene disorders suggest that transplanted bone marrow cells may "fuse" with cells in the "target" organ and, hence, complement any gene deficiency.

ADULT STEM CELL PLASTICITY

Recently, a number of significant papers on the plasticity of bone marrow progenitor cells have been published that focus on their ability to contribute to the different proliferative cell compartments of the intestinal epithelium (and other cell types of the intestinal mucosa). In an elegant series of experiments, bone marrow progenitor cells from a male donor were injected into female recipient mice, which were subsequently subject to induction of mucosal injury (Brittan *et al.*, 2005). Alternatively, bone marrow progenitors were injected into lethally irradiated female mice (Brittan *et al.*, 2002; Brittan and Wright, 2004). *In situ* hybridization was able to demonstrate cells with a male Y chromosome within the regenerating endothelium and within the peri-cryptal fibroblast cell population (the cells that surround the crypt epithelial cells on the basolateral side), but not within the colonic epithelium. The fibroblasts with male chromosomes appeared as columns of cells along the crypt axis within the tissue sections examined, suggesting that bone marrow-derived cells had established a clonal lineage within the proliferative fibroblast cell population.

Other studies have suggested that bone marrow progenitors can contribute to all the different cell lineages within the intestinal epithelium (Rizvi *et al.*, 2006). In mice, injection of labeled female bone marrow progenitor cells into lethally irradiated male mice resulted in cells with female-specific label being found within the proliferative cells of the small intestinal crypts and within the various differentiated cell populations of the villi (columnar, goblet, enteroendocrine, Paneth). The conclusion was that female bone marrow progenitors "fused" with stem or early progenitor cells of the intestinal epithelium of the male recipient. Fusion rather than transdifferentiation was suggested as a mechanism in this study, based on the

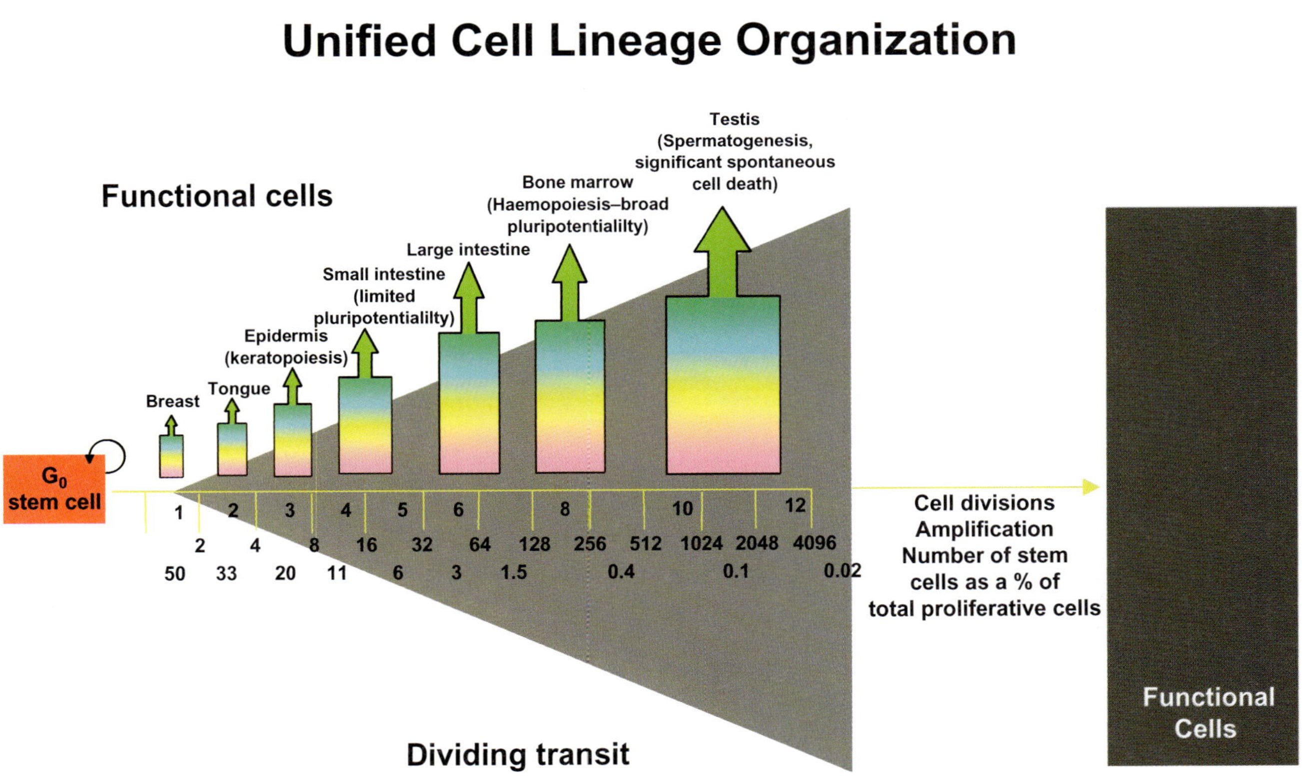

FIGURE 3-8 A diagrammatic representation of a stem cell-derived cell lineage, showing the approximate positions for the number of cell generations in the dividing transit population for a range of murine tissues. Stratified keratinizing epithelia, such as the breast, tongue, and epidermis, tend to have the shortest lineages; the bone marrow and the testis tend to have the longest lineages. Also shown is the degree of theoretical amplification that the dividing transit lineage provides for each stem cell division and the inverse relationship between the degree of amplification and the proportion of the proliferative compartment that the stem cells occupy.

high frequency of cells that were positive for the female-specific marker and the male Y chromosome. In humans, cells from male donors have been found throughout the gastrointestinal tract epithelium of female recipients (Okamoto *et al.*, 2002). It was suggested that the donor cells were able to incorporate into the epithelium and differentiate into mature epithelial cells but did not integrate into the stem cell compartment.

In addition to the regenerating epithelium postinjury, high levels of intestinal epithelial cell proliferation are to be found within tumors of epithelial origin. Further experiments examining the incorporation of female bone marrow progenitors into intestinal tumors of male *Min*/+ mice revealed a large number of epithelial cells within the tumors, which were double positive for male- and female-specific markers (Rizvi *et al.*, 2006). These results paralleled those from the lethal-irradiation experiments and suggest that cell fusion between bone marrow progenitors and epithelial cells is common (within the intestinal epithelium at least) in situations when cell proliferation rates are elevated. It also raises the possibility that bone marrow progenitors can contribute to tumor progression.

This last point is rather controversial, and, to date, only one study has suggested that bone marrow progenitors are directly responsible for tumor development within an epithelial tissue (Houghton *et al.*, 2004). Mice had their bone marrow ablated by irradiation and received a bone marrow transplant from ROSA26 transgenic mice. They were then infected with *Helicobacter felis*, in order to induce gastric carcinogenesis. The mice showed classic progression from metaplasia to dysplasia to carcinoma, over the period of one year, with most dysplastic gastric glands and all neoplasia being derived from the bone marrow progenitors, as assessed by their expression of the ROSA26 transgene. Experiments examining male donor/female recipient combinations failed to show colocalization of male- and female-specific markers in isolated gastric epithelial cells, which suggests that stable fusion events between the bone marrow progenitor cells and gastric epithelial cells do not take place in this experimental system. As evidence for transdifferentiation of bone marrow cells, the authors were able to present evidence of epithelial cell-specific gene expression in bone marrow-derived mesenchymal stem cell cultures exposed to conditioned medium from gastric epithelial cell primary cultures.

It can be clearly seen that different experimental models offer different perspectives on the relative merits of stem cell transdifferentiation and fusion and their ability to contribute to different cell populations within epithelial tissues. There is the possibility that in a highly proliferative tissue like the gastrointestinal epithelium, bone marrow transplantation may introduce into the tissue a population of cells that are a "low-threshold" target for induction of tumorigenesis; however, this still requires further evidence to show whether this is a real phenomenon or just a peculiarity of the experimental model. This aside, adult stem cell transplantation clearly has exciting potential for promoting tissue regeneration following injury.

REFERENCES

Al Barwari, S. E., and Potten, C. S. (1976). Regeneration and dose–response characteristics of irradiated mouse dorsal epidermal cells. *Int. J. Radiat. Biol. Relat. Stud. Phys. Chem. Med. 30*, 201–216.

Brittan, M., and Wright, N. A. (2004). Stem cells in gastrointestinal structure and neoplastic development. *Gut 53*, 899–910.

Brittan, M., Hunt, T., Jeffery, R., Poulsom, R., Forbes, S. J., Hodivala-Dilke, K., Goldman, J., Alison, M. R., and Wright, N. A. (2002). Bone marrow derivation of pericryptal myofibroblasts in the mouse and human small intestine and colon. *Gut 50*, 752–757.

Brittan, M., Chance, V., Elia, G., Poulsom, R., Alison, M. R., MacDonald, T. T., and Wright, N. A. (2005). A regenerative role for bone marrow following experimental colitis: contribution to neovasculogenesis and myofibroblasts. *Gastroenterology 128*, 1984–1995.

Cai, W. B., Roberts, S. A., Bowley, E., Hendry, J. H., and Potten, C. S. (1997). Differential survival of murine small and large intestinal crypts following ionizing radiation. *Int. J. Radiat. Biol. 71*, 145–155.

Cairns, J. (1975). Mutation selection and the natural history of cancer. *Nature 255*, 197–200.

Clarke, R. B., Spence, K., Anderson, E., Howell, A., Okano, H., and Potten, C. S. (2005). A putative human breast stem cell population is enriched for steroid receptor-positive cells. *Dev. Biol. 277*, 443–456.

Ghazizadeh, S., and Taichman, L. B. (2005). Organisation of stem cells and their progeny in human epidermis. *J. Invest. Dermatol. 124*, 367–372.

Hendry, J. H., Potten, C. S., Chadwick, C., and Bianchi, M. (1982) Cell death (apoptosis) in the mouse small intestine after low doses: effects of dose rate, 14.7-MeV neutrons, and 600-MeV (maximum energy) neutrons: *Int. J. Radiat. Biol. Relat. Stud. Phys. Chem. Med. 42*, 611–620.

Houghton, J., Stoicov, C., Nomura, S., Rogers, A. B., Carlson, J., Li, H., Cai, X., Fox, J. G., Goldenring, J. R., and Wang, T. C. (2004). Gastric cancer originating from bone marrow-derived cells. *Science 306*, 1568–1571.

Hume, W. J., and Potten, C. S. (1976). The ordered columnar structure of mouse filiform papillae. *J. Cell Sci. 22*, 149–160.

Karpowicz, P., Morshead, C., Kam, A., Jervis, E., Ramunas, J., Cheng, V., and van der Kooy, D. (2005). Support for the immortal strand hypothesis: neural stem cells partition DNA asymmetrically *in vitro*. *J. Cell Biol. 170*, 721–732.

MacKenzie, I. C. (1997). Retroviral transduction of murine epidermal stem cells demonstrates clonal units of epidermal structure. *J. Invest. Dermatol. 109*, 377–383.

Marshman, E., Booth, C., and Potten, C. S. (2002). The intestinal epithelial stem cell. *Bioessays 24*, 91–98.

Merok, J. R., Lansita, J. A., Tunstead, J. R., and Sherley, J. L. (2002). Cosegregation of chromosomes containing immortal DNA strands in cells that cycle with asymmetric stem cell kinetics. *Cancer Res. 62*, 6791–6795.

Okamoto, R., Yajima, T., Yamazaki, M., Kanai, T., Mukai, M., Okamoto, S., Ikeda, Y., Hibi, T., Inazawa, J., and Watanabe, M. (2002). Damaged epithelia regenerated by bone marrow-derived cells in the human gastrointestinal tract. *Nat. Med. 8*, 1011–1017.

Pare, J. F., and Sherley, J. L. (2006). Biological principles for *ex vivo* adult stem cell expansion. *Curr. Top. Dev. Biol. 73*, 141–171.

Potten, C. S. (1974). The epidermal proliferative unit: the possible role of the central basal cell. *Cell Tissue Kinet. 7*, 77–88.

Potten, C. S. (1977). Extreme sensitivity of some intestinal crypt cells to X and gamma irradiation. *Nature 269*, 518–521.

Potten, C. S. (1998). Stem cells in gastrointestinal epithelium: numbers, characteristics and death. *Philos. Trans. R. Soc. London, Sev. B, Biol. Sci. 353*, 821–830.

Potten, C. S., and Hendry, J. H. (1985). The micro-colony assay in mouse small intestine. *In* "Cell Clones: Manual of Mammalian Cell Techniques." (C. S. Potten and J. H. Hendry, eds.), pp. 50–60. Edinburgh, Churchill Livingstone.

Potten, C. S., and Loeffler, M. (1990). Stem cells: attributes, cycles, spirals, pitfalls and uncertainties. Lessons for and from the crypt. *Development 110*, 1001–1020.

Potten, C. S., and Grant, H. K. (1998). The relationship between ionizing radiation-induced apoptosis and stem cells in the small and large intestine. *Br. J. Cancer 78*, 993–1003.

Potten, C. S., and Booth, C. (2002). Keratinocyte stem cells: a commentary. *J. Invest. Dermatol. 119*, 888–899.

Potten, C. S., Al Barwari, S. E., Hume, W. J., and Searle, J. (1977a). Circadian rhythms of presumptive stem cells in three different epithelia of the mouse. *Cell Tissue Kinet. 10*, 557–568.

Potten, C. S., Booth, C., and Pritchard, D. M. (1997b). The intestinal epithelial stem cell: the mucosal governor. *Int. J. Exp. Pathol. 78*, 219–243.

Potten, C. S., Owen, G., and Booth, D. (2002). Intestinal stem cells protect their genome by selective segregation of template DNA strands. *J. Cell Sci. 115*, 2381–2388.

Potten, C. S., Booth, C., Tudor, G. L., Booth, D., Brady, G., Hurley, P., Ashton, G., Clarke, R., Sakakibara, S., and Okano, H. (2003). Identification of a putative intestinal stem cell and early lineage marker; musashi-1. *Differentiation 71*, 28–41.

Rizvi, A. Z., Swain, J. R., Davies, P. S., Bailey, A. S., Decker, A. D., Willenbring, H., Grompe, M., Fleming, W. H., and Wong, M. H. (2006). Bone marrow-derived cells fuse with normal and transformed intestinal stem cells. *Proc. Natl. Acad. Sci. U.S.A. 103*, 6321–6325.

Shinin, V., Gayraud-Morel, B., Gomes, D., and Tajbakhsh, S. (2006). Asymmetric division and cosegregation of template DNA strands in adult muscle satellite cells. *Nat. Cell Biol. 8*, 677–687.

Till, J. E., and McCulloch, E. A. (1961). A direct measurement of the radiation sensitivity of normal bone marrow cells. *Rad. Res. 14*, 213.

Withers, H. R. (1967). The dose–survival relationship for irradiation of epithelial cells of mouse skin. *Brit. J. Radiolo. 40*, 187.

Withers, H. R., and Elkind, M. M. (1970). Micro-colony survival assay for cells of mouse intestinal mucosa exposed to radiation. *Int. J. Rad. Biol. 17*, 261.

Chapter 4

Mesenchymal Stem Cells

Arnold I. Caplan

INTRODUCTION

In the mid-1980s, my laboratory started experiments to purify mesenchymal stem cells (MSCs) from adult human bone marrow. There were several reasons for initiating our approach, and for our success. It is important to briefly review these events because they established the basis for the current state of the field and the basis for interesting refinements, new observations, and new directions.

From 1968 to 1972, my laboratory developed and refined the methods for liberating and culturing undifferentiated mesenchymal cells from the developing limb buds of 4.5-day-old chick embryos. These embryonic chick limb bud cells multiplied in culture, and depending on the culture conditions and the initial seeding densities, they differentiated into muscle, cartilage, bone, and various connective tissues. At intermediate plating densities, bone-forming cells called osteoblasts were observed to develop, and at a 2.5-fold increase in initial plating density of these same cells, the plates were filled with cartilage with little or no evidence of osteoblast differentiation. The chondrocytes, which formed the cartilage tissue, were shown to continuously modulate the synthesis of the major cartilage water structuring molecule, a proteoglycan called aggrecan; these same changes have also been observed for human articular cartilage. Thus, these cell cultures mimic the full range of cartilage changes from young to old, from bouncy water-filled shock absorbers to the leathery, tough coverings of the joint surface.

In 1965, Professor Marshall R. Urist published his landmark paper on the purification of factors from demineralized bovine bone that induced cartilage and subsequent bone formation in various *in vivo* implantation systems and *in vitro*. The basis for Urist's assays was the observation that demineralized bone chips, when implanted into muscle or subcutaneous pouches in test animals, caused the formation of ectopic bone. Urist coined the term bone morphogenetic proteins (BMPs) to describe the presumptive inductive molecules that stimulated this tissue outcome.

In 1978, Glenn Syftestad joined my laboratory as a postdoc following a stay in Urist's laboratory. Syftestad was determined to purify BMPs from bovine bone, and the embryonic chick limb bud mesenchymal cell system was an obvious way to test for the chondrogenic stimulating activity (CSA) within the extract of demineralized bovine bone. The intermediate plating density where cartilage was never observed became the assay system for identifying and purifying chondrogenic inductive molecules from extracts of adult bone by observing chondrogenic differentiation under conditions in which no cartilage formed normally. Several papers and patents were published on these efforts and centered on a 31-kDa protein. We can now surmise that CSA was a heterodimer of BMPs; unfortunately, we, like Urist and his colleagues, were never able to sequence the proteins or to clone the genes. John Wozney and his collaborators at the Genetics Institute in Cambridge, Massachusetts (now part of Wyeth Pharmaceutical Company) were the first to do this.

In the course of studying CSA, we did many *in vivo* implantations, and observed ectopic cartilage and bone formation. During the course of these studies in the mid-1980s, it occurred to me that cells like the multipotent embryonic chick limb mesenchymal cells must reside in, or must come to, the implantation site in the adult tissue in order for cartilage and bone to form. I call these adult cells MSCs. By this time, several investigators had proposed that osteogenic (bone-forming) progenitor cells reside in bone marrow. This seemed obvious, since all bone is in contact with marrow elements. The experiments of Maureen Owen—and, in particular, her review article in 1988 that summarized A. J. Friedenstein's early work and included the possibility that fat cell progenitors were also in marrow—opened the door for speculation that all of the mesenchymal phenotypes could arise from this same class of progenitors in pathways analogous to those deduced for the blood-forming cells (hematopoiesis) (Figure 4-1).

In this context, several years earlier, my laboratory was deeply involved in the study of bone formation in the

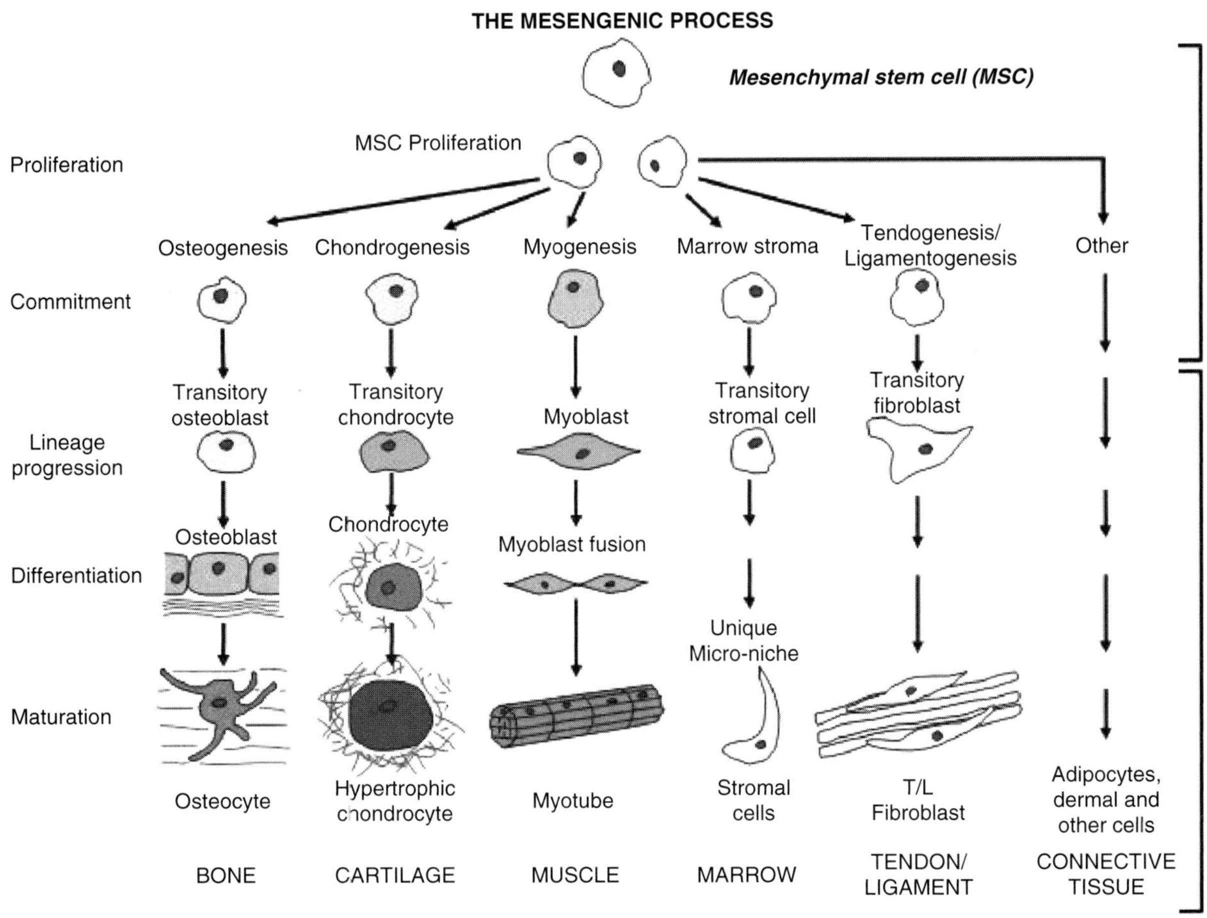

FIGURE 4-1 The mesengenic process depicts mesenchymal progenitor cells entering distinct lineage pathways that contribute to mature tissues.

embryonic chick limb bud *in vivo*, and in an attempt to compare it to *in vitro* events, a young and talented graduate student, Philip Osdoby, had wasted 18 months working with this limb bud mesenchymal cell culture system, because we followed the dogma of the day that cartilage was the progenitor of bone, as evidenced by endochondral bone formation and growth plate dynamics. We studied this by plating embryonic chick limb cells at high density to encourage them to form mounds of cartilage. No matter how we varied the culture medium or its additives, we could not get these cultures to form bone. Osdoby made a key observation in his superb electron micrographs of these cartilage cultures that when rare, mineralized deposits (osteoid) were observed, they were never located in the cartilage nodules, but always in the fibrous tissue between these massive mounds of cartilage. We then realized that osteoblasts formed independently of cartilage and away from it. By plating the cells at a density where cartilage never formed, we could observe many osteoblasts.

When we went back to the whole developing embryonic limb bud and looked at the location of first bone formation, it was outside and away from the already formed cartilaginous rods. Later, based on this early work of Osdoby and the subsequent morphological descriptions of first bone formation, Scott Bruder raised monoclonal antibodies to the tediously dissected first bone collars of the embryonic tibia and used these to establish the lineage transitions of mesenchymal progenitors to osteoblasts and osteocytes.

Thus, from this early work on embryonic mesenchymal cell differentiation, we had the osteogenic and chondrogenic lineages. Others had worked out many of the details of myogenic lineages, and quite interestingly, the suggestion had been made that hematopoietic support tissue (the marrow stroma) also contained cellular elements that provided differential support for separate arms of the hematopoietic pathway. The summary diagram in Figure 4-1 is therefore a composite of many individuals' efforts and was fashioned to mimic the fundamental aspects of the hematopoietic pathway; I called this multilineage pathway the "mesengenic process." I named the multipotent progenitor the mesenchymal stem cell, because all of the lineage pathways depicted in Figure 4-1 were mesenchymal in origin. The reasons for calling this cell a stem cell were rather loose, but were primarily inferred from the observation that the cells we isolated could be induced into different lineage pathways. Thus, these "stemmed" from a common source.

Adult mesenchymal tissues all have unique turnover dynamics—some relatively rapid, like bone, and others relatively slow, like cartilage. In all cases, turnover means that differentiated cells in the tissue die and are replaced by newly created cells. For bone, this death and birth can take place when large sheets of osteoblasts are fabricating sheets of osteoid that will eventually mineralize. Several full generations of cells may come and go while a sector of bone is being fabricated.

Location of Mesenchymal Stem Cells

By the preceding logic, MSCs must occur everywhere mesenchymal tissue turns over. To date, mesenchymal progenitors have been isolated from marrow, muscle, fat, skin, cartilage, and bone. Moreover, every blood vessel in the body has mesenchymal cells on the tissue side of the vessel (some of these are MSCs), although names like pericytes have been given to these multipotent cells.

Test for Mesenchymal Stem Cells

The tests for MSCs are straightforward: populations of cells are placed in medium conditions that include exposure to strong phenotypic inducers. Following phenotypic induction, the expressional activity and morphology of the cells are monitored, and a judgment is made that a specific phenotype is present, based on the analysis of indicator molecules, morphologies, or both. Of course, *in vivo*, we have no unique markers for MSCs, although Stephen Haynesworth and I long ago isolated three monoclonal antibodies (SH2, SH3, and SH4) that preferentially bind to MSCs compared to hematopoietic cells.

Plasticity

One recent experimental complication is that we can take mesenchymal cells of a known phenotype, for example, an adipocyte with fat droplets inside, and expose these cells to inductive medium for another phenotype, for example, osteoblasts. These newly induced cells stop producing fat-specific molecules and start producing bone-specific molecules in osteogenic inductive medium. The capacity to be induced from one lineage into a new lineage is referred to as "plasticity." Once this is recognized, the functional test for an MSC becomes less stringent, because non-MSCs, like the adipocytes cited previously, and MSCs can be induced to become osteoblasts. For emphasis, this is an experimental situation and may not be relevant to *in vivo* events. However, a cell's capacity for plasticity clouds the issue of identifying MSCs by these functional assays for multiple mesenchymal phenotypes.

Mesenchymal Stem Cells Decrease With Age

The body's need for MSCs changes dramatically, depending on the age of the individual. In embryos, mesenchymal tissues form from a relatively high local density of progenitors within a very loose, watery extracellular matrix. The massive differentiation of these groupings of progenitors into differentiated cells, and thus into tissues, involves the substantial up-regulation of synthesis of complex and tissue-specific extracellular matrices. The cell-to-matrix ratios go from high to low during these differentiation events. Moreover, the matrix adds to the many-fold increase in the gross size of the tissues being formed. In this process, many of the progenitor cells are directly converted to differentiated cells, and the relative MSC numbers dramatically decrease in the embryo. Similarly, from birth to the teens, the increase in body-part sizes is many-fold; this also must involve the direct conversion of some of the MSCs into differentiated phenotypes. In this regard, it seems reasonable to expect a mesenchymal tissue injury of a child below the age of 5 years to regeneratively repair because of relatively high MSC levels; in a 20-year-old, the same tissue might only repair with fibrous scar tissue. To take this comparison even further, it would seem reasonable to expect that 50-year-olds would have fewer MSCs locally than 20-year-olds to repair mesenchymal tissue injuries.

Since we have no unique probes for MSCs, we cannot determine the whole-body titers of MSCs. As an estimate, we can measure the colony-forming units—fibroblasts from marrow aspirates or marrow biopsies. Even in this assay, the differences in marrow aspiration techniques, medium additives, and growth-medium serum levels (and prescreening batches of serum) preclude establishing an experimental consensus on the absolute numbers of MSCs in marrow. What everyone seems to agree upon is the *decrease* of MSCs in marrow with age.

Based on this information, I propose six periods of differing MSC titers. Each period would be expected to have a different capacity for tissue expression and regenerative repair. The different periods are: (1) embryology, fertilization to birth; (2) neonatology, birth to 6 years; (3) teens, 7 to 20 years; (4) peak skeletal performance, 20 to 30 years; (5) midlife, 35 to 55 years; and (6) late life, 60 to 90 years. The important distinction is that, although HSC titers and potency are relatively constant from birth to late life, MSC titers change by one or two orders of magnitude. This highlights the fundamental issue that both the definitions and the functional biology of MSCs and HSCs are expected to be different.

TISSUE-ENGINEERED REGENERATION

With the preceding information as a base logic, it follows that the regenerative repair of certain mesenchymal tissues will require the augmentation of MSC numbers at the repair sites in post-teen adults. This can be accomplished in several ways depending on the age and health status of the individual. For years surgeons have used freshly harvested whole marrow to augment local skeletal tissue healing, and methods to concentrate and enrich the marrow are being introduced commercially. Key to this latter use, and to the use of culture-expanded MSCs, are the type and properties of the scaffolds used to isolate, concentrate, and/or deliver the cells to the repair site. Such composite grafts form the backbone of the new scientific discipline of tissue engineering.

ISOLATING AND CULTURING MESENCHYMAL STEM CELLS

MSCs have been isolated from several species and tissues, but the most well characterized, and probably the purest preparation, is from human bone marrow. The studies to purify MSCs from human marrow were initiated in the mid-1980s by Stephen Haynesworth in my laboratory. The key to our initial and rapid success was the lot of fetal bovine serum used to initiate these studies. The original batch of serum was selected from several lots by a selection assay using embryonic chick limb bud mesodermal cells in culture. At that time, we would prepare such cultures weekly from 200 dozen fertile eggs; batches of serum varied greatly from the same commercial supplier and from many suppliers. Based on color alone, many batches, by experience, were not suitable. (A dark brown or greenish tint indicated hemolyzed blood, and batches with these colors were not suitable.) The serum accounts for three key aspects of the MSC purification–expansion protocol: (1) selective attachment to tissue culture plastic to provide discrete colonies of MSCs; (2) optimal mitotic expansion of the cells; and (3) maintenance of the undifferentiated or stem cell state of the MSCs.

Today we use a laborious derivative of our initial, published screening protocol. In this regard, it is more difficult to screen batches of serum in that only one in 20 to 30 lots is suitable for human MSCs. This is because some suppliers blend batches to optimize their use in hybridoma cultures, making it more difficult to screen for a suitable lot for MSCs. Moreover, every animal preparation requires its own lot of serum, with the batch for human being unsuitable for rat, rabbit, or mouse marrow MSCs. (In some cases, batches of serum optimal for human MSCs are toxic to mouse MSCs.) Thus, if an unscreened batch of serum is used, or if an agent such as dexamethasone is added to the basic growth medium, results for colony counts and for cellular response profiles to other bioactive agents will differ from those obtained in my laboratory. In evaluating other reports using MSCs, information pertaining to the serum and other medium components is crucial.

In the early 1990s, Haynesworth, our colleagues, and I organized the transfer of the MSC technology from Case Western Reserve University to a newly formed company called Osiris Therapeutics, Inc. (OTI, now in Baltimore, Maryland). The scientists at OTI have repeated and refined our basic technology and have added to the database of human MSCs. They showed that by second passage the MSC preparation was homogeneous by fluorescence-activated cell-sorting criteria using more than 100 cell-surface antibodies and observed that clones could be obtained that showed multipotency for different phenotypic lineages. Our efforts to use mouse, rat, rabbit, dog, goat, and sheep marrow MSCs for tissue engineering were also transferred to OTI, where various preclinical animal models extended the results first obtained through my laboratory. OTI expanded the safety studies and initial human clinical trials started at Case Western Reserve University and University Hospitals of Cleveland.

TISSUE-ENGINEERING USES OF MESENCHYMAL STEM CELLS

In mature adults, marrow MSCs are used to supply progenitors for normal turnover and for repair–regeneration of tissue damage. The basis for this statement is the obvious involvement of MSCs in marrow stroma in support of normally active hematopoiesis, and of MSCs in normal bone turnover, to both directly support the formation of osteoblasts and indirectly support the formation of osteoclasts through two different lineage pathways (mesengenesis and hematogenesis, respectively). Thus, the use of culture-expanded MSCs to massively reform or repair certain tissues far exceeds their normal use and availability. In this case, we have adopted the logic that, to accomplish reconstruction of tissues that cannot normally themselves provide suitable numbers of progenitors, we must provide these progenitors from exogenous sources. Concomitantly, we must attempt to mimic selected principles of embryonic development and recapitulate these events in adults. The most important aspect of embryonic mesengenesis is that the progenitor cell to matrix ratio is very high just prior to lineage pathway entrance, and that the exact limits or edges of the neotissue are defined. Moreover, we now recognize that each site and each tissue have a unique sequence of inductive events and boundary conditions. Next, I briefly describe the use of MSCs for specific tissue regeneration, taking this embryonic recapitulation logic into account as depicted in Figure 4-2.

Bone Repair

The first and most obvious use of MSCs is in the area of bone regeneration in sites where the body cannot organize this activity, such as in nonunions. Critical size defects in

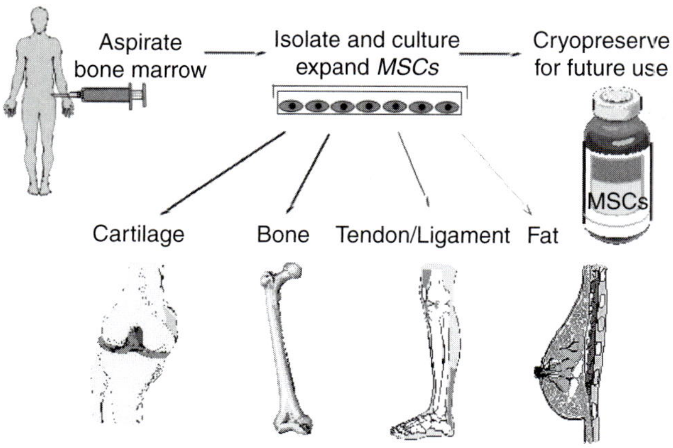

FIGURE 4-2 From an iliac bone marrow aspirate, MSCs could be used for tissue-engineered repair of cartilage, bone, or tendon or for plastic-surgical implant of fat from fresh or frozen stored cells.

nonunion models have shown that culture-expanded marrow MSCs in a porous, calcium phosphate, ceramic delivery vehicle are capable of regenerating structurally sound bone, where whole marrow or the vehicle alone cannot satisfactorily accomplish this repair. These preclinical models include the use of human MSCs in a femoral nonunion model in an immuno-compromised rat that will not reject the human cells.

Cartilage Repair

Cartilage is an avascular tissue incapable of regeneration–repair of even small defects in adults. Although chondrocytes have been used to attempt to repair large cartilage defects, it is difficult to integrate the neotissue with that of the host. In this case, we have pioneered the use of hyaluronan (HA) scaffolds, because of the high content of HA in the embryonic mesenchyme of precartilaginous tissues. These HA-based scaffolds are being tested in human clinical trials in Europe using autologous chondrocytes. These scaffolds provide an inductive microenvironment for MSCs to enter the chondrogenic lineage, and HA-scaffold breakdown products (oligomers) appear to facilitate the integration of neotissue with that of the host.

Marrow Regeneration

Injected MSCs make it back to the bone marrow to refabricate injured marrow stroma. This observation is the basis for a multi-location clinical trial to add back autologous MSCs to chemotherapy–radiation patients receiving bone marrow transplants.

Muscle Regeneration

We were the first to document that congenic (i.e., normal) MSCs could be injected into a specific muscle of the muscular dystrophy (*mdx*) mouse to cure it by providing newly synthesized dystrophin to the affected myotubes. In this case, the donor MSCs differentiated into skeletal myoblasts, fused with the host myotubes, and caused the synthesis and distribution of dystrophin. Moreover, labeled MSCs injected into injured (infarct model) rat or pig heart appear to differentiate into cardiac myocytes.

Fat

MSCs have been induced into the adipocyte pathway and can massively accumulate fat droplets. Although not used in tissue engineering models, we propose that bags of autologous fat are more appropriate for some plastic surgeries than bags of saline or silicone.

Tendon Repair

Studies initiated in my laboratory, and later completed at OTI and the University of Cincinnati, clearly establish the use of autologous MSCs for repair of tendons. The innovative design of an MSC-contracted collagen gel around a restorable suture held in tension allows tendon defects to be spanned by these composite constructs under normal loads. In this case, the same MSC able to differentiate into cartilage or bone develops into appropriate tendon tissue at Achilles or patellar tendon sites.

Gene Therapy

The studies with the *mdx* mouse cited previously establish the capacity of allo-MSCs to cure genetic defects. This principle has been translated to the clinical use of allogeneic bone marrow transplantation with or without additional culture-expanded MSCs to attempt to cure polysaccharide storage diseases or *Osteogenesis imperfecta* with substantial success. In addition, we long ago showed that human MSCs could be transfected with a retroviral construct without affecting their differentiation capacity. Molecules coded for by these gene inserts could be shown to be present many months after introduction into the *in vivo* model. Thus, MSCs hold the potential for a variety of gene therapy applications.

Vascular Support

The relationship between vascular endothelial cells and MSCs is not well studied. It is clear that multipotent mesenchymal progenitor cells are present on blood vessels (pericytes). More important are the details of the interaction between these vascular support cells and angiogenesis–vasculogenesis. Certainly, in embryonic angiogenesis, MSCs play a strong role. In adults, for example, it is now clear that vascular endothelial cells in bone make skeletal inductive molecules, such as BMPs. In addition, MSCs make vascular endothelial growth factor. Based on these and other observations, we predict that, although nearly all tissues contain blood vessels, each tissue has specifically differentiated endothelial cells, and perhaps even unique MSC-derived vascular support cells.

SUMMARY

Adult bone marrow contains mesenchymal progenitor cells that may be called MSCs. Whether these preparations are composed of a pure population of stem cells or a spectrum of cells ranging from pluripotent stem cells to lineage-committed progenitors, it seems clear that these culture-expanded cells have the impressive potential to be used in tissue-engineered regeneration and repair of a variety of tissues. In some cases, as in bone nonunions, these cells are capable of differentiating directly into osteogenic cells. In the cases where they are used in cardiac (infarct) or brain (stroke) therapy, these cells appear to provide instructive and inductive microenvironments, inhibit scarring, and stimulate blood vessel regeneration for facilitated repair of critical lesions. That mesenchymal cells appear to be unusually plastic in their developmental capabilities opens the door for their use in several cell-based therapies.

ACKNOWLEDGMENTS

My thanks to my colleagues and collaborators at Case Western Reserve University and elsewhere for their assistance and encouragement. The National Institutes of Health generously supported much of the experimentation reviewed here.

KEY WORDS

Mesenchymal stem cells Cells that have the capacity to form different mesenchymal tissues given exposure to different inductive agents.

Mesenchyme The middle layer of an embryo, the mesoderm and the tissues that form (bone, cartilage, muscle, fat, etc.).

Osteoblast A cell formed by a multi-step differentiation sequence that becomes capable of fabricating bone.

Tissue engineering A new science that encompasses aspects of the natural and physical sciences that focus on the repair or regeneration of various tissues.

FURTHER READING

Caplan, A. I. (1984). Cartilage. *Sci. Am., 251*, 84–94.

Caplan, A. I. (1991). Mesenchymal stem cells. *J. Ortho. Res., 9*, 641–650.

Caplan, A. I. (2003). Embryonic development and the principles of tissue engineering. In *Novartis Foundation: Tissue Engineering of Cartilage and Bone*. London: John Wiley & Sons.

Goldberg, V. M., & Caplan, A. I. (2004). Principles of tissue engineering and regeneration of skeletal tissues. In V. M. Goldberg & A. I. Caplan (Eds), *Orthopedic Tissue Engineering Basic Science and Practice*. New York: Marcel Dekker.

Horwitz, E., Prockop, D., Fitzpatrick, L., Koo, W., Gordon, P., Neel, M., Sussman, M., Orchard, P., Marx, J., Pyeritz, R., & Brenner, M. (1999). Transplantability and therapeutic effects of bone marrow-derived mesenchymal cells in children with osteogenesis imperfecta. *Nature Medi., 5*, 309–313.

Urist, M. R. (1994). The search for and discovery of bone morphogenetic protein (BMP). In M. R. Urist et al. (Eds), *Bone Grafts Derivatives and Substitutes*, pp. 315–362. London: Butterworth Heinemann.

Chapter 5

Stem Cells, Plasticity, and Regeneration

Bruce M. Carlson

> *The history of science makes it certain that the knowledge derived from research on stem cells will eventually lead to enormous benefits for human health, even if they are unpredictable.*
>
> —Bruce Alberts (2006)

Since the late 1990s, a sea change has been seen in the attitude of the research community toward cellular plasticity, stem cells, and transdifferentiation. Earlier, the possibility of dedifferentiation was denied by most researchers, and transdifferentiation was generally regarded as a biologic curiosity that occurred only in odd systems, such as urodele lens or retinal regeneration. If a role of hematogenous stem cells in regeneration was even mentioned in the literature, it was most often brought up to be refuted by existing lines of evidence. Yet, within just a few years after stem cells burst onto the scene, they have often been viewed as a panacea that can cure an amazing variety of ills. In this early stage in the history of stem cell biology, there is a need to examine the relative roles of traditional regenerative and healing mechanisms and stem cells as they apply to both natural and engineered regenerative phenomena.

Limb regeneration was one of the first fields in which the origin of regenerating cells was discussed and experimentally tested. In response to early discussions of a hematogenous origin of blastema cells, Butler (1935) conducted experiments involving x-radiation showing that a blastema does not form if the distal part of a limb stump is irradiated, but that a blastema does form if that part of the limb stump is protected while the rest of the body is irradiated (see Figure 5-1). This experiment clearly showed the local origin of the regeneration blastema, but it did not exclude the possibility that cells beyond the distal limb stump could, to some extent, participate in the regenerative process. This latter possibility was never seriously tested.

In mammalian tissues, two main models of regeneration took shape. In one model, well exemplified by the liver, parenchymal cells constitute the principal source of regenerating cells, but if there is a deficiency of hepatocytes, then oval cells from the ducts could become activated and compensate for the deficiency in the normal primary cellular source.

The other main model of tissue regeneration requires the presence of a population of tissue-specific progenitor cells that could be called into play after damage to that tissue. Examples of this are the periosteum and endosteum of bone and the satellite cells of skeletal muscle. In these tissues, it was assumed that the primary tissue-specific cells (osteocytes of bone and muscle fibers in muscle) do not survive the injury, and that the progenitor cells have the capacity to restore any tissue deficit. Some of the classic models of pancreatic regeneration involve almost complete destruction of the parenchyma and subsequent regeneration from ductal cells. The origin of these progenitor cells has not been easy to pin down. Whether the progenitor cells come from embryonic rests that did not differentiate together with the rest of the tissue or whether they arise from separate lineages continues to be discussed.

How do stem cells fit into these schemes? A principle articulated decades ago in the limb regeneration literature can serve as a framework. This is the principle of *double assurance,* another term for compensation. Experiments first conducted a century ago involved removing the bones from a limb stump; researchers observed that the regenerated limb contained a completely normal skeleton despite the absence of bones in the stump. With these and other experiments that were conducted at the tissue level, it became apparent that when the first-line regenerative response was denied, a secondary mechanism was called into effect to produce the missing tissue. In a more contemporary context, many of the early single-gene knockout experiments produced disappointingly minor developmental disturbances. It was not until double knockout experiments were tried and resulted in major developmental disturbances that the principle of compensation by a closely related molecule also became apparent at the genetic level. The level of genetic redundancy is such that, for many important genes, a molecular backup system is available to compensate for a single mutation. A similar situation also appears to apply at the cellular level. Many damaged tissues can normally regenerate adequately with local cellular resources, but if for some reason these are deficient, a cellular backup system can be brought into play. In the case of the liver, for example, the backup system can act at several levels. The first level is the parenchyma. The second level is the oval cells, which represent a local progenitor population. A third level of compensation may be stem cells originating from outside the organ itself.

WHAT IS A STEM CELL?

Much of our knowledge of stem cells comes from the numerous studies on the hematopoietic system, which serves as a model, although not necessarily the only one, for understanding the properties and capabilities of stem cells (Körbling and Estrov, 2003; Weissman, 2000). Even in this well-studied system, many major questions remain. A generic definition for stem cells is "self-renewing populations of cells that undergo symmetric and asymmetric divisions to self-renew or differentiate into multiple kinds of differentiated progeny" (Cai *et al.*, 2004, p. 585).

Many different types of stem cells currently have been identified. The first dichotomy in a classification of stem cells is between embryonic (ES) and adult stem cells. ES cells are derived from the inner cell mass of blastocysts. These have been described as totipotent because a single cell from the inner cell mass can, in theory, go on to form any of the differentiated cell types in the body. Permanent cell lines derived from single ES cells can be perpetuated *in vitro*. From such founder lines, subsets of cells can be exposed to defined culture conditions and steered to differentiate along certain specific pathways before being implanted into damaged or deficient tissues of organs. This seemingly straightforward approach shows great promise, but it is fraught with technical difficulties. In addition to understanding how to direct the ES cells toward differentiation in a desired direction, other important issues are the potential developmental lability of these cells and their antigenicity. Ensuring that implanted cells do not form teratomas is a significant challenge. Because descendants of ES cells are likely to be immunologically different from the host in which they are implanted, options are to match donor and host, reduce the antigenicity of the donor cells, or suppress the immune response of the host. Superimposed on all of these scientific problems are political and ethical issues involving obtainment of human ES cells.

Adult stem cells are cells in the postnatal body that retain stem-cell properties throughout life. Although the existence of hematopoietic stem cells in the bone marrow has been both discussed and demonstrated for many decades, the recognition of the presence of stem cells in other parts of the body is much more recent. Several levels of "stemness" are possible. True stem cells must meet the criteria of the definition given at the beginning of this section. Although much of the time stem cells are mitotically quiescent, when a stem cell does divide, several outcomes are possible (Figure 5-1). The division could be a symmetric one in which both daughter cells remain stem cells. Another option is for the division to be an asymmetric one, in which one daughter cell remains a stem cell, but the other one becomes a more restricted intermediate type of cell. Yet, a third option is for each daughter cell in a symmetric division to become an intermediate type of cell.

If one looks at developmental potential, the primary stem cell, sometimes called a *long-term stem cell,* has the greatest ability to create a wide variety of differentiated cell types (Figure 5-2). Some of their descendants, which remain mitotically active, become more restricted in their developmental fates so that they can no longer produce as many cell types as could the primary stem cell. These have been called *short-term stem cells.* Further down the line are transit-amplifying cells. These cells may be restricted to producing only one type of progeny, and they produce the bulk of the cells needed to repair a damaged organ. Finally, their progeny leave the mitotic cycle and differentiate into mature parenchymal cells of the organ in question.

To complicate matters further, a given tissue or organ may house more than one type of stem cell. For example, adult bone marrow is reported to contain at least four types of stem cells: hematopoietic stem cells, mesenchymal stem cells, multipotent adult progenitor cells, and endothelial precursor cells (Körbling and Estrov, 2003). These stem cells have different antigenic properties and have the potential to produce different sets of progeny (Figure 5-3). At one time, it was thought that adult stem cells could be distinguished from tissue-specific progenitor cells, such as satellite cells of muscle or basal cells of epidermis, that were thought to be limited to producing only one differentiated type of progeny, but it now appears that even these progenitor cells have sufficient inherent plasticity to create other cell types when exposed to different environmental conditions.

WHERE ARE ADULT STEM CELLS FOUND?

At one time, only bone marrow was thought to contain adult stem cells, but since the 1990s, adult stem cells have been identified in a wide variety of tissues (Table 5-1). Stem cells are extremely difficult to identify by morphology alone. Marrow-derived stem cells can be identified by unique

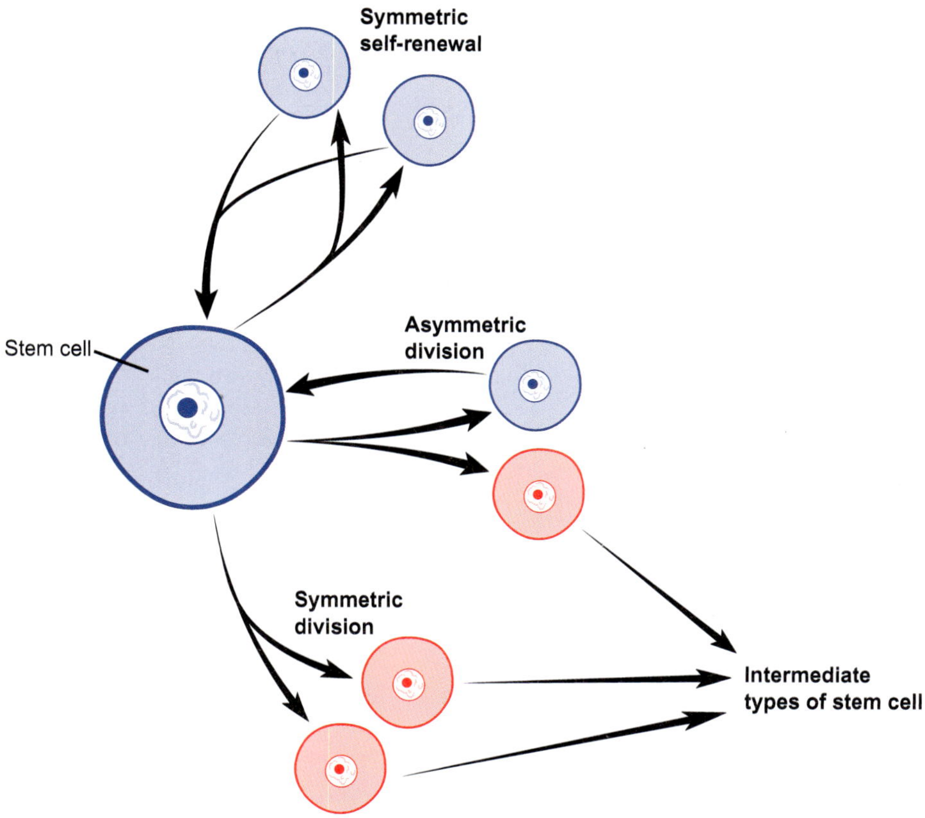

FIGURE 5-1 Types of cell divisions possible for stem cells.

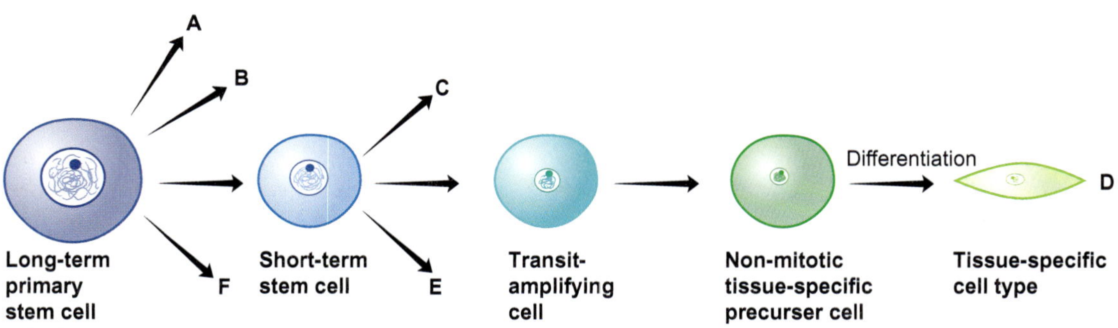

FIGURE 5-2 Scheme illustrating restriction in developmental potential of stem cells and their progeny.

combinations of antigenic markers, but even with these it is easiest to identify them through fluorescence-activated cell sorting rather than in histologic preparations. As a result, it is not easy to localize stem cells in many tissues. There are exceptions, however. In the central nervous system (CNS), stem cells have been identified in regions such as the subventricular zone through their mitotic activity (Gage, 1998). At an even more specific level, stem cells associated with hairs are located in the small epithelial bulge region alongside the hair follicle (Sieber-Blum *et al.*, 2004).

The situation is much less clear in other tissues, such as skeletal muscle. In muscle, tissue-specific satellite cells are located alongside individual muscle fibers, but there appears to be another population of stem cells in muscles that have been defined principally as coming from the "side population" capable of Hoechst 33342 efflux as identified by cell sorting (Deasy *et al.*, 2004). To further complicate the situation, Dreyfus and colleagues (2004) have localized marrow-derived stem cells in muscle in both the connective tissue and beneath the basal laminae of the muscle fibers, the latter being a location thought to be reserved exclusively for conventional satellite cells.

An emerging concept is that of specific microenvironmental niches for the production and maintenance of stem cells (Moore and Lemischka, 2006). In certain situations, such as the intestinal crypt and the bulge region in hair follicles, the topography of the stem cell niche is reasonably well defined, but in others, it is much less clear. Interactions between stem cells and surrounding mesenchymal cells, involving both Wnt and bone morphogenetic

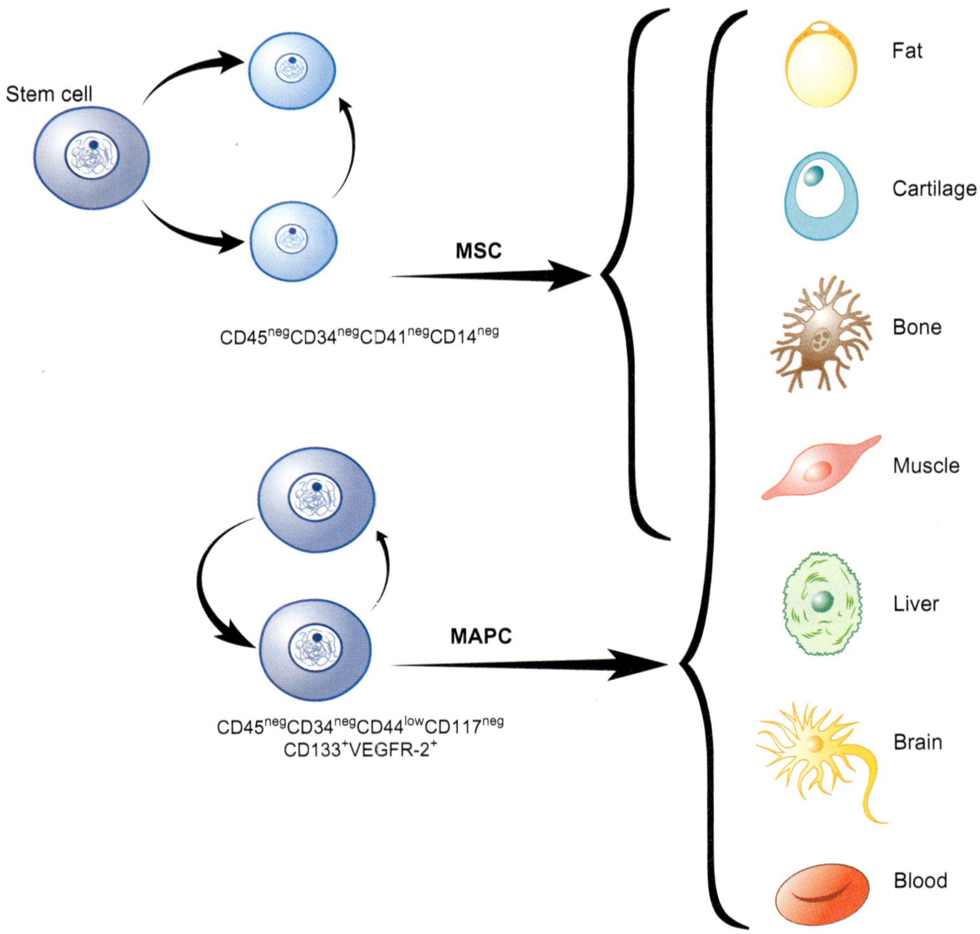

FIGURE 5-3 Types of mesenchymal stem cells present in bone marrow and the progeny that each can produce. MAPC, multipotential adult progenitor cell; MSC, mesenchymal stem cell. (Based on Bunting and Hawley's experiments [2003].)

protein (BMP) pathways and working through nuclear β-catenins in the stem cells, appear to be a common element in maintaining "stemness" in the niches and also in stimulating their production of daughter cells.

PROPERTIES OF ADULT STEM CELLS

In most cases, adult stem cells are best defined by the cell types that they can produce, rather than by intrinsic properties of the cells themselves. Currently, there is intense activity to identify antigens, cellular properties, or gene products that are common to all varieties of stem cells (Cai *et al.*, 2004), but to date, only the most general of characterizations can be made. One important property of stem cells is the absence of specific lineage markers that characterize individual types of differentiated cells. Equally important in the identification of the differentiated progeny of stem cells is the expression of lineage markers, such as type II collagen for chondrocytes and alkaline phosphatase for osteocytes.

Adult stem cells, although quiescent for long periods, must be able to enter the cell cycle when stimulated. Expansion of a stem cell population is triggered by external events, which are often topographically removed from the stem cells themselves. This means that the initiating event must be communicated to the stem cells by some means, probably chemical in most cases. It is likely that the initiating signal is different for most types of stem cells, although the existence of some common element has not been ruled out. Then the stem cell must have the proper receptors to receive the signal and to activate the mitotic apparatus.

Stem cells are often long lived, and they must be protected from DNA damage, which means that they must possess a functional system for DNA repair. Because they may need to divide many times over the course of a lifetime, human stem cells must also have higher telomerase activity than most somatic cells to maintain telomere length toward the end of life. Given these properties for cellular maintenance, stem cells must be able to respond to an activating signal by entering the mitotic cycle. Entry into mitosis by typical somatic cells is regulated at the G1/S checkpoint, but stem cells spend more time in the S-phase, and the checkpoint for them is at the S/G2 junction (rev. Cai *et al.*, 2004). One of the least understood aspects of stem cell division is what determines whether the daughter cells will be channeled into a self-replication

TABLE 5-1 Types of Adult Stem Cells and Locations Where They Can Be Found

Ectodermal	**Endodermal**
Epidermis	Intestinal crypt
Hair follicle	Gastric gland
Neural	Pancreas
Neural crest	Liver
Retina	**Germ Cell**
Dental pulp	Primordial germ cells
Mesodermal	
Cardiac muscle	
Skeletal muscle	
Hematopoietic	
Mesenchyme	
Multipotent adult	
Endothelial	
Umbilical cord blood	
Umbilical cord matrix	

Adapted from Cai, J., et al. 2004. In search of "stemness." Exp Hematol 32:585–598, by permission.

cycle or will enter an asymmetric division, with one remaining a stem cell and the other becoming a more restricted cell heading toward a specific cell lineage. Wnt signaling maintains stem cells in a condition of "stemness" and prevents them from differentiating (Kléber and Sommer, 2004).

Recent research has clarified considerably the molecular basis for maintaining the state of stemness (Bernstein *et al.*, 2006; Lee *et al.*, 2006). This research has focused on chromatin (histone)-based repression of gene activity. Members of the polycomb group proteins catalyze the methylation of histone H3 at specific lysine loci, and this complex silences the regulatory genes, whose proteins turn on more than 200 key developmental genes that are involved in cellular differentiation. The juxtaposition of a silencing site with one that has a stimulating function in what is called a *bivalent domain* allows for greater flexibility and control of gene expression in both stem cells and early ES cells. Like undifferentiated cells in the cleaving embryo, true stem cells express the transcription factor *Oct-4*.

Another important but poorly understood property is translocation. The final progeny of most stem cells mature and settle down in a location other than that of the original stem cell, but what causes them to translocate and how they do it are, for the most part, unknown. The pathway of translocation is known in at least one system. In the CNS, the stem cells in the subventricular zone send their progeny via a rostral migratory stream to the olfactory cortex (Figure 5-4, A), where they differentiate into neurons or glial cells (Lois and Alvarez-Buylla, 1994). Ciliary currents created by the ependymal cells set up a chemorepulsive gradient (Slit-2), which guides neuroblast migration into the olfactory area (Sawamoto *et al.*, 2006).

Once stem cell progeny reach the area in which they will engraft, they must deal with the issue of integration with the other cells of that tissue. This is a tissue-specific phenomenon that is likely dealt with in a specific way by each tissue in the body. Definitive differentiation of stem cell progeny may well involve interactions with other cells or the matrix of that tissue. A final critical property of stem cell progeny is avoidance of neoplastic transformation and overall maintenance of genomic stability.

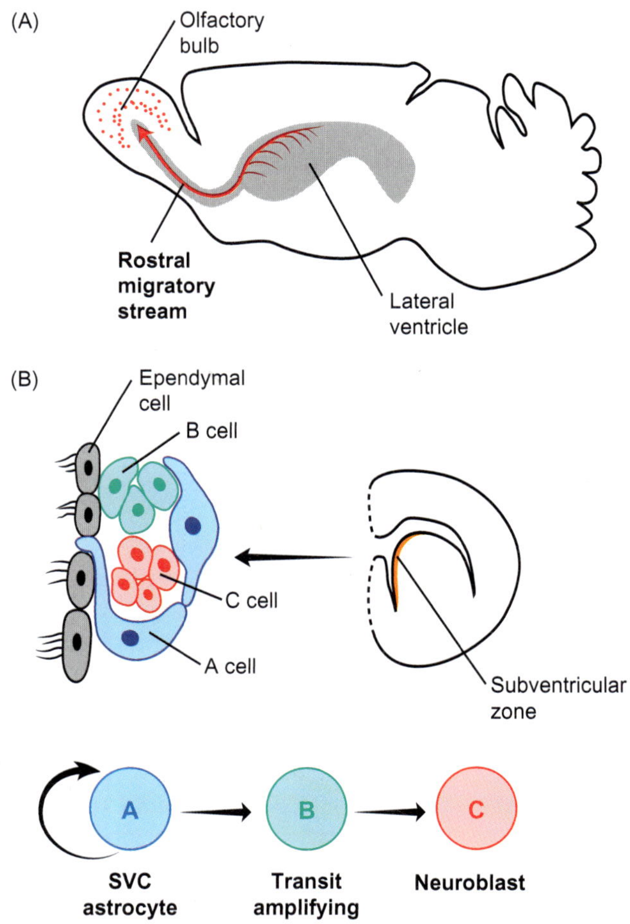

FIGURE 5-4 (A) Representation of the rostral migratory stream, which carries cells generated in the subventricular zone along the lateral ventricle into the region of the olfactory bulb in the rat. (B) Organizational details of cell arrangements in the subventricular zone as seen in cross section. The sequence below shows the pathway of cell transformation. A cells represent subventricular zone (SVC) astrocytes; B cells represent rapidly dividing transit-amplifying cells; C cells represent neuroblasts. (Adapted from Bottai *et al.* [2003]; Alvarez-Buylla and Lim [2004].)

STEM CELLS IN REGENERATION AND TISSUE RECONSTRUCTION

Stem cell progeny are known to populate a wide variety of human tissues and organs. This has been demonstrated in cases of sex-mismatched bone marrow or organ transplants by the use of Y-chromosomal markers. The analyzable combinations include female individuals who have had bone marrow transplants from male donors or male individuals who have received organ transplants from female donors. In the latter case, the presence of cells containing the Y chromosome in the transplant shows that host cells have populated the transplant. In the former case, the presence of cells bearing the Y chromosome in any nonmarrow tissue indicates a cellular mosaic between host tissue and grafted marrow cells. According to one report, after a single, male, bone marrow stem cell was injected intravenously into lethally irradiated mice, progeny of that cell were found in the skin, the kidneys, the liver, and the epithelial lining of lung and small intestine (Krause *et al.*, 2001).

Careful analysis in these studies is essential because the Y chromosome-bearing cells could be macrophages or some other bloodborne cell, rather than a parenchymal cell or a permanent component of the stroma. In transplant situations, male-derived cells have been found in the liver, heart, kidney, endothelium of blood vessels, bone, epithelia of gut, buccal cavity, and skin (rev. Körbling and Estrov, 2003). The reported percentages of immigrant cells vary considerably, from less than 1% to more than 20%. One important observation from the clinical studies is that the percentage of immigrant cells in an organ is typically greater if rejection or some pathologic process is occurring in that organ. This confirms laboratory data suggesting that stem cells are more likely to be called into play when an organ is damaged.

A significant issue in the interpretation of experiments or natural phenomena involving stem cell populations of tissues is the demonstration that under some circumstances stem cells can fuse with other cells (Terada *et al.*, 2002; Ying *et al.*, 2002). The original demonstrations of stem cell fusion were made *in vitro*. It has been more difficult to verify the presence or absence of fusion *in vivo*. Especially in cases in which analytic capabilities are somewhat limited, as in the human situations described earlier, claims of engraftment and conversion of stem cells to organ-specific parenchyma may not be completely substantiated if fusion or the definitive identification of the phenotype of the stem cell derivative is in question.

Because of the difficulty of analysis in human cases, most research on stem cells in regeneration has been conducted on experimental animals. One of the most important reasons for using laboratory animals is the ability to track the fate of implanted stem cells. Three commonly used methods for marking stem cells involve: (1) the use of green fluorescent protein (GFP)-marking of cells, (2) the detection of Y chromosomes by *in situ* hybridization, and (3) the detection of β-galactosidase in transgenic cells. Such tracking, combined with tissue-specific differentiation markers, is one of the most effective ways of determining where stem cells are going and what they turn into. At this stage in the development of stem cell biology, much of the research effort is concentrated on what amounts to lineage tracing, but as stem cell pathways become clearer, effort is turning toward understanding what makes a stem cell a stem cell and how to maximize its participation in regenerative or restorative processes.

EXAMPLES OF STEM CELL PARTICIPATION IN REGENERATION

The promise of stem cell therapy for the treatment of many currently intractable pathologic conditions has led to considerable research on both the presence of stem cells in organs and the ability of stem cells to populate a variety of organs. Organ-specific studies have shown a number of common issues that are generic to stem cell therapy, as well as those that are unique to the organ in question. This section summarizes what has been learned and what remains to be sorted out in three intensively investigated model systems: heart, CNS, and skeletal muscle.

Heart and Cardiac Muscle

One of the real surprises to those who pioneered the isolation and maintenance of ES cells was how often the resulting embryoid bodies contained groups of spontaneously beating cardiac muscle cells. Subsequent studies have shown that the cardiomyocytes differentiating from stem cells go through an orderly sequence of developmental gene expression that closely mirrors that which occurs during *in vivo* cardiac embryogenesis, and that these cells develop the physiologic and pharmacologic properties of terminally differentiated cardiomyocytes (Bohler *et al.*, 2002; Sauer *et al.*, 2004).

A major possibility of using stem cell therapy in the regenerative reconstruction of the heart is based on the pathology of a myocardial infarct. In a typical infarct, a segment of the ventricular wall is deprived of oxygen because of blockage of a coronary vessel, and the cardiomyocytes in that area die. This necrotic area becomes invaded by macrophages and fibroblasts, which ultimately transform into myofibroblasts and produce scar tissue. Between the necrotic area and the relatively undisturbed heart wall is a perinecrotic zone, in which some cellular reparative activity occurs. Over time, the scar tissue of the necrotic zone expands (ventricular remodeling) because of the continuous intraventricular blood pressure, and the pathology of congestive heart failure develops.

A number of strategies have been designed to deal with this problem. One of the more common ones is transplantation of a flap of skeletal muscle (commonly *latissimus dorsi*) over the region of the infarct. Another has been to inject skeletal muscle myoblasts into the area of the infarct or into vessels that lead into the infarcted region (Horackova *et al.*, 2004). Little evidence exists of a transformation of skeletal muscle into cardiac muscle, and one of the significant issues is whether any skeletal muscle that grows or persists on the heart can adapt itself to the continuous beating function required of cardiac muscle. Implanted skeletal myoblasts do not become coupled electrically with neighboring cardiomyocytes (Leobon *et al.*, 2003). They contract independently of the myocardium, which could lead to the development of cardiac arrhythmias. An issue with the injection of myoblasts is the disappearance of more than 90% of the myoblasts from the site of injection within a few hours. The same phenomenon of disappearance of the cellular injectant occurs when myoblasts are injected into dystrophic muscles. One explanation is the removal of the foreign cells by natural killer cells.

The demonstration of the presence of host cells marked by Y chromosomes within transplanted hearts derived from female individuals (Quaini *et al.*, 2002) has stimulated considerable research and even clinical trials (Schächinger *et al.*, 2004; Strauer *et al.*, 2002) on stem cell therapy for the treatment of certain types of heart disease. A strategy that has been touted as a successor to the above treatment is the introduction of stem cells into areas of heart damage. This has been attempted on a number of occasions, especially with marrow-derived adult stem cells. The stem cells have been introduced by direct injection into the infarcted region or into nearby coronary vessels. According to a number of reports, significant numbers of the stem cells have become engrafted into the ventricular wall as new cardiomyocytes (Orlic *et al.*, 2001a,b). In contradiction to these claims, two later reports (Balsam *et al.*, 2004; Murry *et al.*, 2004) have found no or little evidence of conversion of either directly injected hematopoietic stem cells or derivatives of transplanted bone marrow (in irradiated mice) into cardiac muscle cells. Other studies (e.g., Jackson *et al.*, 2001) have found a low percentage (e.g., 0.2%–0.3%) of donor-derived cardiomyocytes and a higher percentage (several percent) of donor-derived endothelial cells in the perinecrotic region.

These contradictory results or interpretations point out some important issues in research on any type of stem cell-based regenerative therapy. First, the donor cells must be clearly labeled, so that later they or their progeny can be clearly identified (Thiele *et al.*, 2004). In addition, it is important to know whether the donor population is homogeneous or whether it possibly includes more than one type of stem cell. The site of introduction is significant (e.g., into the organ itself or into a distant site, such as bone marrow). Once one gets past qualitative issues, such as whether conversion of the stem cells into the desired phenotype actually occurs, then quantitative issues (e.g., yield) become important. On the other side of the equation, it is critically important to precisely identify the phenotype of the stem cell derivative once it is situated in the desired tissue. For research purposes, unequivocal markers (genetic or antigenic) of the phenotype of the regenerative product must exist. For example, in the case of cardiac muscle, the antigen must be unique to cardiac muscle cells, and the antibody demonstrating it should not cross-react with any other type of cell. In addition, the possibility of autofluorescence must be carefully controlled. Most careful laboratory studies use a definitive marker (e.g., Y chromosome, GFP, or an inserted gene) for the introduced stem cells and then one or more specific phenotypic markers for the tissue to be regenerated. Ideally, one should be able to demonstrate the lineage marker and the phenotypic marker on the same cell. Unfortunately, for clinical studies, this is often not always possible. A recent study, based on Cre/lox recombination between graft cells and host cardiomyocytes in mice, showed that fusion is the predominant mechanism of cardiomyocyte generation from bone marrow-derived cellular grafts (Andrade *et al.*, 2005).

The current status of the field of stem cell therapy for damaged cardiac muscle is still very much fluid. The outstanding question remains whether stem cells from the bone marrow or some other location within or outside of the heart can populate a region of myocardial damage and undergo conversion into tissue-specific cells. In fact, it remains unknown whether some form of stem or progenitor cell exists in the normal adult heart. Parmacek and Epstein (2005) have reviewed several recent studies that suggest the presence of some form of stem cell in the heart. One of the more promising is a population of cells derived from the secondary (anterior) heart field in the embryo and that express the islet-1 antigen (Laugwitz *et al.*, 2005). These cells have been isolated from very young animal and human hearts, but they are rare or, in humans, apparently nonexistent in adult hearts.

One group (rev. Sussman and Anversa, 2004) has reported the identification of a population of undifferentiated cells with stem cell characteristics in the heart of the adult rat. These cells express the surface antigens, c-kit, MDR1, and Sca-1, which are found in hematogenous stem cells. It was suggested that groups of these cells are stored in niches, principally in the atria and apex, where they are exposed to low levels of mechanical stress. To be functional in repair, these cells would then have to be able to migrate toward areas of tissue pathology.

A significant consideration in the case of cardiac infarcts is what types of stem cells to introduce into the heart and whether they will actually engraft in the area around the infarct. One strategy is to introduce both endothelial stem cells, to provide a vascular base, and cells that would have the potential of differentiating into cardiomyocytes

(Itescu, 2004). Another issue under active investigation is how best to deliver the stem cells to the heart. Despite the uncertainty concerning what cells or structures the introduced stem cells produce, there still appears to be clinical evidence of improved cardiac function (Couzin and Vogel, 2004). Much of the research on stem cell therapy for the diseased heart is still occupied with technical issues, such as proper identification of cells and delivery methods (Müller *et al.*, 2005). These aside, the fundamental biologic questions remain those that confront all areas of stem cell biology (Table 5-2).

Central Nervous System

For most of the twentieth century, the CNS was considered to be a terminally differentiated tissue with no capacity to replace injured or destroyed cells. The series of elegant autoradiographic studies by Altman during the 1960s (Altman and Das, 1965, 1967) that showed areas of DNA-synthesizing cells in the mammalian brain was largely ignored or sidestepped by the neuroscience community because the results did not fit into the dominant paradigm of the times. Later, Goldman and Nottebohm (1983) showed that adult neurogenesis occurred in a nucleus that controls song in the brain of canaries. Widespread acknowledgment of adult neurogenesis in the brain did not occur until after Luskin (1993) and Lois and Alvarez-Buylla (1993) demonstrated that phenomenon in the subventricular zone of the mammalian brain.

The subventricular zone (see Figure 5-4, B) is a region just beneath the ependyma that lines the lateral ventricle of the brain (Alvarez-Buylla and Lim, 2004; Bottai *et al.*, 2003). Within a specialized extracellular matrix is a group of cells designed to provide replacements for the olfactory system. The subventricular stem cell is a modified astrocyte that can divide to produce more of its own type or generate an intermediate type of rapidly dividing transit-amplifying cell, which ultimately produces type A neuroblasts (rev. Doetsch, 2003). These neuroblasts migrate to the olfactory bulb in a defined rostral migratory stream, and

TABLE 5-2 Important Biologic Questions on Stem Cell-Based Tissue Regeneration

Stem cells themselves
What is the appropriate source (e.g., the organ itself, bone marrow, other organs)?
Can the stem cells used differentiate into the desired cell type?
How can one prevent stem cells from differentiating into undesirable cell types?
Into what other types of cells can they differentiate?
Are the stem cells immunologically compatible with the host?
Are the stem cells capable of expanding their numbers to meet the requirements of the host organ?
Will the stem cells divide symmetrically or asymmetrically?
How many stem cells need to be introduced into the host?
Is the age of the donor of the stem cells relevant?
Is the sex of the donor of the stem cells relevant?
Is there a role for the genetic modification of the stem cells?
Introduction into the organ and engraftment
What is the best route for introduction (e.g., direct injection, injection into the local vasculature, transplantation into the marrow)?
If introduction of stem cells is done outside the desired host organ, how can one ensure that they enter that organ and not other organs?
If stem cells are introduced directly into the organ, should they be placed in the region of pathology or near it? When should stem cells be introduced in relation to specific acute pathologies (e.g., myocardial infarcts)? Does a diseased organ give off stem cell homing signals at specific stages of its pathology? What is the basis for the massive loss of introduced cells within a few hours of injection?
Should introduction of stem cells be accompanied by introduction of an artificial or natural substrate or preceded by its introduction?
Differentiation and functional integration
What are the substrate requirements for the desired differentiation in the host organ?
At what stage of restriction or differentiation should the cells be when they are introduced into the host?
How can differentiation be channeled into the desired phenotype?
What is the role of natural or added growth factors and/or substrates in guiding differentiation of stem cells?
How can differentiation into undesirable phenotypes be avoided?
How can expansion of the stem cell population be controlled once the cells are engrafted in the host organ?
How can the engrafted stem cells ultimately replace existing pathology (e.g., scar tissue)?
How can the stem cells form new histologic functional units (e.g., renal tubules or liver lobules) or integrate into existing ones?
Do introduced stem cells stimulate regeneration or induce cellular responses from the host organ?

within the olfactory bulb they differentiate into neurons. Another major zone of proliferative activity within the adult brain is the sub-ependymal region associated with the dentate gyrus of the hippocampus. Adult neural stem cells *in situ* are located in clusters close to small blood vessels. Shen *et al.*, (2004) have shown that this relation is a functional one, and that products from the endothelial cells stimulate the self-renewal, inhibit differentiation, and enhance neuron production in these stem cells.

Since the demonstration of the pockets of neural stem cells in the subventricular and hippocampal regions in the early 1990s, there has been a veritable explosion of experimental results showing the ability of a wide variety of stem cell types to produce neurons or neuron-like progeny. In addition to sources within the adult CNS, neurons have been produced from ES and fetal stem cells, and in the postnatal body, from sources as diverse as neural crest derivatives, bone marrow, umbilical cord blood, dermis, and hair follicles (Table 5-3). Although neurons or neuron-like cells have been produced from a variety of tissue sources, many produce only a low yield of neurons; therefore, their potential efficacy in stem cell regenerative therapy is questionable.

Stem cell therapy has been proposed for the treatment of a wide variety of pathologic conditions in the CNS, but at this point, a small number of possibilities have emerged as prime targets for potential trials. Parkinson's disease has received the most attention. Parkinson's disease involves the degeneration of a specific neuronal subpopulation, namely, the dopaminergic nigrostriatal fibers. There have already been numerous cases of treatment of this disease with transplants of dopaminergic neurons obtained from human fetuses (Lindvall *et al.*, 1990). Neural stem cell treatment has also been proposed for multiple sclerosis, strokes, and spinal cord transection lesions. In the case of multiple sclerosis, the object of stem cell therapy would be to introduce cells that would home in on demyelinated axons and differentiate into myelin-producing cells, presumably oligodendrocytes. For strokes, the strategy would be similar to that outlined earlier for treatment of myocardial infarcts. In the case of spinal cord trauma, the transplanted stem cells would have to provide any of a variety of functions, from being a source of axons to providing a substrate suitable for axonal regeneration from residual spinal neurons.

TABLE 5-3 Sources of Stem Cells That Can Differentiate into Neurons or Neuron-like Cells

1. Embryonic stem cells
2. Fetal cells
3. Adult cells
   - A. Nervous system
     1. Subventricular zone along lateral ventricle
     2. Hippocampus—dentate gyrus area
     3. Substantia nigra
     4. Cortex—temporal and frontal
     5. Septum
     6. Glia
     7. Optic nerve
     8. Spinal cord
   - B. Neural crest derivatives
   - C. Bone marrow
     1. Bone marrow-derived stem cells
     2. Mesenchymal stem cells
   - D. Umbilical cord blood
   - E. Dermis
   - F. Hair follicles

For any stem cell therapy in the nervous system, a variety of specific issues must be addressed. One of the first is the function to be restored and the desired differentiated cellular outcome, specifically whether the need is for neurons, glial cells, or both, and then what specific type of cell within the category. Once that is decided, the best source of stem cells for introduction needs to be determined. Then the issue becomes whether to try to obtain a certain degree of differentiation toward the desired cell type before transplantation or to let the environment surrounding the engrafted cells determine the differentiative outcome. One of the final issues is how to introduce the cells into the body. For specific localized lesions, direct introduction into or near the lesion may be best, whereas for diffuse lesions, such as in multiple sclerosis, injections into the bloodstream may be more effective. Intraventricular injections have also proved to be effective for the distribution of neural stem cells in a number of animal experiments.

At a purely biologic level, most of the issues confronting those working with neural stem cells are quite similar to those seen in other systems. Once introduced into the host, the neural stem cells must survive long enough to become engrafted. If introduced systemically, they must be able to find their way selectively to the appropriate region in the CNS with a minimal number of cells settling in other locations. In the brain and spinal cord, this also means penetrating the blood–brain barrier. Once inside the blood–brain barrier, neural stem cells appear to be attracted to degenerating neural tissue (Aboody *et al.*, 2000). This mutual affinity is a convenient one for those interested in applying neural stem cells to regenerative situations. The control of proliferation *in situ* is a function of both the intrinsic properties of the implanted stem cells and the microenvironment surrounding them. The importance of growth factors is illustrated by the dramatic expansion of the stem cell population in the subventricular zone after epidermal or fibroblast growth factors have been infused into the brain ventricles (Gage, 2000).

In addition to proliferating, the implanted stem cells must differentiate along appropriate lines. This, too, is largely a function of the local environment. When neural stem cells were introduced into neurogenic regions of the brain, they differentiated into cells appropriate for that environment (Suhonen *et al.*, 1996). However, when neuronal stem cells were grafted into nonneurogenic parts of the brain, neuronal differentiation did not occur. The presence of injury appears to change the environment so that neural stem cells are able to differentiate in areas of the brain that would not normally support such differentiation (Magavi *et al.*, 2000; Nakatomi *et al.*, 2002). Certain molecular environments, such as BMPs together with leukemia inhibitory factor, suppress differentiation in stem cells while promoting proliferation. In the adult neural stem cell environment *in vivo,* important developmental signaling pathways, such as Notch, sonic hedgehog, BMPs, and Eph/ephedrins, remain active. Such pathways regulate the transition from proliferative to differentiative functions in the stem cells. Whittemore and colleagues (1999) exposed cultured neural stem cells to a wide variety of growth factor environments. They found that cells exposed to epidermal growth factor differentiated toward the astrocyte lineage, whereas those exposed to FGF-2 were more likely to differentiate into neurons. However, in this experiment, it was not possible to determine whether the growth factors selected for expansion-specific subpopulations with a bias for neuronal differentiation, or whether they altered the course of differentiation of individual cells. Conditions similar to this must be present at sites of implanted neural stem cells. Ultimately, the stem cells will have to differentiate into neurons or the appropriate glia. During development, this is accomplished by exposure of the cells to different combinations of growth factors. For example, stem cells in the subventricular zone preferentially differentiate into astrocytes rather than into oligodendrocytes or neurons if exposed to BMP signaling (Alvarez-Buylla and Lim, 2004).

Not only differentiation, but connections and homing are important components of a successful neural stem-cell strategy. Neurons derived from stem cells must produce dopamine to be effective in the treatment of Parkinson's disease, and neurons, in general, must make proper synaptic connections to be part of a functional pathway. Stem cell-derived oligodendrocytes, on the other hand, must be able to home to areas of demyelination to be effective. Functional synaptic connections of newly generated neurons in both the olfactory bulb and the dentate gyrus have been demonstrated (Carleton *et al.*, 2003; Kempermann *et al.*, 2004). Much remains to be learned about the nature of "settling down" factors that determine the site of cellular landing in the nervous system.

Ultimately, complete integration with the host is the ideal outcome of neural stem cell therapy. One of the real unknowns in neural stem cell therapy is the extent to which implanted stem cells may themselves provide a new microenvironment that stimulates host cells to do things that they ordinarily would not do.

Skeletal Muscle

As has been the case with a number of other tissues and organs, research on skeletal muscle has gone through what now appears to be a typical sequence of events concerning the recognition of the potential for stem cell participation in tissue regeneration. For many years after the discovery of satellite cells (Mauro, 1961), it was assumed that these local progenitor cells were the main cellular source of regenerating muscle, if not the only one. Two reports (Ferrari *et al.*, 1998; Gussoni *et al.*, 2002) challenged this assumption by demonstrating in laboratory animals and in humans with bone marrow transplants that cells derived from the transplanted marrow could be found in regenerating muscle. These reports cast doubt on the long-standing axiom that satellite cells are the principal source of new material for regenerating skeletal muscle. They also served as the basis for a wave of enthusiasm concerning the possibility of stem cell therapy for muscular dystrophy and some other difficult-to-treat muscle diseases (Deasy *et al.*, 2004; Ferrari and Mavilio, 2002; Jankowski *et al.*, 2002).

Further experimentation showed that adult stem cells derived from a variety of sources in addition to bone marrow have the intrinsic capacity to form skeletal muscle (rev. Goldring *et al.*, 2002). These include neural stem cells and cells from the dermis, adipose tissue, and synovial membrane. Stem cells derived from these sources are also capable of differentiating into cells with the characteristics of bone, cartilage, adipose tissue, and blood. It was shown that nonsatellite cells derived from a population of muscle cells (the "side population" from fluorescence-activated cell sorting) can reconstitute the marrow in irradiated animals. These cells with stem cell properties were shown to be of hematopoietic origin (McKinney-Freeman *et al.*, 2002). The formation of ectopic skeletal tissue in muscle under experimental circumstances (Figure 5-5) or in

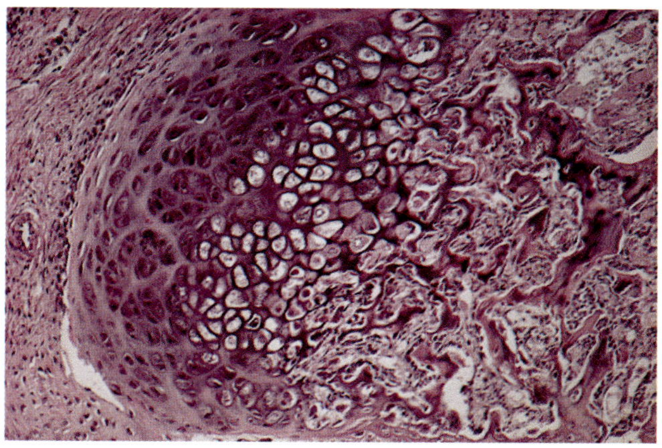

FIGURE 5-5 Nodule containing bone, cartilage, and bone marrow that formed in a rat's minced gastrocnemius muscle regenerate. In addition to these tissue types, the nodule even shows the major structural elements of an epiphyseal plate.

certain pathologic situations (e.g., fibrodysplasia ossificans progressiva) is evidence of the presence of multipotential stem cells in skeletal muscle.

Experiments involving the injection of cloned, single, marrow-derived stem cells into host mice showed that such cells can reconstitute the complete hematopoietic system, but that they can also fuse with skeletal muscle fibers (Camargo *et al.*, 2003; Corbel *et al.*, 2003). Some investigators postulated a natural progression from marrow-derived stem cells into muscle as tissue-specific progenitor cells (satellite cells), which could then be activated and participate in regeneration after some form of injury (Jankowski *et al.*, 2002; LaBarge and Blau, 2002). Several theoretical possibilities exist (Rudnicki, 2003). First, marrow-derived stem cells could migrate to muscle directly and settle down as satellite cells. Second, stem cells resident in muscle could pass through the basal laminae and become traditional satellite cells. Third, inflammatory cells arising in bone marrow could directly fuse with existing muscle fibers. One of the complicating factors in sorting through these options is that there now appears to be heterogeneity among the cells that occupy the traditional satellite cell niche alongside a muscle fiber (Sherwood *et al.*, 2004).

Recent research involving the transplantation of normal muscles into immunosuppressed host LC3F nlacZ-E mice showed that less than 0.01% of the nuclei in the regenerated fibers of the muscle grafts were host-derived (Washabaugh *et al.*, 2004). Sherwood and colleagues (2004) carefully investigated various populations of fiber-associated and interstitial cells isolated from muscle by flow cytometry and determined that only a certain population of muscle-fiber-associated cells (characterized as follows: $CD45^-$, $Sca\text{-}1^-$, $Mac\text{-}1^-$, $CXCR4^+$, β1-integrin[+, $CD34^+$]) readily differentiated into muscle cells either *in vitro* or *in vivo*. These cells reside in the classic anatomic position of satellite cells, between the muscle fiber and the overlying basal lamina. Other cells in the same location and also marrow-derived interstitial cells did not typically form muscle derivatives unless cocultured with myogenic cells or associated with muscle damage *in vivo*. Much remains to be learned about the normal biologic functions of the marrow-derived stem cells that are found in normal muscle or that immigrate into an area of muscle damage. Their role as mediators of inflammation, immunity, or both needs to be separated from their role in reconstructing muscle fibers themselves.

Such research illustrates the complexity of populations of progenitor cells that previously were thought to be homogeneous. These studies also illustrate the limited ability of marrow-derived stem cells to differentiate into specific tissue types. Although they can do so, the percentage of conversion is extremely small. This represents a real challenge for those desiring to use adult stem cells in regenerative therapy.

INTEGRATING STEM CELL BIOLOGY WITH NATURAL REGENERATION

Although the field of stem cell biology is still very young, a few major lessons have been learned. Probably most important is that there are significantly more stem cells than had been imagined previously. The two biggest questions are: (1) What do these cells normally do in the body? and (2) How does one get them to participate effectively in tissue reconstruction and regeneration? It is becoming clear that stem cells are part of a complex hierarchy of cells that are capable of participating in reparative processes throughout the body. They may, at the cellular level, represent part of the double assurance or backup mechanism that exists to compensate for defects at levels from proteins to entire structures.

Based on much recent research, several general principles are becoming clear. One is that bloodborne cells with remarkable powers of differentiation do enter tissues, especially after damage. What they do once they have entered the tissues remains largely a mystery. One option is differentiation into a phenotype appropriate for the parenchyma of that tissue. Another is fusion with cells of the tissue. The latter has been clearly demonstrated (Terada *et al.*, 2002; Ying *et al.*, 2002). A major open question is whether the fused stem cell takes on the phenotypic characteristics of the cell with which it fused or whether it fulfills some other function. Another important question is why stem cells fuse with parenchymal cells in some instances and not in others. Do different types of stem cells react differently when they enter a tissue?

One of the most intriguing issues in stem cell biology is the proclivity of stem cells to home to damaged tissues in much greater numbers than to normal tissue. This has been demonstrated in a number of different systems. Many of these experimental systems have involved irradiation followed by marrow transplantation. These procedures themselves introduce a number of variables that need to be sorted out. However, some models that involve less trauma still indicate a tendency for selective homing to stressed or damaged tissues (Palermo *et al.*, 2005). Because marrow-derived stem cells have been shown to home to a variety of different damaged tissues, one of the most intriguing questions is whether all damaged tissues emit a common type of signal that is received by bloodborne stem cells or whether tissue-specific signals are generated.

The factors that direct the differentiation of stem cells that have penetrated into damaged tissues are also poorly understood. Researchers studying virtually all systems that involve stem cell participation in regeneration stress the importance of the microenvironment in influencing the differentiation of stem cells. One of the most clear-cut examples is the importance of stem cells being associated with differentiating muscle cells in order to differentiate into muscle themselves. Similarly, in the case of bone, the inductive influence of the bone matrix is a critical factor in directing the differentiation of stem cells in this tissue. In addition, the role of the local vasculature and the resulting tissue oxygen concentration acts like a switch in determining whether stem cells will differentiate into osteocytes or chondrocytes.

A significant issue is the relation between aging and the function of stem cells in organ regeneration (van Zant and Liang, 2003). Although there is evidence from irradiation studies on dogs that hematopoietic stem cells in old individuals do not show an apparent decrement in function with age (Zaucha *et al.*, 2001), nevertheless, serial bone marrow transplantation studies have shown that ultimately the stem cell population becomes depleted and fails to support the production of an adequate number of blood cells (Ogden and Mickelm, 1976). Investigators van Zant and Liang (2003) stress the importance of qualitative changes (e.g., restrictions in transdifferentiation potential and impaired homing) as opposed to purely quantitative changes. Old stem cell populations tend to show increased apoptosis after stress, and a larger proportion of the remaining stem cells are engaged in proliferation.

SUMMARY

Stem cells have burst on the landscape of regeneration research. We now recognize that stem cells can be identified in a wide array of tissues and organs, and that many populations of stem cells have not only the capacity to replace tissue-specific defects of the organs in question, but to differentiate into other types of cells as well. Stem cells are resident in tissues and circulate in the blood, and the relative contribution of these populations to the repair of any tissues continues to be debated. Significant questions are: What characterizes "stemness"? How are stem cell populations maintained throughout life? How are stem cells activated? What controls their differentiation?

At this stage in the development of stem cell biology, there are far more questions than answers. Some of the questions have arisen because of overextrapolation of research results. Many others reflect important biologic issues. One of the major lessons learned to date in the stem cell field is the importance of proper markers both for identifying the type and source of stem cells and for determining the cell types into which they transform. One of the fascinating, almost unexplored questions is the potential role of stem cells in facilitating or stimulating regenerative processes in a damaged tissue rather than becoming part of the regenerated tissue themselves.

FURTHER READING

Aboody, K.S., et al. 2000. Neural stem cells display extensive tropism for pathology in adult brain: Evidence from intracranial gliomas. *Proc Natl Acad Sci USA* 97:12846–12851.

Andrade, J., et al. 2005. Predominant fusion of bone marrow-derived cardiomyocytes. *CardiovascRes* 68:387–393.

Altman, J., and G.D. Das. 1965. Autoradiographic and histological evidence of postnatal hippocampal neurogenesis in rats. *J Comp Neurol* 124:319–335.

Altman, J., and G.D. Das. 1967. Postnatal neurogenesis in the guinea-pig. *Nature* 214:1098–1101.

Alvarez-Buylla, A., and D.A. Lim. 2004. For the long run: Maintaining germinal niches in the adult brain. *Neuron* 41:683–686.

Balsam, L.B., et al. 2004. Haemopoietic stem cells adopt mature haematopoietic fates in ischaemic myocardium. *Nature* 428:668–673.

Bernstein, B.E., et al. 2006. A bivalent chromatin structure marks key developmental genes in embryonic stem cells. *Cell* 125:315–326.

Bohler, K.R., et al. 2002. Differentiation of pluripotent embryonic stem cells into cardiomyocytes. *Circ Res* 91:189–201.

Bottai, D., et al. 2003. Neural stem cells in the adult nervous system. *J Hematother Stem Cell Res* 12:655–670.

Bunting, K.D., and R.G. Hawley. 2003. Integrative molecular and developmental biology of adult stem cells. *Biol Cell* 95:563–578.

Butler, E.G. 1935. Studies on limb regeneration in x-rayed *Amblystoma* larvae. *Anat Rec* 62:295-307.

Cai, J., et al. 2004. In search of "stemness." *Exp Hematol* 32:585–598.

Camargo, F.D., et al. 2003. Single hematopoietic stem cells generate skeletal muscle through myeloid intermediates. *Nat Med* 9:1520–1527.

Carleton, A., et al. 2003. Becoming a new neuron in the adult olfactory bulb. *Nat Neurosci* 6:507–518.

Corbel, S.Y., et al. 2003. Contribution of hematopoietic stem cells to skeletal muscle. *Nat Med* 9:1528–1532.

Couzin, J., and G. Vogel. 2004. Renovating the heart. *Science* 304:192–194.

Deasy, B.M., et al. 2004. Tissue engineering with muscle-derived stem cells. *Curr Opin Biotechnol* 15:419–423.

Doetsch, F. 2003. The glial identity of neural stem cells. *Nat Neurosci* 6:1127–1134.

Dreyfus, P.A., et al. 2004. Adult bone marrow-derived stem cells in muscle connective tissue and satellite cell niches. *Am J Pathol* 164:773–779.

Ferrari, G., et al. 1998. Muscle regeneration by bone marrow-derived myogenic precursors. *Science* 279:1528–1530.

Ferrari, G., and F. Mavilio. 2002. Myogenic stem cells from the bone marrow: A therapeutic alternative for muscular dystrophy? *Neuromuscul Disord* 12:S7–S10.

Gage, F.H. 2000. Mammalian neural stem cells. *Science* 287:1433–1438.

Goldman, S.A., and F. Nottebohm. 1983. Neuronal production, migration and differentiation in a vocal control nucleus of the adult female canary brain. *Proc Natl Acad Sci USA* 80:2390–2394.

Goldring, K., et al. 2002. Muscle stem cells. *J Pathol* 197:457–467.

Gussoni, E., et al. 2002. Long-term persistence of donor nuclei in a Duchenne muscular dystrophy patient receiving bone marrow transplantation. *J Clin Invest* 110:807–814.

Horackova, M., et al. 2004. Cell transplantation for treatment of acute myocardial infarction: Unique capacity for repair by skeletal muscle satellite cells. *Am J Physiol Heart Circ Physiol* 287:H1599–H1608.

Jackson, K.A., et al. 2001. Regeneration of ischemic cardiac muscle and vascular endothelium by adult stem cells. *J Clin Invest* 107:1395–1402.

Jankowski, R.J., et al. 2002. Muscle-derived stem cells. *Gene Ther* 9:642–647.

Kempermann, G., et al. 2004. Functional signifi cance of adult neurogenesis. *Curr Opin Neurobiol* 14:186–191.

Kléber, M., and L. Sommer. 2004. Wnt signaling and the regulation of stem cell function. *Curr Opin Cell Biol* 16:681–687.

Körbling, M., and Z. Estrov. 2003. Adult stem cells for tissue repair: A new therapeutic concept? *N Engl J Med* 349:570–582.

Krause, D.S., et al. 2001. Multi-organ, multi-lineage engraftment by a single bone marrowderived stem cell. *Cell* 105:369–377.

LaBarge, M.A., and H.M. Blau. 2002. Biological progression from adult bone marrow to mononucleate muscle stem cell to multinucleate muscle fiber in response to injury. *Cell* 111:589–601.

Laugwitz K.-L., et al. 2005. Postnatal islicardioblasts enter fully differentiated cardiomyocyte lineages. *Nature* 433:647–653.

Lee, T.I., et al. 2006. Control of developmental regulators by polycomb in human embryonic stem cells. *Cell* 125:301–313.

Leobon, B., et al. 2003. Myoblasts transplanted into rat infarcted myocardium are functionally isolated from their host. *Proc Natl Acad Sci USA* 100:7808–7811.

Lindvall, O., et al. 1990. Neural transplantation in Parkinson's disease: The Swedish experience. *Prog Brain Res* 82:729–734.

Lois, C., and A. Alvarez-Buylla. 1993. Proliferating subventricular zone cells in the adult mammalian forebrain can differentiate into neurons and glia. *Proc Natl Acad Sci USA* 90:2074–2077.

Lois, C., and A. Alvarez-Buylla. 1994. Long-distance neuronal migration in the adult mammalian brain. *Science* 264:1145–1148.

Luskin, M.B. 1993. Restricted proliferation and migration of postnatally generated neurons derived from the forebrain subventricular zone. *Neuron* 11:173–189.

Magavi, S.S., et al. 2000. Induction of neurogenesis in the neocortex of adult mice. *Nature* 405:951–955.

Mauro, A. 1961. Satellite cell of skeletal muscle fibers. *J Biophys Biochem Cytol* 9:493–495.

McKinney-Freeman, S.L., et al. 2002. Muscle-derived hematopoietic stem cells are hemapoietic in origin. *Proc Natl Acad Sci USA* 99:1341–1346.

Müller, P., et al. 2005. Myocardial regeneration by endogenous adult progenitor cells. *J Mol Cell Cardiol* 39:377–387.

Murry, C.E., et al. 2004. Haematopoietic stem cells do not transdifferentiate into cardiac myocytes inmyocardial infarcts. *Nature* 428:664–668.

Moore, K.A., and I.R. Lemischka. 2006. Stem cells and their niches. *Science* 311:1880–1885.

Nakatomi, H., et al. 2002. Regeneration of hippocampal pyramidal neurons after ischemic brain injury by recruitment of endogenous neural progenitors. *Cell* 110:429–441.

Orlic, D., et al. 2001a. Bone marrow cells regenerate infarcted myocardium. *Nature* 410:701–705.

Orlic, D., et al. 2001b. Mobilized bone marrow cells repair the infarcted heart, improving function and survival. *Proc Natl Acad Sci USA* 98:10344–10349.

Ogden, D.A., and H.S. Mickelm. 1976. The fate of serially transplanted bone marrow populations from young and old donors. *Transplantation* 22:287–293.

Palermo, A.T., et al. 2005. Bone marrow contribution to skeletal muscle: A physiological response to stress. *Dev Biol* 279:336–344.

Parmacek, M.S., and J.A. Epstein. 2005. Pursuing cardiac progenitors: Regeneration redux. *Cell* 120:295–298.

Quaini, F., et al. 2002. Chimerism of the transplanted heart. *N Engl J Med* 346:5–15.

Rudnicki, M.A. 2003. Marrow to muscle, fission vs. fusion. *Nat Med* 9:1461–1462

Sauer, H., et al. 2004. Development of the cardiovascular system in embryoid bodies derived from embryonic stem cells. In: S. Sell, ed. *Stem cell handbook*. Totowa, NJ: Humana Press. 229–238.

Sawamoto, K., et al. 2006. New neurons follow the flow of cerebrospinal fluid in the adult brain. *Science* 311:629–632.

Schächinger, V., et al. 2004. Transplantation of progenitor cells and regeneration enhancement in acute myocardial infarction. *J Am Coll Cardiol* 44:1690–1699.

Shen, Q., et al. 2004. Endothelial cells stimulate self-renewal and expand neurogenesis of neural stem cells. *Science* 304:1338–1340.

Sherwood, R.I., et al. 2004. Isolation of adult mouse myogenic progenitors: Functional heterogeneity of cells within and engrafting skeletal muscle. *Cell* 119:543–554.

Sieber-Blum, M., et al. 2004. Pluripotential neural crest stem cells in the adult hair follicle. *Dev Dyn* 231:258–269.

Strauer, B.E., et al. 2002. Repair of infarcted myocardium by autologous intracoronary mononuclear bone marrow cell transplantation in humans. *Circulation* 106:1913–1918.

Suhonen, J.O., et al. 1996. Differentiation of adult hippocampus-derived progenitors into olfactory neurons in vivo. *Nature* 383:624–627.

Sussman, M.A., and P. Anversa. 2004. Myocardial aging and senescence: Where have the stem cells gone? *Annu Rev Physiol* 66:29–48.

Terada, N., et al. 2002. Bone marrow cells adopt the phenotype of other cells by spontaneous cell fusion. *Nature* 416:542–545.

Thiele, J., et al. 2004. Regeneration of heart muscle tissue: Quantification of chimeric cardiomyocytes and endothelial cells following transplantation. *Histol Histopathol* 19:201–209.

Washabaugh, C.H., et al. 2004. Nonmuscle stem cells fail to significantly contribute to regeneration of normal muscle. *Gene Ther* 11:1724–1728.

Weissman, I.L. 2000. Stem cells: Units of development, units of regeneration, and units in evolution. *Cell* 100:157–168.

Whittemore, S.R., et al. 1999. Mitogen and substrate differentially affect the lineage restriction of adult rat subventricular zone neural precursor cell populations. *Exp Cell Res* 252:75–95.

Ying, Q.-L., et al. 2002. Changing potency by spontaneous fusion. *Nature* 416:545–548.

Zammaretti, P., and M. Jaconi. 2004. Cardiac tissue engineering: regeneration of the wounded heart. *Curr Opin Biotechnol* 15:430–434.

van Zant, G., and Y. Liang. 2003. The role of stem cells in aging. *Exp Hematol* 31:659–672.

Part II

Methods for Preparing Embryonic or Pluripotent Stem Cells

Chapter 6

Setting up a Facility for Human Embryonic Stem Cell Research

Ian Lyons, David Tan, Philip H. Schwartz, and Mahendra Rao

INTRODUCTION

Here we have distilled our collective experience of setting up over ten stem cell laboratories on three continents, with the intention of lessening the pain of others who are setting up such laboratories for the first time. We are sure of two things: first, the effort extended in setting up systems and processes for a laboratory will be repaid many fold in future efficiencies, and, second, the effort to set systems in place from the very beginning is much less than retrofitting them. So if you are fortunate to be setting up a laboratory from scratch, our advice is to try to get it right, right from the beginning.

OVERVIEW

The recommendations here are for a small-to-medium scale academic or biotech research laboratory; those setting up manufacturing or therapeutic laboratories will have a different set of regulatory and logistic constraints. Most of the design features for a human embryonic stem cell (hESC) laboratory are the same as for any other cell culture laboratory. The physical laboratory space needs to be set up in a way that efficiently reflects the expected workflows, taking into account the numbers of workers who will be sharing the space and the equipment and the expected throughput. Like any well-run cell culture facility, materials and supplies need to be monitored to ensure that quality and amounts match the demands, and waste is disposed of appropriately. Similarly, the ancillary activities for molecular analysis of cell phenotype are not unique to hESCs. However, given the need for a regular supply of cells with defined characteristics and uncompromised differentiation potential, a careful system of cell banking is even more critical for hESC than for other cell types.

The critical issue in setting up the laboratory is planning for long-term success. This will rest on the functional design and utilization of the space and the appropriate equipment. Most important is the implementation of quality control systems that allow a continuous supply of validated cells and reagents. This book provides most of the protocols and methods for these systems, and other resources are also available; however, wherever possible, we recommend that key personnel receive specific training in hESC manipulation, and that the laboratory develop a network of communication and collaboration with other hESC groups.

LABORATORY

General Considerations

- hESC culture laboratories are similar to other cell culture research facilities. Space is required for tissue culture, standard biochemistry and molecular biology, and cell imaging. Incoming material and waste disposal logistics need to be accommodated.
- When setting up a laboratory, we typically define two distinct work areas: a cell culture area with rooms that can be dedicated for different types of work (Figure 6-1) and a molecular biology area with laboratory benches and appropriate equipment organized for common work flows (Figure 6-2). If possible, construction should consist of a modular framework that allows the laboratory to be reconfigured easily, if needed. A centralized CO_2 gas system with all CO_2 gas cylinders located outside the laboratories saves valuable laboratory space and simplifies the monitoring of gas levels, while the exchange of cylinders does not compromise cleanliness of the culture space. Drop lines carry CO_2 gas to equipment in each tissue culture room and in the fluorescence activated cell sorting (FACS) and microscopy rooms, if required for live cell imaging.
- Chemical fume cabinets are located within the molecular biology laboratory and are vented to the external environment as appropriate under local regulations.

FIGURE 6-1 Compartmentalized cell culture facility showing adjacent but separate rooms that are used for culturing different types of stem cells (adult versus embryonic, human versus animal). Arrow shows second biological safety cabinet in adjacent room.

FIGURE 6-2 Arrow shows the molecular biology area that is adjacent to but outside of the cell culture facility, allowing easy access but maintaining the integrity of the cell culture area.

While not a high demand facility, access is needed for tissue perfusion, for fixation procedures, and for organic extraction of nucleic acids.

- High-quality water for the laboratory is supplied through a laboratory-specific Milli-Q or similar system and regular testing procedures for the quality of the water is needed, as water quality is critical to all cell culture.

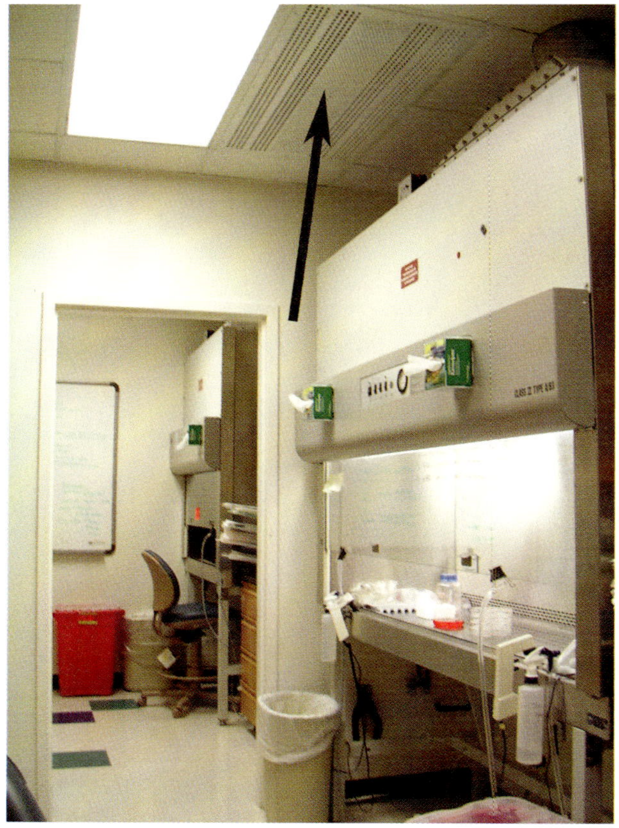

FIGURE 6-3 The air handling system of the cell culture rooms are HEPA-filtered (arrow) and maintain positive pressure with respect to adjacent non-cell culture areas.

- In order to facilitate cleaning and maintenance, vinyl flooring, gloss paint (low in volatile organic compounds), and bench tops of impermeable materials should be used as appropriate. Ideally, the rooms should be lit with fluorescent light and ambient temperature should be constant (typically 18–22°C), while circulating air should be passed through 0.22 μm HEPA filters (Figure 6-3). Cell culture laboratories should be under positive pressure, while molecular biology areas or those used for viral work should be under negative pressure.
- Standard safety requirements for laboratories, conforming to safety standards in the local jurisdiction, need to be implemented, but should include eye-washes, safety showers, first aid kits, fire extinguishers, and chemical spill kits at appropriate sites along the corridors. Sinks are typically located at the end of each workbench and, ideally, the taps will be foot activated (Figure 6-4).
- Common work areas for procedures such as media preparation will be conveniently located in the middle of the laboratory. Computers, internet connection points, and telephones are provided at the end of designated workbenches for communal use.
- When the layout of the laboratory has been determined and equipment locations decided, electrical wiring needs can be planned. Depending on local supply voltage

FIGURE 6-4 All sinks in the cell culture areas are foot-pedal controlled (arrow), allowing hand-washing without contamination of or by fixtures.

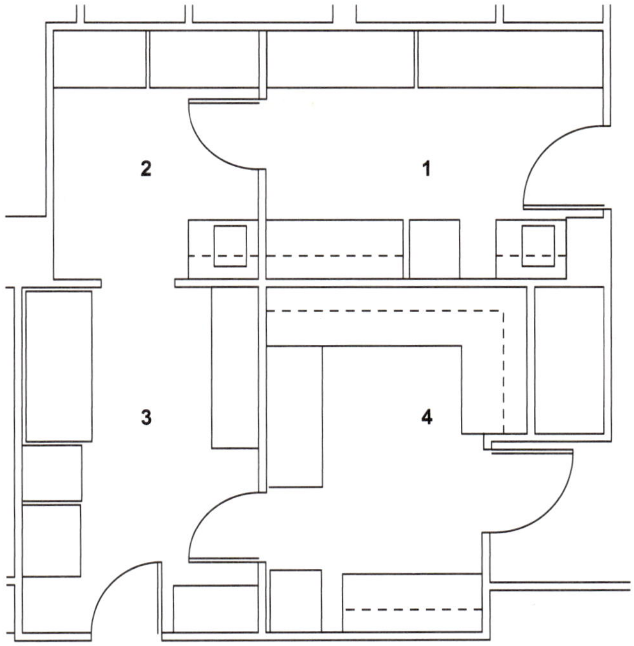

FIGURE 6-5 Example of adjacent cell culture rooms (1, 2, 3, 4) that allow efficient use of space and allow safe culture of disparate cell types in the same facility.

and supply type, special circuits will probably be needed to supply higher voltage or three-phase power to some of the large equipment. In addition, certain equipment, such as ice flake machines and autoclaves require easy access to floor drains and water inlet points.

- We recommend that all storage freezers and incubators be connected to an emergency back-up supply and that a centralized alarm system, linked to key equipment such as freezers and incubators, should be considered.

Cell Culture Area

- Depending on available space, we recommend that cell culture areas be functionally separated into dedicated rooms for hESCs (Figure 6-5), for nonhuman cells (e.g. mouse feeder cells), and for viral work. Ideally, each would have automatic doors such as elbow-activated glass sliding doors for easy access, and cell culture rooms should be under positive air pressure, while the room for viral work should be under negative air pressure.
- We have found it useful to equip each tissue culture room with independent workstations that consist of a biological safety cabinet (BSC), a CO_2 incubator, vacuum aspiration system, a complete set of personal equipment such as pipetting aids, micropipettes (2–1000 μL), and a drawer unit containing a selection of serological pipettes, filters, syringes, needles, culture vessels (flasks, plates, dishes), and cryopreservation supplies. From these workstations, there is easy access to shared equipment such as sinks, water baths, vortex mixers, centrifuges, and inverted microscopes for bright field/phase contrast visualization of cells in culture. Microscopes should be physically isolated from centrifuges and other vibrating equipment.
- For hESC culture, we have each BSC equipped with a heated base and a stereomicroscope to facilitate colony picking or other manipulations, and also the more usual germicidal ultraviolet lamps (under timer control), outlets for CO_2 and vacuum, 2 L collection flasks, and power supply for equipment such as pipetting aids.
- Fitting a zoom dissecting microscope to a class II BSC is not trivial. hESC colony picking requires a relatively large working distance and a heated stage is useful for the extended time the cultures are manipulated. A low microscope base with sloping sides allows easier transfer of the cell culture dish from the worktop to the stereomicroscope with minimal movement, and reduces chances of spillage or contamination.
- We have found the Gelman BH Class II Series (type A/B3) BSCs coupled with the Leica MZ6 to be a good combination. The microscope can be fitted with extended eyepiece tubes, and the BSC is modified to allow the extendable eyepieces to protrude from the sash toward the user (Figure 6-6). A pneumatic window sash is available and swings forward and back, allowing access to the BSC interior for equipment transfer and cleaning without having to move the stereomicroscope.
- An alternative and very cost effective approach, which is not recommended by the BSC manufacturer but has been used successfully by one of this book's editors, is the following: raise the BSC sash to its fully raised position, block all of the rear air vents with tissue paper (i.e. Kimwipes), and place a standard inverted microscope

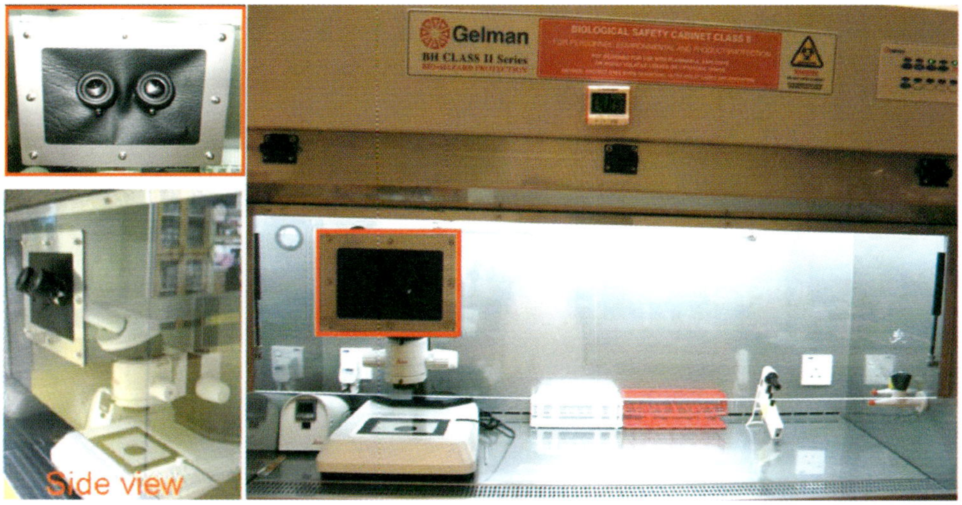

FIGURE 6-6 Biological safety cabinet with fitted dissecting microscope.

FIGURE 6-7 Standard biological safety cabinet modified to permit manual dissection of hESC colonies on an inverted phase-contrast microscope (Olympus CK-2 in this case). Alternatively, a dissecting microscope can be used. The sash is raised to full height (long black arrow) resulting in the sash alarm being activated (white arrowhead), the rear vents are "blocked" with lab tissue paper (Kimwipes, black arrowheads), and the microscope is placed inside the cabinet. Most of the HEPA-filtered airflow, therefore, is directed toward the front vents (black arrows), creating a sterile air barrier near the front of the hood. After a 1–2 min equilibration on the microscope stage, dishes or plates (white arrow) can be uncovered and manual dissection can take place under direct microscopic observation (long white arrow).

into the BSC. Most of the air flow then is into the front vents, creating a sterile curtain behind which culture manipulation on the microscope stage can take place (Figure 6-7) and in front of which the operator can view the process through the microscope's objectives. It should be noted that, in this arrangement, the user may not be adequately protected from the sample; thus, only samples that have been proven negative for potential human pathogens should be used. Dedicated laminar flow hoods, which are relatively inexpensive, can also be used for this purpose.

- The CO_2 incubators are more conveniently situated beside the BSCs instead of behind the user, minimizing movement of culture dishes from BSC to incubator and reducing the chance of accidental spills. Electric outlets in some of the incubators allow temporary installation of equipment inside the incubator (for example an orbital shaker for embryoid body formation).
- There are many CO_2 incubator options available. Important criteria include: readily cleaned internal surfaces without cracks or seams, monitored CO_2 level and temperature, with tight tolerances, and alarms. We like the Binder CB 150 with a data recording system which allows CO_2 levels and changes in temperature levels to be recorded for up to 14 days. Installation of tri-gas control incubators should be considered if the effects of manipulation of oxygen tension are likely to be investigated at some stage. The extra cost is incremental, since these incubators can also be used conventionally at ambient oxygen.

Molecular Biology Area

The organization of the molecular biology area will be determined by both the space available and the numbers of workers employed. For greatest efficiency in high-density laboratory areas, the space can be organized into functional areas for specific workflows. In this system, all the equipment and reagents are readily at hand for specific processes and workers can come to a dedicated and shared area and find all that is needed for their task. Examples of this would be an area for immunostaining, with all equipment and antibodies easily at hand. Another example is a separate bench (preferably a laminar flow cabinet) for RNA preparation and for polymerase chain reaction (PCR) set-up which is physically remote from the area for PCR product manipulation and analysis. The latter has the advantage of separating the PCR set-up from analysis, reducing the chances of reagent contamination with PCR products.

If space is less limited, a separate bench or area can be assigned for each worker. Each person will have access to a basic set of tools including micropipettes (2–1000 μL), a PCR machine, centrifuge, vortex mixer and an under-the-counter refrigerator and freezer.

In addition to the functional areas described above, we have found it convenient to have "communal areas" with more generally used equipment such as analytical balances, centrifuges, spectrophotometers, gel imaging equipment, pH meters, plate readers, ultracentrifuges, and power supplies for electrophoresis.

Additional Considerations

In addition to laboratory areas dedicated to cell culture and to molecular cell analysis, there are a number of other activities that might require dedicated areas. These may be part of the hESC culture facility itself or may be part of core institute facilities; but if access is needed, it should be taken into account during the planning phase.

Dedicated rooms for microscopy, image analysis and flow cytometry equipment will reduce dust and traffic flow. Depending on the equipment used, purpose-built benches with vibration isolation are a good investment.

Sufficient storage space should be available to the laboratory. In particular, disposable plastic cell culture supplies are bulky and having a separate storage room will help avoid bringing potentially dusty cartons into the laboratory. Storage for flammable liquids should be available in fireproof cabinets. Fridges, freezers, and ultra-low freezer units should be located carefully. Apart from under bench fridges and freezers, we recommend that these units be housed separately wherever possible, since they are a major source of ambient noise in laboratories.

Long-term cell storage has traditionally been done in liquid nitrogen freezers. If these are to be used, they should be housed separately from the laboratory in a well-ventilated space with low oxygen alarms to reduce dangers of asphyxiation. This could be in the same area used to store consumables. We also recommend all critical derivations and lines be stored in two different freezers, preferably with one at another site. Ultra-low (−152°C) freezers are an alternative to liquid nitrogen storage; they are reported to have become increasingly reliable and do not need a constant supply of liquid nitrogen, the regular replenishment of which brings dirt into the laboratory and is also expensive.

Hazardous biological waste needs to be disposed of according to local regulations. In most regions this will require access to an autoclave, and this should be taken into account when the laboratory is being planned.

RECORD KEEPING

General Procedures

Good book keeping is essential; paperwork ranging from quotations, purchase orders, delivery orders, invoices, certificates of analysis (C of A), material safety data sheets (MSDS), material transfer agreements (MTAs), research agreements, and licenses should be filed and organized in an easily understood system. This is particularly important in the case of hESCs, where regulations can be stringent in some countries. Careful documentation from the start will save considerable time and stress in the future.

Reagent and Consumable Tracking

A reliable supply of validated reagents is critical to hESC research quality. In general, reagents from major suppliers can be used with confidence; however, some are known to be subject to lot-to-lot variation. There are a number of reasons for this, including the unfortunate fact that some reagents are quality tested against cell types less fastidious than hESCs. An example of this is KnockOut™ serum replacement (KSR; Invitrogen catalog no. 10828-028), a product that is reliable for the cells it was designed for, mouse embryonic stem cells (mESCs), but for which lot-to-lot variation has been described for hESCs. To maintain a continuity of supply of reliable reagents, it is important to have a system that:

- ensures that the laboratory does not exhaust any reagent, and
- provides a system that allows timely testing of incoming reagents to ensure comparability with those being replaced.

For hESC culture, some of the critical reagents that should be tested well in advance of being brought into general use include fetal bovine serum (FBS), KSR, and mouse embryonic fibroblasts (MEFs). Approaches to testing are described below. Other reagents are less variable and are usually used untested.

Stock and Reagent Inventory

The levels of stocks of reagents that need to be held to ensure continuity of supply will depend on the rate of usage by the laboratory, on the delay between ordering and delivery, the probability of an exceptional delay, and the risk aversion of the laboratory director. For a guideline, we recommend that stocks of all reagents be maintained at a minimum of three times the delivery lead time. So for reagents that typically can be delivered one week after ordering, we recommend that a minimum of three weeks normal usage should be maintained. If these reagents are being tested to validate lots, then the holding time should be increased to account for the time it takes to test. Maximum holding times will depend on the storage capacity of the facility and the shelf life of the reagents.

For all reagents, it is important to keep track of the manufacturers' stock keeping units (SKUs) and lot numbers. These should be maintained as part of the laboratory record keeping and inventory records, and each worker should record those details in their laboratory notebook for each experiment or process undertaken. It is relatively rare to have a lot fail, especially if rigorous lot matching and testing is undertaken systematically; however, rare, sporadic lot failure will occur, and recovery from a cell culture "crash" or unexplained failure will be more efficient if all reagents can be traced. The savings in time and money will justify the upfront effort to establish such a system.

Lab Index System

In addition to labeling the outsides of drawers and cabinets with their contents, we have found it useful to number each of the laboratory drawers and cabinets and to keep a database of their contents. This is particularly useful in a laboratory that is likely to have frequent visitors or changes of staff. This also facilitates stocktaking and auditing. The laboratory manager should ensure that these lists are updated regularly so that finding irregularly used items becomes quick and easy.

Freezer Inventory Systems

Keeping track of the stored cell stocks is critical for the smooth operation of the facility on a day-to-day basis, and also allows for the longer term planning. Whether cell stocks are stored in an ultra-low temperature mechanical freezer or in liquid- or vapor-phase nitrogen, a system of cell labeling and record keeping is essential to allow retrieval of cells, or to keep track of the rates of cell stock usage. The cell freezer will typically come with a system of racks to store the ampoules and allow them to be quickly located. This must be backed up by an inventory system that keeps track of each ampoule. This can be done electronically, using a purpose-designed system or a more generic database software, or it can be done manually using a dedicated laboratory notebook or ring binder. In either case, the data need to be regularly backed up. Each freezing vial must be unambiguously labeled either using printable cryoresistant printer labels, or with a cryoresistant marker pen. Details should include cell type and passage number, date, technician, and laboratory notebook reference. Like any system, this will fail if there is not compliance, and every vial added or removed should be noted immediately. Proper adherence by the users is essential to the smooth functioning of the system, so creating an environment where users have ownership over their work is important.

Examples of cryostorage record systems can be found at http://www.cryotrack.com/, http://www.cclims.com/, http://www.nalgenelabware.com/techdata/technical/cryo.pdf.

EQUIPMENT MAINTENANCE

Equipment maintenance is essential for the smooth operation of the laboratory and for the generation of reproducible data. We recommend providing all equipment and accessories with maintenance schedules and a local and alternate contact person for sales and service. All major equipment must be tested and commissioned before use, and maintained on a regular basis by the vendor or reputable service organization. It is important to keep track of dates and records of these procedures so as to prolong the lifespan of each piece of equipment and ensure their proper use. Usage of common large equipment such as centrifuges and spectrophotometers should be tracked using a logbook. For regular maintenance of equipment, there should be clear ownership and visibility.

- *Incubators* should be checked on a weekly basis using a digital thermometer and a Fyrite device. The readings obtained can be recorded on the door of the incubator so that all users are aware of any fluctuations.
- *Water baths* are a common source of contamination. To minimize this, the water should be changed at least weekly. Locating the water baths adjacent to the sinks simplifies the changing of water and cleaning.
- *Microscopes* are often used while not optimally aligned. They should be adjusted regularly, and all scientists and staff using them should be trained in their correct use.
- *BSCs* should have their interior wiped down after every use with 70% ethanol. The germicidal ultraviolet lamps should be switched on at the end of each day. For this reason, it is important that the equipment kept inside the BSC is able to withstand exposure to UV light; remove all vulnerable equipment and supplies before operating the UV light.

SAFETY

Staff should be trained in the safe handling of all equipment, reagents, and biological materials and in the use of first aid and safety equipment. Laboratory gowns and safety glasses should be supplied and be worn at all times. Personal protective equipment (PPE) should be available for all pertinent laboratory work. In particular, full-face shields should be worn when removing and thawing ampoules from cryostorage.

Biological Hazards

Human and other primate cells should be handled using Biosafety Level 2 practices and containment appropriate for the institute and local authorities. All work should be performed in a BSC with appropriate PPE (gloves, glasses, laboratory coats), and all waste material should be decontaminated by autoclaving or disinfection before discarding. For a discussion of recommendations and requirements refer to "Biosafety in Microbiological and Biomedical Laboratories" at http://www.cdc.gov/od/ohs/biosfty/bmbl4/bmbl4toc.htm.

QUALITY CONTROL

The aim of the laboratory set-up is to have systems in place that lead to reliable and reproducible experimental outcomes (Figure 6-8). This is dependent on four critical factors:

- *Reliable techniques* that yield a minimum of variability between processes. Use of methods developed by others and shown to be reliable is a good way to establish a laboratory, and to have the results of that laboratory comparable to the work of others. Refer to other chapters of this book for these methods.
- *Validated reagents*. As this is critical, we describe below a stock keeping and validation system for laboratory reagents.
- The *quality of the hESCs*. We also describe a method for generating a structured stock of hESCs that have been validated molecularly and functionally, and which has sufficient size to cover the demands of the laboratory.
- A *system of testing* and standards that confirms comparability of hESCs being used. They need to be free of contamination, genetically stable, to express defined markers, and capable of differentiation into representatives of the three germ lineages. Much of this is described in other chapters of this book, but below we describe techniques for monitoring of and recovery from contamination.

Reagent Testing

Because lot-to-lot variation has been described for FBS, KSR, and MEFs, we recommend that a regimen be introduced to test lots before they are brought into general use. Other reagents are less variable and are usually used untested.

Lot Testing FBS for MEFs

Almost all FBS lots are able to support MEFs, but they still need to be tested in advance, because the cost may be significant. We confirm FBS lot suitability on the basis of cell expansion over two passages, but do not monitor biological activity. A positive control in this assay is either an FBS sample that has been shown previously to support MEFs, or if unavailable, a commercially available FBS validated for support of mESCs (e.g. Invitrogen catalog no. 10439-016). mESC-qualified FBS could be used routinely

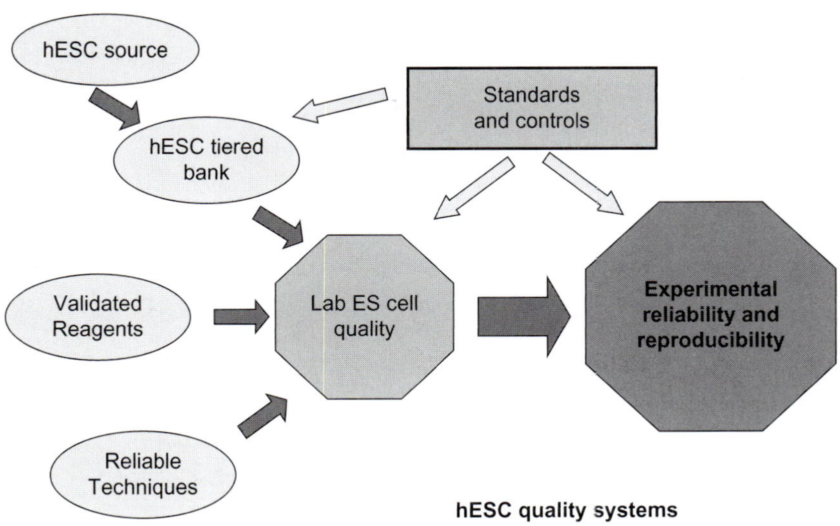

FIGURE 6-8 hESC quality systems.

for MEF propagation, but is significantly more expensive than unvalidated FBS.

Using the protocols described in the chapter on preparing MEFs:

1. Make medium with each of the test FBS samples, and with the control.
2. Seed MEFs at usual density in triplicate 10cm dishes in medium.
3. Harvest cells and count after 5 days (but before the control reaches confluence).
4. Count cells and replate a second passage in each medium.
5. Again harvest cells and count after 5 days (but before the control reaches confluence).

Suitable batches of FBS will have cell yields in excess of 80% of that of the control lot.

Testing FBS and KSR for hESCs

When testing lots of serum or KSR for use with hESCs, we passage cells four times before confirming morphology and marker expression patterns, since many medium components do not have an obvious effect until at least three passages. Ideally, one would use a positive control serum or KSR batch, but these are not commercially available. Typically most laboratories would use a batch that has been previously used successfully. A new laboratory that is being set up will not have that option, though may be able to obtain samples from an established laboratory.

Using the protocols described in Chapter 1 on hESC culture:

1. Make medium with the test serum or KSR batches and with a positive control, if available.
2. Passage cells in the usual manner for three passages in the test media.
3. Passage into appropriate culture systems for immunocytochemistry for SSEA-4, POU5F1/OCT4, and alkaline phosphatase staining, and for expansion to extract RNA for microarray or PCR analysis.
4. Analysis: Serum or KSR batches are acceptable if they support cells as well as the control lot that has been used successfully and/or have these minimal performance characteristics:
   - Immunocytochemistry: >85% of cells express SSEA-4 and POU5F1/OCT4.
   - RT-PCR: We use conventional and quantitative PCR to assess expression of genes that are markers of undifferentiated cells and to confirm absence of markers of differentiation for each of the three germ lineages (see below).
   - Morphology and alkaline phosphatase: more than 95% of colonies express alkaline phosphatase and are morphologically undifferentiated.

Lot Testing MEFs

During preparation, MEFs should be tested for microbiological contamination; testing biological function of each new lot should not be necessary given that the appropriate strain and embryological stage of fetus were used and that methods are reproducibly followed. Testing new batches of MEFs before they are brought into general use, however, is a wise precaution, at least until several lots have been generated successfully. The testing procedures and readouts are essentially the same as those described above for serum and KSR. Cells are maintained on test and preferably control MEFs for four passages and then marker expression is analyzed as described above in the section on Testing FBS and KSR for hESCs.

CELL BANKING

Rationale

A well-structured cell bank is important for successful cell culture in general, but is critical for ESCs, with their inherent karyotypic instability and tendency to differentiate. A tiered or master cell banking system brings validated and reproducible cells to each experiment and procedure. Cells will be:

- At defined range of passage number for all experiments.
- Of normal karyotype.
- Free of microbiological contamination.
- Capable of differentiation to desired lineages.
- Of undifferentiated phenotype with expression of appropriate molecular markers.

Establishment of a tiered banking system

A tiered banking system typically consists of three levels of cryopreserved cells: the earliest "mother stock," the critical "master cell bank," and the "working cell bank" (Figure 6-9). The number of vials of each depends on the cell type and projected usage, and on the extent of the testing required for that cell type. For hESCs, the testing recommended is extensive and relatively expensive, and so we suggest that large banks of cells be frozen.

At each level of banking, we recommend the cells be tested for:

- Karyotype.
- Genotype.
- Differentiation capacity.
- Marker expression (both presence of markers of undifferentiated cells and absence of markers of differentiation).

Other aspects of cell status can also be tested since the cells have been shown to drift with extended culture

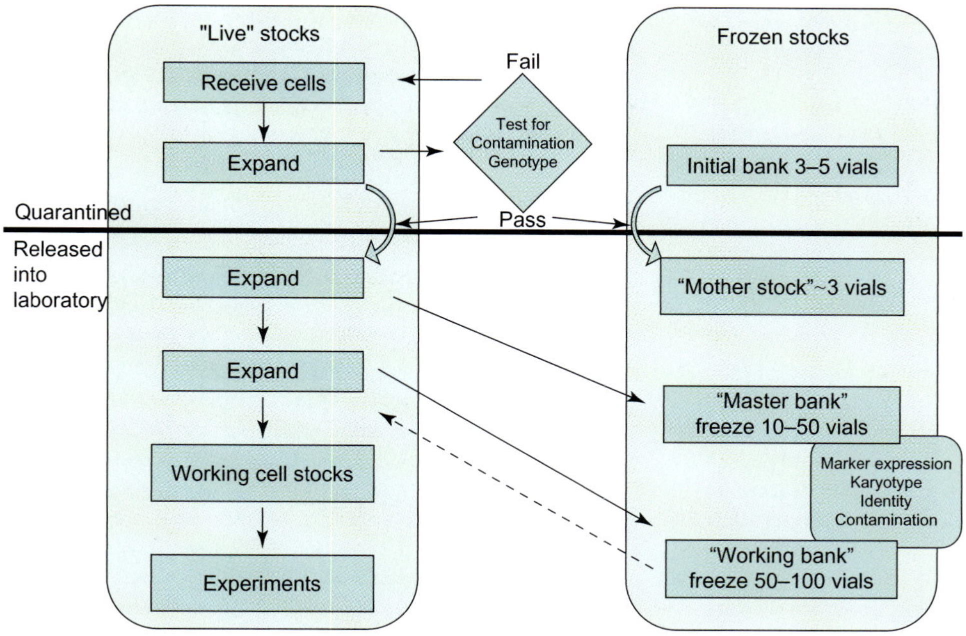

FIGURE 6-9 hESC workflow and tiered cell bank.

(e.g. imprinting status, etc.); however, that drift has not been shown to be associated with systematic loss of pluripotency and so may be seen as less critical.

Incoming Cells

When any cells first arrive, it is prudent to treat them as if they were microbiologically contaminated. Preferably, they should be handled in a separate quarantine facility, though in usual practice this is not always possible and they are handled carefully in the same laboratory, with care to avoid aerosols that would be capable of distributing contaminant microorganisms. Biosafety cabinets are diligently swabbed after handling new cells, and if a separate incubator is not available, the culture vessels are placed in an uncovered, or vented, plastic box to contain any inadvertent spills.

Thawing and Plating Cells

If the cells arrive on dry ice (or more preferably in $N_{2(l)}$), transfer to $N_{2(l)}$ storage immediately. Some distributors recommend immediate thawing of samples shipped on dry ice rather than transfer to $N_{2(l)}$. Cells can then be recovered after all reagents have been assembled and validated as appropriate.

Initial cell recovery and maintenance should be done using the instructions of the cell providers, since troubleshooting by the providers is difficult if they are not familiar with your methods. In general, cells should be recovered as gently as possible, and patience must be exercised. Cryopreserved vials typically contain a few to a few tens of pieces of viable hESC colonies and these can easily be lost. Further, these colony pieces are slow to recover, and often do not appear for up to two weeks (or even after a passage is required due to the exhaustion of the MEFs).

Recovery of Cells Shipped "Live"

Although cells are typically shipped frozen, and preferably in liquid nitrogen in a "dry shipper," it is also possible to ship them as living cultures in a flask at ambient temperatures. Cells sent this way will usually arrive in a "T25" flask completely filled with equilibrated medium and closed tightly give the flask greater compressive strength and to reduce swirling that might dislodge cells. Cells in a flask, however, are much more difficult to passage mechanically.

Swab the flask with 70% ethanol, transfer to the BSC, remove all but 6–8 mL of medium, and inspect the cultures microscopically. If they are not ready to be passaged, incubate and feed, as per usual, until they are ready to passage. When the cells are ready to be passaged, carefully use a heated knife or scalpel blade to cut an aperture in the top of the flask and passage the cells using the microdissection protocol described in Chapter 1. Alternatively, all of the colonies in the flask may be dislodged with a cell scraper and the colonies may be broken up with trituration prior to re-plating. This technique, however, will generally lead to the plating of colonies that are larger than desired and may lead to greater than normal levels of differentiation.

Initial Expansion: "Mother Stock"

The aim of this phase is to establish the earliest bank of frozen cells as soon as possible after receipt, and to confirm that they are not contaminated with microorganisms.

Expansion and Cryopreservation

About three vials should be frozen within two or three passages of the cells' arrival. For this phase, the cells should be maintained using the protocols provided by the supplier of the cells. Cells will be maintained in culture and further expanded to generate the next level of stocks, while simultaneously some cells will be tested for contamination and genotyped to confirm the identity of the cells.

Cells can be cryopreserved using variations on commonly used methods; however, we recommend that, for the first freeze, the methods recommended by the supplier of the cells be followed. Again, this will allow more efficient troubleshooting, if necessary. One way to confirm cryopreservation viability of another cryopreservation method is to thaw one of the vials from the first freezing to confirm recovery before the freezing of additional cells for the master cell bank.

Testing

Microbial contamination

We routinely test for mycoplasma, bacteria, and fungus. The most frequent contaminants of hESC cultures are mycoplasma, which are small prokaryotes without cell walls. Since most antibiotics do not kill mycoplasma, it often persists at low levels without developing the turbidity that indicates bacterial contamination, and may induce a variety of deleterious effects, ranging from changing cell metabolism to chromosomal changes. If the cells are contaminated at this early stage, abandon the expansion and get fresh cells.

Genotype

Cells should be genotyped for positive identification because, although rare, confusion can occur. There are a number of methods available, including single nucleotide polymorphism (SNP), human leukocyte antigen (HLA), and short tandem repeat (STR). These can be carried out in house or by a contract organization such as a paternity testing lab.

Secondary and Tertiary Expansion: "Master Cell Bank" and "Working Cell Bank"

The master and working cell banks are the two critical levels of the tiered cell bank and utmost care should be taken during the expansion of the cells for them. Diligently change medium as recommended, passage using best practise manual dissection, and dissect away any regions of differentiation as described in the supplier's protocols and in Part 1 of this book. Limits are set to the amount of time that cells are serially passaged during normal laboratory activities, and cells must, therefore, be recovered from the working cell bank. As the working cell bank is depleted, it is replaced by expansion of cells from the master cell bank. At each banking step, the cells are extensively validated as described below.

The number of dishes of cells required for this freeze will depend on the size of the cell bank planned, but as a rule of thumb we aim for 20–40 colony pieces per freezing vial, and we usually get four vials per 60 mm dish or 12 vials per six-well plate. So, for a bank of 20 vials, five 60 mm dishes at moderate density are required.

These cells can be cryopreserved by a variety of methods. We recommend thawing a vial of cells previously frozen by your laboratory to ensure that appropriate recovery rates can be achieved, and expanding these cells to undertake the battery of quality control (QC) tests, including microbial contamination, genotype (genetic identity and karyotype), and phenotype (molecular marker expression and differentiation capacity). These tests are described below in the section on QC standards and in protocols found elsewhere in this book.

QC Standards: Routine Monitoring of Cell Status

Although hESC lines are highly similar to each other in their expression of cell surface antigens and markers characteristic of the ESC state and are relatively stable over time in continuous culture, like all other cell types ESCs can undergo changes in culture. Most dramatic are changes in the chromosomal composition of the cells, but other forms of genetic and epigenetic drift have also been detected in hESCs after extended culture. These changes include accumulation of DNA damage, oxidative damage, erosion of telomeres, acquisition of mitochondrial mutations, and loss of imprinting.

The potential for these changes makes it important to monitor cells regularly to ensure maintenance of baseline hESC characteristics. There are many aspects of cell status that could be meaningfully tested; however, to carry out all of these tests while freezing regularly through extended passage would become prohibitively expensive and time consuming. The choice of tests should be most complete at the time of laying down each layer of the tiered frozen cell bank, and less extensive at the time of regular tests through the period of expansion of the cells in culture. We recommend that these regular tests be carried out at least every ten passages and ideally at every fifth. The cost of the tests is currently about US$2000 when performed extensively for a frozen bank and about US$500–800 at every tenth passage. While expensive, it is well worth the savings in time and money that are involved in rescuing a line or retesting interesting results (or, as a worst case scenario, starting all over). Using this strategy, we have been able to maintain five different lines over several years in culture without losing more than an occasional vial to karyotypic or other abnormalities.

Basic Set of Tests for Monitoring Passaged Cells

Karyotype

We recommend using SNP analysis or G-banding (both are described in detail elsewhere in this book). Although fluorescence *in situ* hybridization (FISH), spectral karyotyping (SKY), and comparative genome hybridization (CGH) are also possibilities, these require longer time periods, larger cell numbers or cost much more, so are not appropriate for routine monitoring. They nevertheless remain important second stage tests for verification or more precise definition of an abnormality detected by SNP or G-banding.

Marker Expression: RT-PCR, Immunocytochemical Analysis, and Telomerase

We recommend a combination of PCR and quantitative PCR (qPCR) to assess cell populations. PCR is used to confirm expression of markers of undifferentiated cells and to demonstrate the absence of markers of differentiation and the lack of contaminating feeder populations. If these are detected, we then use a qPCR protocol to assess the degree of contamination by feeders and the extent of differentiation.

While PCR or qPCR give an impression of the total cell population, immunocytochemistry allows assessment of the status of individual cells. Usually we use a double labeling technique to look at, first, POU5F1/OCT4 or SOX2 and, second, a cell surface epitope such as SSEA-3 or, instead, alkaline phosphatase fluorescent histochemistry. We recommend the *in situ* staining of cells in culture to layer the staining information onto the morphology and appearance of the colonies.

While flow cytometry may be a more quantitative way to assess the proportion of cells carrying different cell surface epitopes, we have found that obtaining accurate numbers is difficult as undifferentiated hESCs attach to each other by tight junctions and so obtaining a representative single cell suspension is problematic. Moreover, the total amount of cell sample required is much higher than that required for other tests with equivalent or greater sensitivity.

Telomerase activity is a defining characteristic of hESCs and levels should be assessed regularly.

Additional Tests for Frozen Cell Banks

To the tests described above for routine assessment of cells in culture, we add the following to confirm the identity and characteristics of the cells frozen at each level of the tiered cell bank.

Identity

At each major freezing step or before a critical series of experiments, the identity of the cells should be confirmed by STR or HLA typing. This is particularly important when more than one line is being maintained in the laboratory. Mix-ups are inevitable and are most often attributable to human error. However, such mistakes can be readily determined if a genomic fingerprint of the cells is on file for comparison. STR testing can be readily done using off-the-shelf kits, as can HLA typing. Both are also offered as services by forensic and pathology labs. The cost is easily justified by the alternative greater cost of the consequences of working with misidentified cells.

Demonstration of Differentiation Potential

In addition to the cells being relatively uncontaminated by differentiated cells, it is critical that they retain the ability to differentiate to representative cells of the three germ lineages (endoderm, ectoderm, and mesoderm) and to trophoblast lineages. This can be ascertained most readily by *in vitro* differentiation to embryoid bodies (Chapter 4), followed by molecular analysis (RT-PCR; Chapter 10, or immunocytochemical; Chapter 9). Alternatively differentiation capacity can be demonstrated by the *in vivo* induction of teratomas in immunocompromised mice followed by histological analysis.

Other Tests

Whenever a new line is derived, it should be characterized even more stringently. In addition to the tests described above, mitochondrial sequences, miRNA expression, global methylation profile, and methylation-specific PCR to assess key developmental and imprinted genes should be considered. Our data suggest that although these aspects of cell biology might be important when cell lines are developed for clinical use, they do not appear to affect most cell culture differentiation protocols, and so these may not need to be included in the routine battery of tests.

Standards

The ideal standard to compare gene expression patterns of hESC cultures may be a perfectly defined culture of hESCs with no differentiation at all, but this is difficult to generate in sufficient quantities to allow the use of the same standard sample across time. Although there is no perfect cellular control for the validation of hESC cultures, there are cells that may be used for this purpose because they approximate the expression patterns of hESCs, but are more stable in culture. We recommend two cell lines available from American Type Culture Collection (ATCC): the germ cell tumorderived cell line NTERA-2 cl.D1, and the aneuploid hESC line BG01v. RNA from either of these cell lines can be used as a standard for the comparison of gene expression profiles by qPCR or by microarray. The BG01v cells are useful for immunohistochemistry, since they express the key surface epitopes of undifferentiated euploid hESCs and also are capable of differentiation to the three germ cell lineages. In addition to acting as standards within a laboratory, these cells are very useful in the development of many of the assays and protocols described in this manual and may also be useful for the initial development of new protocols.

Microbial Contamination

Mycoplasma Testing

Mycoplasma is a common contaminant in mammalian cell culture, and can be very insidious. Cultures can be contaminated with mycoplasma from the reagents or from the technician, and once contaminated, low levels may persist undetected.

Fungal and Bacterial Sterility Testing

This is usually done by classical culture techniques that use fluid thioglycollate medium (Sigma catalog no. 90404) for detection of aerobic and anaerobic bacteria and soya bean/casein broth (Sigma catalog no. S1674) for detection of aerobes, facultative anaerobes and fungi. Samples are inoculated into the broths in duplicate, with one being cultured at 22°C and the other at 32°C for 14 days, after which they are examined for turbidity.

Recovery of Normal Populations

While it is possible to recover cells that have become contaminated with microorganisms, the treatments are often harsh, and we strongly recommend going back to an uncontaminated stock if possible. If there has been a significant amount of work invested in the generation of a clone or subline, for example, then the following methods might be attempted; however, once the cells are free of contamination, they should be QC tested to ensure that the relevant properties have been retained.

When using multiwell plates, contamination is generally restricted to only one or a few wells, unless the contamination results from a severe compromise of the medium used to feed all the wells. One option, therefore, is simply to aspirate the contaminated well(s), fill the now-empty well(s) with full-strength bleach, aspirate the bleach, wash the well(s) with 70% ethanol, and aspirate the well(s) to dryness. The remaining uncontaminated well(s) are then likely to remain so. This consideration, by the way, highlights one main advantage of using multiple wells of multiwell plates rather than the equivalent surface area in a single dish—the entire culture may not be lost in the former situation while it is much more likely to be lost in the latter.

It is also important to remember that, unlike bleach and ethanol, antibiotics do not kill microorganisms, they merely retard or stop their proliferation. In the intact animal this allows the immune system time to do its job of actually eradicating the invading microorganisms. In cell culture there is no immune system, so complete eradication of contamination is very difficult and sometimes, at best, all one can hope for is to save the culture with continuous antibiotic supplementation, keeping the contamination at a very low level—it may not ever be possible to go back to an antibiotic-free medium. While this may be acceptable for research purposes, it is clearly unacceptable for clinical purposes, so in this latter case complete clearing of the contamination must be proven by an extended period of continuous antibiotic-free culture.

Mycoplasma Eradication

Mycoplasma infection is a serious threat in hESC laboratories; if a mycoplasma infection is discovered, the best course is to destroy the cultures, then test all the other cultures in the lab. Mycoplasma can be eradicated by treatment with antibiotic cocktails such as BM-cyclin (Roche), MRA (ICN), or Plasmocin (Invivogen), which are all reported to be effective, although with some cytotoxic side effects; however, their successful use with hESCs has not been reported. This is an action of last resort. If possible, go back to frozen mycoplasma-free stocks because the influence of these antibiotics (much less the mycoplasma) on cell karyotype or genotype is not known.

Recovery from Bacterial Contamination

Antibiotics that target bacteria are routinely used in many cell culture situations, and some laboratories use them routinely in hESC culture. Bacterial contamination is usually first recognized by cloudiness in the medium. Contaminated cultures should be immediately isolated from other cultures, remembering that bacteria cannot jump from culture to culture, but rather are spread by spills and aerosols. It is recommended that the cells be handled in separate incubators, and kept in a disposable outer container such as a plastic box. Media and reagents for these cells should be entirely separate from those used with other cells; whenever the contaminated cells are handled, the BSC should be swabbed thoroughly with 70% ethanol before other cells are handled there. We recommend a regimen in which these cells are manipulated last thing at night and, after swabbing the work surfaces with 70% ethanol, the UV lights are left on overnight in the BSC.

1. Working as carefully as possible to prevent aerosols, wash away as much of the bacteria as possible.
2. Aspirate the medium, and replace with (Ca/Mg-free) D-PBS. Aspirate and replace the Ca/Mg-free D-PBS six times. This must be done gently as the Ca/Mg-free buffer loosens the attachment of the hESC colonies to the substrate.
3. Aspirate and replace with hESC medium containing penicillin (100 U/mL) and streptomycin (100 μg/mL). Aspirate and replace with hESC medium containing penicillin (100 U/mL) and streptomycin (100 μg/mL) a second time.
4. If possible, passage the colonies by manual dissection to a fresh dish in order to reduce the burden of bacteria further.
5. Put cultures into the incubator and change the medium twice daily for 2 days and then daily until ready to passage.
6. Keep the cells in antibiotic-containing medium for at least two passages before attempting to wean them from the antibiotics, then monitor carefully by microscope, and, if bacteria reappear, repeat this entire process. It may be prudent to use sterile-filtered MEF-conditioned medium for these latter feeding steps.

Eradication of Yeast and Other Fungi

Option 1

There are a number of antimycotics available, including amphotericin B ("Fungizone") and mycostatin ("Nystatin") (Invitrogen) but these drugs should only be used as a last resort. We are not aware of any groups having successfully recovered fully functional hESC cultures after yeast or fungus contamination and most laboratories simply destroy yeast- or fungus-contaminated cultures.

Contamination is usually recognized microscopically as chains of yeast or filaments of fungus in the culture. Contaminated cultures should be immediately isolated from other cultures, remembering that pre-sporulating fungi and yeast are spread by spills and aerosols. It is recommended that the cells be handled in separate incubators, and kept in a disposable outer container such as a plastic box. Media and reagents for these cells should be entirely separate from those used with other cells, and whenever they are handled, the BSC should be swabbed thoroughly with 70% before other cells are handled there. We, again, recommend a regimen in which these cells are manipulated only at the end of the day and, after swabbing the work surfaces with 70% ethanol, the UV lights are left on in the BSC overnight.

1. In the case of fungal contamination, first carefully aspirate the visible fungal colonies with a Pasteur pipette.
2. Then, working as carefully as possible to prevent aerosols, wash away as much of the fungus as possible. Aspirate the medium, and replace with (Ca/Mg-free) D-PBS. Aspirate and replace the Ca/Mg-free D-PBS six times. This must be done gently as the Ca/Mg-free buffer loosens the attachment of the hESC colonies to the substrate.
3. Aspirate and replace with hESC medium containing the commercial antimycotic. Aspirate and replace with hESC medium containing the antimycotic a second time. If possible, passage the colonies by manual dissection to a fresh dish in order to further reduce the burden of fungi.
4. It is possible to use 2–3 times the usual concentrations of antimycotics during this process but antibiotics must not be used simultaneously with the higher antimycotic concentrations as the toxicity of the antimycotics to the cultured cells is greatly increased in their presence.
5. Put cultures into the incubator and change the medium twice daily for 2 days and then daily until passage.
6. Keep the cells in antimycotic-containing medium for at least two passages before attempting to wean them from the antimycotics, then monitor carefully by microscope, and, if yeast or fungi reappear, repeat this entire process. It may be prudent to use sterile-filtered MEF-conditioned medium for these latter feeding steps.

Option 2

It is also possible, but very difficult, to recover cultures contaminated with fungus without the use of antimycotics. Visible fungal colonies are removed and the cultures are washed repeatedly as described above, first with PBS and then with medium to remove the bulk of the contamination. The colonies are then passaged by microdissection under a high-powered dissecting microscope, taking care to remove pieces of colony without bringing along fungal cells. Those skilled with a pipette can even "clean off" fungal cells from pieces of colony. Individual pieces of colony are transferred to small culture wells (such as Nunc 4 well; SKU 167063). As mentioned above, splitting the colonies up into multiple wells allows for the possibility that separate uncontaminated and contaminated cultures may be generated.

Cultures are monitored daily for contamination. Using this method we have been able to recover about a quarter of the transferred colony pieces. A proportion will carry the fungi, however, and can be discarded, or treated with the antimycotic agent as described above in option 1.

Recovery of Karyotypically Normal Cells

hESC populations are prone to accumulation of karyotypic abnormalities over time. This does not happen spontaneously for all the cells of a given culture, but, presumably, by the appearance of cells with abnormal karyotype which have a growth advantage. The appearance of karyotypic abnormalities has been hypothesized to be associated with the method of passaging; passage by microdissection appears to be less prone to abnormalities than enzymatic passage. The appearance of karyotypic abnormalities is not sudden, but rather there is accumulation of abnormal cells and, by the time the abnormality is recognized, abnormal cells will represent a significant proportion of the culture. A carefully constructed tiered cell bank, as described above, can reduce the impact of aneuploid cells by allowing the researcher, with relatively little cost, to go back to a validated cell population with normal karyotype and with known differentiation capacity.

Sometimes, however, the amount of effort that has been invested in the generation and characterization of a population of cells makes it cost effective to clone out normal cells from the population.

This process is not very efficient, however, and with normal culture conditions a plating efficiency of less than 1% is expected. Cloning is generally done by FACS sorting directly into 96-well trays and it may be made more efficient using medium supplemented with neurotrophins, which are reported to allow cloning of cells with 15% efficiency. In this way, it is quite feasible to clone and expand 10–50 clones and to identify those with normal karyotype.

READING LIST

General

European Collection of Cell Cultures (2001). *The Fundamentals of Cell Culture—A Laboratory Handbook.* http://www.ecacc.org.uk/default.asp?itemid=29673

Overview of General Cell Culture Methods

Freshney R Ian (1999). *Freshney's Culture of Animal Cells: A Multimedia Guide.* New York: John Wiley & Sons.

General Cell Culture Methods and Laboratory Set-up

Centers for Disease Control and Prevention and National Institutes of Health (1999). *Biosafety in Microbiological and Biomedical Laboratories (BMBL)*, 4th edn. http://www.cdc.gov/od/ohs/ biosfty/bmbl4/bmbl4toc.htm

Safe Handling of Human Cells and Tissues

Protocols and manuals published on the internet:
ATCC stem cell methods:
http://www.atcc.org/common/catalog/cellBiology/stem-cells/technicalInfo/protocols.cfm
BresaGen hESC methods:
http://stemcells.nih.gov/research/registry/ESI manual and other Singapore protocols:
http://www.stemcell.edu.sg/resources/methodsProtocols.php
Geron hESC methods:
http://www.geron.com/showpage.asp?code=prodstprot
Melton Laboratory hESC methods:
http://mcb.harvard.edu/melton/HuES/WiCell hESC protocols:
http://www.wicell.org/forresearchers/index.jsp?catid=12&subcatid=20

SPECIFIC REFERENCES

Amit M, Carpenter MK, Inokuma MS, Chiu CP, Harris CP, Waknitz MA, Itskovitz, Eldor J, Thomson JA (2000). Clonally derived human embryonic stem cell lines maintain pluripotency and proliferative potential for prolonged periods of culture. *Dev Biol* 227: 271–278.

Bhattacharya B, Miura T, Brandenberger R, Mejido J, Luo Y, Yang AX, Joshi BH, Ginis I, Thies RS, Amit M, Lyons I, Condie BG, Itskovitz-Eldor J, Rao MS, Puri RK (2004). Gene expression in human embryonic stem cell lines: unique molecular signature. *Blood* 103: 2956–2964.

Brandenberger R, Khrebtukova I, Thies RS, Miura T, Jingli C, Puri R, Vasicek T, Lebkowski J, Rao M (2004). MPSS profiling of human embryonic stem cells. *BMC Dev Biol* 4: 10.

Buzzard JJ, Gough NM, Crook JM, Colman A (2004). Karyotype of human ES cells during extended culture. *Nat Biotechnol* 22: 381–382; author reply 382.

Cai J, Chen J, Liu Y, Miura T, Luo Y, Loring JF, Freed WJ, Rao MS, Zeng X (2006). Assessing self-renewal and differentiation in human embryonic stem cell lines. *Stem Cells* 24: 516–530.

Carpenter MK, Rosler ES, Fisk GJ, Brandenberger R, Ares X, Miura T, Lucero M, Rao MS (2004). Properties of four human embryonic stem cell lines maintained in a feeder-free culture system. *Dev Dyn* 229: 243–258.

Cervantes RB, Stringer JR, Shao C, Tischfield JA, Stambrook PJ (2002). Embryonic stem cells and somatic cells differ in mutation frequency and type. *Proc Natl Acad Sci USA* 99: 3586–3590.

Department of Health and Human Services (2001). Stem cells: scientific progress and future research directions. http://stemcells.nih.gov/info/scireport.

Draper JS, Smith K, Gokhale P, Moore HD, Maltby E, Johnson J, Meisner L, Zwaka TP, Thomson JA, Andrews PW (2004). Recurrent gain of chromosomes 17q and 12 in cultured human embryonic stem cells. *Nat Biotechnol* 22: 53–54.

Ludwig TE, Levenstein ME, Jones JM, Berggren WT, Mitchen ER, Frane JL, Crandall LJ, Daigh CA, Conard KR, Piekarczyk MS, Llanas RA, Thomson JA (2006). Derivation of human embryonic stem cells in defined conditions. *Nat Biotechnol* 24: 185–187.

Maitra A, Arking DE, Shivapurkar N, Ikeda M, Stastny V, Kassauei K, Sui G, Cutler DJ, Liu Y, Brimble SN, Noaksson K, Hyllner J, Schulz TC, Zeng X, Freed WJ, Crook J, Abraham S, Colman A, Sartipy P, Matsui S, Carpenter M, Gazdar AF, Rao M, Chakravarti A (2005). Genomic alterations in cultured human embryonic stem cells. *Nat Genet* 37: 1099–1103.

Miura T, Luo Y, Khrebtukova I, Brandenberger R, Zhou D, Thies RS, Vasicek T, Young H, Lebkowski J, Carpenter MK, Rao MS (2004). Monitoring early differentiation events in human embryonic stem cells by massively parallel signature sequencing and expressed sequence tag scan. *Stem Cells Dev* 13: 694–715.

Noaksson K, Zoric N, Zeng X, Rao MS, Hyllner J, Semb H, Kubista M, Sartipy P (2005). Monitoring differentiation of human embryonic stem cells using real-time PCR. *Stem Cells* 23: 1460–1467.

Pyle AD, Lock LF, Donovan PJ (2006). Neurotrophins mediate human embryonic stem cell survival. *Nat Biotechnol* 24: 344–350.

Rosler ES, Fisk GJ, Ares X, Irving J, Miura T, Rao MS, Carpenter MK (2004). Long-term culture of human embryonic stem cells in feeder-free conditions. *Dev Dyn* 229: 259–274.

Rugg-Gunn PJ, Ferguson-Smith AC, Pedersen RA (2005). Epigenetic status of human embryonic stem cells. *Nat Genet* 37: 585–587.

Sidhu KS, Tuch BE (2006). Derivation of three clones from human embryonic stem cell lines by FACS sorting and their characterization. *Stem Cells Dev* 15: 61–69.

Chapter 7

Approaches for Derivation and Maintenance of Human Embryonic Stem Cells: Detailed Procedures and Alternatives

Irina Klimanskaya and Jill McMahon

INTRODUCTION

Although derivation of mouse ES cell lines has become a common procedure, the limited number and quality of the currently available hES cell lines is testament to the difficulties encountered at various stages of their derivation and maintenance. The current techniques used for the maintenance of hES cells require scrupulous attention to the details of cell culture systems including: quality control of the reagents; monitoring of growth conditions; and an ongoing assessment of the pluripotent status of the cells in order to promote consistency in the cell culture system. Distributors of the available hES cell lines frequently recommend attending special training courses before working with the hES cell lines they provide.

The authors have experience in derivation and culture of over 30 hES cell lines, including hES cells derived without feeder cells, and lines established from single blastomeres (Cowan *et al.*, 2004; Klimanskaya *et al.*, 2005, 2006, 2007; Chung *et al.*, 2008). All our lines have been adapted to trypsinization, which makes passaging of the cells less labor-intensive and generates cells in sufficient numbers to permit experimentation. All of these lines can be successfully frozen and thawed, using very simple procedures, with a recovery rate of 10–30% or higher.

Techniques for deriving and maintaining pluripotent ES cells in culture have been described by a variety of laboratories, and there are similarities and vital differences. For the derivation and maintenance of hES cell lines in our laboratory we adapted previously published methods, and developed an approach which has consistently produced new cell lines and has been successfully taught to many researchers around the world.

The goal of this chapter is to describe aspects of derivation and maintenance of hES cells that we found to be critical for generating hES cell lines. We respond to questions received from other researchers with regard to: equipment, preparation, and quality control of media and reagents; cell passaging techniques; and other aspects of hES cell morphology and behavior.

SETTING UP THE LABORATORY

Equipment

Initial steps in the derivation process are conducted under a dissecting microscope. We make an effort to keep the embryos and the dishes containing the early mechanically passaged dispersions at 37°C. Dishes brought out of the incubators are set on 37°C slide warmers, or viewed using microscopes fitted with heated stages. The mechanical dispersions necessitate having the dishes open for extended periods, and since the cultures are vulnerable to contamination at this time, we have the dissecting microscope within a bench top laminar flow hood. A high quality stereo microscope with a wide range zoom is essential for the mechanical dispersion of colonies, including the ICM outgrowth; it permits an overall assessment of each plate, and evaluation of the morphology of each colony when doing the mechanical passaging.
The equipment that is used in our laboratory is as follows:

- Stereomicroscope for microdissection: Nikon SMZ-1500 with the magnification range of 10–100×. This works well, with its easy zoom and the positioning mirror that

regulates the depth and contrast of the image. A whole 35 mm dish can be scanned for colony morphology, and the zoom permits selection of the parts of the colony that are the best for dissection with precision.
- Inverted microscope: A Nikon TE 300, or any similar microscope, set up with phase contrast and objectives 4×, 10×, 20×, 40×, and Hoffman 20×, 40×. Recommended for fine observation of hES cells, and HMC is required for embryo evaluation.
- Heated microscope stage for both stereo and inverted microscope: (Nikon).
- Slide warmer keeps plates at 37°C during observation and mechanical dispersions.
- Bench-top laminar flow hood with a HEPA filter. Vertical hoods by Terra Universal (Anaheim, Calfornia) are a good inexpensive choice. Mid-Atlantic Diagnostics (Mount Laurel, New Jersey) manufactures workstations that include a biosafety cabinet and a built-in dissecting microscope; these models are more expensive but meet the GMP requirements, if necessary. We found these vertical hoods were tall enough to accommodate a dissecting microscope, and were very convenient and reliable. Horizontal models sometimes produce too much vibration.
- Reliable tissue culture incubator. All parameters (CO_2 concentration, humidity, and temperature) need to be checked daily with external monitoring equipment.

External monitoring equipment:

- Surface thermometer;
- Mercury liquid-immersed thermometer;
- Hygrometer;
- CO_2 monitor and gas calibration kit (GD 444 SR-B by CEA Instruments, Emerson, New Jersey).

Quality Assurance of Equipment

Consistency in growth conditions is very important for the development of the embryos and growth of hES cells. The checklist of parameters which are monitored daily includes the percentage of CO_2 (5.0%), temperature (37°C), and humidity of the incubators (>90%). A checklist with daily readings is very helpful for timely recalibration if any undesirable trend is noticed. Warming stages are constantly monitored with surface thermometers.

The incubators are checked prior to any new derivation round by growing mouse embryos from the two-cell stage to blastocyst; a passing score requires 50% to go to blastocyst. Ideally, until the newly-derived lines can be frozen, the cultures should be split between two incubators, to prevent the loss of a line in the case of equipment failure. If hES cells need to be derived and maintained under the conditions of good manufacturing practices (GMP), GMP protocols for equipment and reagent validation would have to be followed.

Sterility

There are certain aspects of hES cell line derivation which put the associated work under more stringent sterility requirements than any typical cell culture laboratory. Among these is the nature of human embryos, each being unique and very valuable. Because of the labor-intensive nature of derivation and expansion, the new lines usually require a team effort until they are safely frozen. Experiments may require long periods in culture, such as in the course of cell differentiation into embryoid bodies, which often takes several weeks. In addition, because hES cells are prone to spontaneous differentiation under unfavorable conditions, many laboratories prefer not to use antibiotics in cell culture media.

The reagents should either come sterile from the manufacturer, or be filter-sterilized in the laboratory. Most of the cell culture supplies can be bought sterile. However, everything that is in-house sterilized by autoclaving, gas, or dry heat needs to be quality-controlled. We use biological indicators (spore strips from Steris, Mentor, Ohio). If it comes to the worst, a triple-action drug Normocin (active against gram-negative and gram-positive bacteria, fungi, and micoplasma) appears to be well-tolerated by hES cells, without significant changes in their pluripotency or growth rate, and often permits the rescue of contaminated cultures.

PREPARING AND SCREENING REAGENTS

Media Components

- KO-DMEM (Invitrogen cat #10829).
- DMEM high glucose (Invitrogen cat #11960-044).
- Knockout Serum Replacement™ (KSR, Invitrogen cat #10828). Each lot needs to be tested; but, as a guide, we found the lots with osmolarity higher than 470 mOsm/kg and endotoxicity lower than 0.9 EU/ml were the best. Upon thawing, make single-use aliquots and freeze.
- Plasmanate (Talecric, Research Triangle Park, North Carolina). Each lot needs to be tested.
- FBS (Hyclone cat # SH30070.02). Each lot needs to be tested. Heat-inactivate at 55°C for 30 minutes, if desired, and freeze in aliquots.
- Beta-mercaptoethanol, 1000 × solution (Invitrogen cat # 21985-023).
- Non-essential amino acids (NEAA), 100 × solution (Invitrogen cat #11140050).
- Penicillin/streptomycin, 100 × solution (Invitrogen cat #15070-063).
- Glutamax-I, 100 × solution, a stable dipeptide of L-glutamine and L-alanyl, a glutamine substitute (Invitrogen cat #35050-061).

- Penicillin/streptomycin and Glutamax-I are kept in frozen single-use aliquots.
- bFGF (Invitrogen cat #13256-029). Add 1.25 ml of protein-containing medium (we use basal medium, supplemented with Plasmanate and Serum Replacement, as described below) to a vial containing 10 μg of bFGF. This makes 8 μg/ml of stock solution. Increasing final bFGF concentration to 8–20 ng/ml can be beneficial for the cells, especially at early stages of derivation, after thawing or when the cells are grown at low density. Make 120 μl aliquots and freeze.
- Human LIF (Millipore International cat # LIF1010).
- 0.05% trypsin/0.53 mM EDTA (Invitrogen cat #25300-054).
- Gelatin from porcine skin (Sigma, cat. # G1880).
- PBS, Ca^{2+}, Mg^{2+}-free (Invitrogen cat #14190-144).
- Normocin, antibiotic active against Gram +/− bacteria, also has anti-mycoplasma and anti-fungi activity (Invivogen, San Diego, California; cat # ant-nr-2; comes as a 500 × solution).

Media Recipes

Bottles of media that are opened frequently become alkaline rapidly, we suggest making smaller quantities that will last approximately a week.

Primary Mouse Embryo Fibroblast Growth Medium

To a 500 ml bottle of high glucose DMEM add:

- 6 ml penicillin/streptomycin;
- 6 ml Glutamax-1;
- 50 ml FBS.

Human Embryonic Stem Cell Basal Medium (BM)

To a 500 ml bottle of KO-DMEM add:

- 6 ml penicillin/streptomycin;
- 6 ml Glutamax-1;
- 6 ml NEAA;
- 0.6 ml beta-mercaptoethanol. (NOTE: This volume pertains to the 1000 × diluted solution (see the list of reagents), not to 100% beta-mercaptoethanol).

Human Embryonic Stem Cell Derivation Medium

Use this medium at early stages of ICM outgrowth. It has higher LIF and bFGF concentration, and contains fetal bovine serum (FBS). You can switch to hESC growth medium when a steady growth of colonies has been reached (usually, passage 2–4).

To 100 ml of basal medium (bM) add:

- 5 ml Plasmanate;
- 5 ml KSR;
- 5 ml FBS;
- 240 μl of human LIF (final concentration 20 ng/ml);
- 120 μl of bFGF stock solution (final concentration 8 ng/ml) or more (up to 20 ng/ml);
- Sterilize by 0.22 μm filtration.

Alternatively, two media can be prepared: one containing Serum Replacement and Plasmanate only, the other one containing only FBS (10–15%). Mix them 1:1 or 2:1 for the earliest stages of derivation. FBS may help to stimulate the hES cell colony growth, but may reduce its content if any undesirable spontaneous differentiation is observed.

We also supplement derivation medium with hES-cell-conditioned medium. To produce it, derivation medium without FBS is added to an almost confluent culture of hES cells of very good morphology (see below) incubated overnight, collected, filtered through 0.22um filters and used at 25–30%.

Human Embryonic Stem Cell Growth Medium (hESCM)

To 200 ml of basal medium add:

- 20 ml Plasmanate;
- 20 ml KSR;
- 240 μl of human LIF for 10 ng/ml or 480 μl for 20 ng/ml;
- 120 μl of bFGF stock solution (final concentration 4 ng/ml) for 1 × bFGF or more, if a higher concentration is desired;
- Sterilize by 0.22 um filtration.

Gelatin:

Dissolve 0.5 g of gelatin in 500 ml of warm (50–60°C) Milli-Q water. Cool down to room temperature, sterilize by 0.22 μm filtration; makes 0.1% solution. Treat dishes by flooding plates and leaving solution at room temperature for 30 minutes, aspirate solution prior to use.

Mitomycin C:

Add 2 ml of sterile Milli-Q water to a vial (2 mg) of lyophilized Mitomycin C (Sigma, cat. # M 0503); makes 1 mg/ml stock solution. The solution is light-sensitive and is good for one week at 4°C.

Screening Media Components

It is important to be consistent with the screening, aliquoting, and storage of the components of the media. Various lots of KSR, Plasmanate, and FBS should be screened, preferably on hES cells. The screening of Serum Replacement lots should be done prior to the first batch running out, so that an evaluation of its qualities can be compared side-by-side to the previous lot. Some newly-derived hES cells will completely die out with a change in lots at the initial stages.

Screening of Fetal Bovine Serum, Plasmanate, or Knockout Serum Replacement

This test is based on a published procedure for screening FBS lots for mouse ES cell work (Robertson, 1987, p. 74). This approach can be used for screening any combination of reagents or finding the best concentrations for media supplements. The quality of the reagents is assessed by counting the number of colonies, evaluating the morphology of the hES cells, and by staining for alkaline phosphatase activity as detected with a Vector Red Kit (Vector Laboratories, Burlingame, California).

1. Prepare 12-well plates with PMEF. For each lot tested, you will need at least 12 wells to vary the concentration of the component being tested, from the working concentration to high enough concentrations to evaluate toxicity: 8%, 10%, 20%, and 30%. Each concentration is done in triplicate and compared to known media components at working concentrations.
2. Split hES cells onto 12-well plates with a ratio of 1:6–1:10. The difference in reagents will be more noticeable when the cells are started at low density, however, some cell lines grow very slowly and differentiate when they are kept at a low density, therefore adjust the splitting ratio to the specific hES cell line. Resuspend the cells in a small volume of basal medium, and add equal volumes of cell suspension to each well of the test plate; pipette up and down in each well or slowly move the plate in perpendicular directions for even distribution of the cells. Do not rotate, as doing so will move most of the freshly added cells to center.
3. Change the medium daily, and evaluate colony morphology under the microscope. Human ES cells grow in flat, tightly-packed colonies with sharp refractory borders. The colonies appear deep red when stained for alkaline phosphatase activity. In differentiating colonies, the cells are more loosely packed with diffuse borders, and are pinker in comparison. Usually, the difference between the conditions being tested becomes more obvious as colonies grow bigger; however, as the colonies grow larger they can begin to touch each other, and they have a tendency to differentiate. Staining one of the triplicate wells for alkaline phosphatase activity prior to seeing signs of differentiation is advisable. Continue with the other wells in each set for another day or two before staining (see Figure 7-1 for a sample test). This test can also be used to compare performance of different media formulations or new products offered commercially; 8% KSR/8% Plasmanate medium can be used as a control. Adequate record-keeping

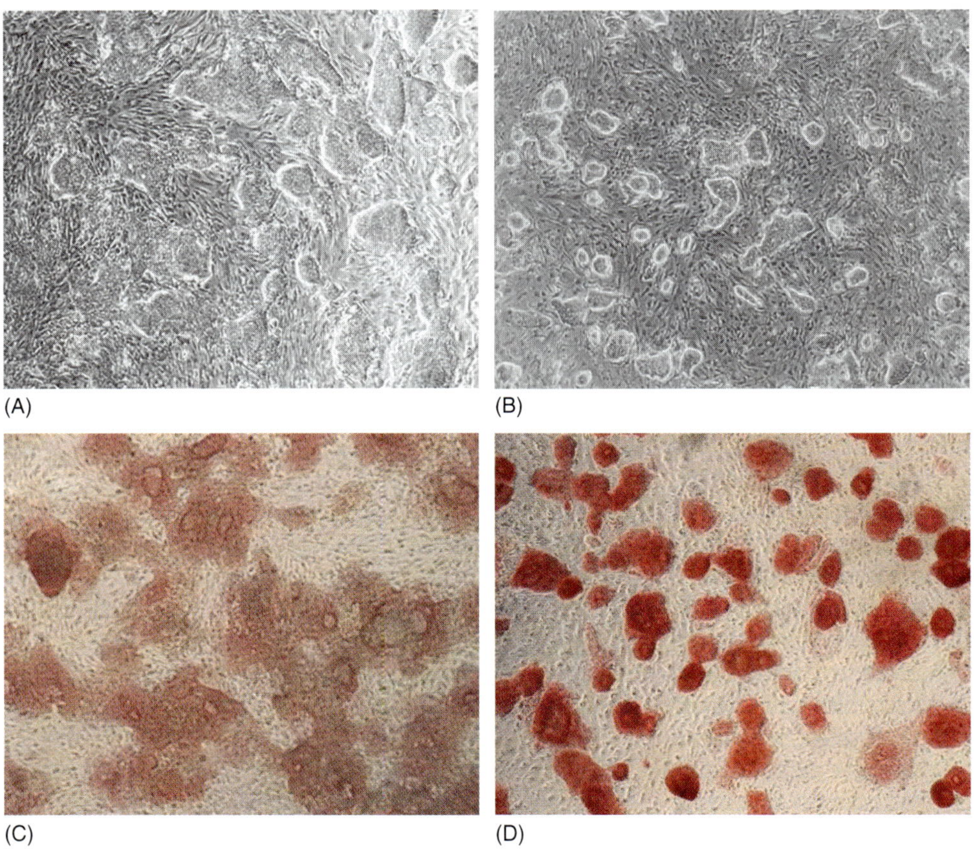

FIGURE 7-1 Media testing. Comparison of 16% Serum Replacement (A, C), with 8% Serum Replacement; 8% Plasmanate (B, D). Quality of the media supplements is assessed by the morphology of the colonies in bright field (A, B), as described in the text, followed by staining for alkaline phosphatase activity (C, D). Note that although the morphology of the colonies in (A) and (B) are comparable, the activity of alkaline phosphatase is higher in medium with both Serum Replacement and Plasmanate (D). Magnification × 40.

for all commercial and in-house prepared lots of reagents is essential for troubleshooting, should the performance of hES cells begin to fail.

PREPARING PRIMARY MOUSE EMBRYO FIBROBLAST (PMEF) FEEDERS

We grow all of our hES cells on PMEF feeders that have been Mitomycin C treated to generate stable monolayers. The PMEF are made by standard procedures using 12.5 dpc ICR mouse embryos (Robertson, 1987, p. 77). The 12.5 dpc embryos are eviscerated, but the heads are left on during tissue disruption in trypsin; plating density is 1.5 embryos per 150 mm plate. PMEF are expanded once after the initial plating (1:5 split), and then frozen (P1). The growth rate of PMEF and their performance as feeders decreases as they go through multiple passages; therefore, thawed PMEF are only passaged once (P2) for expansion purposes prior to Mitomycin C treatment, at which point a new vial of PMEF would be thawed.

Mitomycin C Treatment and Plating

Mitomycin C is added to the medium of a confluent plate of PMEF at a concentration of 10 μg/ml and incubated at 37°C for three hours. The cells are harvested by trypsinization, and plated in PMEF medium onto gelatin-coated plates. In serum-free hESC growth media, the PMEF may appear less confluent due to the spindle-like form the cells adopt; to ensure a confluent monolayer we recommend a plating density of 50–60 000 cells/cm^2. We replace the PMEF medium with hESC growth medium the day after plating, and use plates of PMEF no longer than three to four days after Mitomycin C treatment. We have noticed that some lots of Mitomycin C may be inconsistent—the reagent forms an insoluble precipitate, and the characteristic purple color of the solution lightens dramatically. Reduced concentration of Mitomycin C causes proliferation of "treated" PMEF which results in their detachment in a few days and/or formation of gaps in the monolayer where the drops of medium hit the plate. Correctly inactivated PMEF do not form such gaps, and can remain intact on the plate for up to three to four weeks. Such old PMEF are not good for hES cell culture, are but useful for preparation of extracellular matrix plates for feeder-free culture (Klimanskaya *et al.*, 2005).

MECHANICAL PASSAGING OF HUMAN EMBRYONIC STEM CELL COLONIES

Many established hES cell lines are passaged with collagenase, or by dispase in conjunction with mechanical dispersion. Mechanical dispersion can provide colonies of "perfect" morphology, as it permits one to selectively pick undifferentiated colonies, or even undifferentiated parts from differentiated and overgrown cultures, but it is time-consuming and doesn't yield large numbers of cells, thus limiting expansion of the hES cell lines and experimental design. Nevertheless, this procedure is invaluable at the early stages of derivation, or as a means of producing more homogeneously undifferentiated plates of cells for expansion or for adaptation to trypsin. It is also a tool for a "rescue operation" in critical situations, when the success of salvaging a few colonies means saving a new hES cell line. When hES cell lines recover from freezing with low rates (some providers can only ensure the viability of 1% or less), or show aneuploidy (less than 100%), they can be "rederived" using mechanical passaging.

In new derivations or established cultures it is better to disperse colonies before they grow in contact with each other and prior to signs of differentiation, such as becoming multi-layered (Figure 7-2). The dispersed colonies may be left in the same well or moved to a new well. If there are only a few colony pieces (1–5) they should be placed near to each other, but with enough space to permit growth. When the cells are grown on a 35 mm dish, it is usually sufficient to disperse 50–100 average-sized colonies to populate a new 35 mm plate. In one to two days it may be necessary to disperse some of the larger colonies, leaving the pieces in the same well. Usually, mechanical passaging needs to be done every five to six days, but several larger colonies may need to be dispersed daily.

Materials Needed

Flame-pulled Thin Capillaries

We use an alcohol or gas burner to pull capillaries from glass Pasteur pipettes (pre-sterilized). The capillaries are broken by hand into angled tips (the shape of a hypodermic needle). The diameter of the capillaries may vary, but the best results are achieved when it is about 10–100 ES cells diameter, this is how large the colony pieces are going to be. The choice of a diameter depends on the operation: for instance, to do initial dispersion of an ICM outgrowth or to target undifferentiated parts of a colony, a diameter of 10–30 cells would be used, and to quickly harvest a large number of colonies from a good quality culture, a 50–100 cell capillary diameter would work the best.

Mouth-controlled Suction Device

Similar to a mouth pipette used for embryo transfer, this device consists of a mouth piece (Meditech International, cat. # 15601 P), rubber tubing, and a 0.22 μm syringe filter with a rubber tubing adapter for the capillary. It provides high precision in all manipulations for colony dispersions. Alternatively, a "Stem Cell Cutting Tool" (available from Vitrolife, Sweden) for picking colonies can be used instead of a mouth-controlled device and glass capillaries.

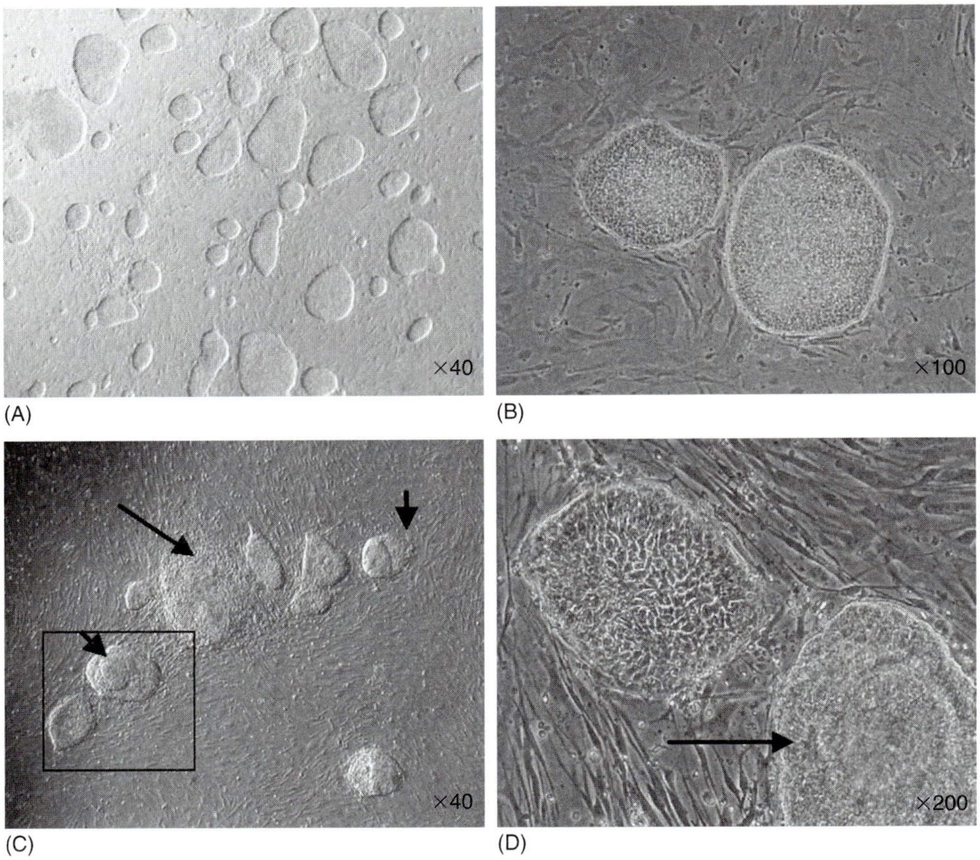

FIGURE 7-2 Morphology of mechanically dispersed hES cells. Colonies on panels (A, B) show no signs of differentiation. Note that the cells are small, tightly-packed, and the colonies have sharp borders. In (C) long arrow points to a very-differentiated colony that should not be dispersed, short arrows show partially-differentiated colonies which can be dispersed, and may produce undifferentiated colonies at the next passage. (D) Higher magnification of the framed colonies on (C), shifted 90°. The arrow points to a partially-differentiated colony, which has become multi-layered in the middle. The other colony is undifferentiated and is similar to colonies shown on (A, B).

Preparation of Primary Mouse Embryo Fibroblast Plate Receiving Mechanically Dispersed Colonies

1. When dispersing the initial ICM outgrowth, change the medium on a 4-well plate of Mitomycin C treated PMEF to hESC derivation medium the night before the derivation day, to let it become conditioned by the PMEF, final volume of 250 µl. Alternatively, instead of conditioning it overnight, the derivation medium can be supplemented with 30% of hESC-conditioned medium. To collect hESC-conditioned medium, add medium to a sub-confluent culture of hES cells with good morphology (see Figure 7-2A for an example of colony density and morphology), leave for 24 hours, collect medium, filter and store for two to three days at 4°C. At early derivation stages, when the colony growth is slow, change ⅔ of the medium every two to three days to keep it conditioned at all times. As more colonies appear, change ⅔ of the medium daily and increase volume to 500 µl/well of a 4-well plate.
2. For established cultures, remove ⅓ of the medium from growing culture, put on new PMEF dish and add ⅔ volume of fresh medium. Change ⅔ of the medium daily.

Mechanical Dispersion

The procedure is similar to vacuuming. Gentle dispersion of the colonies is achieved by simultaneously cutting off the pieces with the angled end of the capillary, very lightly moving them off and sucking them in. With the opening of the capillary positioned horizontally to the bottom of the dish, begin moving from the sides in towards the colony center, chopping off and gently sucking in each piece. The light suction helps to detach the colony parts, and is applied at all times, as you move from the periphery of the colony, collecting the colony parts. If the whole colony is coming away from the monolayer in one piece, it is probably differentiated and should be discarded.

When the desired number of colonies is dispersed, blow out the pieces into the same plate (like for the first ICM dispersion), or into a freshly prepared plate. To avoid having all the colonies stuck to each other in the center of the plate, move the plate gently from side to side, do not swirl. (Figure 7-2 shows examples of mechanically passaged cultures, pointing out parts of colonies that have differentiated and should be avoided when passaging the culture).

DERIVATION OF HUMAN EMBRYONIC STEM CELLS

It is clear there are many factors which are not fully understood, that influence whether an isolated ICM will produce an hES cell line. Some of the factors to consider are: at what stage the embryo was frozen and by what procedure; the length of time the embryo must be in culture to generate a blastocyst; the culture conditions; and the quality of both the ICM and the trophectoderm. The determination of when an embryo is ready for immunosurgery must be determined empirically, usually occurring between d5 and d7. Any embryo that has undergone cavitation and has relatively intact trophectoderm is a candidate for immunosurgery.

Immunosurgery

The process of immunosurgery was performed essentially as described in Solter and Knowles, 1975. It involves removing the zona pellucida with Acid Tyrode's solution, incubating the embryo in an antibody that binds to the trophectoderm and preferably not to the ICM cells (especially important for the embryos with non-intact trophectoderm), and then lysing the trophectoderm cells with complement. The dead cells that surround the ICM are removed by passing the ICM through a narrow capillary. The isolated ICM is put on a prepared PMEF (in microdrops or 4-well plates) for further growth and dispersion.

Materials Needed

- Acid Tyrode's (Specialty Media, cat # MR-004.D).
- Rabbit anti-human red blood cell (RBC) antibody (purified IgG fraction, Inter-Cell Technologies, Hopewell, New Jersey, cat. # AG 28840).
- Aliquoted and stored at −80°C, freshly diluted 1:10 in hES C Derivation Medium.
- Complement (Sigma, cat # S1639).
- Aliquoted and stored at −80°C, freshly diluted 1:10 in Derivation Medium.
- Capillaries for embryo transfer.
- Thinly drawn capillaries (approximately the diameter of the ICM) for the trophectoderm removal,

Prepared Mitomycin C Treated Primary Mouse Embryo Fibroblast Plates

For the initial ICM outgrowth, change the medium on a 4-well plate of Mitomycin C treated PMEF to hESC derivation medium the night before the immunosurgery, to let it become conditioned by the PMEF, final volume of 250 μl.

Alternatively, instead of conditioning it overnight, the derivation medium can be supplemented with 30% of hESC-conditioned medium. To collect hESC-conditioned medium, add medium to a near-confluent culture of hES cells with good morphology (see Figure 7-2A for an example of colony density and morphology), leave for 24 hours, collect medium, filter, and store for two to three days at 4°C.

Microdrops under oil can also be used to minimize the volume of the medium conditioned by the colony growing from an ICM well. To prepare a plate with microdrops first, make gelatin drops on a 60 mm tissue culture dish, and incubate for 30 minutes in a cell culture incubator. Aspirate each drop and immediately add to "footprints" PMEF suspension, 1 million cells/ml. Overlay with embryo-quality light mineral oil. Change the medium the next day to derivation medium, by carefully aspirating the medium from each drop under the dissecting microscope, and replacing with growth medium (a P200 pipette can be used); repeat once or twice to get rid of all PMEF medium.

Immunosurgery Procedure

1. Each embryo is processed separately. A dish is prepared with a series of 30 μl microdrops, three for each step: Acid Tyrodes, anti-human RBC antibody, complement, and three drops of derivation medium for each of the washes between the steps. The drops are covered with embryo-tested mineral oil, and are equilibrated in the CO_2 incubator for 60 minutes.
2. Under the dissecting microscope, transfer the embryo into the first Acid Tyrode's drop for a quick (1–2 seconds) wash, and then move into the second. Watch the embryo closely, and as soon as the zona pellucida thins and is nearly digested, move the embryo into the next series of hES cell media drops. Move the embryos through the first antibody drop into the second and third drops, and place the dish into the incubator for 30 minutes.
3. Transfer the embryos through three drops of Derivation Medium, and through three drops of the complement solution as described above; incubate in the last drop of complement in the CO_2 incubator for 15 minutes, and check for any "bubbling" trophoblast cells. If no cells show signs of lysis or if only a few cells are "bubbling," continue the incubation and recheck in five minutes. The embryo should be transferred to the drop of Derivation Medium as soon as all trophoblast cells are lysed or if no new "bubbling" cells appear after rechecking; the total incubation in complements should not exceed 30 minutes.
4. Gently pass the embryo through the opening of a thinly drawn capillary (about the diameter of the ICM), the lysed trophectoderm cells should detach after 1–2 passes.
5. Wash the ICM in the drops with Derivation Medium, and place into the prepared well of a 4-well plate. The ICM should attach within 24 hours.

Inner Cell Mass Dispersion

At the early stages of derivation, we recommend doing the first dispersion as soon as at least 2–3 colony pieces can be

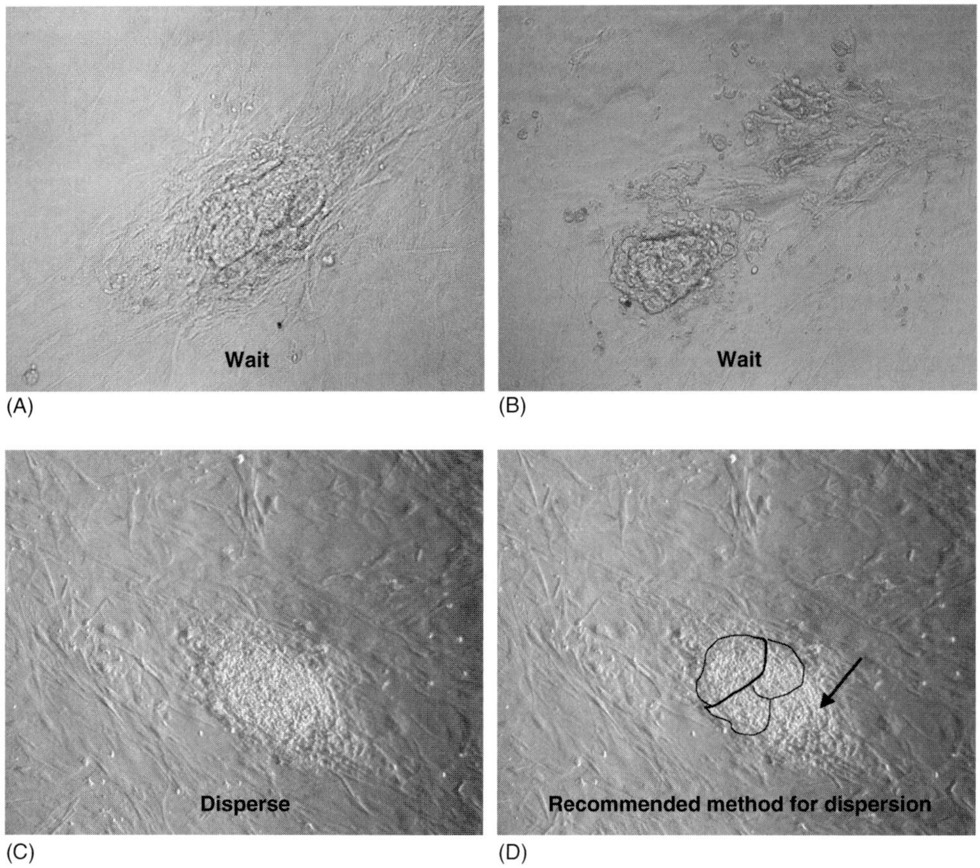

FIGURE 7-3 Initial ICM outgrowth. The initial outgrowth of the ICM rarely shows typical ES colony morphology, and often includes many differentiated-looking cells. When no obvious ES cell-like colonies large enough for dispersion can be located, more time is required before the first colony dispersion can be done. (A, B) The initial ICM outgrowth of two future hES lines. At this stage, when the dispersion is attempted, the outgrowth and the PMEF monolayer come up together, so it is better to wait longer before dispersing. (C) The ICM is ready for dispersion, when a colony of ES-like cells is formed which is large enough for dispersion into several pieces, leaving 20–50% of the outgrowth on the original place for future regrowth. (D) Dotted lines transcribe the number of pieces recommended for the dispersion of this colony. A narrow capillary is used and a small part of colony is left untouched (arrow). Magnification × 100.

obtained from the initial outgrowth (see Figure 7-3). The dispersed colonies may be left in the same well or moved to a new well. If there are only a few colony pieces (1–5) they should be placed close to each other, but with enough space to permit growth. It is better to disperse colonies before they grow in contact with each other and prior to signs of differentiation, such as becoming multilayered. When the colony growth is slow, change ⅔ of the medium every two to three days to keep it conditioned at all times. As more colonies appear, change ⅔ of the medium daily and increase volume to 500 μl/well of a 4-well plate.

Even when an original colony looks differentiated or comes off as a single piece, when replated, it usually gives an outgrowth of hES cells. When doing the initial dispersion, a part of the original colony should be left untouched as a back-up, especially if the picked pieces are transferred into a new well. Expect it to grow back in 1–2 days, and the new outgrowth can be picked and recombined with previously picked cells. Multiple harvests can be obtained from the initial outgrowth. It is critical at this stage to expand the number of colonies slowly and steadily. See Figure 7-4 as an example of the length of time between dispersions and appearance of the cultures during the process of derivation. In this case, no immunosurgery was done because the trophectoderm was not sufficiently intact.

MAINTENANCE OF ESTABLISHED HUMAN EMBRYONIC STEM CELL CULTURES

Usually, once a steady growth of colonies is reached, use of the hESC Derivation Medium is discontinued and the cultures are maintained in hESC Growth Medium. For established cultures, replace ⅔ of the medium volume with fresh medium daily; do not allow the medium to turn yellow. Cultures should be expanded gradually, by the progression from relatively sparsely populated 4-well dishes to confluent 4-wells to 35 mm dishes. Throughout the process, the cultures should be observed daily, differentiated colonies removed and undifferentiated colonies dispersed as necessary

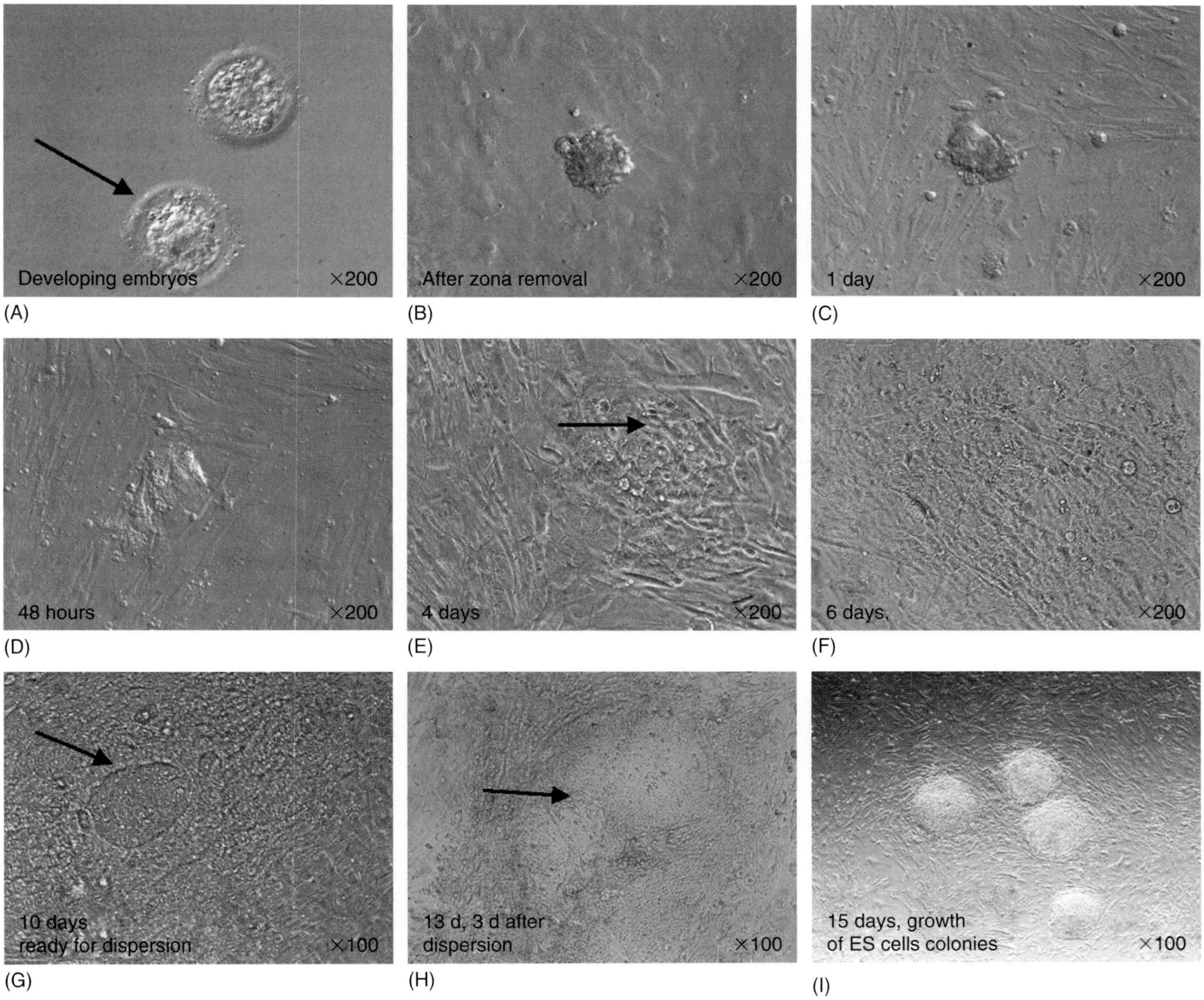

FIGURE 7-4 Early stages of hES cell derivation. A blastocyst of poor quality, grade 3, arrow in (A) underwent the removal of the zona pellucida (B), and next day after being plated on the PMEF monolayer has attached (C). In 48 hours after plating the ICM appears smaller, possibly due to some cell death (D), and in four days after plating an outgrowing group of cells is visible with some small cells in the middle (E, arrow) which become less visible two days later (F). Ten days after plating (G) a small colony of ES-like cells (arrow) has formed within the large group of differentiated-looking cells, and is now large enough for dispersion. Original ICM outgrowth (H) two days after it was dispersed; note regrowth of small ES-like cells in the cleared area (arrow) which are ready for a further dispersion. (I) Formation of ES cell colonies from the recombined first and second dispersion.

(see Figure 7-5 for examples of approaches to be used for dispersion of colonies with different morphologies).

By the time the cells are growing on a 35 mm dish, it is usually sufficient to disperse 50–100 average-sized colonies to populate a new well. In one to two days it may be necessary to disperse some of the larger colonies, leaving the pieces in the same well. Usually, mechanical passaging needs to be done every five to six days, but several larger colonies may need to be dispersed daily.

Adaptation of Human Embryonic Stem Cells to Trypsin

Our experience with trypsinization of over 30 hES cell lines demonstrates that, after the initial adaptation of the lines to trypsin, this procedure can be robust and yield large quantities of hES cells which exhibit all the properties of pluripotential cells. Trypsinized cells retain undifferentiated colony morphology, express characteristic molecular markers, i.e., Oct-4, alkaline phosphatase, SSEA-3, SSEA-4, TRA-1-60, and TRA-1-81, differentiate into three germ layers *in vitro*, and into teratomas *in vivo*, and maintain normal karyotypes.

Newly-derived hES cells may be successfully passaged with trypsin as early as passage 2–3 from a 4-well plate. However, trypsinization is not always successful, and several attempts may be necessary before the cells are adapted to trypsin. We recommend always keeping a back-up well of mechanically passaged cells. It has been shown (Draper *et al.*, 2004; Mitalipova *et al.*, 2005; Chan *et al.*, 2008)

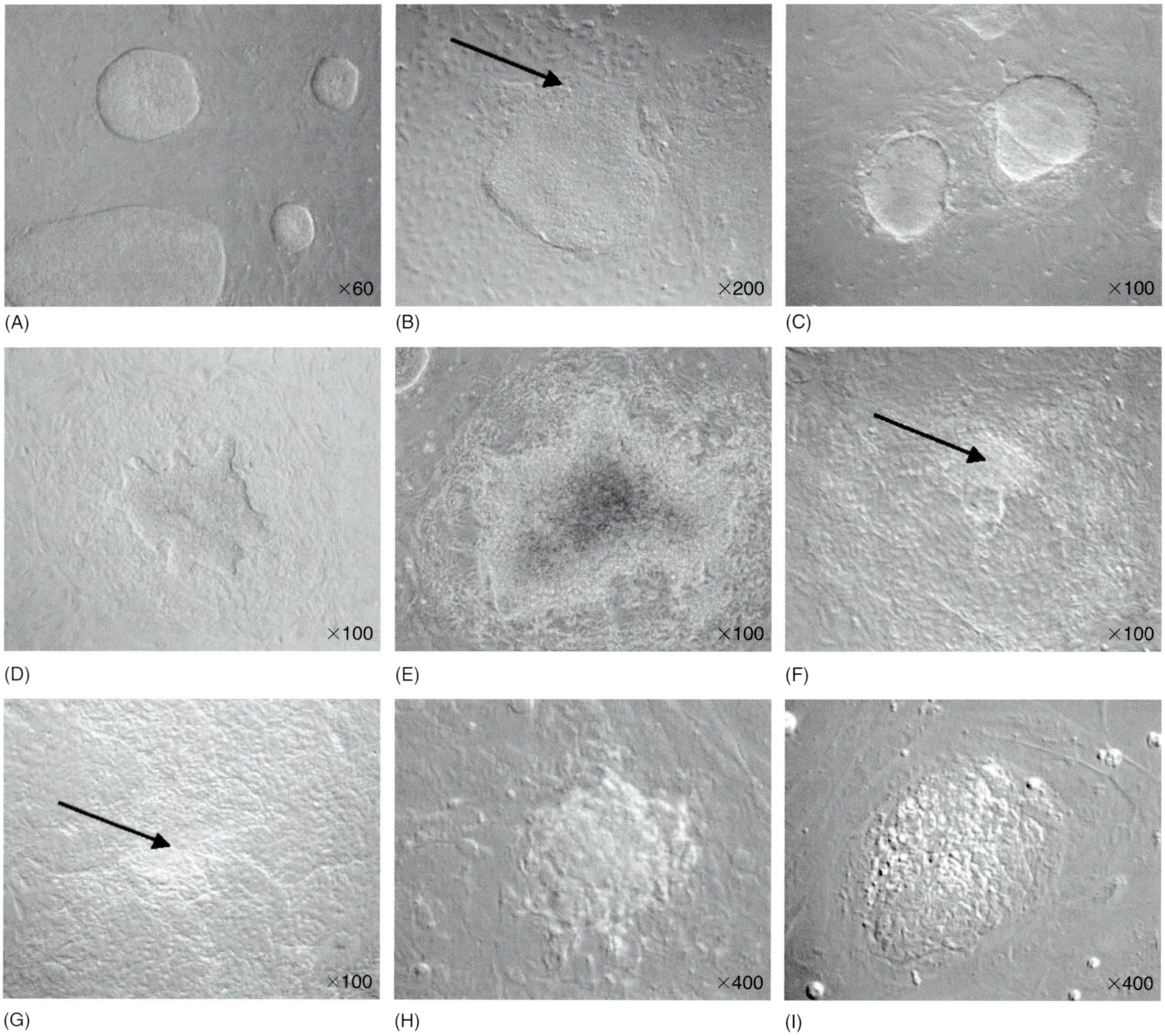

FIGURE 7-5 Evaluation of hES cell colony morphology. Various types of morphology of hES cell colonies encountered at early stages of derivation when mechanical dispersion can be the tool of choice to selectively pick undifferentiated colony parts. (A) All colonies are undifferentiated and can be mechanically passaged. (B) Colony which has a few signs of differentiation (arrow) and is surrounded by differentiated cells; the undifferentiated part is easily separated from surrounding differentiated cells. (C) Partially-differentiated, multi-layered colonies where the centers are thickened and yellowish in color can be mechanically dispersed into several small pieces, but may result in both differentiated and undifferentiated colonies. (D, E, F) All these colonies are more extensively-differentiated, a thin layer of differentiated cells covers them like a veil. They can be cut into pieces through the top cell layer and passaged, may yield undifferentiated colonies. (G, H, I) These colonies are badly-differentiated; the arrow on (G) shows a group of undifferentiated cells within the large differentiated area. If this one must be saved, one should wait for a few days for this group to increase in size.

that aneuploidy can develop in long-term cultures, passaged enzymatically, therefore we recommend periodical karyotyping, and if aneuploidy is apparent, an earlier frozen stock should be expanded, or the line can be re-established by mechanical passaging if only a fraction of the cells appear aneuploid.

The safest approach is to begin with a sub-confluent 35 mm plate of colonies with good morphology. Mechanically pick 50–100 colonies and transfer into a new well as a back-up. Differentiated colonies may be removed mechanically prior to trypsinization. Trypsinize the remaining colonies in the original well, and plate into the same diameter well. The cells should be ready for the next split in five to seven days (Figure 7-6). For second trypsinization, split 1:3. After this step the cells can usually be trypsinized routinely without problems, but a mechanical back-up should always be maintained until the cells are frozen, and the test vials are successfully thawed.

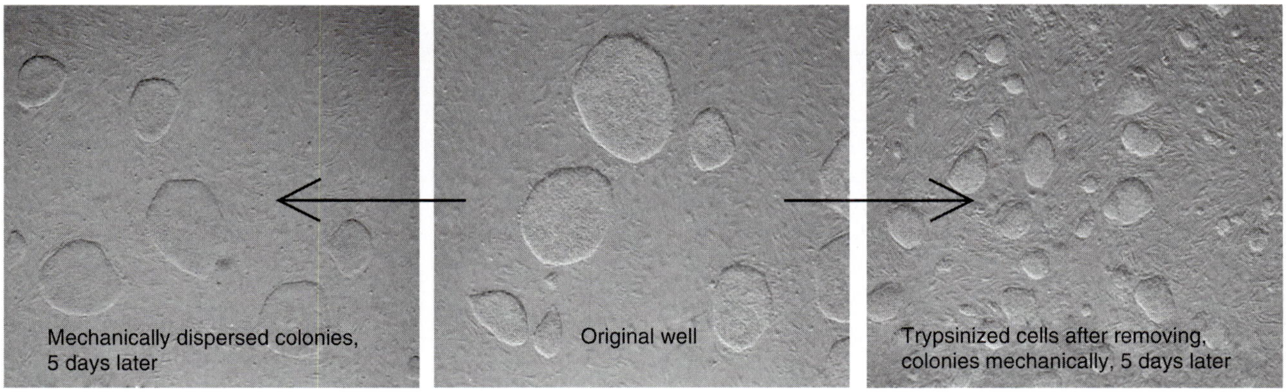

(A) (B) (C)

FIGURE 7-6 Adaptation of mechanically passaged hES cells to trypsin. After five days of growth, colonies were mechanically dispersed and transferred to a fresh plate of PMEF (A). Morphologically they are very similar to the original plate (B). The remaining colonies in the original plate, which were passaged with trypsin and plated onto the same diameter plate (C) show actively-growing colonies, and will probably be ready for passaging in one to two days.

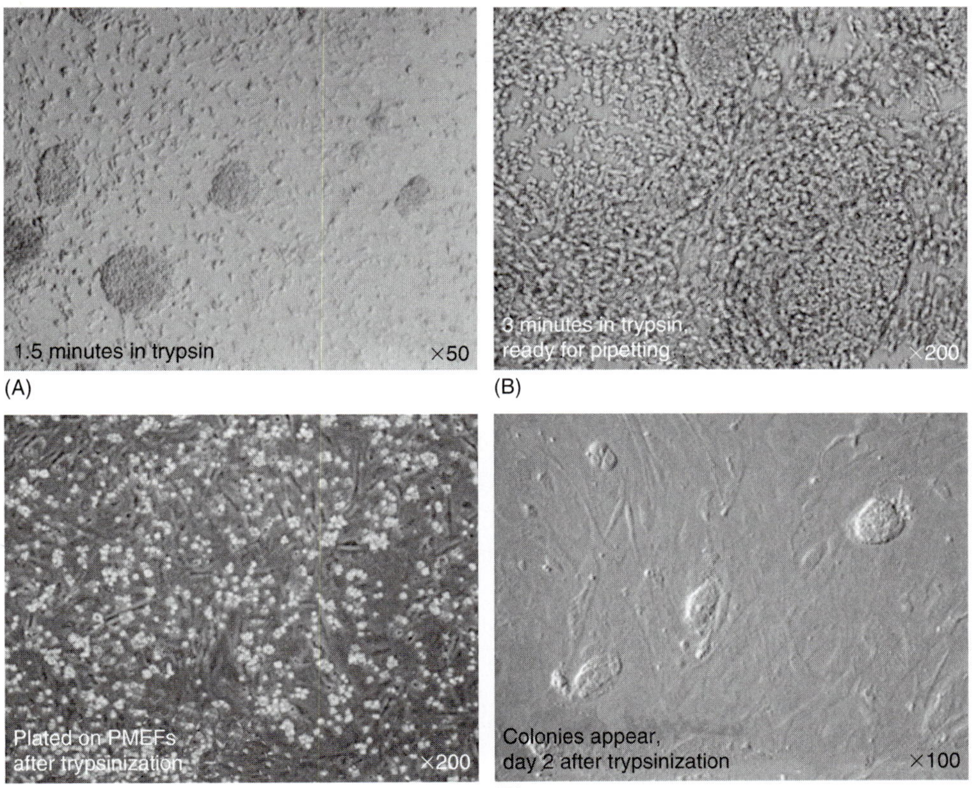

(A) (B) (C) (D)

FIGURE 7-7 Passaging by trypsinization. (A) 1.5 minutes in trypsin, the PMEF look shrunken and the hES colonies have loosened a little. (B) 3 minutes in trypsin at higher magnification: the colonies compacted; this is a good time to begin pipetting. Please, note that depending on the density of the colonies, days after passaging, and the degree of differentiation, the time required to reach this stage may vary and needs to be worked out empirically. (C) Suspension of hES cells after replating onto fresh PMEF. Note the majority of the cells are in small cell aggregates. (D) Small colonies begin to appear on day 2 after plating (at 1:3 ratio). The time when the first colonies are seen may vary, depending on the splitting ratio and the size of the cell aggregates.

Trypsinization

Generally, hES cells recover from trypsinization better when not dispersed to single cells, but remain as small cell clumps of approximately 5–20 cells. The procedure works best when hES cell colonies are dispersed by a combination of enzymatic digestion and pipetting; we do the pipetting before the PMEF monolayer and the colonies turn into a single-cell suspension. The time in trypsin required for the cells to detach varies, depending on the hES cell density, degree of differentiation, age of the culture, temperature of trypsin, etc. Therefore, instead of providing a fixed incubation time in trypsin, we recommend checking the appearance of the hES culture under the microscope, and empirically working out the best incubation time for each plate (see Figure 7-7).

1. Warm trypsin in a 37°C water bath; keep warm until ready for procedure.

2. Rinse the cells with PBS, Ca^{2+}-, Mg^{2+}-free, twice (1–2 ml per 35 mm dish).
3. Add 1 ml of trypsin to each 35 mm dish. Incubate in the hood at room temperature for several minutes, usually 2–5, frequently checking the cells under the microscope. The cells are ready for mechanical dispersion when the PMEF begin to shrink; the colonies should round up but remain attached. Some cells may begin to detach and float (Figure 7-7A, B).
4. Prepare a centrifuge tube with 10 ml of warm PMEF medium. (NOTE: It is necessary to use PMEF medium to inactivate the trypsin, because our hESC medium is serum-free.) Tilt the plate and begin to gently pipette the trypsin solution up and down with a 1 ml pipetteman (Gilson type), pouring it over the cell monolayer at an angle. Properly-digested cells should detach easily, leaving visible clear gaps in the monolayer, where the trypsin solution was poured. If no such gaps appear, leave it for another 1–2 minutes and test again. Expect the monolayer to detach after several repetitions. Usually, on cell cultures less than five-days-old, you should be able to completely disperse the monolayer, but if the culture is older or very dense, there may be some undigested material which can be discarded. Usually, it takes 5–10 pipetting strokes to break the colonies into small clumps of cells (see Figure 7-7C for approximate clump size). Extensive pipetting should be avoided. Using a cell culture pipette in place of an automatic pipettor will result in larger cell clumps, and thus may require longer time in trypsin.
5. Transfer the trypsinized cell suspension into the prepared centrifuge tube; centrifuge for five minutes at 160 g.
6. Aspirate the medium, and resuspend the pellet in hESC medium, again avoiding extensive pipetting to preserve small cell aggregates, and replate at the desired ratio. The colonies should become visible in one to two days, depending on the splitting ratio and the clump size (Figure 7-7D).

Although hES cultures passaged with trypsin can be maintained in an undifferentiated state, if conditions are unfavorable due to changes in media/reagents quality, too high or low splitting ratio, or problems with PMEF quality, the cultures can have a degree of differentiation which should be evaluated prior to the next trypsinization (see Figure 7-8).

Freezing Human Embryonic Stem Cells

Many of the established hES cells have low recovery rates upon thawing, as low as 0.1–1%. This may be due to the method of passaging the cells. Mechanical picking or using collagenase dispersion usually results in large cell aggregates, which presumably do not get cryopreserved as effectively as smaller clumps. Trypsinized cells in our

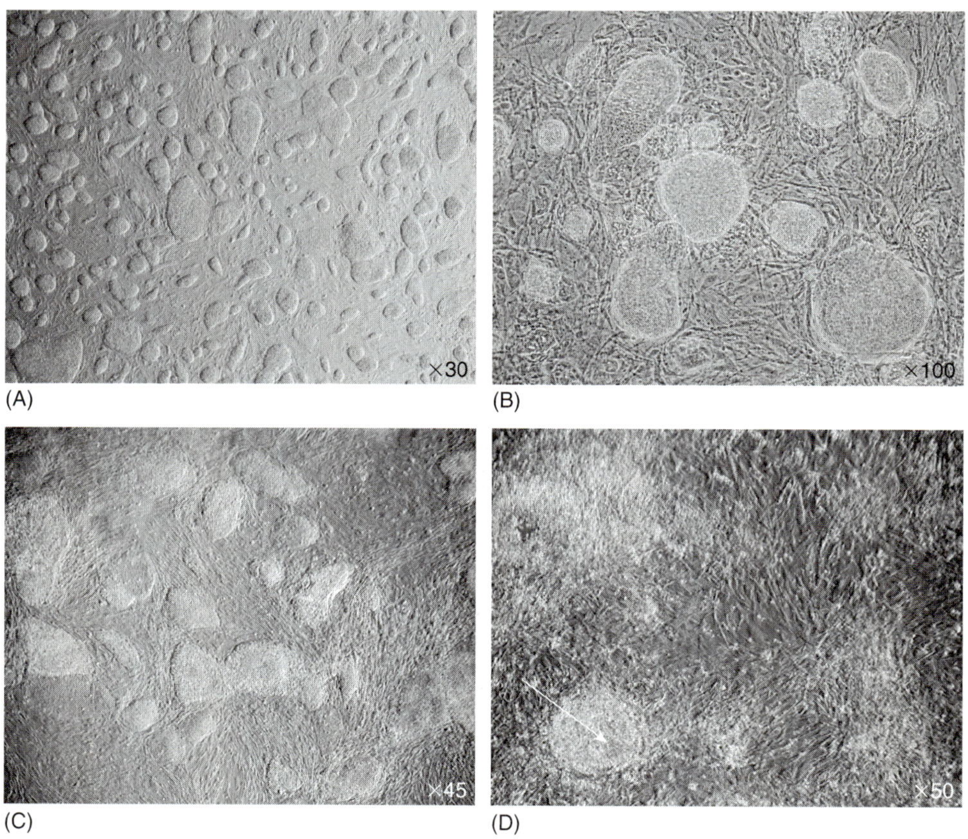

FIGURE 7-8 Evaluation of hES cell culture in preparation for the next trypsinization. (A, B) Mostly undifferentiated hES cells ready for trypsinization. (C) The colonies are a little overgrown and show signs of differentiation, but can still be safely passaged with trypsin. (D) Badly differentiated hES cells, an arrow points to a colony, which can still be rescued by mechanically picking colony pieces and passaging.

laboratory have a recovery rate of about 10–20% or higher, and do not require more complicated procedures like vitrification (Figure 7-9).

Freezing Medium

The best recovery rate was observed in freezing medium consisting of 90% FCS, 10% DMSO. However, Oct-4 expression in the thawed cells was lower than in cells frozen in hESC Growth Medium with 10% DMSO. Nevertheless, by the next passage, the expression and distribution of Oct-4 and other markers of undifferentiated cells were indistinguishable between these two freezing conditions. We routinely use the 90% FBS, 10% DMSO medium.

Freezing Protocol

Select a high-quality confluent culture with good morphology for freezing. We also recommend taking a picture of a sample field, and staining for molecular markers characteristic of undifferentiated hES cells for future reference.

Materials Needed

- Freezing medium: 90% FBS, 10% DMSO.
- Cryovials, labeled with the line, passage number, date.
- Cryovial rack (rack with ice reservoir by Corning).
- Styrofoam rack from packaging for 15 ml centrifuge tubes.
- −80°C freezer.

1. Trypsinize the cells; centrifuge in PMEF medium (see above).
2. Resuspend the pellet in cold freezing medium. We recommend freezing one confluent 35 mm plate per vial in 0.5 ml of freezing medium. Work quickly, and keep the cells on ice after addition of the freezing medium.
3. Aliquot cell suspension into pre-chilled freezing vials and sandwich the vials between two Styrofoam racks; tape to prevent the two racks from separating and transfer to a −80°C freezer overnight. Transfer cryovials to liquid nitrogen for long term storage.

Thawing Human Embryonic Stem Cells

Thawing hES cells is a relatively simple procedure. The main rule to follow is to do everything quickly.

Preparation

1. Prepare Mitomycin C-treated PMEF a day before thaw.
2. Make Thawing Medium. We use 70% hESC Growth Medium, supplemented with 20 ng/ml hLIF and 8 ng/ml bFGF with 30% hESC or PMEF-conditioned medium.

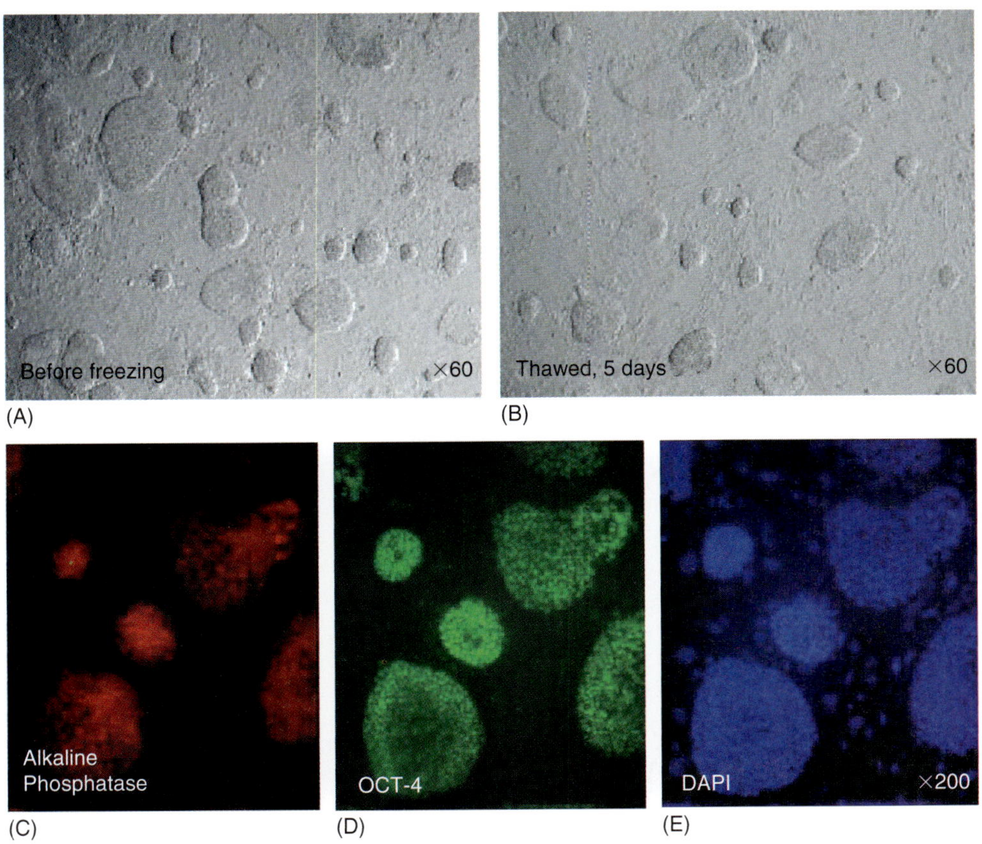

FIGURE 7-9 Freezing and thawing of hES cells. The top panel shows the approximate density and morphology of the colonies of hES cells at freezing (A). When thawed onto the same diameter plate, the colonies should be ready for the next split in five to seven days (B). Thawed hES cells show high expression of Oct-4 and alkaline phosphatase (C, D).

3. Change the medium on the PMEF plate to the Thawing Medium; equilibrate in the CO_2 incubator for one hour. For 35 mm plates use 1.5 ml, for 4-well plates use 0.5 ml medium per well.
4. Prepare a 50 ml conical tube with 10–15 ml of warm hESC growth medium.

Thawing

1. Thaw the vial(s) in a 37°C water bath, constantly agitating, while ensuring that the neck of the vial is above the water level. Check the content of the vial in about 40 seconds, and after that at 10 second intervals, until only a small piece of ice remains.
2. Quickly spray the vials with 70% isopropanol then, using a 1 ml pipetteman, add warm hESC medium to the contents of the vial drop-wise with gentle agitation. Do it quickly, but very gently. Immediately transfer the contents into the prepared 50 ml tube with warm hESC medium centrifuge at 160 g for five minutes.
3. Remove the medium completely, not touching the pellet.
4. Add 0.5 ml hESC Thawing Medium; gently resuspend the cells using a 1 ml pipetteman (2–4 repetitions), transfer to prepared PMEF plates with equilibrated hESC Thawing Medium. Spread the cells evenly throughout the well by moving the plate several times in two directions, at 90° to each other, avoid swirling.
5. Check the cells the next day; if there are many dead cells or the medium has changed color, change ⅔ of the medium, otherwise, don't change it for another day.
6. The colonies usually begin to appear in three to four days, and can be ready for splitting in five to ten days (Figure 7-7).

Human Embryonic Stem Cell Quality Control

Although morphology of hES cells is often used for evaluating the quality of the culture and its readiness for passaging or freezing, this criteria alone cannot be used for assessment of the ES cells pluripotency. Staining for the expression of Oct-4 or alkaline phosphatase, even in colonies of "perfect" morphology, can result in one or both of these markers appearing in the cells only at the periphery of the colony. It is important, therefore, to regularly assess the cells by analyzing the expression of markers of pluripotent cells. We look at Oct-4, Nanog, SSEA-3, SSEA-4, TRA 1-60, and TRA 1-81 by immunostaining, or perform an enzyme assay for alkaline phosphatase. The procedures for such assays and available antibodies are described elsewhere (Xu, 2006). Routine karyotyping performed either in-house or commercially provides an assessment of the degree of aneuploidy in hES cultures, and testing for differentiation potential is another important parameter to assess. In our laboratory we routinely perform differentiation of hES cells into retinal pigment epithelium and hemangioblast as a part of other ongoing projects, and if abnormal cell behavior, such as a low differentiation rate, is observed in differentiation assays, the cultures are discarded and fresh stock is thawed. Other differentiation models for hES cells can be used for such routine assessment of the quality of the culture as determined by the research goals.

GLOSSARY TERMS

Conditioned medium Medium left in contact with cultured cells, usually for a prolonged period of time.

Immunosurgery A method of removing the trophectoderm of a blastocyst using antibodies bound to the surface antigens of the trophectoderm and complement.

PMEF (primary mouse embryonic fibroblasts) A mixed population of cells derived from dispersed mouse embryos and cultured for a limited number of passages under conditions favoring the growth of fibroblasts.

FURTHER READING

Amit, M., Carpenter, M. K., Inokuma, M. S., Chiu, C. P., Harris, C. P., Waknitz, M. A., Itskovitz-Eldor, J., & Thomson, J. A. (2000). Clonally derived human embryonic stem cell lines maintain pluripotency and proliferative potential for prolonged periods of culture. *Dev. Biol.*, *227*(2), 271–278.

Chan, E. M., Yates, F., Boyer, L. F., Schlaeger, T. M., & Daley, G. Q. (2008). Enhanced plating efficiency of trypsin-adapted human embryonic stem cells is reversible and independent of trisomy 12/17. *Cloning Stem Cells*, *10*(1), 107–118.

Chung, Y., Klimanskaya, I., Becker, S., Li, T., Maserati, M., Lu, S. J., Zdravkovic, T., Ilic, D., Genbacev, O., Fisher, S., Krtolica, A., & Lanza, R. (2008). Human embryonic stem cell lines generated without embryo destruction. *Cell Stem Cell*, *2*(2), 113–117.

Cowan, C. A., Klimanskaya, I., McMahon, J., Atienza, J., Witmyer, J., Zucker, J. P., Wang, S., Morton, C. C., McMahon, A. P., Powers, D., & Melton, D. A. (2004). Derivation of embryonic stem-cell lines from human blastocysts. *N. Engl. J. Med.*, *25*, 350(13), 1353–1356. Epub 3 Mar. 2004.

Draper, J. S., Smith, K., Gokhale, P., Moore, H. D., Maltby, E., Johnson, J., Meisner, L., Zwaka, T. P., Thomson, J. A., & Andrews, P. W. (2004). Recurrent gain of chromosomes 17q and 12 in cultured human embryonic stem cells. *Nat. Biotechnol.*, *22*(1), 53–54.

Klimanskaya, I., Chung, Y., Meisner, L., Johnson, J., West, M. D., & Lanza, R. (2005). Human embryonic stem cells derived without feeder cells. *Lancet*, *365*(9471), 1636–1641.

Klimanskaya, I., Chung, Y., Becker, S., Lu, S. J., & Lanza, R. (2006). Human embryonic stem cell lines derived from single blastomeres. *Nature*, *444*(7118), 481–485. Epub 23 Aug. 2006. Erratum in: *Nature*, *444*(7118), 512, *Nature*, *446*(7133), 342.

Klimanskaya, I., Chung, Y., Becker, S., Lu, S. J., & Lanza, R. (2007). Derivation of human embryonic stem cells from single blastomeres. *Nat. Protoc.*, *2*(8), 1963–1972.

Lanzendorf, S. E., Boyd, C. A., Wright, D. L., Muasher, S., Oehninger, S., & Hodgen, G. D. (2001). Use of human gametes obtained from anonymous donors for the production of human embryonic stem cell lines. *Fertil. Steril.*, *76*(1), 132–137.

Lebkowski, J. S., Gold, J., Xu, C., Funk, W., Chiu, C. P., & Carpenter, M. K. (2001). Human embryonic stem cells: culture, differentiation, and genetic modification for regenerative medicine applications. *Cancer J.*, *7* Suppl 2, S83–S93.

Mitalipova, M. M., Rao, R. R., Hoyer, D. M., Johnson, J. A., Meisner, L. F., Jones, K. L., Dalton, S., & Stice, S. L. (2005). Preserving the genetic integrity of human embryonic stem cells. *Nat. Biotechnol.*, *23*(1), 19–20.

Reubinoff, B. E., Pera, M. F., Fong, C. Y., Trounson, A., & Bongso, A. (2000). Embryonic stem cell lines from human blastocysts: somatic differentiation *in vitro*. *Nat. Biotechnol.*, *18*(4), 399–404. Erratum in: *Nat. Biotechnol.*, *18*(5), 559.

Richards, M., Fong, C. Y., Chan, W. K., Wong, P. C., & Bongso, A. (2002). Human feeders support prolonged undifferentiated growth of human inner cell masses and embryonic stem cells. *Nat. Biotechnol.*, *20*(9), 933–936. Epub 5 Aug. 2002.

Robertson, E. J. (Ed.) (1987). *Teratocarcinomas and Embryonic Stem Cells: A Practical Approach*. Oxford: IRL Press.

Thomson, J. A., Itskovitz-Eldor, J., Shapiro, S. S., Waknitz, M. A., Swiergiel, J. J., Marshall, V. S., & Jones, J. M. (1998). Embryonic stem cell lines derived from human blastocysts. *Science*, *282*(5391), 1145–1147. Erratum in: *Science*, *282*(5395), 1827.

Solter, D., & Knowles, B. B. (1975). Immunosurgery of mouse blastocyst. *PNAS*, *72*, 5099–5102.

Thomson, J. A., Kalishman, J., Golos, T. G., Durning, M., Harris, C. P., Becker, R. A., & Hearn, J. P. (1995). Isolation of a primate embryonic stem cell line. *Proc. Nat. Acad. Sci. USA*, *92*(17), 7844–7848.

Xu, C., Inokuma, M. S., Denham, J., Golds, K., Kundu, P., Gold, J. D., & Carpenter, M. K. (2001). Feeder-free growth of undifferentiated human embryonic stem cells. *Nat. Biotechnol.*, *19*(10), 971–974.

Xu, C. (2006). Characterization and evaluation of human embryonic stem cells. *Methods Enzymol.*, *420*, 18–37.

Chapter 8

Derivation and Differentiation of Human Embryonic Germ Cells

Michael J. Shamblott, Candace L. Kerr, Joyce Axelman, John W. Littlefield, Gregory O. Clark, Ethan S. Patterson, Russell C. Addis, Jennifer N. Kraszewski, Kathleen C. Kent, and John D. Gearhart

Embryonic germ (EG) cells are pluripotent stem cells derived from primordial germ cells (PGCs) that arise in the late embryonic and early fetal period of development. EG cells have been derived from several species, including mouse, pig, chicken, and human. Mouse, pig, and chicken EG cells have been demonstrated to contribute to experimentally-produced chimeric animals, including germline transmission in the latter two species. Furthermore, germline transmission of mouse and chicken EG cell derivatives have been demonstrated. Mouse and human EG cells can be differentiated *in vitro* to form embryoid bodies (EBs). Like EBs generated from embryonic stem (ES) cells, EG-derived EBs contain differentiated cells representing all three germ layers, as well as mixed-cell populations of less differentiated progenitors and precursors. These human EB-derived (EBD) cells are capable of considerable cell proliferation and express a variety of lineage-specific markers. Human EBD cell cultures have a normal and stable karyotype and normal patterns of genomic imprinting, including X-inactivation. Transplantation studies have demonstrated that human EBD cells can engraft into a variety of rodent tissues and can participate in the recovery of rats following motor neuron injury.

INTRODUCTION

Pluripotent stem cells have been derived from two embryonic sources. Embryonic stem (ES) cells were first derived from the inner cell mass of mouse pre-implantation embryos, and embryonic germ (EG) cells were initially derived from mouse primordial germ cells (PGCs). Subsequently, EG cells have been derived from chicken, pig, and human PGCs. Pig, chicken, and mouse EG cells have been demonstrated to contribute to experimentally-produced chimeric animals, including germline transmission in the latter two species.

Primordial Germ Cells

PGCs are the sole means of genetic transmission between parent and offspring, as they generate eggs and sperm. In many species, such as *C. elegans*, germ cells are segregated very early in development, during the first embryonic cleavages, and are marked by deposition of ribonucleoprotein P-granules. In mammals, the process occurs later in development, and seems to be directed more by extrinsic factors than by preprogrammed intrinsic differences. For example, in mice, cells that generate PGCs are located close to extraembryonic ectoderm during gastrulation. Rather than having a previously determined fate, cells in this location receive external signals to further differentiate into PGCs, as demonstrated by the observation that transplantation of cells from other parts of the epiblast to this region can take on a PGC fate. Several components of this signaling process have been identified. Initially, bone morphogenetic protein 4 (BMP4) and BMP8b are produced by extraembryonic ectoderm program cells from the epiblast to become extraembryonic mesoderm precursors or PGCs. Cells destined to become PGCs express higher levels of membrane protein fragilis than nuclear protein stella.

In the mouse, PGCs are visible as alkaline phosphatase (AP) positive cells at the base of the allantois at 7.5 to 8.0 days postcoitus (dpc). They begin to associate with the endoderm that is invaginating to form the hindgut at 8.5dpc. By 10.5dpc, PGCs are associated with dorsal mesenteries and are translocated to the genital ridges. The migration of PGCs is caused by both cellular migration and association with moving tissues. Throughout this migration, PGCs expand from approximately 130 cells at 8.5dpc to more than 25 000 at 13.5dpc. Once they arrive at the genital ridge, PGCs continue proliferating until they enter prophase of the first meiotic division. In males, entry into meiosis is inhibited by signals from the developing testis, blocking PGCs at G0 until after birth. In the absence of inhibitory signals, female PGCs undergo oogenesis. Although not as thoroughly studied, much is known regarding the migratory path of human PGCs, including their association with gut endoderm and migration into developing genital ridges.

PGCs do not survive well under standard tissue culture conditions, and are not pluripotent stem cells *in vivo* or *in vitro*. Early attempts to use various growth factors and feeder layers succeeded in prolonging their survival, but proliferation was limited. The combination of leukemia inhibitory factor (LIF), basic fibroblast growth factor (bFGF), and c-kit ligand (KL, also known as stem cell factor, mast cell factor, or steel factor) proved to result in an immortal cell population, especially if the KL was presented in the transmembrane form by a layer of "feeder" cells (see the section Feeder Layer). Instead of simply encouraging PGC proliferation, these factors cause the normally solitary PGCs to congregate and proliferate as multicellular colonies, known as EG cells, and to gain pluripotency. Mouse EG cell lines have been derived from PGCs prior to migration around 8.0–8.5dpc, during migration at 9.5dpc, and after entry into the genital ridges between 11.5 and 12.5dpc.

The roles played by KL and the tyrosine kinase receptor for KL, c-Kit, in the *in vitro* derivation of EG cells from PGCs have parallels *in vivo*. c-Kit is expressed in PGCs, and KL is expressed along the PGC migratory pathway and in the genital ridges. The roles of KL and c-Kit in PGC survival were originally characterized through several mutations at their respective loci, *Sl* and *W*, which resulted in subfertile or sterile mice. PGCs are formed in homozygous mutant embryos of *W* and *Sl*, but mitosis is severely impaired, and the few PGCs that reach the gonad do not survive. KL is produced as a membrane-bound growth factor that can undergo proteolytic cleavage to generate a soluble form. Mice lacking the membrane-bound KL, but not the soluble form, maintain low PGC numbers and are sterile, suggesting that the membrane-bound form but not the soluble form is essential for PGC survival. The mechanism involved in KL-induced PGC survival has been shown to involve suppression of apoptosis. The c-kit receptor has also been shown to be involved in the adhesion of mouse PGCs to somatic cells *in vitro*. Other recent studies attempting to identify signaling pathways activated by KL binding to its receptor in mouse PGCs have shown activation of AKT kinase and telomerase.

In contrast to the embryologically early and relatively undifferentiated epiblast, PGCs arise late and have a specialized role during normal development. In this regard, it is somewhat surprising that exposure to three cytokines can convert PGCs to pluripotent stem cells *in vitro*. It is possible that the flexibility provided by extrinsic signaling during PGC specification, rather than intrinsic preprogramming, allows for this conversion.

Comparison to Embryonic Stem Cells

Both mouse ES and EG cells are pluripotent and demonstrate germline transmission in experimentally produced chimeras. Mouse ES and EG cells share several morphological characteristics, such as high levels of intracellular AP and presentation of specific cell-surface glycolipids and glycoproteins. These properties are characteristic of, but not specific for, pluripotent stem cells. Other important characteristics include growth as multicellular colonies, normal and stable karyotypes, the ability to be continuously passaged, and the capability to differentiate into cells derived from all three EG layers: endoderm, ectoderm, and mesoderm.

HUMAN EMBRYONIC GERM CELL DERIVATION

Although many combinations of cytokines and feeder layers have been evaluated, the standard practice for derivation of human EG cells remains similar to techniques developed for the mouse. As of 2003, approximately 140 human EG cultures have been derived in our laboratory using this general technique.

Initial Disaggregation and Plating

Gonadal ridges and mesenteries of week 5–9 postfertilization human embryos (obtained as a result of therapeutic termination of pregnancies) are collected in 1 ml ice-cold growth media and rapidly transported to a sterile work space. The tissues are then soaked in calcium–magnesium-free Dulbecco's phosphate-buffered saline (DPBS) for five minutes, and then transferred to 0.1 ml trypsin–EDTA solution. The concentration of trypsin and EDTA is varied, such that at the earliest developmental stages, a gentler 0.05% trypsin–0.5 mM EDTA is used, and at later developmental stages, a stronger 0.25% trypsin–0.5 mM EDTA solution is used. The tissue is mechanically disaggregated thoroughly using a fine forceps and iris scissors. This process is carried out for 5 to 10 minutes at room temperature and then incubated at 37°C for 5 to 10 minutes. This disaggregation process often results in a single-cell suspension and large pieces of undigested tissue. To stop the digestion, serum-containing growth media is added. The digested tissue is transferred to wells of a 96-well tissue culture plate that has been previously prepared with feeder layer (see the sections Plating an STO Feeder Layer Prior to Inactivation and Plating an STO Feeder Later after Inactivation). Usually, the initial plating occupies four to 10 wells of the 96-well plate. The plate is incubated at 37°C in 5% CO_2 with 95% humidity for seven days. Approximately 90% of the growth media is removed each day, and the plate is replenished with fresh growth media.

Subsequent Passage of Embryonic Germ Cell Cultures

In the first seven days of derivation (passage 0), most human EG cultures do not produce visible EG colonies. Staining for AP activity demonstrates the presence of solitary PGCs, with either stationary or migratory morphology (Figure 8-1A, B). Often, colonies of cells that do not stain AP+ are seen (Figure 8-1D, E), as are small clumps of tissue remaining from the initial disaggregation (Figure 8-1C). After seven days, the media is removed, and the wells are rinsed twice with calcium–magnesium-free DPBS for a total of five minutes. Then, 40 ml of freshly thawed trypsin solution is added to each well, and the plate is incubated on a heated platform or in a tissue culture incubator for five minutes at 37°C. As previously described, the trypsin solution is 0.05% trypsin–0.5 mM EDTA, 0.25% trypsin–0.5 mM EDTA, or a mixture of these two solutions. The important point at this stage is to facilitate the complete disaggregation of the STO cell feeder layer (see the section Feeder Layer), which can be a significant challenge. After the incubation in trypsin, the edge of the 96-well culture plate is hit firmly against a solid surface until the STO cells have completely lifted off the growth surface. This process can be aided by scraping the well and gently triturated. After the STO cells have been loosened, fresh growth media is added to each well, and the contents are triturated. This phase is critical to successful disaggregation of STO feeder layer and EG cells.

All subsequent passages are done as described. After 14 to 21 days (during passage 1 or 2), large and recognizable EG colonies will arise in some of the wells (Figure 8-2). At this point, wells that do not have EG colonies are discarded.

Several common problems occur during the passage of human EG cells. One observation is that the STO feeder layer will sometimes not fully disaggregate. This can be observed by the presence of large cell aggregates immediately following disaggregation. If this occurs with regularity, it is a sign that the trypsinization method is insufficient, or that the STO cells have become less contact inhibited and have overgrown during the seven days of culture. Another common occurrence is that the EG colonies do not fully disaggregate. The consequences of poor disaggregation

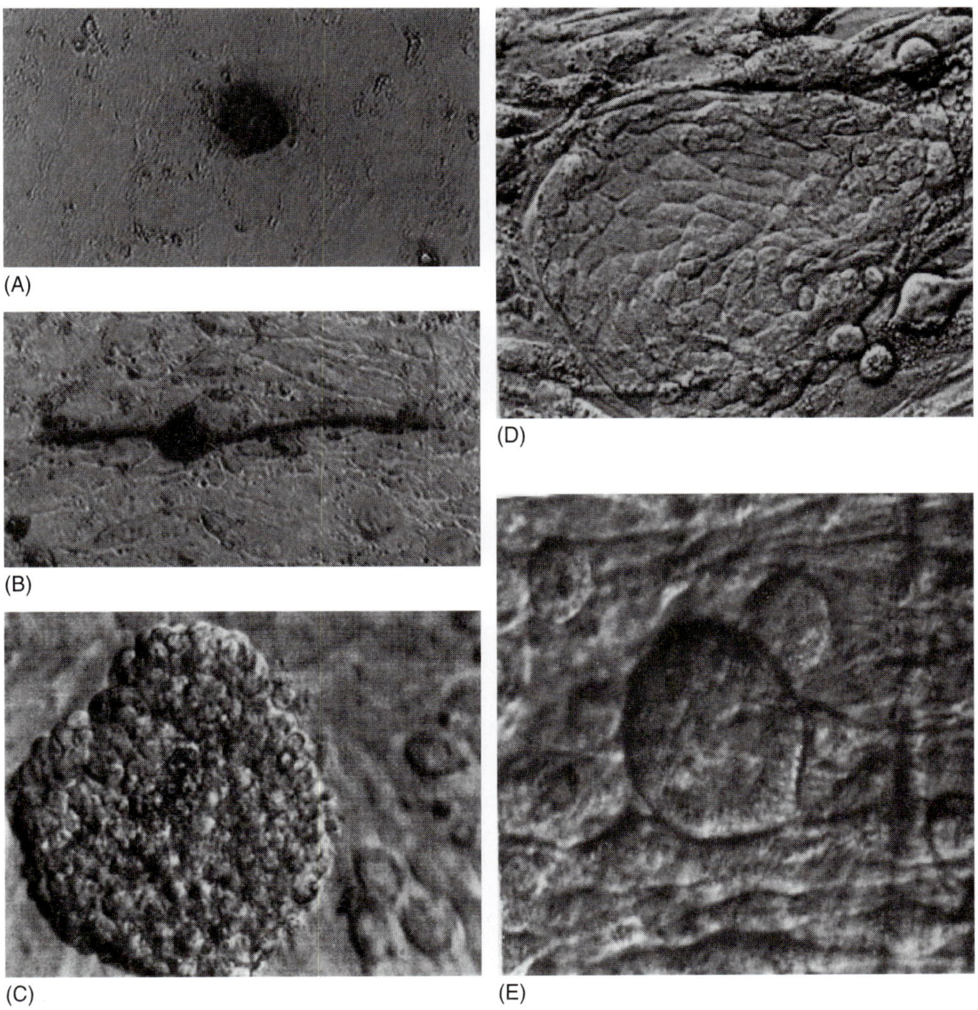

FIGURE 8-1 Cell morphologies seen in early passage human embryonic germ culture. (A, B) Alkaline phosphatase-positive (AP+) stationary and migrating primordial germ cells. (C) Multicellular piece of undisaggregated gonadal tissue. (D, E) Flat and round cell colonies that do not lead to human EG cells.

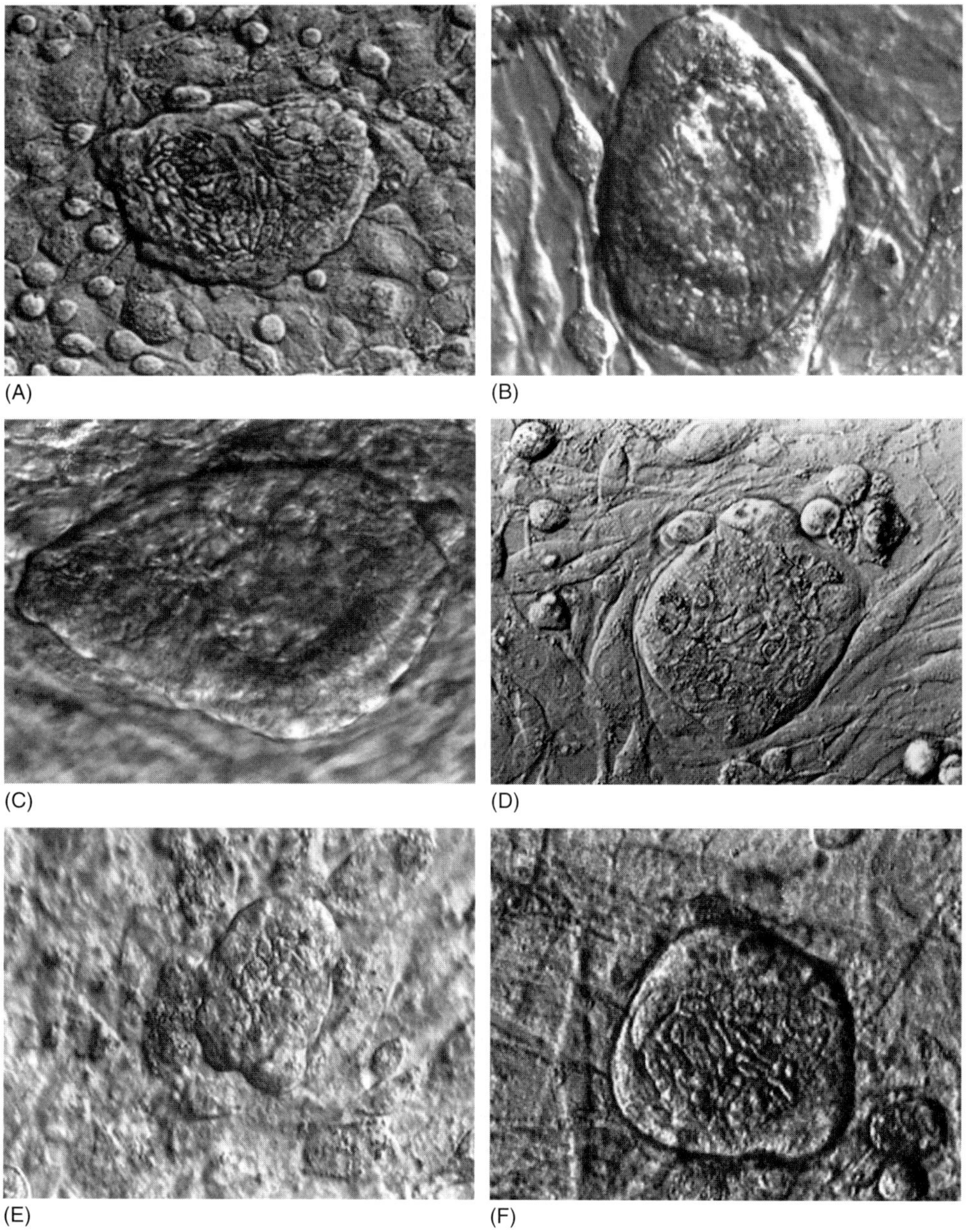

FIGURE 8-2 Human EG colonies growing on an STO cell feeder layer.

are that the large pieces differentiate or die, and fewer EG cells are available for continued culture expansion. Although much effort has been expended to find a solution to this problem, it remains the most difficult aspect of, and challenging hurdle to, human EG cell biology.

To gain some insight into this problem, a series of electron microscopic images were taken to compare the cell–cell interactions found in mouse ES, mouse EG, and human EG cell colonies. It is evident from these images that cells within the human EG colonies adhere more completely to each other than cells within mouse ES and EG colonies (Figure 8-3). It is possible that this tight association within the colony limits the access of disaggregation reagents. At this time, neither the nature of the cell–cell interactions nor an effective solution to this problem is evident.

Because of incomplete disaggregation and other intrinsic or extrinsic signals, many human EG colonies (10–30% per passage) differentiate to form three-dimensional structures termed EBs (see the section EB Formation and Analysis) or flatten into structures that are AP− and do no continue to proliferate (Figure 8-4). EG colonies that are more fully disaggregated go on to produce new EG colonies, and under the best circumstances, EG cultures can be expanded continuously for many months, and have routinely exceeded 20 passages. Inevitably, as large EG colonies are removed from the culture as a result of EB formation, the cultures become sparse, and are discontinued for practical considerations. Efforts employing standard dimethyl sulfoxide (DMSO) cryopreservation techniques have so far been unsuccessful.

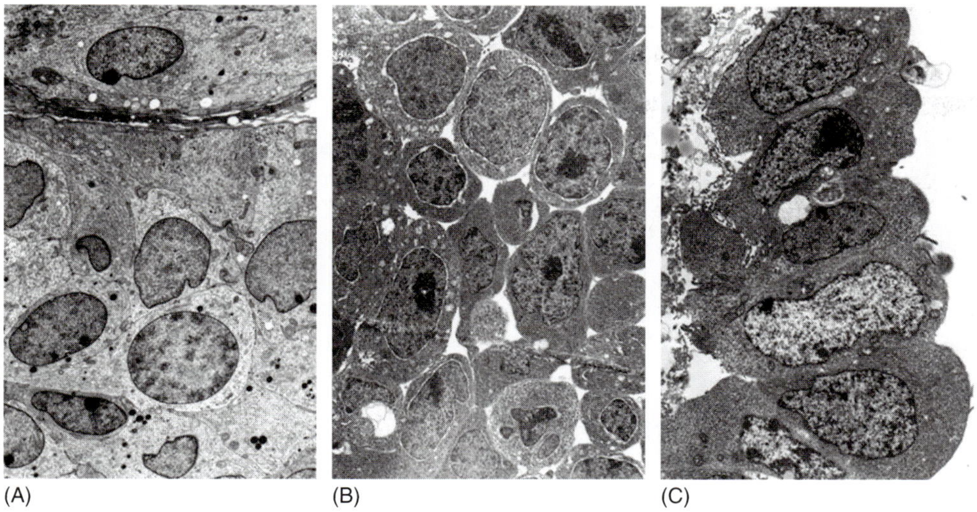

FIGURE 8-3 Electron micrograph of EG and ES colonies. (A) Human EG colony; (B) mouse EG colony; and (C) mouse ES colony.

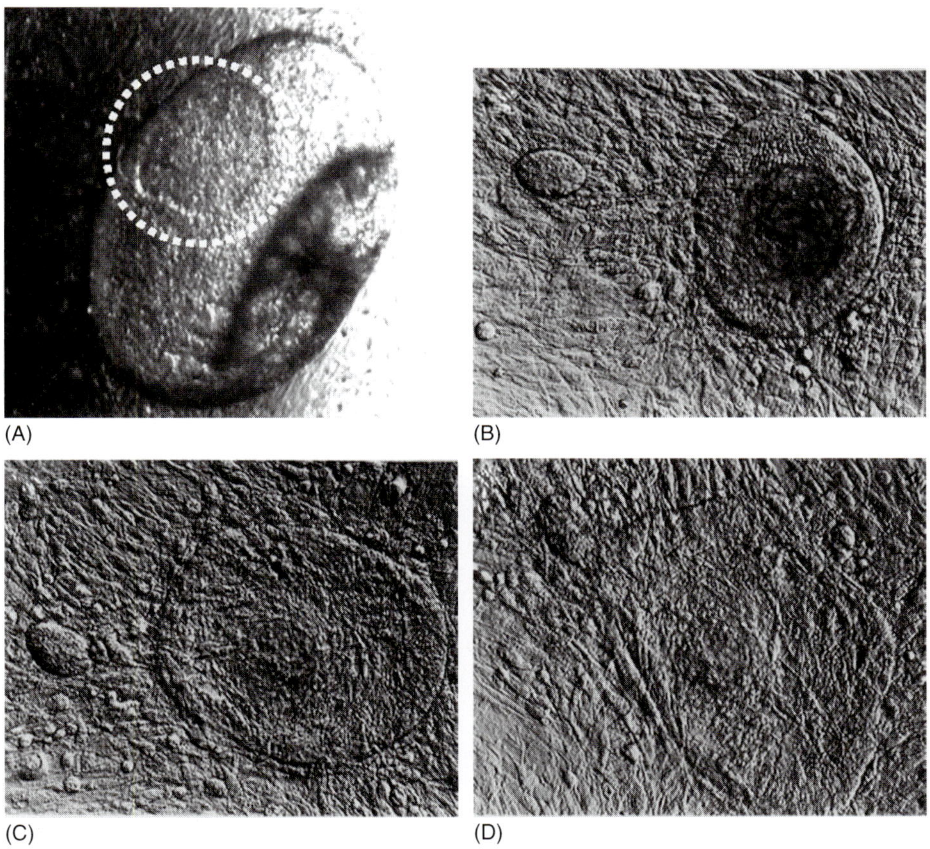

FIGURE 8-4 Differentiating human EG colonies. (A) Stereomicrograph of an EG colony with attached embryoid body (EB). The EG colony is circled. The diameter of the EB is approximately 0.5 mm. (B–D) Time-lapse study of EG cell flattening. (B) Four days after disaggregation, the small EG colony on the left results from more complete disaggregation. The large colony on the right is composed of a dark residual colony surrounded by new EG cell growth. (C) Five days after disaggregation, the large EG colony has begun to flatten. (D) Seven days later, the large EG colony has completely flattened and did not survive as a recognizable EG cell colony after disaggregation and replating. In a separate experiment, EBs and these flattened structures are shown to be largely AP−.

Feeder Layer

Unlike ES cells, EG cell derivation is highly dependent on a specific type of feeder layer. Mouse STO fibroblasts are a spontaneously transformed cell line from the Sandoz inbred mouse. They are thioguanine- and ouabain-resistant, features of historical interest, but not used in this context. STO cells have been used to derive mouse embryonal carcinoma (EC), ES, and EG cells, and to date, they are the only cell type capable of generating human

EG cells. The factor or factors provided exclusively by this cell type are not fully understood. However, the transmembrane form of KL is present on STO fibroblasts, but not on most other cell types evaluated in our laboratory.

Although STO is a clonal cell line, individual isolates vary greatly in their ability to support human EG derivation. This is further complicated by the known phenotypic variation of STO cells in continuous culture. Given the very limited supply of human tissue, it is prudent to screen STO cells for suitability prior to use. The most reliable screening method is to produce a number of clonal STO lines (by limiting dilution or cloning cylinder), and to evaluate them for their ability to support the derivation of mouse EG cells. The growth of existing mouse EG lines is not a sufficient method, as most mouse EG lines become feeder layer independent after derivation. Derivation of mouse EG cells is not a simple undertaking. In an effort to screen STO cells more rapidly, and to investigate the role of transmembrane KL, we have begun to screen STO cell lines by using an immunocytochemical method to detect KL. Once a supportive STO fibroblast line is identified, it should be immediately cryopreserved in several low-passage aliquots. One of these aliquots can then be expanded to provide many medium passage aliquots, which are thawed and used with limited further expansion. Continuous passage of STO fibroblasts without frequent screening should be avoided.

Plating an STO Feeder Layer Prior to Inactivation

Two different methods can be used to prepare an STO feeder layer: plate-then-irradiate and irradiate-then-plate. Most human EG cultures are derived using the former method. This is practical when a large g-irradiator is available. STO cells are passaged for short periods (not continuously) in PGC growth media without LIF, bFGF, or forskolin, and are disaggregated using 0.05% trypsin–EDTA solution. One day prior to use, a 96-well culture dish is coated with 0.1% gelatin for 30 minutes. The gelatin is withdrawn, and 5×10^4 STO cells are plated per well in PGC growth media without the LIF, bFGF, or forskolin. Similar cell densities ($\sim 1.5 \times 10^5$ cells/cm^2) can be achieved in other well configurations. The cells are grown overnight and then exposed to 5000 rads (1 rad = 0.01 Gy) g-radiation or X-ray. The cells are then returned to the tissue culture incubator until required. Prior to use, the growth media is removed, 0.1 ml PGC growth medium with added factors is added to each well (or half of the required well volume), and the dish is returned to the tissue culture incubator.

Plating an STO Feeder Layer after Inactivation

This method of STO cell preparation is used when a large g-radiation unit is not available, when large amounts of cells are required, or if better control of STO cell density is required. STO cells are grown as described previously, trypsinized, counted, and resuspended in PGC growth media without added factors. The cells are placed into one or more 50 ml conical tubes, and placed into the g-irradiator or X-ray device. Cells are exposed to 5000 rads g-radiation or X-ray. Following exposure, cells are adjusted to a convenient concentration with PGC growth media without added factors, counted, and plated into tissue culture dishes that have been previously coated with 0.1% gelatin for 30 minutes. Cells are allowed to adhere overnight; then the media is replaced with half the final volume of PGC growth media, including factors prior to use.

Primordial Germ Cell Growth Media Components

Human EG cells are derived and maintained in Dulbecco's modified Eagle's medium (Gibco BRL) supplemented with 15% fetal bovine serum (FBS, Hyclone), 0.1 mM nonessential amino acids (Gibco BRL), 0.1 mM 2-mercaptoethanol (Sigma), 2 mM glutamine (Gibco BRL), 1 mM sodium pyruvate (Gibco BRL), 100 μ/ml penicillin (Gibco BRL), 100 mg/ml streptomycin (Gibco BRL), 1000 μ/ml human recombinant LIF (hrLIF, Chemicon), 1–2 ng/ml human recombinant bFGF (hrbFGF, R&D systems), and 10 mM forskolin (Sigma) prepared in DMSO.

Evaluation of Embryonic Germ Cell Cultures

Of 150 human PGC cultures initiated, 142 (~95%) demonstrated morphological, biochemical, and/or immunocytochemical characteristics consistent with previously-characterized pluripotent stem cell lines. The easiest and most consistent method for evaluating EG cultures is to note the presence of tightly-compacted multicellular colonies resembling early-passage mouse ES and EG cell colonies (Figure 8-2), rather than the flattened and more loosely associated colonies seen in human EC and rhesus ES cells. Under the best circumstances, the number of colonies should double or triple over a one-week passage. The trend is usually to start with small colonies, and to end with larger colonies that result from incomplete colony disaggregation (see the earlier section Subsequent Passage of EG Cultures).

High levels of AP activity are associated with human EG cells. Under standard culture conditions, human EG colonies are >70–90% AP+. As colonies differentiate, one can observe a lower staining percentage and weaker staining, sometimes restricted to the periphery of the colony.

Human EG cells have been further characterized by a bank of five monoclonal antibodies: SSEA-1; SSEA-3; SSEA-4; TRA-1-60; and TRA-1-81. Colonies stain strongly for four of the five antibodies. The antibody recognizing SSEA-3 antigen stains the cells inconsistently and weakly. As with the results of AP staining, the percentage of cells within a colony that stain positive is variable.

The histological profile of human EG cells (AP+, SSEA-1+, SSEA-3+, SSEA-4+, TRA-1-60+, and TRA-1-81+) differs from undifferentiated human EC and rhesus ES cells, which are SSEA-1−. The fact that differentiation of the human EC line NTERA2 leads to increased expression of SSEA-1 may suggest that this is indicative of differentiation in the human EG cultures. However, NTERA2 differentiation is accompanied by the loss of the other markers, which are not observed in these cultures.

Karyotypic analyses carried out at passage 8–10 (60–70 days in culture) indicated apparently normal human chromosomes at the 300-band level of resolution. Both XX and XY cultures have been derived. Other markers of pluripotency, such as mRNA expression of the human ortholog of mouse Oct3/4 and telomerase enzyme activity, have been examined. Results differ greatly, depending on the status of the EG colony, culture, or both. In general, relatively undifferentiated EG colonies are Oct4 mRNA+ as detected by RTPCR and have detectable levels of telomerase.

Alkaline Phosphatase and Immunocytochemical Staining

Cells are fixed for detection of AP activity in 66% acetone–3% formaldehyde and then stained with naphthol/FRV-alkaline AP substrate (Sigma). For immunocytochemistry, cells are fixed in 3% paraformaldehyde in DPBS. Cell-surface glycolipid- and glycoprotein-specific monoclonal antibodies are used at 1:15 to 1:50 dilution. MC480 (SSEA-1), MC631 (SSEA-3), and MC813-70 (SSEA-4) antibodies were supplied by the Developmental Studies Hybridoma Bank (University of Iowa). TRA-1-60 and TRA-1-81 were a gift from Dr Peter Andrews (University of Sheffield, UK). Antibodies are detected by using biotinylated antimouse secondary antibody, streptavidin-conjugated horseradish peroxidase, and AEC chromogen (BioGenex).

Embryoid Body Formation and Analysis

Embryoid bodies (Ebs) form spontaneously in human EG cultures. Although this represents a loss of pluripotent EG cells from the culture, EBs provide evidence for the pluripotent status of the culture, and provide cellular material for subsequent culture and experimentation (see the section EBD Cells). Initially, EBs provided the only direct evidence that human EG cultures were pluripotent, as all attempts to form teratomas in mice from human EG cells failed. To this day, there is no evidence of teratoma formation from human EG cells or their derivatives.

Embryoid Body Embedding and Immunohistochemistry

The constituent cells of EBs can be identified most reliably by embedding them in paraffin and staining sections with a bank of well-characterized antibodies. This process avoids the significant problem of antibody trapping that complicates analyses of large three-dimensional structures when direct staining is attempted. EBs are collected from cultures and placed into a small drop of molten 2% low melting point agarose (FMC), prepared in DPBS, and cooled to 42°C. Solidified agarose-containing EBs are then fixed in 3% paraformaldehyde in DPBS and embedded in paraffin. Individual 6 mm sections are placed on microscope slides (ProbeOn Plus, Fisher Scientific). Routinely, immunohistochemical analysis is carried out by using a BioTek-Tech Mate 1000 automated stainer (Ventana-BioTek Solutions). Manual staining is also carried out using standard immunohistochemical methods; however, antigen retrieval is required for some of the antibodies employed. Cryosections of EBs generally produce less satisfactory cell morphology and have not been used extensively.

Antibodies used on paraffin sections include HHF35 (muscle specific actin, Dako), M 760 (desmin, Dako), CD34 (Immunotech), Z311, (S-100, Dako), sm311 (pan-neurofilament, Sternberger Monoclonals), A008 (a-1-fetoprotein), CKERAE1/AE3 (pancytokeratin, Boehringer Mannheim), OV-TL 12/30 (cytokeratin 7, Dako), and K_S20.8 (cytokeratin 20, Dako). Primary antibodies are detected by using biotinylated antirabbit or antimouse secondary antibody, streptavidin-conjugated horseradish peroxidase, and DAB chromogen (Ventana-BioTek Solutions). Slides are counterstained with hematoxylin. Using these and other antibodies it can be demonstrated that, when human EG cells differentiate, they form EBs comprised of endodermal, ectodermal, and mesodermal derivatives.

EMBRYOID BODY-DERIVED CELLS

Although a compelling demonstration of the potential of human EG cells, the limited growth characteristics of differentiated cells within EBs, and the difficulties associated with their isolation, make extensive experimental manipulation difficult and limit their use in future cellular transplantation therapies. At least two possibilities exist to explain the presence of the observed differentiated cell types. These cells could be generated directly from the pluripotent EG cell, or they could proceed through a series of precursor or progenitor cell types, prior to the acquisition of a mature phenotype. It seems unlikely that cells within EBs would bypass normal pathways of differentiation, so efforts have been made to isolate and expand these populations. The hypothesis was that progenitor–precursor cells would have desirable proliferation characteristics, and could be recognized by the expression of molecules known to mark progenitor–precursor populations, as well as by the simultaneous expression of markers normally considered part of a mature cellular expression repertoire. Simultaneous expression of neuronal and glial markers by neural progenitors and the expression of a variety of lineage-affiliated transcription factors and cytokine receptors by multipotent hematopoietic progenitors provide some basis for this hypothesis. In this model of differentiation, multilineage gene expression by precursor or progenitor cells defines a ground state from which cell-extrinsic and -intrinsic signals work to continuously define a differentiated expression pattern and phenotype, resulting in the developmental plasticity observed after the differentiation of bone marrow and central nervous system stem cells.

The method used to isolate cell populations from EBs is conceptually similar to microbiological selective media experiments. EBs are disaggregated and plated into several different cellular growth environments. These environments consist of combinations of a growth media and a matrix. Although many combinations were evaluated, most EBD cell cultures have been derived from one of six environments, formed by combinations of two growth media and three plating surfaces. The growth media are an RPMI 1640 media supplemented with 15% FBS, and a low (5%) FBS media supplemented with bFGF, epidermal growth factor (EGF), insulin-like growth factor 1 (IGF1), and vascular endothelial growth factor (VEGF). The plating surfaces are bovine type I collagen, human extracellular matrix extract, and tissue culture-treated plastic. These are not intended to be highly selective environments. Instead, they favor several basic themes: cells thriving in high serum and elevated glucose (10 mM) conditions versus cells proliferating in low glucose (5 mM) under the control of four mitogens. Surfaces included binding to type I collagen, a biomatrix often thought to favor undifferentiated proliferation versus human extracellular matrix, a more complex mixture of laminin, collagen, and fibronectin. The initial assay is to determine conditions that favor extensive cell proliferation, with the hypothesis that this condition will favor undifferentiated cell populations, or at least provide numerical disadvantage to the terminally-differentiated constituents of the EB.

Embryoid Body-derived Growth and Expression Characteristics

Cell populations capable of long-term and robust proliferation can be isolated in this way from human EG-derived EBs. Embryoid body-derived is the generic term used to describe cells derived in this way *and* capable of extensive further proliferation. In general, the type I collagen and human extracellular matrices combined with the low serum media, provide the most rapid and extensive cell proliferation. EBD cell lines and cultures are routinely maintained in the environment in which they were derived. The EBD naming convention aids this process. The first two letters of the name refer to the EG culture from which it was derived. The second letter indicates the growth media (E for EGM2MV and R for RPMI1640), and the last letter indicates the matrix (C for collagen, E for human extracellular matrix extract, and P for plastic). For example, the EBD culture SDEC was derived from EG culture SD in EGM2MV medium on type I collagen.

To distinguish EBD cells from a simple population of cells, rapidly proliferating but presumably uninteresting in terms of molecules expressed, it is important to establish a robust expression profile assay. The assay should use redundant measures when possible and must combine breadth, sensitivity, specificity, and speed. A series of 24 RTPCRs detecting products from five cell lineages (neuronal, glial, muscle, hematopoietic–vascular endothelia, and endoderm) combined with immunocytochemical staining provides a rapid measure of cell expression. Inevitably, molecular markers are not as definitively specific as desired, so multiple markers for each lineage are advisable.

Using mRNA and antibody expression profiling, we can demonstrate that most rapidly proliferating EBD cell cultures simultaneously express a wide array of mRNA and protein markers normally associated with distinct developmental lineages. This is not a surprising property considering that EBD cells are, at least during the derivation stage, a mixed-cell population. More remarkable is the finding that most (11 of 13) EBD cell lines isolated by dilution cloning also exhibit a broad multi-lineage gene expression profile. It can also be demonstrated that the expression profile for a given EBD culture remains stable throughout the lifespan of the culture. This normally exceeds 70 population doublings, but is not unlimited since EBD cells are not immortal.

More than 100 EBD cell cultures and clonally isolated cell lines have been derived and characterized as described here. Most of these cultures share the properties of rapid and robust proliferation, and broad multi-lineage gene expression. Less than 10% of the EBD cultures derived have a narrow expression profile, with one extreme case expressing only nestin, vimentin, and a-1-fetoprotein mRNA. Other general trends in EBD expression are that many cultures appear to be neurally biased, with strong expression of neuronal, glial, and neural progenitor markers, and relatively weak in expression of muscle markers.

Other general characteristics of EBD cells are the relative ease with which they can be genetically manipulated, using lipofection and electroporation, as well as retroviral, adenoviral, and lentiviral vectors. Adenoviral and lentiviral vectors are capable of nearly 100% transduction efficiency. These techniques have been used to generate EBD lines that constitutively and tissue-specifically express enhanced GFP and contain many different genetic selection vectors. EBD cells can be immortalized by retroviral-based expression of the telomerase RNA subunit (pBABE). Interestingly, after several hundred population doublings, these lines often become genetically unstable, generating at least two rearrangements: [47,XX,−1,+del(1)(q12),+i(1)(q10)] and [46,XX, del(4)(p14)]. In addition, these EBD lines tend to have narrow expression profiles.

The imprinting pattern of several EBD cultures has been examined. In one study, expression levels of four imprinted genes (TSSC5, H19, SNRPN, and IGF2) were determined in five EBD cultures. Three of these genes (TSSC5, H19, and SNRPN) had normal monoallelic expression levels, and IGF2 had a partially-relaxed imprinting pattern comparable to levels found in normal somatic cells. This study also determined that the imprinting control region that regulates H19 and IGF2 imprinting had a normal pattern of DNA methylation. A second study determined that two XX EBD cultures had a normal pattern of X-inactivation.

The proliferation and expression characteristics of EBD cells suggest that they may be useful in the study of human cell differentiation, and as a resource for cellular transplantation therapies. One important property in this regard is that no tumor of human origin has arisen in any animal receiving EBD cells, although hundreds of mice, rats, and African green monkeys have received EBD transplants in a variety of anatomical locations, often consisting of more than one million cells injected. This is in contrast to the infrequent, yet significant, number of teratocarcinomas that have arisen following transplantation of cells produced through neural and hematopoietic differentiation of mouse ES cells.

Embryoid Body-derived Cell Transplantation

Transplantation of EBD cells into animal models of human disease constitutes an active and promising research avenue. Studies with EBD cells and many other cell types have suggested that tissue injury can be highly instructive to transplanted cells. This provides a powerful method to test the potential of cells to differentiate, without an initial understanding of the underlying mechanisms. These studies also suggest possibilities for the eventual treatment of patients suffering from these diseases.

One example of EBD cell transplantation is the use of an EBD culture named SDEC. This culture was initially selected for further study because of its strong neural expression bias. SDEC cells were introduced into the cerebrospinal fluid of normal rats and rats exposed to the neuroadapted Sindbis virus. This virus specifically targets spinal cord motor neurons, and infection results in permanent hind limb paralysis. SDEC cells transplanted into virally injured rats engrafted extensively the length of spinal cord and migrated into the cord parenchyma. Substantial engraftment was not observed in uninjured animals receiving SDEC cells. Engrafted SDEC cells took on expression characteristics of mature neurons and astrocytes. Remarkably, albeit at low frequency, engrafted SDEC cells became immunoreactive to choline acetyltransferase, and sent axons into the sciatic nerve. Even more remarkably, after 12 and 24 weeks, paralyzed animals receiving SDEC cells partially recovered hind limb function. In this experiment, the frequency and total number of neurons generated from SDEC cells was not sufficient to easily explain the significant recovery of function, which is not surprising since the transplanted SDEC cells are a mixed-cell population rather than a line or culture grown or differentiated to promote neural outcome. The mechanism proposed to explain the functional recovery involves EBD cells protecting host motor neurons from death, and facilitating host motor neuron reafferentation, possibly through the secretion of transforming growth factor-a and brain-derived neurotrophic factor.

This example illustrates several important points. The engraftment (and possibly one or more steps in cellular differentiation) of SDEC cells was promoted by an injury signal following viral infection. Once engrafted, cells within the SDEC population were capable of differentiation *in vivo* into mature astrocytes and neurons, some of which sent processes along the correct pathway to the sciatic nerve and were then capable of retrograde transport. It is difficult to see how this elaborate and spatially precise differentiation could be carried out *ex vivo* and then introduced into an animal or patient. The use of a mixed-cell population may have allowed the variety of cellular responses that ultimately resulted in functional recovery. Some cells within the population were capable of forming new neural cells, and others took on supportive and protective roles. Although, in this experimental model, the multiple roles carried out by use of a mixed-cell population may have resulted in functional recovery, future experiments will need to focus on isolation of sub-populations to increase the efficiency of differentiation into required cell types, and to address issues of safety. Lastly, the rats in all treatment groups received immunosuppressive drugs to discourage rejection of the human EBD cells. In the near term, this will likely be a feature of all EBD-based cellular transplantation experiments and therapies.

Embryoid Body-derived Derivation, Growth, and Cryopreservation Methods

EBs formed in the presence of PGC growth media are harvested in groups of 10 or more and are dissociated by digestion in 1 mg/ml Collagenase/Dispase (Roche) for 30 minutes to one hour at 37°C. Cells are then spun at 1000 rpm for five minutes and resuspended in various growth media–matrix environments. These include RPMI growth media (RPMI 1640 (LTI), 15% FCS, 0.1 mM nonessential amino acids, 2 mM L-glutamine, 100 μ/ml penicillin, and 100 mg/ml streptomycin) and EGM2MV media (Clonetics) (5% FCS, hydrocortisone, hbFGF, hVEGF, R^3-IGF1, ascorbic acid, hEGF, heparin, gentamycin, and amphotericin B). Matrices are bovine collagen I (Collaborative Biomedical, 10 mg/cm^2), human extracellular matrix (Collaborative Biomedical, 5 mg/cm^2), and tissue culture plastic. EBD cells are cultured at 37°C, 5% CO_2, 95% humidity, and are routinely passaged 1:10 to 1:40 using 0.025% trypsin–0.01% EDTA (Clonetics) for five minutes at 37°C. Low serum cultures are treated with trypsin inhibitor (Clonetics) and then spun down and resuspended in growth media. EBD cells are cryopreserved in the presence of 50% FCS, 10% DMSO, in a controlled rate freezing vessel and stored in liquid nitrogen.

SUMMARY

Human EG cells can be derived from PGCs by using methods similar to those used to derive mouse EG cultures. Like mouse ES and EG cells, human EG cells require LIF for proliferation as undifferentiated stem cells. Unlike mouse EG cells, however, human EG cells do not readily lose their dependence on exogenous cytokines and factors supplied by the feeder layer, and they have a higher frequency of spontaneous differentiation into EBs. Although EBs are a loss to the pluripotent stem cell population, they are a source of cells expressing markers of mature cellular phenotypes, as well as their presumed progenitors and precursors. Cells that retain a high capacity for cell proliferation and express makers of multiple lineages can be isolated from EBs, and can be used in a variety of *in vitro* and *in vivo* differentiation paradigms. The current challenges are to match individual EBD cultures to desired endpoints, and to enrich or purify populations of cells within EBD cultures to more specifically address biological requirements.

KEY WORDS

Embryoid body A multicellular structure comprised of multiple-cell types formed during the *in vitro* and *in vivo* differentiation of embryonic stem cells.

Embryonic germ cell A pluripotent stem cell derived from primordial germ cells.

Primordial germ cell A cell type that colonizes the developing gonad and turns into egg or sperm.

FURTHER READING

Kerr, D. A., Llado, J., Shamblott, M. J., Maragakis, N. J., Irani, D. N., Crawford, T. O., Krishan, C., Dike, S., Gearhart, J. D., & Rothstein, J. D. (2003). Human embryonic germ cell derivatives facilitate motor recovery of rats with diffuse motor neuron injury. *J. Neurosci.*, *23*, 5131–5140.

Matsui, Y., Toksoz, D., Nishikawa, S., Nishikawa, S., Williams, D., Zsebo, K., & Hogan, B. L. (1991). Effect of steel factor and leukemia inhibitory factor on murine primordial germ cells in culture. *Nature*, *353*, 750–752.

Matsui, Y., Zsebo, K., & Hogan, B. L. (1992). Derivation of pluripotential embryonic stem cells from murine primordial germ cells in culture. *Cell*, *70*, 841–847.

McLaren, A. (2000). Establishment of the germ cell lineage in mammals. *J. Cell Physiol.*, *182*, 141–143.

Resnick, J. L., Bixler, L. S., Cheng, L., & Donovan, P. J. (1992). Long-term proliferation of mouse primordial germ cells in culture. *Nature*, *359*, 550–551.

Shamblott, M., Axelman, J., Littlefield, J., Blumenthal, P., Huggins, G., Cui, Y., Cheng, L., & Gearhart, J. (2001). Human embryonic germ cell derivatives express a broad range of developmentally distinct markers and proliferate extensively *in vitro*. *Proc. Nat. Acad. Sci. USA*, *98*, 113–118.

Shamblott, M. J., Axelman, J., Wang, S., Bugg, E. M., Littlefield, J. W., Donovan, P. J., Blumenthal, P. D., Huggins, G. R., & Gearhart, J. D. (1998). Derivation of pluripotent stem cells from cultured human primordial germ cells. *Proc. Nat. Acad. Sci. USA*, *95*, 13726–13731.

Tam, P. P., & Snow, M. H. (1981). Proliferation and migration of primordial germ cells during compensatory growth in mouse embryos. *J. Embryol. Exp. Morphol.*, *64*, 133–147.

Chapter 9

Genetic Manipulation of Human Embryonic Stem Cells

Yoav Mayshar and Nissim Benvenisty

Genetic manipulation, the process of inducing changes in gene expression and the expression of novel genes, has proven to be an indispensable tool in recent genetic research. The implementation of increasingly powerful genetic tools to mouse embryonic stem (ES) cells has led to an explosion of data concerning the specific properties of an extremely large array of genes. Current available techniques allow for the specific elimination of target gene expression, tissue specific induction of reporter gene expression, the overexpression of cellular genes, and more. It is not surprising, therefore, that a short time after the derivation of human ES cells, several studies describing genetic manipulation of these cells have been reported. This chapter focuses on the recent advances in the genetic manipulation of human ES cells, the methods currently available, and their possible utilization. A brief discussion of advanced genetic manipulation techniques not yet demonstrated in these cells is also presented.

INTRODUCTION

The new technology of establishing human embryonic stem (ES) cell lines has raised great hopes for breaking new ground in basic and clinical research. Clearly, genetic manipulation will play a pivotal role in applying this technology to biological research, as well as for the specific needs of transplantation medicine.

Genetic manipulation has proved to be a key experimental procedure in the field of mouse ES cell research since their initial isolation, over 20 years ago. Many effective techniques have since been developed for bringing about specific genetic modifications. Some of these methodologies have recently been adapted for the manipulation of human ES cells. These include both transfection and infection protocols, as well as overexpression and homologous recombination procedures. The newly-acquired availability of these techniques should make genetic engineering of human ES cells routine, and enable great advances in the field of human ES cell research.

By introduction of genetic modifications to mouse ES cells, a great number of genetically modified animals have been produced, which have proved to be an extremely valuable tool for research. Obviously, no equivalent research can be conducted in humans. Thus, despite the great potential attributed to human ES cells, much of the basic developmental research using these cells may be limited to *in vitro* studies. Researchers into human ES cells are, therefore, forced to adopt developmental models such as embryoid bodies (EBs) for studying early embryogenesis. This chapter will present the principle aspects regarding the *in vitro* genetic manipulation of human ES cells, and the possible uses of the variety of methods currently available.

METHODS OF GENETIC MANIPULATION

A variety of methods are currently used for manipulating mammalian gene expression. Many of these have been used in mouse ES cell research. Of these, several gene delivery methods have been shown to be effective in human ES cells. These include transfection by various chemical reagents, electroporation, and viral infection. Transfection is probably the most commonly used method for genetic manipulation. The system is straightforward, and relatively easy to calibrate. It provides sufficient numbers of cells for clonal expansion, and allows for insertion of constructs of virtually unlimited size. Recently, long-term transgene expression in ES cells was demonstrated using infection by lentiviral vectors. This technique has been shown to be highly effective, and could therefore emerge as a popular tool in human ES cell genetic engineering.

Transfection

The method used for gene delivery is one of the major variables that directly affect the efficiency of genetic manipulation. The choice of the gene delivery method should be based, first and foremost, on the nature of experiment to be performed. The duration of the genetic modification is one factor that will dictate the method of choice. For short-term induction, transient expression can be achieved through the introduction of supercoiled plasmid DNA, transcription from which has been shown to be more efficient than from the linear form. Transcription from transiently introduced plasmids usually peaks at around 48 hours post-transfection. For the most part, positive cells demonstrate relatively high expression levels, as a result of a high copy number of plasmids being introduced into transfected cells. Stable transfection is accomplished by selecting for cells where the vector DNA has integrated into the genome. In this type of experiment, it is important to linearize the vector, as this significantly raises integration and targeting efficiencies. A positive selection marker under regulation of a strong constitutive promoter is frequently included in the insert when the gene target is non-selectable. Selection should not be carried out immediately following transfection. Delaying the selection gives the cells time to recuperate and express the resistance marker. Given that the various selection reagents are toxic to all mammalian cells, the feeder layer supporting the cells must also be resistant to the selection drug. To this end, lines of transgenic mice, carrying various genes conferring resistance to selection drugs, are currently available for the derivation of resistant feeder cells.

Transfection Reagents

In the case of chemical-based transfection, carrier molecules bind the foreign nucleic acids, and introduce them into the cells through the plasma membrane. In general, the cellular uptake of the exogenous nucleic acids is thought to occur either through endocytosis, or in the case of lipid-based reagents, through fusion of a lipidic vesicle to the plasma membrane. This method can routinely produce transient transfection rates of approximately 10%, and stable transfection efficiency of 10^{-5}–10^{-6}. The main advantages of chemical transfection are, notably, the relative simplicity of the procedure, and the fact that it can be performed on an adherent cell culture.

Electroporation

Transfection by electroporation was adopted early on as the method of choice for gene-targeting in mouse ES cells. This process involves applying a brief high voltage shock to cells in suspension, in the presence of DNA molecules. The electric shock is thought to cause transient pores to open up in the cell membrane, enabling DNA to enter. A recent study performed by Zwaka and Thomson demonstrated the efficiency of electroporation for the production of transgenic HESC clones. Importantly, the mouse electroporation protocol that proved inadequate for HESCs was customized for these cells. Using electroporation, a stable integration rate of $\sim 10^{-5}$ was achieved and homologous recombination efficiencies were reported to be between 2–40%, subject to the properties of the vector employed. A substantial number of homologous recombination clones have thus been generated for a number of different vector constructs, demonstrating the feasibility of achieving efficient gene-targeting in human ES cells.

Infection

Unlike transfection methods, viral vectors can produce a very high percentage of infected cells. Viral vectors also allow easy determination of the vector copy number, both for transient and stable expression. Several groups have recently reported transgene expression in human ES cells using lentiviral vectors. Lentiviral infection provides a very high proportion of stable integrants, although integration cannot be targeted, and vector size may be an additional limiting factor. Because of its efficiency, this method could prove useful for bypassing the need for selection and time-consuming clonal expansion, as well as for experiments intended to induce random insertional mutagenesis or gene trap.

GENETIC MODIFICATION APPROACHES

One of the major roles of genetic engineering will be research into the early stages of human development, and the genes involved in this process. The roles of many of these genes have been elucidated using genetic manipulation techniques, mostly with the aid of genetically-modified animals. Several studies using EB models have shown that specific early embryonic developmental processes can, to some extent, be recapitulated *in vitro*. Human ES cells thus provide researchers with a unique system to investigate the early stages of embryonic development through the implementation of genetic manipulation techniques. These techniques can also be used to determine gene function at the cellular level.

Several approaches can be used to discover the activity of specific genes in ES cells through genetic manipulation. These include overexpression of cellular or foreign genes, use of reporter genes, and gene silencing. In this section we attempt to summarize the major techniques currently used in human ES cells, together with a number of other techniques that are likely to be implemented in human ES cell research in the near future (see Figure 9-1).

Exogenous Gene Expression

The concept of directed differentiation is an example of the possible utility of introducing exogenous properties into human ES cells. Extensive studies are currently being carried out on directed differentiation in human ES cells. The addition of specific growth factors to EBs during the differentiation process has been shown to modify their

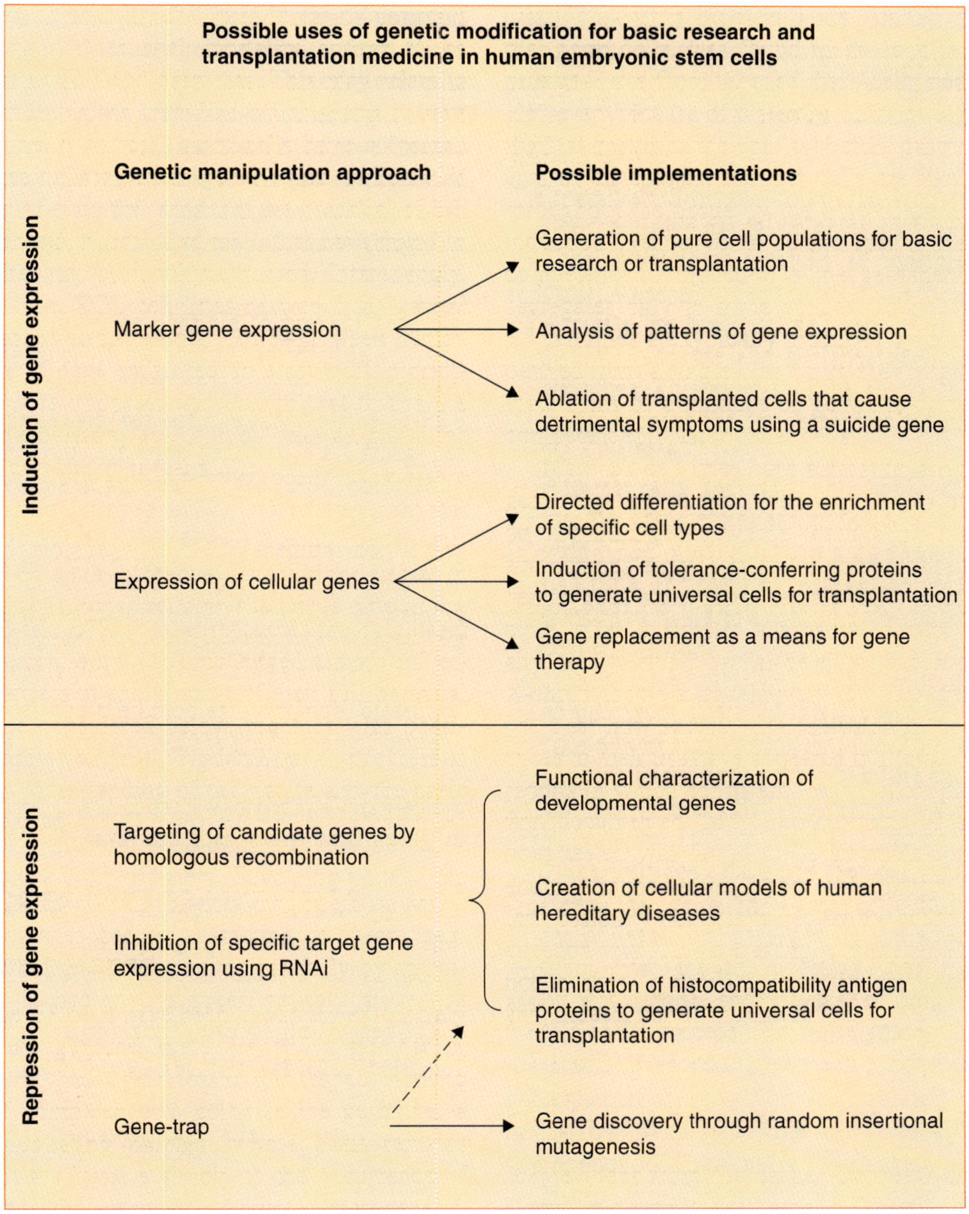

FIGURE 9-1

differentiation potential. Use of various enrichment protocols has made it possible to obtain several different cell types, such as neurons, hematopoietic cells, and cardiomyocytes. While creation of substantial amounts of certain cell types has been reported, the outcome of some differentiation protocols are often inconsistent, and are currently unable to yield the pure cell populations required for transplantation. Genetic manipulation could complement these methods, and facilitate the practical application of human ES cells in transplantation medicine.

Genetic Labeling

Introduction of constitutively expressed reporter genes, such as the gene encoding for green fluorescent protein (GFP) into the cells, allows their detection in a mixed population. This method was applied in a transplantation experiment where undifferentiated human ES cells were introduced into an early stage chick embryo. By using the same reporter gene regulated by a promoter of an ES cell-enriched gene (*Rex1*), it was possible to follow undifferentiated cell populations during differentiation *in vitro*, at the same time screening out differentiated cells with a fluorescence-activated cell sorter (FACS). Accordingly, Zwaka and Thomson created knockin cell lines using GFP or neomycin resistance promoterless constructs. Through this method, in the homologous recombination events that were selected, GFP expression was regulated by the endogenous OCT4 promoter. By placing the reporter under the regulation of the native gene, possible position effects that could confound the interpretation of the results are avoided, and the entire set of *cis*-acting elements is utilized.

Alternatively, the introduction of a conditional suicide gene could solve one of the principle concerns facing cellular transplantation. Adverse effects from transplants may occur as a result of malfunction or uncontrolled proliferation of grafted cells. Selective ablation of human ES cells expressing herpes simplex virus thymidine kinase (HSV-*tk*) has been demonstrated in an animal model. Growth of teratomas formed by transplanting PGK-HSV-*tk* transfected cells was successfully arrested by oral administration of the prodrug gancyclovir. Using this technique, it will be possible to transplant "fail-safe" cells into patients. These cells could then be eliminated on the onset of harmful symptoms.

Probably the most direct way for the enrichment of a specific cell type using genetic manipulation is to isolate it, using a tissue specific promoter driving a reporter or resistance gene. This method has been demonstrated for the enrichment of undifferentiated human ES cells, but can also be applied to isolate specific populations of differentiated cells. In mouse ES cells, through expression of a neomycin resistance gene regulated by the endogenous promoter of *sox2*, a neuronal progenitor specific gene, Li *et al.* (1998) produced highly-enriched neural precursors from a heterogeneous population following selection. Similarly, mouse cardiac cells were isolated by using a reporter gene driven by the cardiac alpha actin promoter. Of course, the efficiency of such direct selection methods could be improved under culture conditions that allow greater quantities of the desired cell type to be produced.

Bacterial Artificial Chromosomes and Recombineering

One of the major problems in reporter gene expression is to faithfully recapitulate the endogenous gene. Conventional DNA constructs are of limited size, and are therefore constrained by the maximal size of control elements that can be included. This can be overcome, as has been previously described, by a knockin approach—even though this is a difficult procedure and also disrupts one of the copies of the gene. Bacterial artificial chromosome (BAC)-based constructs containing cloned human DNA have been combined with homologous recombination in *E. coli* (recombineering) to facilitate in solving this problem. The basis of the method is the use of publicly and commercially available BACs containing cloned genomic regions of 100–200-kb, together spanning practically the entire human genome. Thus, it is possible to obtain a BAC containing virtually any gene of interest, along with extensive upstream and downstream sequences. These flanking sequences could be critically important for faithfully recapitulating gene expression by long-distance *cis* regulatory elements. Utilization of the BAC for genetic manipulation is facilitated by the use of recombineering. This procedure takes advantage of phage λ recombination proteins that induce homologous recombination within *E. coli*, using only very short homologous flanking sequences. Thus, BACs have been used to modify mouse ES cells, and transgenic animals have been established. In these experiments, reporter genes were the main use of the constructs, although targeting vectors for genomic modification were also demonstrated.

Over-expression of Cellular Genes as Dominant Effectors of Cell Fate

The overexpression of developmentally important genes is currently being used as a means of achieving directed differentiation. Several regulatory genes have been introduced in various mouse experiments. Hepatocyte nuclear factor 3 (HNF3) transcription factors -α and -β induced differentiation of mouse ES cells towards the early endodermal lineage. Ectopic expression of HoxB4, a homeotic selector gene, has enabled the use of mouse ES cells for the repopulation of myeloid and lymphoid lineages in immune-deficient mice. Similarly, the expression of the nuclear receptor related-1 (Nurr1) transcription factor generated a highly-enriched population of midbrain neural stem cells that have been shown to ameliorate the phenotype of a mouse model of Parkinson's disease.

As yet no single key factor or gene has been found that could cause undifferentiated ES cells to differentiate *en masse* to a specific lineage or cell type. Therefore, a combination of genetic manipulation, and selective culture conditions, would probably be necessary to generate differentiated transplant ready cells from undifferentiated ES cells.

Gene Silencing

As opposed to the exogenous expression of various genes, gene silencing assesses the functionality of a particular gene of interest through its elimination. In the mouse, gene silencing is commonly achieved through homologous recombination, in which an exogenous sequence is inserted within the gene's coding region, thus disrupting it. Gene trap is an additional method through which a wide range of genomic sequences can be randomly mutated. These techniques that generate knockout alleles are, however, of more limited use in human ES cells, as will be discussed later. Thus, in these cells the most common emerging method for gene silencing is post-transcriptional RNA interference—also known as knockdown.

Gene Targeting

Genetic modification has, for many years, been considered a cornerstone of mammalian biology research. Two major advances in genetic manipulation of the mammalian genome were the isolation of mouse ES cells, and the development of targeting vectors. Together, these two methods have enabled the relatively straightforward production of genetically modified mice, carrying predesigned alterations to their genome. The subsequent proliferation of data has shed light on the functional characteristics of a very large number of genes, and has led to the creation of various models for human disease. Although a number of mouse models have been created, some have proved inadequate in recapitulating some or all aspects of the human phenotype. Research is often limited by lack of an adequate mouse model, and working with primary cultures can often be problematic. In light of these difficulties, the creation of a cellular model of human disease, using human ES cells, will be very valuable.

Mutation of the hypoxanthine guanine phosphoribosyltransferase gene (HPRT) in humans is the basis for Lesch–Nyhan syndrome, a severe genetic disease. This syndrome is manifested in a variety of problems related to high uric acid levels. These include gout-like arthritis and renal failure. There are also neurological disorders associated with this disease, but the exact causes are still unknown. Mouse models for HPRT deficiency failed to fully-recapitulate the human disease, possibly because uric acid metabolism is different in humans. Homologous recombination was recently used to disrupt HPRT using positive negative selection. Analysis of mutant human ES cell metabolism showed that, unlike their mouse counterparts, human cells mimic the Lesch–Nyhan phenotype.

HPRT is ideal for analyzing homologous recombination, as it is X-linked. Thus, mutation of the single allele in male cells may produce a phenotype. Furthermore, cells lacking HPRT activity can be selected with 6-thioguanine (6TG). By culturing these cells in HAT medium, they can be selectively eliminated. In many cases, the rate-limiting step towards the obtaining of targeted recombinants is the high number of nonhomologous integrations, making screening for homologous recombinants very time-consuming. The additional negative selection is useful, in order to confirm successful homologous recombination events and eliminate false positives. Thus, by using an HPRT– cell line as a base for further genetic manipulation, an HPRT minigene can be used as an additional positive and negative selection, by correcting the HPRT deficiency. These advantages have made HPRT very useful in the initial study of homologous recombination. In cases other than HPRT targeting, a vector is usually constructed that contains a positive selection cassette (usually the neomycin-resistance gene), flanked by sequences homologous to the endogenous target gene. A negative selection marker, usually HSV-*tk*, is often placed in such a way that, following linearization of the vector, it is located outside the region of homology. Thus, in those cases in which homologous recombination has successfully occurred, the transfected cells will not contain the HSV-*tk* gene, and will be resistant to the drug gancyclovir.

When targeting expressed genes, promoterless constructs can be employed, in order to increase the efficiency of homologous recombination by reducing the selectability of most random integrations. This method requires careful calibration of the concentration of the selection reagent, subject to the strength of endogenous promoter. Zwaka and Thomson (2003) targeted the gene encoding OCT4 (*POU5F1*) using this approach, and achieved homologous recombination rates of up to 40%.

As opposed to gene-targeting, random gene disruption on a wide scale is currently being performed by gene trap. In this method, a reporter gene lacking an essential regulatory element is randomly inserted into the genome. Thus, in cases in which the reporter gene is expressed, often this is due to the use of an endogenous regulatory sequence. A large-scale disruption of genes is therefore made possible, allowing for the discovery of new genes, and the creation of banks for a wide variety of mutations.

Induction of Homozygosity in Mutant Alleles

Several methods have been developed for the homozygotization of mutated alleles in mouse ES cells. When mutating the mouse genome, there is often an option of generating homozygosity through breeding two heterozygotes. However, research with human models will require *in vitro* manipulation. Two strategies can be employed to this end: (1) targeting the remaining allele by a second round of homologous recombination using distinct selection markers; and (2) growing transfected cells in high concentrations of the selection reagent, thus selecting for spontaneous homozygotization of the mutated allele containing the resistance gene.

The Effect of Extent and Nature of Homology on Targeting Efficiency

The length of the homologous regions has been shown to be a critical factor in determining the effectiveness of targeted recombination. By increasing the length of homology from 2-kb to 10-kb, an exponential increase in targeting efficiency is achieved. A less dramatic rate of increase is observed with up to ~14-kb of homology. The length of the short arm of homology should not be less than approximately 0.5-kb. Use of isogenic, rather than non-isogenic, DNA has also been shown to increase the rate of targeted recombination dramatically, although the need for isogenic DNA for human somatic cell lines and ES cells has recently been disputed.

Advanced Methods of Gene Targeting

It has been shown that the strategy of inserting a selection cassette to generate a knockout phenotype can confound the interpretation of these experiments, as the exogenous promoter may disrupt endogenous gene expression at distances of over 100-kb from the point of insertion. This disruption of the local regulation elements can produce an artifact phenotype where none exists. This is demonstrated by distinct phenotypic effects that may be caused by different constructs used to target the same gene. Removal of the selection cassette is therefore recommended. Clean mutations can be achieved, either by the "hit-and-run" or "tag-and-exchange" methods, or by application of an exogenous recombinase.

"Hit-and-Run" and "Tag-and-Exchange" Methods

Several strategies have been demonstrated that can create specific substitutions. These strategies create subtle mutations in the gene, as opposed to disrupting it with a large insert. The "hit-and-run" method is based on selection of intrachromosomal recombination events that excise the vector sequence. In another strategy, termed "tag-and-exchange" (double replacement), a series of targeting procedures is performed. In the first stage, the gene is targeted with a vector that contains both positive and negative selections in the insert. In the second stage, a vector containing the mutation of choice is inserted by homologous recombination, selected by the loss of the negative selection marker.

Exogenous Recombinases

By introducing highly-efficient and specific exogenous recombination systems, both efficiency and tight control over the recombination event are achieved. The system most widely-used is the Cre recombinase from phage P1. Cre identifies a highly-specific 34 bp target sequence (*lox P*), and promotes reciprocal recombination between two such sites. The versatility of the system is such that, depending on the location of the *lox P* sequences, it can promote either reversible integration or excision, while retaining the recombination sites' integrity. Thus the Cre-*lox* system has been used for a variety of ends. These include the induction of chromosomal rearrangements, spatial or temporal inactivation of cellular genes by deletion, or activation of reporter genes. They also include multiple insertions into the same genomic site, and the creation of clean mutations devoid of exogenous regulatory sequences.

RNA Interference

While gene-targeting remains the most powerful method for the silencing of endogenous gene expression, post-transcriptional gene silencing has lately been characterized as a potent method of down-regulating specific genes. Double stranded RNA (dsRNA) molecules have been shown to initiate RNA interference (RNAi), causing sequence-specific mRNA degradation. Because RNAi does not entail any modification of the targeted gene itself, it is relatively simple to generate gene silencing using this method. Thus, RNAi inducing molecules can be directly introduced by transfection, or synthesized within the cells using expression vectors, either transiently or by stable integration.

The evolutionarily conserved machinery that underlies the phenomenon of homology-dependent gene silencing involves endonuclease (Dicer) processing of dsRNA into small interfering RNA (siRNA). siRNAs are 21- to 23-nt long fragments that, in turn, are incorporated into RISC, a multi-component nuclease, that uses the siRNA as a guide for the degradation of mRNA substrate. While the use of long dsRNA has been demonstrated to be effective, it is well-established that dsRNA within mammalian cells may induce a cellular response by interferon, thus generating non-specific effects. Direct use of siRNA has been shown to circumvent this problem.

In recent years, the RNAi technology has been extensively utilized for both mouse and human ES cells. The relative ease of designing siRNA molecules against specific gene targets, along with the straightforwardness of the knockdown experiment, is especially important in human ES cells for which double allele knockouts have not yet been achieved. The major hindering factor in the application of RNAi in human ES cells is the relatively low rate of transient transfection achieved, using standard methods, of up to 70%. Viral vectors can be used to achieve more efficient rates of genetic modification, and it has recently been proposed that nucleofection may also be at least as efficient. Nevertheless, siRNA is currently the most common method used by researchers wishing to down-regulate specific genes in human ES cells.

Conditional Genetic Modification

For many experimental applications, it could be worthwhile to perform conditional or controlled genetic modification. The means to achieve such modification are mainly through an inducible system, such as the tetracycline transactivator system, or by site-specific recombination, such as the previously described Cre-*lox*P system. Keeping the transgene from being expressed during the process of clone establishment can be essential in the case of deleterious or toxic genes. This is specifically important when transient transfection is problematic, either due to the relatively low transfection rates in human ES cells, or when long-term effects need to be examined. Additional applications could be the reversal of a phenotype conferred by a transgene—by its recombination-mediated removal or by the shutdown of the conditional system—or the conditional deletion of a genomic sequence.

CONCLUSIONS

One of the advantages of ES cells over other stem cell types currently under research is their accessibility for genetic manipulation. Progress so far indicates that genetic manipulation in human ES cells closely resembles that of their veteran mouse counterparts. Thus, most methods that have been used for the genetic manipulation of mouse ES cells have also been demonstrated in human ES cells. One handicap that human ES cells have is the problem of homozygotization of null alleles, which in the mouse is usually solved by breeding heterozygotes. Thus, it would seem that most of the genetic loss-of-function experiments in human ES cells will use RNAi. Possible uses for genetically-modified human ES cells could be the creation of cellular models for disease, studying various aspects of early human development, transplantation medicine, and many more. Thus, in the near future, a large number of studies using genetic manipulation techniques in human ES cells are foreseeable.

KEY WORDS

DNA transfection The introduction of foreign DNA molecules into cultured cells.

Gene-targeting The modification of a gene in a specific and directed manner through homologous recombination with a DNA segment. Possible applications include the elimination of gene expression, or correction of a mutation in a gene.

Homologous recombination The substitution of one DNA segment by another, based on exact or near identity between them.

Negative selection The application of growth conditions eliminating those cells expressing the selection marker.

Positive selection The application of specific growth conditions causing selective elimination of cells that do not contain an exogenous genetic element (selection marker) conferring resistance.

RNA interference (RNAi) The process of post-transcriptional down-regulation of gene function through the use of non-coding RNA molecules, such as small interfering RNA (siRNA).

Stable transfection Transfection in which foreign DNA is incorporated into the host genome, generating a long-term genetic modification.

Transient transfection Transfection in which plasmid DNA is not physically incorporated into the genome, and is retained as an episome, generating short-term gene expression of the foreign DNA.

FURTHER READING

Eiges, R., Schuldiner, M., Drukker, M., Yanuka, O., Itskovitz-Eldor, J., & Benvenisty, N. (2001). Establishment of human embryonic stem cell-transfected clones carrying a marker for undifferentiated cells. *Curr. Biol.*, *11*, 514–518.

Gu, H., Marth, J. D., Orban, P. C., Mossmann, H., & Rajewsky, K. (1994). Deletion of a DNA polymerase beta gene segment in T cells using cell type-specific gene targeting. *Science*, *265*, 103–106.

Hannon, G. J. (2002). RNA interference. *Nature*, *418*, 244–251.

Li, M., Pevny, L., Lovell-Badge, R., & Smith, A. (1998). Generation of purified neural precursors from embryonic stem cells by lineage selection. *Curr. Biol.*, *8*, 971–974.

Pfeifer, A., Ikawa, M., Dayn, Y., & Verma, I. M. (2002). Transgenesis by lentiviral vectors: lack of gene silencing in mammalian embryonic stem cells and preimplantation embryos. *Proc. Nat. Acad. Sci. USA*, *99*, 2140–2145.

Rideout, W. M., 3rd, Hochedlinger, K., Kyba, M., Daley, G. Q., & Jaenisch, R. (2002). Correction of a genetic defect by nuclear transplantation and combined cell and gene therapy. *Cell*, *109*, 17–27.

Smithies, O., Gregg, R. G., Boggs, S. S., Koralewski, M. A., & Kucherlapati, R. S. (1985). Insertion of DNA sequences into the human chromosomal beta-globin locus by homologous recombination. *Nature*, *317*, 230–234.

Stanford, W. L., Cohn, J. B., & Cordes, S. P. (2001). Gene-trap mutagenesis: past, present and beyond. *Nat. Rev. Genet.*, *2*, 756–768.

Thomas, K. R. & Capecchi, M. R. (1987). Site-directed mutagenesis by gene targeting in mouse embryo-derived stem cells. *Cell*, *51*, 503–512.

Urbach, A., Schuldiner, M., & Benvenisty, N. (2004). Modeling for Lesch–Nyhan disease by gene targeting in human embryonic stem cells. *Stem Cells*, *22*, 635–641.

Zwaka, T. P., & Thomson, J. A. (2003). Homologous recombination in human embryonic stem cells. *Nat. Biotechnol.*, *21*, 319–321.

Chapter 10

Induced Pluripotent Stem Cell Derivation

Junying Yu and James A. Thomson

INTRODUCTION

Derivation of induced pluripotent stem cells directly from differentiated human somatic cells offers an easy and efficient alternative to derive patient-specific pluripotent stem cells (Takahashi *et al.*, 2007; Yu *et al.*, 2007; Lowry *et al.*, 2008; Park *et al.*, 2008). Through screening genes with enriched expression in human embryonic stem cells, we found that overexpression of *OCT4*, *SOX2*, *NANOG* and *LIN28* alone was sufficient to induce pluripotent stem cells directly from fetal, newborn, and adult human fibroblasts. These induced pluripotent stem cells are morphologically similar to human embryonic stem cells, express typical human embryonic stem cell-specific cell surface antigens and endogenous genes, and are able to differentiate into multiple lineages *in vitro*. More importantly, when injected into immunocompromised mice, they form teratomas containing differentiated derivatives of all three primary germ layers (Yu *et al.*, 2007). In this chapter, we describe the detailed procedure for the derivation of induced pluripotent stem cells, using newborn human foreskin fibroblasts as an example.

MATERIALS

Reagents

- DMEM basal medium (Invitrogen, Cat. No. 11965).
- DMEM/F12 basal medium (Invitrogen, Cat. No. 11330-057).
- IMEM basal medium (Invitrogen, Cat. No. 12440).
- Fetal bovine serum (FBS) (Hyclone, Cat. No. SV30014.03).
- Knockout serum replacer (Invitrogen, Cat. No. 10828-028).
- Non-essential amino acids (Invitrogen, Cat. No. 11140-050).
- GlutaMAX (Invitrogen, Cat. No. 35050-061).
- Sodium pyruvate (Invitrogen, Cat. No. 11360-070).
- β-mercaptoethanol (Sigma, Cat. No. M7522).
- DMSO (Sigma, Cat. No. D-2650).
- Trypsin/EDTA solution (Invitrogen, Cat. No. 25300).
- Collagenase Type IV (Invitrogen, Cat. No. 17104-019).
- PBS without Ca^{2+} and Mg^{2+} (Invitrogen, Cat. No. 14190-250).
- Geneticin (Invitrogen, Cat. No. 10131035).
- Gelatin (Sigma, Cat. No. G1393).
- SuperFect transfection reagent (Qiagen, Cat. No. 301305).
- bFGF (zebrafish) is produced as previously described (Ludwig *et al.*, 2006).
- RNeasy® Mini Kit (Qiagen, Cat. No. 74106).
- QIA shredder (Qiagen, Cat. No. 79656).
- RNase-Free DNase set (Qiagen, Cat. No. 79254).
- SuperScript® III First-Strand Synthesis System (Invitrogen, Cat. No. 18080-051).
- SYBR green QPCR mastermix (Applied Biosystems, Cat. No. 4309155).
- Polybrene (Hexadimethrine bromide) (Sigma, Cat. No. H9268).
- Clorox® regular bleach

Cell Stocks

- 293FT cells (Invitrogen, # R700-07).
- Newborn human foreskin fibroblasts (ATCC, Cat. No. CRL-2097).
- Mouse embryonic fibroblast (MEF) feeder cells are prepared from day 13 CF1 pregnant mouse, and irradiated as previously described (Xu *et al.*, 2004).

Stem Cell Anthology

Solutions and Culture Medium

- Polybrene solution (6 mg/ml). Add 60 mg of polybrene to 10 ml water. Filter to sterilize. Aliquot (100 µl each) and store at −20°C.
- 0.1% gelatin solution. Add 0.5 gram of gelatin powder to 500 ml of water. Autoclave to sterilize.
- 1 mg/ml collagenase solution. Add 50 mg of collagenase type IV to 50 ml of DMEM/F12 basal medium. Filter to sterilize. Store at 4°C for no longer than one week.
- 293FT cell culture medium. Add 100 ml of heat-inactivated FBS, 10 ml of 200 mM glutamax stock, and 10 ml of 10 mM non-essential amino acid stock to 880 ml of DMEM basal medium to make up the 1 liter volume. Filter to sterilize. Store at 4°C for no longer than two weeks.
- MEF feeder cell culture medium. Add 100 ml of heat-inactivated FBS and 10 ml of 10 mM non-essential amino acid stock to 890 ml of DMEM basal medium to make up the 1 liter volume. Filter to sterilize. Store at 4°C for no longer than two weeks.
- Human ES cell culture medium without bFGF. Add 200 ml of Knockout Serum Replacer, 5 ml of 200 mM glutamax stock, 10 ml of 10 mM non-essential amino acid stock and 7 µl of β-mercaptoethanol to 785 ml of DMEM/F12 basal medium to make up the 1 liter volume. Filter to sterilize. Store at 4°C for no longer than two weeks.
- Human ES cell culture medium with bFGF. Add 100 µg zebrafish bFGF to 1 liter of human ES cell culture medium without bFGF to obtain a final bFGF concentration of 100 ng/ml. Filter to sterilize. Store at 4°C for no longer than two weeks.

Supplies and Equipment

- T75 tissue culture treated flasks (Fisher, Cat. No. 10-126-7).
- T175 tissue culture treated flasks (Fisher, Cat. No. 12-565-54).
- 10 cm tissue culture treated dishes (Fisher, Cat. No. 08-772E).
- 6-well tissue culture treated plates (Fisher, Cat. No. 14-832-11).
- 48-well tissue culture treated plates (Fisher, Cat. No. 12-565-322).
- Cell culture filters–50 ml (Fisher, Cat. No. SE1M179M6).
- Cell culture filters–150 ml (Fisher, Cat. No. SCGPU01RE).
- Cell culture filters–250 ml (Fisher, Cat. No. SCGPU02RE).
- Cell culture filters–500 ml (Fisher, Cat. No. SCGPU05RE).
- Cell culture filters–1000 ml (Fisher, Cat. No. SCGPU11RE).
- 115 ml low-protein binding cellulose acetate filters (0.45 µM) (Corning, Cat. No. 430945).
- 15 ml conical tubes (Fisher, Cat. No. 05-538-53D).
- 50 ml conical tubes (Fisher, Cat. No. 05-538-55).
- Incubators: all cells are maintained in a humidified incubator in a 5% CO2/95% air atmosphere at 37°C.
- Optima™ L-80 XP ultracentrifuge (Beckman, PN # 392051).
- 45Ti rotor for ultracentrifuge (Beckman, PN # 339160).
- 70 ml centrifugation bottles (Beckman, PN # 355622).
- Horizontal shaker (Fisher, Cat. No. 11+67152Q).
- Thermocyler (Applied Biosystems, Cat. No. 4375786).
- Real-time PCR system (Applied Biosystems, Cat. No. 4351106).
- 96-well PCR plate (Fisher, Cat. No. NC9726586).
- Optimal Adhesive Film (Fisher, Cat. No. NC9745301).

Plasmids

- pSIN4-EF2-EGFP-IRES-Puro
- pSIN4-EF2-OCT4-IRES-Puro
- pSIN4-EF2-SOX2-IRES-Puro
- pSIN4-EF2-NANOG-IRES-Puro
- pSIN4-EF2-LIN28-IRES-Puro
- psPAX2
- pMD.G

PREPARING TRANSGENE-EXPRESSING LENTIVIRUS

Packaging Lentivirus

1. Maintain 293FT cells in 293FT cell culture medium supplemented with 500 µg/ml geneticin according to the product manual (Invitrogen).
2. For the lentiviral packaging, 293FT cells are expanded in T175 tissue culture treated flasks in 293FT cell culture medium supplemented with 500 µg/ml geneticin. The number of T175 flasks needed is approximately twice the number of constructs to be packaged. Gelatin-coating of T175 flasks is not needed.
3. Day 1 (~11 am–12 pm): when 293FT cells reach approximately 80–90% confluency, remove culture medium, and wash cells with 8 ml/T175 flask of PBS without Ca^{2+} and Mg^{2+}.
4. Add 6 ml/T175 flask of Trypsin/EDTA solution, and incubate at room temperature for three to five minutes.
5. Add 12 ml/T175 flask of 293FT cell culture medium to neutralize the trypsin activity.
6. Transfer 293 FT cells to 50 ml conical tubes, and pellet cells by centrifugation at 200 × g for five minutes.

7. Remove supernatant, and gently resuspend 293FT cells in 293FT cell culture medium without geneticin (approximately 30 ml for each T175 flask of 293FT cells). The cell density is usually between $1 - 1.5 \times 10^6$ cells/ml.
8. For each construct, prepare 6 × T75 tissue culture treated flasks. Gelatin-coating of T75 flasks is not needed.
9. Aliquot 10 ml of 293FT cell suspension to each T75 flask.
10. Prepare transfection mixture as steps 11 to 14 (Table 10-1).
11. Centrifuge plasmids pMD.G, psPAX2 and transgene-expressing pSIN vectors at maximum speed (~20000 × g) for 10 minutes, to remove debris and possible bacterial contamination. The concentration of all plasmids should be at least 0.5 μg/μl, to avoid dilution when they are combined with transfection reagents.
12. Mix the corresponding amount of each plasmid in room temperature serum-free IMDM basal medium, as in Table 10-1. Prepare enough plasmids in 15 ml conical tubes for 6 × T75 flasks for each transgene-expressing construct.
13. Add the corresponding amount of Superfect to the DNA solution, and mix with brief vortexing.
14. Briefly spin down to collect liquid at the bottom of the tube, and incubate the DNA/Superfect mixture at room temperature for 10 minutes.

Transfection of 293FT Cells

1. To transfect 293FT cells, add fresh 293 FT cell culture medium to the DNA/Superfect mixture (~3.4 ml for 6 × T75 flasks), and mix by pipetting up and down three times. Avoid bubble formation.
2. Add DNA/Superfect mixture dropwise to the 293FT cell suspension (~1 ml/T75 flask).
3. Gently mix 293FT cells, and incubate at 37°C overnight.
4. Day 2 (~9 am–10 am): replace the transfection medium with fresh 293FT cell culture medium, supplemented with 1 mM sodium pyruvate (~8 ml/T75 flask).
5. Day 4 (~9 am): collect lentivirus (~68 to 72 hours post-transfection) by pouring the supernatant from 6 × T75 flasks for each transgene-expressing construct directly to a 50 ml conical tube.
6. Remove cell debris by centrifugation at 1000 × g for five minutes.
7. Filter the supernatant through 115 ml low-protein binding cellulose acetate filters (0.45 μM).
8. Concentrate lentivirus through ultracentrifugation of the supernatant in 70 ml sterilized tubes at 50000 × g for 2.5 hours at 4°C. The 70 ml centrifugation tubes can be sterilized by Clorox® regular bleach, followed by three washes with sterile water.
9. Carefully remove the supernatant without disturbing the pellet.
10. Add ~500 μl of PBS without Ca^{2+} and Mg^{2+}, and resuspend the pellet by gently shaking the centrifuge tubes at 4°C for about three to six hours on a horizontal rocker.
11. Transfer the resuspended lentivirus to 2 ml eppendorf tubes pre-chilled at 4°C.
12. Remove any residual debris from the lentiviral preparations by centrifugation at 1000 × g for 10 minutes at 4°C.
13. Aliquot (~50 μl each) the lentivirus-containing supernatant to 0.5 ml eppendorf tubes pre-chilled at −20°C, and store at −80°C.

TABLE 10-1 Transfection of 293FT Cells

	per T75 flask	6 × T75 flasks
pMD.G	5 μg	30 μg
psPAX2	10 μg	60 μg
pSIN vector	5 μg	30 μg
Superfect	40 μl	240 μl
IMDM	400 μl	2.4 ml

Titering Lentivirus

Determine the Titer for EGFP-expressing Lentivirus Through Transduction

1. A 6-well tissue culture treated plate is coated with gelatin by adding 1.5 ml/well of 0.1% sterile gelatin solution, and incubated at 37°C for at least one hour. Remove the gelatin solution immediately prior to use.
2. Grow 1 × T75 flask of 293 FT cells in 293 FT cell culture medium supplemented with 500 μg/ml geneticin.
3. When 293 FT cells reach approximately 80–90% confluency, remove culture medium, and wash cells with 6 ml of PBS without Ca^{2+} and Mg^{2+}.
4. Add 3 ml of Trypsin/EDTA solution to the T75 flask, and incubate at room temperature for three to five minutes.
5. Add 6 ml of 293 FT cell culture medium to the T75 flask to neutralize the trypsin activity.
6. Transfer 293 FT cells to a 15 ml conical tube, and pellet cells by centrifugation at 200 × g for five minutes.
7. Remove the supernatant, and gently resuspend 293 FT cells in 10 ml of 293 FT cell culture medium without geneticin.
8. Determine and adjust the cell density to about 1×10^6 cells/ml.
9. Add 1 ml/well of 293 FT cells to the gelatin-coated 6-well plate (1×10^6 cells /well).

10. Add 1 μl of polybrene (6 mg/ml stock) to each well.
11. Add the following amount of the concentrated EGFP-expressing lentivirus to each well (μl): 0, 1, 2, 5, 10, 20.
12. Mix 293 FT cells and incubate at 37°C overnight.
13. The next day, replace the transduction mixture with 3 ml of fresh 293 FT cell culture medium.
14. After 48–72 hours post-transduction, determine the percentage of EGFP positive cells in each well using flow cytometry analysis.
15. Use the percentages of non-lentivirus-saturating wells to calculate the titer for the EGFP-expressing lentivirus. For example, if the percentage of the 2 μl-well is "a", the percentage of the 5 μl-well is "b", and both "a" and "b" are less than 90%, the titer for the EGFP-expressing lentivirus is the following: $[(a \times 10^6/2 + b \times 10^6/5)/2 \times 10^3$ viral particles (vp)/ml, i.e., $[(a/2 + b/5)/2] \times 10^9$ vp/ml. The titer for the EGFP-expressing lentivirus is generally greater than 1×10^8 vp/ml using the packaging protocol above.

Determine the Titers for OCT4, SOX2, NANOG and LIN28-expressing Lentivirus through Real-time RT-PCR

1. Extract total RNA from 10 μl of the concentrated lentivirus expressing EGFP (as a relative standard), OCT4, SOX2, NANOG, and LIN28, as described in the RNeasy® Mini Kit (Qiagen). To remove DNA contamination, on-column DNA digestion is carried out with RNase-free Dnase, as suggested in the RNeasy® Mini Kit. To reduce experimental errors, RNA is extracted from triplicates of each transgene-expressing lentivirus.
2. Elute RNA in 50 μl of sterile water.
3. Take 8 μl of each eluate for reverse transcription (RT) reactions with random hexamers as described in the Superscript III First-Strand Synthesis System (Invitrogen) (Table 10-2).

1. Real-time PCR is carried out with the following primer set (100 bp): Titer-F (5′ gacaggcccgaaggaatag 3′) and Titer-R (5′ actgccatttgtggcgatac 3′).
2. 10 μl of cDNAs from EGFP-expressing lentivirus (triplicate) is diluted with the following dilution factors: 1, 2^1, 2^2, 2^3, 2^4, 2^5, 2^6, 2^7, which are used as a relative standard for real-time PCR.
3. cDNAs from the OCT4, SOX2, NANOG and LIN28-expressing lentivirus are diluted eight-fold, e.g., 5 μl of cDNA to 35 μl of sterile water.
4. Real-time PCR is carried out with SYBR green QPCR Master Mix (Applied Biosystems) (Table 10-3).

1. Calculate the titers for the OCT4, SOX2, NANOG and LIN28-expressing lentivirus as the following:
   a. Plot $-\log_2$ (dilution factor) (y) against $Ct_{ave.}$ (x), using results from the EGFP-expressing lentivirus standard in Microsoft Excel. Ct_{ave} is the average of triplicate Ct values for the EGFP-expressing lentivirus. Add the linear trendline to the data, and obtain equation y = a*x + b.
   b. Use the following equation to calculate the titer for each transgene-expressing lentivirus: $[(2^{-(a \cdot Ct1+b)} + 2^{-(a \cdot Ct2+b)} + 2^{-(a \cdot Ct3+b)})/3] * 3 *$ titer (EGFP)

TABLE 10-2 First Strand cDNA Synthesis

Reagents	Volume (μl)
RNA	8
Random hexamers (50 ng/μl)	1
Incubate at 70°C for 10 minutes Immediately transfer to ice for 2 minutes	
10× RT buffer	2
25 mM $MgCl_2$	4
DTT (0.1 M)	2
dNTPs (10 mM)	1
RNaseOUT (40 U/μl)	1
Superscript RT III (200 U/μl)	1
Mix and incubate at room temperature for 10 minutes	
Incubate at 50°C for 50 minutes	
Incubate at 85°C for 5 minutes to inactivate the enzyme	
Chill on ice and proceed to real-time PCR or store at −20°C	

TABLE 10-3 Real-time PCR Set-up

Components	Volume (μl)
PCR Master Mix (2×)	12.5
20 μM Titer-F	0.25
20 μM Titer-R	0.25
Template	1
Water	11
Total	25
95°C: 10′ 40 × [95°C: 30″, 60°C: 1′] 72°C: 3′	

vp/ml. The titer (EGFP) refers to the infectious titer determined through transduction of 293FT cells as shown above.

6. The MEF dishes or plates can be used to support reprogramming culture or culture of human ES cells and human induced pluripotent stem (iPS) cells for up to two weeks.

REPROGRAMMING HUMAN FORESKIN FIBROBLASTS

Culturing Human Foreskin Fibroblasts

1. Perform a quick thaw of a vial of frozen human foreskin fibroblasts (ATCC, Cat. No. CRL-2097) in a 37°C water bath. To avoid contamination, do not immerse the vial above the level of the cap.
2. When only a small crystal of ice remains in the tube, take the vial out of the water bath, and wipe dry.
3. Decontaminate the vial exterior with 70% ethanol.
4. Transfer the content of the vial to a 15 ml conical tube.
5. Slowly add 10 ml pre-warmed 293FT cell culture medium to the tube dropwise to reduce osmotic shock.
6. Pellet the cells by centrifugation at 200 × g for five minutes.
7. Remove the supernatant, and gently resuspend the foreskin fibroblasts in 10 ml 293FT cell culture medium.
8. Transfer the cells to a T75 tissue culture treated flask. No gelatin coating is needed.
9. Replace the medium the next day, and thereafter every three days.
10. When fibroblasts reach around 90% confluency, split the cells at a ratio of 1:6 according to the product manual (ATCC). Foreskin fibroblasts can be frozen down with 95% 293FT cell culture medium supplemented with 5% DMSO as suggested in the product manual.

Preparing Mouse Embryonic Fibroblast Dishes and Plates

1. 10 cm tissue culture treated dishes, 6-well and 48-well tissue culture treated plates are coated with gelatin by adding 0.1% sterile gelatin solution (6 ml/10 cm dish, 1.5 ml/well of 6-well plates and 0.5 ml/well of 48-well plates), and incubated at 37°C for at least 1 hour. Remove the gelatin solution immediately prior to use.
2. Mouse embryonic fibroblast (MEF) feeder cells are prepared from day 13 CF1 pregnant mouse and irradiated as previously described (Xu *et al.*, 2004).
3. Irradiated MEF are resuspended in MEF feeder cell culture medium at a density of 0.75×10^5 cells/ml.
4. Add the following amount of MEF to each gelatin-coated culture dish or plate: 15 ml/10 cm dish, 2.5 ml/well of 6-well plates and 0.3 ml/well of 48-well plates.
5. Incubate at 37°C overnight. Remove the MEF feeder cell culture medium immediately prior to use.

Preparing Conditioned Medium

1. T75 flasks are coated with gelatin by adding 6 ml of 0.1% sterile gelatin solution, and incubated at 37°C for at least one hour. Remove the gelatin solution immediately prior to use.
2. Irradiated MEF are resuspended in MEF feeder cell culture medium at a density of 2.12×10^5 cells/ml.
3. Add 20 ml of irradiated MEF to each gelatin-coated T75 flask, and incubate at 37°C overnight.
4. Remove the MEF feeder cell culture medium the next day, and wash the cells once with 8 ml/ T75 flask of PBS without Ca^{2+} and Mg^{2+}.
5. Add 20 ml fresh human ES cell culture medium without bFGF to each T75 MEF flask and incubate at 37°C for about 24 hours.
6. Collect the conditioned human ES cell culture medium without bFGF (CM). The CM can be stored at −80°C. To be used in reprogramming culture, the CM is sterilized by filtration following the addition of bFGF at a final concentration of 100 ng/ml, which can be stored at 4°C for up to two weeks.
7. Repeat step 5 and 6 for five days with 20 ml/ T75 flask of human ES cell culture medium without bFGF, and eight additional days with 15 ml/ T75 flask of human ES cell culture medium without bFGF.

Reprogramming Human Foreskin Fibroblasts

1. A 6-well tissue culture treated plate is coated with gelatin by adding 1.5 ml/well of 0.1% sterile gelatin solution, and incubated at 37°C for at least one hour. Remove the gelatin solution immediately prior to use.
2. Prepare 12 × 10 cm MEF dishes.
3. Grow human foreskin fibroblasts in T75 flasks in 293FT cell culture medium.
4. When human foreskin fibroblasts reach around 80–90% confluency, remove culture medium, and wash cells with 6 ml of PBS without Ca^{2+} and Mg^{2+}.
5. Add 3 ml of Trypsin/EDTA solution to the T75 flask, and incubate at 37°C for three to five minutes.
6. Add 6 ml of 293FT cell culture medium to the T75 flask to neutralize the trypsin activity.
7. Transfer the individualized foreskin fibroblasts to a 15 ml conical tube, and pellet cells by centrifugation at 200 × g for five minutes.
8. Remove the supernatant, and gently resuspend 293FT cells in 4 ml of 293FT cell culture medium.
9. Determine and adjust the cell density to about 1×10^6 cells/ml.

10. Add 2 ml/well of fresh 293FT cell culture medium to two wells of the gelatin-coated 6-well plate.
11. Add 0.15 ml/well of the resuspended human foreskin fibroblasts to the two wells prefilled with fresh 293FT cell culture medium ($\sim$1.5 $\times$ 10^5 cells/well), and incubate at 37°C.
12. The next day, when fibroblasts reach $\sim$70–80% confluency, perform the lentiviral transduction by adding 2.5 μl of polybrene (6 mg/ml stock) and the following amount of lentivirus: OCT4 ($\sim$2 $\times$ 10^6 vp), SOX2 ($\sim$8 $\times$ 10^5 vp), NANOG ($\sim$2 $\times$ 10^6 vp), and LIN28 ($\sim$4 $\times$ 10^5 vp) to one well. The other well of fibroblasts is used as a non-transduced control.
13. Incubate the fibroblasts with the lentivirus overnight at 37°C.
14. The next day (day 1 post-transduction), remove the lentivirus-containing culture medium, and wash cells with 2 ml/well of PBS without Ca^{2+} and Mg^{2+}.
15. Add 1 ml/well of Trypsin/EDTA solution to the fibroblasts, and incubate at 37°C for three to five minutes.
16. Add 5 ml/well of fresh 293FT cell culture medium to neutralize the trypsin activity.
17. Prepare the 12 $\times$ 10 cm MEF dishes by replacing the MEF feeder cell culture medium with 15 ml/dish of fresh 293FT cell culture medium.
18. Aliquot 1 ml/dish of the individualized control and lentivirus-transduced fibroblasts to the 10 cm MEF dishes, and incubate at 37°C.
19. Two days following fibroblast transfer to the 10 cm MEF dishes (day 3 post-transduction), the 293FT cell culture medium is replaced with 10 ml/dish of human ES cell culture medium with bFGF.
20. The culture medium is exchanged every two days.
21. When a significant number of MEF die off, which usually occurs around day 10 post-transduction, the human ES cell culture medium with bFGF is replaced with 10 ml/dish of the CM supplemented with bFGF.
22. The CM with bFGF is exchanged every two days.
23. The earliest human foreskin iPS cell colonies are generally visible around day 14 post-transduction. Between day 20 and 25 post-transduction, on average, there will be around 5 to 15 iPS cell colonies on each 10 cm dish, about one third of which can be picked for expansion (see below). The morphology of various iPS cell colonies and a typical non-iPS-cell colony at day 22 post-transduction are shown in Figure 10-1.

Picking iPS Cell Clones

1. Prepare a 48-well MEF plate prior to iPS cell colony picking.
2. Immediately before iPS cell colony picking, add 0.5 ml/well of human ES cell culture medium with bFGF to an empty 48-well tissue culture treated plate.

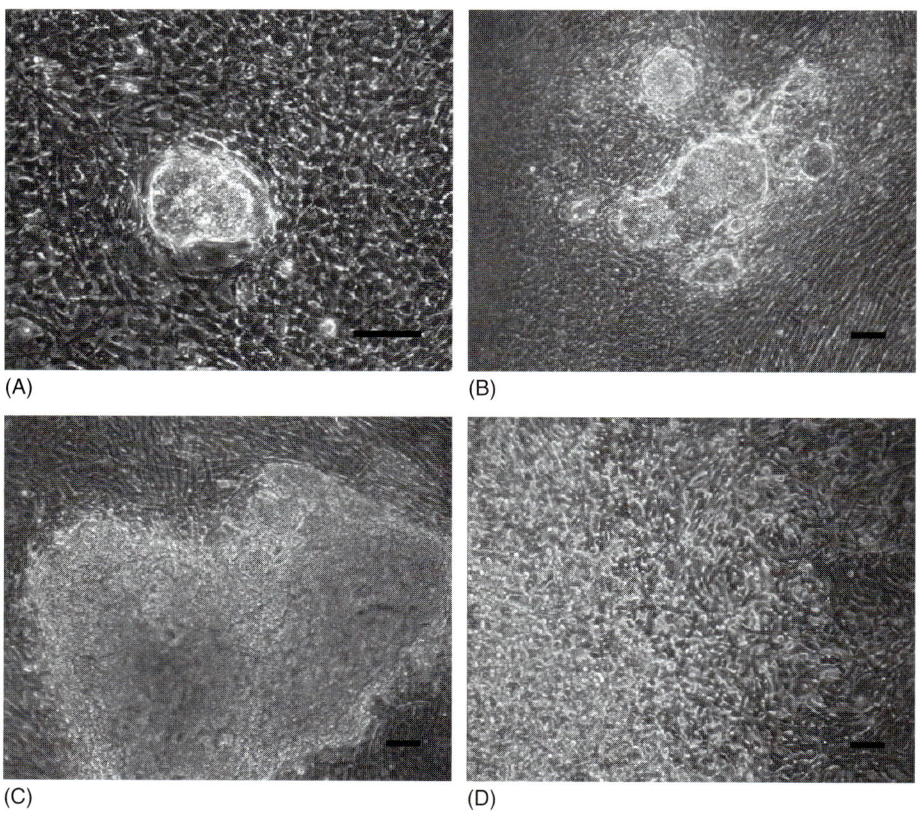

FIGURE 10-1 Morphology of typical iPS cell and non-iPS-cell colonies on day 22 post-transduction of human foreskin fibroblasts. (A) A single late-arising iPS cell colony. (B) A cluster of iPS cell colonies, which most likely originates from a single transduced fibroblast. (C) An overgrown iPS cell colony. (D) A non-iPS-cell colony. Scale bar: 0.1 mm.

3. At day 20 to 25 post-transduction, use a 20 μl pipette (set to 10 μl) to pick each iPS cell colony to each well of the 48-well plate prefilled with human ES cell culture medium with bFGF under a dissection microscope.
4. Use a 1000 μl pipette (set to 600 μl) to mechanically break up the picked iPS cell colonies by pipetting up and down three times.
5. Remove the MEF feeder cell culture medium from the 48-well MEF plate.
6. Transfer the picked iPS cell colonies to the 48-well MEF plate, and incubate at 37°C.
7. After two days of incubation, replace the culture medium with fresh human ES cell culture medium with bFGF, followed by daily exchange of culture medium.

Expanding Induced Pluripotent Stem Cell Clones

1. Prepare the appropriate number of 6-well MEF plates.
2. At day 7 to 10 following colony picking when iPS cell clones are large enough to split, remove culture medium from the iPS cells.
3. Add 0.5 ml/well of collagenase solution (1 mg/ml) to the 48-well plate.
4. Incubate at 37°C for 6–8 minutes. iPS cell colonies should curl up around the edge, but do not come off the plate.
5. Remove the collagenase solution.
6. Wash the iPS cells three times with 0.5 ml/well of DMEM/F12 basal medium.
7. Add 0.5 ml/well of fresh human ES cell culture medium with bFGF.
8. Remove MEF feeder cell culture medium from the 6-well MEF plates.
9. Add 2 ml/well of fresh human ES cell culture medium with bFGF to the 6-well MEF plates.
10. Use a 1000 μl pipette (set to 600 μl) to mechanically break up the collagenase-treated iPS cells by pipetting. DO NOT BREAK UP THE iPS CELL COLONIES TOO MUCH AT THIS STEP.
11. Transfer each iPS cell clones to each well of 6-well MEF plates, and incubate at 37°C.
12. After two days of incubation, replace the culture medium with fresh human ES cell culture medium with bFGF, followed by daily exchange of culture medium.
13. From this step on, the culture and characterization of human iPS cells can be carried out as that for normal human ES cells as previously described (Xu *et al.*, 2004).

ACKNOWLEDGMENTS

We thank Deborah Faupel for critical reading of this manuscript.

KEY WORDS

Differentiation Process of specialization wherein a cell, through changes in gene expression, acquires the correct shape, polarity, and orientation with respect to its neighbors and the extracellular matrix, and an appropriate set of organelles and proteins to enable it to carry out metabolic, signaling, transport, contractile, structural, or other specific functions in a particular tissue.

Fibroblast A cell in connective tissue that produces collagen and other fibers.

Induced pluripotent stem cells A type of pluripotent stem cell artificially derived from a non-pluripotent cell, typically an adult somatic cell, by inducing a "forced" expression of certain genes.

Lentiviral vector A type of retrovirus that can infect both dividing and nondividing cells, because their preintegration complex (virus "shell") can get through the intact membrane of the nucleus of the target cell.

Pluripotent Able to give rise to differentiated cells of all three germ layers. Cells of the inner cell mass and ES cells are pluripotent.

Teratoma A tumor composed of tissues of multiple germ layer lineages.

FURTHER READING

Lowry, W. E., Richter, L., Yachechko, R. *et al.* (2008). Generation of human induced pluripotent stem cells from dermal fibroblasts. *Proc. Nat. Acad. Sci. USA*, *105*(8), 2883–2888.

Ludwig, T. E., Bergendahl, V., Levenstein, M. E., Yu, J., Probasco, M. D., & Thomson, J. A. (2006). Feeder-independent culture of human embryonic stem cells. *Nat. Methods*, *3*(8), 637–646.

Park, I. H., Zhao, R., West, J. A. *et al.* (2008). Reprogramming of human somatic cells to pluripotency with defined factors. *Nature*, *451*(7175), 141–146.

Takahashi, K., Tanabe, K., Ohnuki, M. *et al.* (2007). Induction of pluripotent stem cells from adult human fibroblasts by defined factors. *Cell*, *131*(5), 861–872.

Yu, J., Vodyanik, M. A., Smuga-Otto, K. *et al.* (2007). Induced pluripotent stem cell lines derived from human somatic cells. *Science*, *318*(5858), 1917–1920.

Xu, C., & Carpenter, M. K. (2004). Feeder-free Culture. *Handbook of Stem Cells* – Vol. 1 Embryonic Stem Cells, *1*, 535–542.

Part III

Types and Properties of Stem Cells

Chapter 11

Molecular Bases of Pluripotency

Fatima Cavaleri and Hans Schöler

INTRODUCTION

Early mammalian embryogenesis is characterized by a gradual restriction in the developmental potential of the cells that constitute the embryo. The zygote and single blastomeres from a two to four cell embryo are totipotent. As the embryo continues to cleave, the blastomeres lose the potential to differentiate into all lineages. The blastocyst is the first landmark of the embryo in which lineage restriction is apparent. At this stage, the outer cells of the embryo compact into the trophectoderm, from which the placenta will derive. The inner cells, termed inner cell mass (ICM), will give rise to all cell lineages of the embryo proper, but cannot contribute to the trophoectoderm, and thus are considered pluripotent. Once isolated and cultured *in vitro* under permissive conditions, the ICM may be propagated as an embryonic stem (ES) cell line. As a matter of fact, these cells are the *in vitro* substitutes for embryos in the search for the genetic switches and molecular mechanisms required to ensure pluripotency. Mutations affecting the ability of ES cells to self-renew or differentiate and contribute to distinct cell lineages provide the necessary tools to unravel the molecular network underlying pluripotency.

In an attempt to define the molecular basis underlying pluripotency, we will focus on four main areas:

1. Influence of the IL6 family of cytokines on ES cell pluripotency and self-renewal;
2. Extrinsic determinants of pluripotency;
3. Intrinsic determinants (i.e., transcription factors) of pluripotency;
4. Epigenetic configuration of pluripotent cells.

CELLULAR MODELS OF PLURIPOTENCY

Three different types of pluripotent cell lines are currently available as cellular models for the study of mechanisms of pluripotency: EC, ES and EG.

EC

Embryonal carcinoma (EC) cells are historically the first pluripotent stem cells derived from mouse embryonic or fetal tissues. EC cells are self-renewing undifferentiated cells derived from teratocarcinomas—malignant tumors containing undifferentiated cells mixed with various differentiated tissues of the three primary germ layers. Teratocarcinomas, first observed in some strains of mice as naturally gonadal developing tumors, arise from pre-implantation/pre-gastrulating embryos or primordial germ cells (PGCs) that had been ectopically grafted *in vivo*. EC cells can re-form teratocarcinomas *in vivo* and differentiate *in vitro* when cultured in suspension. They provide a suitable model system to study cellular commitment and differentiation, but have a major disadvantage: EC cells are tumor cells and consequently they are typically aneuploid. Although they could integrate into a developing embryo and contribute to adult tissues, their contribution was extremely poor and lacked consistency and reproducibility. Most importantly, EC cell contribution to the germline was shown to be a very rare event.

ES

EC cells set the technical stage for derivation and handling of ES cells a few years later. In 1981 two different scientists, Evans and Martin, reported the establishment of pluripotent cell lines from blastocyst stage mouse embryos. The pluripotentiality of these cells was considered to be proven by their capability to form teratocarcinoma upon subcutaneous injection into syngenic mice, and to differentiate from embryoid bodies into tissues of all germ layers *in vitro*. Although prone to discard the Y chromosome, ES cells are in general euploid and constitute an ideal, direct *in vitro* link to and from the embryo. In 1984 Bradley showed that pluripotent ES cells could be efficiently used to alter the mouse body composition and germline throughout generations. A few years later, the basis of modern mouse developmental genetics was laid down with the generation of genetically modified mice derived from ES cells that had been manipulated by retroviral or homologous recombination methods.

EG Cells

In addition to EC and ES cells, a third type of pluripotent stem cells, called embryonic germ or EG cells, has been isolated from mouse. Migratory PGCs form colonies that are morphologically indistinguishable from ES colonies when grown on feeder cells in the presence of serum and a cocktail of growth factors, namely LIF (leukemia inhibitory factor), bFGF (basic fibroblast growth factor), and SCF (stem cell factor). Like ES cells, EG cells show full developmental capacity, being able to differentiate into derivatives of all germ layers *in vitro*, form teratocarcinoma *in vivo*, and contribute to all tissues of chimeric mice, including germline, upon injection into host blastocysts.

Human ES and EG Cells

In a way similar to mouse, pluripotent embryonic stem and germ cells have been derived from human blastocysts and PGCs. They are capable of differentiating into cells of all three germ layers *in vitro* and of forming teratocarcinoma when transplanted under the skin of a recipient mouse. Unlike their murine counterpart, human ES cells can even give rise to cells of the trophoectoderm lineage when cultured in the appropriate conditions.

In summary, pluripotent murine embryo-derived ES and EG cells exhibit unique properties which make them an extremely powerful model system to unveil the molecular basis of pluripotency both *in vivo* and in *vitro*:

- Unlimited self-renewal: they can be grown in large numbers and indefinitely passaged *in vitro*;
- Stable karyotype;
- Refractoriness to senescence, i.e., they are virtually immortal;
- Highly efficient and reproducible differentiation potential: they can give rise to derivatives of all three embryonic germ layers *in vitro* and *in vivo*;
- Germline colonization;
- Clonogenicity: they grow as separate colonies that can be expanded as independent subclones following genetic manipulations;
- High versatility to genetic manipulation without loss of pluripotency, as introduction of foreign DNA does not affect their ability to be fully integrated into the founder tissue of a host embryo or to colonize its germline.

THE EMBRYONIC STEM CELL ENVIRONMENT: CYTOKINES AND PLURIPOTENCY

The Stem Cell Niche

In many cases the culture of stem cells has been complicated by the lack of knowledge of their cellular environment or niche. Hematopoietic stem cells, for example, were identified more than 40 years ago, but conditions still have to be established to ensure their maintenance *in vitro*. Additionally, they might require complicated three-dimensional arrangements of specific stromal cells in order to proliferate.

The ES Cell Niche

In contrast, ES cells are relatively easy to be handled technically. The first murine embryonic stem lines were established by culturing early embryos on a layer of mitotically inactivated mouse embryonic fibroblasts. Without such a "feeder" layer, cultured embryonic cells would not remain pluripotent, *suggesting that fibroblasts either promote self-renewal or suppress differentiation, or both*.

LIF and Other Cytokines

Embryonic fibroblasts maintain pluripotency of ES cells by secreting a factor, which was identified as leukemia-inhibiting factor, also known as differentiation inhibiting activity (DIA). LIF is a member of the interleukin-6 family of cytokines, including IL-6, oncostatin M (OSM), ciliary neurotrophic factor (CNTF) and cardiotrophin-1 (CT-1). The IL-6 family cytokines are structurally and functionally related. They act on a variety of cells, i.e., they are pleiotropic, and can mediate proliferation or differentiation or both, according to the target cell types. For example, LIF, OSM and IL-6 are competent to induce myeloma growth and to inhibit macrophage differentiation of M1 cells. The redundancy in biological function is mainly due to the structural similarity of the receptor complex involved in signal transduction (see below).

Redundant Cytokine Functions

The absence of a developmental phenotype in IL-6-, LIF-, and CTNF-null mice confirms that IL-6-related cytokines are indeed functionally redundant. However, LIF mutant females are infertile, as the interaction between embryo and uterine wall (decidual reaction) strictly depends on a surge of estrogen on the fourth day of gestation, which coincides also with a surge in LIF production by the uterus. As a consequence, LIF−/− females fail to support embryo implantation. However LIF−/− embryos can implant and develop to term in a normal uterus.

ES self-renewal dependence on cytokine supply can be attributed to several factors. LIF may influence the rate of cell proliferation or cell cycle progression and act on the stem cell phenotype by activating a signaling cascade which operates on the up- or/and down-regulation of genes that are exclusively expressed in "pluripotent" and differentiated cells, respectively. A complete and systematic analysis of the target genes lying downstream of the LIF-induced signaling pathways is necessary to clarify the cytokine *modus operandi* on the ES cell phenotype.

CYTOKINE-RECEPTOR BINDING ON ES CELLS: MULTIPLE RELAY STATIONS

Receptor Structure

The receptors involved in the IL-6 family cytokine signaling cascade belong to the cytokine receptor class I family. The extra-cellular domain of all members of cytokine receptor class I family is composed of a variable number of fibronectin type III modules. Two of the fibronectin modules are conserved among all members of the family and constitute the cytokine-binding module. The cytoplasmatic domain of the receptor contains three conserved motifs, called box 1, box 2, and box 3 in a membrane proximal to distal order, and lacks intrinsic kinase activity. These three subdomains are responsible for transmitting the extracellular signal into the cytoplasm.

Receptor Dimerization

Binding of the IL-6 family cytokine to their cognate receptors leads to homodimerization of gp130 or heterodimerization of gp130 with the cytokine cognate receptor. For example, IL-6/IL6R and IL-11/IL-11-R complexes induce gp130 homodimerization. Both LIF and CT-1 bind to LIFR and induce LIFR/gp130 heterodimerization. CTNF engages LIFR/gp130 as a signaling competent complex through an association with CTNF/CTNFR, whereas OSM engages the LIFR/gp130 or OSMR/gp130 by binding to the gp130 portion of the heterodimers (see Figure 11-1). All the described receptor complexes share gp130 as the common component critical for signal transduction, which explain the observed redundancy in cytokine functions.

Expression of Cytokine Receptors

Although gp130 is widely expressed in various tissues, ligand-specific receptor components display a more restricted expression. In the mouse blastocyst LIF and LIFR have a reciprocal pattern of expression: the cytokine is expressed in the trophectoderm and the receptor in the ICM.

LIFR, OSMR, and CTNFR are expressed in ES cells, consequently CT-1, OSM, CNTF, and LIF are interchangeable in supporting ES cell derivation and maintenance of ES cells in culture. IL-6 and IL-11 cannot substitute for CT-1, OSM, and CNTF as the IL-6 and IL-11 receptors are not expressed in ES cells. However, IL-6 can prevent ES differentiation if delivered in conjunction with a soluble form of the IL-6 receptor, which retains ligand binding activity and capability to induce gp130 homodimerization (see Figure 11-1).

Cytokine Receptor Function *In Vivo*

LIFR-null embryos die shortly after birth, and exhibit reduced bone mass and profound loss of motoneurons. Embryos homozygous for the gp130 mutation die between 12 and 18 days post coitum (dpc), because of placental, myocardial, hematological, and neurological disorders. CTFR deficient mice exhibit perinatal death and display profound motor neuron deficits.

Gp130 and LIFR Function in Diapause and ES Cells

The late embryonic lethality of gp130−/− fetuses is in conflict with gp130 requirement for ES cell self-renewal

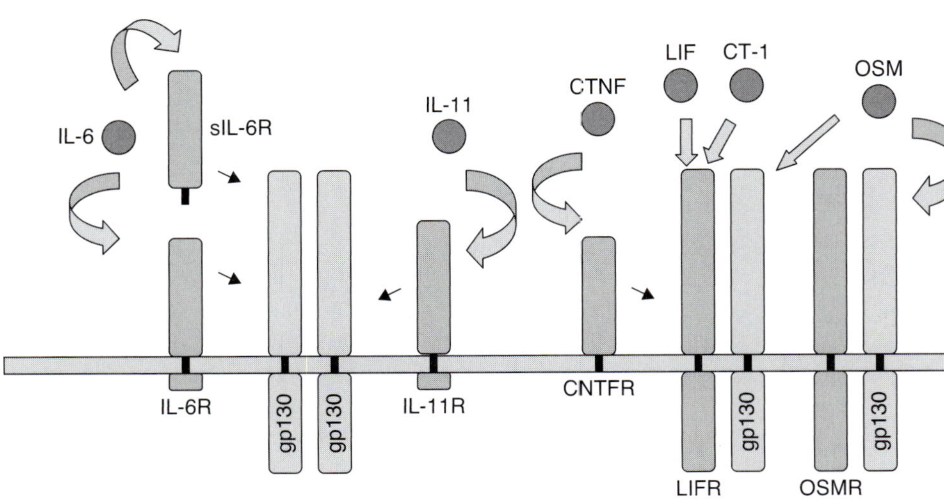

FIGURE 11-1 Schematic structure of the cytokine receptor complexes sharing gp130 as a common subunit. IL-6/IL6R and IL-11/IL-11-R complexes induce gp130 homodimerization. Both LIF and CT-1 bind to LIFR and induce LIFR/gp130 heterodimerization. CTNF engages LIFR/gp130 as a signaling competent complex through an association with CTNF/CTNFR, whereas OSM engages the LIFR/gp130 or OSMR/gp130 by binding to the gp130 portion of the heterodimers.

in vitro. Since delayed or quiescent blastocysts were used in the initial experiments of ES cell derivation, the possibility that cytokine receptors may have an embryonic function in the quiescence embryo state was investigated. Lactating females can conceive while still nursing their pups, but cannot support blastocyst implantation because they do not produce estrogen at the fourth day of gestation. Consequently, embryonic development is arrested and resumes under favorable conditions for optimal development. This phenomenon, termed diapause, can be artificially induced by ovariectomy after fertilization. The embryos reach the blastocyst stage, hatch from the zona pellucida and float *in uterus* in a quiescent status for up to four weeks. In this scenario, the epiblast, which normally preserves its pluripotent status for about three days (from 3.5 dpc, when it forms, up to 6.5 dpc, when gastrulation starts), can be maintained for longer time periods and resumes development when an estrogen-rich environment is established.

LIFR−/− and gp130−/− delayed embryos are unable to resume embryogenesis after twelve and six days of diapause, respectively. The number of cells that constitute the ICM of delayed gp130−/− blastocysts gets gradually reduced by apoptosis during a six-day period of diapause. Moreover, ICMs isolated from delayed gp130 null blastocysts cannot form a pluripotent outgrowth *in vitro*, as they differentiate exclusively into parietal endoderm.

Thus, it appears that maintenance of the epiblast during diapause is temporally dependent on different cytokines and that gp130 plays a more critical role than LIFR in this process. Two models may explain why epiblast cells enter apoptosis in the absence of gp130 signaling:

1. gp130 relays a cell survival signal from the extracellular compartment to the nucleus. This model is supported by the anti-apoptotic activity of the transcription factor STAT3 in a variety of cells.
2. gp130 suppresses epiblast differentiation. In the absence of gp130 signal, epiblast cells may differentiate inappropriately (as shown by the endoderm formation solely), and consequently die.

SIGNAL TRANSDUCTION: CASCADES TO THE STEM CELL NUCLEUS

Homo- or heterodimerization of gp130 results in the activation of receptor-associated kinases of the Janus family Jak1, Jak2 and Tyk2. These tyrosine kinases constitutively interact with the conserved regions box 1 and box 2 of gp130. The receptor complex is inactive until ligand-induced receptor dimerization brings the associated Jak kinases within sufficient proximity to allow *trans*phosphorylation and activation of the kinase catalytic domain. Activated Jaks phosphorylate specific tyrosines on the intracellular domain of gp130, creating docking sites for the recruitment of SH2 proteins to the activated receptor complex. When gp130 is phosphorylated, several signaling pathways, involving STAT1, STAT3, the SH2-domain containing tyrosine phosphatase (SHP2), ERK1/ERK2 (extra-cellular signal receptor kinases) or mitogen activated kinases (MAPK), and phosphatidylinositol-3 kinase (PI3K), are activated.

STATS: LATENT TRANSCRIPTION FACTORS TRANSMITTING SIGNALS

The STAT Family

STAT proteins belong to a group of latent cytoplasmic transcription factors that play a central role in transmitting signals from the cytoplasmic membrane to the nucleus, hence their name (signal transducers and activator of transcription). Seven major STAT proteins have thus far been identified in mouse (STAT1 to STAT6, including STAT5a and STAT5b). With the exception of STAT4, which is restricted to myeloid cells and testis, STAT factors are ubiquitously expressed. They are activated in many cell types by a broad range of cytokines, growth factors and interferons (IFNs) and they are substrates for tyrosine kinases of the Src and Jak families.

Structure

STAT proteins share several conserved structural and functional domains. A tetramerization and a leucine zipper-like domain are located at the amino terminus, followed by a DNA binding domain, a Src homology domain 3-like region (SH3, proline rich motif binding domain), a Src homology domain 2 (SH2), a critical site of tyrosine phosphorylation (Y705 in STAT3), and a carboxy-terminal transactivation domain. No evidence has emerged so far to suggest SH3 function.

The SH2 domain plays three important roles:

- Recruitment to activated receptor complexes
- Interaction with Jaks
- STAT dimerization and DNA binding.

Regulation

The regulation of STAT signaling is mostly post-translational and involves both tyrosine and serine phoshorylation. Phosphorylation of the conserved tyrosine (Y701 for STAT1, Y705 for STAT3) results in dimerization of STAT1/STAT3 through the intermolecular interaction of the SH2 domains and the domain containing the phosphorylated tyrosine. STAT1 and STAT3 homo and heterodimers translocate to the nucleus, where they activate gene transcription by binding to specific DNA sequences. Consistent with the requirement for dimerization-induced activation, the consensus binding sites are symmetrical dyad sequences.

STAT3

STAT3 was originally identified as an acute response factor which, upon IL-6-family cytokine stimulation, induces the expression of a variety of genes, referred to as acute response genes, whose expression dramatically increases with tissue injury and inflammation.

Targeted disruption of STAT3 gene *in vivo* leads to embryonic lethality. STAT3 null embryos develop into elongated egg cylinders, but degenerate at around E7.0. At this stage wild-type embryos start to express STAT3 in the visceral endoderm. Embryonic lethality is then explained as a consequence of the failure in establishing metabolic exchanges between embryo and maternal blood. Thus, STAT3 plays a crucial role during embryonic development that cannot be compensated by other members of the STAT family.

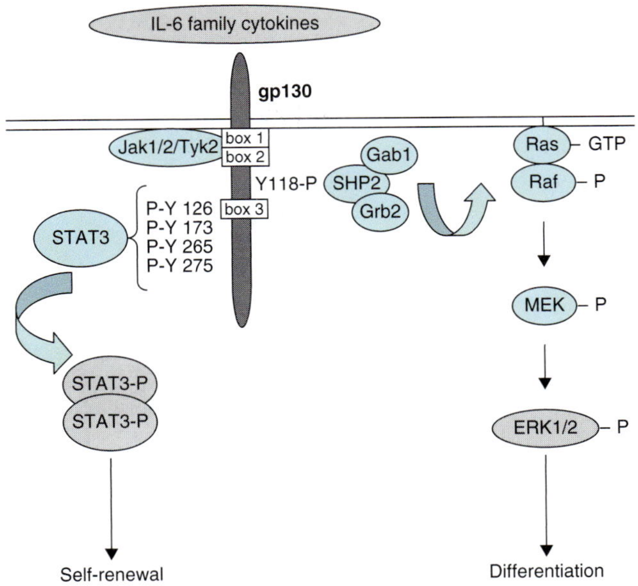

FIGURE 11-2 Schematic description of the signaling pathways induced by the IL-6 family cytokines. Following gp130 hetero- or homodimerization, activated JAKs phosphorylate the intracellular domain of gp130 on Y126, 173, 265, 275, and 118. STAT3 association with gp-130 phosphorylated Y126-275 leads to STAT3 phosphorylation, dimerization and translocation to the stem cell nucleus. Association of SHP-2 with phosphorylated gp-130 Y118 leads via adaptor proteins to activation of the Ras pathway and translocation of ERK1/2 to nucleus. STAT3 activation induces ES self-renewal, whereas ERK activation causes cell differentiation.

LIF and STAT3

The initial studies conducted on LIF-dependent transcriptional activation in embryonic stem cells showed induction of a DNA binding activity that correlated with LIF treatment. Steady state levels of STAT1, 3, 5, and 6 were unaffected by LIF treatment. In contrast, co-immunoprecipitation experiments indicated STAT3 as being a component of the tyrosine-phosphorylated complex formed upon LIF induction. In addition, tyrosine kinase inhibitors were shown to impair formation of the activated STAT3 complex and to alter undifferentiated ES cell morphology versus a differentiated phenotype.

STAT3, ES Cell Renewal and the Undifferentiated State

Identification of STAT3 as a key determinant of ES renewal came from the elegant studies conducted in Smith's and Yokota's laboratory at the end of the 1990s. Granulocyte colony-stimulating factor receptor (G-CSF-R) belongs to the class I cytokine receptor family and is not expressed in ES cells. In order to characterize the functional role of the gp130/LIFR intracellular domains or residues in signal transduction, chimeric receptors constituted of the extra-cellular domain of G-CSF-R fused to the transmembrane and cytoplasmic region of gp130 or LIFR were engineered. G-CSFR/gp130 and G-CSF-R receptors can support ES self-renewal in the presence of G-CSF at high and low efficiency, respectively, whereas the G-CSF-R/LIF-R chimera does not support formation of stem cell colonies. This result suggests that gp130 is an essential component in signaling self-renewal in ES cells and that G-CSF-R and gp130 intracellular domains activate common signaling pathways.

The intracellular domain of gp130 harbors four consensus motifs, **YXXQ**, whose phosphorylated tyrosines were shown to function as docking sites for STAT3. Single mutation of these tyrosine residues (Y126, 173, 265, and 275, enumeration starting from the transmembrane domain) did not affect self-renewal appreciably (see Figure 11-2). Rather, mutations of Y265/275 completely abolished STAT3-induced binding activity and formation of undifferentiated colonies. These data demonstrate that STAT3 docking sites are essential in mediating transmission of the signal from gp130 to STAT3 in self-renewing ES cells, although the tyrosine residues are not functionally equivalent. Moreover, conditional expression of STAT3F, containing a tyrosine 705 into phenylalanine substitution, caused complete differentiation of ES cells even in the presence of LIF.

In order to uncouple STAT activation from any signaling pathway induced at a membrane receptor by extra-cellular factors, Matsuda and collegues constructed a fusion protein composed of the STAT3 entire coding region and the ligand-binding domain of the estrogen receptor. The STAT3ER chimeric protein specifically translocates into the nucleus in the presence of the synthetic steroid ligand, 4-hydroxytamoxifen (4OHT). Parental ES and ES expressing STAT3ER were grown in the presence of LIF or 4OHT. The compact colonies formed by STAT3ER ES in the presence of either LIF or 4HT clearly demonstrated that STAT3 activation is sufficient to maintain the undifferentiated phenotype of

ES cells. Most importantly, besides being morphologically undifferentiated, cells grown for one month in 4OHT widely contributed to all tissues of chimeric mice.

STAT3 Independent ES Cell Renewal

The striking paradox created by STAT3- or gp130-null embryo phenotypes and the strict dependence on gp130-STAT3 signaling pathways for maintenance of pluripotent ES cells is suggestive of the existence of alternative pathways governing pluripotency *in vivo*. Dani and colleagues presented some evidence of STAT3-independent signaling pathways operating in self-renewing ES cells. When ES cells are grown at high density in the absence of LIF, the newly differentiated cells start synthesizing LIF, which in turn allows the expansion of undifferentiated cells. In order to eliminate the LIF-dependent pathways of self-renewal, a LIF deficient ES cell line was generated. Strikingly, LIF null ES cells can still produce undifferentiated colonies, though at lower efficiency than wild type or heterozygous lines when induced to differentiate. A differentiated, LIF-deficient, parietal endoderm-like cell line was derived and shown to support ES cell propagation via production of a soluble activity named ES cell renewal factor (ESRF). ES cells grown in the presence of ESRF retain their pluripotent character, as they can still contribute to any tissues upon blastocyst injection. ESRF does not operate via LIFR or gp130, as it is effective on LIFR deficient ES cells and is not blocked by anti-gp130 antibodies. Additionally, STAT3 is not induced in the presence of ESRF.

SHP-2/ERK SIGNALING

STAT 1 and 3 are only two of the downstream effector molecules induced by cytokine signaling via gp130. LIF treatment of ES cells increases MAP kinase activity and induces phosphorylation of ERK1 and ERK2. The bridging factor between cytokine receptor and MAP kinase is the widely expressed tyrosine phosphatase SHP-2.

SHP-2 Structure

SHP-2 contains two N-terminal SH2 domains and a C-terminal catalytic domain. SHP-2 interacts with the intracellular domain of Gp130 through phosphorylated tyrosine 118, located inside the consensus **YSTV** sequence. Recruitment to the activated receptor induces SHP-2 phosphorylation, which leads to an increase in phosphatase activity by preventing an intramolecular interaction between the SH2 and the catalytic domain of SHP2. In BAF-B03 cells, mutation of gp130 tyrosine 118 into phenylalanine was shown to block SHP-2 phosphorylation and induction of ERK2 activation.

SHP-2-ERK Signaling and ES Cells

G-CSF responsive transgenic ES cells provided an excellent tool for establishing the role played by the SHP-2-ERK signaling in ES cell propagation.

A mutant G-CSFR-gp130 chimeric receptor that cannot be phoshorylated on Y118 (and that consequently cannot induce SHP-2 phosphorylation) does not impair the self-renewal of ES cells (see Figure 11-2). In contrast, the mutated chimera makes ES more sensitive to G-CSF, as they can be maintained at a lower concentration of G-CSF (1000-fold less) than the one required by ES cells expressing the unmodified receptor. Moreover, the signal started at the phosphorylated Y118 mediates attenuation of activated STAT3, as cells expressing the G-CSFR-gp-130 mutant chimeric receptor show a slower decay of phosphorylated STAT3.

Enhanced ES self-renewal is observed when either a catalytically inactive SHP-2 is overexpressed or ERK phosphorylation is chemically blocked. These results indicate that SHP-2/ERK activation is not required for the maintenance of self-renewal signaling, they rather inhibit it. This conclusion is confirmed by enhanced LIF sensitivity and increased proliferation rates observed in ES cells, and embryoid bodies derived thereof, expressing a ΔSH2-SHP-2 protein. However, cardiac/epithelial differentiation of SHP-2 mutant cells is inhibited and delayed, indicating that SHP-2 plays a positive role in ES differentiation. In conclusion, ES cell self-renewal is a consequence of the precise balance of antagonistic signaling pathways.

SHP-2 Communicates With ERKs

Activated ERKs undergo nuclear translocation, which enables them to modulate the activities of transcription factors that govern proliferation, differentiation, and cell survival. SHP-2 interaction with two adaptor molecules, Grb2 and Gab1, provides two alternative routes that couple cytokines to the activation of ERK pathways. SHP-2/Grb2 association leads to Ras activation through the GTP-GDP exchange protein SOS. SHP-2 has also been shown to associate with the scaffold Grb2-associated binder protein Gab1 and PI3-kinase and activate ERK MAP kinases via a Ras dependent pathway.

Grb2 and Gab1 Adaptor Proteins

Like most adaptor proteins, Grb-2 contains SH2 and SH3 domains. The SH2 domain mediates Grb2 binding to SHP-2 while the SH3 domain mediates interaction with SOS. Tyrosine phosphorylation of Gab1 following stimulation with cytokines depends on the gp130 site (Y118) docking SHP-2, but is independent from the gp130 C-terminal domain interacting with STAT3. Gp130-mediated ERK2 activation is enhanced by Gab1 expression and inhibited by a dominant negative Ras.

Functional Ablation of Grb2 in Embryos and ES Cells

Null mutation of the Grb2 gene leads to embryonic lethality at around 7.5 dpc. The differentiation potential of Grb2-null ICM cells is compromised as cultured blastocysts lack visceral and parietal endodermal cells. The ability of Grb2 to support endoderm differentiation is abrogated by mutation in either the SH2 or SH3 domain. Interestingly, transformation of Grb2-deficient ES cells with an activated variant of Ras restores endoderm differentiation, indicating that the Grb2-Ras-MAPK pathway is essential for early specification of the endoderm tissues.

FGF4 AND ERK

Fibroblast growth factor 4 (FGF4) is produced by ICM cells, and was initially postulated to function in an autocrine fashion to promote the proliferation and expansion of the ICM. This hypothesis derived from two different observations:

1. FGF4- and fibroblast growth factor receptor 2(FGFR2)-null embryos cannot form an egg cylinder and die soon after implantation.
2. When cultured *in vivo*, both FGF4- and FGFR2-null ICMs fail to expand and eventually degenerate.

In contrast to the embryo ICM or epiblast, FGF4−/− ES cells are able to proliferate in the absence of the growth factor, suggesting that *in vitro* FGF4 is not required for ES cell proliferation.

ES stimulation with FGF4 results in massive increase in Erk1/2 phosphorylation, whereas ES culture in the presence the FGFR receptor inhibitor SU5402 elicits a strong decrease in phospho-Erk1/2 levels, indicating that FGF4 constitutes the major stimulus for the activation of ERk signaling in mouse ES cells.

Since FGF4 is produced in an autocrine fashion by undifferentiated ES cells in order to establish the role of FGF4 autocrine production on ES self-renewal and commitment, FGF4 signaling was abrogated by treatment with chemical inhibitors of the FGF receptor or by genetic ablation of either FGF4 or Erk2. In all three cases, interference with the FGF signaling did not affect propagation of undifferentiated ES cells. However, Fgf4- and Erk2-null ES cells are unable to differentiate along the neural or mesodermal lineage, and display sustained expression of *Oct4* and *Nanog*. Addition of FGF4 to FGF4-null ES cells abrogates the blockade in neural/mesoderm differentiation, indicating that FGF/Erk stimulation is directly responsible for forcing ES cells to exit self-renewal and for inducing neural/mesoderm lineage commitment.

WNT SIGNALING

The canonical Wnt pathway is initiated upon Wnt protein binding to the Frizzled receptor on the cell membrane. Activation of the pathway leads to inhibition of Glycogen Synthase Kinase 3 (GSK3), which functions in inducing the proteasome-mediated degradation of cytoplasmic β-catenin. Thus, GSK3 inhibition results in increased nuclear accumulation of β-catenin, which activates the transcription of target genes in association with T-cell-specific factors (Tcfs). Although LIF/STAT3 signaling is crucial for mouse ES maintenance, it does not appear to have a role in hES propagation. Sato and collegues were able to demonstrate that activation of the Wnt pathway is sufficient to maintain self-renewal of both mouse and human ES cells. They used 6-bromoindirubin-3′-oxime (BIO), a pharmacological intracellular inhibitor of GSK3, to activate the Wnt pathway in both mouse and human ES cells. BIO-treated ES cells display nuclear accumulation of β-catenin, which demonstrates that ES cells are capable of transducing Wnt signaling. Additionally, BIO-activated Wnt signaling results in sustained expression of the pluripotent markers Oct-4, *Nanog*, and Zfp-42 *Rex1*. Notably, in short-term cultures, BIO can substitute for LIF and MEF-conditioned medium in the propagation of pluripotent mouse and human ES cells respectively.

LIF and Wnt Signaling Converge on a Common Target

Stat3 is required for the gp-130 dependent activation of the *c-Myc* gene in pro-B cell lines. Catwright and collegues showed that *c-Myc* is a STAT3 target gene in ES cells as well. STAT3 binding to *c-Myc* promoter depends on LIF. However, MYC protein is degraded upon GSK3-dependent phosphorylation of Threonin 58 (T58). ES cells that conditionally express a T58A mutant protein can be propagated in the absence of LIF, while cells expressing a dominant negative MYC protein with decreased activity are prone to differentiation. These results indicate a requirement for a stable form of MYC, and consequently for inhibition of T58 phophorylation during ES self-renewal, which is consistent with the observation that differentiating ES cells display an increase in GSK3 activity as early as one day after LIF removal. LIF/STAT3 and Wnt signaling converge thus on *c-Myc*, by ensuring respectively gene transcriptional activation and stabilization protein.

THE TGFβ SIGNALING PATHWAY

The TGFβ superfamily of ligands includes nearly 30 proteins in mammals, e.g., TGFβ, Activins, Nodal, GDF, and BMPs. Smad1/5/8 transduce signals from Bmps and GDFs ligands, whereas Smad2/3 transduce signals from TGFβ,

Activins, and Nodal. Upon activation by phosphorylation and association with a common Smad4, the receptor-activated Smads translocate to the nucleus to regulate gene expression in concert with other transcription factors. Smad6 and Smad7 are inhibitory SMADs.

BMP4

As described before, LIF is sufficient to guarantee mouse ES self-renewal in a serum-enriched growth environment. However, when ES cells are cultivated in serum-free medium containing LIF, they tend to spontaneously undergo neural differentiation. This indicates the existence of one or multiple growth factors contained in fetal calf serum (FCS), which cooperate with LIF in the maintenance of ES cell pluripotency. Ying and collegues identified BMP4 as the growth factor that acts in combination with LIF to sustain self-renewal, and to prevent/block neural differentiation of ES cells in serum-free cultures. This was an unexpected result, as Bmp4 alone promotes differentiation into non-neural-fates. Bmp4 induces phosphorylation of Smad1/5/8 which, in association with Smad4, translocates into the nucleus to activate expression of the Inhibitor of differentiation (Id) protein. This in turn blocks neural differentiation by functionally antagonizing neurogenic bHLH factors (Figure 11-3). Forced expression of Id eliminates BMP or serum dependence and allows self-renewal in the presence of LIF alone. However, Id-expressing ES cells cannot differentiate into the neural lineages, further confirming the Bmp-4 induced lineage-specific block. When added to hES, BMP4 triggers differentiation into primitive endoderm and trophoblast lineages.

Activin/Nodal

In an alternative study, Ogawa *et al.* showed that in serum-free culture exogenous activin or Nodal, but not Bmp or Tgfβ, significantly enhance ES propagation without affecting pluripotency. Over-expression of Smad7, a specific inhibitor of both activin-Nodal and Bmp signaling, decreases ES proliferation, whereas Smad6 overexpression, which specifically inhibits Bmp signaling, does not elicit such an effect. Although the function of Smad7 in mouse embryogenesis is still unknown, these data are supported by the analysis of Nodal function *in vivo*. Nodal ablation leads, in fact, to a substantial reduction of the embryo epiblast size and to low *Oct4* expression at 6.5 dpc. The results presented by Ogawa *et al.* are in contradiction with those reported by Ying *et al.* This inconsistency is probably due to the fact that different serum-free culture conditions were used in the two studies.

Activation of TGFβ/activin/nodal signaling via SMAD2/3 was also shown to be required for maintenance of the undifferentiated state of human ES cells. Additionally, hES differentiation is accompanied by a decrease in SMAD2/3 signaling and by activation of SMAD1/5 signaling, which is instead inhibited in undifferentiated cells. Activin, MEF conditioned medium and BIO all sustain high levels of Smad2/3 phosphorylation and *Oct4* expression in hES cells, indicating that Smad2/3 activation is the common denominator to the three signaling pathways in ensuring hES cell identity.

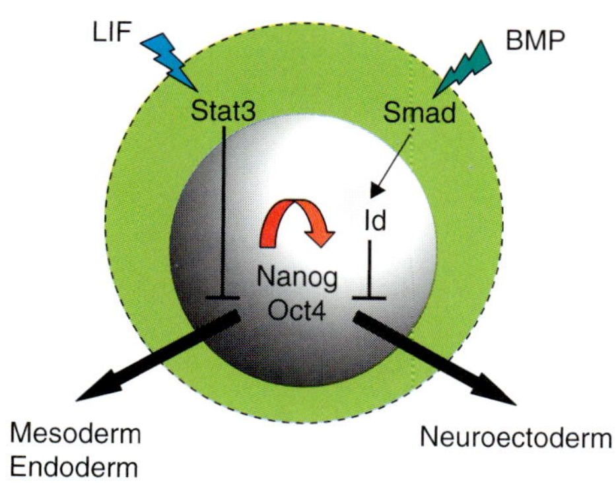

FIGURE 11-3 In serum-free conditions BMP and LIF collaborate in maintaining ES cell self-renewal by suppressing multi-lineage differentiation. Bmp accomplish this role by inducing Id gene expression via the Smad pathway. Adapted from Ying *et al.*, 2003.

bFGF and TGFβ Signaling in hES Cells

Human ES cells can be maintained in an undifferentiated status in the presence of basic fibroblast growth factor (bFGF) when grown on fibroblast feeder layers or in fibroblast-conditioned medium. They can be cultured in feeder-free conditions as well, but only in the presence of a combination of bFGF and Noggin, an inhibitor of Bmp4. Strikingly, in order to condition medium for hES growth/self-renewal, MEF must also be cultured with bFGF. Genome-wide expression profiling of supportive (grown with bFGF) and non-supportive MEF cells (grown without bFGF) revealed that bFGF stimulation regulates the expression of key members of the TGF-β pathway, including Inhba, Tgfb1, Grem1 (a Bmp4 inhibitor), and Bmp4. In particular, Tgfb1, Inhba, and Grem1 are up-regulated, whereas Bmp4 is down-regulated in bFGF-treated MEF. As these same genes are also targets of bFGF regulation in hES cells, these results suggest that FGF and TGF-β signaling operate in a concerted manner to achieve hES cell self-renewal.

THE PI3K/AKT SIGNALING PATHWAY PARTICIPATES IN ES CELL PROLIFERATION, SURVIVAL, AND PLURIPOTENCY

PI3K phosphorylate phosphatidylinositol-3,4-diphosphate (PIP2) into the second messanger PIP3, which signals to downstream effectors including AKT/PKB, a serine/threonine kinase implicated in the regulation of cell cycle progression, cell death, and tumorigenesis. The lipid phosphatase and tumor suppressor PTEN antagonizes AKT activation by dephosphorylating PIP3 into PIP2. Involvement of the PI3K signaling pathway in ES cell propagation was first proposed through the analysis of ES cell behavior after PTEN deletion. PTEN-null ES cells display increased proliferation rate/cell survival and tumorigenicity. They also have enhanced AKT/PKB activity, and consequently elevated amounts of PIP3. In contrast, double null pten/AKT ES cells revert to the proliferation rates of wild type ES cells, indicating that *Akt* up-regulation is the main cause of the tumorigenic/proliferation phenotype observed in PTEN-null ES cells. Treatment of ES cells with the PI3K inhibitor LY294002 inhibits proliferation and induces cell cycle arrest at G0/G1. Additionally, genetic deletion of the small GTPase Eras, which is an activator of PI3K/AKT signaling, reduces the endogenous proliferative and tumorigenic potential of ES cells.

The PI3K signaling pathway is activated and/or induced by LIF treatment, as assessed by measurements of PI3K-dependent modifications, e.g., phosphorylation of serine 473 on PKB and of serine 21/9 on GSK3β. Chemical inhibition of PI3K and AKT induces differentiation of, respectively, mouse and human ES cells by enhancing activation of the MAP kinases ERK1/2. This result suggests that, besides having an impact on ES proliferation, the PI3K/AKT signaling cascade may have a potential role in the maintenance of ES cell pluripotency. Watanabe and collegues showed that a myristoylated active form of AKT is capable of maintaining mouse and primate ES cells undifferentiated in the absence of LIF and MEF cells, respectively. AKT/PKB inhibits GSK3 through the phosphorylation of Ser-9. However, AKT signaling maintains cells in the undifferentiated state independently of Wnt/β catenin and STAT3/ERK signaling, as myr-AKT expressing ES cells show neither nuclear β-catenin accumulation nor increased Stat3 phophorylation. AKT promotes a G1 to S phase transition by regulating cyclin D expression, and by facilitating formation of cyclin/CDK complexes. ES cells are unique for going through a very short G1 phase, owing to absence of the cell cycle inhibition mediated by the CDK inhibitor Ink4 and tumor suppressor RB. Since cell cycle inhibitory mechanisms are acquired during differentiation, then AKT may exert its role on pluripotency by blocking the ability to gain those inhibitory mechanisms. However, cyclin D overexpression does not maintain ES cells in an undifferentiated state, indicating that regulation of cell cycle alone cannot explain the effect of AKT on ES pluripotency.

THE NEW PARADIGM OF EMBRYONIC STEM CELL RENEWAL

Unlimited proliferation of mouse ES cells *in vitro* is guaranteed by the presence of LIF and BMP or serum. Interestingly, ICM-derived cells remain pluripotent in the embryo *in vivo*, even when the LIF/Stat3 signaling is genetically inactivated. The explanation of this apparent paradox was finally given by the investigations conducted in Smith's laboratory. According to the model proposed in the study published in May 2008, LIF and serum or BMP do not have an instructive role in stem cell renewal, but rather they function by imposing a block on differentiation-inducing signals.

Since differentiation is characterized by auto-inductive stimulation of the MAPK/ERK pathway by FGF4, SU5402 and PD184352 were used to inhibit FGF receptor tyrosine kinases and the ERK cascade, respectively. Undifferentiated ES cells can be passaged in the presence of both inhibitors, although occasional neuronal differentiation is observed. Low levels of phospho-ERK are detected when cells are grown with SU5402 and PD184352. However, the two inhibitors compromise the viability and growth rate of ES cells, as indicated by poor survival at clonal density and after passage. Alone, CHIR99021, a selective inhibitor of GSK3, enhances survival at low density, but also induces non-neural differentiation. Crucially, the combination of the three inhibitors supports *de novo* ES derivation and efficient propagation of undifferentiated ES cells even at low cell density. Additional validation to the role that FGF signaling has on stem cell renewal is provided by fgf4-null ES cell propagation in CHIR99021 alone. Thus, at the ground state, ES cells are doomed to spontaneous differentiation by auto-inductive stimulation of the ERK pathway. Self-renewal is ensured by elimination of ERK signaling. SU5402 and PD184352 enable self-renewal by inhibiting ERK activation; LIF and BMP counteract signaling emanating from activated ERK1/2 (see Figure 11-4).

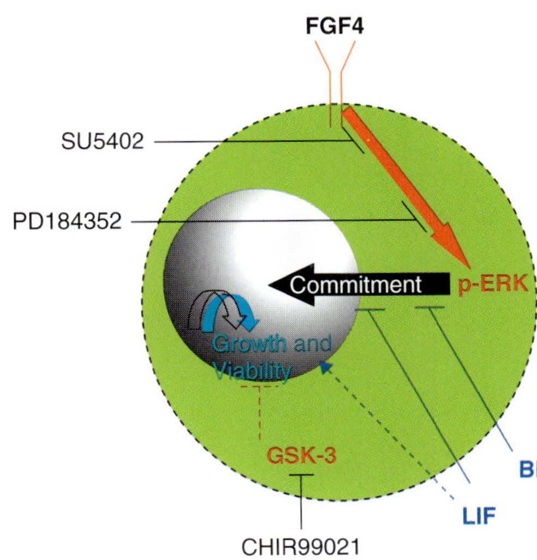

FIGURE 11-4 Schematic representation of the replicative pluripotent state in different culture conditions. Phospho-ERKs are the source of differentiation-inducing signals. Self-renewal results from either the upstream chemical inhibition of ERK activation by SU5402 and PD184352 or from LIF/Bmp-mediated inhibition of signaling emanating from phospho-ERKs. In the first scenario inhibition of GSK3 is necessary in order to augment ES cell growth and viability. Adapted from Ying *et al.*, 2008.

INTRINSIC DETERMINANTS OF PLURIPOTENCY

The phenotype of a particular cell type is the culmination of a well-defined and unique gene expression pattern, characterized by the activation of some genes and repression of others. Therefore, elucidation of the molecular basis of pluripotency is based primarily on the identification of the key transcription factors involved in regulating gene expression at the pluripotent "state." In a very simplistic model, genes that promote and maintain an undifferentiated cellular state would be expressed in pluripotent cells, while those that are activated during stem cell differentiation would be repressed. Conversely, in terminally differentiated cells, pluripotency-related genes would be silenced, whereas those required for the differentiated cell state would be expressed. Elements that influence the transcriptional regulation of gene expression include epigenetic modifications of the genome, such as methylation and histone deacetylation, which determine what types of interactions are allowed between DNA and transcription factors and modulating cofactors. A combination of cell culture and gene ablation studies in mouse embryos has led to the identification of the intracellular determinants of ICM and ES pluripotency.

OCT4 IS A KEY TRANSCRIPTION FACTOR FOR PLURIPOTENCY

Oct4 is a transcription factor belonging to the class V of POU factors. The POU family of transcription factors binds to the octamer motif ATGCA/TAAT found in the regulatory domains of cell type-specific as well as ubiquitous genes (ATGCAAAT). POU factors have a common conserved DNA binding domain, called the POU domain, which was originally identified in the transcription factors *Pit1*, *Oct1*, *Oct2*, and Unc86. The POU domain is comprised of two structurally independent subdomains—the POU specific domain (POU_S) and the homeodomain (POU_H)—connected by a flexible linker of variable length. The POU domain of *Oct4* is characterized by several properties that confer the OCT4 protein an impressive versatility in the operational mode of transcriptional regulation:

- Flexible amino acid-base interactions allowing recognition of moderately variable cognate DNA elements;
- Variable orientation, spacing and positioning of DNA-tethered POU subdomains relative to each other on the DNA, as evidenced by the different arrangement on the Palindromic Oct factor Recognition Element (PORE);
- Cooperative binding of the two subdomains to the DNA;
- Interaction with other transcription factors and regulatory modulators;
- Post-translational modifications modulating OCT4 transactivation in various cell types.

The POU domain binds to DNA via interaction of the third "recognition" helix of the POU_H with bases in the DNA major groove at the 3′A/TTTA portion of the octamer site. The POU_H domain is structurally similar to other homeodomains. The POU_S domain exhibits site-specific, high-affinity DNA binding and bending capability. Both the POU_S and POU_H subdomains function as structurally independent units with cooperative high-affinity DNA-binding specificity. Functional cooperation between the two subdomains may occur indirectly via the DNA by overlapping base contacts from the two subdomains.

In addition to the DNA binding function, both the POU_H and POU_S subdomains can participate in protein–protein interactions. In ES cells, OCT4 activates gene transcription irrespective of the octamer motif distance from the transcriptional initiation site. However, OCT4 can transactivate only from a proximal location in differentiated cells. In this scenario, interaction between the adenovirus E1A or human papillomavirus E7 protein and OCT4 POU domain is sufficient for OCT4 to elicit transcriptional activation from remote binding sites. E1A and E7 proteins would therefore mimic unidentified embryonic stem cell specific coactivators that serve a similar function in pluripotent cells.

Two domains spanning the N- and C-terminal portion of Oct4 protein define the transactivation capacity of the POU transcription factor. The N-terminal region (N domain) is a proline- and acid residue-rich region, whereas the C-terminal domain (C domain) is a region rich in proline, serine, and threonine residues. The N-terminal domain can function as an activation domain in heterologous cell systems. However, the C-terminal domain of OCT4 exhibits a POU domain mediated cell type-specific function. Intramolecular interactions between the POU domain and C-terminal domain may lead to cell type-specific interactions with different cofactors or kinases. An interesting regulation model of the C-domain predicts that association between the OCT4 POU domain and other factors can alter the phosphorylation status of the protein and ultimately modulate the activity of the C-domain.

OCT4 EXPRESSION

In the Mouse

Oct4, otherwise designated *Oct3* or *POU5F1*, is a maternally-inherited transcript that is developmentally regulated in mice. It is expressed at low levels in all blastomeres until the four-cell stage, at which time the gene undergoes zygotic activation resulting in high OCT4 protein levels in the nuclei of all blastomeres until compaction. After cavitation, *Oct4* is maintained only in the inner cells (ICM) of the blastocyst and is down-regulated in the differentiated trophectoderm (TE). Following implantation, *Oct4* expression is restricted to the primitive ectoderm (epiblast), although it is transiently expressed at high levels in cells of the forming hypoblast (primitive endoderm). During gastrulation, *Oct4* expression becomes down-regulated in the epiblast in an anterior to posterior manner. From 8.5 dpc on, *Oct4* becomes restricted to precursors of the gametes or PGCs. *Oct4* is also expressed in undifferentiated mouse ES, EG, and EC cell lines. ES and EC cell treatment with the differentiation-inducing agent retinoic acid (RA) induce rapid *Oct4* down-regulation in both cell types. *Oct4* is not expressed in differentiated tissues. Recently, a very low amount of *Oct4* messenger was detected in multipotent adult precursor cells (MAPCs) and human breast cancer cells.

In Humans

Oct4 is highly expressed in ICM cells relative to TE cells in discarded human embryos. It is also expressed in pluripotent human EC and ES/EG lines.

OCT4 ACTIVITY IN ASSAYS

Gene inactivation experiments clearly indicate that *Oct4* plays a determinant role in the specification of mouse pluripotent cells. However, *Oct4* function is not confined to pluripotent cells. Extrapolation of the data obtained by manipulating *Pou5f1* expression in ES cells indicates that *Oct4* is a regulator of cell fate in all three tissues of a preimplantation embryo.

Embryos With No Oct4

Oct4-deficient embryos die at the peri-implantation stage and form implantation sites that contain trophoblastic cells but are devoid of yolk sac or embryonic structures. The embryos develop up to the 3.5 dpc blastocyst stage, with an unaltered number of cells in the ICM or trophoectoderm as compared to wild type embryos, suggesting that in the absence of OCT4 protein, cell proliferation is not affected during early stages of development. *In vitro* cultures of ICM cells isolated from the *Oct4* null blastocysts are devoid of pluripotent and extra-embryonic endoderm cells, but do contain trophoblastic giant cells.

These results indicate that the ICM of *Oct4*-deficient blastocyst is defined only by the stereo arrangement of non-pluripotent cells. These findings lead to two main conclusions:

1. *Oct4* is essential in the establishment of the ICM pluripotency;
2. In the absence of *Oct4*, interior cells of the pre-implantation embryo are diverted to a trophoectoderm fate.

ES Cells With Low and High Levels of Oct4

Based on *Oct4* expression levels during the first two events of lineage commitment in the early embryo, it is not entirely surprising that a critical amount of *Oct4* has recently been reported to be crucial for the maintenance of ES cell self-renewal.

ES cells containing a bacterial genetic system for the conditional expression and repression of *Oct4* transgene allowed researchers to determine the cellular phenotype linked to variable levels of OCT4 protein. Briefly, ES cells in which one or both endogenous *Oct4* alleles had been inactivated were transfected with constructs for a tetracycline-regulated transactivator and a responsive *Oct4* transgene, whose precise level of expression was modulated by variable amounts of the antibiotic.

A 50% decrease in the endogenous *Oct4* levels relative to that of undifferentiated ES cells resulted in the commitment of ES cells to trophoectoderm lineages, containing both proliferating and endo-duplicating giant cells based on the culture conditions. This result is consistent with the phenotype of *Oct4* deficient embryos and the observation that trophoectoderm differentiation of outer cells of the morula is accompanied by *Oct4* down-regulation (Okamoto, 1990; Rosner, 1990; Schöler, 1990a).

However, a 50% increase beyond the threshold level of *Oct4* leads to the concomitant differentiation of ES cells into extra-embryonic endoderm and mesoderm. Interestingly, LIF withdrawal leads to the specification of the same lineages. Smaller changes in *Oct4* level (whether increases or decreases) do not affect ES cell self-renewal. In conclusion, the precise level of OCT4 protein governs commitment of embryonic cells along three distinct cell fates (self-renewal, trophectoderm, or extra-embryonic endoderm and mesoderm) (see Figure 11-5).

The conditional *Oct4*-null ES line proved to be a useful tool to establish which domains of the OCT4 protein are sufficient to maintain ES cells in complementation assays. The complementation assay was based on the ability of a proper *Oct4* molecule to rescue the self-renewal capability of cells that would otherwise differentiate due to Oct4 down-regulation. OCT4 is the only POU protein that has the ability to rescue the self-renewing phenotype, as OCT2 or OCT6

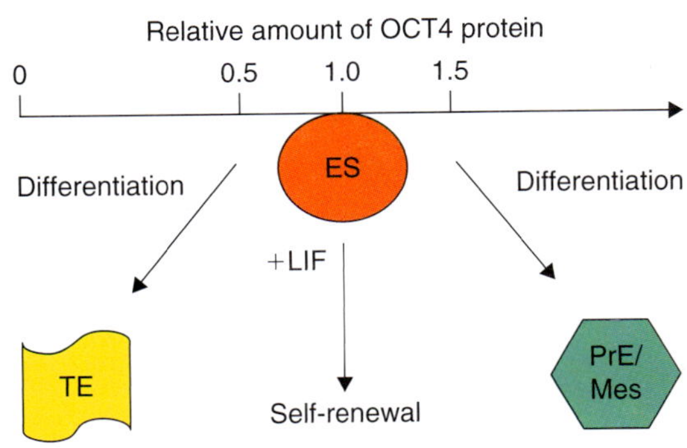

FIGURE 11-5 Model depicting the consequences of OCT4 variable protein concentration on ES cell fate. TE stands for trophoectoderm; PrE/Mes stands for primitive endoderm/mesoderm.

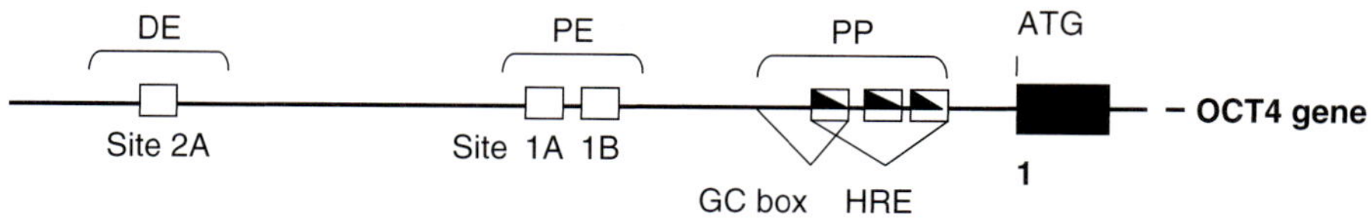

Distal enhancer	Proximal enhancer
Pre-implantation embryo Oocyte through morula stage Post-implantation embryo: Primordial germ cells	Post-implantation embryo: Embryonic ectoderm
Embryonic stem cells Embryonic germ cells F9 embryonal carcinoma cells	P19 embryonal carcinoma cells

FIGURE 11-6 Representation of the OCT4 gene. Open and partially filled boxes indicate conserved regions included within the proximal/distal enhancer and the proximal promoter. The black box indicates the first exon of the gene. Cell and tissue specificity conferred by the two enhancers is also described.

has no effect on cell fate in this system. A truncated OCT4 protein containing the OCT4 POU domain and either the N- or C-terminal domain can support ES cell self-renewal as the wild-type counterpart. Furthermore, gene expression analysis revealed that OCT4 transactivation domains, although equivalent in sustaining the undifferentiated stem cell phenotype, elicit activation of different target genes. It would be interesting to determine the consequences of either the N- or C- domain deletion on embryo development *in vivo*.

REGULATION OF OCT4 EXPRESSION

Use of *Oct-4/LacZ* transgenes has allowed the identification of two distinct enhancer elements, which are driving the cell type-specific expression of *Oct4* in a reciprocal fashion. The distal enhancer or DE, located approximately 5 Kb upstream of the promoter, regulates *Oct4* expression in pre-implantation embryos (morula and ICM), PGCs, and in ES, F9 EC, and EG cells. However, the DE is inactive in cells of the epiblast. The proximal enhancer or PE, located approximately 1.2 Kb upstream, directs *Oct4* expression in the epiblast, including *Oct4* down-regulation in the anterior to posterior direction observed during gastrulation, and in P19 EC cells (see Figure 11-6). Two similar inverted elements, the 2A site within the DE and the 1A site within the PE, are bound by transcription factors in undifferentiated ES and EC cells *in vivo*. Upon RA-induced differentiation, protection of these two areas is lost. These data suggest that these elements are involved in regulating *Oct4* transcription probably through local signals and/or cell type-specific factors.

The *Oct4* gene comprises a TATA-less promoter, called PP, containing a GC-rich Sp1-like sequence and three hormone response element half sites—(HRE) R1, R2 and R3. Several nuclear receptors, retinoid acid receptors and the orphan receptors COUP-TFI/II, SF1, LRH1 and GCNF, have been shown to be involved in the regulation of Oct4 expression *in vitro*. However, knockout mouse models have substantiated such a role only for *LRH1 and gcnf*.

ChIP analysis was employed to demonstrate that binding of LRH1 to the PE and PP regions of the Oct4 gene occurs in undifferentiated ES/P19 cells and is lost upon RA induced differentiation. LRH1 binding sites drive the expression of *Oct4*-PE-PP-luciferase constructs in undifferentiated P19 cells. Additionally, *lrh1*-null ES cells, display a much quicker loss of *Oct4* expression upon LIF removal than is observed for wt ES cells. *In vivo*, *LRH1* is expressed in the ICM and visceral endoderm of the blastocyst, and at the egg cylinder stage throughout the whole embryo. *LRH1* deficient embryos die at 6.5 dpc and display no detectable Oct4 mRNA, which indicates that *LRH1* is specifically required to maintain *Oct4* in the embryo epiblast.

During RA-induced differentiation of P19 cells, Oct4 down-regulation is temporally followed by *gcnf* up-regulation. GCNF binding to the promoter R2 repeat represses transcription driven by the proximal enhancer–promoter (PEP) in P19 cells. Analysis of *gcnf*-deficient embryos has emphasized the impact of *gcnf*-induced repression of the *Oct4* gene. Low *gcnf* expression is detected in the whole mouse embryo at 6.5 dpc. At 7.5 dpc, increasing *gcnf* and decreasing *Oct4* mRNA levels are observed in neural folds and at the posterior of the embryo. As mentioned above, at the end of gastrulation *Oct4* is maintained only in PGCs at the base of the allantois. In 8.0–8.5 dpc *gcnf*-deficient embryos, Oct4 mRNA is detected in the putative hindbrain region and posterior of the embryo, which indicates that *gcnf* ablation leads to loss of *Oct4* repression in somatic cells or loss of GCNF-induced *Oct4* restriction in the germline.

The same phenotype is observed in *gcnf*-deficient mice containing a targeted deletion of the DNA binding domain of GCNF. *gcnf*-null ES cells maintain expression of *Oct4* (and pluripotent genes) during a six-day time course of RA treatment. Taken together, these findings strongly point towards GCNF as an essential regulator of *Oct4* repression both *in vivo* and *in vitro*.

DNA methylation constitutes an important mechanism of gene expression regulation during embryogenesis. A global loss of DNA methylation occurs during cleavage prior to the 16-cell stage. Following implantation, a wave of *de novo* DNA methylation occurs in all genes, except those containing CpG islands. *In vivo* analysis of the PE-P region of the *Oct4* gene reveals that *Oct4* is not methylated from the blastocyst stage up to 6.25 dpc, a time during which other genes undergo *de novo* methylation. Interestingly, a cis-specific demethylation element exists within the *Oct4* PE. Mutational analysis shows that site 1A within PE is involved in preventing methylation. The presence of this cis-specific demodification element causes demethylation of sequences that had been methylated prior to their introduction into EC cells, and protects from *de novo* methylation *in vivo*. Interestingly, methylated PE sequences induce a decrease in transcriptional activation both of a reporter gene and the endogenous *Oct4* in P19 cells. According to the model proposed by Gidekel and Bergman, binding of trans-acting factors to the PE element offers protection against *de novo* methylation, which can begin soon after gastrulation, when these factors are down-regulated. At the end of gastrulation, these factors would be present only in the germ cell lineage, the sole lineage to express *Oct4* by 8.5 dpc.

As for the PE, the *Oct4* promoter region spanning between −562 and the putative ATG was shown to contribute to the epigenetic regulation of *Oct4* transcription. The region contains 16 CpG sites, which are unmethylated in undifferentiated ES cells and display increased methylation status upon RA treatment. In the absence of GCNF, *Oct4* promoter is hypomethylated both in differentiating ES cells and in 8.5 dpc embryos. Using a two-hybrid assay, Gu *et al.* found the methyl DNA binding domain proteins MBD3 and MBD2 as GCNF-interacting factors whose GCNF-dependent recruitment to *Oct4* promoter is necessary for Oct4 repression and methylation. *Mbd3* and *Mbd2* belong to the MeCP1 and NurD (Nucleosome remodeling and histone deacetylation) transcriptional repression complexes. In agreements with these results, *Mbd3*-deficient ES cells display LIF-independent self-renewal, and a compromised *in vivo* differentiation potential. *In vitro* they can only differentiate upon RA treatment.

OCT4 TARGET GENES

Human chorionic gonadropin (hCG) is required for implantation and maintenance of pregnancy. It is secreted by the trophectoderm of peri-implantation blastocyst. *Oct4* has been shown to silence the expression of α and β subunit genes of hCG in human choriocarcinoma cells. *Oct4* binding to a unique octamer motif (ACAATAATCA) in the hCGβ −305/−249 promoter region considerably reduces both hCGβ mRNA and protein levels in JAr cells. Although Oct4 is a potent inhibitor of hCGα expression, no octamer binding site has been identified in the promoter region of hCGα.

Like CG in humans, tau interferon (IFN-τ) is needed to prevent regression of the maternal *corpus luteum* during the early stages of pregnancy in ruminant species. In cattle, members of the multigene IFN-τ family are expressed in the TE from the blastocyst stage up to the beginning of the placentation process. IFN-τ expression is promoted by an Ets-2 enhancer, which is located in the promoter region of the gene.

Unlike *Oct1* or *Oct2*, *Oct4* specifically interferes with Ets-2-induced activation of *IFN-τ* promoter in JAr cells. These data, taken together with the observation that reduced Oct4 protein level triggers differentiation of murine ES cells into trophoectoderm, suggest that silencing of *Oct4* in the TE is a prerequisite for up-regulation of *hCGs* and *IFN- τ*.

The *Rex1* gene, which encodes a zinc finger protein, is expressed at high levels in ES and EC cells, and is down-regulated upon RA treatment. Both processes are mediated by the octamer motif located in the promoter region of *Rex1*. Low levels of OCT4 protein are sufficient to activate *Rex1* promoter in P19 RA-differentiated cells, whereas high OCT4 or OCT6 level inhibit transcription in F9 cells. Distinct Oct4 protein domains elicit the observed effects, suggesting different molecular mechanisms of *Oct4* mediated transcriptional activation and repression.

Expression of the platelet-derived growth factor α receptor (PDGFαR) in undifferentiated human EC Tera2 cells depends on a canonical octamer motif within the gene promoter. Mutation of the octamer site has been shown to decrease the promoter activity and *PDGFαR* expression.

OCT4 DOES NOT PLAY *SOLO:* SOX2

Sox2 belongs to the Sox (*S*ry-related HMG b*ox*-containing) family of proteins that bind to DNA through the 79-amino acid HMG (high mobility group) domain. In contrast to most DNA-binding proteins, which access to DNA through the major groove, the HMG box interacts with the minor groove of the DNA helix and, as a consequence, induces a dramatic bend in the DNA molecule. As a result of this energetically high cost interaction, Sox proteins bind to DNA with a high dissociation constant.

Sox2 Expression

Sox2 is co-expressed with *Oct4* in the ICM of preimplantation embryos, ES cells, EC cells, and germ cells. At the egg cylinder stage, *Sox2* is expressed both in the epiblast and in the adjacent extra-embryonic ectoderm (ExE). In 7.5 dpc embryos *Sox2* mRNA marks the chorion and the presumptive neuroectoderm.

Sox2 Function *In Vivo*

Analysis of *Sox2*-deficient embryos supported the hypothesis that functional association between OCT4 and SOX2, previously observed in cultured cells, constitutes a new paradigm of gene activation in the early embryo as well (see following paragraph). *Sox2*-deficient embryos die shortly after implantation at around 6.0 dpc, with implants lacking *Oct4*-positive epiblast cells and having disorganized extra-embryonic tissues. Post-implantation lethality cannot be attributed to lack of *Sox2* expression in ExE, as wild-type ES cell can rescue development of *Sox2*-deficient embryos up to E7.5–7.75.

Immunosurgically isolated *Sox2*-null ICMs are diverted to a trophoectodermal and endodermal fate. In *Sox2*-deficient blastocysts, very low *Oct4* expression can be detected by RT-PCR, indicating that a pluripotent ICM can be specified but not maintained (see Figure 11-7). However, complete knockout of the *Sox2* gene is not practicable during early development, as maternal SOX2 protein persists throughout pre-implantation development. Failures in establishing *Sox2*-null ES cell lines confirm *Sox2* role in the maintenance of pluripotent stem cells.

Sox2 and OCT4 Partnership

Regulation of the *fgf*4 and *utf1* genes provided insight into an OCT4-induced gene activation model based on parternship with the transcription factor *Sox2*.

Distinct regulatory elements govern *fgf4* gene expression in the mouse ICM, myotomes and developing limb.

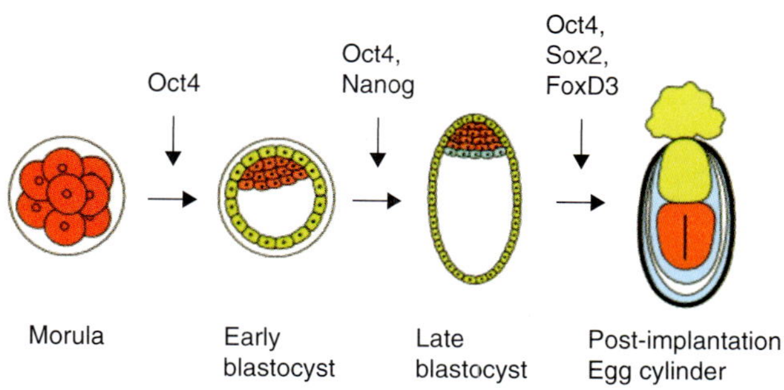

FIGURE 11-7 Model relative to the roles played by Oct4, Nanog, Sox2, and FoxD3 during early mouse development. Oct4 is essential in the first embryonic lineage specification, whereas Nanog function is to prevent endoderm differentiation of the ICM cells. Sox2 and FoxD3 are essential in the maintenance of a pluripotent epiblast. Depicted in red are the pluripotent compartments of the embryo, e.g., ICM and epiblast. Blastocyst trophoectoderm and trophoectoderm-derivatives of the post-implantation embryo are depicted in yellow. Blastocyst primitive endoderm and visceral/parietal endoderm of the post-implantation embryo are depicted in light blue.

fgf4 expression in ICM and in ES/EC cells is conferred by a distal enhancer localized in the 3′ UTR of the gene. *Oct4* (or *Oct1*) and SOX2 bind to the respective cognate sites in the "embryonic" enhancer and form a unique ternary complex that elicits strong synergistic transcriptional activation. Formation of an active ternary complex is highly dependent on the spatial arrangement of the adjacent Sox and Oct binding sites on the DNA, since insertion of three base pairs between the Sox and Oct recognition sequences of the *fgf4* gene severely impairs the enhancer function. The C-terminal domains of both SOX2 and OCT4 contribute to the functional activity of Sox2/OCT4 complex. The synergistic action of SOX2 and OCT4 results from two distinct yet concerted events. The first event involves cooperative binding of SOX2 and OCT4 to the DNA via their respective DNA binding domains. The tethering of each factor to the enhancer region on fgf4 ameliorates the intrinsic activity of the activation domains of each protein. Upon formation of the ternary SOX2/OCT4 complex, novel DNA–protein and protein–protein interactions induce conformational changes that may lead to activation of latent domains and constitute a new, distinct platform for the recruitment of other co-activators. In line with this model, synergistic activation of *fgf*4 transcription by SOX2 and OCT4 in HeLa cells is mediated by p300, a potential bridging factor, which should promote enhancer–promoter interactions. Notably, the embryonic fgf4 enhancer sequences are conserved in the mouse and human genes.

Embryonic stem cell-specific expression of *utf1* is also regulated by the synergistic action of SOX2 and OCT4. The binding sites for these two factors are arranged with no intervening spacing in the second intron of the *utf1* gene. Interestingly, one base difference in the canonical octamer binding sequence enables the recruitment of an active OCT4/SOX2 complex and prevents binding of a transcriptionally inactive complex containing OCT1 or OCT6. The sequence ACTAGCAT (canonical sequence: ATTA/TGCAT) is recognized specifically by OCT4, while OCT1 and OCT6 can bind to it only by exploiting half of the 3′ adjacent Sox site (AACAATG). The OCT4 POU_H domain is essential in ensuring unique DNA binding ability on the *utf1* octamer consensus sequence.

The F-box containing protein 15(*Fbx*15) is expressed predominantly in ES cells. Embryonic expression of *Fbx15* results from the cooperative binding of OCT4 and SOX2 to cognate sites juxtaposed in the *Fbx15* ES specific enhancer.

Zfp206, a transcription factor that is highly expressed in mouse and human ES and down-regulated during ES differentiation, is another example of OCT4/SOX2 target genes.

OCT4–OCT4 Dimers and Sox2

The structural flexibility of the linker sequence between the two independent POU subdomains endows OCT4 with the ability to form homodimers on specific DNA sequences. A novel PORE, composed of an inverted pair of homeodomain-binding sites and separated by five base pairs (ATTTGaaatgCAAAT), has been identified in first intron of the EC/ES cell-specific *opn* gene enhancer.

OPN is an extra-cellular phosphoprotein, which modulates cell migration and adhesion of the primitive endoderm cells by binding to specific integrins. In the mouse embryo, *opn* is co-expressed with *Oct4* in ICM, forms primitive endoderm at 4.0 dpc, and is down-regulated in the epithelial hypoblast of E 4.5 embryos. In EC and ES cells, *opn* expression is inversely correlated with SOX2 protein levels.

The POU transcription factors OCT1, OCT6, and OCT4 bind to the DNA PORE element as monomers, homo-, and heterodimers. Oct4 monomer does not elicit transcriptional activation via the PORE in EC cells. Rather, activation of gene transcription is highly dependent on OCT4 dimerization on the PORE. In fact, mutation in the palindromic element of either home domain-binding sites drastically impairs dimerization and transcriptional activation. Interestingly, SOX2 represses OCT4 mediated *opn* gene transactivation by binding to the cognate site adjacent to the PORE in the *opn* intron. SOX2 may interfere with OCT4 dimer formation.

Dimerization on the PORE creates a specific conformational surface that may be suitable for interaction with unidentified co-activators. In fact, OCT1/ or OCT2/PORE dimers can interact and synergize in transcriptional activation with the lymphoid specific co-activator OBF-1. According to a model of the PORE OCT4 dimer structure, Isoleucine 21 (I21) of the POU_S and Serine 107 (S107) of the POU_H of two distinct OCT4 molecules make specific contact in the PORE dimer interface. Mutation of either residue impairs dimerization on the PORE element. S107 corresponds to a conserved phosphorylation site identified in OCT1 and PIT1. Phosphorylation of this serine has been shown to influence OCT1 binding to DNA and may provide an additional level of regulation of OCT4 dimer activity. Mutation of S107 into glutamate might structurally mimic a phosphorylated serine. To date, *opn* is the only known *Oct4* target gene to be transcriptionally regulated by OCT4 dimers. Mutant mice with impaired dimerization of the OCT4 protein might be valuable in identifying new Oct4 dimer target genes and their function during early development.

FOXD3

FoxD3, a transcription factor belonging to the forkhead family, is expressed in mouse and human ES cells, and during mouse embryogenesis in the epiblast and neural crest cells. FoxD3-deficient blastocysts present a regular pattern of *Oct4*, *Sox2*, and *fgf4* expression before implantation. Nevertheless, mutant mouse embryos die at 6.5 dpc and display a reduction of the epiblast, extension of the proximal extra-embryonic endoderm and ectoderm tissues, and lack of primitive streak. In addition, FoxD3-deficient ICMs fail to expand after prolonged culture *in vitro*. These results suggest that FoxD3 is essential for epiblast maintenance, but is not required for differentiation of extra-embryonic tissues (Figure 11-7).

NANOG

In 2003 Chambers and Matsui reported on the identification of *Nanog* as a new member of the embryonic stem cell stage. Like OCT4, NANOG protein contains a homeodomain, which shares a maximum of only 50% amino acid identity with members of the NK2 family of homeoproteins. This makes NANOG a unique divergent homeodomain protein.

Nanog Expression

Nanog is found to be expressed in a quite restricted pattern. *In vivo*, *Nanog* expression is first detected in the interior cells of compacted morulae and then in the ICM and the epiblast of a pre-implantation blastocyst, but it declines prior to implantation. Following implantation, *Nanog* transcripts can be found in a subset of epiblast cells and at E11.5 in post-migratory germ cells (PGCs). *In vitro*, *Nanog* marks all pluripotent cell lines, from ES (both murine and human) to EG and EC. Hence the name Nanog, after the Celtic land of the ever-young Tir nan Og.

Nanog Function *In Vivo* and *In Vitro*

Nanog-deficient embryos die soon after implantation due to a failure in the specification of the pluripotent epiblast, which is diverted to an endodermal fate. Similarly, *Nanog* ablation in ES cells causes differentiation into parietal/visceral endoderm lineages. These data demonstrate that Nanog is essential for maintenance of a pluripotent status both *in vivo* and *in vitro*, and that *Nanog* function is to prevent endoderm differentiation (see Figure 11-7).

ES cells that express elevated levels of Nanog can self-renew in the absence of LIF, or even in the presence of LIF antagonists. These results indicate that *Nanog* is capable of rendering ES cells independent from LIF/STAT3 stimulation for self-renewal. However, *Nanog* and STAT3 exert their function independently from each other, as *Nanog* overexpression does not result in the activation of the STAT3 signaling pathway. In addition, Nanog overexpressing ES cells grown at clonal density in the presence of LIF and form increased numbers of purely undifferentiated colonies than do non-transgenic ES cells, indicating that LIF and *Nanog* have an additive effect on ES cell self-renewal. *Nanog* and *Oct4* work in concert to support stem cell potency and

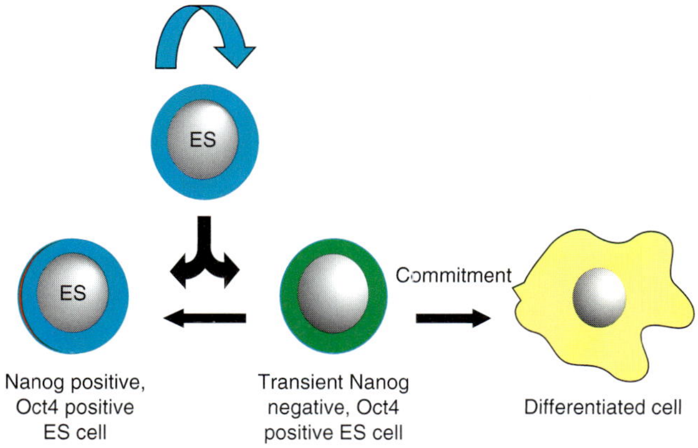

FIGURE 11-8 Nanog functions as a safeguard of pluripotency. A Nanog positive ES cell can either self-renew or give rise to a Nanog negative ES cell, which is predisposed towards differentiation while also being able to give rise to a Nanog positive ES cell. Adapted from Chambers *et al.*, 2007.

self-renewal. This conclusion is based on the two observations: first, *Nanog* is expressed in *Oct4*-deficient embryos; and second, *Nanog* overexpression cannot reverse the differentiation program induced in ES cells by *Oct4* down-regulation.

Chambers *et al.* further investigated the role of *Nanog* in ES self-renewal, and found that a fraction of ES cells that express *Oct4*, but do not have detectable *Nanog*, can still give rise to *Nanog*- and *Oct4*-positive cells upon trypsinization. Surprisingly, he found that permanent genetic ablation of *Nanog* does not abolish pluripotency, but rather reduces self-renewal capability by generating a transient state which is reversibly poised to both self-renewal and differentiation commitment (Figure 11-8). However, *Nanog*-deficient ES cells form PGCs that can colonize the genital ridge, but are unable to survive beyond E11.5. The conclusion of this study is that *Nanog in vivo* is essential for ICM determination and germ cell development. *In vitro* it works/acts by preventing progression to differentiation, and can thus be considered as a safeguard of pluripotency (Figure 11-8).

Nanog Regulation

Nanog proximal promoter (−289+117 relative to the ATG) contains a composite element of adjacent Sox and Oct binding sites, which are fully conserved among eutherian mammals. OCT4 and SOX2 control *Nanog* transcription by binding to this element in both mouse and human ES cells. The same Oct binding site was shown to mediate *Nanog* biphasic regulation by OCT4. Sub-steady state levels of *Oct4* expression elicit a positive effect on *Nanog* transcription, whereas steady state or higher doses exert a negative effect. This could explain why *Nanog* depletion and *Oct4* overexpression both lead to endoderm differentiation.

Nanog proximal promoter also mediates gene regulation via epigenetic mechanism. The −327 to −10 region is a tissue-dependent and differentially-methylated region or T-DMR. It contains six CpG dinucletides, which are found to be hypomethylated in ES cells and hypermethylated in trophoblastic stem and 3T3 cells.

Nanog and p53

Somatic cells respond to DNA damage by activating p53, which causes cell cycle arrest or apoptosis. However, ES cells lack p53-G1/S checkpoint and undergo apoptosis only in response to high levels of DNA damage. When low levels of DNA damage are introduced, ES cells tend instead to differentiate into a cell type, which is more prone to undergo p53-dependent cell cycle arrest and apoptosis. Two p53 binding motifs, located at −871 to −849 and −611 to −585 relative to the ATG, have been identified within the *Nanog* promoter region. P53 is known to interact with the corepressor mSin3a in order to repress transcription by recruiting histone deacetyleases. Upon DNA damage, p53 binds to *Nanog* promoter and suppresses Nanog transcription. P53 transcriptional activity and *Nanog* expression inversely correlate during RA-induced ES differentiation. However, p53 activity and *Nanog* suppression depend on p53 phosphorylated serine 315, which is essential for efficient recruitment of mSin3a to the Nanog promoter.

Nanog–Nanog Interaction

As NANOG is a homeobox protein, the possibility that it may homodimerize in order to exert target gene regulation was investigated. Co-immunoprecipitation experiments of tagged NANOG and deletion mutants of NANOG protein showed that NANOG can assemble into homodimers, and that dimerization is mediated by the tryptophan-rich C-terminal domain. NANOG homodimers are able to bind to *Nanog* consensus with high efficiency *in vitro*. Most importantly, Nanog dimers are responsible for directing ES cell LIF-independent self-renewal, as enforced expression of a Nanog mutant that does not contain the C-terminal domain leads to ES differentiation upon LIF removal.

PI3K and Nanog

Pharmacological inhibition of PI3K signaling with LY294002 results in reduction of *Nanog* mRNA and protein levels. In contrast, *Oct4* protein and messenger remain unaltered, indicating that the observed change in *Nanog* expression does not result from LY294002-induced global differentiation. Chemical inhibition of GSK3 activity reverses the effects of LY294002 treatment on *Nanog*. Since PI3K signaling leads to the AKT/PKB mediated inactivation of GSK3, these results suggest that the PI3K–GSK3 signaling axis is directly involved in the regulation of *Nanog* expression.

The *Oct4/FoxD3/Nanog* Regulatory Loop

FOXD3 was demonstrated to activate *Nanog* transcription in F9 EC cells through a binding site located between nt−544 and −269 of *Nanog* promoter. FOXD3 and NANOG in turn strongly activate a 2 Kb *Oct4* promoter, which interestingly is also repressed by *Oct4* itself. We mentioned earlier that OCT4 also acts as a biphasic regulator of *Nanog* expression. Since OCT4 concentration must be kept at precise levels in order to maintain stem cell pluripotency, these data point towards the existence of an *Oct4, Nanog*, and *FoxD3* regulatory circuit that would ensure the proper level of *Oct4* expression in ES cells.

THE CORE TRANSCRIPTIONAL REGULATORY CIRCUITRY CONTROLLING ES CELL IDENTITY

As mentioned before, in mouse ES cells expression of *utf1*, *fgf4*, *Rex1*, *Fbx15*, and *Nanog* is controlled by both SOX2 and OCT4. Remarkably, *Oct4* and *Sox2* genes are also regulated by the OCT4/SOX2 complex. OCT4 and SOX2 specifically mediate *Oct4* expression in ES cells by binding to site 2B of the Oct4 DE, whereas a non-canonical binding site for OCT4 and SOX2 within the 3′ Regulatory Region of the *Sox2* gene, called SRR2, is involved in *Sox2* expression in ES/EC cells. SOX2 and OCT4 or OCT6 can bind simultaneously, but not cooperatively, to SRR2. OCT6 or OCT4 (but not OCT1) have been shown to slightly increase *Sox2*-dependent transcription of the *Sox2* gene.

All these data suggest not only that *Oct4* and *Sox2* expression is auto-maintained, but also that pluripotency may be controlled by a restricted group of master regulators that could determine ES identity by regulating themselves together with crucial developmental factors. This hypothesis was confirmed to hold true both for mouse and human ES cells. Oct4/Sox2/Nanog binding sites along the human genome were mapped by "ChIP-on-chip" technology. The results of this genome-wide occupancy analysis revealed that *Oct4, Nanog*, and *Sox2* in the hES collaborate in forming a transcriptional regulatory circuitry consisting of both feedback autoregulatory and feed-forward loops. Remarkably, the three factors co-occupy a good portion of their target genes, with target co-occupancy leading to either target transcription activation or repression. ES cell-specific transcription factors and components of the Wnt/TGF-β signaling pathways belong to the group of genes that are activated by the OCT4/SOX2/NANOG, whereas development-related transcription factors like Neurogenin, Pax6, and Myf5 are among the negatively-regulated target genes. Oct4/Sox2/Nanog in hES team up as the main catalysts of ES identity by positively regulating signaling pathways that are essential to ES self-renewal, and by repressing genes required for developmental processes to occur. Genome-wide analysis of OCT4/NANOG binding sites also showed co-occupancy of the downstream target genes in mES cells. The tight functional association between OCT4 and NANOG is further supported by the discovery that the two proteins interact with each other. In addition, they associate with components (Mta1/2 and Hdca1/2) of the nuclear remodeling and histone deacetylation NurD and Sin3A complexes to co-occupy some of the known *Nanog* target genes, including *FoxD3*, *Oct4*, and *Nanog* itself.

Chromatin immunoprecipitation of the kruppel-like factors KLF2, KLF4, and KLF5 revealed that they bind and auto-regulate their own genes; they bind to *Nanog* enhancer and coregulate some downstream targets with NANOG. *Tcf3*, a terminal component of the Wnt canonical pathway, seems to be another integral component of the core regulatory circuitry of ES cells, as it was found to occupy, with OCT4 and NANOG, common ES and differentiation genes.

THE ROLE OF MICRO-RNAS IN PLURIPOTENCY

Small non-coding RNAs can be divided into two different classes: micro-RNA (miRNAs); and short interfering RNAs (siRNAs). Micro-RNAs control the stability and translation of target mRMAs and are involved in a wide variety of biological processes, including cell-cycle regulation, apoptosis, and cell differentiation. The double-stranded RNA-binding protein DGCR8 and the RNAse III enzyme Drosha process in the nucleus long primary miRNAs (pri-miRNAs) into short hairpins called precursor miRNAs (pre-miRNAs). The resulting hairpins are exported into the cytoplasm and processed by DICER into mature effector RNA molecules. DICER is also required for the maturation of short interfering RNAs, which regulate gene transcription by inducing cleavage of target mRNA.

Dicer

In vivo, Dicer is essential for mouse development and stem cell maintenance in the embryo. *Dicer*-deficient ES cells, which can be derived by gene targeting, express ES-cell specific markers but fail to differentiate both *in vivo* and *in vitro*. These ES cells also have high levels of double-stranded centromeric RNA transcripts that cannot be processed in the absence of *Dicer*. In addition, they display reduced levels of genomic DNA methylation and of histone H3-K9 di- and trimethylation at centromeric satellite repeats. These data indicate that *Dicer* is essential for stem cell differentiation, maintenance of centromeric heterochromatin structure and epigenetic centromeric silencing.

DGCR8

Significant reduction in the expression of the majority of known pre-miRNAs observed in *Dgcr8* knockout ES cells confirms that DGCR8 is essential for micro-RNA biogenesis. Although *dgcr8*-null ES cells fail to fully down-regulate pluripotency markers when induced to differentiate in the presence of retinoic acid, they do express some differentiation markers, albeit at lower levels than wild type cells.

This phenotype being different from the one described for *Dicer*-null ES cells, two conclusions can be drawn:

1. *Dicer* has miRNA-independent roles;
2. miRNA are essential for efficient silencing of the ES cell program during differentiation.

REST and miR-21

The neuronal repressor *Rest* is expressed at high levels in ES cells and down-regulated during differentiation. Both heterozygous deletion of *Rest* and its knockdown in ES cells cause a considerable loss of self-renewal and lead to the activation of markers specific for all lineages. Remarkably, decreased levels of *Oct4*, *Nanog*, *Sox2*, and *c-Myc* correspond to *Rest* transcript knocking down. Examination of miRNA profiles of *Rest+/−* and wt ES cells revealed a significant up-regulation of miR-21, -106a, -106b and -124a in heterozygous cells. However, only exogenous addition of precursor miR-21 to ES cells results in a decrease of self-renewal and in reduced expression of *Oct4, Nanog*, SOX2, and *c-Myc*. As REST binds to the miR−21 gene chromatin, it is proposed that *Rest* maintains self-renewal and pluripotency of mouse ES cells by suppressing miR−21 expression.

Notably, bioinformatics predicted miR−21 binding sites are present in *Sox2* and *Nanog* mRNAs. Additionally OCT4, NANOG, and SOX2 were found to co-occupy Rest promoter. REST was also found as part of the NANOG-OCT4 complex. All this data strongly point at Rest as a major component of the regulatory network regulating self-renewal and pluripotency of mouse ES cells.

THE CHROMATIN STATUS DEFINING STEM CELL IDENTITY

Embryonic stem cell identity depends on the combined expression of pluripotent genes and silencing of differentiation genes. However, both methylation of CpG dinucleotides and histone modifications have critical impact on transcription regulation by modulating global chromatin structure and accessibility. Therefore, gene profiling studies alone, while giving important information about the relative abundance of known transcription factors, are not sufficient to define cell identity. Thanks to the advent of ChIP-on-chip technology, a great amount of data resulting from genome-wide gene target, methylation, and chromatin analysis are now converging into the emergence of a global and comprehensive picture of stem cell configuration.

Histone H3 lysine 4 and lysine 27 methylation, respectively, mediate transcription activation and repression. Methylation of H3 lysine 4, which is catalyzed by the trithorax group of proteins (trxG), induces histone acetylation and nucleosome remodeling. Methylation of H3 lysine-27, which is catalyzed by the Polycomb Repressive Complex (PRC) PRC2, is required for PRC1-induced chromatin condensation, and consequently gene silencing. Polycomb group proteins (PcG) are the constituents of multiple PRC complexes and are conserved from Drososphila to humans. EZH2 is the methyltransferase component of the PRC2 complex.

Sequence analysis of 56 highly-conserved non-coding regions within the mouse genome showed significant enrichment for genes involved in embryonic development. So far, the distribution of repressive and activating histone modifications was thought to be non-overlapping. Surprisingly, genome-wide mapping of Histone H3 lysine 4 and 27 methylation within those 56 conserved non-coding regions revealed the existence of bivalent chromatin domains containing a combination of both repressive and active marks, i.e., H3 Lys4- and Lys27-methylation. Bivalent opposing marks are associated with low levels of expression in pluripotent cells, but these are resolved into a monovalent methylation mark during ES differentiation. Remarkably, genes with bivalent domains are mostly transcription factors, which are essential for differentiation to occur, hence the idea that the bivalent mark silences genes in ES cells but poises/primes them for activation upon differentiation. The significance of this new chromatin code, or signature of pluripotent cells, is further strengthened by the fact that similar bivalent marks were found in lineage-specific genes within the hES genome.

Another step forward in deciphering the chromatin code of pluripotency was taken with the studies conducted on the Polycomb group of proteins. The fact that PcG proteins may be important in dictating ES identity came from the observation that genetic ablation of *Suz12* and *ezh2*, both components of the PRC2 complex, results in early post-implantation lethality. Furthermore, ES lines cannot be derived from EZh2-deficient blastocysts.

Mapping of SUZ12/RNA Pol II binding and of H3K27 trimethylation across the nonrepeat portion of hES genome by ChIP-on-chip analysis revealed that SUZ12 occupies and silences key genes involved in a variety of developmental processes (such as members of the Fox, Gata, and Tbx family). These developmental regulators are preferentially activated during differentiation, indicating that PcG-mediated silencing is associated with pluripotency and that loss of PcG occupancy correlates with differentiation (Figure 11-9). Similar results were obtained with mouse ES cells. However, a large number of genes, which are

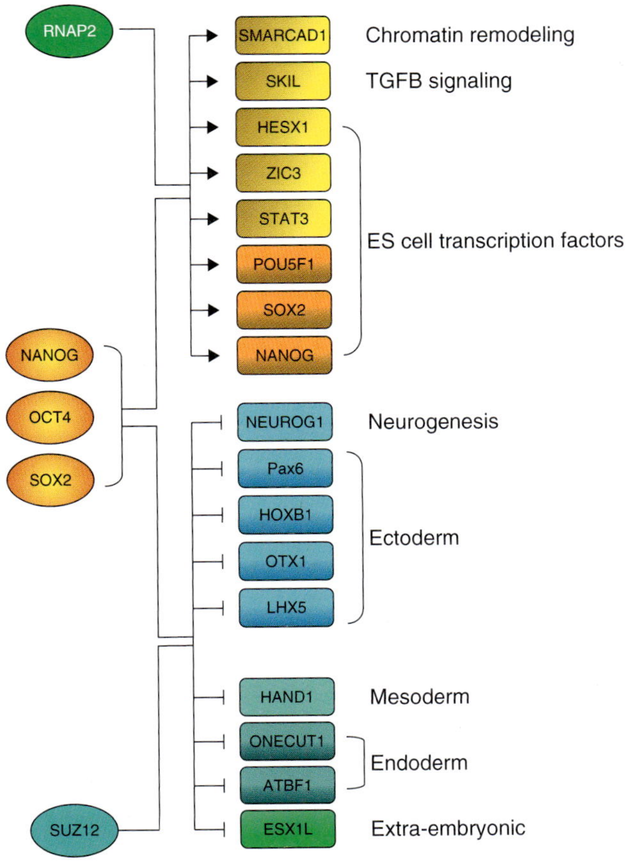

FIGURE 11-9 Examples of pluripotency associated genes co-occupied by Nanog/Oct4/Sox2 and RNAP2, and repressed developmental regulators co-occupied by Nanog/Oct4/sox2 and SUZ12 in hES cells. Adapted by Lee *et al.*, 2006.

not marked by either histone modification or Polycomb protein association, are silenced in ES cells.

As previously stated, DNA methylation constitutes an important mechanism of gene expression regulation during embryogenesis. The two DNA methyltransferases *Dnmt3a* and 3b are essential for *de novo* methylation, and their ablation leads to early embryonic lethality. Triple mutant ES cells lacking *Dnmt3a*, 3b, and *Dnmt1* (a methyl DNA maintenance enzyme) can self-renew, but undergo massive apoptosis when stimulated to differentiate, indicating that DNA methylation likely contributes to the "pluripotent" configuration.

Genome-wide mapping of DNA methylation patterns at proximal promoters in mouse ES cells revealed that 87% of genes with neither H3K4me3 nor H3K27me3 are CpG methylated and mostly repressed, and that 71.3% of genes repressed by PcG complex are unmethylated. These results indicate that Polycomb association, histone modifications, and DNA methylation constitute distinct epigenetic regulators that work in a non-overlapping manner.

Importantly, a direct link exists between epigenetic control/chromatin modulation of ES cell identity and the pluripotency master gene *Oct4*. By ChIP analysis and transfection essays, OCT4 was found to bind and to activate the expression of Jmjd1a and Jmjd2c, which are respectively H3K9 Me2 and H3K9Me3 demethylases. *Tcl1* and *Nanog* are in turn downstream effectors of, respectively, Jmjd1a and Jmjd2c and specific knockdown of either one of the two demethylases leads to ES differentiation. Thus *Oct4*, *Tcl1*, and *Nanog* bridge the ES cell transcription circuitry to H3K9-mediated chromatin modulation.

REPROGRAMMING OF SOMATIC CELLS TO A PLURIPOTENT STATE: IPS VERSUS ES

Is pluripotency imposed by one single master gene or by a group of genes? Discovery of such a master gene(s) would constitute the panacea of modern human regenerative medicine, as it would obviate the need for human cloning, with all its ethical implications. From what was discussed in the previous paragraph, it appears that pluripotency may most likely be achieved through the combination of properly sequenced processes that modify chromatin accessibility or inaccessibility, and ultimately end with the activation and repression of pluripotency and differentiation genes, respectively.

Up to two years ago, reprogramming of a somatic cell to a pluripotent state could be accomplished only either by fusion with a pluripotent cell or by nuclear transfer into oocytes. In August 2006 Takahashi and Yamanaka demonstrated that pluripotent ES-like stem cells could be generated from both embryonic and adult mouse fibroblasts by retroviral transduction of four genes, i.e., *Oct4*, Sox2, Klf4, and c-Myc. The fibroblasts used in the study carried a β-geo knockin construct within the *Fbx15* alleles, which allowed isolation of reprogrammed fibroblasts by growth selection in G418. The induced pluripotent stem cells, named iPS, displayed morphology and growth properties typical of ES cells. Global gene-expression profiling of iPS revealed that these cells cluster more closely with ES cells than with fibroblasts. Although the microarray analysis showed genes commonly up-regulated in ES and iPS, it also highlighted genes which were expressed at higher levels in ES cells than in iPS. In addition, *Oct4* and *Nanog* promoters were found partially methylated in iPS cells, i.e., incompletely reprogrammed. When the pluripotent potential of the cells was tested, it turned out that, even though iPS were endowed with multi-lineage differentiation capability *in vitro*, *in vivo* they could contribute to fetal but not to adult mouse development. This was probably due to the fact that expression of all four factors was driven by constitutively active promoters, which are not able to mediate transgene down-regulation during differentiation. Altogether, these data indicated that *Fbx15*-iPS were similar, but not identical to ES cells.

iPS isolation from drug selected fibroblast carrying a neomycin gene inserted within the Oct4 or *Nanog* locus resulted instead in the generation of iPS with gene expression, chromatin, and DNA methylation characteristics, alias epigenetic state of pluripotent ES cells. Remarkably, *Nanog- and Oct4-neo* iPS displayed increased developmental potency, which was identical to the one showed by ES cells, as they were even germline competent. Expression analysis of transduced and endogenous *Oct4/Sox2/Klf4/c-Myc* genes showed that retroviral transgenes are silenced during the induction of *Oct4*- or *Nanog*-iPS. This result indicates that exogenous *Oct4, Sox2*, Klf4, and *c-Myc* are essential for the induction of ES-like characteristics, but that the activity of the endogenous genes must be engaged in order to achieve pluripotency and full differentiation potential. However, iPS-derived animals developed tumors, probably due to the reactivation of the *c-Myc* transgene. Successful reprogramming of fibroblasts was shown to be accomplished without *c-Myc*, but evidence that iPS derived by *Oct4*, *Sox2*, and *Klf4* transduction will not induce tumor formation in adult mice is still missing.

iPS can be isolated from non-transgenic fibroblasts by using exclusively morphological criteria for the recognition, and hence picking, of ES-like colonies after viral transduction. This method gives a rate of reprogramming efficiency that is 5–10-fold higher than that observed with the aid of drug selection. Importantly, fetal and adult fibroblasts are not the only differentiated somatic cell type that can give rise to iPS. Non-terminally differentiated B-lymphocytes were shown to be reprogrammed to a pluripotent state by inducible expression of the four factors, whereas mature B-cell reprogramming required additional genetic manipulation, leading to the suppression of the B-cell transcriptome, i.e., ectopic expression of the myeloid TF CCAAT/enhance-binding protein-alfa or knock-down of the B-cell specific transcription factor *Pax5*.

The derivation of iPS has finally brought into the stem cell field the sacred formula for the conversion of any cell into the ever young embryonic stem cell, but a mechanistic insight of the *in fieri* conversion is still fundamental for the understanding, and hence controlling, developmental potency. An attempt to fill up the gap came from two different studies. When the process of mouse iPS derivation was monitored at different time-points for pluripotency marker gene expression, alkaline phosphatase first, SSEA1, and finally *Nanog* and *Oct4* were consistently found to be switched on in a sequential temporal order. These data would point towards the existence of an *iter* of gene reprogramming which is defined and gradual, rather than chaotic. Further confirmation of this hypothesis was given by a comprehensive genomic/integrative genomic characterization of cells during iPS generation. Cells were transduced with inducible lentiviral vectors encoding for the four factors and sampled at day 4, 8, 12, and 16 of initial induction for expression profiling. The most immediate response to the four-factor induction consisted in the up-regulation of proliferation genes, probably mediated by *c-Myc*, and in the down-regulation of MEF-specific mesenchymal genes. Remarkably, a strong up-regulation of kidney neural epidermal lineage specific genes and of stress-induced antiproliferative genes (*Klf4* and *c-Myc* targets) was also observed, indicating that the majority of cells are prevented from reaching an embryonic status because they mostly tend to stop uncontrollable proliferation or/and activate a differentiation program which reflects *Sox2* and *Klf4* function in tissues other than the embryonic one. Fully reprogrammed cells express all known pluripotent genes and are epigenetically similar to ES cells, whereas partially reprogrammed cell lines, like the *Fbx15* iPS, reactivate only a subset of pluripotent genes, because of high levels of promoter methylation at pluripotent loci and incomplete down-regulation of lineage-specific genes. That DNA methylation status has a critical role in cellular reprogramming was also proved by the observed improvement in the reprogramming efficiency of fibroblasts treated with DNA methyltransferase inhibitors.

A further advancement in the understanding of the mechanism of pluripotency reacquisition occurred in July 2008 with the report of Kim and collegues. Neural stem cells (NSCs) express *Sox2* and *c-Myc* at levels that are about two-fold and ten-fold higher than in ES cells, respectively whereas *Klf4* expression is eight-fold lower. Exogenous Oct4, in combination with either *Klf4* or *c-Myc*, are sufficient to generate iPS cells from NSCs. These two-factor iPS are similar to ES cells at the molecular level, they form chimeras and are able to contribute to the developing germline. These data strongly support the hypothesis that the minimum number of factors required to fully revert somatic into pluripotent embryonic cells strictly depends on the level of expression of endogenous *Oct4, Sox2, Klf4*, and *c-Myc* relative to ES cells. Following the same hypothesis, if *Oct4* and *c-Myc* perform like *Oct4* and *Klf4*, then it is conceivable that even *Klf4* might be omissible in the reprogramming of NSCs, and that *Oct4* alone could fulfill the process. This was eventually found to be the case as Kim and co-workers succeeded in producing 1-factor iPS in February 2009, demonstrating that *Oct4* alone is required and sufficient to directly reprogram NSCs to pluripotency.

CONCLUSIONS

The therapeutic potential of induced pluripotent cells was tested in a sickle cell anemia mouse model. Autologous iPS were generated from skin cells of a diseased animal, targeted for correction of the endogenous human sickle hemoglobin gene and induced to differentiate into hematopoietic progenitors. The efficient treatment of sickle cell anemia obtained by transplantation of these host specific cells into the diseased mouse provides a proof of principle for the clinical achievements that combined gene and cell therapy can lead to. Importantly, induced pluripotent cells have also been derived from human fetal, neonatal, and adult fibroblasts by ectopic expression of the same four-factor cocktail that yielded mouse iPS. Unfortunately, genetic manipulation of cells brings inevitable drawbacks: ectopic expression of *c-Myc* causes tumorigenicity, and random insertion of the viral sequences within the genome may produce unwanted mutagenesis events. None-the-less, an essential step forward in coupled gene repair plus cell therapy has been made. We have learned that the fundamental transcriptional network governing pluripotency in humans and mice is conserved, regardless of the differences between the two species for growth factor requirements. As of today, the whole picture eventually appears in its marvelous completeness, thanks to Zhou and colleagues who apply the missing piece of the puzzle and obtained iPS by the 'simple' protein transduction of the 4 factors.

KEY TERMS

Pluripotent Able to give rise to differentiated derivatives of all three germ layers. Cells of the inner cell mass and ES cells are pluripotent.

Totipotent Able to give rise to all type of differentiated cells, including cells of the trophoectoderm lineage.

FURTHER READING

Avilion, A. A., Nicolis, S. K., Pevny, L. H., Perez, L., Vivian, N., & Lovell-Badge, R. (2003). Multipotent cell lineages in early mouse development depend on Sox2 function. *Genes Dev.*, *17*, 126–140.

Bernstein, B. E., Mikkelsen, T. S., Xie, X., Kamal, M., Huebert, D. J., Cuff, J., Fry, B., Meissner, A., Wernig, M., Plath, K., Jeanisch, R., Wagschal, A., Feil, R., Schreiber, S. L., & Lander, E. S. (2006). A bivalent chromatin structure marks key developmental genes in embryonic stem cells. *Cell*, *125*, 315–326.

Boef, H., Hauss, C., Graeve, F. D., Baran N., & Kedinger, C. (1997). Leukemia inhibitory factor-dependent transcriptional activation in embryonic stem cells. *J. Cell Biol.*, *138*, 1207–1217.

Boyer, L. A., Lee, T. I., Cole, M. F., Johnstone, S. E., Levine, S. S., Boyer, L. A., Plath, K., Zeitlinger, J., Brambrik, T., Medeiros, L. A., Lee, T. I., Levine, S. S., Wernig. M., Tajonar, A., Ray, M. K., Bell, G. W., Otte, A. P., Vidal, M., Gifford, D. K., Young, R. A., & Jaenisch, R. (2006). Polycomb complexes repress developmental regulators in murine embryonic stem cells. *Nature*, *441*, 349–353.

Bradley, A., Evans, M., Kaufman, M. H., & Robertson, E. (1984). Formation of germ line chimeras from embryo-derived teratocarcinoma cell lines. *Nature*, *309*, 255–256.

Burdon, T., Chambers, I., Stracey, C., Niwa, H., & Smith, A. (1999). Signaling mechanism regulating self-renewal and differentiation of pluripotent embryonic stem cells. *Cells Tissues Organs*,*165*, 131–143.

Cavaleri, F., & Schoeler, H. R. (2003). Nanog: a new recruit to the embryonic stem cell orchestra. *Cell*, *113*, 551–552.

Chambers, I., Silva, J., Colby, D., Nichols, J., Nijmeijer, B., Robertson, M., Vrana, J., Jones, K., Grotewold, L., & Smith, A. (2007). Nanog safeguards pluripotency and mediates germline development. *Nature*, *450*, 1230–1234.

Dani, C., Chambers, I., Johnston, S., Robertson, M., Ebrahimi, B., Saito, M., Taga, T., Li, M., Burdon, T., Nichols, J., & Smith, A. G. (1998). Paracrine induction of stem cell renewal by LIF-deficient cells: a new ES regulatory pathway. *Dev. Biol.*, *203*, 149–162.

Evans M. J., Kaufman M. H. Establishment in culture of pluripotential cells from mouse embryos. Nature. 1981 Jul 9;292(5819):154–6.

Furhman, G., Chung, C. K., Jackson, K. J., Hummelke, G., Baniahmad, A., Sutter, J., Sylvester, I., Schoeler, H. R., & Cooney, A. J. (2001). Mouse germline restriction of Oct4 expression by germ cell nuclear factor. *Dev. Cell.*, *1*, 377–387.

Gidekel, S., & Bergman, Y. (2002). A unique developmental pattern of Oct-3/4 DNA methylation is controlled by a cis-demodification element. *J. Biol. Chem.*, *277*, 34521–34530.

Hanna, L. A., Foreman, R. K., Tarasenko, E. A., Kessler, D. S., & Labonsky, P. A. (2002). Requirement for FoxD3 in maintaining pluripotent cells of the early mouse embryo. *Genes Dev.*, *16*, 2650–2661.

Heinrich, P. C., Behrmann, I., Muller-Newen, G., Schaper, F., & Graeve, L. (1998). Interleukin-6-type cytokine signaling through the gp130/Jak/STAT pathway. *Biochemical Journal*, *334*, 297–314.

James, D., Levine, A. J., Besser, D., & Hemmati-Brivanlou, A. (2005). TGFβactivin/nodal signaling is necessary for the maintenance of pluripotency in human embryonic stem cells. *Dev.*, *132*, 1273–1282.

Kim J. B, Sebastiano V., Wu G., Araúzo-Bravo M. J., Sasse P., Gentile L., Ko K., Ruau D., Ehrich M., van den Boom D., Meyer J., Hübner K., Bernemann C., Ortmeier C., Zenke M., Fleischmann B. K., Zaehres H., Schöler H. R. Oct4-induced pluripotency in adult neural stem cells. *Cell. 136*:411–9.

Kim, J. B., Zahres, H., Wu, G., Gentile, L., Ko, K., Sebastiano, V., Arauzo-Bravo, M. J., Ruau, D., Han, D. W., Zenke, M., & Schoeler, H. R. (2008). Pluripotent stem cells induced from adult neural stem cells by reprogramming with two factors. *Nature*, *454*, 646–650.

Kunath, T., Saba-El-Leil, M. K., Almousailleakh, M., Wray, J., Meloche, S., & Smith, A. (2007). FGF stimulation of the Erk1/2 signalling cascade triggers transition of pluripotent embryonic stem cells from self-renewal to lineage commitment. *Dev.*, *134*, 2895–2902.

Lee, T. I., Jenner, R. G., Boyer, L. A., Guenther, M. G., Levine, S. S., Kumar, R. M., Chevalier, B., Johnstone, S. E., Cole, M. F., Isono, K., Koseki, H., Fuchikami, T., Abe, K., Murray, H. L., Zucker, J. P., Yuan, B., Bell, G. W., Herbolsheimer, E., Hannett, N. M., Sun, K., Odom, D. T., Otte, A. P., Volkert, T. L., Bartel, D. P., Melton, D. A., Gifford, D. K., Jaenisch, R., & Young, R. A. (2006). Control of developmental regulators by Polycomb in human embryonic stem cells. *Cell*, *125*, 301–313.

Loh, Y. H., *et al.* (2006). The Oct4 and Nanog transcription network regulates pluripotency in mouse embryonic stem cells. *Nature Genet.*, *38*, 431–440.

Matsuda T., Nakamura T., Nakao K., Arai T., Katsuki M., Heike T., & Yokota T. (1999). STAT3 activation is sufficient to maintain an undifferentiated state of mouse embryonic stem cells. EMBO J. Aug 2;18(15):4261–4269.

Martin G (December 1981). "Isolation of a pluripotent cell line from early mouse embryos cultured in medium conditioned by teratocarcinoma stem cells". Proc Natl Acad Sci USA 78(12): 7634–7638. doi:10.1073/pnas.78.12.7634. PMID 6950406.

Nichols, J., Zevnik, B., Konstantinos Anastassiadis, K., Niwa, H., Klewe-Nebenius, D., Chambers, I., Schoeler, H. R., & Smith, A. (1998). Formation of pluripotent stem cells in the mammalian enbryo depends on the POU transcription factor Oct4. *Cell*, *95*, 379–391.

Niwa, H., Burdon, T., Chambers, I., & Smith, A. G. (1998). Self-renewal of pluripotent embryonic stem cells is mediated via activation of STAT3. *Genes Dev.*, *12*, 2048–2060.

Niwa, H., Miyazaki, J., & Smith, A. G. (2000). Quantitative expression of Oct-3/4 defines differentiation, de-differentiation or self-renewal of ES cells. *Nature Genet.*, *24*, 372–376.

Ogawa K, Saito A, Matsui H, Suzuki H, Ohtsuka S, Shimosato D, Morishita Y, Watabe T, Niwa H, Miyazono K. (2007) Activin-Nodal signaling is involved in propagation of mouse embryonic stem cells. *J Cell Sci.*, Jan 1;120(Pt 1):55–65.

Pesce, M., & Schoeler, H. R. (2001). *Oct-4*: Gatekeeper in the Beginnings of Mammalian Development. *Stem Cells*, *19*, 271–278.

Sato, N., Meijer, L., Skaltsounis, L., Greengard, P., & Brivanlou, A. H. (2004). Maintenance of pluripotency in human and mouse embryonic stem cells through activation of Wnt signaling by a pharmalogical GSK-3 specific inhibitor. *Nature Med.*, *10*, 55–63.

Smith, A. G. (2001). Embryo-derived stem cells: of mice and mice. *Annu. Rev. Cell. Biol.*, *17*, 435–462.

Taga, T., & Kishimoto, T. (1997). Gp130 and the interleukin-6 family of cytokines. *Ann. Rev. Immunol.*, *15*, 797–819.

Takahashi, K., & Yamanaka, S. (2006). Induction of pluripotent stem cells from mouse embryonic and adult fibroblast cultures by defined factors. *Cell*, *126*, 663–676.

Takahashi, K., Tanabe, K., Ohnuki, M., Narita, M., Ichisaka, T., Tomoda, K., & Yamanaka, S. (2007). Induction of pluripotent stem cells from adult human fibroblasts by defined factors. *Cell*, *131*, 861–872.

Wang, J., Rao, S., Chu, J., Shen, X., Lavasseur, D. N., Theunisen, T. W., & Orkin, S. H. (2006). A protein interaction network for pluripotency of embryonic stem cells. *Nature*, *444*, 364–368.

Watanabe, S., Umehara, H., Murayama, K., Okabe, K., Kimura, T., & Nakano, T. (2006). Activation of AKT signaling is sufficient to maitain pluripotency in mouse and primate embryonic stem cells. *Oncogene*, *25*, 2697–2707.

Wernig, M., Meissner, A., Foreman, R., Brambrik, T., Ku, M., Hochedlinger, K., Bernstein, B. E., & Jaenisch, R. (2007). *In vitro* reprogramming of fibroblasts into a pluripotent ES-cell-like state. *Nature*, *448*, 318–324.

Ying, Q., Wray, J., Nichols, J., Battle-Morera, L., Doble, B., Woodgett, J., Cohen, P., & Smith, A. (2008). The ground state of embryonic stem cell renewal. *Nature*, *453*, 519–523.

Ying, Q. L., Nichols, J., Chambers, I., & Smith, A. (2003). BMP induction of Id proteins suppresses differentiation and sustains embryonic stem cell self-renewal in collaboration with STAT3. *Cell*, *115*, 281–292.

Zhou H, Wu S, Joo JY, Zhu S, Han DW, Lin T, Trauger S, Bien G, Yao S, Zhu Y, Siuzdak G, Schöler HR, Duan L, Ding S. (2009). Generation of Induced Pluripotent Stem Cells Using Recombinant Proteins. *Cell Stem Cell*. Apr 22. [Epub ahead of print]

Zucker, J. P., Guenther, M. G., Kumar, R. M., Murray, H. L., Jenner, R. G., Gifford, D. K., Melton, D. A., Jaenisch, R., & Young, R. A. (2005). Core transcriptional regulatory circuitry in human embryonic stem cells. *Cell*, *122*, 947–956.

Chapter 12

Characteristics and Characterization of Human Pluripotent Stem Cells

Anne G. Bang and Melissa K. Carpenter

INTRODUCTION

Pluripotent stem cells are defined as cells that have the capacity to differentiate into all three germ layers, specifically: ectoderm; mesoderm; and endoderm, which can give rise to all cell types of the body. These cell populations can be derived from embryonic tissue sources such as the inner cell mass, the epiblast, and primordial germ cells or they can be induced by reprogramming using somatic cell nuclear transfer (SCNT), or the introduction of pluripotency genes into somatic cells. These different pluripotent stem cell populations have similarities and differences, as discussed below.

Embryonic Cell Sources

Embryonic Carcinoma Cells

Embryonic carcinoma (EC) cells were the first pluripotent cells identified. These cells were derived from malignant germ cell tumors called teratocarcinomas. These tumors are characterized by the presence of cell types from all three germ layers, as well as undifferentiated cells. Mouse EC cells were initially derived from testicular tumors, and were utilized as *in vitro* models of development. Human EC cells were subsequently derived, and these cells show surface marker expression that is different from mouse EC cells. For example, mouse EC cells express SSEA-1, while human EC cells express SSEA3, SSEA-4, TRA-1-60, and TRA-1-81. It remains unclear whether these differences in surface marker expression are correlated with functional differences between the cell lines. Another difference between mouse and human cells is their cytogenetic stability. While mouse EC cells are generally euploid, human EC cells are aneuploid, which limits their utility.

Embryonic Stem Cells

Mouse embryonic stem cells (ESCs) were first isolated, over 20 years ago, by removing the inner cell mass (ICM) from mouse pre-implantation embryos and serially passaging these cells on inactivated mouse feeder layers. The resulting cell lines showed unrestricted proliferative potential, and were pluripotent. These ESCs differentiated into derivatives of all three germ layers and, on injection into blastocysts, they contributed to all the tissue types in the resulting chimera. Because these cell lines can be genetically modified, they have been largely used for the generation of transgenic or knockout mice, but they also provide a unique tool for studying the early development of the embryo.

The first human ESC (hESC) lines were derived by Thomson *et al.* from the ICM of surplus human embryos resulting from assisted reproductive procedures. Since the first report in 1998, hESC lines have been derived in multiple laboratories throughout the world (see, for example, the NIH registry website at http://escr.nih.gov).

To date, ESCs have been derived from mouse, rat, and non-human primate and human embryos. While ESCs from these different species share many characteristics, there are also some differences. The defining features of ESCs include: derivation from preimplantation embryos; pluripotency; capacity for prolonged proliferation and self-renewal; a normal euploid karyotype; and the expression of distinctive markers. Importantly, not only the derivation methods, but also the culture conditions and handling of ESC lines determine their successful maintenance. Existing hESC lines were derived in different laboratories using different procedures, which may result in cell lines with different qualities. Although some of the lines have been characterized extensively, most cell lines have not been available for long enough to allow full characterization; therefore, further study will be required to definitively determine their differences and similarities.

Similar to EC cells, mouse, rat, and human ESCs show species specific differences in surface marker expression, but all express characteristic markers of pluripotency, including OCT4, SOX2, and NANOG (Table 12-1). Although all three species express common pluripotency markers, the rodent cell lines appear to be fundamentally different from

Stem Cell Anthology

TABLE 12-1 Comparison of Mouse, Rat, and Human Pluripotent Cells

	Mouse EC	Human EC	Mouse ES	Rat ES	Human ES	Mouse epiblast	Mouse EGCs	EGCs	Mouse iPSCs	Human iPSCs
Alkaline Phosphatase	+	+	+	+	+	–	+	+	+	–
SSEA-1	+	–	+	+	–	+	+	+	+	–
SSEA-3	–	+	–	–	+	–	–	+	–	+
SSEA-4	–	+	–	–	+	–	–	+	–	+
TRA-1-60	–	+	–	+	+	–	–	+	–	+
TRA-1-81	–	+	–	+	+	–	–	+	–	+
OCT 3/4	+	+	+	+	+	+	+	+	+	+
NANOG	+	+	+	+	+	+	+	+	+	+
SOX 2	+	+	+	+	+	+	+	N/A	+	+
REX 1	N/A	+	+	+	+	–	N/A	N/A	+	+
TERT	+	+	+	N/A	+	N/A	+	+	+	+
FGF4	+	+	+	+	+	N/A	N/A	N/A	+	+
UTF-1	+	+	+	N/A	+	N/A	N/A	N/A	+	N/A
CRIPTO	+	+	+	N/A	+	+	+	N/A	+	+

the human lines in the signaling pathways responsible for self-renewal. Mouse and rat ESC self-renewal is regulated via the LIF/STAT3 signaling pathway, while self-renewal in human ESCs appears to be regulated via the bFGF signaling pathway. It is unclear whether these differences are indicative of other fundamental differences between species.

ESCs can also be derived from parthenogenetic embryos. In this case, oocytes arrested at the second meiotic metaphase are activated in the presence of cytochalasin, such that extrusion of the second polar body is prevented. The resulting parthenogenetic zygote develops into a blastocyst from which diploid ESC lines that are genetically identical to the egg donor can be isolated (reviewed in Cibelli *et al.*, 2006). Using this or similar processes, ESCs have been derived from mouse, non-human primate, and human parthenogenetic blastocysts. Although this work has been very intriguing, several groups have found abnormalities in these ESCs which may impact their ability to function normally. More recently, it has been shown that ESCs derived from mouse parthenogenetic embryos can generate teratomas, and can contribute to chimeric mice after injection into the blastocyst, although no germline transmission was observed. Thus, ESCs derived from parthenogenetic embryos are capable of differentiation to many lineages; however, it is still unclear whether homozygosity in some regions of the genome and/or lack of paternal imprinting might disrupt cell function.

Embryonic Germ Cells

Primordial germ cells have been shown to be capable of generating ESC-like pluripotent stem cells called embryonic germ cells (EGCs). Mouse EGCs show the fundamental characteristics of pluripotent cells; they can generate teratomas, and can contribute to chimeric mice after injection into a blastocyst. Human EGCs were isolated in 1998, but to date, the proliferation of these cells appears quite limited (Turnpenny *et al.*, 2006). These cells can differentiate into various cell types *in vitro*, but teratoma formation has not been demonstrated. More recently, pluripotent stem cells have been derived from both mouse and human testis. It remains unclear whether these cells represent an endogenous cell population, or are a product of the *in vitro* culture system.

Epiblast Cells

As the early embryo develops, the cells transition from the totipotent zygote to the pluripotent ICM and epiblast. Recently, pluripotent cells have also been isolated from the mouse epiblast. These cells show expression of the pluripotency transcription factors OCT4, SOX2, and NANOG, differentiate *in vitro*, and generate teratomas, but do not contribute to chimeras. Similar to ESCs, these cells show extensive self-renewal; however, they require the presence of FGF and activin for proliferation, rather than LIF. Therefore, it has been speculated that human ESCs are more similar to mouse epiblast cells than mouse ESCs, and therefore actually represent human epiblast cells (Rossant, 2008). As these different cell populations are more extensively characterized, these distinctions should become clear.

Reprogrammed Cells

Pluripotency can be achieved by reprogramming. Reprogramming induces the dedifferentiation of somatic cells, which is thought to occur through the erasure of epigenetic signatures of the cell, resulting in an increase in potency. Most of the work in the reprogramming field has been done using nuclear transfer techniques (reviewed in Gurdon and Melton, 2008; Jaenisch and Young, 2008). More recently, several groups have demonstrated that reprogramming can be induced by the introduction of pluripotency genes. Successful reprogramming of somatic cells essentially "dedifferentiates" these cells to a pluripotent state, and allows the subsequent derivation of pluripotent cell lines. The generation of pluripotent cells using reprogramming allows generation of cell lines which are genetically identical to the donor.

Somatic Cell Nuclear Transfer

Somatic cell nuclear transfer (SCNT) takes advantage of a unique property of the oocyte cytoplasm that allows somatic nuclei to be reprogrammed to a pluripotent state. In this case, the nucleus of a somatic cell is transferred into an enucleated oocyte. The somatic nucleus is then reprogrammed, and partial development to the ICM stage can occur in culture, followed by either transplantation into a prepared uterus in order to generate cloned animals, or harvesting the ICM to generate ESC lines (reviewed in Yang *et al.*, 2007; Gurdon and Melton, 2008). In 2005, a group in South Korea reported the generation of human ESCs from patient specific blastocysts created using SCNT. Unfortunately, this work was later shown to be fraudulent, and to date, the generation of human ESC lines using SCNT has not been reported.

Induced Pluripotent Cells

More recently, it has been shown that pluripotency can be achieved by reprogramming adult and fetal somatic cells by the introduction of four genes, OCT4, SOX2, KLF4, c-MYC or OCT4, SOX2, NANOG, and Lin28, into mouse and human fibroblasts. The resulting induced pluripotent stem cells (iPSCs) show many of the characteristics of ESCs, including similar marker expression (Table 12-1), the ability to differentiate into all three germ layers and, in the case of mouse cells, to generate germline chimeras.

CHARACTERIZATION OF PLURIPOTENT CELLS

Markers

Identification of reliable markers for the characterization of pluripotent cells is of great importance in order to exploit their potential. To date, human pluripotent cell lines have been characterized using markers previously used to characterize other pluripotent stem cell populations, such as human embryonal carcinoma (EC) cells and mouse ESCs. The comparison of expression profiles of the different pluripotent cell populations presents similarities and dissimilarities between species and cell type. As shown in Table 12-1, expression of surface markers varies between these different cell populations. Similar to mouse ESCs, the hESCs express alkaline phosphatase-related antigens. The undifferentiated hESCs express the globoseries glycolipid antigens designated stage-specific embryonic antigen (SSEA-3 and SSEA-4), which are expressed by human EC cells, but not by mouse ESCs. In contrast to the mouse ESCs, the undifferentiated hESCs lack expression of a lactoseries oligosaccharide antigen SSEA-1. The keratin sulfate-related antigens, TRA-1-60 and TRA-1-81, expressed by human EC cells are also expressed by hESCs. Although the expression of surface markers has been a very useful tool in characterizing the different pluripotent cell populations, none of these markers has been functionally associated with pluripotency. For instance, the depletion of SSEA-4 from human ESCs does not impact the ability to proliferate or differentiate into all three germ layers. Therefore, there is considerable interest in finding markers which are predictive of the pluripotent state.

A number of transcription factors that play a critical role in maintaining stem cell self-renewal have now been identified, and analysis of their expression is also used to characterize pluripotent cells. The POU transcription factor, OCT4, is expressed in pluripotent cell populations, such as early embryo cells, blastomeres, and germ cells. *In vitro,* OCT4 is expressed in pluripotent cells, such as EC, ES, and EGCs, and is down-regulated on differentiation. Furthermore, it has been demonstrated that OCT4 is necessary to retain pluripotency in murine embryonic cells, and that the level of expression can influence the status of pluripotency and direct the cells toward specific phenotypes. Several target genes for OCT4 have been identified, including UTF1, REX1, PDGFαR, OTX2, Lefty-1, and NANOG, but their exact roles in pluripotency and self-renewal are unclear. Other transcriptional regulators such as SOX2 act cooperatively with OCT4. The molecular mechanisms through which OCT4 and various co-factors act to regulate ESC pluripotency and self-renewal are complex, and are discussed in greater detail in a reviews by Niwa (2001) and Jaenisch and Young (2008).

Expression profiling has been utilized to identify functional markers associated with pluripotency. Initial experimentation focused on the gold standard pluripotent cell, ESCs. Transcriptional profiles comparing mouse and human ESCs lines identified a set of 918 genes enriched in undifferentiated hESCs, compared to hESCs that had undergone differentiation. This set of genes included ligand–receptor pairs for the FGF, TGF-β–BMP, and Wnt signaling pathways. Some of these components corresponded to genes previously identified as "stemness" genes, indicating that a "core molecular program" may exist. In contrast, others have found a number of differences between expression profiles of mouse ESCs and human ESC lines using immunocytochemistry, reverse transcription-polymerase chain reaction (RT-PCR), and microarray analysis (Ginis *et al.*, 2004). As summarized in Table 12-1, ESCs from both species expressed transcription factors associated with pluripotent cells, such as OCT4, SOX2, TERT, UTF1, and REX1. Using RT-PCR and immunocytochemistry, significant differences were found in expression of vimentin, β-tubulin, α-fetoprotein, eomesodermin, HEB, ARNT, FOXD3, and the LIF receptor. Focused microarrays revealed profound differences in cell cycle regulation, control of apoptosis, and cytokine expression. These findings confirm that there are many differences between mouse and human ESCs, and demonstrate the need to use multiple methodologies to characterize these differences.

In order to identify a specific molecular signature associated with pluripotency, various groups have compared different pluripotent cell populations. Several groups have evaluated multiple hESC lines using microarrays, SAGE analysis, and transcriptional profiling to define "stemness". Rao and colleagues (Bhattacharya *et al.* 2004) proposed a set of pluripotency genes of around one hundred, based on transcriptional profiling. More recently, with the availability of increasing numbers of human pluripotent stem cell populations, researchers have begun defining regulatory networks and molecular signatures which are specific to pluripotent, multipotent, and unipotent stem cells. This has led to an unbiased metanalysis of multiple cell populations, and identified a set of genes associated with pluripotent cell populations termed PluriNet (Muller *et al.*, 2008). These data are intriguing, and demonstrate progress toward the identification pluripotent cell markers; however, broader application of this marker set is required to establish its predictive value.

Cytogenetic Analysis

Pluripotent cells have the capacity for prolonged cell proliferation *in vitro*, and the cytogenetic stability of these cells over time in culture is critical to their utility in both research and clinical applications. Currently, the most information about cytogenetic stability has been obtained from assessment of human ESCs. In most laboratories, this has been measured by G-banding, with most hESCs exhibiting a normal complement of chromosomes. There are now several reports in which maintenance of normal karyotype over long-term culture has been demonstrated. In contrast, some investigators have reported abnormal karyotypes in hESCs that are similar to those abnormalities observed in human ECs, such as trisomy 12 and 17. Further experimentation has suggested that, although hESCs maintain an apparently stable phenotype and karyotype over long-term culture, they are like other cells and can acquire mutations. Most recently, two separate groups have identified recurrent chromosomal abnormalities in hESCs maintained over long-term culture, including amplification at 20q11.21, which is associated with oncogenic transformation. It is important to note that abnormal cultures retain expression of standard hESC markers and pluripotency. It is unclear whether the presence of atypical cells in human ESC cultures will alter the fundamental characteristics of the cells or affect growth rate, cell cycle regulation, and capacity to differentiate; however, the occasional identification of these cells, albeit at a low frequency, highlights the importance of careful monitoring.

Differentiation

The fundamental characteristic of pluripotent cells is the ability to differentiate into cells of all three germ layers, which can be measured in several ways. Perhaps the most definitive test of pluripotency is the formation of chimeric mice. In this paradigm, ESCs are injected into the blastocyst, and the contribution of the ESCs to the resulting chimera is assessed to determine the differentiative capacity of the injected cells. Although this approach can assess mouse cells, this assay is clearly not suitable for human cells. Therefore, embryoid body (EB) formation *in vitro* and teratoma formation after injection into immunocompromised mice are currently used to measure pluripotency in the human system. This is accomplished by scoring the presence/absence of ectoderm, endoderm, and mesoderm in EBs and teratomas. However, quantitation of these assays has not yet been accomplished, making any quantitative evaluation of pluripotency in human cells particularly difficult. In order to develop more quantitative assays for differentiation of pluripotent cells, many groups are working on techniques to direct differentiation toward specific cell lineages. These "directed differentiation" techniques may ultimately allow for the assessment of the capacity to form specific cell types.

SUMMARY

Currently there are a number of human pluripotent cell populations that can be generated both from embryonic cell sources, and via reprogramming techniques. These different cell populations show similarities in expression of surface markers and transcription factors associated with pluripotency. To date, it remains unclear whether these markers will be predictive of the differentiative capacity of the pluripotent cells. Current efforts are directed towards identifying a molecular signature which can be used prospectively to identify pluripotent cell populations. Continued characterization of the existing and newly-created pluripotent cell populations is required to understand the mechanisms involved in retaining the undifferentiated state of these cells, as well as the mechanisms involved in differentiation.

KEY WORDS

Aneuploidy Any deviation from the normal haploid number of chromosomes.

Karyotype A standard arrangement of the chromosome complement. A normal human female karyotype would have each of the 22 pairs of autosomes (non-sex chromosomes) arranged in numerical order together with the two X chromosomes.

Pluripotent The capacity to give rise to cells representing all germ layers.

FURTHER READING

Bhattacharya, B., Miura, T., Brandenberger, R., Mejido, J., Luo, Y., Yang, A., Joshi, B., Ginis, I., Thies, R., Amit, M., Lyons, I., Condie, B., Itskovitz-Eldor, J., Rao, M.S., & Puri, R. (2004). Gene expression in human embryonic stem cell lines: unique molecular signature. *Blood, 103*, 2956–2964.

Cibelli, J. B., Cunniff, K., & Vrana, K. E. (2006). Embryonic stem cells from parthenotes. *Methods Enzymol., 418*, 117–135.

Ginis, I., Luo, Y., Miura, T., Thies, R., Brandenburg, R., Gerecht-Nir, S., Amit, M., Hoke, A., Carpenter, M., Itskovitz-Eldor, J., & Rao, M. (2004). Differences between human and mouse embryonic stem cells. *Developmental Biology, 269*, 360–380.

Gurdon, J. B., & Melton, D. A. (2008). Nuclear reprogramming in cells. *Science, 322*, 1811–1815.

Jaenisch, R., & Young, R. (2008). Stem cells, the molecular circuitry of pluripotency and nuclear reprogramming. *Cell, 132*, 567–582.

Muller, F. J., Laurent, L. C., Kostka, D., Ulitsky, I., Williams, R., Lu, C., Park, I. H., Rao, M. S., Shamir, R., Schwartz, P. H., Schmidt, N. O., & Loring, J. F. (2008). Regulatory networks define phenotypic classes of human stem cell lines. *Nature, 455*, 401–405.

Niwa, H. (2001). Molecular mechanism to maintain stem cell renewal of ES cells. *Cell Struct. Funct., 26*, 137–148.

Rossant, J. (2008). Stem cells and early lineage development. *Cell, 132*, 527–531.

Thomson, J. A., Itskovitz-Eldor, J., Shapiro, S. S., Waknitz, M. A., Swiergiel, J. J., Marshall, V. S., Jones, J. M. (1998). Embryonic stem cell lines derived from human biastocysts. Science, 282(5391): 1145–1147.

Turnpenny, L., Spalluto, C. M., Perrett, R. M., O'Shea, M., Hanley, K. P., Cameron, I. T., Wilson, D. I., & Hanley, N. A. (2006). Evaluating human embryonic germ cells: concord and conflict as pluripotent stem cells. *Stem Cells, 24*, 212–220.

Yang, X., Smith, S. L., Tian, X. C., Lewin, H. A., Renard, J. P., & Wakayama, T. (2007). Nuclear reprogramming of cloned embryos and its implications for therapeutic cloning. *Nat. Genet., 39*, 295–302.

Chapter 13

Multipotent Adult Progenitor Cells

Catherine M. Verfaillie and Annelies Crabbe

INTRODUCTION

In this chapter, we want to provide updated information regarding a rare cell population we have named multipotent adult progenitor cells, or MAPC. In 2001–2002, we published a series of papers demonstrating that, while attempting to select and culture mesenchymal stem cells (MSC) from human and subsequently mouse and rat bone marrow (BM), we accidentally identified a rare population of cells that has characteristics unlike most adult somatic stem cells, in that they appear to proliferate without senescence, and have broad differentiation ability *in vitro* and *in vivo* (Reyes *et al.*, 2001; Jiang *et al.*, 2002).

ISOLATION AND PHENOTYPIC CHARACTERIZATION OF MULTIPOTENT ADULT PROGENITOR CELLS

MAPC can be cultured from mouse, rat, swine, and human bone marrow (BM). MAPC from all species are significantly smaller than their mesenchymal stem cell (MSC) counterpart, and differ phenotypically from MSC. We have phenotyped murine MAPC most extensively. Unlike MSC, murine MAPC do not express major histocompatibiliy (MHC)-class I antigens, do not express or express only low levels of the CD44 antigen, and do not express CD34 (Jiang *et al.*, 2002; Breyer *et al.*, 2006; Ulloya-Montoya *et al.*, 2007). Unlike hematopoietic stem cells (HSC), murine MAPC do not express CD45 and CD34, but like HSC, mouse MAPC express high levels of cKit. Unlike MSC, MAPC also do not express the PDGF-Rβ, but express PDGF-Rα (Ulloya-Montoya *et al.*, 2007). Human MAPC are HLA-class one low, do not express CD45 and CD34, but express low levels of CD133 (Aranguren *et al.*, 2007).

Unlike most adult somatic stem cells, MAPC proliferate without obvious signs of senescence, and have active telomerase. In humans, the length of MAPC telomeres is 3–5-kB longer than in neutrophils and lymphocytes, and telomere length is no different when MAPC are derived from young or old donors. This suggests that MAPC are derived from a population of cells that either has active telomerase *in vivo*, or that is highly quiescent *in vivo*, and therefore have not yet incurred telomere shortening *in vivo*. Human MAPC that are expanded to >70 population doublings, and hence well beyond the Hayflick point (Shay and Wright, 2000), remain cytogenetically normal. Likewise, MAPC isolated from swine bone marrow can be expanded to more than 80–100 population doublings without cytogenetic abnormalities, even though some cytogenetic abnormalities are seen when cells continue to be expanded for a longer period of time (Zeng *et al.*, 2006). Rodent MAPC can be culture-expanded for several hundreds of population doublings. However, after many rounds of cryopreservation and expansion, several populations of mouse and rat MAPC have acquired numerical, as well as structural, chromosomal abnormalities (manuscript in preparation). Such abnormalities are, however, not present in early passage cells (manuscript in preparation). We have not found significant cytogenetic abnormalities in human cell populations. The phenomenon of acquisition of cytogenetic abnormalities, more pronounced perhaps in rodent compared to larger mammalian cell populations, is a common problem seen when stem cells are expanded long-term *in vitro* (Rosler *et al.*, 2004; Maitra *et al.*, 2005; Mitalipova *et al.*, 2005; Bochkov *et al.*, 2007) and is a reason why many believe that safe clinical application of *ex vivo* expanded stem cells will likely require earlier passage cells to be used.

Although isolation of MAPC remains technically demanding, the maintenance of established cell lines in their undifferentiated form has become significantly less demanding. The challenge in MAPC isolation is that cells with morphological and phenotypic features of MAPC can only be detected in bone marrow cultures after several population doublings. As will be discussed below, we still have not yet determined whether MAPC exist as such in the freshly isolated bone marrow, or whether this cell phenotype

Stem Cell Anthology

is acquired via the culture process. Culture conditions that affect isolation of MAPC include cell density, CO_2, and O_2 among others. An update on the culture conditions and quality control of rodent MAPC can be found in the manuscript by Breyer *et al.* (2006). A similar report will be forthcoming for the isolation and maintenance of human cells. Because of these optimizations, new investigators unfamiliar with the technology can now maintain the different cell lines with minimal instruction.

We have evidence that, under the same culture conditions, clonal populations can be isolated that differ in relation to cell phenotypic and expressed gene profile. These differences correlate with the breath of differentiation that can be observed for the different cell populations, as discussed in Ulloa *et al.* (2007). For instance, we have described that multiple murine cell populations isolated from a single bone marrow, or from different bone marrow samples, differ in the level of Oct4 expression, ranging from no Oct4 at all (mClone-1) (Ulloa *et al.*, 2007), to levels of Oct4 as high as 10% of that of murine ESC (qRT-PCR compared with the mRNA level of GAPDH: some clones express Oct4 at a delta CT of 4–8, with ESC expressing Oct4 at a delta CT of 2–3 (Ulloa *et al.*, 2007)). Likewise, when isolating clonal populations of rat MAPC, we have found cell populations without Oct4 (rClone-1) (Ulloa *et al.*, 2007), and populations with Oct4 mRNA at a delta CT compared with the household gene GAPDH of 2 (rMAPC-1 (Ulloa *et al.*, 2007), and several newer isolations).

Cell populations that do not express Oct4 cluster more closely by transcriptome analysis to MSC, whereas cell populations expressing higher levels of Oct4 cluster more closely with ESC. The expression level of Oct4 is also associated with differences in differentiation potential of the clonal cell lines: cells that do not express Oct4 differentiate significantly less well into cells with endothelial, endodermal, and neuroectodermal characteristics.

MAPC that express Oct4 also express a number of other transcription factors known to be necessary for the pluripotency of ESC, such as Sall-4 (Zhang *et al.*, 2006), Rex-1 (Cenki *et al.*, 2003), Tbx-3 (Rowley *et al.*, 2004), among others, or necessary for the induction of a pluripotent state in adult cells, such as Lin28 (Viswananthan *et al.*, 2008) or nMYC (Cartwright *et al.*, 2005). Noteworthy, MAPC that express Oct4 do not express Sox2 (Lewitzky and Yamanaka, 2007) and Nanog (Pan and Thomson, 2007), the two other key transcription factors involved in a recently well-characterized feed-back transcription system that maintains ESC undifferentiated (Chew *et al.*, 2005; Okita and Yamanaka, 2006). However, MAPC express several transcription factors involved in mesodermal and endodermal differentiation, including Gata-4 and Gata-6 (Fletcher *et al.*, 2006), myocardin (Wang and Olson, 2004), Pitx2 (Franco and Campione, 2003), Foxa2 (Friedman and Kaestner, 2006), and Sox17 (Sinner *et al.*, 2004), among others. This may suggest that MAPC can be placed developmentally between the epiblast stage and a primitive streak/mesendoderm committed cell type, which is perhaps consistent with the differentiation ability they exhibit (Figure 13-1).

IN VITRO DIFFERENTIATION POTENTIAL OF MULTIPOTENT ADULT PROGENITOR CELLS

MAPC from all species can be successfully differentiated into typical mesenchymal lineage cells, including osteoblasts, chondroblasts, and adipocytes. In contrast to MSC, human, swine, mouse, and rat MAPC can be induced to differentiate into cells with morphological, phenotypic and functional characteristics of endothelial cells (EC) (Jiang *et al.*, 2002; Ross *et al.*, 2006). Moreover, like ESC but unlike cord blood or bone marrow-derived CD133 cells in human,

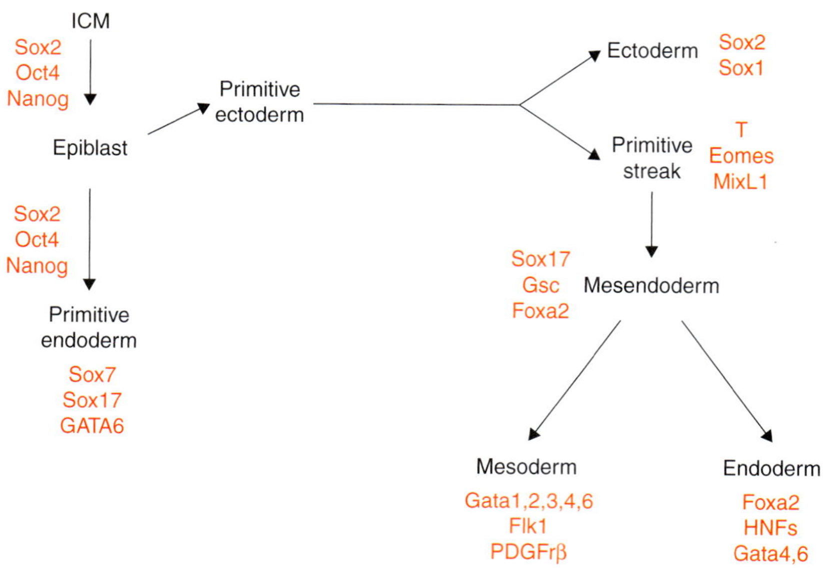

FIGURE 13-1 MAPC appear to have a phenotype consistent with a developmental stage between epiblast and mesendoderm/endoderm-mesoderm.

we have demonstrated that human MAPC treated with vascular endothelial growth factor (VEGF) start to express arterial and venous endothelium specific transcripts and proteins, and that addition of the Notch ligand, delta-like-4 (Dll4) and sonic hedgehog (Shh) inhibits venous, while enhancing arterial, endothelium specification (Aranguren *et al.*, 2007). This is true not only *in vitro*, but also when human MAPC are implanted in Matrigel *in vivo*.

MAPC from all species can be coaxed to differentiate into cells with phenotypic and functional characteristics of smooth muscle cells (SMC) (Ross *et al.*, 2006). Like primary aorta SMC from newborn rats, MAPC-SMC exhibit functional calcium channels. When entrapped in fibrin, in circular vascular molds, MAPC-SMC circumferentially align to generate ring-shaped constructs. Like constructs generated with primary SMC, MAPC-SMC derived constructs generate force in response to KCl, as well as a number of other agonists, including the L-type channel opener FLP64176, serotonin or endothelin-1, and complete relaxation when Rho-Kinase is inhibited.

We have published that human, mouse, rat, and swine MAPC can be committed down an endodermal path, more specifically giving rise to cells that have morphological, phenotypic and functional characteristics of hepatocytes (Jiang *et al.*, 2002; Zeng *et al.*, 2006; Ross *et al.*, 2006). In the initial studies we demonstrated that this could be accomplished by addition of high concentrations of dexamethasone and the cytokines, hepatocyte growth factor (HGF), and fibroblast growth factor-4 (FGF-4). Subsequent studies by Snykers *et al.* (2006) demonstrated that when the cytokines are added sequentially rather than simultaneously, more robust differentiation of MAPC to hepatocyte-like cells could be found. The same cytokines, HGF and FGF4, also induce human ESC towards cells with characteristics of hepatocyte-like cells (Schwartz *et al.*, 2005), although the percentage of albumin positive cells was low, and differentiation required initial culture in embryoid bodies (EB) (Schwartz *et al.*, 2005). Since the initial description of hepatic-like cell differentiation from cells with greater potency, we have further optimized the differentiation schema, and have developed a four-step protocol that allows not only ESC, but also MAPC to differentiate sequentially to mesendoderm/definitive endoderm—and this following expression of genes found during primitive streak formation, hepatoblasts, and ultimately hepatocyte-like cells. For ESC, differentiation can be obtained without intermediate EB formation. Progeny from such differentiation cultures is still impure, with 10–20% of cells expressing mature hepatocyte proteins, with persistent immature hepatoblasts, as well as continued presence of mesodermal cells, and undifferentiated ESC and MAPC. Nevertheless, functional studies have demonstrated that even such a mixed cell population contains a significant number of cells that display many functional features of mature hepatocytes (manuscript in preparation).

Finally, in 2003 we published that rodent MAPC can be induced down a neuroectodermal path, generating neurons expressing dopamine and with functional voltage-gated sodium channels. Since then we have continued to evaluate whether, like ESC, MAPC could be induced to commit to a neural stem cell (NSC) fate using methods described for ESC. Preliminary studies demonstrate that a neural stem cell phenotype (based on transcriptome and protein analysis) can be obtained from MAPC with culture conditions described for ESC (Conti *et al.*, 2005).

Interestingly, the lineage that continues to be elusive is cardiac myoblasts, a differentiation path that is almost a default differentiation pathway for mouse ESC. Although we have shown that expression of the transcription factors Nkx2.5, Tbx5, Mef2C, more mature cardiac structural proteins such as troponins, and myosin heavy chain mRNA and proteins can be induced, we have not observed spontaneous rhythmic contracting or beating cell populations. As we have solid evidence that MAPC can generate many types of mesodermal cells, as well as hepatocyte-like cells, following well-known developmental intermediates, the lack of differentiation to cardiomyocytes remains unexplained.

IN VIVO ENGRAFTMENT OF MULTIPOTENT ADULT PROGENITOR CELLS

In the initial description of murine MAPC in 2002, we demonstrated that when undifferentiated murine MAPC were transplanted intravenously in minimally irradiated murine hosts, low levels of engraftment within the hematopoietic system could be detected, and some engraftment was seen in the liver, lung and gastrointestinal epithelium (Jiang *et al.*, 2002). Since these initial studies, many more studies have been done to assess the ability of undifferentiated rodent, swine, and human MAPC to contribute to tissue formation/repair *in vivo*.

MULTIPOTENT ADULT PROGENITOR CELLS IN ISCHEMIA MODELS

Peripheral Vascular Disease

MAPC were injected in a mouse model of limb ischemia (Aranguren *et al.*, 2008). Murine undifferentiated MAPC, or a combination of MAPC-derived EC and MAPC-derived SMC, were compared with mouse bone marrow cells. Human MAPC were compared with human CD133+ cells from bone marrow or cord blood. Significant and persistent improvement in limb perfusion and muscle function were seen with undifferentiated MAPC, which outclassed human CD133+ cells and murine MAPC-EC/MAPC-MSC. Of note, murine BM cells after an initial improvement in perfusion and limb function, caused deterioration in the vascular status, as well as function of the limb, for reasons that are not fully understood. The mechanisms underlying the effect of MAPC are likely mixed; the undifferentiated MAPC contribute, albeit relatively modestly, to endothelium, smooth muscle, and skeletal muscle. However, the number and size of host blood vessels was significantly increased following MAPC therapy, with decreased fibrosis, necrosis and significant increase in host myogenesis.

Cardiac Ischemia

In a second series of studies, rodent and swine MAPC were injected in the peri-infarct zone of the heart following left anterior descendent (LAD) artery ligation. Compared to BM mononuclear cells and fibroblasts, murine MAPC improved LV contractile function significantly for at least eight weeks post-transplantation, with a significantly greater decrease in scar size. MAPC, bone marrow cells or fibroblasts were present beyond one week; therefore, the effect of MAPC was not by differentiation and direct contribution to the vascular or cardiomyocyte compartment (Pelacho *et al.*, 2007). However, there was significantly greater vascularity eight weeks post-transplantation, possibly due to the production of monocyte chemoattractant protein-1 (MCP)-1, VEGF, platelet derived growth factor (PDGF)-BB, and transforming growth factor (TGF)-β_1, secreted by MAPC. Tolar *et al.* (2007) also grafted MAPC in the peri-infarct zone of murine hearts, using both immunocompetent, syngeneic C67Bl/6 mice and Rag2/IL-2Rγc−/− immunodeficient animals. Significantly more MAPC could be detected in the heart of Rag2/IL-2Rγc−/− than C57Bl/6 mice. However, left ventricular contractile function was significantly better in C57BL/6 than Rag2/IL-2Rγc−/− recipients, likely due to increased vascular density in infarct and peri-infarct zones. Thus, the immune competence of the recipient appears to modulate the therapeutic impact of MAPC cells in myocardial infarcts, and a more vigorous immune response may be advantageous for therapeutic myocardial repair. In a third rodent study, Agbulut *et al.* (2006) grafted human MAPC in a rat model of chronic myocardial infarction. At one month the left ventricular function of untreated rats deteriorated significantly. Although the ejection fraction of cell treated animals remained more stable, the difference did not reach statistical significance. While MAPCs could be identified early post-transplant, no evidence of engraftment was further observed at one month by immunohistochemistry, electron microscopy, or PCR. Hence, in this model, MAPC did not improve global pump function, and although some of these cells expressed endothelial markers during the early post-transplant period, MAPC did not differentiate into cardiomyocytes.

In another study, swine MAPC were grafted in the peri-infarct zone of the heart of allogeneic swine. Some animals were treated with cyclosporine, whereas others did not receive immunosuppression (Zeng *et al.*, 2007). The beneficial effect of the MAPC was seen irrespective of the immunosuppressive therapy. Again, few cells persisted in the heart, but they induced significant neovascularization and cardiomyocyte regeneration that stabilized bioenergetic and contractile function in the infarct zone and border zone. Magnetic resonance imaging (MRI) demonstrated that the infarct zone systolic thickening fraction improved significantly from systolic "bulging" in untreated animals to active thickening, while the left ventricular ejection fraction improved significantly. This study thus demonstrates a possible role for allogeneic MAPC in the treatment of ischemic disease.

Stroke

A similar trophic effect as shown in peripheral vascular disease models, as well as cardiac ischemia models, was also seen when human MAPC were grafted in the penumbra of stroke in cyclosporine treated rats that underwent a middle cerebral artery occlusion (MCAO) seven days earlier. Although significant functional improvement was seen on week six, no significant contribution of the grafted cells to the vasculature or the brain parenchyma was seen; whether cyclosporine is needed to detect the beneficial effect was not tested (Zhao *et al.*, 2003).

In a second study, Yasuhara *et al.* (2006) tested the effect of syngeneic or allogeneic rat MAPC injected in the hippocampus of seven-day-old rats subjected to unilateral hypoxic-ischemic brain injury by ligation of the common carotid artery and subsequent exposure to hypoxic environment. All animals received immunosuppression. There was a significant reduction in motor asymmetry in MAPC-treated animals, and MAPC-treated animals could stay significantly longer on the rotarod, whether syngeneic or allogeneic cells were grafted. Although some cells persisted, again no significant differentiation towards neurons was seen. In a subsequent study, similar results were also seen when MAPC were transplanted systemically, compared with direct intra-brain transplantation (Yasuhara *et al.*, 2008).

SYSTEMIC TRANSPLANTATION OF MULTIPOTENT ADULT PROGENITOR CELLS

Tolar *et al.* (2006) demonstrated that MAPC which express low levels of major histocompatibility (MHC)-class I molecules do not stimulate T-cell alloresponses, and are targets of NK cytolysis. Following transplantation *in vivo*, MAPC thus persist best in animals devoid of NK cells. Based on this data, many of the studies, including transplantation of rodent MAPC in the ischemic limb model and myocardial infarct model described above, as well as the studies evaluating the differentiation of MAPC after systemic transplantation, were done using animals in whom NK cells were eliminated by anti-asialo-GM1 antibodies.

When murine MAPC from CD45.2/GFP donor mice were intravenously injected into sub-lethally irradiated NOD–SCID mice also treated with anti-asialo-GM1 antibodies, MAPC-derived CD45.2+ donor cells could be detected in the blood, bone marrow, and spleen at 10–20 weeks after transplantation (Serafini *et al.*, 2007). However, MAPC could not rescue animals from radiation-induced pancytopenia. MAPC generated 1–2% ckit+ Sca1+ Lin− (KSL cells) cells, indicating that they can generate LT-HSC. This was confirmed by transplantation of Lin− cells, as well as KSL cells, in lethally-irradiated secondary recipients, where they could re-establish long-term hematopoiesis.

Chimera Formation

Finally, we have shown that transfer of 10–12 mouse MAPC into mouse blastocysts results in the generation of chimeric mice, with the degree of chimerism varying between 1–40% in about 50% of mice (Jiang *et al.*, 2002). When more recently derived MAPC lines were used, significantly less chimerism was seen, with a maximal 2% degree of chimerism. The reason for the differences between lines is not known. Also not known is whether the initial high levels of chimerism seen in rare animals is due to fusion between MAPC and cells within the inner cell mass, or via direct contribution. Also not yet known is whether MAPC contribute to the germline when injected in the blastocyst.

OTHER ADULT STEM CELLS WITH GREATER DIFFERENTIATION POTENTIAL

Since our publication of 2002, many other groups have isolated cells from bone marrow, cord blood, and amniotic fluid with characteristics similar to those of MAPC. Specifically bone marrow-derived multipotent stem cells (BMSC) (Yoon *et al.*, 2005), marrow-isolated adult multilineage inducible cells (MIAMI) (D'Ippolito *et al.*, 2004), pre-mesenchymal stem cells (Anjos-Afonso and Bonnet, 2007), very small embryonic-like cells (VSEL) (Kucia *et al.*, 2006), and multipotent adult stem cells (MASC) (Beltrami *et al.*, 2007) have been isolated from bone marrow of mouse and man, amniotic fluid derived stem (AFS) cells are isolated from amniotic fluid (De Coppi *et al.*, 2007), and unrestricted somatic stem cells (USSC) come from umbilical cord blood (Kogler *et al.*, 2006). All these cell populations are CD45−ve cells with some degree of variation in expression of cell surface antigens. They can be cultured for extended periods, and differentiate to cells of at least two different germlayers. Some of these cell populations may express the ESC-specific transcription factors, Oct4, Sox2, and Nanog (Lewitzy and Yamanaka, 2007; Pan and Thomson, 2007; Niwa *et al.*, 2000). It should, however, be noted that for human cells, care needs to be taken when assessing Oct4 and Nanog expression, as multiple pseudogenes exist for these transcription factors, making RT-PCR evaluation technically difficult (Liedtke *et al.*, 2008). An overview of the different adult stem cells with greater differentiation potential can be seen in Table 13-1.

POSSIBLE MECHANISMS UNDERLYING THE GREATER POTENCY OF MULTIPOTENT ADULT PROGENITOR CELLS

Currently we do not fully understand the mechanism(s) underlying the greater potency of MAPC. We have definitive data to demonstrate that the greater potency of MAPC is not due to presence of multiple stem cells within the cultured MAPC population. Using retroviral marking studies, we have shown that a single cell can differentiate *in vitro* to cells of mesoderm, both mesenchymal and non-mesenchymal, neuroectoderm, and hepatocyte-like cells (Jiang *et al.*, 2002).

A second possibility for the greater degree of differentiation potential would be that cells undergo fusion and acquire greater pluripotency via this mechanism. Fusion has been shown to be responsible for apparent ES characteristics of marrow and neural stem cells that had been co-cultured with ES cells *in vitro*, and for the apparent lineage switch of bone marrow cells to hepatocytes, neural cells, and cardiac and skeletal muscle cells. In the ESC-neural/hematopoietic cell fusions occurring *in vitro* (Ying *et al.*, 2002; Terada *et al.*, 2002), the majority of genes expressed in the marrow or neural cells that fused with ESC were silenced, and the majority of the genes expressed in ESC were persistently expressed. Likewise for the bone marrow-hepatocyte fusion occurring when hematopoietic cells were infused in animals with hereditary tyrosinemia due to lack of the fumarylacetoacetate hydroxylase (FAH) gene, the majority of genes expressed normally in hematopoietic cells (except the FAH gene) were silenced, whereas gene expressed hepatocytes predominated (Dorrell and Grompe, 2005; Camargo *et al.*, 2004; Wang *et al.*, 2003). The fused cells were in general tetraploid or aneuploid. It is unlikely that this phenomenon underlies the greater potency of MAPC, as cultivation and differentiation *in vitro* does not require that MAPC are co-cultured with other cells, making the likelihood that MAPC are the result of fusion very low. Smith *et al.* suggested in a commentary that MAPC could be caused by fusion of multiple cell types early on during culture leading to reprogramming of the genetic information and pluripotency. However, we have no evidence that MAPC, even early during culture, are tetraploid or aneuploid, making this possibility less likely.

TABLE 13-1 An Overview of Different Adult Stem Cells with Greater Differentiation Potential

Name	Source	Phenotype	ES transcription factors	Differentiation capacity
MAPC (Jiang *et al.*, 2002)	Bone marrow	Mouse: CD45−, PDGFRα+, ckit+, Sca1^{low} Rat: CD45−, CD44^{dim}, CD31+ Human: CD45−, CD44+, CD105−, CD90+, MHC−I^{dim} Swine: CD45− CD45−, MHC−	Oct4+, Nanog− Sox2−	Mesoderm, endoderm, and neuroectoderm-like cells Some mouse lines contribute to chimeric animals
BMSC (Yoon *et al.*, 2005)	Human bone marrow	CD29−, CD44−, CD73−, CD105−, CD90−, CD117−	Oct4−	Mesoderm, endoderm, and neuroectoderm-like cells
MIAMI (D'Ippolito *et al.*, 2004)	Bone marrow	CD34−, CD36−, CD45−, CD117−, HLA−, CD29+, CD63+, CD81+, CD122+, CD164+	Oct4+, Rex1+	Mesoderm, endoderm, and neuroectoderm-like cells
Pre-MSC (Anjos-Afonso & Bonnet, 2007)	Murine bone marrow	Lin−, CD45−, CD31− or CD45−, CD11− and SSEA1+	Oct4+, Nanog+, Sox2+	Mesoderm, endoderm, and neuroectoderm-like cells
VSEL Kucia *et al.*, 2006)	Human and mouse bone marrow/umbilical cord blood	Sca1+, lin−, CD45−	Oct4+, Nanog+, Sox2+	Mesoderm, endoderm, and neuroectoderm-like cells
AFS (De Coppi *et al.*, 2007)	Human amniotic fluid	CD45−, CD34−, CD133+, CD29+, CD44+, CD73+, CD90+, CD106+, MHCI+	Oct4+	Mesoderm, endoderm, and neuroectoderm-like cells
USSC (Kogler *et al.*, 2006)	Human umbilical cord blood	CD14−, CD33−, CD34−, CD45−, CD49−, CD50−, CD62−, CD106−, CD117−, CD13+, CD29+, CD44+, CD90+, CD105+	Oct4−, Nanog−	Mesoderm, endoderm, and neuroectoderm-like cells
MASC (Beltrami *et al.*, 2007)	Human bone marrow Liver Heart	CD13+, CD49b+, CD90+, CD73−, CD44−, HLA−, CD29−, CD105−, CD49a−	Oct4+, Rex1+, Nanog+	Mesoderm, endoderm, and neuroectoderm-like cells
Germ cells (Kanatsu-Shinohara *et al.*, 2004)	Mouse testis	SSEA1+, EE2+, CD9+, integrin+, EpCAM+, ALP+	Oct4+, Rex1+, Nanog+	Embryoid bodies and teratomas
MASC (Seandel *et al.*, 2007)	Mouse testis	GPR125+, ckit+, PLZF−, DAZL−	Oct4+, Nanog+, Sox2+	Embryoid bodies, teratomas, and partial chimeric mice
maGSC (Guan *et al.*, 2006)	Testis	Sca1^{dim}, Ter119−, CD34−	Oct4+, Nanog+, Rex1+	Mesoderm, endoderm, and neuroectoderm-like cells Embryoid body formation Teratoma formation Chimeric mice, including germline

The *in vivo* studies were not set up to rule out this possibility fully. As indicated above, it remains possible that the chimera formation seen with some of the MAPC lines might be due to fusion of MAPC with cells in the inner cell mass (Jiang *et al.*, 2002). The contribution of MAPC to skeletal muscle in the limb ischemia model is certainly due to fusion, as muscle fibers are the result of fusion events (Aranguren *et al.*, 2008). Whether MAPC also generate satellite cells, the muscle fiber progenitor, was not addressed. We do, however, have evidence that skeletal muscle can be generated from mouse and human MAPC *in vitro*, without the need for cell fusion (Araguren *et al.*, 2008; Muguruma *et al.*, 2003). When injected in ischemic swine hearts, rare cardiac muscle cells, apparently derived from donor MAPC, were seen (Zeng *et al.*, 2007). Likewise, when murine MAPC were grafted intravenously in sublethally irradiated mice, rare apparently donor-derived cardiomyocytes were seen (Serafini *et al.*, 2007). However, the studies did not allow us to determine whether this is due to fusion or not. The contribution of MAPC to the hematopoietic system is, however, unlikely due to fusion, as no cells could be found that co-expressed markers from host and donor origin, and cytogenetic analysis demonstrated that the MAPC-derived hematopoietic cells were euploid (Serafini *et al.*, 2007).

Currently, we do not know whether MAPC exist *in vivo*, or are a culture-induced phenomenon. Several approaches have been used to isolate cells from murine or rat bone marrow or brain with a phenotype of culture established MAPC, including expression of SSEA1, c-Kit, CD31, PDGFRα, but absence of CD45 or Sca1, and presence of Oct4 (using Oct4-GFP transgenic mice obtained from The Jackson Laboratory or Oct4-knockin add-on GFP mice, a kind gift from Dr R. Jaenisch) (Lengner *et al.*, 2007). No cells with the phenotype or morphology of MAPCs could be isolated. During MAPC isolation, cells are maintained for three to four weeks prior to subcloning at five cells following column depletion of CD45+/Ter119+ cells. If MAPC are present in fresh bone marrow, cells proliferating from the five-cell subcloning step would contain 20% MAPC. However, we see the appearance of small cells that express Oct4 and that proliferate relatively rapidly out-growing the rest of the culture only after six to eight weeks following subcloning. This suggests strongly that MAPC may not exist in rodent tissues, but are in part generated in culture. This notion is consistent with the data from Afonso *et al.* (2007), who demonstrated that rare cells in the bone marrow of mice expressing SSEA1 and very low levels of Oct4 mRNA can, under MAPC culture conditions, be expanded *in vitro* with a 100–10 000-fold further induction of Oct4 transcript levels following culture. It should be noted that the low levels of Oct4 expression in the freshly-isolated cells was quickly lost when cells were cultured under MSC conditions, and that cells that did not express SSEA1 or Oct4 could not be induced to express these markers, even when cultured under MAPC conditions. These studies may thus suggest that MAPC are generated from a rare sub-population of cells within bone marrow under the specific culture conditions used to isolate and expand the cell population.

Consistent with the notion that culture may induce a more immature phenotype are studies from three groups demonstrating that, when spermatogonial stem cells are cultured *ex vivo*, cell populations are established that have nearly all or all characteristics of ESC, such as expression of Nanog and Sox2 after two weeks to two to three months, aside from Oct4, differentiation to embryoid bodies, and for one of the cell populations contribution to all tissues, including the germline, in chimeric mice. These cells were named multipotent adult stem cells or MASC, and multipotent/adult germline stem cells or maGSC (Beltrami *et al.*, 2007; Kanatsu-Shinohara *et al.*, 2004; Guan *et al.*, 2006).

This notion is obviously also supported by the recent findings by, initially, the group of Dr S. Yamanaka (Takahashi and Yamanaka, 2006), and since then by many other groups (Okita *et al.*, 2007; Wernig *et al.*, 2007; Yu *et al.*, 2007; Lowry *et al.*, 2008; Park *et al.*, 2008), that it is possible to reprogram mouse and human terminally differentiated cells to an embryonic-like cell, called induced pluripotent stem cells, following the overexpression of Oct4 and Sox2, alone or combined with either Klf4, c-Myc, Nanog, and/or Lin28. There is already evidence that transduction with transcription factors may not be necessary if the same signaling pathways can be activated using small molecules (Shi *et al.*, 2008). Hence, we hypothesize that culture of bone marrow cells, under stringent MAPC conditions, may lead to the reprogramming of certain cells in the marrow to a MAPC-like state. It will obviously be of importance to understand the mechanism(s) underlying the putative spontaneous reprogramming to enhance the efficiency of this phenomenon.

KEY WORDS

Adult stem cell Cell derived from postembryonic stage tissue with self-renewal ability, and ability to generate all cell types of the tissue where it has been derived from, but not other tissues.

MAPC (multipotent adult progenitor cell) Cell isolated from murine bone marrow and created in culture, able to self-renew and generate certain cell types (e.g., endoderm, mesoderm).

Neural stem cell A type of an adult stem cell located in the brain. When there is an injury, these cells can move towards the site of injury, where some can replace a mature neuron, astrocyte or oligodendrocyte.

Trophic effect When stem cells are transplanted in a damaged tissue, they release certain growth factors and chemokines in order to stimulate the endogenous stem cells to replace the lost cells.

FURTHER READING

Agbulut, O., *et al.* (2006). Can bone marrow-derived multipotent adult progenitor cells regenerate infarcted myocardium? *Cardiovasc. Res.*, *72*, 175–13.

Anjos-Afonso, F., & Bonnet, D. (2007). Nonhematopoietic/endothelial SSEA-1+ cells define the most primitive progenitors in the adult murine bone marrow mesenchymal compartment. *Blood*, *109*, 1298–1306.

Aranguren, X. L., *et al.* (2007). *In vitro* and *in vivo* arterial differentiation of human multipotent adult progenitor cells. *Blood*, *109*, 2634–2642.

Aranguren, X. L., *et al.* (2008). Multipotent adult progenitor cells sustain function of ischemic limbs in mice. *J. Clin. Invest.*, *118*, 505–514.

Beltrami, A. P., *et al.* (2007). Multipotent cells can be generated *in vitro* from several adult human organs (heart, liver, and bone marrow). *Blood*, *110*, 3438–3446.

Bochkov, N. P., *et al.* (2007). Chromosome variability of human multipotent mesenchymal stromal cells. *Bull. Exp. Biol. Med.*, *143*, 122–126.

Breyer, A., *et al.* (2006). Multipotent adult progenitor cell isolation and culture procedures. *Exp. Hematol.*, *34*, 1596–1601.

Camargo, F. D., Finegold, M., & Goodell, M. A. (2004). Hematopoietic myelomonocytic cells are the major source of hepatocyte fusion partners. *J. Clin. Invest.*, *113*, 1266–1270.

Cartwright, P., *et al.* (2005). LIF/STAT3 controls ES cell self-renewal and pluripotency by a Myc-dependent mechanism. *Development*, *132*, 885–896.

Cenkci, B., Petersen, J. L., & Small, G. D. (2003). REX1, a novel gene required for DNA repair. *J. Biol. Chem.*, *278*, 22574–22577.

Chew, J. L., *et al.* (2005). Reciprocal transcriptional regulation of Pou5f1 and Sox2 via the Oct4/Sox2 complex in embryonic stem cells. *Mol. Cell Biol.*, *25*, 6031–6046.

Conti, L., *et al.* (2005). Niche-independent symmetrical self-renewal of a mammalian tissue stem cell. *PLoS Biology*, *3*, e283.

D'Ippolito, G., *et al.* (2004). Marrow-isolated adult multilineage inducible (MIAMI) cells, a unique population of postnatal young and old human cells with extensive expansion and differentiation potential. *J. Cell Sci.*, *117*, 2971–2981.

De Coppi, P., *et al.* (2007). Isolation of amniotic stem cell lines with potential for therapy. *Nat Biotechnol.*, *25*, 100–106.

Dorrell, C., & Grompe, M. (2005). Liver repair by intra- and extrahepatic progenitors. *Stem Cell Rev.*, *1*, 61–64.

Fletcher, G., Jones, G. E., Patient, R., & Snape, A. (2006). A role for GATA factors in Xenopus gastrulation movements. *Mech. Dev.*, *123*, 730–745.

Franco, D., & Campione, M. (2003). The role of Pitx2 during cardiac development. Linking left-right signaling and congenital heart diseases. *Trends Cardiovasc. Med.*, *13*, 157–163.

Friedman, J. R., & Kaestner, K. H. (2006). The Foxa family of transcription factors in development and metabolism. *Cell Mol. Life Sci.*, *63*, 2317–2328.

Guan, K., *et al.* (2006). Pluripotency of spermatogonial stem cells from adult mouse testis. *Nature*, *440*, 1199–1203.

Jiang, Y., *et al.* (2002). Pluripotency of mesenchymal stem cells derived from adult marrow. *Nature*, *418*, 41–49.

Kanatsu-Shinohara, M., *et al.* (2004). Generation of pluripotent stem cells from neonatal mouse testis. *Cell*, *119*, 1001–1012.

Kogler, G., Sensken, S., & Wernet, P. (2006). Comparative generation and characterization of pluripotent unrestricted somatic stem cells with mesenchymal stem cells from human cord blood. *Exp. Hematol.*, *34*, 1589–1595.

Kucia, M., *et al.* (2006). A population of very small embryonic-like (VSEL) CXCR4(+)SSEA-1(+)Oct-4+ stem cells identified in adult bone marrow. *Leukemia*, *20*, 857–869.

Lengner, C. J., *et al.* (2007). Oct4 expression is not required for mouse somatic stem cell self-renewal. *Cell Stem Cell*, *1*, 403–415.

Lewitzky, M., & Yamanaka, S. (2007). Reprogramming somatic cells towards pluripotency by defined factors. *Curr. Opin. Biotechnol.*, *18*, 467–473.

Liedtke, S., Stephan, M., & Kogler, G. (2008). Oct4 expression revisited: potential pitfalls for data misinterpretation in stem cell research. *Biol. Chem.*, *389*, 1529–1536.

Lowry, W. E., *et al.* (2008). Generation of human induced pluripotent stem cells from dermal fibroblasts. *Proc. Nat. Acad. Sci. USA*, *105*, 2883–2888.

Maitra, A., *et al.* (2005). Genomic alterations in cultured human embryonic stem cells. *Nat. Genet.*, *37*, 1099–1103.

Mitalipova, M. M., *et al.* (2005). Preserving the genetic integrity of human embryonic stem cells. *Nat. Biotechnol.*, *23*, 19–20.

Muguruma, Y., *et al.* (2003). *In vivo* and *in vitro* differentiation of myocytes from human bone marrow-derived multipotent progenitor cells. *Exp. Hematol.*, *31*, 1323–1330.

Niwa, H., Miyazaki, J., & Smith, A. G. (2000). Quantitative expression of Oct-3/4 defines differentiation, dedifferentiation or self-renewal of ES cells. *Nat. Genet.*, *24*, 372–376.

Okita, K., & Yamanaka, S. (2006). Intracellular signaling pathways regulating pluripotency of embryonic stem cells. *Curr. Stem Cell. Res. Ther.*, *1*, 103–111.

Okita, K., Ichisaka, T., & Yamanaka, S. (2007). Generation of germline-competent induced pluripotent stem cells. *Nature*, *448*, 313–317.

Pan, G., & Thomson, J. A. (2007). Nanog and transcriptional networks in embryonic stem cell pluripotency. *Cell Res.*, *17*, 42–49.

Park, I. H., *et al.* (2008). Reprogramming of human somatic cells to pluripotency with defined factors. *Nature*, *451*, 141–146.

Pelacho, B., *et al.* (2007). Multipotent adult progenitor cell transplantation increases vascularity and improves left ventricular function after myocardial infarction. *J. Tissue Eng. Regen. Med.*, *1*, 51–59.

Reyes, M., *et al.* (2001). Purification and *ex vivo* expansion of postnatal human marrow mesodermal progenitor cells. *Blood*, *98*, 2615–2625.

Rosler, E. S., *et al.* (2004). Long-term culture of human embryonic stem cells in feeder-free conditions. *Dev. Dyn.*, *229*, 259–274.

Ross, J. J., *et al.* (2006). Cytokine induction of functional smooth muscle cells from multipotent adult progenitor cells. *J. Clin. Invest.*, *116*, 3139–3149.

Rowley, M., Grothey, E., & Couch, F. J. (2004). The role of Tbx2 and Tbx3 in mammary development and tumorigenesis. *J. Mammary Gland Biol. Neoplasia*, *9*, 109–118.

Schwartz, R. E., *et al.* (2005). Defined conditions for development of functional hepatic cells from human embryonic stem cells. *Stem Cells Dev.*, *14*, 643–655.

Seandel, M., *et al.* (2007). Generation of functional multipotent adult stem cells from GPR125+ germline progenitors. *Nature*, *449*, 346–350.

Serafini, M., *et al.* (2007). Hematopoietic reconstitution by multipotent adult progenitor cells: precursors to long-term hematopoietic stem cells. *J. Exp. Med.*, *204*, 129–139.

Shay, J. W., & Wright, W. E. (2000). Hayflick, his limit, and cellular ageing. *Nat. Rev. Mol. Cell Biol.*, *1*, 72–76.

Shi, Y., *et al.* (2008). A combined chemical and genetic approach for the generation of induced pluripotent stem cells. *Cell Stem Cell, 2*, 525–528.

Sinner, D., Rankin, S., Lee, M., & Zorn, A. M. (2004). Sox17 and beta-catenin cooperate to regulate the transcription of endodermal genes. *Development, 131*, 3069–3080.

Snykers, S., *et al.* (2006). Robust differentiation of rat MAPCs to hepatocyte-like cells by exposure to cytokines that closely reflects their temporal expression during development. *Toxicol. Sci., 94*, 330–341, Epub 13 July 2006.

Takahashi, K., & Yamanaka, S. (2006). Induction of pluripotent stem cells from mouse embryonic and adult fibroblast cultures by defined factors. *Cell, 126*, 663–676.

Terada, N., *et al.* (2002). Bone marrow cells adopt the phenotype of other cells by spontaneous cell fusion. *Nature, 416*, 542–545.

Tolar, J., *et al.* (2006). Host factors that impact the biodistribution and persistence of multipotent adult progenitor cells. *Blood, 107*, 4182–4188.

Tolar, J., *et al.* (2007). The host immune response is essential for the beneficial effect of adult stem cells after myocardial ischemia. *Exp. Hematol., 35*, 682–60.

Ulloa-Montoya, F., *et al.* (2007). Comparative transcriptome analysis of embryonic and adult stem cells with extended and limited differentiation capacity. *Genome Biol., 8*, R163.

Viswanathan, S. R., Daley, G. Q., & Gregory, R. I. (2008). Selective blockade of microRNA processing by Lin28. *Science, 320*, 97–100.

Wang, D. Z., & Olson, E. N. (2004). Control of smooth muscle development by the myocardin family of transcriptional coactivators. *Curr. Opin. Genet. Dev., 14*, 558–566.

Wang, X., *et al.* (2003). Cell fusion is the principal source of bone-marrow-derived hepatocytes. *Nature, 422*, 897–901.

Wernig, M., *et al.* (2007). *In vitro* reprogramming of fibroblasts into a pluripotent ES-cell-like state. *Nature, 448*, 318–324.

Yasuhara, T., *et al.* (2006). Behavioral and histological characterization of intrahippocampal grafts of human bone marrow-derived multipotent progenitor cells in neonatal rats with hypoxic-ischemic injury. *Cell Transplant., 15*, 231–238.

Yasuhara, T., *et al.* (2008). Intravenous grafts recapitulate the neurorestoration afforded by intracerebrally delivered multipotent adult progenitor cells in neonatal hypoxic-ischemic rats. *J. Cereb. Blood Flow Metab., 28*(11), 1804–1810.

Ying, Q. L., Nichols, J., Evans, E. P., & Smith, A. G. (2002). Changing potency by spontaneous fusion. *Nature, 416*, 545–548.

Yoon, Y. S., *et al.* (2005). Clonally expanded novel multipotent stem cells from human bone marrow regenerate myocardium after myocardial infarction. *J. Clin. Invest., 115*, 326–338.

Yu, J., *et al.* (2007). Induced pluripotent stem cell lines derived from human somatic cells. *Science, 318*, 1917–1920.

Zeng, L., *et al.* (2006). Multipotent adult progenitor cells from swine bone marrow. *Stem Cells, 24*, 2355–2366.

Zeng, L., *et al.* (2007). Bioenergetic and functional consequences of bone marrow-derived multipotent progenitor cell transplantation in hearts with postinfarction left ventricular remodeling. *Circulation, 115*, 1866–1875.

Zhang, J., *et al.* (2006). Sall4 modulates embryonic stem cell pluripotency and early embryonic development by the transcriptional regulation of Pou5f1. *Nat. Cell Biol., 8*, 1114–1123.

Zhao, L. R., Duan, W. M., Reyes, M., Verfaillie, C. M., & Low, W. C. (2003). Immunohistochemical identification of multipotent adult progenitor cells from human bone marrow after transplantation into the rat brain. *Brain Res. Brain Res. Protoc., 11*, 38–45.

Chapter 14

Bone Marrow Stem Cells: Properties and Pluripotency

Munira Xaymardan, Massimo Cimini, Richard D. Weisel, and Ren-Ke Li

INTRODUCTION

The formation of new tissue in animals is generally confined to the embryonic and developmental stages. Regeneration of highly complex tissue does occur in some amphibians and reptiles. In mammals, however, healing of damaged tissue essentially results in the replacement of functional cells by highly fibrotic reparative tissue, which leads to diminished or even lost organ function. In the past 5 years, accumulating evidence has shown that multipotential stem cells are in fact present in many adult tissues. Bone marrow and tissue-specific stem cells can be induced to differentiate into adult cell types previously thought terminally differentiated, including cardiomyocytes, skeletal muscle cells, and neurons.

Stem cells, by definition, have two characteristics: (1) the ability to self-renew and generate more stem cells through cell division and (2) under appropriate induction, the ability to give rise to clonal progency that continue to differentiate into one or more specialized cell types. In adults, bone marrow is a major reservoir for stem cells. Unlike totipotent stem cells (such as a fertilized egg), which can give rise to entire organism, bone marrow stem cells (BMSCs) are multipotent cells, which can give rise to most of the adult cell types, but not yet proven to be able to develop into a fetus (Stocum, 2001). Stem- and progenitor-based therapies are currently being developed for the treatment of cardiovascular diseases, which represent the major cause of death in the Western world (NIH, 2000). A number of clinical trials are underway to test the efficacy of local and systemic delivery of bone marrow-derived stem cells for the replacement of cardiomyocytes and vascular endothelial cells (Britten *et al.*, 2003; Perin and Silva, 2004) (Table 14-1). Preliminary results are suggestive, but their widespread application necessitates a thorough understanding of the mechanisms of cellular replacement in order to optimize the efficient use of BMSCs for vascular repair and cardioprotection.

BONE MARROW STEM CELLS

Bone marrow is hematopoietic tissue that lies within the trabecular bone. The trabecular and the bone marrow stroma are the elements that physically support and physiologically maintain the hematopoietic tissue. In adult humans, bone marrow is the site for production of all hematopoietic cells; the supporting stroma consists of reticular cells, osteocytes, adipocytes, vascular endothelium, and extracellular matrix. And together with the blood vessels, the bone marrow forms a hematopoietic inductive microenvironment that controls adult hematopoiesis, where five billion blood cells are produced every day.

The vascular structure of bone marrow consists of sinusoidal vasculature in which the endothelial cells do not have subsequent encapsulation of other types of cells; this is highly permissive for the emigration and immigration of the bone marrow cells.

Hematopoietic Stem Cells

Hematopoietic stem cells (HSCs) are the stem cells from which all red and white blood cells develop. They are entirely responsible for the development, maintenance, and regeneration of the blood forming tissue for life (Weissman, 2000). Because HSCs can reconstitute and restore the **hematopoietic** system of a myeloablated host, they have been traditionally used for treating hematologic disorders, starting in 1945 (Gengozian and Makinodan, 1956), when donor-derived HSCs were first used to protect a lethally irradiated civilian population.

In adult mouse bone marrow, HSC activity has been shown in a cell population marked by c-kit^{pos}, thy-1^{low}, and sca-1^{pos}(Bradfute *et al.*, 2005). In adult humans, HSCs are marked by c-kit^{pos}, thy-1^{pos}, and CD34^{pos} (Weissman, 2000). HSCs from mice and humans are being isolated, starting with a lineage depletion step in which all the lineage-specific cells (B220, CD3, 4, 8, 11b Mac-1, Gr-1 and Tcr-119

TABLE 14-1 Summary of Clinical Trials

Author	Trial name	Disease	Trial size	Length of follow-up	Cell source and type	Delivery route	Outcome
Perin and Silva (2004)		Ischemic heart failure	21	4 months	BM, mono-nuclear cells	Trans-endocardial	Increase LVEF
Kang *et al.* (2004)	MAGIC; randomized-controlled	>2 days	27	6 months	G-CSF + CPC CPC	Intra-coronary	G-CSF induced restenosis
Schachinger *et al.* (2004)	TOPCARE-AMI	AMI, 4.7 +/−1.7 days	59	6 months and 1 year	BM, CPC (contain CD133; CD34)	Intra-coronary	Improved global and regional contractility decreased MI size
Wollert *et al.* (2004)	BOOST, random-controlled	AMI, 4.5 days	60	6 months	BM, CD34	Intra-coronary	Improved global LVEF
Lunde *et al.* (2005)	ASTAMI Random-controlled	AMI, 5–8 days	100	1 year	BMC	Intra-coronary	No benefit
Cleland *et al.* (2006)	REPAIR-AMI; random-controlled	4 days after AMI	204	18 months plus	BMC	Intra-coronary	Benefit up to 18 months

Cx-non-treatment control; AMI-acute myocardial infarction; BM-bone marrow; CPCs-circulating blood progenitor cells; LV left ventricular; EF-ejection fraction.

for mice and CD10, 14, 15, 16, 19, and 20 in human) are removed (Figure 14-1). The resultant population, referred to as Lin^{neg}, can be enriched 10–100-fold, and is able to repopulate bone marrow of a lethally irradiated host. *In vitro* expansion of HSCs can be achieved by co-culturing them with stromal cells from bone marrow.

Researchers have found several subpopulations within Lin^{neg} HSCs. One homogenous population is characterized as side population (SP) cells based on their unique ability to extrude hoechst dye. When examined by fluorescence activated cell sorter (FACS) analysis, SP cells fall within a separate population to the side of the rest of the cells on a dot plot of emission data. SP cells express the ABCG2 transporter, a transmembrane protein, which allows them to actively exclude hoechst dye and fluoresce in this specific manner. These cells are also able to home rapidly to the bone marrow of a lethally irradiated host (Goodell *et al.*, 1996) and contribute progeny to the lung and liver in irradiated mice, and infiltrate into the infarcted heart (Abe *et al.*, 2003). SP cells are also present in other tissues, including skeletal muscle and skin (Liadaki *et al.*, 2005). Data are conflicting: some suggest that SP cells can be tissue-specific stem cells within these organs, and others suggest that are actually bone marrow-derived SP cells lodged within these tissues.

Another group of highly plastic stem cells isolated from bone marrow are known as **bone marrow-derived stem cells** (Leone *et al.*, 2005). Several recent studies indicate that these cells are highly plastic, exhibiting tremendous differentiation activity in numerous non-hematopoietic organs. It is unclear whether these populations are enriched for pre-hematopoietic cells that maintain greater pluripotentiality than HSCs. An additional possibility is that a differentiated hematopoietic cell, such as a macrophage, may be able to assume the gene expression pattern of a different cell type by fusion (Ozturk *et al.*, 2004).

Mesenchymal Stem Cells

Mesenchymal stem cells (MSCs) are stem cells found in bone marrow, from where they can generate bone, cartilage, fat, and fibrous connective tissue. They are the non-hematopoietic, structural components of bone marrow that support hematopoiesis by providing extracellular matrix components, cytokines, and growth factors. MSCs represent 0.001–0.01% of the bone marrow cell population, and in culture as clonal, plastic adherent cells that assume a spindle cell morphology with a finite life span. Friedenstein first discovered MSCs in the 1970s (1978), when his laboratory was able to culture these cells in media and induce them to differentiate into multilineage cell types, including osteoblasts, chondroblasts, and adipocytes, in response to appropriate stimuli. No specific constellation of surface markers has been agreed upon for these cells, but human MSCs are typically isolated from the mononuclear layer of the bone marrow after separation by discontinuous gradient centrifugation. In some cases, further purification is performed based on MSC markers, such as STRO-1 (Gronthos *et al.*, 1994). Other surface antigens reported to exist on MSCs are: CD13 (aminopeptidase-N), CD49a and CD49b (integrins-alpha), CD29 (integrin-beta), CD44 (hyaluronate), CD71 (transferrin), CD90 (thy-1), CD106 (vascular cell adhesion molecule-1), CD166 (activated leukocyte cell adhesion molecule), CD54 (intercellular adhesion molecule-1), CD55 (decay accelerating factor), and CD124

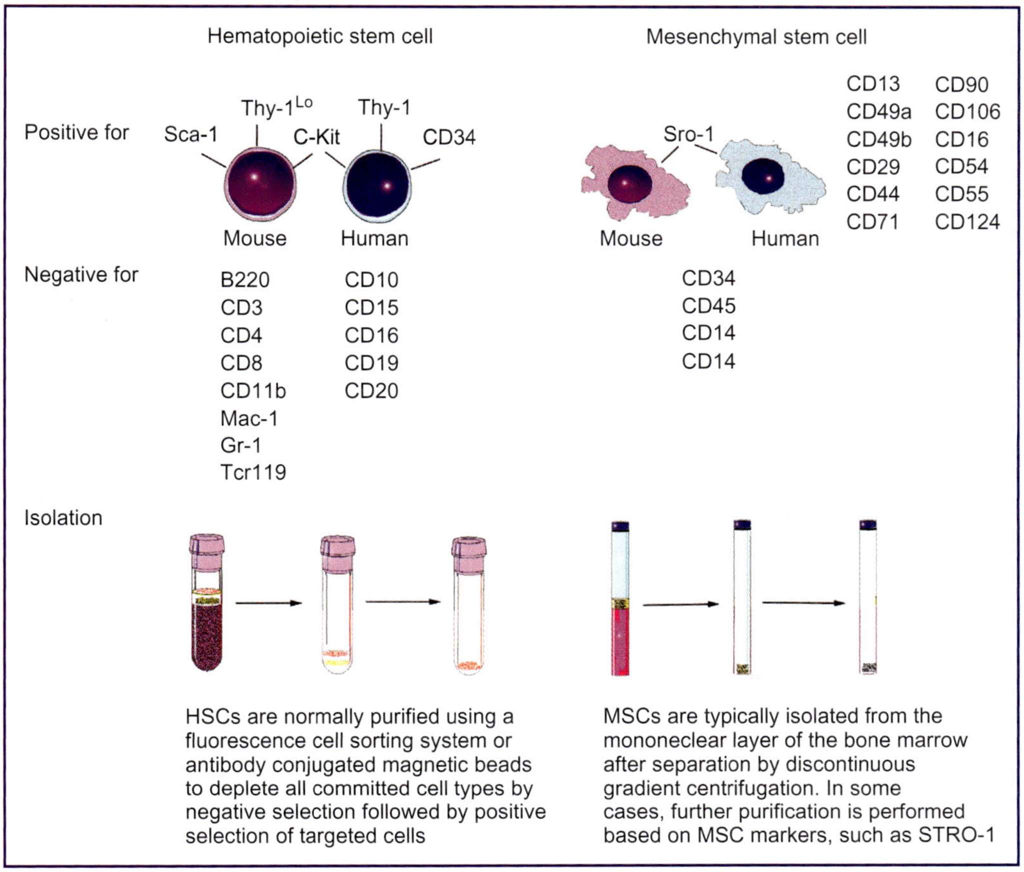

FIGURE 14-1 Isolation of HSCs and MSCs from bone marrow.

(inter-leukin-4 (IL-4) receptor). MSCs uniformly lack antigens CD34, CD45, CD14, and CD31 that typically identify hematopoietic cells (Pittenger and Martin, 2004) (Figure 14-1). A wide array of cytokines, including fibroblast growth factor-2 (FGF-2), FGF-4, platelet-derived growth factor-BB (PDGF-BB), and leukemia inhibitory factor (LIF), have been used to expand MSCs (Gregory *et al.*, 2005).

Because MSCs are easily expandable in culture and differentiate into multiple tissue lineages, there has been much interest in their clinical potential for tissue repair and gene therapy. In particular, a population of highly plastic, adult-derived bone marrow cells, referred to as multipotent adult progenitor cells (MAPCs), can be grown *in vitro* from the postnatal marrow (and other organs) of mice, rats, and humans. These cells co-purify initially with MSCs and grow as adherent cells *in vitro* (Reyes *et al.*, 2001). However, unlike MSCs, MAPCs can be cultured indefinitely in a relatively nutrient-poor medium (Jiang *et al.*, 2002). Specific changes in growth factors induce differentiation of MAPCs into myoblasts, hepatocytes, and even neural tissue (Jiang *et al.*, 2002; Schwartz *et al.*, 2002).

Endothelial Progenitor Cells

Endothelial progenitor cells (EPCs) are a group of non-endothelial cells that can give rise to endothelial cells. Stemness of the cells is not clear, but they can be expanded, and increasing evidence shows that EPCs play a major role in postnatal *neovascularization*. Bone marrow HSCs and MSCs, as well as other tissues (fat, cord blood, and circulating blood), are the sources of the EPCs, of which HSC-derived EPCs are perhaps the best characterized. HSC-derived EPCs are maintained in the BMSC niche and are released upon mobilization with cytokines such as vascular endothelial growth factor (VEGF) or stromal cell-derived factor-1 (SDF-1), which are synthesized by ischemic tissue (Leone *et al.*, 2005).

Indeed, Asahara *et al.* (1999) demonstrated that bone marrow-derived HPCs give rise to endothelial cells and contribute to endothelial recovery and new capillary formation after ischemia. EPCs have been subsequently defined as cells that express HSC markers such as CD34 or CD133, and an endothelial marker protein, VEGF receptor 2 (VEGFR2 or flk-1). Isolated cells express the classic HSC marker protein CD34 or the more immature HSC marker protein CD133. Both cell populations differentiate to endothelial cells *in vitro* under appropriate endothelial differentiation-promoting factors (Gehling *et al.*, 2000). Most importantly, injection of $CD34^{pos}$ or $CD133^{pos}$ cells enhanced neovascularization in animal models after ischemia (Asahara *et al.*, 1999).

Likewise, MSCs can differentiate into endothelial cells (Oswald *et al.*, 2004) and improve neovascularization *in vivo* (Pittenger and Martin, 2004). Because MSCs can release a variety of angiogenic growth factors, this cocktail of growth factors may also act in a paracrine manner to support angiogenesis and arteriogenesis. Verfailliea's group (Reyes *et al.*, 2002) reported that MAPCs that co-purify with MSCs isolated from postnatal human bone marrow can differentiate into cells that express endothelial markers, function *in vitro* as mature endothelial cells, and contribute to *in vivo* neoangiogenesis during tumor angiogenesis and wound healing (Reyes *et al.*, 2002).

Interaction of Bone Marrow Cells and Stem Cell Niches

The stem cells in the bone marrow are not randomly distributed. They reside in specific compartments consisting of support cells known as niche, the microenvironment, which in turn controls the fate of the stem cells.

The concept of a **stem cell niche** was first proposed for the human hematopoietic system in the 1970s (Schofield, 1978). At present, the hematopoietic niche is conceptually divided into two parts: an osteoblastic niche and a vascular niche. The osteoblastic niche located near the trabecular bone is a hypoxic environment that hosts quiescent state HSCs (slow cycling or G0), whereas the vascular niche located near sinusoids is an oxygenic niche, where stem/progenitor cells actively proliferate. The presence of osteoblasts not only sustains the bone but is required for the maintenance and expansion of HSCs through interaction of N-cadherin and betacatenin (Zhang *et al.*, 2003). Other soluble and membrane bound proteins which are required for self-renewing within the niches are: mKirre, the Wnt proteins, stem cell factor (SCF), and bone morphogenic proteins (BMP) such as BMP-4 (Ueno *et al.*, 2003; Zhang *et al.*, 2003; de Boer *et al.*, 2004). The limiting factor of HSC selfrenewal is perhaps the space within the niche (Zhang *et al.*, 2003). As a niche is filled with stem cells, the excess cells are pushed into the adjacent vascular niche, which fosters the maturation of the HSCs and where HSCs finally mature and egress from marrow into the peripheral circulation via the bone marrow sinusoids. Hemotopoiesis in the vascular niche is partly regulated by growth factors (cytokines), particularly by ILs and colonal stimulating factor (CSF) (Barria *et al.*, 2004), both of which stimulate the proliferation and maturation of the HSCs. The growth factor binding ligands are tyrosine kinase receptors such as c-kit, flt-3, and thrombopoietin, and all are expressed on primitive hematopoietic cells. The major inhibiting factors of hematopoiesis are perhaps transforming growth factor-beta (TGF-β), and tumor necrosis factor alpha (TNF-α).

BMSCS AND TISSUE REGENERATION

BMSCs are multipotential in that they not only act as myelo-regenerative and supportive cells, but they also can differentiate into multilineage cell types. HSCs are capable of differentiating into endothelial cells and have also demonstrated an ability to differentiate into liver cells, skeleton muscle cells, and cardiac cells. Increasing evidence indicates that MSCs can differentiate into functional cells and repair damaged tissue. MSCs have been demonstrated to adopt osteoblasts, chondrocytes, and adipocytes *in vitro* (Friedenstein *et al.*, 1978). When implanted *in vivo*, they are able to help repair multiple tissues including blood vessels, heart, liver, kidney, and muscle (Pittenger and Martin, 2004). Their ability to generate almost all the mesenchymal lineages of connective tissues has strengthened the idea that MSCs represent, or at least contain, a population of stem cells from which all mesenchymal lineages originate under the influence of different microenvironments. Understanding the molecular signals that underlie the process of bone marrow cell differentiation, and moreover, controlling the microenvironment, will help advance cell-based therapies (Figure 14-2).

In cases where female animals or female human patients have received a male donor bone marrow transplant, tracing of a bone marrow cell that differentiated into multiple tissue types is achieved through fluorescent *in situ* hybridization (FISH) techniques to detect the Y-chromosome (Deb *et al.*, 2003). Alternatively, wild-type animals may be transplanted with green fluorescence protein cells, which are easily detected using a fluorescence microscopy (Orlic *et al.*, 2001b).

BMSCs and Heart Regeneration

Ventricular remodeling following an acute myocardial infarction leads to ventricular dilatation and progressive heart failure. The remodeling process is characterized by the removal of necrotic **cardiac cells** accompanied by granulation tissue formation with the simultaneous induction of neovascularization in the peri-infarct bed. The latter is a prerequisite for the survival of surrounding hypertrophic but viable cardiomyocytes, and the prevention of further cardiomyocyte loss by apoptosis. Ultimately, the remodeling process culminates in the formation of a non-contractile fibrous scar, which may expand, leading to further cardiac deterioration and heart failure (Chandrashekhar, 2005).

Historically, the adult heart has been viewed as a terminally differentiated organ without the capacity of self-renewal or regeneration. But recent data challenges this doctrine, suggesting the existence of innate mechanisms for myocardial regeneration. Studies have shown evidence of low-level mitotic activity in the normal human myocardium, and proliferation of cardiomyocytes increases in the heart with end-stage ischemic disease (IHD) (Beltrami *et al.*, 2001). The most intriguing finding perhaps is the data from Quaini *et al.* (2002), showing that cardiac regeneration following orthotopic heart transplantation. Using the Y-chromosome as a marker, this study found recipient-derived

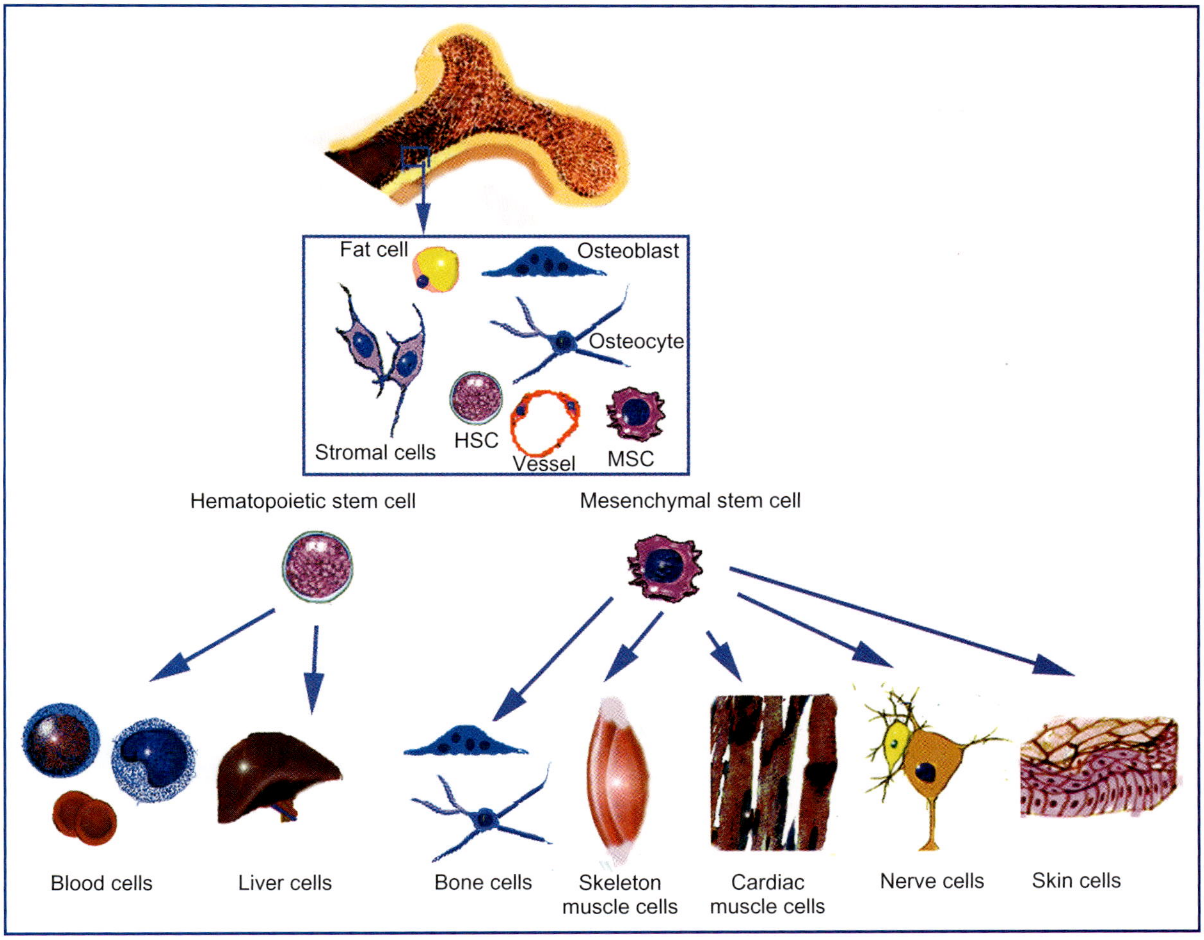

FIGURE 14-2 Pulripotency of BMSCs.

cardiomyocytes and vascular structure within the donor hearts of male patients who had received female donor hearts. BMSCs are considered to be the major contributors to the regeneration of cardiac tissue. Supporting this hypothesis are reports from female patients who received sex-mismatch bone marrow transplantation. The female hearts were examined for Y-chromosome and results confirmed the presence of bone marrow cells within the myocardium (Deb *et al.*, 2003).

Both HSCs and MSCs are reported to have the ability to repair damaged hearts. The establishment of a cardiomyogenic cell line from murine bone marrow MSCs marks a typical example of bone marrow differentiation into cardiomyocyte (Tomita *et al.*, 1999; Fukuda, 2000). Fukuda's group repeatedly passaged bone marrow cells until a single clone of immortalized homogenous fibroblast-like cells was obtained. After prolonged treatment with the DNA demethylating agent 5-azacytidine, the cells formed myotubes connected by intercalated disks, which beat synchronously after 3 weeks of culture. Ultrastructurally, the differentiated myotubes had well-organized sarcomeres, central nuclei, and contained atrial granules and mitochondria.

The cardiomyocyte phenotype was further confirmed by both electrophysiological and cardiac-specific gene expression studies. These included the identification of both sinus node-like action potentials and ventricular cardiomyocyte-like action potentials. Differentiated myotubes expressed atrial natriuretic peptide (ANP), brain natriuretic peptide (BNP), low levels of a-MHC and a-cardiac actin, and high levels of b-MHC and a-skeletal actin, as well as MLC-2v, which was consistent with a fetal ventricular phenotype. The cells have also been shown to express functional adrenergic and muscarinic receptors, which mediate heart rate, conduction velocity, contractility, and cardiac hypertrophy (Fukuda, 2000). 5-xathydidine treated cardiomyocyte cell lines have been shown to establish stable cardiac engraftment and site-specific differentiation in myocardial scar tissue in the rat cryo-injury model of infarction (Tomita *et al.*, 1999). These findings are also supported by rat myocardial infarction and porcine myocardial ischemia models (Bittira *et al.*, 2002; Moscoso *et al.*, 2005). Other experiments have shown that bone marrow cell interaction with neonatal cardiomyocytes or cellular extract may induce cardiomyogenic differentiation of the BMSCs. For example, bone marrow c-kit cells express cardiac markers when co-cultured with neonatal cardiomyocyte (Lagostena *et al.*, 2005). Anversa's group isolated Lin^{neg}/c-kit^{pos} cells from adult bone marrow. When injected into an ischemic heart, these cells reconstituted well-differentiated myocardium formed by blood-carrying new vessels and myocytes with the characteristics of young cells (Beltrami *et al.*, 2003). The results of some studies cast doubt on the ability of HSCs to adopt cardiac myocyte phenotypes *in vivo*. For example, Murry *et al.* isolated HSCs from mice carrying the alpha-cardiac myosin heavy chain promoter driving nuclear-localized enhanced green fluorescence protein (EGFP), and delivered these cells into mice after acute myocardial infarction. Unfortunately, neither systemic delivery nor direct injection of HSCs produced myocyte regeneration. Most studies have shown improved cardiac function by exogenous delivery of HSCs. This implies that an alternative mechanism of the HSC cardiac repair may be due to the paracine system: BMSCs secrete growth factors that augment angiogenesis, which in turn improves the remodeling process associated with **cardiac regeneration**.

BMSCS and Skeletal Muscle Regeneration

BMSCs are reported to differentiate into skeletal myoblasts. Human BMSCs are shown to differentiate into multinucleated myotubes in culture (Bossolasco *et al.*, 2004). Moreover, direct injection of human whole bone marrow into the right tibialis anterior muscle of immunodeficient mice, previously treated with cardiotoxin to induce muscle degeneration, showed a variable but significant level of human cell engraftment (Bossolasco *et al.*, 2004).

BMSCS and Bone Regeneration

The osteogenic lineage is considered a default pathway of *in vitro* differentiation of the bone marrow stromal cells. Indeed, regeneration of the bone tissue has been successfully used in clinical practice. The earliest studies used clonal forming unit fibroblasts (CFU-F) like cells to form bone structures in culture (Friedenstein *et al.*, 1978). The phenomenon is also observed in the stro-1 population of MSCs (Gronthos *et al.*, 1994). The bone forming ability of these cells has also been tested in diffusing wound chambers in rabbit models. When cells are isolated and expanded in the presence of FGF-2, the frequency of clones able to differentiate into the osteogenic, chondrogenic, and adipogenic lineages is greater than in the other lineages.

BMSCS and Liver Regeneration

BMSC engraftment to *hepatocytes* using male-to-female bone marrow transplantation in rats and mice (Fujii *et al.*, 2002) was first demonstrated in response to liver damage, which may promote BMSC-to-hepatocyte transition. In rats, a combination of hepatotoxin, which induces widespread liver damage, and 2-acetylamino-fluorine, which prevents endogenous liver repair, was used. A combination of Y-chromosome FISH and transgene expression was then employed to confirm that BMSCs were the source of the resultant hepatocytes. The effect on other forms of liver damage could be assessed in liver samples from men who received orthotopic liver transplants from female donors. In these patients, the degree of subsequent damage to the transplanted liver correlated with the extent of male (host-derived) hepatocyte engraftment (Theise *et al.*, 2000).

BMSCS and Nerve Cell Regeneration

Two different systems show that bone marrow-derived stem cells can serve as progenitors of non-hematopoietic cells in the murine central nervous system (CNS). In one study, lethally irradiated adult mice that received whole marrow intravenously developed donor-derived brain cells bearing the neuronal antigens NeuN and class 3 b-tubulin (Brazelton *et al.*, 2000). Similarly, adult rat and human BMSCs induced the stromal cells to exhibit a neuronal phenotype, expressing neuron-specific enolase, NeuN, neurofilament-M, and tau (Woodbury *et al.*, 2000).

Bone Marrow to Kidney, Pancreas, Lung, and Gastrointestinal Tract

Similarly, BMSCs can also differentiate *in vivo* into pancreas islet cells, lung clara cells, and GI crypt cells, which are the functional stem cells of the gastrointestinal epithelium.

IMPORTANT FACTORS REGULATING BMSC HOMING AND DIFFERENTIATION

As discussed in the previous sections, there is an accumulating body of evidence suggesting that stem and progenitor cells have the potential to regenerate and revascularize injured tissue. Stem cell-mediated cardiac repair involves three components: (1) the bone marrow as a stem cell reservoir; (2) the injured myocardium as the area where repair is required with the release of mediating factors; and (3) the circulation for transport of the signals and stem cells from the bone marrow to the injured myocardium (Vandervelde *et al.*, 2005). Upon myocardial injury, molecular pathways are upregulated immediately, followed by streams of chemical mediators (**cytokines** and **chemokines**) that are released into circulation to help recruit BMSCs and allow for their homing to the distal injured myocardium (Vandervelde *et al.*, 2005). Alternatively, exogenous BMSCs can be delivered directly to the injured site. Repair may have three different foci: (1) the vasculature, (2) the cardiomyoctes, and (3) the stability of the extracellular matrix. Together, these components orchestrate the signaling, mobilization, homing, incorporation, survival, proliferation, and differentiation of stem cells—a progression that involves a dynamic process of metalloproteinase activity, adhesion molecules, and remodeling of the extracellular matrix. The cytokines and chemokines involved may be classified according to function as mediators of homing and mobilization, inflammation (to aid in incorporation), survival and differentiation of stem cells (Figure 14-3).

Granulocyte colony-stimulating factor (G-CSF) and VEGF are among the best-characterized cytokines for mobilization of BMSCs and EPCs to the site of injury.

Vascular Endothelial Growth Factor

VEGFs are a group of secreted proteins produced by almost every cell type; they appear to be the most prominent protein that guides vascular growth during vasculargenesis and angiogenesis (Carmeliet *et al.*, 1996). However, the angiogenic capabilities of VEGF often overshadow its importance in the mobilization of BMSCs. Patients with high levels of plasma VEGF were found to have an increase in BMSCs in the heart, indicating the ability of VEGF to recruit stem cells post-myocardial infarction (Kamihata *et al.*, 2001). In animal models, infusion of BMSCs was related to reduction in infarct size; the effect was attenuated with neutralizing antibodies to either VEGF and by increasing its soluble receptor VEGF-R1 (Flt-1) (Hiasa *et al.*, 2004). Injection of VEGF plasmid DNA has also been documented to have mitogenic effects on porcine cardiomyocytes (Laguens *et al.*, 2002). Furthermore, naked plasmid DNA directly injected into the ischemic myocardium of symptomatic myocardial ischemia patients led to a reduction in symptoms and improved myocardial perfusion (Laguens *et al.*, 2002). Recent study shows that the VEGF is sufficient for organ homing of BMSCs to the perivascular area (Grunewald *et al.*, 2006). VEGF may therefore enhance homing, mobilization of BMSCs, and augment cardiomyocyte proliferation.

G-CSF

G-CSF is a hematopoietic factor that stimulates neutrophils and BMSC mobilization through cleavage of intercellular adhesion molecule-1, thereby disrupting the homing mechanism of the stem cells in the bone (Levesque *et al.*, 2001). G-CSF is also involved in the proliferation, differentiation, and survival of bone marrow-derived stem and progenitor cells. The mobilization properties of G-CSF are widely utilized in clinical stem cell therapies with promising results seen in most of the trials (Kang *et al.*, 2004; Valgimigli *et al.*, 2005). Indeed, G-CSF has been shown to increase the

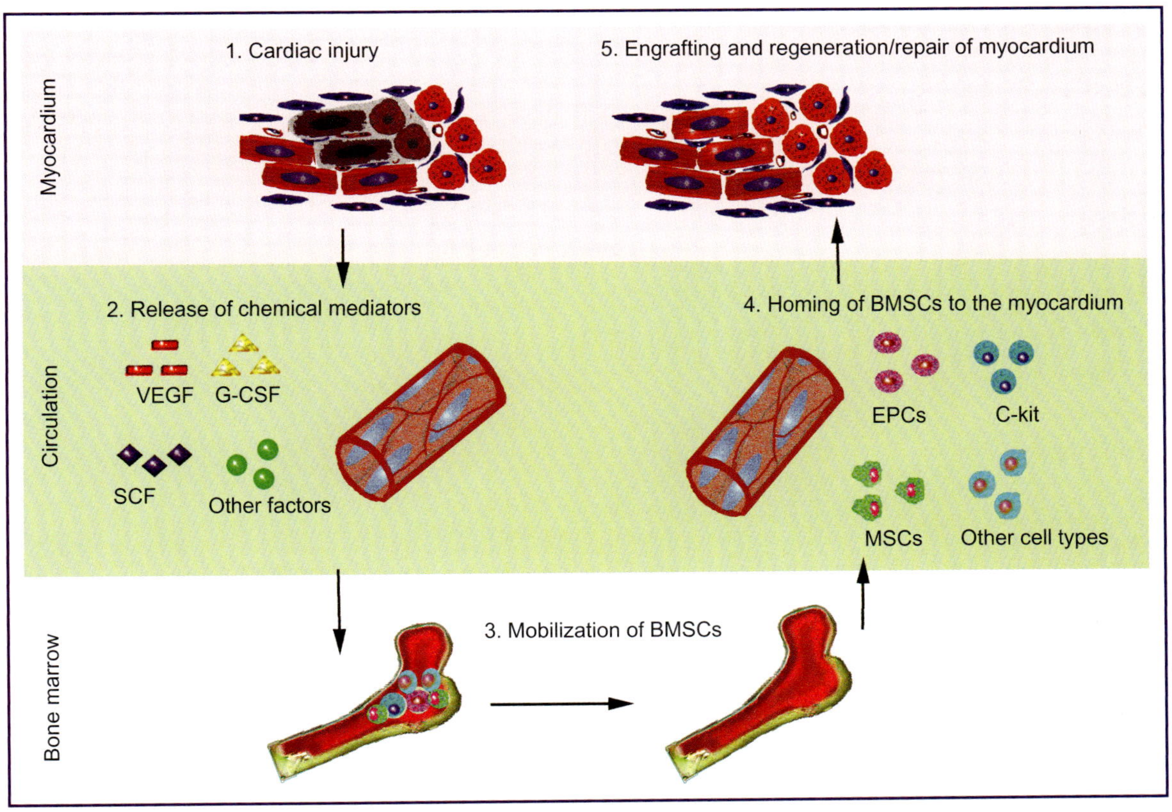

FIGURE 14-3 The stages of BMSC recruitment involved in regeneration of the damaged myocardium.

number of CD34^{pos} cells in the circulation from 5- to 30-fold (Powell *et al.*, 2005). When Kocher *et al.* isolated circulating human CD34^{pos} cells released by G-CSF treatment and injected it into the infarcted hearts of nude rats, they found these CD34^{pos} cells demonstrated phenotypic and functional properties of embryonic hemangioblasts stimulating neoangiogenesis in the infarct vascular bed (Kocher *et al.*, 2001). G-CSF treatment was also found to increase the density of macrophages and neutrophils which may enhance the absorption of necrotic tissue in post-infarct myocardium, and coincide with proliferating cardiomyocytes and improved cardiac function; these results suggest additional pathways for G-CSF treatment (Minatoguchi *et al.*, 2004). Further, G-CSF may act on non-hematopoietic cells that may potentially serve as the origin of the bone marrow-derived cardiomyocytes observed in the mouse myocardial infarction model (Kawada *et al.*, 2004).

STROMAL CELL-DERIVED FACTOR-1

Stromal cell-derived factor-1 (SDF-1) (CXCL12) is produced by bone marrow stromal cells; its function is to promote homing and engraftment of HSCs within the recipient bone marrow (Peled *et al.*, 2000). SDF-1 transcription is partially controlled by hypoxia inducible factor-1 (HIF-1), upregulated by hypoxia during vascular injury (Ceradini and Gurtner, 2005). It therefore seems logical that SDF-1 forms a gradient from the hypoxic to the oxygenic bone marrow compartments. Studies have shown that blockage of SDF-1 binding to its receptor, CXCR4, inhibits stem cell homing to the infarcted heart, strongly suggesting that SDF-1/CXCR4 interactions play a crucial role in the recruitment of BMSCs to the heart after myocardial infarction (Abbott *et al.*, 2004). Interestingly, a non-hematopoietic CXCR4^{pos} population in the bone marrow has been found to also express early cardiac progenitor markers such as Nkx2.5/Csx, GATA-4, and MEF2C. This population can be mobilized into the peripheral blood after experimental myocardial infarction, providing a possible therapeutic target for myocardial regeneration (Kucia *et al.*, 2004).

Stem Cell Factor

SCF, also known as c-kit ligand or steel factor, binds to its receptor c-kit and induces chemotactic properties in stem and progenitor cells (Chute *et al.*, 2005). SCF is abundantly expressed in the normal bone marrow and heart, but it is downregulated following myocardial infarction (Woldbaek *et al.*, 2002). Orlic *et al.* demonstrated that combined SCF and G-CSF treatment synergistically improved mouse cardiac function after myocardial infarction via mobilization of the BMSCs (Orlic *et al.*, 2001); however, other groups failed to reproduce the described effect (Norol *et al.*, 2003; Ohtsuka *et al.*, 2004). Since SCF is produced by infiltrating macrophages in the ischemic myocardium and attracts mast cell precursors, SCF treatment for increased BMSCs may have a detrimental hyper-inflammatory effect (Frangogiannis *et al.*, 1998).

Interleukin-8

IL-8 is a member of the small chemokine CXC family. It is upregulated by pro-inflammatory cytokines, like SCF-1, and is an important factor for stem cell proliferation in the bone marrow niche and promoting rapid mobilization of BMSCs. On the one hand, IL-8 promotes endothelial cell migration to sites of injury duration (Fibbe *et al.*, 2000). On the other hand, it activates neutrophil adhesion to cardiomyocytes, subsequently promoting cardiomyocyte death (Kukielka *et al.*, 1995). Similar to SCF, IL-8 may have detrimental effects on the cardiac tissue.

Transforming Growth Factor-Beta and Bone Morphogenetic Proteins

TGF- β**s** and **bone morphogenetic proteins (BMPs)** constitute a single morphogenic protein super family involved in cardiogenesis (Zaffran and Frash, 2002). Both have been demonstrated to induce embryonic stem cell differentiation into a cardiogenic phenotype (Behfar *et al.*, 2002). Recently, it has been demonstrated that CD117^{pos} cells, partially positive in various fractions for Lin, CD34, and Sca-1, and negative for Lin, were able to undergo cardiomyogenic differentiation when treated with TGF-β_1 (Li *et al.*, 2005). Furthermore, transplantation of these cells into the infarcted region expressed ventricular heavy chain myosin, reduced fibrosis, and improved shortening. Unfortunately, angiogenesis, wall thickness, and LVEDD/LVESD did not significantly differ between the untreated and preprogrammed c-kit transplanted cells. Similar effects have also been reported in MSCs treated with BMP-2 and FGF-4 in a rat model of experimental myocardial infarction (Yoon *et al.*, 2005), and hepatic growth factor (HGF) and platelet-derived growth factor-B (PDGF-B) were reported to stimulate BMSC survival and differentiation in experimental bone marrow transplantation models.

CLINICAL APPLICATIONS OF BMSCS FOR CARDIAC REGENERATION

Based on the *ex vivo* confirmation of bone marrow plasticity and the idea of possible cardiomyocyte regeneration, numerous clinical trials have been initiated to augment this process by transplanting exogenous bone marrow cells into damaged myocardium in patients with acute **myocardial infarct** or ischemic heart failure. Both HSC and MSC populations have been used in patients, with delivery methods including: (1) intra-coronary delivery (Britten *et al.*, 2003; Kang *et al.*, 2004); (2) direct injection into myocardium (Stamm *et al.*, 2003; Perin and Silva, 2004; Pompilio *et al.*, 2004); and (3) G-SCF-mediated BMSC mobilization (Kang *et al.*, 2004). Most of the techniques described in these studies have been combined with conventional treatments, including surgical revascularization like angioplasty and stenting. Although some authors indicated that the cells used in these studies contained CD133 and/or CD34 populations, most of the cells and treatments were not clearly characterized, and their fates are undetermined. The results, however, (summarized in Table 14-1) have demonstrated that **bone marrow transplantation** in ischemic heart disease patients is safe and feasible, with the exception of a single report by Kang *et al.* (2004), which showed increasing restenosis in patients treated with G-CSF.

While G-SCF release of smooth muscle progenitor cells may contribute to increased in-stent restenosis, transplantation of bone marrow cells into the ischemic heart augments angiogenesis and improves cardiac function. Most trials did not find significant risk in patients receiving bone marrow treatments.

Indeed, most of the short-term trials have shown improvements in left ventricular ejection fraction and other functional parameters tested (Britten *et al.*, 2003; Stamm *et al.*, 2003; Kang *et al.*, 2004; Perin and Silva, 2004; Pompilio *et al.*, 2004; Kuethe *et al.*, 2005). However, more recently, randomized, controlled clinical trials have produced controversial results. The REPAIR-AMI study (Cleland *et al.*, 2006) (Germany and Switzerland) randomly assigned 204 patients to infusion of BMSCs or cell-free supernatant an average of 4 days after a myocardial infarction. By 4 months after treatment, left ventricular ejection fraction had improved in both groups, but the improvement was significantly greater in the patients who had received stem cells. In contrast, the ASTAMI study (Lunde *et al.*, 2005) (Norway) randomly assigned 100 patients to stem cell implant or treatment after an acute anterior myocardial infarction. The investigators observed no benefit from stem cell implants, and indeed suggested that at 6 months, left ventricular ejection fraction had increased more in the control group. A recent update of the BOOST study (Wollert *et al.*, 2004) suggested that the benefits of bone marrow transfer post-myocardial infarction were sustained at 18 months. However, there was a further improvement in global left ventricular function in the control group rendered the inter-group comparison non-significant.

CONCLUSION

As discussed in this chapter, BMSCs provide a promising new arena for regenerative medicine. Although the challenging nature of the research raises some skepticism within the field, bone marrow studies still in their infancy are showing great potential for regeneration of various tissues, at least through the delivery of endothelium and paracrine factors improving revascularization and preventing apoptosis. Current debates addressing the therapeutic potential of this fundamental biologic process should encourage collaborative effort in defining the microenvironment that controls BMSC transdifferentiation. The characterization of such a factor could help harness the mechanism by which cellular repair is achieved. With an understanding of the mechanisms involved in BMSC activity, and with solutions to existing technical difficulties—through improved cellular tracking, improved imaging technology and research standards, and increased communication between international research groups—the mist surrounding BMSCs may soon be lifted.

REFERENCES

Abbott, J.D., Huang, Y., Liu, D., Hickey, R., Krause, D.S. and Giordano, F.J. (2004). Stromal cell-derived factor-1alpha plays a critical role in stem cell recruitment to the heart after myocardial infarction but is not sufficient to induce homing in the absence of injury. *Circulation 110*(21): 3300–3305.

Abe, S., Lauby, G., Boyer, C., Rennard, S.I. and Sharp, J.G. (2003). Transplanted BM and BM side population cells contribute progeny to the lung and liver in irradiated mice. *Cytotherapy 5*(6): 523–533.

Achen, M.G. and Stacker, S.A. (1998). The vascular endothelial growth factor family; proteins which guide the development of the vasculature. *Int. J. Exp. Pathol. 79*(5): 255–265.

Asahara, T., Masuda, H., Takahashi, T., Kalka, C., Pastore, C., Silver, M., Kearne, M., Magner, M. and Isner, J.M. (1999). Bone marrow origin of endothelial progenitor cells responsible for postnatal vasculogenesis in physiological and pathological neovascularization. *Circ. Res. 85*(3): 221–228.

Barria, E., Mikels, A. and Haas, M. (2004). Maintenance and self-renewal of long-term reconstituting hematopoietic stem cells supported by amniotic fluid. *Stem Cells Dev. 13*(5): 548–562.

Behfar, A., Zingman, L.V., Hodgson, D.M., Rauzier, J.M., Kane, G.C., Terzic, A. and Puceat, M. (2002). Stem cell differentiation requires a paracrine pathway in the heart. *Faseb J. 16*(12): 1558–1566.

Beltrami, A.P., Urbanek, K., Kajstura, J., Yan, S.M., Finato, N., Bussani, R., Nadal-Ginard, B., Silvestri, F., Leri, A., Beltrami, C.A., *et al.* (2001). Evidence that human cardiac myocytes divide after myocardial infarction. *N. Engl.J.Med. 344*(23): 1750–1757.

Beltrami, A.P., Barlucchi, L., Torella, D., Baker, M., Limana, F., Chimenti, S., Kasahara, H., Rota, M., Musso, E., Urbanek, K., *et al.* (2003). Adult cardiac stem cells are multipotent and support myocardial regeneration. *Cell 114*(6): 763–776.

Bittira, B., Kuang, J.Q., Al-Khaldi, A., Shum-Tim, D. and Chiu, R.C. (2002). *In vitro* preprogramming of marrow stromal cells for myocardial regeneration. *Ann. Thorac. Surg. 74*(4): 1154–1159; discussion 1159–1160.

Bossolasco, P., Corti, S., Strazzer, S., Borsotti, C., Del Bo, R., Fortunato, F., Salani, S., Quirici, N., Bertolini, F., Gobbi, A., *et al.* (2004). Skeletal muscle differentiation potential of human adult bone marrow cells. *Exp. Cell Res. 295*(1): 66–78.

Bradfute, S.B., Graubert, T.A. and Goodell, M.A. (2005). Roles of Sca-1 in hematopoietic stem/progenitor cell function. *Exp. Hematol. 33*(7): 836–843.

Brazelton, T.R., Rossi, F.M., Keshet, G.I. and Blau, H.M. (2000). From marrow to brain: expression of neuronal phenotypes in adult mice. *Science 290*(5497): 1775–1779.

Britten, M.B., Abolmaali, N.D., Assmus, B., Lehmann, R., Honold, J., Schmitt, J., Vogl, T.J., Martin, H., Schachinger, V., Dimmeler, S., *et al.* (2003). Infarct remodeling after intracoronary progenitor cell treatment in patients with acute myocardial infarction (TOPCARE-AMI): mechanistic insights from serial contrast-enhanced magnetic resonance imaging. *Circulation 108*(18): 2212–2218.

Carmeliet, P., Ferreira, V., Breier, G., Pollefeyt, S., Kieckens, L., Gertsenstein, M., Fahrig, M., Vandenhoeck, A., Harpal, K., Eberhardt, C., *et al.* (1996). Abnormal blood vessel development and lethality in embryos lacking a single VEGF allele. *Nature 380*(6573): 435–439.

Ceradini, D.J. and Gurtner, G.C. (2005). Homing to hypoxia: HIF-1 as a mediator of progenitor cell recruitment to injured tissue. *Trends Cardiovasc. Med. 15*(2): 57–63.

Chandrashekhar, Y. (2005). Role of apoptosis in ventricular remodeling. *Curr. Heart Fail. Rep. 2*(1): 18–22.

Chute, J.P., Muramoto, G.G., Dressman, H.K., Wolfe, G., Chao, N.J. and Lin, S. (2006). Molecular profile and partial functional analysis of novel endothelial cell derived growth factors that regulate hematopoiesis. *Stem Cells 24*(5): 1315–1327.

Cleland, J.G., Freemantle, N., Coletta, A.P. and Clark, A.L. (2006). Clinical trials update from the American Heart Association: REPAIR-AMI, ASTAMI, JELIS, MEGA, REVIVE-II, SURVIVE, and PROACTIVE. *Eur. J. Heart Fail. 8*(1): 105–110.

de Boer, J., Siddappa, R., Gaspar, C., van Apeldoorn, A., Fodde, R. and van Blitterswijk, C. (2004). Wnt signaling inhibits osteogenic differentiation of human mesenchymal stem cells. *Bone 34*(5): 818–826.

Deb, A., Wang, S., Skelding, K.A., Miller, D., Simper, D. and Caplice, N.M. (2003). Bone marrow-derived cardiomyocytes are present in adult human heart: a study of gender-mismatched bone marrow transplantation patients. *Circulation 107*(9): 1247–1249.

Driessen, R.L., Johnston, H.M. and Nilsson, S.K. (2003). Membrane-bound stem cell factor is a key regulator in the initial lodgment of stem cells within the endosteal marrow region. *Exp. Hematol. 31*(12): 1284–1291.

Fibbe, W.E., Pruijt, J.F., van Kooyk, Y., Figdor, C.G., Opdenakker, G. and Willemze, R. (2000). The role of metalloproteinases and adhesion molecules in interleukin-8-induced stem-cell mobilization. *Semin. Hematol. 37*(1 Suppl 2): 19–24.

Frangogiannis, N.G., Lindsey, M.L., Michael, L.H., Youker, K.A., Bressler, R.B., Mendoza, L.H., Spengler, R.N., Smith, C.W. and Entman, M.L. (1998). Resident cardiac mast cells degranulate and release preformed TNF-alpha, initiating the cytokine cascade in experimental canine myocardial ischemia/reperfusion. *Circulation 98*(7): 699–710.

Friedenstein, A.J., Ivanov-Smolenski, A.A., Chajlakjan, R.K., Gorskaya, U.F., Kuralesova, A.I., Latzinik, N.W. and Gerasimow, U.W. (1978). Origin of bone marrow stromal mechanocytes in radiochimeras and heterotopic transplants. *Exp. Hematol. 6*(5): 440–444.

Fujii, H., Hirose, T., Oe, S., Yasuchika, K., Azuma, H., Fujikawa, T., Nagao, M. and Yamaoka, Y. (2002). Contribution of bone marrow cells to liver regeneration after partial hepatectomy in mice. *J. Hepatol. 36*(5): 653–659.

Fukuda, K. (2000). Generation of cardiomyocytes from mesenchymal stem cells. *Tanpakushitsu Kakusan Koso 45* (13 Suppl): 2078–2084.

Gehling, U.M., Ergun, S., Schumacher, U., Wagener, C., Pantel, K., Otte, M., Schuch, G., Schafhausen, P., Mende, T., Kilic, N., *et al.* (2000). *In vitro* differentiation of endothelial cells from AC133-positive progenitor cells. *Blood 95*(10): 3106–3112.

Gengozian, N. and Makinodan, T. (1956). Antibody response of lethally x-irradiated mice treated with rat bone marrow. *J. Immunol. 77*(6): 430–436.

Goodell, M.A., Brose, K., Paradis, G., Conner, A.S. and Mulligan, R.C. (1996). Isolation and functional properties of murine hematopoietic stem cells that are replicating *in vivo*. *J. Exp. Med. 183*(4): 1797–1806.

Gregory, C.A., Prockop, D.J. and Spees, J.L. (2005). Non-hematopoietic bone marrow stem cells: molecular control of expansion and differentiation. *Exp. Cell Res. 306*(2): 330–335.

Gronthos, S., Graves, S.E., Ohta, S. and Simmons, P.J. (1994). The STRO-1 + fraction of adult human bone marrow contains the osteogenic precursors. *Blood 84*(12): 4164–4173.

Grunewald, M., Avraham, I., Dor, Y., Bachar-Lustig, E., Itin, A., Yung, S.,Chimenti, S., Landsman, L., Abramovitch, R. and Keshet, E. (2006). VEGF-induced adult neovascularization: recruitment, retention, and role of accessory cells. *Cell 124*(1): 175–189.

Hiasa, K., Egashira, K., Kitamoto, S., Ishibashi, M., Inoue, S., Ni, W., Zhao, Q., Nagata, S., Katoh, M., Sata, M., *et al.* (2004). Bone marrow mononuclear cell therapy limits myocardial infarct size through vascular endothelial growth factor. *Basic Res. Cardiol. 99*(3): 165–172.

Jiang, Y., Jahagirdar, B.N., Reinhardt, R.L., Schwartz, R.E., Keene, C.D., Ortiz-Gonzalez, X.R., Reyes, M., Lenvik, T., Lund, T., Blackstad, M., *et al.* (2002). Pluripotency of mesenchymal stem cells derived from adult marrow. *Nature 418*(6893): 41–49.

Kajstura, J., Leri, A., Finato, N., Di Loreto, C., Beltrami, C.A. and Anversa, P. (1998). Myocyte proliferation in end-stage cardiac failure in humans. *Proc. Natl Acad. Sci. USA 95*(15): 8801–8805.

Kamihata, H., Matsubara, H., Nishiue, T., Fujiyama, S., Tsutsumi, Y., Ozono, R., Masaki, H., Mori, Y., Iba, O., Tateishi, E., *et al.* (2001). Implantation of bone marrow mononuclear cells into ischemic myocardium enhances collateral perfusion and regional function via side supply of angioblasts, angiogenic ligands, and cytokines. *Circulation 104*(9): 1046–1052.

Kang, H.J., Kim, H.S., Zhang, S.Y., Park, K.W., Cho, H.J., Koo, B.K., Kim, Y.J., Soo Lee, D., Sohn, D.W., Han, K.S., *et al.* (2004). Effects of intracoronary infusion of peripheral blood stem-cells mobilised with granulocyte-colony stimulating factor on left ventricular systolic function and restenosis after coronary stenting in myocardial infarction: the MAGIC cell randomised clinical trial. *Lancet 363*(9411): 751–756.

Kawada, H., Fujita, J., Kinjo, K., Matsuzaki, Y., Tsuma, M., Miyatake, H., Muguruma, Y., Tsuboi, K., Itabashi, Y., Ikeda, Y., *et al.* (2004). Nonhematopoietic mesenchymal stem cells can be mobilized and differentiate into cardiomyocytes after myocardial infarction. *Blood 104*(12): 3581–3587.

Kocher, A.A., Schuster, M.D., Szabolcs, M.J., Takuma, S., Burkhoff, D., Wang, J., Homma, S., Edwards, N.M. and Itescu, S. (2001). Neovascularization of ischemic myocardium by human bone-marrow-derived angioblasts prevents cardiomyocyte apoptosis, reduces remodeling and improves cardiac function. *Nat. Med. 7*(4): 430–436.

Kucia, M., Dawn, B., Hunt, G., Guo, Y., Wysoczynski, M., Majka, M., Ratajczak, J., Rezzoug, F., Ildstad, S.T., Bolli, R., *et al.* (2004). Cells expressing early cardiac markers reside in the bone marrow and are mobilized into the peripheral blood after myocardial infarction. *Circ. Res. 95*(12): 1191–1199.

Kuethe, F., Figulla, H.R., Herzau, M., Voth, M., Fritzenwanger, M., Opfermann, T., Pachmann, K., Krack, A., Sayer, H.G., Gottschild, D., *et al.* (2005). Treatment with granulocyte colony-stimulating factor for mobilization of bone marrow cells in patients with acute myocardial infarction. *Am. Heart J. 150*(1): 115.

Kukielka, G.L., Smith, C.W., LaRosa, G.J., Manning, A.M., Mendoza, L.H., Daly, T.J., Hughes, B.J., Youker, K.A., Hawkins, H.K., Michael, L.H., *et al.* (1995). Interleukin-8 gene induction in the myocardium after ischemia and reperfusion *in vivo*. *J. Clin. Invest. 95*(1): 89–103.

Lagostena, L., Avitabile, D., De Falco, E., Orlandi, A., Grassi, F., Iachininoto, M.G., Ragone, G., Fucile, S., Pompilio, G., Eusebi, F., *et al.* (2005). Electrophysiological properties of mouse bone marrow c-kit cells co-cultured onto neonatal cardiac myocytes. *Cardiovasc. Res. 66*(3): 482–492.

Laguens, R., Cabeza Meckert, P., Vera Janavel, G., Del Valle, H., Lascano, E., Negroni, J., Werba, P., Cuniberti, L., Martinez, V., Melo, C., *et al.* (2002). Entrance in mitosis of adult cardiomyocytes in ischemic pig hearts after plasmid-mediated rhVEGF165 gene transfer. *Gene Ther. 9*(24): 1676–1681.

Leone, A.M., Rutella, S., Bonanno, G., Contemi, A.M., de Ritis, D.G., Giannico, M.B., Rebuzzi, A.G., Leone, G. and Crea, F. (2005a). Endogenous G-CSF and CD34() cell mobilization after acute myocardial infarction. *Int. J. Cardiol. 111*(2): 202–208.

Leone, A.M., Rutella, S., Bonanno, G., Abbate, A., Rebuzzi, A.G., Giovannini, S., Lombardi, M., Galiuto, L.,Liuzzo, G., Andreotti, F., *et al.* (2005b). Mobilization of bone marrow-derived stem cells after myocardial infarction and left ventricular function. *Eur. Heart J. 26*(12): 1196–1204.

Levesque, J.P., Takamatsu, Y., Nilsson, S.K., Haylock, D.N. and Simmons, P.J. (2001). Vascular cell adhesion molecule-1 (CD106) is cleaved by neutrophil proteases in the bone marrow following hematopoietic progenitor cell mobilization by granulocyte colony-stimulating factor. *Blood 98*(5): 1289–1297.

Li, T.S., Hayashi, M., Ito, H., Furutani, A., Murata, T., Matsuzaki, M. and Hamano, K. (2005). Regeneration of infarcted myocardium by intramyocardial implantation of *ex vivo* transforming growth factor-beta-preprogrammed bone marrow stem cells. *Circulation 111*(19): 2438–2445.

Liadaki, K., Kho, A.T., Sanoudou, D., Schienda, J., Flint, A., Beggs, A.H., Kohane, I.S. and Kunkel, L.M. (2005). Side population cells isolated from different tissues share transcriptome signatures and express tissue-specific markers. *Exp. Cell Res. 303*(2): 360–374.

Lunde, K., Solheim, S., Aakhus, S., Arnesen, H., Abdelnoor, M. and Forfang, K. (2005). Autologous stem cell transplantation in acute myocardial infarction: The ASTAMI randomized controlled trial. Intracoronary transplantation of autologous mononuclear bone marrow cells, study design and safety aspects. *Scand Cardiovasc. J. 39*(3): 150–158.

Murry, C.E., Soonpaa, M.H., Reinecke, H., Nakajima, H., Nakajima, O.H., Rubart, M. Pasumarthi, K.B., Virag, J.I., Bartelmez, S.H. and Pappa, V. *et al.* (2004). Haematopoietic stem cells do not transdifferentiate into cardiac myocytes in myocardial infarctions. *Nature 428*: 664–668.

Menasche, P. (2003). Cell transplantation in myocardium. *Ann. Thorac. Surg. 75*(6 Suppl): S20–S28.

Minatoguchi, S., Takemura, G., Chen, X.H., Wang, N., Uno, Y., Koda, M., Arai, M., Misao, Y., Lu, C., Suzuki, K., *et al.* (2004). Acceleration of the healing process and myocardial regeneration may be important as a mechanism of improvement of cardiac function and remodeling by postinfarction granulocyte colony-stimulating factor treatment. *Circulation 109*(21): 2572–2580.

Moscoso, I., Centeno, A., Lopez, E., Rodriguez-Barbosa, J.I., Santamarina, I., Filgueira, P., Sanchez, M.J., Dominguez-Perles, R., Penuelas-Rivas, G. and Domenech, N. (2005). Differentiation "*in vitro*" of primary and immortalized porcine mesenchymal stem cells into cardiomyocytes for cell transplantation. *Transplant. Proc. 37*(1): 481–482.

NIH (2000). *National Institutes of Health, Morbidity and Mortality*. Washington, DC: NIH.

Norol, F., Merlet, P., Isnard, R., Sebillon, P., Bonnet, N., Cailliot, C., Carrion, C., Ribeiro, M., Charlotte, F., Pradeau, P., *et al.* (2003). Influence of mobilized stem cells on myocardial infarct repair in a nonhuman primate model. *Blood 102*(13): 4361–4368.

Ohtsuka, M., Takano, H., Zou, Y., Toko, H., Akazawa, H., Qin, Y., Suzuki, M., Hasegawa, H., Nakaya, H. and Komuro, I. (2004). Cytokine therapy prevents left ventricular remodeling and dysfunction after myocardial infarction through neovascularization. *Faseb J. 18*(7): 851–853.

Orlic, D., Kajstura, J., Chimenti, S., Limana, F., Jakoniuk, I., Quaini, F., Nadal-Ginard, B., Bodine, D.M., Leri, A. and Anversa, P. (2001a). Mobilized bone marrow cells repair the infarcted heart, improving function and survival. *Proc. Natl Acad. Sci. USA 98*(18): 10344–10349.

Orlic, D., Kajstura, J., Chimenti, S., Jakoniuk, I., Anderson, S.M., Li, B., Pickel, J., McKay, R., Nadal-Ginard, B., Bodine, D.M., *et al.* (2001b). Bone marrow cells regenerate infarcted myocardium. *Nature 410*(6829): 701–705.

Oswald, J., Boxberger, S., Jorgensen, B., Feldmann, S., Ehninger, G., Bornhauser, M. and Werner, C. (2004). Mesenchymal stem cells can be differentiated into endothelial cells *in vitro*. *Stem Cells 22*(3): 377–384.

Ozturk, M.A., Guven, G.S. and Haznedaroglu, I.C. (2004). How hematopoietic stem cells know and act in cardiac microenvironment for stem cell plasticity? Impact of local renin–angiotensin systems. *Med. Hypotheses 63*(5): 866–874.

Peled, A., Kollet, O., Ponomaryov, T., Petit, I., Franitza, S., Grabovsky, V., Slav, M.M., Nagler, A., Lider, O., Alon, R., *et al.* (2000). The chemokine SDF-1 activates the integrins LFA-1, VLA-4, and VLA-5 on immature human CD34() cells: role in transendothelial/stromal migration and engraftment of NOD/SCID mice. *Blood 95*(11): 3289–3296.

Perin, E.C. and Silva, G.V. (2004). Stem cell therapy for cardiac diseases. *Curr. Opin. Hematol. 11*(6): 399–403.

Pittenger, M.F. and Martin, B.J. (2004). Mesenchymal stem cells and their potential as cardiac therapeutics. *Circ. Res. 95*(1): 9–20.

Pompilio, G., Cannata, A., Peccatori, F., Bertolini, F., Nascimbene, A., Capogrossi, M.C. and Biglioli, P. (2004). Autologous peripheral blood stem cell transplantation for myocardial regeneration: a novel strategy for cell collection and surgical injection. *Ann. Thorac. Surg. 78*(5): 1808–1812.

Powell, T.M., Paul, J.D., Hill, J.M., Thompson, M., Benjamin, M., Rodrigo, M., McCoy, J.P., Read, E.J., Khuu, H.M., Leitman, S.F., *et al.* (2005). Granulocyte colony-stimulating factor mobilizes functional endothelial progenitor cells in patients with coronary artery disease. *Arterioscler. Thromb. Vasc. Biol. 25*(2): 296–301.

Quaini, F., Urbanek, K., Beltrami, A.P., Finato, N., Beltrami, C.A., Nadal-Ginard, B., Kajstura, J., Leri, A. and Anversa, P. (2002). Chimerism of the transplanted heart. *N. Engl.J.Med. 346*(1): 5–15.

Reyes, M., Dudek, A., Jahagirdar, B., Koodie, L., Marker, P.H. and Verfaillie, C.M. (2002). Origin of endothelial progenitors in human postnatal bone marrow. *J. Clin. Invest. 109*(3): 337–346.

Schofield, R. (1978). The relationship between the spleen colony-forming cell and the haemopoietic stem cell. *Blood Cell 4*(1–2): 7–25.

Schwartz, R.E., Reyes, M., Koodie, L., Jiang, Y., Blackstad, M., Lund, T., Lenvik, T., Johnson, S., Hu, W.S. and Verfaillie, C.M. (2002).

Multipotent adult progenitor cells from bone marrow differentiate into functional hepatocyte-like cells. *J. Clin. Invest. 109*(10): 1291–1302.

Stamm, C., Westphal, B., Kleine, H.D., Petzsch, M., Kittner, C., Klinge, H., Schumichen, C., Nienaber, C.A., Freund, M. and Steinhoff, G. (2003). Autologous bone-marrow stem-cell transplantation for myocardial regeneration. *Lancet 361*(9351): 45–46.

Stocum, D.L. (2001). Stem cells in regenerative biology and medicine. *Wound Repair Regen. 9*(6): 429–442.

Theise, N.D., Nimmakayalu, M., Gardner, R., Illei, P.B., Morgan, G., Teperman, L., Henegariu, O. and Krause, D.S. (2000). Liver from bone marrow in humans. *Hepatology 32*(1): 11–16.

Tomita, S., Li, R.K., Weisel, R.D., Mickle, D.A., Kim, E.J., Sakai, T. and Jia, Z.Q. (1999). Autologous transplantation of bone marrow cells improves damaged heart function. *Circulation 100*(19 Suppl): II247–II256.

Ueno, H., Sakita-Ishikawa, M., Morikawa, Y., Nakano, T., Kitamura, T. and Saito, M. (2003). A stromal cell-derived membrane protein that supports hematopoietic stem cells. *Nat. Immunol. 4*(5): 457–463.

Valgimigli, M., Rigolin, G.M., Cittanti, C., Malagutti, P., Curello, S., Percoco, G., Bugli, A.M., Della Porta, M., Bragotti, L.Z., Ansani, L., *et al.* (2005). Use of granulocyte-colony stimulating factor during acute myocardial infarction to enhance bone marrow stem cell mobilization in humans: clinical and angiographic safety profile. *Eur. Heart J. 26*(18): 1838–1845.

Vandervelde, S., van Luyn, M.J., Tio, R.A. and Harmsen, M.C. (2005). Signaling factors in stem cell-mediated repair of infarcted myocardium. *J. Mol. Cell Cardiol. 39*(2): 363–376.

Weissman, I.L. (2000). Translating stem and progenitor cell biology to the clinic: barriers and opportunities. *Science 287*(5457): 1442–1446.

Woldbaek, P.R., Hoen, I.B., Christensen, G. and Tonnessen, T. (2002). Gene expression of colony-stimulating factors and stem cell factor after myocardial infarction in the mouse. *Acta Physiol. Scand. 175*(3): 173–181.

Wollert, K.C., Meyer, G.P., Lotz, J., Ringes-Lichtenberg, S., Lippolt, P., Breidenbach, C., Fichtner, S., Korte, T., Hornig, B., Messinger, D., *et al.* (2004). Intracoronary autologous bone-marrow cell transfer after myocardial infarction: the BOOST randomised controlled clinical trial. *Lancet 364*(9429): 141–148.

Woodbury, D., Schwarz, E.J., Prockop, D.J. and Black, I.B. (2000). Adult rat and human bone marrow stromal cells differentiate into neurons. *J. Neurosci. Res. 61*(4): 364–370.

Yoon, J., Min, B.G., Kim, Y.H., Shim, W.J., Ro, Y.M. and Lim, D.S. (2005). Differentiation, engraftment and functional effects of pre-treated mesenchymal stem cells in a rat myocardial infarct model. *Acta Cardiol. 60*(3): 277–284.

Zaffran, S. and Frasch, M. (2002). Early signals in cardiac development. *Circ. Res. 91*(6): 457–469.

Zhang, J., Niu, C., Ye, L., Huang, H., He, X., Tong, W.G., Ross, J., Haug, J., Johnson, T., Feng, J.Q., *et al.* (2003). Identification of the haematopoietic stem cell niche and control of the niche size. *Nature 425*(6960): 836–841.

Chapter 15

Hematopoietic Stem Cell Properties, Markers, and Therapeutics

S.M. Chambers, William J. Lindblad, and M.A. Goodell

INTRODUCTION

Hematopoietic stem cells (HSCs), which primarily reside in bone marrow (BM), maintain blood formation and replenish themselves throughout the adults' lifespan. The activity of BM HSCs was discovered half a century ago when Ford *et al.* (1956) identified a robust contribution of donor BM cells in lethally irradiated recipient mice. After three decades of work, the contribution of donor hematopoietic cells in recipients had been demonstrated to originate from a few "clones," suggesting the existence of HSCs (Becker *et al.*, 1963; Lemischka *et al.*, 1986). However, the isolation of HSCs was not achieved until 1988 when Weissman *et al.* enriched HSCs from the murine BM using a fluorescent-activated cell sorter (Spangrude *et al.*, 1988). Since these seminal studies, researchers have been able to demonstrate that HSCs possess stem cell properties including the ability to give rise to daughter HSC (self-renewal) as well as to repopulate all of the hematopoietic lineages (differentiation, Figure 15-1). As one of the most investigated tissue stem cells, studies of HSCs have inspired the exploring of various stem cells, and will continuously provide insight of stem cell biology.

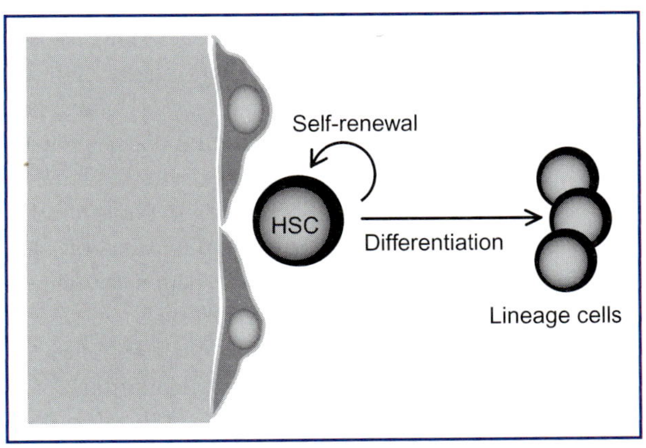

FIGURE 15-1 Self-renewal and differentiation of HSC: The HSC-to-niche interaction influences the two definitive properties of HSC. HSC can expand to create more HSC (self-renew) and they can regenerate the hematopoietic system (differentiate).

DEVELOPMENTAL ORIGIN OF HEMATOPOIESIS

During embryonic development, two waves of hematopoiesis occur: primitive hematopoiesis and definitive hematopoiesis, which, respectively, give rise to embryonic and adult hematopoietic cells. Primitive hematopoiesis begins at day 7 of gestation in the mouse yolk sac and generates embryonic primitive erythroblasts (EryP) (reviewed in Lensch and Daley (2004)). However, the hematopoietic precursors from yolk sac are not able to reconstitute lethally irradiated adult recipients (Medvinsky *et al.*, 1993; Muller *et al.*, 1994), which is the gold standard for demonstrating functional HSC activity. The second wave of hematopoiesis, definitive, or adult hematopoiesis arises around day 10 of gestation in the aorta-gonad-mesonephros (AGM) region (Muller *et al.*, 1994; Medvinsky and Dzierzak, 1996). The definitive embryonic HSCs are able to self-renew and give rise to mature hematopoietic lineages in adults. These cells seed the BM, where HSCs contribute to blood formation throughout the lifespan of the adult (Lensch and Daley, 2004).

Several studies have identified molecules required for primitive and definitive hematopoiesis, bringing insight to HSC ontogeny and shedding light on mechanisms that regulate HSC self-renewal and differentiation (reviewed in Lensch and Daley (2004) and Medvinsky and Dzierzak (1998)). Deficiency in some of these genes is found to cause embryonic anemia due to inefficient hematopoiesis, indicating they are essential for HSC formation. For example, mutants of Runx1, a member of the runt transcription factor family, exhibit normal primitive hematopoiesis in the yolk sac but lack of hematopoietic clusters in the intra-aorta region at E10.5 of the AGM, and are embryonic lethal at E12.5 with anemic fetal liver, a temporary site of primitive hematopoiesis between E11 and E14

(Okuda *et al.*, 1996; North *et al.*, 1999). The evidence indicates that Runx1 is indispensable for the definitive hematopoiesis but not primitive hematopoiesis. Flk-1 (vascular endothelial growth factor receptor-2) null mice are also embryo lethal (at E8.5–E9.5) with defects in forming blood clusters and in developing vascular network in yolk sac region (Shalaby *et al.*, 1995; Sakurai *et al.*, 2005). Likewise, Scl/Tal1 null mice are found to be embryonic lethal (at E9.5–E11.5), and lack yolk sac vitelline vessels and primitive hematopoiesis (Robb *et al.*, 1995; Shivdasani *et al.*, 1995). Scl/Tal1 null cells also fail to contribute to definitive hematopoiesis of both the AGM and fetal liver in chimeric mice (Porcher *et al.*, 1996; Robb *et al.*, 1996), suggesting critical roles of Scl/Tal1 in both primitive and definitive hematopoiesis.

INTRINSIC REGULATORS OF SELF-RENEWAL

Adult HSCs comprise only ~0.02% of whole BM cells but possess abilities to self-renew and replenish the whole hematopoietic system. Self-renewal, a signature process of all stem cells, is the process by which one stem cell is able to give rise to at least one daughter stem cell via cell division. By using retroviral integration to mark HSCs, it has been demonstrated that HSCs undergo clonal expansion while repopulating the hematopoietic system *in vivo* (Lemischka *et al.*, 1986). Moreover, as few as a single HSC is sufficient to establish long-term multi-lineage engraftment (Osawa *et al.*, 1996; Camargo *et al.*, 2005), which would only be possible through a self-renewal process. However, culture conditions have not been fully established to expand HSC, which has hindered our ability to manipulate HSC *in vitro*. Nevertheless, in the attempt to understand the mechanisms that intrinsically underlie HSC self-renewal *in vivo*, gene-targeted mice have been developed. HSCs lacking self-renewal mediators usually bear defects in maintaining homeostasis of stem cell population upon proliferation stimuli. Genes that are involved in intrinsically regulating HSC self-renewal can be roughly divided into several groups: cell cycle regulators, transcription factors, anti-apoptotic molecules, and development pathway regulators (Table 15-1).

The majority of HSCs remain in G0 phase under normal homeostasis but are able to extensively expand upon receiving proliferation cues. p18INK4C, p21cip/waf1, and p27kip1 are G1-phase cyclin-dependent kinase inhibitors (CKIs) that tightly regulate the G0/G1 stage transition of cell cycle. By perturbing these cell cycle regulators, HSCs exhibit distinct self-renewal phenotypes. Surprisingly, although all three gene-targeted HSCs exhibit proliferative phenotypes, p21 null HSCs exhaust over time (Cheng *et al.*, 2000b), whereas p27 and p18 null HSCs possess increased engraftment in transplantation assays (Cheng *et al.*, 2000a; Yuan *et al.*, 2004). More interestingly, the increased multi-lineage engraftment in p18 and p27 null mice originates from different mechanisms; p18 null HSCs possess a higher proportion of self-renewing HSCs (Yuan *et al.*, 2004), whereas p27 null progenitors preferentially outcompete the progenitors of wild type HSC (Cheng *et al.*, 2000a). These data demonstrate how components of cell cycle regulation impinge on HSC cell fate decision. Gfi-1, a zinc-finger proto-oncogene, has been found to be involved in self-renewal with a regulatory role in HSC proliferation. In addition to the defects in generation of neutrophils which will be discussed later, Gfi-1 null HSC exhibited a higher proliferation rate but failed to engraft after serial transplantation, indicating Gfi-1 is a transcriptional regulator that restricts HSC proliferation and prevents exhaustion of HSC pool (Hock *et al.*, 2004).

Molecules that regulate cell apoptosis have been implicated to affect HSC function as well. Overexpression of Bcl-2, an anti-apoptotic factor, leads to, not only an increased HSC number, but also elevated HSC repopulating ability (Domen *et al.*, 2000). In addition, deletion of Mcl-1, a Bcl-2 family member, leads to a loss of HSCs and committed progenitors, and ultimately severe anemia in mice (Opferman *et al.*, 2005).

The largest group of genes that regulate HSC self-renewal are transcription factors. Homeobox genes (HOX) have been found to be involved in both hematopoiesis and leukemia ontogenesis (Abramovich *et al.*, 2005). Wild-type HSCs quickly lose engraftment ability after *in vitro* culture. However, enforced expression of Hoxb4 with a retroviral target vector extends the self-renewing capability of cultured-mouse HSC. Hoxb4-expressing HSC engraft as well as freshly isolated HSC while the control retroviral-transduced HSC showed 5–10% decline in engraftment over the course of cell culture (Sauvageau *et al.*, 1995; Thorsteinsdottir *et al.*, 1999). Similarly, overexpression of Hoxb9 enhances engraftment when they are directly competed against wildtype HSC (Thorsteinsdottir *et al.*, 2002). The findings that Hoxb4 overexpression in HSC leads to a robust expansion of HSC *in vitro* (Antonchuk *et al.*, 2002), but retained HSC under control of homeostasis *in vivo* (Thorsteinsdottir *et al.*, 1999), suggests a therapeutic strategy to expand HSC *in vitro* prior to transplantation. Human HSCs with enforced expression of Hoxb4 have been shown to be capable of repopulating immunodeficient mice. Unfortunately, the repopulation which is derived from Hoxb4-transduced HSC preferentially differentiates into myeloid lineages at a loss of lymphoid cells (Sauvageau *et al.*, 1995; Brun *et al.*, 2003), limiting the clinical utility of this strategy.

Polycomb genes that are involved in epigenetic regulation also regulate HSC self-renewal. Bmi-1-deficient mice exhibit a hypocellular BM phenotype, and their HSCs exhaust 2 months after birth. Fetal liver and adult BM cells from Bmi-1-deficient mice failed to engraft after transplantation, which indicates Bmi-1 is critical for HSC self-renewal (Park *et al.*, 2003). Moreover, enforced expression

TABLE 15-1 Intrinsic factors of HSC self-renewal

Gene		HSC phenotypes	Reference
Cell cycle regulators			
p21cip/waf1	KO	Increased number of proliferating HSC (sensitive to 5FU treatment) Increased HSC cell number (CAFC assay) Decreased self-renewal ability (serial transplantation)	Cheng *et al.* (2000)
p27kip1	KO	Unaltered HSC pool Increased frequency of committed progenitor Increased multi-lineage repopulation due to outgrowth of progenitors (competitive transplantation)	Cheng *et al.* (2000)
p18INK4X	KO	Increased frequency of self-renewing HSC	Yuan *et al.* (2004)
Transcription factors			
Hoxb4	KO	Lower proliferative rate in HSC compartment Decreased repopulation efficiency when competing with wild-type HSC Impaired embryo primitive hematopoiesis Normal hematopoiesis	Bjornsson *et al.* (2003)
	Tg (retroviral)	Extend the self-renewing ability of HSC after short-term in vitro cell culture (transplantation)	Sauvageau *et al.* (1995) and Thorsteinsdottir *et al.* (1999)
Hoxa9	Tg (retroviral)	Expanded number of self-renewing HSC after short-term in vitro cell culture (competitive transplantation)	Thorsteinsdottir *et al.* (2002)
Pbx1	KO	Defects in fetal liver-derived HSC to engraft (competitive transplantation)	DiMartino *et al.* (2001)
c-myc	Tg (retroviral)	Loss of adhesion molecule expression and long-term repopulation ability	Wilson *et al.* (2004)
CBP	KO-Het	CBP+/− HSC pool prematurely exhausted after birth	Kung *et al.* (2000) and Rebel *et al.* (2002)
Bmi-1	KO	HSC prematurely exhausted after birth, mice die from anemia	Park *et al.* (2003)
	Tg (retroviral)	Bmi1 null HSC fail to repopulate Enforced expression of Bmi-1 increased HSC engraftment	Iwama *et al.* (2004)
Rae28	KO	Reduced repopulating activity (competitive transplantation, serial transplantation)	Ohta *et al.* (2002)
Mel-18	KO	Slightly increased repopulation activity of Mel-18 null HSC (competitive repopulation)	Kajiume *et al.* (2004)
	Tg-mice	Slightly decrease in HSC proliferation status Decreased repopulation efficiency Increased frequency of proliferative HSC	
Gfi-1	KO	Increased frequency of proliferative HSC and HSC pool Compromised repopulation ability of Gfi-1 null HSC (transplantation and competitive transplantation)	Hock *et al.* (2004)
Anti-apoptotic molecules			
Bcl2	Tg-mice	Higher frequency of HSC compartment Increased repopulation ability of HSC (competitive transplantation) Decreased frequency of proliferative HSC	Domen *et al.* (2000)
MCL-1	KO	Decreased engraftment after Mcl-1 gene deletion Mcl-1 gene deletion leads to loss of HSC and committed progenitors	Opferman *et al.* (2005)

(*Continued*)

TABLE 15-1 (Continued)

Gene		HSC phenotypes	Reference
Anti-oxidative stress			
ATM	KO	Decreased frequency of BM HSC compartments (KTSL and KSL compartments) Decreased long-term engraftment (competitive transplantation)	Ito *et al.* (2004)
Developmental molecules			
Notch1	Tg (retroviral)	HSCs are able to survive through a long-term cell culture, retain repopulation ability after *in vitro* cell culture	Varnum-Finney *et al.* (2000)
β-catenin	Tg (retroviral)	Enforced expression leads to higher engraftment ability	Reya *et al.* (2003)

These genes have been demonstrated to alter aspects HSC self-renewal; Tg: transgenic, KO: knock-out.

of Bmi-1 by a retrovirus in murine HSC has shown an increased repopulating activity, suggesting a therapeutic role of Bmi-1 (Iwama *et al.*, 2004). Other polycomb genes such as Rae28 (Ohta *et al.*, 2002) and Mel-18 (Kajiume *et al.*, 2004) have also been shown to modulate HSC self-renewal. Interestingly, unlike Rae28 and Bmi-1, Mel-18 acts as a negative regulator of HSC self-renewal. Mel-18 null HSCs possess a higher repopulation activity, whereas Mel-18 transgenic HSCs have decreased ability to engraft into lethally irradiated mice (Kajiume *et al.*, 2004).

Pathways that trigger cell fate decisions during early development of vertebrates and invertebrates have been demonstrated to be involved in dictating cell fate decisions of HSC during cell division. Overexpression of constitutively activated Notch1 in HSC immortalizes HSC in long-term *in vitro* culture. A single Notch1-transduced HSC clone is able to undergo multi-lineage repopulation *in vivo* (Varnum-Finney *et al.*, 2000). The Wnt signaling pathway has also been implicated in HSC self-renewal. Extrinsic stimulation of Wnt3a or overexpression of beta-catenin in combination with the presence of Bcl2 leads to HSC with a high repopulating activity after a long period of *in vitro* cell culture while ectopic expression of Axin, an inhibitor of the Wnt pathway, decreased HSC proliferation *in vitro* and reconstitution function *in vivo*, suggesting a role of Wnt signaling in retaining HSC self-renewal (Reya *et al.*, 2003).

MULTI-LINEAGE REPOPULATION

During adult hematopoiesis, BM-HSCs generate both lymphoid and myeloid cells (Figure 15-2). Lymphoid cells are comprised of primarily T-cells, B-cells, and natural killer cells (NK cells). Myeloid cells include granulocytes, macrophages, megakaryocytes, and erythrocytes. Hematopoiesis is a gradual differentiation process that involves multiple decision points beginning with HSC and ending with terminally differentiated lineages (Figure 15-2). This concept of stepwise hematopoiesis has lead to the identification of several differentiation intermediates. From this concept, Morrison and Weissman described two populations within BM that possess transient engraftment ability when transplanted into lethally irradiated mice. These two populations are considered short-term HSC (ST-HSC, Mac-1^{lo}CD4^{-}) and multipotent progenitor (MPP, Mac-1^{lo}CD4^{lo}), which are distinguished from long-term repopulating HSC (Morrison and Weissman, 1994; Morrison *et al.*, 1997). Weissman *et al.* also first identified common lymphoid progenitors (CLPs) within BM which specifically give rise to lymphoid lineages (Tcells, B-cells, and NK cells) (Kondo *et al.*, 1997), and common myeloid progenitors (CMPs), which give rise to granulocytes/macrophages and megakaryocytes/erythrocytes colonies in methylcellulose cell culture and in lethally irradiated recipients (Akashi *et al.*, 2000). More recently, Adolfsson *et al.* have revised the role of MPPs as lymphoid-primed multipotent progenitors (LMPP). They discovered that MPPs (now LMPP) preferentially differentiate into lymphoid lineages, but retain some myeloid development capacity (dashed line in Figure 15-2), restricted to granulocytes and macrophages (Adolfsson *et al.*, 2005). The CMP is the major generator of all myeloid cells, including megakaryocytes/erythrocytes lineages. In summary, these studies have suggested a hematopoietic hierarchy in which long-term HSCs give rise to ST-HSC that differentiates into CMP and CLP. The CMP and CLP then generate the myeloid and lymphoid lineages (Figure 15-2). Although the differentiation pathway in humans is not as well established as in rodent models, transplantation of HSC has been utilized for decades to treat patients with hematopoietic diseases, demonstrating the repopulation ability of human HSC to reconstitute the entire hematopoietic system.

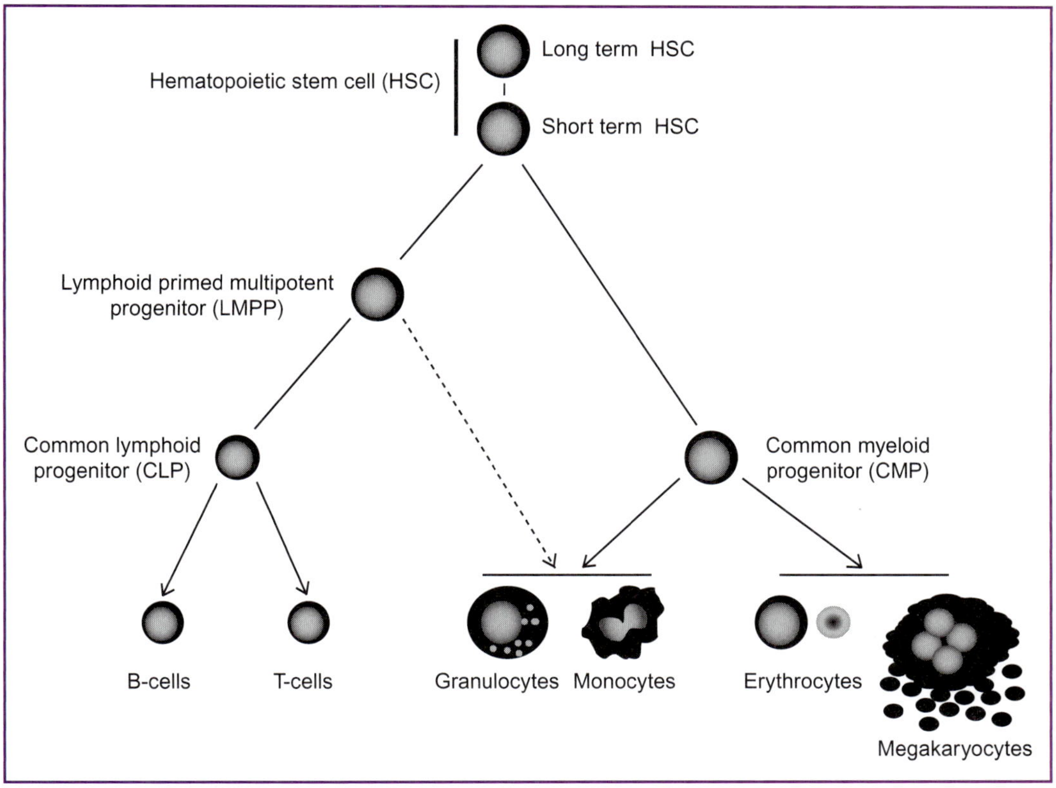

FIGURE 15-2 HSC differentiation: HSC regenerates the hematopoietic system, which is comprised of a myeloid and lymphoid branch, ultimately creating all the cells that comprise the blood. The multipotent-progenitor (MPPs), previously thought of as a bipotential progenitor, is now identified as a lymphoid-primed progenitor (LMPP).

PLAYERS IN HEMATOPOIESIS

Transcription factors that regulate cell fate decisions during adult hematopoiesis have been identified mostly by phenotypes of gene-targeted mice. Deficiencies in genes controlling cell fate lead to defective, definitive hematopoiesis in embryos and lineage skewing in adult hematopoietic system. For example, PU.1, a member of the ETS transcription factor family, has been shown to dictate cell fate at the divergence point of myeloid and lymphoid differentiation in a dosage-dependent manner. Overexpression of PU.1 drives fetal liver-derived hematopoietic precursors to differentiate into macrophages; whereas haploinsufficiency in PU.1+/− cells leads to both pro-B-cells and macrophages (DeKoter and Singh, 2000). In humans, aberrant expression of several transcription factors was shown to cause diseases that result from defects in hematopoietic development or abnormal proliferation of a particular hematopoietic lineage. For example, mutations of a proto-oncogene Gfi-1, a zinc finger transcription factor, result in neutropenia in which patients lack neutrophils (Hock *et al.*, 2003; Person *et al.*, 2003). Chromosome translocation that results in overexpression of Scl/Tal, a basic helix-loop-helix gene that is also critical to murine primitive and definitive hematopoiesis, causes human acute T-lymphocytic leukemia (T-ALL) (Begley *et al.*, 1989; Chen *et al.*, 1990). Runx1-Eto, the fusion protein caused by chromosome translocation t(8;21), and Tel-Runx1, caused by t(12;21), result in AML and B-cell precursor ALL, respectively (reviewed in Izraeli (2004)). In addition, retroviral integration which results in overexpressing LMO2 has been found to cause T-cell malignancy in a retroviral-gene therapy that was originally designed to treat X-linked severe combined immunodeficient (SCID) patients (Hacein-Bey-Abina *et al.*, 2003). These studies have shed light on therapeutic methodology to target hematopoietic disorders as well as defining roles of genes that control homeostasis of the hematopoietic system.

IN VITRO DIFFERENTIATION OF HEMATOPOIETIC LINEAGES

In vitro cell culture of blood precursors has been established to quantify the differentiation ability of hematopoietic precursors. The *in vitro* culture conditions were established with the goal to mimic the *in vivo* growth stimuli and maturation signals. A functional assay of hematopoietic precursors, the colony forming unit in cell culture (CFU-C) was first established in the 1980s and measured generation of myeloid and erythroid cells (Metcalf, 1989). *In vitro* differentiation of lymphocytes was later found to require microenvironments that are provided by co-cultured cells. *In vitro* co-culture of stromal cell line with B-cell precursors (Cumano *et al.*, 1990) and fetal thymic organ culture (FTOC) to generate thymocytes (Robinson and Owen, 1978) have been aimed at recapitulating the *in vivo* environment and identifying regulators of lymphocyte differentiation. Moreover, *in vitro* assays such as cobblestone area forming cells (CAFC) and long-term culture initiating cells (LT-CIC) were developed to detect earlier precursors, including the HSCs. A more comprehensive outline of *in vitro* HSC differentiation assays has been described by Ramos *et al.* (2003). In addition, differentiation of murine and human hematopoietic progenitor cells *in vitro* has been utilized to modulate immune responses. The best example is terminally differentiated dendritic cells (DCs), one of the professional antigen presenting cells in immune system. Antigen (Ag)-pulsed DCs have been utilized to modulate Ag-specific immune response. In clinical trials of immunotherapy, these *in vitro* differentiated DCs are able to stimulate tumor antigen-specific immune responses and to induce tolerance in autoimmune diseases (Figdor *et al.*, 2004).

IN VITRO EXPANSION OF SELF-RENEWING HSCs

Expansion of HSC *in vitro* has been the most difficult challenge for decades due to a decline in repopulation capacity of HSC in long-term *ex vivo* culture. There are currently two main strategies for *ex vivo* HSC expansion: HSC stromal cell co-cultivation and HSC suspension culture. BM stromal cells support HSC maintenance, measured by repopulating ability, in the absence of additional growth factors (Fraser *et al.*, 1990; Fraser *et al.*, 1992). BM stromal cells are thought to mimic the microenvironment of the HSC niche. To identify the molecules in these stromal cells that retain HSC function, genome wide studies of cloned stromal cells have been reported (Moore, 2004), providing an emerging picture that underlies the HSC-to-niche interaction which we will discuss below. For suspension culture, growth factor cocktails have been utilized in an attempt to expand human and murine HSC (Sauvageau *et al.*, 2004), albeit with low recovery rate of repopulating HSC. The differentiation of HSC during *in vitro* culture has drawn into question whether or not HSC self-renewal occurs during cell expansion. To address the question, Glimm and Eaves labeled HSC with a fluorescent membrane-specific dye, carboxyfluorescein diacetate succinimidyl (CFSE), to track the proliferation history. By transplanting cells that had divided from an *in vitro* cell suspension culture, they discovered that human HSCs were still able to give rise to multi-lineage repopulation after a low number of cell divisions (Glimm and Eaves, 1999). Additionally, Nakauchi *et al.* have cultured highly purified murine single-cell HSC to track cell division as well as cell fate. They have been able to generate limited self-renewing HSC under the influence of various combinations of cytokines (Ema *et al.*, 2000). However, after more than two cell divisions *in vitro*, the HSCs greatly lost their repopulating activity.

HSC NICHE

In adults, HSCs reside in the BM cavity, closely associated with surrounding stromal cells. There is mounting evidence suggesting that the most primitive HSCs localize to the interior surface of bone (periosteum/endosteum border) on the basis of colony forming assays (Lord and Hendry, 1972) and Brd-U label retention (Zhang *et al.*, 2003), bringing them within close contact with osteoblasts. Murine osteoblasts have long been thought to provide essential cues for HSC, as they (or their transformed counterparts) express various cytokines known to influence hematopoiesis, including but not limited to G/M/GM-CSFs, Il-1, Il-6, SDF-1, and VEGF (reviewed in Taichman (2005)). In addition to expressing HSC-modulating cytokines, recent genetic evidence from mice has demonstrated that expanding the number of osteoblasts within the BM increases the relative percentage of HSC (Calvi *et al.*, 2003; Zhang *et al.*, 2003), and genetically ablating osteoblasts results in the failure of BM hematopoiesis (Visnjic *et al.*, 2004), suggesting that osteoblasts provide a direct physical niche for the HSC that maintains their self-renewing capacity via various cell surface molecules (Figure 15-3). Researchers have also provided evidence of a second HSC niche provided by sinusoidal endothelial cells resident in the BM (Kiel *et al.*, 2005).

Molecules including N-cadherin, Notch-1, Tie2, and CXCR-4 have all been implicated in the HSC-to-niche interface. N-cadherin, a Ca^{2+}-dependent homophilic adhesion molecule expressed on osteoblasts, exhibits an asymmetrical localization on HSC, as determined by fluorescence microscopy (Zhang *et al.*, 2003), and is found expressed on a fraction (10%) of hematopoietic progenitors that include the HSC, thus suggesting that it may be involved in self-renewal or niche retention. However, it remains to be seen whether or not N-cadherin is required for HSC-to-niche interaction, by examining if there is a functional difference between N-cadherin$^+$ and N-cadherin$^-$ HSC *in vivo*.

Several studies have demonstrated that Notch-1 is expressed on HSC and its activation by incubating with Jagged-1 expressing cells (Varnum-Finney *et al.*, 2000) or constitutive Notch-1 signaling (Varnum-Finney *et al.*, 2000) results in *in vitro* expansion of self-renewing HSC with normal homeostasis while transplanting into lethally irradiated mice. In contrast to the evidence discovered from *in vitro* Notch1 stimulation, targeted disruption of Jagged1 and Notch-1 in mice does not result in reconstitution or self-renewal

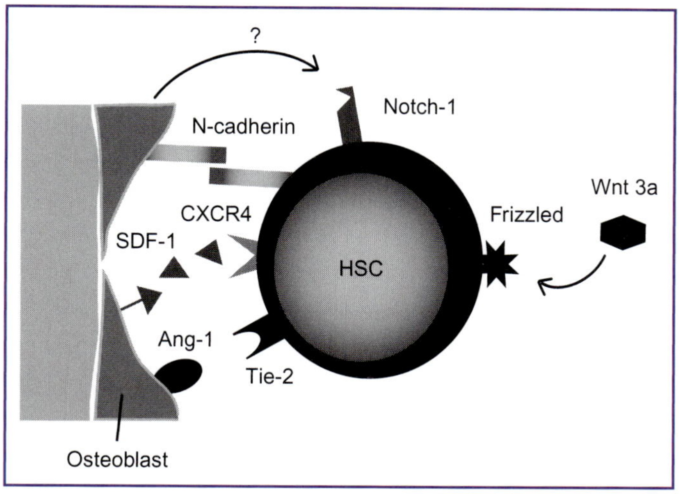

FIGURE 15-3 Hypothetical HSC-to-niche interactions: Through several cell surface molecules (markers) and cytokines, the HSC is thought to directly or indirectly interact with osteoblast cells. These are few of the molecules that provide instructions for self-renewal and differentiation to the HSC.

defects *in vivo* (Mancini *et al.*, 2005). Therefore, Notch-1 stimulation may regulate HSC cell fate decision toward self-renewal during *in vitro* cell culture, although it is not essential for HSC homeostasis *in vivo*. Hence, stimulation of the signaling pathway is of interest in HSC expansion.

A second such receptor–ligand interaction is that of Tie2, a tyrosine kinase receptor that is expressed on HSC, and its ligand Angiopoietin-1 (Ang-1), expressed by osteoblasts. These two molecules were demonstrated to be important retention factors for HSC (Arai *et al.*, 2004). Incubation of HSC in Ang-1 or overexpression of Tie-2 in transduced BM cells resulted in expansion of the quiescent portion of the HSC compartment. A loss of self-renewal or rapid differentiation continues to be a hurdle for *in vitro* expansion of HSC. Exploiting Tie2 signaling with soluble Ang-1 may be a promising method for expanding long-term, quiescent HSC cultures.

One such receptor that has been employed in the clinic is the chemokine receptor, CXCR-4, expressed on both human (Viardot *et al.*, 1998) and mouse (Wright *et al.*, 2002) HSCs. In addition, SDF-1, the ligand for CXCR-4, is a well-established HSC homing factor expressed by osteoblasts and BM fibroblasts (Ponomaryov *et al.*, 2000). Antibodies against SDF-1 or CXCR-4 block human HSC engraftment in the non-obese severe combined immunodeficient (NOD/SCID) mouse model, and SDF-1 enhances transwell migration of human HSC (Peled *et al.*, 1999). Therefore, the CXCR-4/SDF-1 axis is thought to provide a BM homing and retention mechanism for HSC *in vivo*. In a secondary transplantation assay, treatment of stem cell factor (SCF) and IL6 enhanced HSC engraftment, correlating with elevated CXCR4 expression and increased *in vitro* migration activity to SDF-1 (Peled *et al.*, 1999). However, they did not distinguish between a rescued migratory defect and enhanced self-renewal as the cause for increased engraftment *in vivo*. Therefore, it remains less clear whether CXCR-4 plays a role in HSC self-renewal and *in vitro* expansion of HSC.

In summary, over the past decade it has become increasingly clear that components such as N-cadherin, Tie2/Ang-1, Notch-1, and CXCR-4/SDF-1 may be required in order to establish an *ex vivo* niche for HSC expansion and self-renewal, unfortunately, research has yet to elucidate the appropriate cocktail of soluble factors, cytokines, and/or niche support cells needed to stimulate faithful and prolonged *in vitro* HSC self-renewal and expansion.

PURIFICATION AND MOLECULAR SIGNATURE OF HSC

In order to assess HSC function and identify novel molecular components, it is essential that HSCs are distinguished from the heterogeneous cell mixture that comprises the BM niche. Consequently, functional purity has been an important focus in the HSC field, since the purification method can drastically influence engraftment and hematopoietic reconstitution. Murine and human HSCs are purified by a combination of enrichment (by magnetic cell sorting) and fluorescence-activated cell sorting based on cell surface markers or vital dye staining (Table 15.2). Human HSCs have been shown to express CD34 and lack the expression of CD38, forming the basis of all human HSCs purification schemes. Positive selection for CD34 and Thy-1, as well as the removal of differentiated progeny by CD38 and a cocktail of antibodies that recognize lineage-specific cell surface molecules (Lin cocktail), has also been used to purify human hematopoietic progenitors from fetal liver (Muench *et al.*, 2002), umbilical cord blood (Gluckman *et al.*, 1989; Rocha *et al.*, 2004), BM (Thomas, 2000), or patient peripheral blood treated with an HSC mobilizing agent (Ho *et al.*, 1996; Broxmeyer *et al.*, 2005).

In the mouse, HSCs are typically harvested from the medullar cavity of the leg bones (tibias and femurs). Similar

TABLE 15-2 HSC cell surface molecules that allow for purification of specific HSC subpopulation; UK:unknown,

Gene	Function	Homology		Details/discovery	References
		Mouse	Human		
N-Cadherin	Cell surface marker; Wnt signaling	+	UK	Homophilic adhesion molecule found to be expressed on both HSC and osteoblasts	Zhang *et al.* (2003)
Notch-1	Cell surface marker; Notch signaling	+	+	*In vitro* activation results in self-renewal/expansion	Varnum-Finney *et al.* (1998) and Varnum-Finney *et al.* (2000)
Tie2	Cell surface marker	+	UK	Activation with Ang-1 results in enhanced quiescence	Arai *et al.* (2004)
Endoglin	Cell surface marker; TGF-b signaling	+	UK	Discovered by microarray, marks LT-HSC	Chen *et al.* (2002)
CD34	Cell surface marker	ST-HSC	+	Purification marker in humans; ST-HSC marker in mouse	Civin *et al.* (1996)
CXCR-4	Cell surface marker	+	+	Homing and retention of HSC to niche	Peled *et al.* (1999)
CD150 (SLAM)	Cell surface marker	+	UK	Discovered by microarray, marks LT-HSC	Kiel *et al.* (2005)
CD48	Cell surface marker	ST-HSC	UK	Discovered by microarray, marks ST-HSC	Kiel *et al.* (2005)
Bcrp-1	Cell property	+	+	Thought to be responsible for HSC SP phenotype	Zhou *et al.* (2002)
Thy-1	Cell surface marker	+	−	ST-HSC marker	Uchida and Weissman (1992)
Sca-1	Cell surface marker; HSC self-renewal	+	−	Purification marker, null HSC exhibit reduced *in vivo* repopulation (competitive transplantation and serial transplantation)	Ito *et al.* (2003)
c-Kit	Cell surface marker; Tyrosine Kinase Receptor; expansion and self-renewal of HSC pool *in vivo*	+	+	HSC from W/Wv mice fail to engraft (transplantation, parabiosis)	Gardner *et al.* (1988)

HSC cell surface molecules: Both mouse and human HSC express cell surface molecules (markers) that allow for purification of specific HSC subpopulations.

to human HSCs, differentiated cells are excluded from the purification on the basis of expression of lineage-specific cell surface markers (Lineage^{negative} or Lin^{-}). To further purify mouse HSCs, cells that express two canonical HSC markers, stem cell antigen-1 (Sca-1) and the tyrosine receptor kinase c-Kit, are selected. Low expression of the marker Thy-1 is sometimes used to further subdivide this population but its utility is limited to strains bearing the Thy-1.1 allele. This isolation scheme has been collectively referred to as "KTSL" in the literature (c-Kit^{+}, Thy^{low}, Sca-1^{+}, Lin^{-}).

Two purification schemes have used to simplify the bewildering number of antibodies needed in order to obtain highly purified HSC. The first isolation scheme exploits a property of HSC to efflux the fluorescent DNA binding dye Hoechst 33342. Cells that posses this efflux ability are referred to as side population (SP) cells (Goodell *et al.*, 1996) and have been shown to have a KTSL phenotype (Camargo *et al.*, 2005). This efflux property is thought to be conferred by the ABC transporter Bcrp1 (ABCG2) (Zhou *et al.*, 2002). In mouse BM, SP cells comprise approximately 0.03–0.07% of all nucleated cells, express low to no lineage marker expression, and greater than 95% express c-kit and Sca-1 (Figure 15-4). Details about the Hoechst staining procedure can be found in a recent review (Goodell, 2002; Goodell *et al.*, 2005). Finally, a new scheme may simplify the purification scheme further. Antibodies against

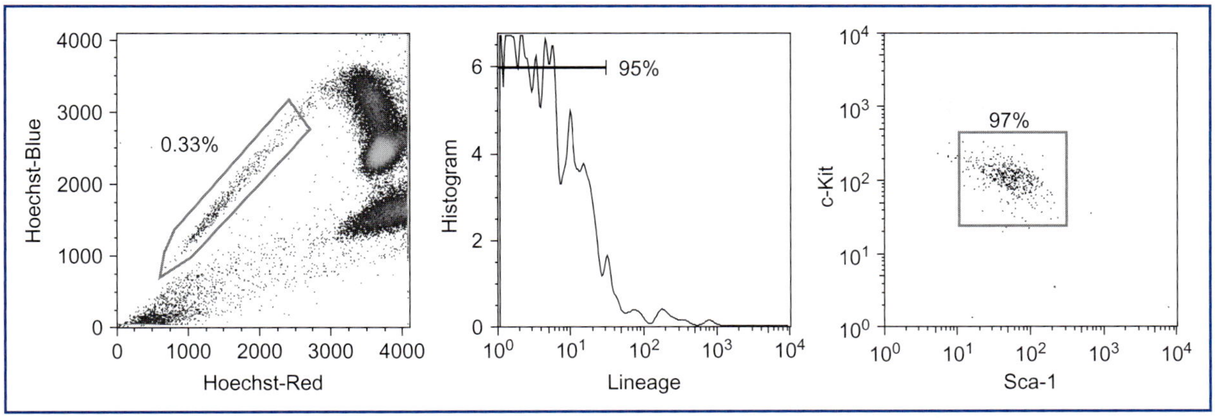

FIGURE 15-4 HSC SP phenotype: Mouse HSC can be purified by their ability to efflux Hoechst dye. When whole BM is excited with a UV laser and viewed by two wavelengths (Hoechst-Red, Hoechst-Blue), HSCs are found in a SP (left). Less than 5% of these SP cells express differentiation markers (lineage, middle), and greater than 95% express the two canonical stem cell markers Sca-1 and c-Kit (right).

a combination of three markers, all members of the SLAM family of proteins, have been demonstrated to recognize high-purity HSC comparable to KTSL (Kiel *et al.*, 2005). Although these results have yet to be verified by other labs, this method reduces the markers needed to purify HSC to three ($CD150^+$, $CD48^-$, and $CD244^-$).

Now that HSCs have been purified to near homogeneity, microarrays have been employed to examine HSC transcriptome under steady state and activated conditions. Initial analyses were aimed at identifying transcriptional similarities shared between embryonic, hematopoietic, and neuronal stem cells ("stemness") (Ivanova *et al.*, 2002; Ramalho-Santos *et al.*, 2002). While some similarities were observed, the implications remain unclear (Evsikov and Solter, 2003; Fortunel *et al.*, 2003; Vogel, 2003). Fortunately, transcriptional profiling of both mouse and human HSCs under steady state conditions has lead to a number of observations about cell surface markers and potential regulators of HSC function. For instance, CD150 was found to be a nearly exclusive marker of LTHSC, after comparing the expression profiles of ST- and LT-HSC (Kiel *et al.*, 2005). The expression profile of mouse HSCs under activated conditions has been studied by chemotherapeutic drug 5-fluorouracil (5-FU) treatment, which destroys the majority of hematopoietic cells prompting hematopoietic regeneration. This stressful, anemic environment stimulates the normally quiescent HSC ($\sim$1% S-phase) into a transient proliferative phase (>20% S-phase) lasting approximately 12 days. The expression profile of HSC was monitored at 2–3 day intervals after 5-FU injection (Venezia *et al.*, 2004). The data not only elucidated a model of HSC activation but identified CD48, a molecule which marks proliferating HSC and has been shown to distinguish ST-HSC from LT-HSC (Kiel *et al.*, 2005). Venezia, *et al.* also employed a refined gene ontology (GO) analysis (Young *et al.*, 2005), which quantified fold enrichment (over random chance) of all GO biological processes, thus conferring more biological relevance to the global microarray results. Microarrays are an extremely powerful emerging technology, with continued bioinformatic developments that can augment and standardize the results, microarrays will become increasingly prevalent in the HSC field as a tool for expression profiling, molecular phenotyping, and screening.

PLASTICITY AND THERAPEUTICS

The central tenet to embryonic development states that HSCs become committed to a multipotent program of differentiation, in that they exclusively create all of the cells of the hematopoietic hierarchy. However, reports began to arise that HSCs could give rise to non-autologous tissues leading to the idea the HSCs may be pluripotent (Ferrari *et al.*, 1998; Brazelton *et al.*, 2000). Although provocative, many of these reported cases of HSC "plasticity" have been called into question, due to inadequate assays (Jackson *et al.*, 2004), cell fusion (Terada *et al.*, 2002; Ying *et al.*, 2002), and extramedullary HSC (McKinney-Freeman *et al.*, 2002). Through stringent genetic studies it has become clear that certain hematopoietic progeny, most likely the fusogenic and invasive macrophage, are able to contribute at subtherapeutic levels to skeletal muscle (Camargo *et al.*, 2003) and liver (Camargo *et al.*, 2004) regeneration.

Clinical trials of HSC transplantation for the purpose of replenishing hematopoiesis have been widely performed for treating leukemia (Armitage, 1994), severe combined immunodeficiency (Fischer *et al.*, 2004), and severe autoimmune disease (Tyndall and Saccardi, 2005). In addition, hematopoietic treatment has been used in adjunct to chemotherapy for non-hematopoietic cancer such as breast cancer, neuroblastoma, and testicular cancer (reviewed in Armitage (1994)). Currently, HSC transplant research is investigating ways to improve the overall success by reducing graft-versus-host-disease, infection during recovery, and accelerating robust hematopoiesis after transplantation.

CONCLUSIONS

The HSC field continues to be at the forefront of regenerative medicine with therapeutic potential to cure a wide range of diseases. Developmental origin, self-renewal, differentiation, molecular signature, and therapeutic potential of HSC are currently being established. The next horizon for the HSC field is to create an *ex vivo* niche. Identifying the essential support cells and their secreted cytokines required for HSC self-renewal and expansion would open the gateway for unfettered genetic modification of HSC to aid the continued efforts in gene therapy. Equally important, HSC expansion could drastically reduce the number of HSCs needed to provide adequate reconstitution in human BM transplant therapy. Therefore, further characterization of components that contribute to regulation of HSC self-renewal is needed in order to co-ordinate HSC expansion.

REFERENCES

Abramovich, C., Pineault, N., Ohta, H. and Humphries, R.K. (2005). Hox genes: from leukemia to hematopoietic stem cell expansion. *Ann. NY Acad. Sci. 1044*: 109–116.

Adolfsson, J., Mansson, R., Buza-Vidas, N., Hultquist, A., Liuba, K., Jensen, C.T., Bryder, D., Yang, L., Borge, O.J., Thoren, L.A., *et al.* (2005). Identification of Flt3+ lympho-myeloid stem cells lacking erythro-megakaryocytic potential a revised road map for adult blood lineage commitment. *Cell 121*: 295–306.

Akashi, K., Traver, D., Miyamoto, T. and Weissman, I.L. (2000). A clonogenic common myeloid progenitor that gives rise to all myeloid lineages. *Nature 404*: 193–197.

Antonchuk, J., Sauvageau, G. and Humphries, R.K. (2002). HOXB4-induced expansion of adult hematopoietic stem cells *ex vivo*. *Cell 109*: 39–45.

Arai, F., Hirao, A., Ohmura, M., Sato, H., Matsuoka, S., Takubo, K., Ito, K., Koh, G.Y. and Suda, T. (2004). Tie2/ angiopoietin-1 signaling regulates hematopoietic stem cell quiescence in the bone marrow niche. *Cell 118*: 149–161.

Armitage, J.O. (1994). Bone marrow transplantation. *N. Engl. J. Med. 330*: 827–838.

Becker, A.J., Mc, C.E. and Till, J.E. (1963). Cytological demonstration of the clonal nature of spleen colonies derived from transplanted mouse marrow cells. *Nature 197*: 452–454.

Begley, C.G., Aplan, P.D., Davey, M.P., Nakahara, K., Tchorz, K., Kurtzberg, J., Hershfield, M.S., Haynes, B.F., Cohen, D.I., Waldmann, T.A., *et al.* (1989). Chromosomal translocation in a human leukemic stem-cell line disrupts the T-cell antigen receptor delta-chain diversity region and results in a previously unreported fusion transcript. *Proc. Natl Acad. Sci. USA 86*: 2031–2035.

Brazelton, T.R., Rossi, F.M., Keshet, G.I. and Blau, H.M. (2000). From marrow to brain: expression of neuronal phenotypes in adult mice. *Science 290*: 1775–1779.

Broxmeyer, H.E., Orschell, C.M., Clapp, D.W., Hangoc, G., Cooper, S., Plett, P.A., Liles, W.C., Li, X., Graham-Evans, B., Campbell, T.B., *et al.* (2005). Rapid mobilization of murine and human hematopoietic stem and progenitor cells with AMD3100, a CXCR4 antagonist. *J. Exp. Med. 201*: 1307–1318.

Brun, A.C., Fan, X., Bjornsson, J.M., Humphries, R.K. and Karlsson, S. (2003). Enforced adenoviral vector-mediated expression of HOXB4 in human umbilical cord blood CD34+ cells promotes myeloid differentiation but not proliferation. *Mol. Ther. 8*: 618–628.

Calvi, L.M., Adams, G.B., Weibrecht, K.W., Weber, J.M., Olson, D.P., Knight, M.C., Martin, R.P., Schipani, E., Divieti, P., Bringhurst, F.R., *et al.* (2003). Osteoblastic cells regulate the haematopoietic stem cell niche. *Nature 425*: 841–846.

Camargo, F.D., Green, R., Capetanaki, Y., Jackson, K.A. and Goodell, M.A. (2003). Single hematopoietic stem cells generate skeletal muscle through myeloid intermediates. *Nat. Med. 9*: 1520–1527.

Camargo, F.D., Finegold, M. and Goodell, M.A. (2004). Hematopoietic myelomonocytic cells are the major source of hepatocyte fusion partners. *J. Clin. Invest. 113*: 1266–1270.

Camargo, F.D., Chambers, S.M., Drew, E., McNagny, K.M. and Goodell, M.A. (2005). Hematopoietic stem cells do not engraft with absolute efficiencies. *Blood 107*: 501–507.

Chen, Q., Cheng, J.T., Tasi, L.H., Schneider, N., Buchanan, G., Carroll, A., Crist, W., Ozanne, B., Siciliano, M.J. and Baer, R. (1990). The tal gene undergoes chromosome translocation in T cell leukemia and potentially encodes a helix-loop-helix protein. *Embo. J. 9*: 415–424.

Cheng, T., Rodrigues, N., Dombkowski, D., Stier, S. and Scadden, D.T. (2000a). Stem cell repopulation efficiency but not pool size is governed by p27(kip1). *Nat. Med. 6*: 1235–1240.

Cheng, T., Rodrigues, N., Shen, H., Yang, Y., Dombkowski, D., Sykes, M. and Scadden, D.T. (2000b). Hematopoietic stem cell quiescence maintained by p21cip1/waf1. *Science 287*: 1804–1808.

Cumano, A., Dorshkind, K., Gillis, S. and Paige, C.J. (1990). The influence of S17 stromal cells and interleukin 7 on B cell development. *Eur. J. Immunol. 20*: 2183–2189.

DeKoter, R.P. and Singh, H. (2000). Regulation of B lymphocyte and macrophage development by graded expression of PU.1. *Science 288*: 1439–1441.

Domen, J., Cheshier, S.H. and Weissman, I.L. (2000). The role of apoptosis in the regulation of hematopoietic stem cells: overexpression of Bcl-2 increases both their number and repopulation potential. *J. Exp. Med. 191*: 253–264.

Ema, H., Takano, H., Sudo, K. and Nakauchi, H. (2000). *In vitro* self-renewal division of hematopoietic stem cells. *J. Exp. Med. 192*: 1281–1288.

Evsikov, A.V. and Solter, D. (2003). Comment on " 'Stemness': transcriptional profiling of embryonic and adult stem cells" and "a stem cell molecular signature". *Science 302*: 393 (author reply 393).

Ferrari, G., Cusella-De Angelis, G., Coletta, M., Paolucci, E., Stornaiuolo, A., Cossu, G. and Mavilio, F. (1998). Muscle regeneration by bone marrow-derived myogenic progenitors. Science *279*: 1528–1530.

Figdor, C.G., de Vries, I.J., Lesterhuis, W.J. and Melief, C.J. (2004). Dendritic cell immunotherapy: mapping the way. *Nat. Med. 10*: 475–480.

Fischer, A., Hacein-Bey-Abina, S. and Cavazzana-Calvo, M. (2004). Gene therapy for immunodeficiency diseases. *Semin. Hematol. 41*: 272–278.

Ford, C.E., Hamerton, J.L., Barnes, D.W. and Loutit, J.F. (1956). Cytological identification of radiation-chimeras. *Nature 177*: 452–454.

Fortunel, N.O., Otu, H.H., Ng, H.H., Chen, J., Mu, X., Chevassut, T., Li, X., Joseph, M., Bailey, C., Hatzfeld, J.A., *et al.* (2003). Comment on " 'Stemness': transcriptional profiling of embryonic and adult stem cells" and "a stem cell molecular signature". *Science 302*: 393 (author reply 393).

Fraser, C.C., Eaves, C.J., Szilvassy, S.J. and Humphries, R.K. (1990). Expansion *in vitro* of retrovirally marked totipotent hematopoietic stem cells. *Blood 76*: 1071–1076.

Fraser, C.C., Szilvassy, S.J., Eaves, C.J. and Humphries, R.K. (1992). Proliferation of totipotent hematopoietic stem cells *in vitro* with retention of long-term competitive *in vivo* reconstituting ability. *Proc. Natl Acad. Sci. USA 89*: 1968–1972.

Glimm, H. and Eaves, C.J. (1999). Direct evidence for multiple self-renewal divisions of human *in vivo* repopulating hematopoietic cells in short-term culture. *Blood 94*: 2161–2168.

Gluckman, E., Broxmeyer, H.A., Auerbach, A.D., Friedman, H.S., Douglas, G.W., Devergie, A., Esperou, H., Thierry, D., Socie, G., Lehn, P., *et al.* (1989). Hematopoietic reconstitution in a patient with Fanconi's anemia by means of umbilicalcord blood from an HLA-identical sibling. *N. Engl. J. Med. 321*: 1174–1178.

Goodell, M.A. (2002). Stem Cell Identification and Sorting Using the Hoechst 33342 Side Population (SP), Vol. 2. New York: Wiley.

Goodell, M.A., Brose, K., Paradis, G., Conner, A.S. and Mulligan, R.C. (1996). Isolation and functional properties of murine hematopoietic stem cells that are replicating *in vivo*. *J. Exp. Med. 183*: 1797–1806.

Goodell, M.A., McKinney-Freeman, S. and Camargo, F.D. (2005). Isolation and characterization of side population cells. *Method. Mol. Biol. 290*: 343–352.

Hacein-Bey-Abina, S., Von Kalle, C., Schmidt, M., McCormack, M.P., Wulffraat, N., Leboulch, P., Lim, A., Osborne, C.S., Pawliuk, R., Morillon, E., *et al.* (2003). LMO2-associated clonal T cell proliferation in two patients after gene therapy for SCID-X1. *Science 302*: 415–419.

Ho, A.D., Young, D., Maruyama, M., Corringham, R.E., Mason, J.R., Thompson, P., Grenier, K., Law, P., Terstappen, L.W. and Lane, T. (1996). Pluripotent and lineage-committed CD34+ subsets in leukapheresis products mobilized by G-CSF, GM-CSF vs. a combination of both. *Exp. Hematol. 24*: 1460–1468.

Hock, H., Hamblen, M.J., Rooke, H.M., Traver, D., Bronson, R.T., Cameron, S. and Orkin, S.H. (2003). Intrinsic requirement for zinc finger transcription factor Gfi-1 in neutrophil differentiation. *Immunity 18*: 109–120.

Hock, H., Hamblen, M.J., Rooke, H.M., Schindler, J.W., Saleque, S., Fujiwara, Y. and Orkin, S.H. (2004). Gfi-1 restricts proliferation and preserves functional integrity of haematopoietic stem cells. *Nature 431*: 1002–1007.

Ivanova, N.B., Dimos, J.T., Schaniel, C., Hackney, J.A., Moore, K.A. and Lemischka, I.R. (2002). A stem cell molecular signature. *Science 298*: 601–604.

Iwama, A., Oguro, H., Negishi, M., Kato, Y., Morita, Y., Tsukui, H., Ema, H., Kamijo, T., Katoh-Fukui, Y., Koseki, H., *et al.* (2004). Enhanced self-renewal of hematopoietic stem cells mediated by the polycomb gene product Bmi-1. *Immunity 21*: 843–851.

Izraeli, S. (2004). Leukaemia—a developmental perspective. *Br. J. Haematol. 126*: 3–10.

Jackson, K.A., Snyder, D.S. and Goodell, M.A. (2004). Skeletal muscle fiber-specific green autofluorescence: potential for stem cell engraftment artifacts. *Stem Cells 22*: 180–187.

Kajiume, T., Ninomiya, Y., Ishihara, H., Kanno, R. and Kanno, M. (2004). Polycomb group gene mel-18 modulates the self-renewal activity and cell cycle status of hematopoietic stem cells. *Exp. Hematol. 32*: 571–578.

Kiel, M.J., Yilmaz, O.H., Iwashita, T., Yilmaz, O.H., Terhorst, C. and Morrison, S.J. (2005). SLAM family receptors distinguish hematopoietic stem and progenitor cells and reveal endothelial niches for stem cells. *Cell 121*: 1109–1121.

Kondo, M., Weissman, I.L. and Akashi, K. (1997). Identification of clonogenic common lymphoid progenitors in mouse bone marrow. *Cell 91*: 661–672.

Lemischka, I.R., Raulet, D.H. and Mulligan, R.C. (1986). Developmental potential and dynamic behavior of hematopoietic stem cells. *Cell 45*: 917–927.

Lensch, M.W. and Daley, G.Q. (2004). Origins of mammalian hematopoiesis: *in vivo* paradigms and *in vitro* models. *Curr. Top. Dev. Biol. 60*: 127–196.

Lord, B.I. and Hendry, J.H. (1972). The distribution of haemopoietic colony-forming units in the mouse femur, and its modification by x rays. *Br. J. Radiol. 45*: 110–115.

Mancini, S.J., Mantei, N., Dumortier, A., Suter, U., Macdonald, H.R. and Radtke, F. (2005). Jagged1-dependent Notch signaling is dispensable for hematopoietic stem cell self-renewal and differentiation. *Blood 105*: 2340–2342.

McKinney-Freeman, S.L., Jackson, K.A., Camargo, F.D., Ferrari, G., Mavilio, F. and Goodell, M.A. (2002). Muscle-derived hematopoietic stem cells are hematopoietic in origin. *Proc. Natl Acad. Sci. USA 99*: 1341–1346.

Medvinsky, A. and Dzierzak, E. (1996). Definitive hematopoiesis is autonomously initiated by the AGM region. *Cell 86*: 897–906.

Medvinsky, A.L. and Dzierzak, E.A. (1998). Development of the definitive hematopoietic hierarchy in the mouse. *Dev. Comp. Immunol. 22*: 289–301.

Medvinsky, A.L., Samoylina, N.L., Muller, A.M. and Dzierzak, E.A. (1993). An early pre-liver intraembryonic source of CFU-S in the developing mouse. *Nature 364*: 64–67.

Metcalf, D. (1989). The molecular control of cell division, differentiation commitment and maturation in haemopoietic cells. *Nature 339*: 27–30.

Moore, K.A. (2004). Recent advances in defining the hematopoietic stem cell niche. *Curr. Opin. Hematol. 11*: 107–111.

Morrison, S.J. and Weissman, I.L. (1994). The long-term repopulating subset of hematopoietic stem cells is deterministic and isolatable by phenotype. *Immunity 1*: 661–673.

Morrison, S.J., Wandycz, A.M., Hemmati, H.D., Wright, D.E. and Weissman, I.L. (1997). Identification of a lineage of multipotent hematopoietic progenitors. *Development 124*: 1929–1939.

Muench, M.O., Suskind, D.L. and Barcena, A. (2002). Isolation, growth and identification of colony-forming cells with erythroid, myeloid, dendritic cell and NK-cell potential from human fetal liver. *Biol. Proceed. Online 4*: 10–23.

Muller, A.M., Medvinsky, A., Strouboulis, J., Grosveld, F. and Dzierzak, E. (1994). Development of hematopoietic stem cell activity in the mouse embryo. *Immunity 1*: 291–301.

North, T., Gu, T.L., Stacy, T., Wang, Q., Howard, L., Binder, M., Marin-Padilla, M. and Speck, N.A. (1999). Cbfa2 is required for the formation of intra-aortic hematopoietic clusters. *Development 126*: 2563–2575.

Ohta, H., Sawada, A., Kim, J.Y., Tokimasa, S., Nishiguchi, S., Humphries, R.K., Hara, J. and Takihara, Y. (2002). Polycomb group gene rae28 is required for sustaining activity of hematopoietic stem cells. *J. Exp. Med. 195*: 759–770.

Okuda, T., van Deursen, J., Hiebert, S.W., Grosveld, G. and Downing, J.R. (1996). AML1, the target of multiple chromosomal translocations in human leukemia, is essential for normal fetal liver hematopoiesis. *Cell 84*: 321–330.

Opferman, J.T., Iwasaki, H., Ong, C.C., Suh, H., Mizuno, S., Akashi, K. and Korsmeyer, S.J. (2005). Obligate role of antiapoptotic MCL-1 in the survival of hematopoietic stem cells. *Science 307*: 1101–1104.

Osawa, M., Hanada, K., Hamada, H. and Nakauchi, H. (1996). Long-term lymphohematopoietic reconstitution by a single CD34-low/negative hematopoietic stem cell. *Science 273*: 242–245.

Park, I.K., Qian, D., Kiel, M., Becker, M.W., Pihalja, M., Weissman, I.L., Morrison, S.J. and Clarke, M.F. (2003). Bmi-1 is required for maintenance of adult self-renewing haematopoietic stem cells. *Nature 423*: 302–305.

Peled, A., Petit, I., Kollet, O., Magid, M., Ponomaryov, T., Byk, T., Nagler, A., Ben-Hur, H., Many, A., Shultz, L., *et al.* (1999). Dependence of human stem cell engraftment and repopulation of NOD/SCID mice on CXCR4. *Science 283*: 845–848.

Person, R.E., Li, F.Q., Duan, Z., Benson, K.F., Wechsler, J., Papadaki, H.A., Eliopoulos, G., Kaufman, C., Bertolone, S.J., Nakamoto, B., *et al.* (2003). Mutations in proto-oncogene GFI1 cause human neutropenia and target ELA2. *Nat. Genet. 34*: 308–312.

Ponomaryov, T., Peled, A., Petit, I., Taichman, R.S., Habler, L., Sandbank, J., Arenzana-Seisdedos, F., Magerus, A., Caruz, A., Fujii, N., *et al.* (2000). Induction of the chemokine stromal-derived factor-1 following DNA damage improves human stem cell function. *J. Clin. Invest. 106*: 1331–1339.

Porcher, C., Swat, W., Rockwell, K., Fujiwara, Y., Alt, F.W. and Orkin, S.H. (1996). The T cell leukemia oncoprotein SCL/tal-1 is essential for development of all hematopoietic lineages. *Cell 86*: 47–57.

Ramalho-Santos, M., Yoon, S., Matsuzaki, Y., Mulligan, R.C. and Melton, D.A. (2002). "Stemness": transcriptional profiling of embryonic and adult stem cells. *Science 298*: 597–600.

Ramos, C.A., Venezia, T.A., Camargo, F.A. and Goodell, M.A. (2003). Techniques for the study of adult stem cells: be fruitful and multiply. *Biotechniques 34*: 572–578, 580–574, 586–591.

Reya, T., Duncan, A.W., Ailles, L., Domen, J., Scherer, D.C., Willert, K., Hintz, L., Nusse, R. and Weissman, I.L. (2003). A role for Wnt signalling in self-renewal of haematopoietic stem cells. *Nature 423*: 409–414.

Robb, L., Lyons, I., Li, R., Hartley, L., Kontgen, F., Harvey, R.P., Metcalf, D. and Begley, C.G. (1995). Absence of yolk sac hematopoiesis from mice with a targeted disruption of the scl gene. *Proc. Natl Acad. Sci. USA 92*: 7075–7079.

Robb, L., Elwood, N.J., Elefanty, A.G., Kontgen, F., Li, R., Barnett, L.D. and Begley, C.G. (1996). The scl gene product is required for the generation of all hematopoietic lineages in the adult mouse. *Embo. J. 15*: 4123–4129.

Robinson, J.H. and Owen, J.J. (1978). Transplantation tolerance induced in foetal mouse thymus *in vitro*. *Nature 271*: 758–760.

Rocha, V., Sanz, G. and Gluckman, E. (2004). Umbilical cord blood transplantation. *Curr. Opin. Hematol. 11*: 375–385.

Sakurai, Y., Ohgimoto, K., Kataoka, Y., Yoshida, N. and Shibuya, M. (2005). Essential role of Flk-1 (VEGF receptor 2) tyrosine residue 1173 in vasculogenesis in mice. *Proc. Natl Acad. Sci. USA 102*: 1076–1081.

Sauvageau, G., Thorsteinsdottir, U., Eaves, C.J., Lawrence, H.J., Largman, C., Lansdorp, P.M. and Humphries, R.K. (1995). Overexpression of HOXB4 in hematopoietic cells causes the selective expansion of more primitive populations *in vitro* and *in vivo*. *Genes Dev. 9*: 1753–1765.

Sauvageau, G., Iscove, N.N. and Humphries, R.K. (2004). *In vitro* and *in vivo* expansion of hematopoietic stem cells. *Oncogene 23*: 7223–7232.

Shalaby, F., Rossant, J., Yamaguchi, T.P., Gertsenstein, M., Wu, X.F., Breitman, M.L. and Schuh, A.C. (1995). Failure of blood-island formation and vasculogenesis in Flk-1-deficient mice. *Nature 376*: 62–66.

Shivdasani, R.A., Mayer, E.L. and Orkin, S.H. (1995). Absence of blood formation in mice lacking the T-cell leukaemia oncoprotein tal-1/SCL. *Nature 373*: 432–434.

Spangrude, G.J., Heimfeld, S. and Weissman, I.L. (1988). Purification and characterization of mouse hematopoietic stem cells. *Science 241*: 58–62.

Taichman, R.S. (2005). Blood and bone: two tissues whose fates are intertwined to create the hematopoietic stem-cell niche. *Blood 105*: 2631–2639.

Terada, N., Hamazaki, T., Oka, M., Hoki, M., Mastalerz, D.M., Nakano, Y., Meyer, E.M., Morel, L., Petersen, B.E. and Scott, E.W. (2002). Bone marrow cells adopt the phenotype of other cells by spontaneous cell fusion. *Nature 416*: 542–545.

Thomas, E.D. (2000). Landmarks in the development of hematopoietic cell transplantation. *World J. Surg. 24*: 815–818.

Thorsteinsdottir, U., Sauvageau, G. and Humphries, R.K. (1999). Enhanced *in vivo* regenerative potential of HOXB4-transduced hematopoietic stem cells with regulation of their pool size. *Blood 94*: 2605–2612.

Thorsteinsdottir, U., Mamo, A., Kroon, E., Jerome, L., Bijl, J., Lawrence, H.J., Humphries, K. and Sauvageau, G. (2002). Overexpression of the myeloid leukemia-associated Hoxa9 gene in bone marrow cells induces stem cell expansion. *Blood 99*: 121–129.

Tyndall, A. and Saccardi, R. (2005). Haematopoietic stem cell transplantation in the treatment of severe autoimmune disease: results from phase I/II studies, prospective randomized trials and future directions. *Clin. Exp. Immunol. 141*: 1–9.

Varnum-Finney, B., Xu, L., Brashem-Stein, C., Nourigat, C., Flowers, D., Bakkour, S., Pear, W.S. and Bernstein, I.D. (2000). Pluripotent, cytokine-dependent, hematopoietic stem cells are immortalized by constitutive Notch1 signaling. *Nat. Med. 6*: 1278–1281.

Venezia, T.A., Merchant, A.A., Ramos, C.A., Whitehouse, N.L., Young, A.S., Shaw, C.A. and Goodell, M.A. (2004). Molecular signatures of proliferation and quiescence in hematopoietic stem cells. *PLoS Biol. 2*: e301.

Viardot, A., Kronenwett, R., Deichmann, M. and Haas, R. (1998). The human immunodeficiency virus (HIV)-type 1 coreceptor CXCR-4 (fusin) is preferentially expressed on the more immature CD34+ hematopoietic stem cells. *Ann. Hematol. 77*: 193–197.

Visnjic, D., Kalajzic, Z., Rowe, D.W., Katavic, V., Lorenzo, J. and Aguila, H.L. (2004). Hematopoiesis is severely altered in mice with an induced osteoblast deficiency. *Blood 103*: 3258–3264.

Vogel, G. (2003). Stem cells. "Stemness" genes still elusive. *Science 302*: 371.

Wright, D.E., Bowman, E.P., Wagers, A.J., Butcher, E.C. and Weissman, I.L. (2002). Hematopoietic stem cells are uniquely selective in their migratory response to chemokines. *J. Exp. Med. 195*: 1145–1154.

Ying, Q.L., Nichols, J., Evans, E.P. and Smith, A.G. (2002). Changing potency by spontaneous fusion. *Nature 416*: 545–548.

Young, A., Whitehouse, N., Cho, J. and Shaw, C. (2005). Ontology Traverser: an R Package for GO Analysis. *Bioinformatics*. 2005 Jan 15; *21*(2): 275–276.

Yuan, Y., Shen, H., Franklin, D.S., Scadden, D.T. and Cheng, T. (2004). *In vivo* self-renewing divisions of haematopoietic stem cells are increased in the absence of the early G1-phase inhibitor, p18INK4C. *Nat. Cell. Biol. 6*: 436–442.

Zhang, J., Niu, C., Ye, L., Huang, H., He, X., Tong, W.G., Ross, J., Haug, J., Johnson, T., Feng, J.Q., *et al.* (2003). Identification of the hematopoietic stem cell niche and control of the niche size. *Nature 425*: 836–841.

Zhou, S., Morris, J.J., Barnes, Y., Lan, L., Schuetz, J.D. and Sorrentino, B.P. (2002). Bcrp1 gene expression is required for normal numbers of side population stem cells in mice, and confers relative protection to mitoxantrone in hematopoietic cells *in vivo*. *Proc. Natl Acad. Sci. USA 99*: 12339–12344.

Chapter 16

Stem Cells in the Gastrointestinal Tract

Stuart A. C. McDonald, Trevor A. Graham, Adam Humphries, Nicholas A. Wright, Sean L. Preston, Mairi Brittan, and Natalie C. Direkze

Turnover of the epithelial cell lineages within the gastrointestinal tract is a constant process, occurring every two to seven days under normal homeostasis and increasing after damage. This process is regulated by multipotent stem cells, which generate all gastrointestinal epithelial cell lineages and can regenerate whole intestinal crypts and gastric glands. The stem cells of the gastrointestinal tract are as yet undefined, although it is generally agreed that they are located within a "niche" in the intestinal crypts and gastric glands. Studies of allophenic, tetraparental chimeric mice and targeted stem cell mutations suggest that a single stem cell undergoes an asymmetrical division to produce an identical daughter cell, thus replicating itself, and a committed progenitor cell, which further differentiates into an adult epithelial cell type. The discovery of stem cell plasticity in many tissues, including the ability of transplanted bone marrow to transdifferentiate into intestinal subepithelial myofibroblasts, provides a potential use of bone marrow cells to deliver therapeutic genes to damaged tissues, for example, in the treatment of mesenchymal diseases in the gastrointestinal tract, such as fibrosis and Crohn's disease. Studies are beginning to identify the molecular pathways that regulate stem cell proliferation and differentiation into adult gastrointestinal cell lineages, such as the Wnt and Notch–Delta signaling pathways, and to discover the importance of mesenchymal–epithelial interactions in normal gastrointestinal epithelium and in development and disease. Finally, despite some dispute, a strong case can be made for intestinal neoplasia arising as a result of a series of mutations in stem cells. The mechanism and direction of the spread of this mutated clone in the gastrointestinal mucosa is hotly disputed, and central to this argument is the position and nature of the gastrointestinal stem cell.

INTRODUCTION

There has been a tremendous increase in interest in stem cell biology and its potential applications in recent years. Although this has been galvanized by the exploitation of research in embryonic stem cells, it is interesting to note that what might be called the "intestinal stem cell community," albeit small, has been working productively for some 40 to 50 years. With stem cells now claiming considerable attention, in retrospect, many of the basic tenets that govern our understanding of organ-specific stem cells have come from studies of the gastrointestinal tract and the haematopoietic system.

In the gastrointestinal tract, there is a large body of evidence that multipotential stem cells are found in specific zones, or niches, within gastric glands and intestinal crypts, composed of and maintained by myofibroblasts in the adjacent lamina propria. In this chapter, we review evidence that these multipotential stem cells generate all gastrointestinal epithelial cell lineages through committed precursor cells housed in the proliferative compartments of intestinal crypts and gastric glands, a concept that has had a long and difficult gestation. Notwithstanding their obvious significance, the gastrointestinal stem cells remain elusive and unidentified, mainly because of a lack of accepted morphological and functional markers at the single-cell level. We also explore concepts of stem cell number, location, and fate, and we touch on the ability of gastrointestinal stem cells to regenerate cell lineages of whole intestinal crypts and villi after damage. The luminal gut shows regional specializations of function—the stomach primarily for absorption and the intestinal mucosa for both absorption and secretion. This is reflected by variation in the adult cell lineages native to each tissue, and it is thereby consistent that stem cell fate within each tissue is also different. We are beginning to understand the mechanisms that govern such variation. Controversial findings regarding stem cell plasticity in the gastrointestinal tract are examined in this chapter. Because of their longevity, putatively the same as that of the organism itself, stem cells are often viewed as the target cells for carcinogens, and the cells of origin for spontaneous tumors. This includes a "top-down" proliferation of mutated stem cells located within intercryptal zones on the mucosal surface, which spread downwards into the adjacent crypts. Contrasting to this is a "bottom-up" theory of the upward proliferation of mutated stem cells in the crypt base, to produce dysplastic crypts that replicate and expand by crypt fission. This brings other facets of gut biology into sharp focus—the mechanisms of crypt reproduction; the clonal architecture of normal and dysplastic gastric glands, intestinal crypts, and their derivative tumors; and the role that stem cells take in these events. Here, we propose that the stem cells accumulate the multiple genetic events leading to tumorigenesis, and we explore the manner by which such mutated clones spread in gastrointestinal epithelia. Many of these concepts are being explored at the level of molecular regulatory pathways, including the signaling pathways of Wnt and transforming growth factor β (TGFβ).

This is quite a brief, since several of the most common tumors originate in the gastrointestinal tract, and with increasing incidence in Barrett's esophagus. Our concepts of stem cell biology impinge considerably on our understanding of how these tumors arise.

GASTROINTESTINAL MUCOSA CONTAINS MULTIPLE LINEAGES

In the small intestine, the epithelial lining forms numerous crypts and larger, finger-shaped projections called villi. In the colon there are many crypts, which vary in size throughout the colon; the shortest is in the ascending colon. Overall, four main epithelial cell lineages exist in the intestinal epithelium. These are the columnar cells, the mucin-secreting cells, the endocrine cells, and Paneth cells in the small intestine. Other less-common cell lineages are also present, such as the caveolated cells, and membranous or microfold cells. Columnar cells, with apical microvilli, are the most abundant epithelial cells, termed enterocytes in the small intestine and colonocytes in the large intestine. "Goblet" cells, containing mucin granules—and thus producing swollen, goblet-shaped cells—are found throughout the colonic epithelium, secreting mucus into the intestinal lumen. Endocrine, "neuroendocrine," or "enteroendocrine" cells form an abundant cell population distributed throughout the intestinal epithelium; these cells secrete peptide hormones in an endocrine or paracrine manner from their contained dense core of neurosecretory granules. Paneth cells are located almost exclusively at the crypt base of the small intestine and ascending colon, they contain large apical secretory granules, and express several proteins—including lysozyme, tumor necrosis factor, and the antibacterial cryptins (low molecular-weight peptides related to defensins).

In the stomach, the epithelial lining forms long, tubular glands divided into foveolus, isthmus, neck, and base regions. Gastric foveolar or surface mucus cells are located on the mucosal surface and in the foveola. They contain tightly-packed mucous granules in the supranuclear cytoplasm, and do not possess a theca. The mucus neck cells are situated within the neck and isthmus of the gastric glands, and contain apical secretory mucin granules. The peptic–chief or zymogenic cells are located in the base of the glands in the fundic and body regions; they secrete pepsinogen from oval zymogenic granules. The parietal or oxyntic, acid-secreting cells are located in the body of the stomach in the base of the glands. These cells have many surface infoldings, or canaliculi, which form a network reaching almost to the base of the gland. Endocrine cell families include the enterochromaffin-like cells in the fundus or body that produce histamine; the gastrin-producing cells are a major component of the antral mucosa.

The intestinal crypts and gastric glands are enclosed within a fenestrated sheath of intestinal subepithelial myofibroblasts (ISEMFs). These cells exist as a syncytium that extends throughout the lamina propria and merges with the pericytes of the blood vessels. The ISEMFs are closely applied to the intestinal epithelium, and play a vital role in epithelial–mesenchymal interactions. ISEMFs secrete hepatocyte growth factor (HGF), TGFβ, and keratinocyte growth factor (KGF), but the receptors for these growth factors are located on the epithelial cells. Thus, the ISEMFs are essential for the regulation of epithelial cell differentiation through the secretion of these and possibly other growth factors. Platelet-derived growth factor-α (PDGF-A), expressed in the intestinal epithelium, acts by paracrine signaling through its mesenchymal receptor, PDGFR-α, to regulate epithelial–mesenchymal interactions during development. Studies of mice with targeted deletions in the PDGF-A or PDGFR-α genes have shown defects in normal proliferation and differentiation of PDGFR-α-positive mesenchymal cells. Typically, ISEMFs are α-smooth muscle actin-positive (αSMA+) and desmin−, but some myofibroblasts also express myosin-heavy chains. ISEMFs undergo proliferation despite this MyoD expression—unlike the skeletal muscle myoblast, which decycles once MyoD is expressed. It has been proposed that these cells form a renewing population, migrating upwards as they accompany the epithelial escalator. Although they appear to proliferate and migrate, they move relatively slowly, and then move off into the lamina propria to become polyploid.

A second myofibroblast population in the intestine is the interstitial cells of Cajal. These cells are located close to neurones in the muscular layers; they act as pacemakers for gastrointestinal smooth muscle activity, propagate electrical events, and modulate neurotransmission. They are said to be αSMA+ and desmin+, and to immunostain for *c-kit* and CD45.

EPITHELIAL CELL LINEAGES ORIGINATE FROM A COMMON PRECURSOR CELL

Little is known of the location and fate of the stem cells within the gastrointestinal tract, because of the lack of distinctive and accepted stem cells markers, although they are usually said to appear undifferentiated and can be identified operationally by their ability to repopulate crypts and glands after damage.

The Unitarian hypothesis states that all the differentiated cell lineages within the gastrointestinal epithelium emanate from a common stem cell origin. Although widely propounded, little definitive evidence exists to underpin this hypothesis. Moreover, Pearse and coworkers famously proposed that gastrointestinal endocrine cells derive from migrating neuroendocrine stem cells in the neural crest, a concept that still has its adherents. Although studies of quail neural crest cells transplanted into chick embryos, or experiments where the neural crest is eradicated, show gut endocrine cells to be of endodermal origin, Pearse subsequently suggested that the endoderm is colonized by "neuroendocrine-programmed stem cells" from the primitive epiblast, which generate gut endocrine cells. This hypothesis was not ruled out by chick–quail chimera experiments; therefore, other models must be used to ascertain the gut endocrine cell origins, such as the chimeric mouse studies described later.

Several lines of evidence suggest that stem cells reside in the base of the crypts of Lieberkuhn in the small intestine, just superior to the Paneth cells (approximately the fourth or fifth cell position in mice). In the large intestine, they are presumed to be located in the midcrypt of the ascending colon, and in the crypt base of the descending colon. However, within the gastric glands, migration of cells is bidirectional from the neck–isthmus region to form the simple mucous epithelium of the foveolus or pit, and cells migrate downwards to form parietal cells and chief cells. Therefore, the stem cells are believed to be within the neck–isthmus region of the gastric gland. The Unitarian hypothesis is now supported by a considerable body of research.

SINGLE INTESTINAL STEM CELLS REGENERATE WHOLE CRYPTS CONTAINING ALL EPITHELIAL LINEAGES

The ability of intestinal stem cells to regenerate epithelial cell populations of entire intestinal crypts and villi following cytotoxic treatment has been demonstrated using the crypt microcolony assay. Four days after irradiation, sterilized crypts undergo apoptosis and disappear, but they can be identified by remaining radio-resistant Paneth cells at the crypt base. At higher radiation dose levels, only single cells survive in each crypt, since a unit increase in radiation leads to a unit reduction in crypt survival. Survival of one or more clonogenic cells in a crypt after radiation ensures crypt persistence, and there is regeneration of all epithelial cell populations of that crypt and, in the small intestine, of the overlying villi. Therefore, following cytotoxic damage, a single surviving stem cell can produce all cell types of the intestinal epithelium to reproduce a crypt.

MOUSE AGGREGATION CHIMERAS SHOW THAT INTESTINAL CRYPTS ARE CLONAL POPULATIONS

Mouse embryo aggregation chimeras are readily made, wherein the two populations can be easily distinguished. The lectin *Dolichos biflorus* agglutinin (DBA) binds to sites on the B6-derived, but not on the SWR-derived, cells in C57BL/6J Lac (B6)↔SWR mouse embryo aggregation chimeras, and it can be used to distinguish the two parental strains in gut epithelium. The intestinal crypts in each chimera studied were either positive or negative for DBA, and there were no mixed crypts in the tens of thousands studied. Therefore, each crypt forms a clonal population. This is the case for Paneth, mucous, and columnar cells, although it was not possible to detect the markers in endocrine cells, because of their inability to bind the lectin on their surface. In neonatal C57BL/6J Lac (B6)↔SWR chimeras, there were mixed (i.e., polyclonal) crypts for the first two weeks after birth, suggesting that multiple stem cells exist during development. However, all crypts ultimately become derived from a single stem cell between birth and postnatal day 14, so-called monoclonal conversion. This apparent cleansing or "purification" of crypts could be caused by the stochastic loss of one stem cell lineage, or by the segregation of lineages because of an extremely active replication of crypts by fission, which occurs at this developmental period. To exclude the possibility that crypts from distinct strains segregate differentially during organogenesis, Griffiths confirmed the findings in mice bearing an X-linked defective gene for glucose-6-phosphate dehydrogenase (*G6PD*) (Figure 16-1A–C).

In the stomach, the situation is similar, although more complex. Epithelial cell lineages in the antral gastric mucosa of the mouse stomach, including the endocrine cells, derive from a common stem cell. Identification of the Y-chromosome by *in situ* hybridization in XX–XY chimeric mice showed that gastric glands were also clonal populations. These findings were confirmed by Tatematsu in 1994 in CH3↔BALB/c chimeric mice, where each gastric gland was composed of either CH3 or BALB/c cells; there were no mixed glands. Thus, we might advance the general hypothesis that gastric glands in the mouse, in addition to the intestinal crypts, are clonally derived. Additionally, by combining immunohistochemistry for gastrin, an endocrine cell marker, with *in situ* hybridization to detect the Y-chromosome, the male regions of the gastric glands were shown to be almost exclusively Y-chromosome positive with gastrin-positive endocrine cells, whereas the female areas in the chimeric stomach were gastrin positive and Y-chromosome negative. These results finally negate the

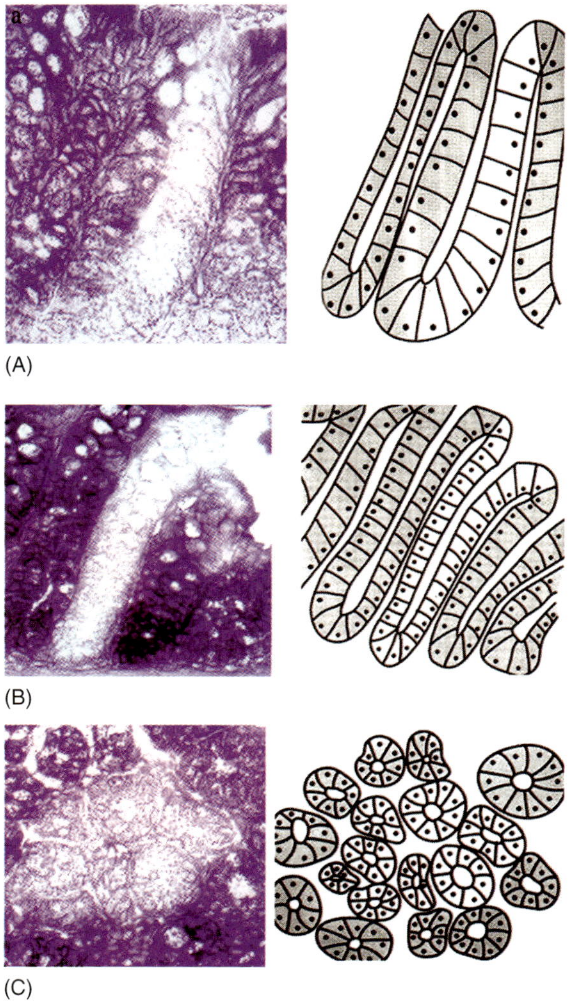

FIGURE 16-1 Gastrointestinal clonality studies. (A-C) G6PD histochemistry in frozen sections of colonic mucosa in an ENU-treated C3H mouse: (A) partially negative crypt, (B) completely negative crypt, and (C) cross section of an eight-crypt patch at 21 weeks. with permission from Park.

Pearse concept that gut endocrine cells originate from a separate stem cell pool.

Nomura used X-inactivation mosaic mice expressing a *lacZ* reporter gene to study clonality of gastric glands in the fundic and pyloric regions of the developing mouse stomach. As in the intestine, most glands are initially polyclonal, with three or four stem cells per gland, but they become monoclonal during the first six weeks of life—again either by purification of the glands, where division of one stem cell eventually overrides all other stem cells, or by gland fission. A population of approximately 5–10% of mixed, polyclonal glands persists into adulthood. The significance of these mixed glands with an increased stem cell number is not known, but it is possible that they do not undergo fission or have reduced fission rates; perhaps they even have an increased stem cell number during development, and therefore maintain a higher number of stem cells after crypt fission.

SOMATIC MUTATIONS IN STEM CELLS REVEAL STEM CELL HIERARCHY AND CLONAL SUCCESSION

Somatic mutations at certain loci allow us to study stem cell hierarchy and clonal succession within the gastrointestinal tract. Mutations in the *DLb-1* on chromosome 11 are one good example of this; C57BL/6J↔SWR F1 chimeric mice show heterozygous expression of a binding site on intestinal epithelial cells for the DBA lectin. This binding site can be abolished when the *Dlb-1* locus becomes mutated, either spontaneously or by the chemical mutagen ethyl nitrosourea (ENU). After ENU treatment, crypts emerge that initially are partially, and then entirely, negative for DBA staining. Perhaps the simplest explanation for this phenomenon is that a mutation occurs at the *Dlb-1* locus in a stem cell within the small intestinal crypt. This mutated cell could expand stochastically to produce a clone of cells that cannot bind DBA, and remain unstained. If this is the case, then a single stem cell can generate all the epithelial lineages within an intestinal crypt of the small intestine.

A "knockin" strategy at the *Dlb-1* locus can also be used to explain the preceding findings. If SWR mice do not express a DBA-binding site on their intestinal epithelial cells, but can be induced to bind DBA by ENU treatment, wholly DBA+ or DBA− intestinal crypts would result. From the use of this model, Bjerknes and Cheng propose that "committed epithelial progenitor" cells exist in mouse intestinal crypts by visualizing the morphology, location, and longevity of mutant clones in crypts and villi of the mouse small intestine. These transitory committed progenitor cells—the columnar cell progenitors (C_o) and the mucus cell progenitors (M_o)—evolve from pluripotential stem cells, and then differentiate further into adult intestinal epithelial cell types.

Not much is known about the mechanisms that regulate the proliferation of these progenitor cells, but administration of proglucagon-derived, glucagon-like peptide 2 (GLP-2) to SWR mice was found to induce intestinal epithelial growth and repair, specifically by stimulating the columnar cell progenitors, resulting in increased crypt and villus size in the normal small intestine. The receptor for GLP-2 (GLP2R) has been shown to be located on enteric neurons, and not on the gut enteroendocrine cells and in the brain as previously thought. GLP-2 activation of enteric neurons produces a rapid induction in *c-fos* expression, which signals growth of columnar epithelial cell progenitors and stem cells that generate adult columnar cell types. There is no stimulatory effect on the mucus cell lineage; instead, it is stimulated by KGF. Thus, the committed progenitor cells are involved in regeneration of damaged epithelia, possibly through a neural regulatory pathway (Figure 16-2B).

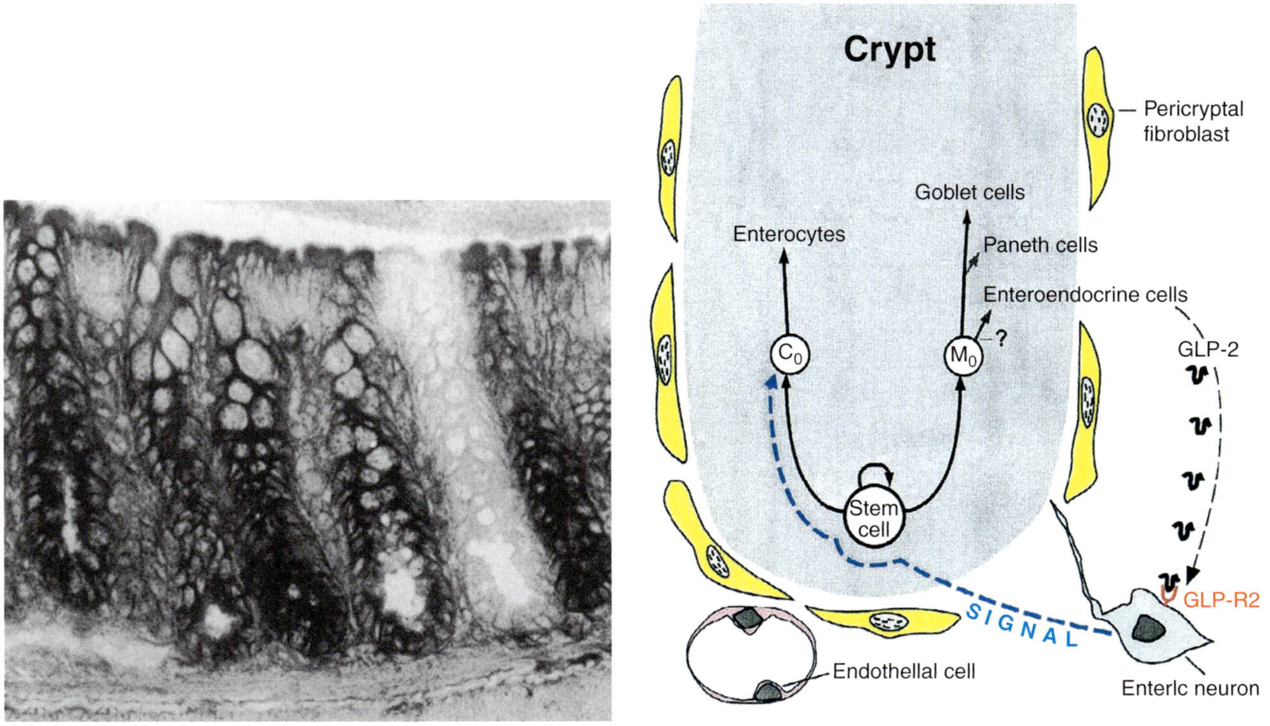

(A) (B)

FIGURE 16-2 Stem cell niche hypothesis. (A) G6PD-stained frozen section of colonic mucosa from a CH3 mouse 14 days after a single dose of DMH, showing a single, wholly mutated crypt in which all cells have been replaced by a mutant phenotype. (B) Diagrammatic representation of the stem cell niche. An active, multipotent stem cell produces a daughter cell (C_0) that generates enterocytic lineages and another that generates goblet, Paneth, and enteroendocrine cells (although it is not known whether all derive from M_0). GLP-2, produced by a subset of enteroendocrine cells, stimulates proliferation of the C_0 daughter through an interaction with enteric nervous system neurons that express the GLP-2 receptor (GLP-2R). Panel A reproduced with permission from Williams and panel B reproduced with permission from Mills.

The possibility remains that cells from different parental strains of chimeric animals segregate independently during development to produce "monophenotypic" crypts and not "monoclonal" crypts, as we assumed earlier in this chapter. Mosaic expression of the electrophoretic isoenzymes PGK-1A and PGK-1B were analyzed in colonic crypts of mice heterozygous for the X-linked alleles *Pgk-1*^{a} and *Pgk-1*^{b}. In agreement with the results from the chimeric studies, no mixed crypts were seen, thus eliminating the possibility that these crypts are monophenotypic, and verifying that they are derived from a single progenitor cell. In a further model, mice that heterozygously express a defective *G6PD* gene have a crypt-restricted pattern of G6PD expression, thus confirming that the intestinal crypts are derived from a single stem cell. Moreover, mice treated with the colon carcinogen dimethylhydrazine (DMH) or ENU also develop crypts that initially are partially, and later are wholly, negative for G6PD. The partially-negative crypts could conceivably result from the mutation of a cell in the dividing transit population of the crypt that lacks stem cell properties. This is supported by the observation that these partially-negative crypts are transient and decrease in frequency, parallel to an increase in wholly-negative crypts. Conversely, such partially-negative crypts could become wholly-negative by stochastic expansion of a mutant stem cell. Wholly-negative crypts would then be a clonal population derived from this mutant stem cell.

After administration of a mutagen, in both the *Dlb-1* and the G6PD models, the time taken for the decrease in partially mutated crypts and the emergence of entirely negative crypts to reach a plateau is approximately four weeks in the small intestine, and up to 12 weeks in the large intestine (Figure 16-2A). This difference is intriguing, and was initially thought to be because of cell cycle time differences between the colon and the small intestine. However, a favored explanation can be found in the "stem cell niche hypothesis." This hypothesis suggests that multiple stem cells occupy a crypt with random cell loss after stem cell division, and was originally formulated as the stem cell zone hypothesis by Bjerknes and Cheng. The numbers of stem cells may be larger in the small intestine than in the large intestine, causing the difference in time taken for phenotypic changes following mutagen treatment as the mutant stem cell expands stochastically (Figure 16-2A).

An alternative hypothesis might lie in crypt fission: the rate of fission at the time of mutagen administration was higher in the colon than in the small intestine. During crypt fission, when crypts divide longitudinally, selective segregation of the two cell populations could occur, "cleansing" the partially mutated crypts by segregating the mutated and nonmutated cells, and by duplicating the wholly-negative crypts to create monoclonal crypts (Figure 16-3). First proposed as a concept by Park, this phenomenon was illustrated by Bjerknes and Cheng.

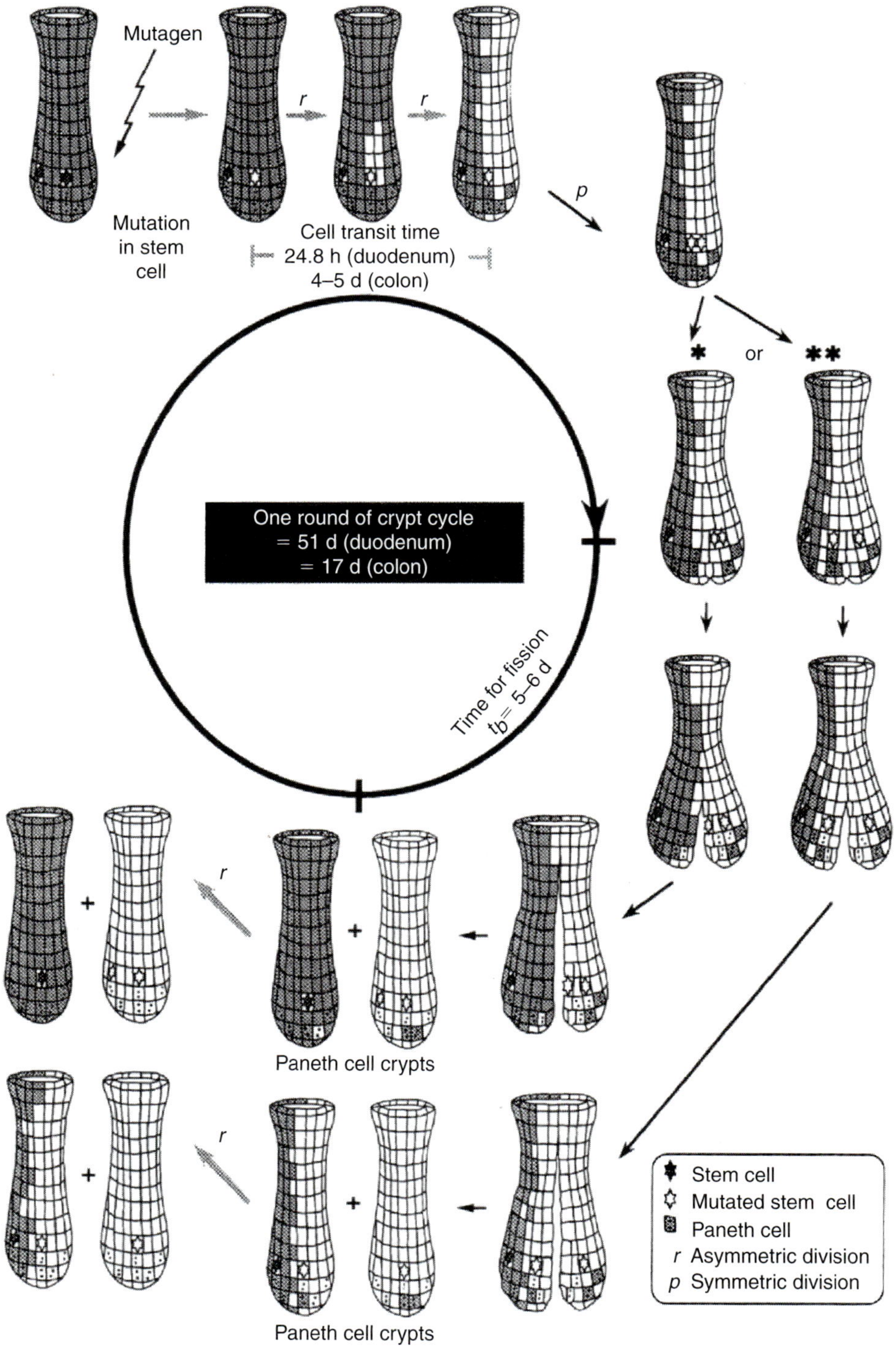

FIGURE 16-3 Crypt cycle. Emergence of transformed crypts after a single stem cell mutation. Crypt division is triggered by symmetric stem cell divisions (p), indicated by light-colored arrows, whereas dark arrows show asymmetric stem cell divisions (r). This is independent of the crypt cycle and enables individual crypts to maintain a constant number of stem cells and to continuously replace differentiated progeny over a crypt cell turnover time, on average. Reproduced with permission from Park.

HUMAN INTESTINAL CRYPTS CONTAIN MULTIPLE EPITHELIAL CELL LINEAGES DERIVED FROM A SINGLE STEM CELL

The investigation of gastrointestinal stem cells in humans has proved difficult. The human colorectal carcinoma cell line HRA19 was derived from a primary adenocarcinoma of the rectum. A colony of single morphology was cloned and grown as a monolayer *in vitro*. When engrafted subcutaneously into *nude* mice, these clones produce tumors histologically-identical to the original tumor and contain columnar, goblet, and neuroendocrine cells (Figure 16-4A–D). These malignant epithelial cells are therefore multipotential, and can produce all differentiated cell types in the human colorectal epithelium. However, these results cannot definitively be applied to normal human gastrointestinal epithelia.

In most of the human population, the colonic goblet cells secrete O-acetylated mucin. However, approximately 9% of the human Caucasian population have a homozygous genetic mutation in the enzyme O-acetyl transferase (*OAT−/−*), and in goblet cells secreting this non-O-acetylated sialic acid; these are then positive when stained with mild periodic acid Schiff (mPAS). In heterozygotes, which comprise approximately 42% of the population (*OAT−/OAT+*), O-acetylation proceeds, and crypts thereby stain negative for mPAS. Loss of the remaining active *OAT* gene converts the genotype to *OAT−/OAT−*, resulting in occasional, apparently randomly located, positive mPAS-stained crypts with uniform staining of goblet cells from base to luminal surface, an effect that increases with age. This could be because of a somatic mutation or nondisjunction in a single crypt stem cell and subsequent colonization of the crypt by the mutated stem cell. The frequency of these events is racially determined and increases after irradiation. This has been interpreted as indicating an increased rate of stem cell mutation. Interestingly, there is no increase in the rate of apparent stem cell mutation in hereditary nonpolyposis carcinoma patients, nor in the background mucosa of left- and right-sided carcinomas. However, just as in the mouse stem cell mutation models, when patients are followed by the mPAS method with time after irradiation, initially there is partial crypt staining, and then whole crypts appear where the goblet cells are mPAS+. A "clonal stabilization time" (defined as the period required for the emergence of most of such wholly-stained crypts) in humans is approximately one year, a process we referred to previously as monoclonal crypt conversion. These results have implications for the origins of goblet cell lineages in the gut: they indicate that they arise from crypt stem cells; however, they say little about the other cell lineages.

Perhaps the best evidence for the clonality of human intestinal crypts, and the stem cell derivation of all contained epithelial cell lineages, comes from studies of the colon of a rare XO–XY patient who received a prophylactic colectomy for familial adenomatous polyposis (FAP). Nonisotopic *in situ* hybridization (NISH) using Y-chromosome-specific probes showed the patient's normal intestinal crypts to be composed almost entirely of either Y-chromosome positive or Y-chromosome negative cells (Figure 16-4E), with about 20% of crypts being XO. Immunostaining for neuroendocrine specific markers and Y-chromosome NISH, used in combination, showed that crypt neuroendocrine cells shared the genotype of other crypt cells (Figure 16-4F). In the small intestine, the villus epithelium was a mixture of XO and XY cells, in keeping with the belief that the villi derive from stem cells of more than one crypt. Of the 12614 crypts examined, only four crypts were composed of XO and XY cells, which could be explained by nondisjunction with a loss of the Y-chromosome in a crypt stem cell. Importantly, there were no mixed crypts at patch boundaries. These observations agree with previous findings using chimeric mice that intestinal crypt epithelial cells, including neuroendocrine cells, are monoclonal and derive from a single multipotential stem cell. Consequently, the hypothesis that enteroendocrine cells and other differentiated cell types within the colorectal epithelium share a common cell of origin (the Unitarian

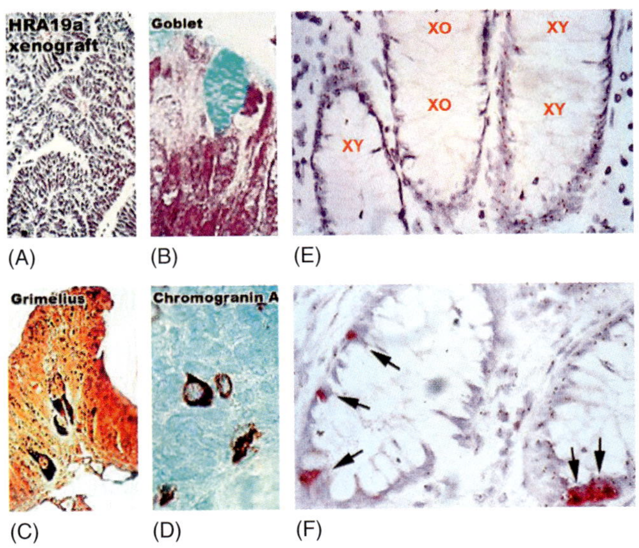

FIGURE 16-4 Monoclonal origin of human clonic crypts. (A) A tumor growing in a nude mouse derived from a single-cell cloned human colorectal carcinoma cell line, stained (B) with Alcian blue to show goblet cells, and with (C) Grimelius and (D) an antibody against chromogranin A to demonstrate endocrine cells. (B) Monoclonal origin of human colonic crypts: normal colonic mucosa in an XO–XY mosaic individual stained by *in situ* hybridization for a Y-chromosome-specific probe showing an XO crypt (central) surrounded by two XY crypts. (C) Endocrine cells, highlighted by their chromogranin A content, also show clonality with the other crypt cell lineages in the same patient. (D) Villi, receiving cells from more than one crypt of different clonal derivation, show a polyclonal pattern in this X0–XY patient. Apart from the occasional Y-chromosome positive inflammatory cell (white arrow), most cells on the right of this villus are XO (black arrows); on the left side, the cells are XY. Panel A–D courtesy of S. Kirkland, panels E and F courtesy of M. Novelli.

hypothesis) appears to apply to both mice and humans. These observations have been confirmed in Sardinian women heterozygous for a defective *G6PD* gene.

Recently it has been shown that colonic stem cells and their progeny contain non-pathogenic mutations in their mitochondrial DNA (mtDNA), specifically mutations affecting the activity of cytochrome *c* oxidase (COX)—a component of complex IV of the respiratory chain, encoded by mtDNA, that are relatively common. The mitochondrial genome is small (16.6 kb) and prone to mutation, due to a lack of protective histones and poor DNA repair mechanisms, in addition to residing in an environment with high oxidation stress. Genetic drift and expansion of a single mutated mtDNA allows mutations to expand stochastically within a cell and over time cells will become either "homoplasmic"—all the mitochondria in the cell are mutated, or "heteroplasmic"—the cell contains a mixture of mutated and wild-type mitochondria. This stochastic expansion is a lengthy process, and COX-deficient cells are rare in young people. For a mutated cellular phenotype to be observed, homoplasmy or a high degree of heteroplasmy must be present, thus stem cells are the only cells that have a sufficient life-span to accumulate these mitochondrial mutations to a level that results in a biochemical deficiency. Two-color enzyme histochemistry can be used to simultaneously detect the mtDNA-encoded cytochrome *c* oxidase and nuclear DNA-encoded succinate dehydrogenase (SDH), a component of complex II of the respiratory chain. Three different types of crypts are observed: wild-type crypts; wholly-mutated crypts; and partially-mutated crypts containing both mutant and wild-type cells. Laser micro-dissection on wholly-mutated crypts, followed by PCR coupled with mtDNA sequencing, confirms the presence of conserved clonal mitochondrial mutations within crypts. This confirms that crypts are clonal populations, as here a crypt has been repopulated from a single mutated stem cell via "niche succession" (the process by which one stem cell takes over the entire stem cell population of a crypt) and "monoclonal conversion" (the process by which the progeny of a single stem cell clone make up the entire crypt). The existence of crypts containing two cell populations, one wild-type and one mutated, demonstrates that human intestinal crypts contain at least two stem cells, and has enabled the dynamic analysis of clonal expansion and identification of the stem cell niche in human colonic tissue. Thus, we now have evidence in human tissue of niche succession and monoclonal conversion "in action," and evidence that multiple stem cells maintain a crypt.

This technology is not only applicable to studying the expansion of clones in the colon, but has also been used as a tool to study the mechanism of field cancerization in the human stomach and duodenum. In particular, the complex structure of the human gastric unit (comprised of multiple glands and a single pit) meant that it has been very difficult to establish the location of the stem cells and the clonal architecture of the human gastric unit. Nomura used X-chromosome-linked inactivation to study fundic and pyloric glands in human female stomachs. Studies using polymorphisms on X-linked genes, such as the androgen receptor (HUMARA) to distinguish between the two X-chromosomes revealed that, although pyloric glands appear homotypic and thus monoclonal, about half of the fundic glands studied were heterotypic for at the HUMARA locus, and were consequently polyclonal. This finding suggests that a more complex situation occurs in humans than the studies of gastric gland clonality in chimeric mice indicate. However, we have seen that some glands in the mouse remain polyclonal throughout life. Using the COX model, the authors were able to demonstrate that human gastric units are clonal and that they are able to divide by gland fission, forming entirely clonal patches of mutated gastric units: this strong evidence to support fission as a mechanism of field cancerization.

It has been proposed that insight into stem cell organization can be gained from the study of the methylation pattern of nonexpressed genes in the colon. In the normal human colon, methylation patterns are somatically-inherited endogenous sequences that randomly change and increase in occurrence with aging. Investigation of methylation patterns is a possible alternative to histological markers to investigate crypt histories and allow fate mapping. Examination of methylation tags of three neutral loci in cells from normal human colon showed variation in sequences between crypts and mosaic methylation patterns within single crypts. Multiple unique sites were present in morphologically identical crypts; for example, one patient had no identical methylation sequences of one gene within any of the crypts studied, even though all sequences were related. This indicates that some normal human colonic crypts are quasi-clonal, with multiple stem cells per crypt. Differences in methylation tags can highlight relationships among cells in a crypt where less closely-related cells show greater sequence alterations, and where closely-related cells have similar methylation patterns. Sequence differences suggest that crypts are maintained by stem cells, which are randomly lost and replaced in a stochastic manner, eventually leading to a "bottleneck" effect, in which all cells within a crypt are closely-related to a single stem cell descendant. This reduction to the most recent common crypt progenitor is predicted to occur several times during life, superficially resembling the clonal succession of tumor progression.

BONE MARROW STEM CELLS CONTRIBUTE TO GUT REPOPULATION AFTER DAMAGE

The haematopoietic bone marrow stem cell is of mesodermal origin, and its functionality and cell surface markers have been well-characterized. When transplanted into lethally irradiated animals and humans, as in clinical bone marrow transplantation, it has long been considered that the haematopoietic stem cell colonizes host tissues to form only new erythroid, granulocyte–macrophage, megakaryocyte, and lymphoid lineages. Although earlier studies suggest that vascular endothelium could derive from transplanted donor marrow, other studies have not only confirmed these earlier proposals concerning endothelial cells, but have also indicated that adult bone marrow stem cells possess a considerable degree of plasticity and can differentiate into different cell types, including hepatocytes, biliary epithelial cells, skeletal muscle fibres, cardiomyocytes, central nervous system cells, and renal tubular epithelial cells. These pathways can be bidirectional, as muscle and neuronal stem cells can also apparently form bone marrow. Furthermore, it appears that selection pressure induced by target organ damage can intensify the efficacy of this process as bone marrow stem cells differentiate into cardiomyocytes, endothelial cells, and smooth muscle cells in mice, with ischemic cell death following myocardial infarction and coronary artery occlusion. Bone marrow stem cells have also been shown to differentiate into pancreatic β-cells, and possibly more persuasively, fully-differentiated cells can transdifferentiate into other adult cell types without undergoing cell division; for example, exocrine pancreatic cells can differentiate into hepatocytes *in vitro*. Furthermore, isolated potential hepatic stem cells from fetal mouse livers, which differentiate into hepatocytes and cholangiocytes when transplanted into recipient animals, can form pancreatic ductal acinar cells and intestinal epithelial cells when transplanted directly into the pancreas or duodenal wall. Thus, the conventional view that bone marrow stem cells generate cell types of a single lineage (i.e., all formed elements in the peripheral blood) has been rectified in favor of findings that adult bone marrow stem cells are highly plastic, and can differentiate into many cell types within various organs.

These observations have raised the possibility of regeneration of a failing organ by transplanting an individual's own bone marrow stem cells to colonize and repopulate the diseased tissue, thus avoiding the allograft reaction. In apparent proof-of-principle, fumarylacetoacetate hydrolase (FAH)-deficient mice, which resemble Type 1 tyrosinemia in humans, are rescued from liver failure by transplantation of purified hematopoietic stem cells that become morphologically normal hepatocytes, express the FAH enzyme, and therefore are functionally normal.

There are now several reports that bone marrow cells can repopulate both epithelial and mesenchymal lineages in the gut. Brittan analyzed colons and small intestines of female mice that received a bone marrow transplant from male mice donors, and gastrointestinal biopsies from female patients with graft-versus-host disease following bone marrow transplant from male donors. Bone marrow cells frequently engraft into the mouse small intestine and colon, and differentiate to form ISEMFs within the lamina propria. *In situ* hybridization confirmed the presence of Y-chromosomes in these cells; their positive immunostaining for αSMA and negativity for desmin, the mouse macrophage marker F4/80, and the hematopoietic precursor marker CD34 determined their phenotype as pericryptal myofibroblasts in the lamina propria derived from transplanted bone marrow. This engraftment and transdifferentiation occurred as early as one week after bone marrow transplantation; almost 60% of ISEMFs were bone marrow-derived six weeks after the transplantation, indicating that transplanted bone marrow cells are capable of a sustained turnover of the ISEMF cells in the lamina propria (Figure 16-5). Y-chromosome-positive ISEMFs were also seen in the human intestinal biopsy material. Lethally-irradiated female mice given a male bone marrow transplant and a subsequent foreign-body peritoneal implant formed granulation tissue capsules containing myofibroblast cells derived from the hematopoietic stem cells of the transplanted bone marrow. This suggests that myofibroblasts may generally derive from bone marrow cells.

There are a growing number of reports that bone marrow cells can repopulate gastrointestinal epithelial cells in animals and man. Krause found bone marrow-derived epithelial cells in the lung, gastrointestinal tract, and skin 11 months after transplantation of a single hematopoietic bone marrow stem cell in the mouse. In the gastrointestinal

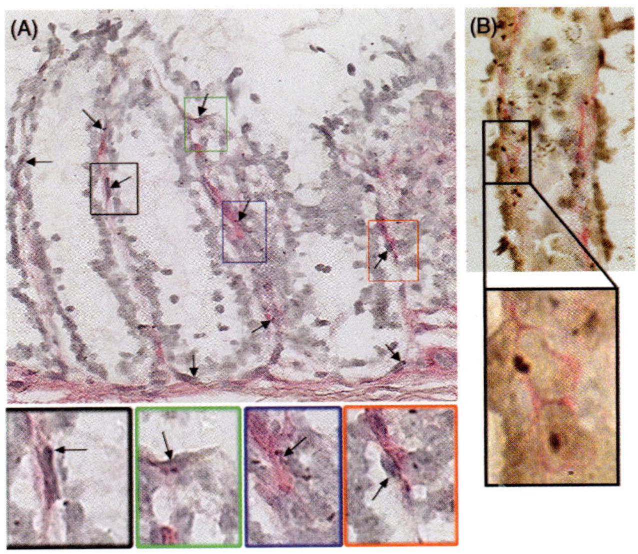

FIGURE 16-5 Bone marrow-derived ISEMFs in the lamina propria of the mouse colon. Female mouse colon following a male whole bone marrow transplant. Bone marrow-derived ISEMFs are present as Y-chromosome-positive cells, immunoreactive for α-SMA (A) two weeks after transplant (arrows and high-power inserts) and as columns reaching from crypt base to tip (B) six weeks after transplant (arrows and high-power inserts). The Y-chromosome is seen as a brown–black punctate density, and red cytoplasmic staining indicates immunoreactivity for α-SMA. Courtesy of M. Brittan.

tract, engrafted cells were present as columnar epithelial cells in the esophageal lining, a small intestinal villus, colonic crypts, and gastric foveola. No apparent engraftment into the pericryptal myofibroblast sheath was reported, and as only a single hematopoietic stem cell was transplanted, it is possible that the ISEMFs derive from mesenchymal stem cells within transplanted whole bone marrow. It is, however, generally believed that stromal cell populations do not survive following bone marrow transplantation—although if an empty niche exists, as after irradiation in the gut, engraftment might occur.

In biopsies from female patients who had undergone sex-mismatched hematopoietic bone marrow transplantation, *in situ* hybridization for a Y-chromosome-specific probe with immunohistochemical staining for cytokeratins demonstrated mucosal cells of donor origin in the gastric cardia. Moreover, Okamoto reported four long-term bone marrow transplant survivors with multiple engraftment of esophageal, gastric, small intestinal, and colonic epithelial cells by donor bone marrow cells up to eight years after transplantation, emphasizing the long-term nature of this transdifferentiation.

It would be impossible to finish even such a short section on this topic without mentioning the mechanisms and significance of such phenomena. At first considered to be caused by transdifferentiation, or lack of lineage fidelity—commonly seen in invertebrates during gastrulation, or during organogenesis—it is becoming clear that such changes are neither simple nor readily reproducible. Some laboratories have been unable to reproduce earlier findings, and there are claims that adult tissues are contaminated with bone marrow precursors. Finally, the fusion of a transplanted bone marrow cell with an indigenous adult cell has been proposed as the mechanism by which bone marrow stem cells acquire the phenotype of target cell lineages. Initially, cell fusion was seen as a rare event, occurring only *in vitro*, and under circumstances of extreme selection. In the FAH model described previously, cell fusion has been shown to be common, and it cannot be ruled out as the main mechanism by which bone marrow stem cells transform to functional hepatocytes. However, the genetic mechanisms whereby gene expression is switched off in the recipient cells—followed by clonal expansion to repopulate large areas of the liver, for example—are as yet unclear. We should also recall that several tissues in the mouse are polyploid, such as the liver and the acinar cells of the exocrine pancreas. Other studies in which mixed-sex bone marrow transplants have been used to show plasticity have not reported evidence of cell fusion in animals or man: for example, bone marrow-engrafted cells in the human stomach, intestine, buccal mucosa, and pancreatic islet cells showed a normal complement of X- and Y-chromosomes. Whatever the mechanism, it is clear that the most important criterion for altered lineage commitment, that of function, has only been fulfilled in a few models, such as the FAH model and possibly postinfarction cardiomyocyte engraftment. The future in this field will prove interesting.

In inflammatory bowel diseases such as Crohn's disease, intestinal myofibroblasts are activated to proliferate and synthesize an extracellular matrix, and excessive collagen deposition causes fibrosis and postinflammatory scarring in the lamina propria and muscularis layers of the gut wall. Intestinal inflammation is believed to be mediated by luminal bacteria and bacterial wall polymers, and tumor necrosis factor (TNF) plays a key role in the pathogenesis of intestinal inflammatory disease, since mice with targeted deletion of the AU-rich elements in TNF develop chronic ileitis resembling a Crohn's disease-like phenotype (Figure 16-6A–D). Moreover, a single dose of anti-TNF antibody to Crohn's disease patients can dramatically alleviate inflammation. Other cytokines, including interleukin-10 and TGFβ3 are implicated in the development of fibrosis in inflammatory bowel disease.

It has previously been concluded that fibrotic reactions, which occur in inflammatory bowel disease and other diseases, are caused by the local proliferation of myofibroblasts and fibroblasts. Our data showing that transplanted bone marrow contributes to the intestinal subepithelial myofibroblast population implicates that extraintestinal cells may have a role in fibrosis. In this respect, we have shown that cells with a fibroblast–fibrocyte phenotype can derive from transplanted bone marrow, and contribute to fibrotic reactions in and around the intestinal wall (Figure 16-6E). The concept that cells of the lamina propria, other than the lymphoid and myeloid lineages, exist in equilibrium with bone marrow precursors is an interesting one and could provide an opportunity for therapeutic delivery of cytokines to the intestine to prevent the development of fibrosis, or even to treat it. Further data show that many of the myofibroblasts and fibroblasts recruited to, or expanding in, the lamina propria and submucosa of mice with induced TNBS colitis are bone marrow-derived (Figure 16-6F).

The IL-10 knockout mouse is a well-established model of colitis. Bamba transplanted female IL-10−/− mice with either male IL-10 bone marrow or wild-type bone marrow. In the former group, a florid colitis developed with up to 45% of myofibroblasts being bone marrow-derived. However, IL-10−/− mice displayed an ameliorated course of disease after wild-type bone marrow transplant with up to 30% of myofibroblasts being of donor origin. Subsequently, work in patients with Crohn's disease has shown an improvement in disease course with CD34 enriched and hematopoietic stem cells, as well as mobilized peripheral blood stem cells. In these cases the therapeutic effect is postulated to be due to the immunosuppressive effects of the conditioning regimen required for stem cell transplant. The autologous stem cell transplant serves to shorten the neutropenic interval, rather than reconstitute the myofibroblast component of the colonic niche—a mechanism that has not as yet been proven.

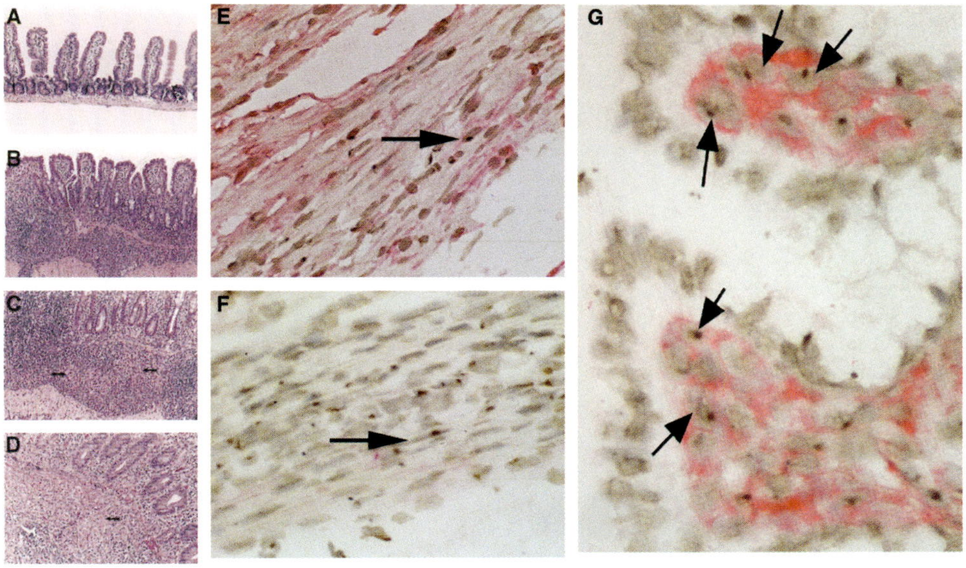

FIGURE 16-6 Murine models of fibrosis. Inflammatory bowel disease in $TNF^{\Delta ARE}$ and SAMP/Yit mouse ilea resembling human Crohn's disease. (A) Normal ileal morphology in an 8-week control mouse. (B) $TNF^{\Delta ARE}$ homozygous knockout at 7 weeks with blunt, distorted villi and increased inflammatory infiltrate. (C) $TNF^{\Delta ARE}$ heterozygote at 16 weeks with villus blunting and chronic inflammation. An ill-defined noncaseating granuloma is in the submucosa (arrows). (D) Ileocecal region from a human patient with Crohn's disease demonstrating similar location and composition of granuloma compared with mouse model (arrow). (E) A fibrotic reaction in the serosal tissues of a paracetmol-treated male–female radiation chimeric mouse, showing numerous spindle-shaped vimentin-positive cells with Y-chromosomes revealing their bone marrow origin, which are also (F) smooth muscle actin negative, confirming their fibroblast lineage. (G) Massive expansion of myofibroblasts of bone marrow origin in a male–female chimeric mouse with colitis induced by TNBS. Panel D reproduced with permission from Kontoyiannis, panels E–F courtesy of N. Direkze, and panel G courtesy of M. Brittan.

GASTROINTESTINAL STEM CELLS OCCUPY A NICHE MAINTAINED BY INTESTINAL SUBEPITHELIAL MYOFIBROBLASTS IN THE LAMINA PROPRIA

Stem cells within many tissues are thought to reside within a niche formed by a group of surrounding cells and their extracellular matrices, which provide an optimal microenvironment for the stem cells to function. The identification of a niche within any tissue involves knowledge of the location of the stem cells; as we have seen, this has proved problematical in the gastrointestinal tract. According to Spradling, to prove that a niche is present, the stem cells must be removed and subsequently replaced while the niche persists, providing support to the remaining exogenous cells. Although this has been accomplished in *Drosophila*, such manipulations have not yet been possible in mammals. In this context, the survival of a single epithelial cell following cytotoxic damage to intestinal crypts in the microcolony assay is interesting, as many of the intestinal subepithelial myofibroblasts are also lost after irradiation, although sufficient numbers may remain or be replaced by local proliferation or migration from the bone marrow to provide a supportive niche for the surviving stem cell or cells. The ISEMFs surround the base of the crypt and the neck–isthmus of the gastric gland, a commonly proposed location for the intestinal and gastric stem cell niches, respectively. It is proposed that ISEMFs influence epithelial cell proliferation and regeneration through epithelial–mesenchymal cross talk, and that they ultimately determine epithelial cell fate.

There has been a long quest for markers of stem cells in the intestine. The neural RNA-binding protein marker Musashi-1 (Msi-1) is a mammalian homolog of a *Drosophila* protein evidently required for asymmetrical division of sensory neural precursors. In the mouse, Msi-1 is expressed in neural stem cells and has been proposed as the first intestinal stem cell marker, because of its expression in developing intestinal crypts and specifically within the stem cell region of adult small intestinal crypts. This is further substantiated by its expanded expression throughout the entire clonogenic region in the small intestine after irradiation.

The regulatory mechanisms of stem cell division within the niche to produce, on average, one stem cell and one cell committed to differentiation is as yet unknown, although there is no shortage of potential models. In the stem cell zone hypothesis, the bottom few cell positions of the small intestinal crypt are occupied by a mixture of cell types: Paneth, goblet, and endocrine. The migration vector is towards the bottom of the crypt. Above cell position 5, cells migrate upwards, although only the cells that divide in the stem cell zone beneath are stem cells. Other models envisage the stem cells occupying a ring immediately above the Paneth cells, although there is little experimental basis for such an assertion since "undifferentiated" cells of similar appearance are seen among the Paneth cells in thin sections. Moreover, there is no difference in the expression of Msi-1 and Hes1—a transcriptional factor regulated by the Notch signaling pathway also required for neural stem cell renewal and neuronal lineage commitment—in undifferentiated cells located in either the stem cell zone or immediately above the Paneth cells. This suggests that both populations may have the same potential as putative stem cells.

The number of stem cells in a crypt or a gland is presently unknown. Initially, all proliferating cells were believed to be stem cells. Although clonal regeneration experiments using the microcolony assay indicated that intestinal crypts contained a multiplicity of stem cells, it was clear that this was less than the proliferative cellularity. Proposed stem cell numbers have varied from a single stem cell to 16 or more. Others have proposed that the number of stem cells per crypt varies throughout the crypt cycle, with the attainment of a "threshold number" of stem cells per crypt being the signal for fission to occur. There is little experimental evidence to support these proposals. Although all cells in a crypt are initially derived from a single cell, as shown by the X-inactivation experiments discussed previously, mutagenesis studies such as those shown in Figure 16-1A argue strongly for more than one stem cell per crypt, with stochastic clonal expansion of a mutant clone. A three stem cell colonic crypt has been suggested by Williams on this basis.

In organisms such as *Drosophila* and *Caenorhabditis elegans*, stem cell divisions are known to be asymmetric. We have no such firm concept in the mammalian gut, although there is some evidence to support the proposal. By labeling DNA template strands in intestinal stem cells with tritiated thymidine during development or tissue regeneration, and by labeling newly-synthesized daughter strands with bromodeoxyuridine, segregation of the two markers can be studied. The template DNA strand labeled with tritiated thymidine is retained, but the newly-synthesized strands labeled with bromodeoxyuridine become lost after the second division of the stem cell. This indicates not only that asymmetric stem cell division occurs, but also that by discarding the newly-synthesized DNA which is prone to mutation into the daughter cell destined to differentiate, a mechanism of stem cell genome protection is afforded.

When a stem cell divides, the possible outcomes are that two stem cells (P) are produced, that two daughter cells destined to differentiate cells (Q) are produced, or that there could be an asymmetric division resulting in one P and one Q cell. These are sometimes called p, q, and r divisions, or p and q divisions (Figure 16-7). If $p = 1$ and $q = 0$, then regardless of the number of stem cells per crypt, the cells are immortal and there will be no drift in the niche with time. Such a situation is called "deterministic." However, if $p < 1$ and $Q > 0$ (i.e., a stochastic model), there will be eventual extinction of some stem cell lines, and a drift towards a common stem cell from which all other cells derive. We previously described the variation in methylation patterns or "tags" that occur in human colonic crypts, and explained that crypts apparently show several unique tags. The variance in these unique tags was compared with those expected using a variety of models, including no drift with aging (the deterministic model), drift with immortal stem cells with divergence (the numbers of unique tags are proportional to the stem cell number), drift with one stem cell per crypt, and a stem cell niche with more divergence (with loss of stem cells occurring proportional to the time since divergence). Multiple unique tags were found in some crypts, and the number of unique tags increased with the number of markers counted,

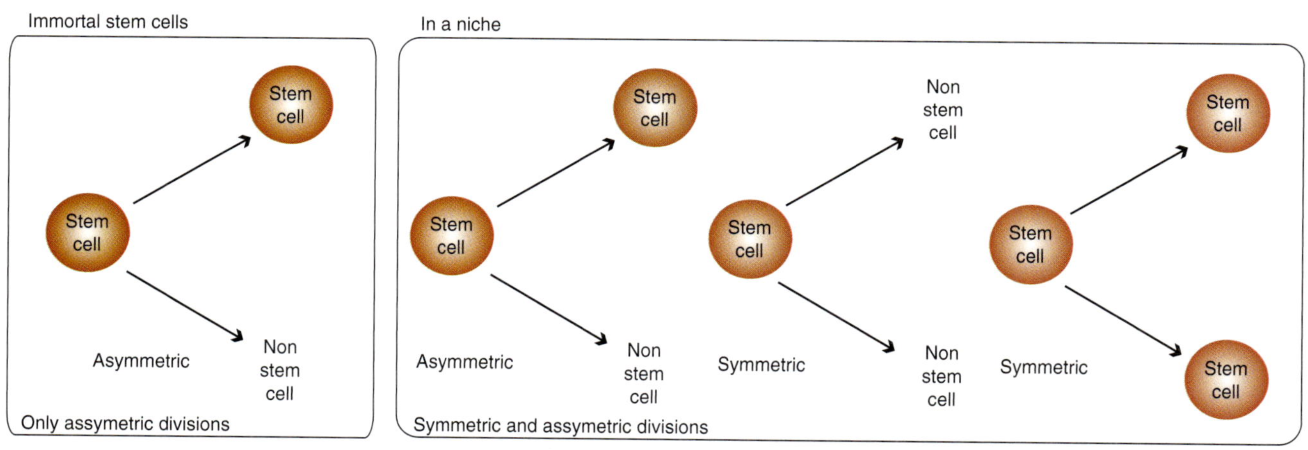

FIGURE 16-7 How stem cells divide in a niche. Classically, stem cells always replace themselves by asymmetrical division; therefore, their lineages never become extinct. However, although the stem cell number remains constant in a niche, both asymmetrical and symmetrical divisions occur. Symmetrical division can lead to one stem cell lineage becoming extinct as both daughters leave the niche, but to maintain a constant stem cell number in the niche, this loss is balanced by stem cell expansion, where both daughters remain in the niche as stem cells. Such stem cell loss with ensuing replacement can lead to the extinction of all lineages bar one, or *clonal succession.* Reproduced with permission from Kim.

which favors random tag drift and multiple stem cells per crypt. The variances were consistent with drift in immortal stem cells, where N (the number of stem cells) $= 2$, but favored a model where $0.75 < P < 0.95$, and $N < 512$. Thus, the data supported a stochastic model with multiple stem cells per crypt. However, as in many such attempts, there are several major assumptions necessary, such as a constant stem cell number. It is clear that variation in both P and N occur in this model. However, this analysis is consistent with the results of Campbell, who deduced that the time taken for monoclonal conversion, or the "clonal stabilization time," of OAT+/− individuals to convert to OAT−/− cells following irradiation was found to be about one year. Assuming 64 stem cells per niche, and $P = 0.95$, the mean time for conversion should be some 220 days. The same assumptions suggest a bottleneck, where all stem cells are related to the most recent common ancestral cell, occurring every 8.2 years.

MANY SIGNALING PATHWAYS REGULATE GASTROINTESTINAL DEVELOPMENT, PROLIFERATION, AND DIFFERENTIATION

Although the molecular mechanisms by which pluripotent stem cells of the gastrointestinal tract produce differentiated cell types are not entirely understood, considerable progress has been made in the last few years. An increasing number of genes and growth factors have been identified that regulate development, proliferation, and differentiation, as well as the development of tumors. The Wnt/β-catenin pathway appears to be a crucial mechanism in regulating proliferation, and constitutionally active Wnt signaling is sufficient to initiate tumor growth. Other pathways, including the Notch pathway and bone morphogenic protein (BMP) pathways and Hedgehog signaling, have also been shown to influence intestinal crypt cell proliferation and differentiation. The intestinal mesenchymal and epithelial cells themselves express these molecules. Details of these pathways are described in the following sections.

WNT/β-CATENIN SIGNALING PATHWAY MAINTAINS THE PROLIFERATIVE COMPARTMENT OF INTESTINAL CRYPTS

The Wnt family of signaling proteins is critical during embryonic development and organogenesis in many species. There are 19 known murine *Wnt* genes, which bind to receptors of the frizzled (*Fz*) family, 10 of which have been identified in mice, and to the low-density lipoprotein receptor-related proteins (*LRP*). Canonical Wnt signaling is mediated by the multifunctional protein β-catenin. β-catenin normally interacts with the so-called "destruction complex," a complex of glycogen synthase kinase 3-β (GSK3-β), axin, and the adenomatous polyposis coli (APC) tumor-suppressor protein. Subsequent serine phosphorylation of cytosolic β-catenin by GSK3-β leads to its ubiquitination, and to its proteasomal degradation, thereby maintaining low levels of cytosolic and nuclear β-catenin. Wnt ligand binding to its cell surface receptors activates the cytoplasmic phosphoprotein "disheveled," which in turn initiates a signaling cascade resulting in the inhibition of the destruction complex and increased cytosolic levels of β-catenin. The free β-catenin then translocates to the cell nucleus, where it forms a transcriptional activator by combining with members of the T-cell factor/lymphocyte enhancer factor (Tcf/LEF) DNA-binding protein family. This activates transcription of specific genes, including *c-myc*, *cyclin D1*, *CD44*, *c-Jun*, *Fra-1*, and the putative stem cell marker *Lgr5*, and results in cell proliferation (see Figure 16-8).

β-catenin/Tcf signaling is required for maintenance of the proliferative compartment of the intestinal crypt. Mice with targeted disruption of *Tcf-4* have no proliferating cells within their small intestinal crypts, and lack a functional stem cell compartment. Similarly, conditional ectopic expression of Dickkopf-1 in the intestine of adult mice, a potent Wnt inhibitor, has similar effects. In colorectal cancer cells, disruption of β-catenin/TCF signaling causes rapid cell cycle arrest. Furthermore, induced deletion of the Wnt target gene *c-myc* in adult mice causes rapid loss of intestinal crypts. Gregorieff, using *in situ* hybridization, showed the distribution of wnts, and their receptors and antagonists, along the major axis of murine intestinal crypts. A number of Wnts, and their receptors and antagonists, show differential expression along the crypt axis: in the small intestine for example, Wnt-3, Wnt-6, and Wnt-9b are expressed predominantly at the crypt base. As well as maintaining a proliferative compartment, Wnt signals also appear to be involved in inducing cell differentiation: conditional deletion of the *Frizzled-5* receptor gene in mice resulted in Paneth cells failing to home to the crypt base.

As well as being required to maintain homeostasis in the healthy intestine, the Wnt/β-catenin pathway plays a key role in malignant transformation. Mutations of the *APC* tumor suppression gene, which typically result in a truncated protein product, are present in up to 80% of human sporadic colorectal tumors. This mutation inhibits cytosolic β-catenin degradation by the GSK3β/axin/APC complex. This results in increased nuclear β-*catenin*, and a subsequent increase in β-catenin-induced Tcf/LEF transcription. Thus, one of the main tumor suppressing functions of *APC* is the destabilization of β-catenin. Indeed, accumulation of nuclear β-catenin is one of the earliest events, or perhaps even *the* initiating event, in human and mouse intestinal tumorigenesis. This majority of the tumorigenic Wnt signaling may be mediated by the Wnt target

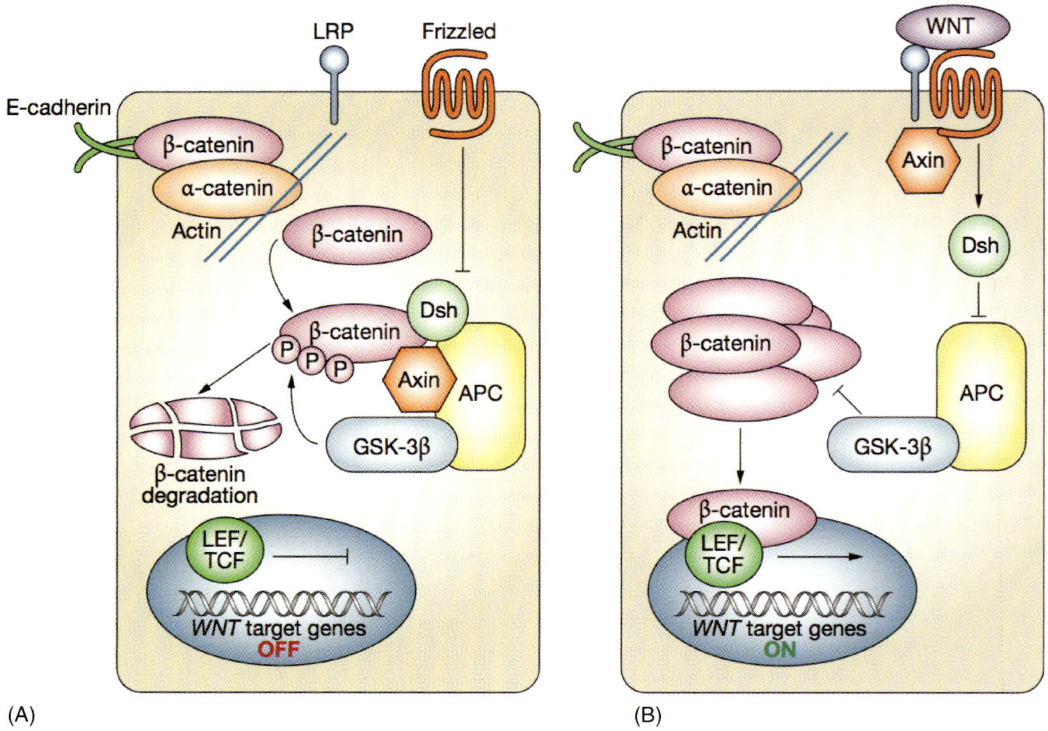

FIGURE 16-8 The Wnt signalling pathway. (A) In the absence of WNT bonding to frizzled, dishevelled remains unactivated and B-catenin is targeted for degradation via the adenomatous polyposis coli/axin/glycogen-synthase-kinase 3Bpathway. (B) When Wnt binds to frizzled (right), dishevelled is activated and uncouples B-catenin from the adenomatous polyposis coli complex: B-catenin can then associate with transcription factors and mediate transcription of the target gene. APC, adenomatous polyposis coli; Dsh, dishevelled; GSK-3B, glycogen synthase kinase 3B; LEF, lymphoid enhancer-binding factor; LRP, LDL-receptor related protein; P, phosphorylation; TCF, T cell factor.

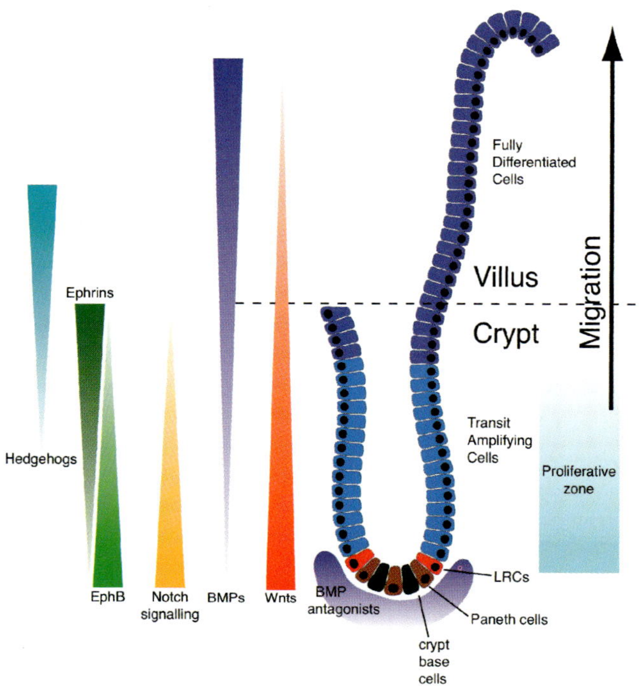

FIGURE 16-9 Schematic of signaling pathways in intestinal crypts. Proliferation tends to be restricted to the crypt base, where Wnt signals are most strongly expressed. BMP signaling increases away from the crypt base, whereas the BMP-antagonists, which include Germlins and Noggin, are principally expressed in the ISEMFs at the crypt base. The hedgehog molecules, antagonists of Wnt signaling, are chiefly expressed at the crypt tops.

c-myc, as simultaneous disruption of both *APC* and *c-myc*, restricts the tumor growth induced by *APC* disruption. Interestingly, the level of Wnt signaling required for tumor cell growth appears to be finely-tuned: tumor cells appear to select for "just-right" levels of β-catenin/TCF signaling. Tumorigenic APC mutations tend to result in a truncated protein product, which tends to retain one, or less frequently, two or three, of the 20 amino acid β-catenin-binding domains. The truncated APC protein can still regulate β-catenin activity, suggesting that there is selection for an optimal level of Wnt signaling during tumor growth. Furthermore, Segditsas and colleagues recently reported "third-hits" in APC in human colorectal cancers, suggesting that the Wnt pathway may be subjected to repeated targeting during cancer development. These third-hits may be selected for on the basis that they provide a new optimal level of Wnt-signaling for the cancer cells, which may be different to the Wnt-signal that provided an optimal proliferation signal in the early lesion.

WNT TARGETS EPHB DETERMINE CELL POSITIONING ALONG THE CRYPT AXIS

The Wnt targets, the receptor tyrosine kinases *EphB2* and *EphB3* are required for the correct positioning of epithelial cells along the crypt axis. In EphB3−/− knockout mice, for

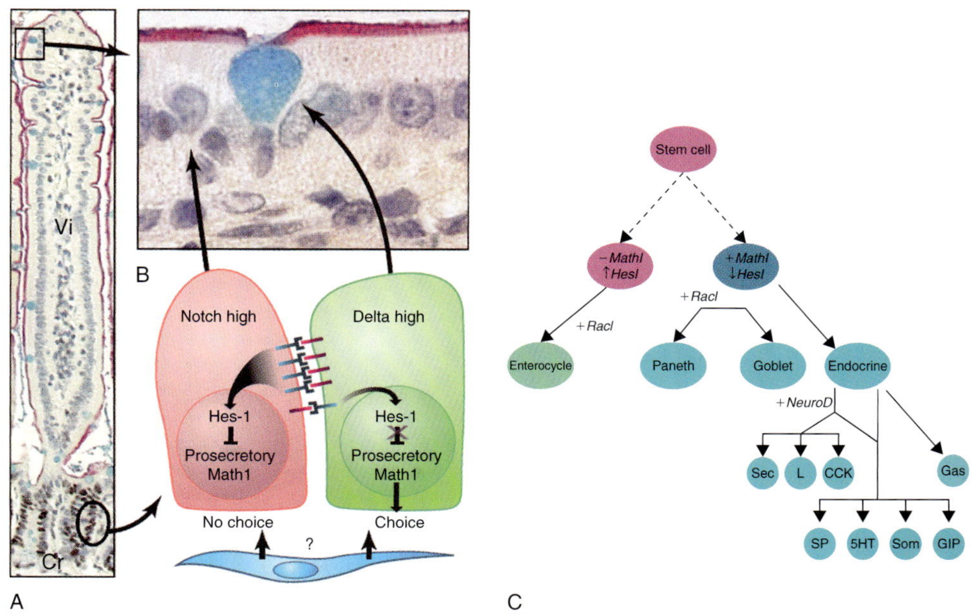

FIGURE 16-10 Math1 signaling pathway. (A) Low-power section of adult murine small intestine. Precursor cells are stained for cyclin proliferating cell nuclear antigen, enterocytes express intestinal alkaline phosphatase, and goblet cells secrete mucins. Inset shows high-power image of small intestinal enterocytes and goblet cells. (B) Math1, a component of the Notch signaling pathway, influences intestinal epithelial cell fate decisions. In crypt progenitor stem cells that express high levels of Notch, the Hes1 transcription factor is switched on, and the expression of Math1 and of other "prosecretory" genes is blocked. The result is that the precursor cells become enterocytes. In cells expressing low amounts of Notch, levels of Delta are high, production of Hes1 is blocked, and Math1 expression is induced. Production of the Math1 helix–loop–helix transcription factor allows precursor cells to make a choice: whether to become goblet cells, Paneth cells, or enteroendocrine cells. (C) Math1 is essential for secretory cells. Whether Math1-expressing cells descend directly from stem cells or an intermediate progenitor remains unknown. (Vi: villus, Cr: crypt, Sec: secretin, L: glucagon–peptide YY, CCK: cholecystokinin, SP: substance P, 5HT: serotonin, Som: somatostatin, GIP: gastric inhibitory peptide, and Gas: gastrin). Panels A and B reproduced with permission from van Den Brink and panel C reproduced with permission from Yang.

example, Paneth cells do not migrate downwards to their normal position at the bottom of the crypt, but instead scatter along crypt and villus. EphB2 and EphB3 receptor expression is strongest at the crypt base, whereas their ligand ephrin-B1 is expressed most strongly at the crypt top (Figure 16-10). The expression of EphB receptors is increased in response to elevated Wnt-signaling. Interestingly, EphB receptor expression is lost in human colorectal cancers, and inducible deletion of *EphB* in *ApcMin/+* mice accelerates tumorigenesis and leads to the development of invasive carcinomas. This suggests that EphB mediate a pathway crucial in permitting tumor cells to become invasive.

NOTCH SIGNALING IS ALSO TO MAINTAIN THE PROLIFERATIVE COMPARTMENT OF INTESTINAL CRYPTS, AND DETERMINES CELL FATE

Notch receptors, the ligand delta, and the associated Hes transcription factors are chiefly expressed at the base of intestinal crypts. Binding of the ligand to the Notch receptor causes the cleavage of part of the Notch receptor. The resulting free intra-cellular domain of the receptor translocates to the nucleus, where it binds to the transcription factor CSL/RBP-J and activates target gene transcription. Constitutive Notch signaling is required to maintain the proliferative compartment of intestinal crypts: intraperitoneal injection of mice with a γ-secretase inhibitor, which prevents the cleavage of the activated Notch receptor, rapidly transformed the proliferative cells in the crypt to post-mitotic Goblet cells. Similar results were observed following conditional removal of the CSL/RBP-J transcription factor in adult mice. The requirement for Notch signaling in order to maintain cell proliferation in intestinal crypts is maintained during intestinal tumorigenesis: γ-secretase inhibition in *ApcMin/+* mice caused widespread Goblet cell conversion within adenomas.

In mice, deletion of the *Math1* gene, a basic helix–loop–helix transcription factor and downstream component of the *Notch* signaling pathway, depletes goblet, Paneth, and enteroendocrine cell lineages in the small intestine. Math1− progenitors merely become enterocytes. This indicates that *Math1*, and indeed Notch signaling, is essential for determining stem cell fate (Figure 16-10). High levels of Notch switch on the Hes1 transcriptional repressor, which in turn blocks expression of Math1 so that cells remain progenitors and ultimately become enterocytes. Conversely, low Notch expression increases levels of its ligand Delta, which induces Math1 expression by blocking

Hes1, causing cells to become goblet cells, Paneth cells, or enteroendocrine cells. Correspondingly, Hes1-null mice have elevated Math1 expression, with increased enteroendocrine and goblet cells and fewer enterocytes. Conditional activation of a constitutionally active form of Notch causes a rapid and temporary increase in the number of Goblet cells in the crypt.

BONE MORPHOGENIC PATHWAY SIGNALING REGULATES PROLIFERATION ALONG THE INTESTINAL CRYPT AXIS

The bone morphogenic pathway (BMP) is part of the transforming growth factor β superfamily, and involves many ligands and receptors. BMPs bind to their receptors on the cell surface, causing a signal to be transduced to the nucleus via Smad proteins (crucially Smad4), where they regulate gene expression. Expression of BMPs is highest at the top (luminal end) of intestinal crypts, primarily in the mesenchyme, whereas BMP antagonists, including Gremlin 1 and 2 and Noggin, are expressed strongly at the crypt base (Figure 16-9).

BMPs appear to antagonize Wnt signaling, via the PTEN tumor-suppressor protein, and so permit differentiation and restrict proliferation. Conditional inactivation of the BMPR1A receptor in the mouse intestine resulted in increased cell proliferation. Inhibition of BMP signaling by conditional ectopic expression of the BMP-antagonist Noggin also increased proliferation, and led to *de novo* crypt formation and eventual polyposis. Gremlin 1, a BMP antagonist, was shown to partially inhibit cell differentiation *in vitro* and promote proliferation. Conversely, stimulation of colon cell lines by BMP2 inhibited growth. BMP signaling is also implicated in intestinal tumorigenesis, further illustrating the likely importance of the pathway in the regulation of proliferation in intestinal crypts. The rare inherited syndrome juvenile polyposis is associated with both germline mutations in the *BMPR1A* receptor, and with mutations in the *Smad4* gene. A recent genome-wide association study found that common genetic variants in BMP4 and upstream of BMP2, and in SMAD7, are associated with increased risk of colon cancer.

HEDGEHOG SIGNALING PATTERNS THE CRYPT AXIS

Hedgehog (Hh) signaling appears to work in the opposite direction to the other pathways discussed above. Expression of the two hedgehog ligands Indian and Sonic hedgehog tends to be restricted to the epithelium, and is predominantly in the mature colonocytes at the crypt top, whereas Hh receptors are expressed by the adjacent mesenchyme. Hedgehog signaling is an antagonist of Wnt signaling: transfection of Indian hh (Ihh) into colon cancer cells down-regulates TCF4 expression, whereas conversely, inhibition of TCF4 into colon cancer cells with constitutionally active Wnt-signaling restores Ihh expression. Thus, the differential expression of Hh along the crypt axis appears to restrict Wnt signaling to the crypt base. This localization of Hh is crucial for the correct positioning and maturation of colonocytes: reduction in Hedgehog signaling by expression of a pan-Hh inhibitor in transgenic mice causes epithelial remodeling, and mislocation of the subepithelial myofibroblasts. Differentiation was also disrupted when rats were treated with the pan-Hh inhibitor.

HOX GENES DEFINE REGIONAL GUT SPECIFICATION

Mammalian homeobox genes *Cdx-1* and *Cdx-2* display specific regional expression in developing and mature colon and small intestine. During embryogenesis, *Cdx-1* localizes to the proliferating cells of the crypts, and maintains this expression during adulthood. The Tcf-4 knockout mouse does not express *Cdx-1* in the small intestinal epithelium, thus the Wnt/β-catenin complex appears to induce *Cdx-1* transcription in association with Tcf-4 during the development of intestinal crypts. Mice heterozygous for a *Cdx-2* mutation develop colonic polyps composed of squamous, body, and antral gastric mucosa with small intestinal tissue. Proliferation of *Cdx-2*−colonic cells with low *Cdx-2* levels can produce clones of cells phenotypically similar to epithelial cells of the stomach or small intestine. This could indicate a possible homeotic shift in stem cell phenotype. Regional specificity of genes such as *Cdx-1*, *Cdx-2*, and *Tcf-4* appear to influence the morphological features of different regions of the intestinal epithelium, and regulate the proliferation and differentiation of the stem cells.

GASTROINTESTINAL NEOPLASMS ORIGINATE IN STEM CELL POPULATIONS

We can use the development of colorectal carcinoma as a paradigm. The concept of the adenoma–carcinoma sequence, whereby adenomas develop into carcinomas, is now widely-accepted, and most colorectal carcinomas are believed to originate in adenomas. The initial genetic change in the development of most colorectal adenomas is thought to be somatic mutations at the *APC* locus—mutations in *APC* can be identified in up to 80% of sporadic colorectal carcinomas—and the molecular events associated with these stages are clear: a second hit in the *APC* gene is sufficient to give microadenoma development, at least in FAP. There are two models for adenoma histogenesis, both of which closely involve basic concepts

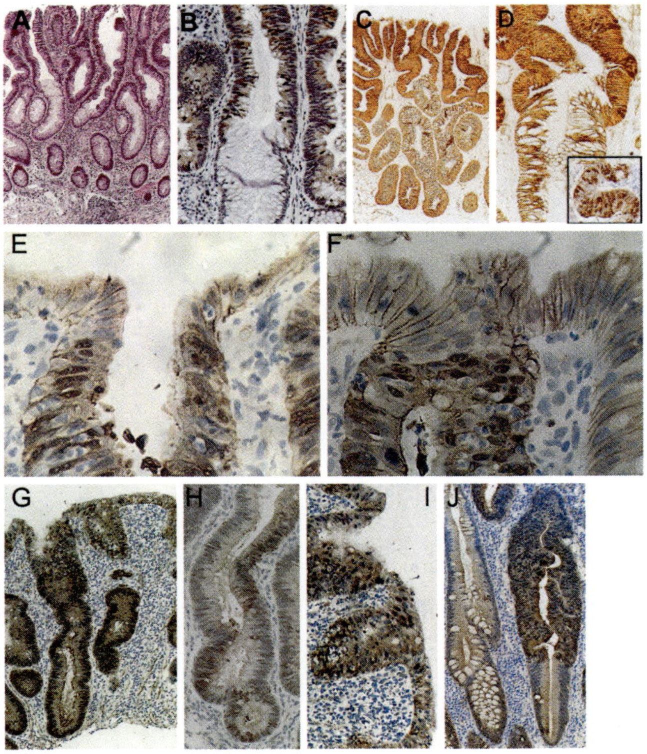

FIGURE 16-11 Contrasting theories for the morphogenesis of adenomas and the part played by stem cells. (A) Hematoxylin- and eosin-stained sections of a small tubular adenoma. Dysplastic epithelium is superficial within the crypts, with histologically normal underlying epithelium. (B) Abrupt transition between dysplastic and normal-appearing epithelial cells at the midpoint of this crypt. Proliferative activity assessed with the Ki-67 antibody, distributed throughout the dysplastic epithelium at the top of the crypts. (C) Nuclear β-catenin is highly expressed and distributed throughout the dysplastic epithelium at the top of the crypts but not in the crypt bases. (D) β-catenin in the nuclei of adenomatous crypts from a tiny tubular adenoma. (E) Nuclear β-catenin extends to the bottom of crypts in early adenomas, including the very bases of the crypts. (F) β-catenin staining in nuclei of budding crypts. (G) Junction between early adenomatous crypts, showing a sharp junction on the surface with accumulation of nuclear β-catenin giving way sharply to membranous staining in the normal surface cells. (H) High-power serial sections, demonstrating the sharp junction between nuclear staining in the adenomatous cells and membranous staining in normal surface epithelial cells. (I) Surface continuity between crypts showing nuclear β-catenin staining. (J) Crypts from a larger adenoma stained for β-catenin, showing invasion of adjacent crypt territories in a top-down fashion. Panels A–D reproduced with permission from Shih and panels E–J reproduced with permission from Preston.

of stem cell biology in the colon: in the "top-down" model, mutant cells appear in the "intracryptal zone" between crypt orifices and, as the clone expands, the cells migrate laterally and downwards, displacing the normal epithelium of adjacent crypts. A modification of this proposal is that a mutant cell in the crypt base, classically the site of the stem cell compartment, migrates to the crypt apex where it expands. These proposals are based on findings in some early non-FAP adenomas, where dysplastic cells were seen exclusively at the orifices and on the luminal surface of colonic crypts; measurement of loss of heterozygosity (LOH) for *APC*, and nucleotide sequence analysis of the mutation cluster region of the *APC* gene carried out on microdissected, well-orientated histological sections of these adenomas showed that half the sample had LOH in the upper portion of the crypts, most with truncating *APC* mutations. Only these superficial cells showed prominent proliferative activity, with nuclear localization of β-catenin indicating an *APC* mutation only in these apical cells (Figure 16-11C). Earlier morphological studies have drawn attention to the same appearances. The top-down morphogenesis has wide implications for established concepts of stem cell biology in the gut. It is clear that most evidence indicates that crypt stem cells are found at the origin of the cell flux, near the crypt base. These proposals, however, seek to either re-establish the stem cell compartment in the intracryptal zone, or make the intracryptal zone a favored locus where stem cells, having acquired a second hit, clonally expand.

An alternative hypothesis—the bottom-up theory—proposes that the earliest lesion is the "unicryptal" or "monocryptal" adenoma, where dysplastic cells occupy an entire single crypt with surrounding normal epithelium, with clonal expansion by crypt fission being the predominant mode of spread of an adenomatous crypt. A stem cell acquires a second hit in *APC*, then it expands stochastically, or more likely because of a selective advantage, to colonize the whole crypt and form a monocryptal adenoma. These lesions are the earliest detecable precursor lesions of adenoma development, and are commonly found in FAP; although they are rare in non-FAP patients they have been described. As would be expected from the experimental data demonstrating the clonal nature of crypts, such monocryptal lesions are indeed clonal. Similar crypt-restricted expansion of mutated stem cells has been well-documented in mice after ENU treatment and in humans heterozygous for the *OAT* gene, where after LOH, initially half, and then the whole crypt is colonized by the progeny of the mutant stem cell. Interestingly, *OAT+/OAT−* individuals with FAP show increased rates of stem cell mutation with clustering of mutated crypts. Thus, the mutated clone expands, not by lateral migration, but by "crypt fission," whereby the crypt divides, usually symmetrically at the base (Figure 16-12A) or by budding (Figure 16-12B).

Several studies have shown that fission of adenomatous crypts is the main mode of adenoma progression—predominantly in FAP, where such events are readily evaluated,

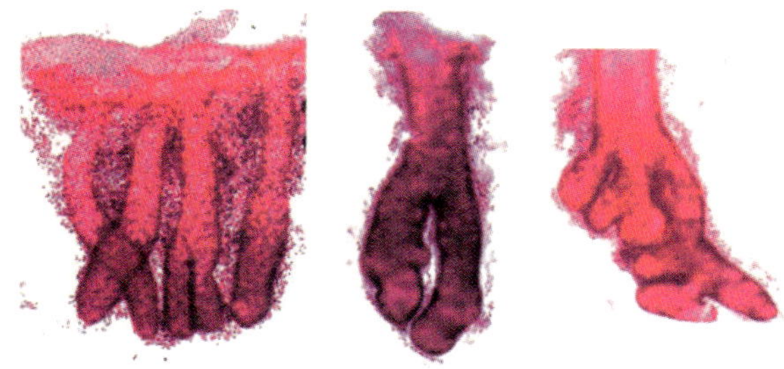

FIGURE 16-12 Microdissected crypts from normal colonic mucosae and adenomas. (A) Symmetrical fission of normal colonic crypts. (B) Isolated adenomatous crypt showing frequent crypt fission with atypical and asymmetrical branching. (C) Another crypt from an adenoma with a bizarre shape, asymmetrical branching, and multiple budding. Reproduced with permission from Preston.

but also in sporadic adenomas. The nonadenomatous mucosa in FAP, with only one *APC* mutation, not only shows a large increase in crypt fission rates, but also expansion of the crypt base stem cell population—a proposed mechanism for increased crypt fission. At the bottom of the crypt, progenitor cells accumulate nuclear β-catenin and express *Tcf* target genes, as a result of *Wnt* stimulation from surrounding basal pericryptal myofibroblasts. In normal crypts, cells that reach the mid-crypt region down-regulate β-catenin/*Tcf*, resulting in cell cycle arrest and differentiation. Cells that bear a β-catenin, or an *APC*, mutation do not respond to signals controlling β-catenin/*Tcf* activity, and these cells continue to behave as crypt progenitor cells in the surface epithelium, generating microadenomas. Boman used computer modeling to suggest that an expansion in the crypt stem cell population explained the putative proliferative abnormality in FAP, namely an upward shift in the proliferative compartment towards the top of the crypt. Simulation of labeling index distribution curves from FAP crypts using a single mechanistic design was able to fit the data from both control and FAP crypts, indicating that the proliferative abnormality does not alter the rate of cell cycle proliferation, differentiation, or apoptosis of proliferating crypt cells. Instead, it suggests an expansion in the crypt stem cell population sufficient to explain the observed proliferative abnormality in FAP. Thus, β-catenin signaling in the colonic crypt controls the number of stem cells. The stem cell population is expanded in FAP crypts, because of a germline APC mutation activating Tcf-4. Any increase in the size of the stem cell population might be expected to result in an increase in the rate of crypt fission. Crypt fission is therefore an essential event in the expansion of mutated clones in adenomas. Although the morphology of this process is distinct, the molecular mechanisms that govern it are far from clear.

Aberrant crypt foci (ACF) are clusters of small numbers of abnormal crypts, thought to be a further early step in the development of adenomas, and have been shown to grow by crypt fission, as do hyperplastic polyps. The bottom-up concept does not exclude the possibility that the clone later expands by lateral migration, with downward spread into adjacent crypts, but with the initial lesion the monocryptal adenoma, this model of tumorigenesis is conceptually very different.

Work supporting the bottom-up spread of colorectal adenomas looked at a number of small (<3 mm) tubular adenomas. Here, nuclear accumulation of β-catenin (Figure 16-11G) was seen, indicating loss of function of one of the genes in the *Wnt* pathway, most likely *APC*, with subsequent translocation of β-catenin to the nucleus. Serial sections (Figure 16-11E, F) showed that the β-catenin nuclear staining extended to the bottom of the crypts, and was present in crypts in the process of crypt fission (Figure 16-11H). β-catenin expression was particularly marked in the nuclei of buds. At the surface, there was a sharp cut-off between the adenomatous cells in the crypt that showed nuclear β-catenin, and those surface cells that did not (Figure 16-11E, F). The adjacent crypts were filled with dysplastic cells containing nuclear β-catenin, which were not confined to the upper portions of the crypts. In larger adenomas, there was unequivocal evidence of surface cells growing down and replacing the epithelium of normal-looking crypts (Figure 16-11J). Crypt fission was rare in normal and non-involved mucosa, and usually began with basal bifurcation at the base of the gland (Figure 16-12A), whereas in adenomas, fission was commonly asymmetrical and disordered, with budding from the superficial and mid-crypt (Figure 16-12B). Multiple fission events were frequently observed in adenomas (Figure 16-12C). The "crypt fission index" (the proportion of crypts in fission) in adenomas has been demonstrated to be significantly greater than that in non-involved mucosa.

It is often stated that adenomas do not display a stem cell architecture: observations on possibly older adenomas have suggested that maximum proliferative activity is found towards the top of the crypts, indicating that migration kinetics are reversed, with cells flowing toward the bottom of the crypt, this is corroborated by the report of increased apoptosis at the bottom of adenomatous crypts. However, more recent experimental data from both the human and mouse demonstrates crypts in early adenomas show similarities to normal crypts in the distribution of their proliferative activity, with an upward expansion of the proliferative compartment towards the surface in adenomas, and maintenance of a stem cell/progenitor hierachy. Tsao investigated the likely microstructure of colorectal adenomas and carcinomas from hereditary nonpolyposis colon cancer, by sampling microsatellite mutations and then using experimental modeling to recreate these measurements; they concluded that the discrete stem cell architecture of normal epithelium is likely to persist in adenomas. This has further been supported

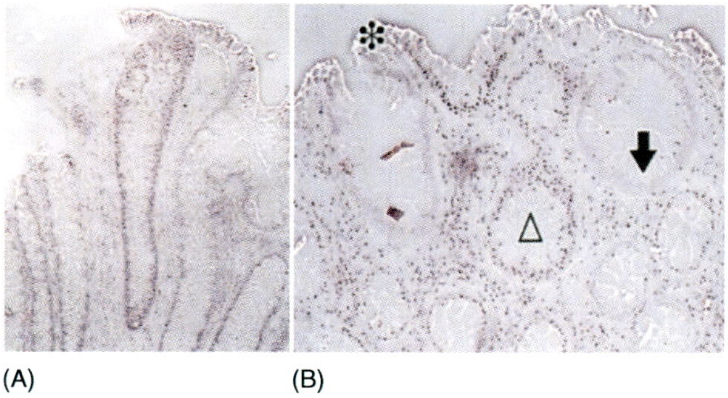

FIGURE 16-13 Adenomas from an XO–XY mosaic individual with FAP, stained by *in situ* hybridization for a Y-chromosome-specific probe. (A) Clonal monocryptal adenoma. (B) Polyclonal adenoma, with a mixture of XO (arrow) and XY (triangle) crypts. Note that there is a sharp margin between the territories of XO and XY adenomatous crypts (*) at the surface with no evidence of invasion. Reproduced with permission from Preston.

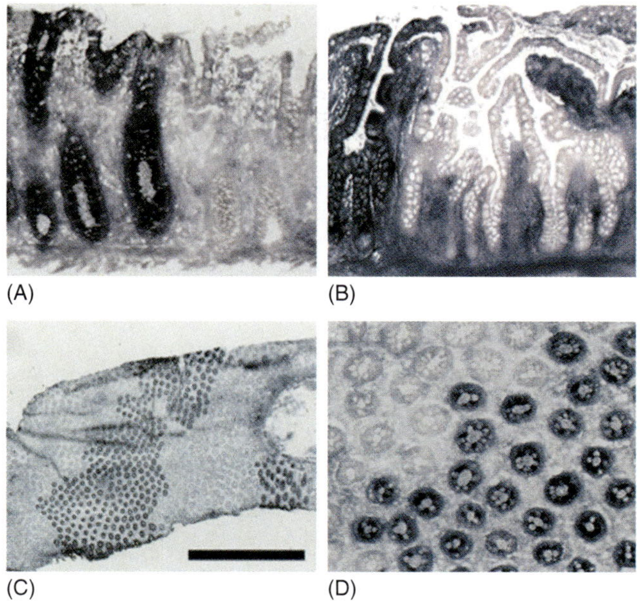

FIGURE 16-14 Visualization of X-inactivation patches directly by enzyme histochemistry in surgical resections from Sardinian females heterozygous for the G6PD Mediterranean mutation (563 C 3 T), previously shown to have reduced G6PD enzyme activity. Heterozygosity for the G6PD Mediterranean mutation was confirmed by polymerase chain reaction analysis of genomic DNA, followed by MBOII restriction endonuclease. (A) G6PD staining in longitudinal sections of colonic crypts. In individual crypts, all epithelial cells show a similar staining pattern. (B) Longitudinal sections of crypts and villi in small intestinal mucosa stained for G6PD activity. Epithelial cells in individual crypts show a similar staining pattern, but the villous epithelium shows areas of positive and negative staining, confirming its polyclonal derivation. (C) G6PD staining in colonic patches in low power. (D) High-power view of large patches of crypts with irregular patch borders. Bar indicates 2 mm. Reproduced with permission from Novelli.

by Barker who, using lgr5 as a stem cell marker, demonstrated in the mouse that deletion of APC in crypt stem cells allows rapid dysplastic transformation and development of microadenomas. The microadenomas continued to grow and formed macroscopic adenomas in three to five weeks, importantly the stem-cell hierarchy was maintained in the adenomas and deleting APC in non stem cells did not result in sustained adenoma growth—strengthening the concept that it is mutations in stem cells that initiate gastrointestinal cancers.

Examination of the adenomas of an XO–XY individual with FAP showed none exceeded 2.5 mm in diameter. The monocryptal adenomas showed either the XO or the XY genotype, and hence are clonal proliferations, as would be expected from the observation that crypts are clonal (Figure 16-13A). Many microadenomas showed the XY genotype, and none of these early lesions showed a mixture of XO and XY nuclei occupying the same crypt (Figure 16-13B). It was shown that 76% of the microadenomas were polyclonal. There were also sharp boundaries at the surface between adjacent adenomatous crypt territories.

What are the implications of these considerations on our concepts of the single (stem) cell origin or clonality in colorectal adenomas? We have seen that crypts are clonal units; thus, these lesions would be polyclonal because of the mixture of clonal crypts and clonal adenoma—although in this instance, they have different clonal derivation. A study of both sporadic and FAP adenomas, using X-linked restriction fragment length polymorphisms, showed that such lesions were apparently monoclonal in origin. On the other hand, the X-linked patch in the colon is large—and can be in excess of 450 crypts in diameter (Figure 16-14). So, unless an adenoma grows over a patch boundary and involves crypts on either side of that boundary, X-inactivation analysis will always show that such lesions are monoclonal.

On the "bottom-up" proposals, expansion of a clonal monocryptal adenoma by crypt fission would inevitably lead to a monoclonal microadenoma, and thus adenoma. However, as mentioned before, studies on the XO–XY individual, who has a mean patch size of 1.48 crypts, indicated that some 76% of adenomas were polyclonal, supported by observations in Rosa26–Min chimeric mice. Explanations for this are that random collisions occur between independently-transformed neighboring crypts, as there was a high tumor load in this study. Alternatively, active cooperation or clonal interaction is required between dysplastic clones in order for the continued survival and growth of an adenoma.

So, what is happening to the stem cells in the development of these early, monocryptal lesions? Bjerknes showed that adenomatous crypts in FAP contained two stem cell lines, *APC*−/+ and *APC* −/−, and interpreted this to mean that the *APC*−/− cells were expanding stochastically within the same crypt. There is good evidence that such monoclonal conversion occurs as a stochastic process. It has previously been shown that a quantitative analysis of age-related methylation suggests that crypts are maintained by niches containing multiple stem cells. Random stem cell loss with replacement suggests that all niche lineages except one will become extinct. It is fairly clear that clonal succession is related to tumor progression. Although we have mentioned previously that Shibata has argued for clonal succession occurring in a stem cell niche without mutation, genetic alterations aiding clonal expansion of a mutant clone might confer growth advantages. Several mouse genotypes exist in which crypt morphology—and therefore implicitly stem cell proliferative behavior—is normal, such as APC+/−, TP53+/−, Trp53−/−, Apc+/−, Trp53−/−, MLH1−/0, Apc+/−, Mlh1−/−, and Tgfbr2+/−. However, more subtle alterations, such as an increased incidence of crypt fission that may be dependent upon stem cell behavior, might well be missed.

We have argued that gastrointestinal cancers originate in stem cells; not only are these the only cells that would appear to have a sufficient lifespan required to accumulate all the mutations necessary for malignant transformation, but we now have experimental evidence that only dysplastic transformation of stem cells can drive the formation of adenomas. Recent studies have now shown that a subpopulation of cancer-initiating cells is found in colonic carcinomas. These so-called cancer stem cells were identified by the cell surface marker CD133, and only this CD133 population of cancer cells was able to initiate new tumors in immunodeficent mice. It is proposed, therefore, that cancers themselves are maintained by a population of cancer stem cells, and it is only these cells that are able to migrate and form distant metastases.

Let us now summarize the situation in a normal individual, where stem cells are wild-type (APC+/+), and one APC allele is lost in a single stem cell. This APC+/− stem cell, unless it possesses a growth advantage, could be lost by stochastic ejection from the niche—indeed, it could be argued that this will happen in the over-riding majority of instances, but the APC+/ − cell could survive and the niche will be populated by progeny of this APC+/− cell, which will resemble an FAP crypt. A further hit in APC in this crypt will lead to the formation of a dysplastic crypt—the monocryptal adenoma. Initial growth occurs through crypt fission, with spread into adjacent crypt territories a secondary event, resulting in an aberrant crypt focus, or microadenoma, that in turn expands further to form an established adenoma. We have argued that adenomas are polyclonal, with possible transformation of non-involved crypts by transformed stem cells, and with clonal interaction needed for the survival of the early adenoma. The stem cell hierarchy is maintained in the adenoma, but with an upward expansion in the stem cell compartment, and hugely increased rates of disordered crypt fission. Over time, the adenoma accumulates further mutations in critical genes, resulting in malignant transformation to an adenocarcinoma that, possibly, is also maintained by a cancer stem cell population.

SUMMARY

The cells of the gastrointestinal tract undergo constant renewal, and respond to damage by regeneration and repopulation. Each region of the gastrointestinal tract is morphologically distinct, with its own repertoire of cell types. Although the stem cells are the most important cells of the gastrointestinal tract, responsible for the production of every other cell type in the gastrointestinal mucosa, they have not yet been closely characterized. There is, initially at any rate, a single stem cell in every intestinal crypt or gastric gland that indirectly generates a clone containing further stem cells, transit amplifying and differentiated cells, through the production of committed progenitor cells. This cell also produces new crypts by crypt fission, repairs entire crypts and villi when damaged, and generates gastrointestinal tumors. The stem cell or cells occupy a niche, formed by mesenchymal cells, such as the ISEMFs, and extracellular matrix molecules, which regulate epithelial stem cells through mesenchymal–epithelial crosstalk. The molecular events that regulate the development of the gastrointestinal tract, and epithelial cell turnover in the normal tissue and in formation of carcinomas are beginning to be identified. It is clear that the Wnt/β-catenin signaling pathway and downstream molecules such as APC, Tcf-4, Fkh-6, Cdx-1, and Cdx-2 are vital for normal gastrointestinal stem cell function. We are beginning to identify molecular pathways that determine further proliferation of committed progenitor cells into specific epithelial cell lineages: the Notch–Delta signaling pathways involving Hes1 and Math1 transcription factors regulate differentiation of goblet, Paneth, and enteroendocrine cells in the small intestine. Factors secreted and expressed by the mesenchymal cells (KGF, HGF, etc.) that regulate gastrointestinal mucosal development and epithelial proliferation are rapidly becoming identified. Epithelial and intestinal subepithelial myofibroblast lineages, apparently derived from bone marrow, present the possibility of delivering therapeutic genes to the damaged lamina propria—for example, in diseases that cause fibrosis, such as Crohn's disease—and even repopulating the damaged gut. Finally, we conclude that the colonic stem cell is pivotal in understanding mechanisms of tumorigenesis in the colon. The isolation and characterization of gastrointestinal stem cells is a priority in gut biology.

KEY WORDS

Crypt clonality Intestinal crypts, wherein all cell lineages are originally derived from a single cell during embryonic development.

Gastrointestinal stem cells Founder cells capable of limitless self-renewal throughout the lifetime of their host. Gastrointestinal stem cells are also responsible for the maintenance of tissue homeostasis, by producing daughter cells committed to multipotent differentiation to form all adult lineages within the gastrointestinal tract.

Gastrointestinal stem cell signaling Elucidation of the components involved in molecular pathways, such as Wnt/b-catenin, which regulate cell proliferation and differentiation in the gastrointestinal tract, will provide an insight into the mechanisms of normal epithelial cell turnover and neoplastic development.

Gastrointestinal tumor morphogenesis The mode of dysplastic growth within the gastrointestinal tract is a topic of heated debate, with two main conflicting theories. The first "bottom-up" theory proposes that neoplastic lesions arise in the base of the crypt, the "stem cell zone" and proliferate upwards. The alternative, "top-down" theory states that lesions arise within an "intercryptal" zone, and spread laterally and down into adjacent crypts.

Stem cell plasticity (in the gastrointestinal tract) Evidence suggests that a population of stem cells in the adult bone marrow can engraft within the gastrointestinal tract and form both epithelial and mesenchymal lineages, with an increased propensity in damaged tissue. However, the mechanisms by which bone marrow cells form gastrointestinal lineages are unresolved.

FURTHER READING

Alison, M. R., Poulsom, R., Jeffery, R., Dhillon, A. P., Quaglia, A., Jacob, J. *et al.* (2000). Hepatocytes from non-hepatic adult stem cells. *Nature*, *406*(6793), 257.

Barker, N., & Clevers, H. (2007). Tracking down the stem cells of the intestine: strategies to identify adult stem cells. *Gastroenterology*, *133*(6), 1755–1760.

Greaves, L. C., Preston, S. L., Tadrous, P. J., Taylor, R. W., Barron, M. J., Oukrif, D. *et al.* (2006). Mitochondrial DNA mutations are established in human colonic stem cells, and mutated clones expand by crypt fission. *Proc. Nat. Acad. Sci. USA*, Jan. 17, *103*(3), 714–719.

Hardwick, J. C., Kodach, L. L., Offerhaus, G. J., & van den Brink, G. R. (2008). Bone morphogenetic protein signaling in colorectal cancer. *Nature Reviews Cancer*, *8*(10), 806–812.

Klaus, A., & Birchmeier, W. (2008). Wnt signaling and its impact on development and cancer. *Nature Reviews Cancer*, *8*(5), 387–398.

Preston, S. L., Wong, W. M., Chan, A. O., Poulsom, R., Jeffery, R., Goodlad, R. A. *et al.* (2003). Bottom-up histogenesis of colorectal adenomas: origin in the monocryptal adenoma and initial expansion by crypt fission. *Cancer Research*, *63*(13), 3819–3825.

Ricci-Vitiani, L., Lombardi, D. G., Pilozzi, E., Biffoni, M., Todaro, M., Peschle, C. *et al.* (2007). Identification and expansion of human colon-cancer-initiating cells. *Nature*, *445*(7123), 111–115.

Rowan, A. J., Lamlum, H., Ilyas, M., Wheeler, J., Straub, J., Papadopoulou, A. *et al.* (2000). APC mutations in sporadic colorectal tumors: A mutational "hotspot" and interdependence of the "two hits." *Proc. Nat. Acad. Sci. USA*, *97*(7), 3352–3357.

Wang, X., Willenbring, H., Akkari, Y., Torimaru, Y., Foster, M., Al-Dhalimy, M. *et al.* (2003). Cell fusion is the principal source of bone-marrow-derived hepatocytes. *Nature*, *422*(6934), 897–901.

Wright, N. A. (2000). Epithelial stem cell repertoire in the gut: clues to the origin of cell lineages, proliferative units and cancer. *Int. J. Exp. Pathol.*, *81*(2), 117–143.

Chapter 17

Stem Cells Derived from Amniotic Fluid and Placenta

Paolo De Coppi, Shay Soker, and Anthony Atala

INTRODUCTION

Amniotic fluid and placenta have been recently taken into consideration as potential sources of progenitor cells. Amniocentesis and chorionic villus sampling (CVS) are widely accepted methods for prenatal diagnosis. Minimal or no ethical concerns would be present if embryonic and fetal stem cells would be taken from amniotic fluid and placenta before or at birth. In the last few years our and other groups have described the presence of stem cell with various differentiative and proliferative potential in the amniotic fluid and placenta. We will briefly describe the techniques in use for amniocentesis and CVS, and we will examine the different progenitors that have been described.

CVS AND AMNIOCENTESIS

The first reported amniocentesis took place in 1930 when attempts were being made to correlate the cytologic examination of cell concentration, count and phenotypes in the amniotic fluid to the sex and the health of the baby. Since then, the development technique of karyotype and the discovery of reliable diagnostic markers, such as alpha-fetoprotein, as well as the development of ultrasound-guided amniocentesis, have greatly increased the reliability of the procedure as a valid diagnostic tool as well as the safety of the procedure (Milunsky, 1979; Hoehn and Salk, 1982; Gosden, 1983; Crane and Cheung, 1988). One of the primary uses of amniocentesis is as a safe method of isolating cells from the fetus that can be karyotyped and examined for chromosomal abnormalities. In general, the protocol consists of acquiring 10–20 ml of fluid using a transabdominal approach. Amniotic fluid samples are then centrifuged, and the cell supernatant is resuspended in culture medium. Approximately 10^4 cells are seeded on 22 × 22 mm cover slips. Cultures are grown to confluence for 3–4 weeks in 5% CO_2 at 37°C, and the chromosomes are characterized from mitotic phase cells (Brace and Resnik, 1999). Amniocentesis is performed typically around 16 weeks of gestation, although in some cases it may be performed as early as 14 weeks when the amnion fuses with the chorion and the risk of bursting the amniotic sac by needle puncture is minimized. Amniocentesis can be performed as late as term. The amniotic sac is usually noticed first by ultrasound around the 10-week gestational time point.

With the introduction of CVS in the 1980s, first-trimester diagnosis became a reality. A small sample of chorionic villi (tissue from the developing placenta) is obtained from the mother's uterus under ultrasound guidance, either transvaginally or transabdominally. Sampling of chorionic villi from the fetus is performed from 10 weeks of gestation. The biopsy is usually taken under ultrasound guidance via a transabdominal approach. Alternatively, the cervical approach may be utilized. Each biopsy yields 5–30 mg of tissue that can be used for fetal sexing, karyotyping, biochemical studies and DNA analysis. A direct fetal chromosomal analysis on cultured cells is possible within 24 h. However, given the problem with mosaicism in CVS samples, this should always be followed by chromosomal analysis on cultured cells from the sample 2–3 weeks later. An advantage of CVS is that termination can be completed in the first trimester when it is technically easier.

DIFFERENTIATED CELLS FROM AMNIOTIC FLUID AND PLACENTA

Amniotic fluid cell culture consists of a heterogeneous cell population displaying a range of morphologies and behaviors. Studies on these cells have characterized them into many shapes and sizes varying from 6 μm to 50 μmm in diameter and from round to squamous in shape. Most cells in the fluid are terminally differentiated along epithelial lineages and have limited proliferative and differentiation capabilities. Previous studies have noted an interesting composition of the fluid consisting of a heterogeneous cell population expressing markers from all three germ layers. (Sarkar *et al.*, 1980; Cousineau *et al.*, 1982; Medina-Gomez and Johnston, 1982; von Koskull *et al.*, 1984). The source of these cells and of the fluid itself underwent a great deal of research. Current theories suggest that the fluid is largely derived from the urine and pulmonary secretion from the fetus as well as from some ultra filtrate from the plasma of the mother entering though the placenta. The cells in the fluid have been shown to be overwhelmingly from the fetus and are thought to be mostly cells sloughed off the epithelium and digestive and urinary tracts of the fetus as well as the amnion (Lotgering and Wallenburg, 1986; Underwood *et al.*, 2005).

MESENCHYMAL CELL FROM PLACENTA AND AMNIOTIC FLUID

Preliminary studies have been published a few years ago describing very simple protocols for the isolation of a non-specific population of cells with "mesenchymal" characteristics from amniotic fluid and placenta (Haigh *et al.*, 1999; Kaviani *et al.*, 2001, 2002, 2003). These cells were able to proliferate *in vitro*, to be engineered in a three-dimensional structure and used *in vivo* to repair a tissue defect (Kaviani *et al.*, 2003). A few years later In't Anker *et al.* were able to prove for the first time that both amniotic fluid and placenta were abundant sources of fetal mesenchymal stem cells (MSCs) that exhibit a phenotype and multilineage differentiation potential similar to that of postnatal bone marrow (BM)-derived MSCs (In't Anker *et al.*, 2003). They described a simple and repeatable protocol for their isolation and expansion. Briefly, amniotic fluid samples were centrifuged for 10 min at 1,283 rpm. Pellets were resuspended in Iscove's modified Dulbecco's medium containing 2% fetal calf serum (FCS) and antibiotics (defined as washing medium). Similarly, for the placenta, approximately 1 cm^3 was washed in phosphate-buffered saline (PBS) and single-cell suspensions were made by mincing and flushing the tissue parts through a 100 μm nylon filter with washing medium. Single-cell suspensions of amniotic fluid and placenta were plated in six-well plates and cultured in M199 supplemented with 10% FCS, 20 μg/ml endothelial cell growth factor, heparin (8U/ml), and antibiotics. After 7 days, non-adherent cells were removed and the medium was refreshed. When grown to confluence, adherent cells were detached with trypsin/EDTA and expanded in culture flasks pre-coated with 1% gelatin and kept in a humidified atmosphere at 37°C. The expansion potency of fetal MSCs was higher compared with adult BM-derived MSCs. As a result, they were able to expand amniotic fluid MSCs to about 180×10^6 cells within 4 weeks (three passages). The phenotype of the culture-expanded amniotic fluid-derived cells was similar to that reported for MSCs derived from second-trimester fetal tissues and adult BM. They were able to show that amniotic fluid-derived MSCs showed multilineage differentiation potential into fibroblasts, adipocytes, and osteocytes (In't Anker *et al.*, 2004). Furthermore, amniotic fluid-derived MSCs were successfully isolated, cultured, and enriched without interfering with the routine process of fetal karyotyping. Flow cytometry analyses showed that they were positive for SH2, SH3, SH4, CD29 and CD44, low positive for CD90 and CD105, but negative for CD10, CD11b, CD14, CD34, CD117, and EMA (Tsai *et al.*, 2004). Most importantly, immuno-phenotypic analyses demonstrated that these cells expressed HLA-ABC, class I major histocompatibility complex (MHC-I), but they did not express HLA-DR, DP, and DQ (MHC-II molecules) (Li *et al.*, 2005a). Li et al. have extensively investigated their immunological role and described that when mononucleated cells recovered from placentas by density gradient fractionation were added to umbilical cord blood (UCB) lymphocytes stimulated by human adult lymphocytes or potent T-cell mitogen phytohemagglutinin, a significant reduction in lymphocyte proliferation was observed. This immunoregulatory feature strongly implies that they may have potential application in allograft transplantation. As it is possible to obtain placenta and UCB from the same donor, they suggested the placenta as an attractive source of MSCs for co-transplantation in conjunction with UCB-derived hematopoietic stem cells (HSCs) to reduce the potential graft-versus-host disease (GVHD) in recipients (Li *et al.*, 2005b). Other methods have been described for the isolation of mesenchymal cell from human placenta. Dissection and proteinase digestion are alternative techniques to harvest high numbers of viable mononuclear cells from human placenta at term, and a mesenchymal cell population with characteristic expression of CD9, CD29, and CD73 could be obtained in culture. The *in vitro* growth behavior of such placenta-derived mesengenic cells was similar to that of human BM mesengenic progenitor cells. Differentiation experiments showed differentiation potential along osteogenic, chondrogenic, adipogenic, and myogenic lineages (Figure 17-1). However, after *in vitro* propagation for more than three passages, the cells were exclusively of maternal origin (Wulf *et al.*, 2004). Similar cells isolated from term placenta were described by Yen *et al.* They exhibited many markers common to mesenchymal stem cells – including CD105/endoglin/SH-2, SH-3, and SH-4 – and they lack hematopoietic-, endothelial-, and trophoblastic-specific cell markers. In addition, they exhibit embryonic stem (ES) cell surface markers of SSEA-4, TRA-1-61, and TRA-1-80. Adipogenic, osteogenic, and neurogenic differentiation were achieved after culturing under the appropriate conditions (Yen *et al.*, 2005). Mesenchymal cells were also isolated from placentas collected after neonatal delivery (38–40 weeks of gestation). The cells expressed CD13, CD44, CD73, CD90, CDIO5, and HLA class I as surface epitopes, but not CD31, CD34, CD45, and HLA-DR, differentiated into osteocytes, chondrocytes, and adipocytes under specific culture conditions, and were also induced to form neural-like cells (Fukuchi *et al.*, 2004; Igura *et al.*, 2004). Different types of tissue were obtained by *in vivo*

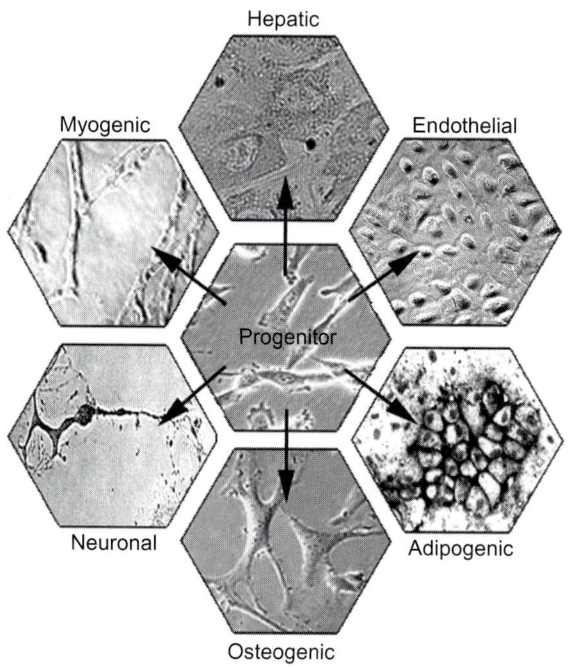

FIGURE 17-1 The isolated progenitor cells were capable of differentiation into multiple cell types, including muscle, liver, endothelial cells, adipocytes, osteoblasts, and neurons.

implantation of the cells. Several studies suggested that human amniotic fluid and placenta-derived mesenchymal cells can be one of the possible allogeneic cell sources for tissue engineering of cartilage. In particular, Zhang *et al.* (2005) showed the possibility to make a cartilage-like tissue embedding mesenchymal stem cells derived from human placenta, into an atelocollagen gel with chondrogenic induction media. The *in vitro* pre-induced cells were implanted into nude mice and also into nude rats with osteochondral defect, and they were able to form chondrogenic structures. Ovine mesenchymal amniocytes have also been cultured and engineered into a collagen hydrogel in order to replace partial diaphragmatic loss or absences. The authors showed that diaphragmatic repair with an autologous tendon engineered from mesenchymal amniocytes leads to improved mechanical and functional outcomes when compared with an equivalent acellular bioprosthetic repair, depending on scaffold composition (Fuchs *et al.*, 2004). Different groups have claimed that mesenchymal cells from placenta and amniotic fluid could have more plasticity than that initially thought. Phenotypic and gene expression studies indicated mesenchymal stem cell-like profiles in both amnion and chorion cells that were positive for neuronal, pulmonary, adhesion, and migration markers. In addition, transplantation in neonatal swine and rats resulted in human microchimerism in various organs and tissues, suggesting that amnion and chorion cells may represent an advantageous source of progenitor cells with potential applications in a variety of cell therapy and transplantation procedures (Bailo *et al.*, 2004). Similarly, Zhao *et al.* have reported that human amniotic mesenchymal cells (hAMC) may also be a suitable cell source for cardiomyocytes. He showed that freshly isolated hAMC expressed cardiac-specific transcription factor GATA4, cardiac-specific genes, such as myosin light chain (MLC)-2a, MLC-2v, cTnI, and cTnT, and the alpha-subunits of the cardiac-specific L-type calcium channel (alpha1c). After stimulation with basic fibroblast growth factor (bFGF) or activin A, hAMC expressed Nkx2.5, a specific transcription factor for the cardiomyocyte and cardiac-specific marker atrial natriuretic peptide. In addition, the cardiac-specific gene alpha-myosin heavy chain was detected after treatment with activin A. Co-culture experiments confirmed that hAMC were able to both integrate into cardiac tissues and differentiate into cardiomyocyte-like cells. After transplantation into the myocardial infarcts (AMI) in rat hearts, hAMC survived in the scar tissue for at least 2 months and differentiated into cardiomyocyte-like cells (Zhao *et al.*, 2005). However, we have recently shown that this potential does not belong to mesenchymal progenitor cells in bigger animals, such as pigs. Amniotic fluid-derived mesenchymal cells (AFC) autotransplanted in a porcine model of AMI were able to transdifferentiate to cells of vascular cell lineages but failed to give origin to cardiomyocytes (Sankar and Muthusamy, 2003; Sartore *et al.*, 2005). Regarding neuron regeneration, it has been shown that rat amniotic epithelial (RAE) cells were positive *in vitro* for both neuronal and neural stem cell markers, neurofilament microtubule-associated protein 2, and nestin. RT-PCR revealed that these cells expressed nestin mRNA. The RAE cells were also transplanted into the hippocampus of adult gerbils that were subjected to temporal occlusion of bilateral carotid arteries. Five weeks after transplantation, grafted cells migrated into the CA1 pyramidal layer that showed selective neuronal death, and survived in a manner similar to CA1 pyramidal neurons (Okawa *et al.*, 2001). Different reports suggest that human amniotic epithelial cells (HAEC) also possess certain properties similar to that of neural and glial cells (Tsai *et al.*, 2005). When transplanted into the transection cavities in the spinal cord of bonnet monkeys, HAEC were able to survive, support the growth of host axons through them, prevent the formation of glial scar at the cut ends and may prevent death in axotomized cells or attract the growth of new collateral sprouting (Okawa *et al.*, 2001). Amniotic epithelial cells isolated from human term placenta express surface markers normally present on ES and germ cells. In addition, they express the pluripotent stem cell-specific transcription factors octamer-binding protein 4 (Oct-4) and nanog. Under certain culture conditions, amniotic epithelial cells form spheroid structures that retain stem cell characteristics. Amniotic epithelial cells did not require other cell-derived feeder layers to maintain Oct-4 expression, did not express telomerase, and are non-tumorigenic upon transplantation. Based on immunohistochemical and genetic analysis, amniotic epithelial cells had the potential to differentiate to all three germ layers—endoderm (liver, pancreas), mesoderm (cardiomyocyte), and ectoderm (neural cells) *in vitro* (Miki *et al.*, 2005). Sarkar *et al.* (2003) have also shown that HAEC obtained from human placenta were able to survive into the transection cavities in the spinal cord of bonnet monkeys, to support the growth of host axons through them, to prevent the formation of glial scar at the cut ends and may prevent death in axotomized cells or attract the growth of new collateral sprouting. They speculated that HAEC may be having certain properties equal to the beneficial effects of neural tissue in repairing spinal cord injury. Apart from this speculation, there are two more reasons for why HAEC transplantation studies are warranted to understand the long-term effects of such transplantations. First, there was no evidence of immunological rejection probably due to the non-antigenic nature of the HAEC. Second, unlike neural tissue, procurement of HAEC does not involve many legal or ethical problems (Sakuragawa *et al.*, 1996, 2000; Elwan and Sakuragawa, 1997; Takahashi *et al.*, 2002).

PLURIPOTENT STEM CELLS FROM PLACENTA AND AMNIOTIC FLUID

In the midgestation murine embryo, several major vascular tissues contain pluripotent stem cells, mainly defined for their HSC activity. These include the aorta–gonad–mesonephros (AGM) region, yolk sac, and fetal liver. Recently, different reports have shown that the mouse placenta functions as a hematopoietic organ that harbors a large pool of pluripotent HSCs during midgestation (Alvarez-Silva *et al.*, 2003; Ottersbach and Dzierzak, 2005). The onset of HSC activity in the placenta parallels that of the AGM region starting at E10.5–E11.0. However, the placental HSC pool expands until E12.5–E13.5 and contains >15-fold more HSCs than the AGM (Gekas *et al.*, 2005). Placental HSC activity starts before HSCs are found in circulation or have colonized the fetal liver. Moreover, hematopoietic cells in midgestation mouse placenta are not instructed for differentiation along the myeloerythroid lineage, as in the fetal liver. These findings suggest that the placenta provides a supportive niche where the definitive HSC pool can be temporarily established during development. Furthermore, if the stem cell-promoting properties of the placental niche can be harnessed *in vitro* to support HSC formation, maturation, and/or expansion in culture, these assets may greatly improve HSC-based therapies in the future (Mikkola *et al.*, 2005). A part of HSCs, the presence of pluripotent stem cells, similar to ES cells, have been pointed out by us and others. Oct-4 is a marker for pluripotent human stem cells so far known to be expressed in embryonal carcinoma cells, ES cells, and embryonic germ cells. Performing RT-PCR, Western blot, and immunocytochemical analyses, it has been evident that, in human amniotic fluid, in the background of Oct-4-negative cells, a distinct population of cells can be found, which express Oct-4 in the nucleus. Oct-4-positive amniotic fluid cell samples also expressed stem cell factor, vimentin, and alkaline phosphatase mRNA. The Oct-4-positive amniotic fluid cells were actively dividing, proven by the detection of cyclin A expression. They suggested that human amniotic fluid could represent a new source for the isolation of human Oct-4-positive stem cells without raising the ethical concerns associated with human embryonic research (De Coppi *et al.*, 2001, 2002; Prusa *et al.*, 2003; Karlmark *et al.*, 2005). Established cell lines derived from human placenta by cloning technique using alpha-MEM culture medium containing 10ng/ml of EGF (epidermal growth factor), 10 ng/ml of hLIF, and 10% FBS (fetal bovine serum) appeared to maintain a normal karyotype indefinitely *in vitro* and expressed markers characteristic of stem cells from mice and human, namely alkaline phosphatase. These cells contributed to the formation of chimeric mouse embryoid bodies and gave rise to cells of all germ layers *in vitro* (Tamagawa *et al.*, 2004). Koegler *et al.* have also described a new pluripotent human somatic stem cell, CD45-negative population from human cord blood, termed unrestricted somatic stem cells (USSCs). This rare population grows adherently and can be expanded to 10^{15} cells without losing pluripotency. *In vitro* USSCs showed homogeneous differentiation into osteoblasts, chondroblasts, adipocytes, and hematopoietic and neural cells including astrocytes and neurons that express neurofilament, sodium channel protein, and various neurotransmitter phenotypes. Stereotactic implantation of USSCs into intact adult rat brain revealed that human Tau-positive cells persisted for up to 3 months and showed migratory activity and a typical neuron-like morphology. *In vivo* differentiation of USSCs along mesodermal and endodermal pathways was demonstrated in animal models. Bony reconstitution was observed after transplantation of USSC-loaded calcium phosphate cylinders in nude rat femurs. Chondrogenesis occurred after transplanting cell-loaded gelfoam sponges into nude mice. Transplantation of USSCs in a non-injury model, the pre-immune fetal sheep, resulted in up to 5% human hematopoietic engraftment. More than 20% albumin-producing human parenchymal hepatic cells with absence of cell fusion and substantial numbers of human cardiomyocytes in both atria and ventricles of the sheep heart were detected many months after USSC transplantation. No tumor formation was observed in any of these animals (Kogler *et al.*, 2004).

We have recently described a pluripotent population of cells derived from both amniotic fluid and placenta. We will describe in the following paragraphs in *detaiòs* their isolation, characterization, and differentiation *in vitro* into different lineages (De Coppi *et al.*, 2007).

Isolation and Characterization of Chorionic Villi and Amniotic-Derived Stem Cells

Chorionic villi samples and human amniotic fluid were obtained under informed consent at 12–18 weeks of pregnancy from a total of 300 women between 23 and 42 years of age. In all cases the karyotype evaluated from the cultured chorionic villi and amniotic fluid cells was normal. Samples were seeded in a 22 × 22 mm cover slip in a volume of 2 ml and grown to confluence for 3–4 weeks at 95% humidity and 37°C. Fresh medium was applied after 5 days of culture and every third day thereafter. The culture medium consisted of alpha-MEM (GIBCO/BRL, Grand Island, NY), 18% Chang medium B (Irvine Scientific, Santa Ana, CA), 2% Chang C (Irvine Scientific, Santa Ana, CA) with 15% ES cell-certified FBS (ES-FBS, GIBCO/BRL, Grand Island, NY), 1% antibiotics (GIBCO/BRL, Grand Island, NY), and L-glutamine (Sigma-Aldrich, St. Louis, MO). The cells were subcultured using 0.25% trypsin containing 1mM EDTA for 5min at 37°C.

In order to test the hypothesis that placenta and amniotic fluid could contain stem cells that would be able to differentiate into multiple lineages, cell colonies derived from single cells were expanded.

The cells were successfully isolated from 300 fetuses and maintained in culture in Chang medium. The presence of cells of maternal origin in placenta and amniotic fluid is extremely low. In order to evaluate for the presence of maternal cells, the studies were performed using cells from male fetuses. Karyotypic analyses of the $ckit^{pos}$ cells showed an *xy* phenotype in all the cells. Female fetuses were used as controls and they did not show any difference in their pluripotential ability. Cytofluorimetric analysis and immunocytochemistry showed that most of the amniotic cells were epithelial and stained positive for cytokeratins. Most of the stromal cells stained for alpha-actin, and only a few cells were positive for desmin or myosin. Fluorescence-activated cell sorter (FACS) analyses showed that between 18% and 21% of the cells expressed CD105, while approximately the same proportion of cells (between 0.8% and 3%) expressed *ckit* and CD34. The $ckit^{pos}$ cells were successfully isolated and maintained in culture in Chang medium. The $ckit^{pos}$ cells were shown to be pluripotent. They maintained a round shape when cultured in bacterial plates for almost a week while they had a very low proliferative capability. After the first week the cells started to adhere to the plate and changed their morphology, becoming more elongated, and they started to proliferate. The medium was changed every 3 days and they were passed whenever they reached confluence. If the cells were not passaged, they aggregated, forming embryoid-shaped tissue-like structures measuring 1 × 5 mm^3. Serial sections of these structures showed specific markers for the three embryonic germ layers immunohistochemically. The embryo-shaped tissue, if disaggregated, was still able to differentiate into different lineages under appropriate growth conditions. The CD105, CD90, and CD34 immunoseparated cells, and the remaining non-immunoseparated cells did not show any pluripotential ability. No feeder layers were required, either for maintenance or expansion (Takeda *et al.*, 1992; Mosquera *et al.*, 1999).

Telomerase Activity

Telomerase activity was evaluated using the telomerase repeat amplification protocol (TRAP) assay, and the presence of telomerase was analyzed immunocytochemically. No telomerase activity could be detected with the TRAP assay, either in the $ckit^{pos}$ cells (lanes1) or in the control BM stem cells (lane 2). In contrast, the prostate cancer cell line PC3 and an epithelial tumor cell line (HeLa), as a control, showed high telomerase activity (lanes 4 and 5). Anti-telomerase antibodies positively stained the amniotic $ckit^{pos}$ cells, suggesting that the cells may express telomerase protein, but the levels were too low to be detected by the TRAP assay.

Differentiation Potential Induction of Osteogenic Phenotype (Figure 17-2a)

Light microscopy analysis showed that $ckit^{pos}$ cells, within 4 days in osteogenic medium, developed an osteoblastic-like appearance with finger-like excavations into the cells' cytoplasm (Karsenty, 2000; Olmsted-Davis *et al.*, 2003). At 16 days the cells aggregated in the typical lamellar bone-like structures and increased their expression of alkaline phosphatase. Ca accumulation was evident after 1 week and increased over time. To confirm the cytochemical findings, AP activity was measured using a quantitative assay, which measured *p*-nitrophenol, equivalent to AP production. The $ckit^{pos}$ cells showed more than a two hundred time increase in AP production in the osteogenic-inducing medium compared to cells grown in control medium at days 16 and 24. After that time the levels of AP decreased. No AP production was detected in $ckit^{neg}$ amniotic cells cultured in osteogenic medium at any time point. AP expression was confirmed at the RNA level. No activation of the AP gene was detected at 8, 16, 24, and 32 days in the $ckit^{pos}$ cells grown in control medium. In contrast, $ckit^{pos}$ cells grown in osteogenic medium showed an activation of the AP gene at each time point. Expression of cbfa1, a transcription factor specifically expressed in osteoblasts and hypertrophic **chondrocytes**, was highest in cells grown in osteogenic-inducing medium **at day 8**, and decreased slightly at **days** 16, 24, and 32. The expression of cbfa1 in the controls was significantly lower at each time point. Osteocalcin was expressed only in the $ckit^{pos}$ cells in osteogenic conditions at 8 days. No expression of ostecalcin was detectable in the $ckit^{pos}$ cells in the control medium and in the $ckit^{neg}$

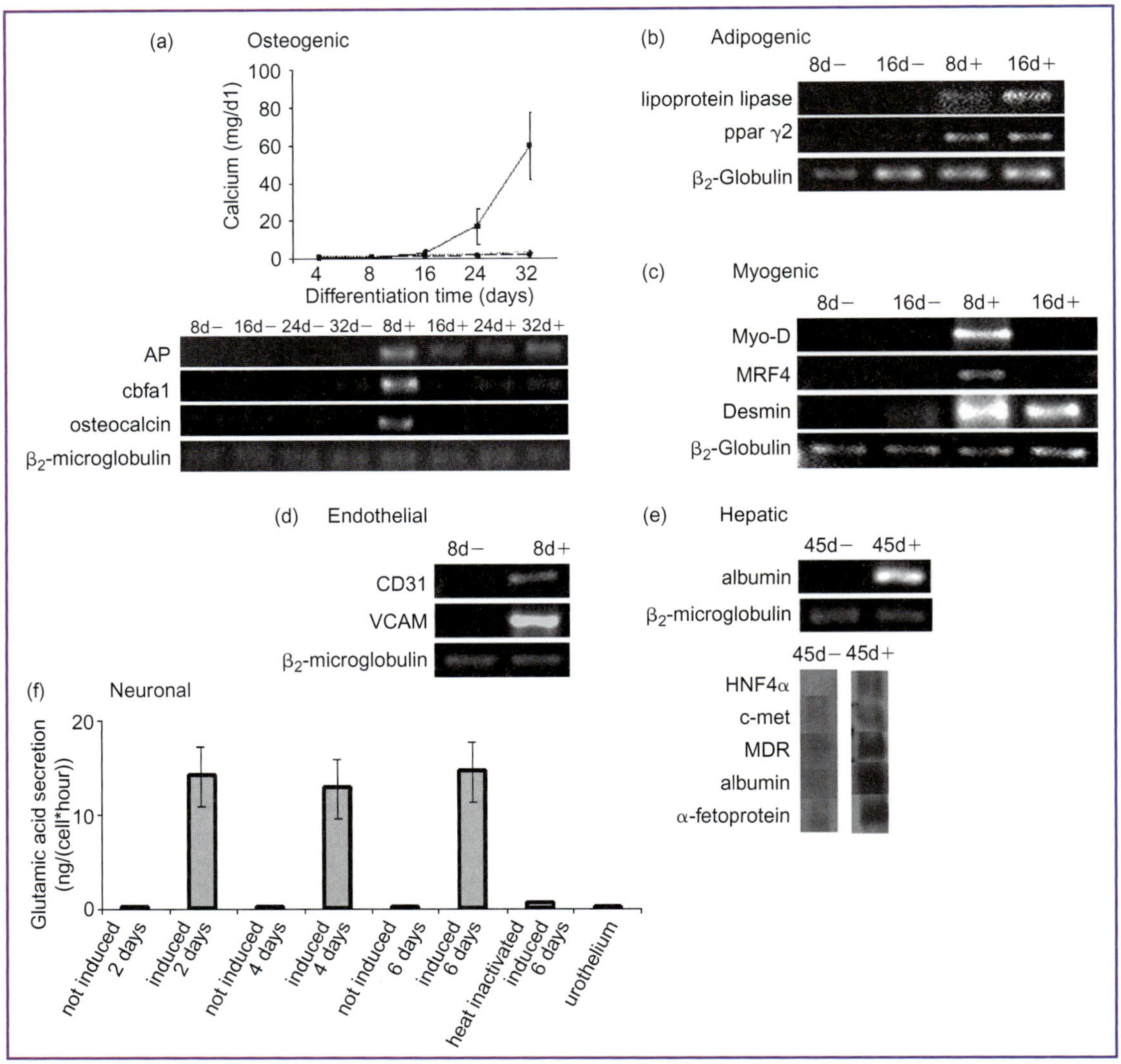

FIGURE 17-2 The differentiated cell types expressed functional and biochemical characteristics of the target tissue. (a) Osteogenic-induced progenitor cells showed a significant increase of calcium deposition starting at day 16 (solid line). No calcium deposition was detected in the progenitor cells grown in control medium or in the negative control cells grown in osteogenic conditions (dashed line). RT-PCR showed presence of *cbfa1* and *osteocalcin* at day 8 and confirmed the expression of *AP* in the osteogenic-induced cells. (b) Gene expression of *pparγ* and *lipoprotein lipase* in cells grown in adipogenic-inducing medium was noted at days 8 and 16 (lanes 3 and 4). (c) Myogenic-induced cells showed a strong expression of *desmin* expression at day 16 (lane 4). *MyoD* and *MRF4* were induced with myogenic treatment at day 8 (lane 3). Specific PCR-amplified DNA fragments of *MyoD*, *MRF4*, and desmin could not be detected in the control cells at days 8 and 16 (lanes 1 and 2). (d) RT-PCR of progenitor cells induced in endothelial medium (lane 2) showed the expression of *CD31* and *VCAM*. (e) RT-PCR revealed an upregulation of albumin gene expression. Western blot analyses of cell lysate showed the presence of the hepatic lineage-related proteins HNF-4a, c-met, MDR, albumin, and alpha-fetoprotein. Undifferentiated cells were used as negative control. (f) Only the progenitor cells cultured under neurogenic conditions showed the secretion of glutamic acid in the collected medium. The secretion of glutamic acid could be induced (20 min in 50-mM KCl buffer).

cells in the osteogenic conditions at any time point. A major feature of osteogenic differentiation is the ability of the cells to precipitate calcium. Cell-associated mineralization can be analyzed using von Kossa staining and by measuring the calcium content of cells in culture. Von Kossa staining of cells grown in the osteogenic medium showed enhanced silver nitrate precipitations by day 16, indicating high levels of calcium. Calcium precipitation continued to increase exponentially at 24 and 32 days. In contrast, cells in the control medium did not form silver nitrate precipitations after 32 days. Microscopic examination of stained cells showed no calcification in the osteogenic treated cells at day 4 or 8, but strong black silver nitrate precipitates were noticed in osteogenic-induced cells after 16, 24, and 32 days in culture. In cells cultured in control medium, no precipitates were noticed over the 32-day time period. Calcium deposition by the cells was also measured with a quantitative chemical assay, which measures calcium–cresolophthalein complexes. Cells undergoing osteogenic induction showed a significant increase in calcium precipitation after 16 days (up to 4 mg/dl). The precipitation of calcium increased up to 70 mg/dl at 32 days. In contrast, cells grown in control medium did not show any increase in calcium precipitation (1.6 mg/dl) by day 32.

Induction of Adipogenic Lineage (Figure17-2b)

Ckit^{pos} cells cultured in a medium containing dexametasone, insulin, indomethacin, and 3-isobutyl-1-methylxanthine, within 8 days, changed their morphology from elongated to round (Kim *et al.*, 1998). This coincided with the accumulation of intracellular triglyceride droplets. After 16 days in culture, more than 95% of the cells had their cytoplasm completely filled with lipid-rich vacuoles, which stained positively with Oil-O-Red. The amniotic *ckit*^{neg} cells that were induced with the same medium and the *ckit*^{pos} cells cultured in control medium did not show any phenotypic change of adipogenic differentiation and did not stain with Oil-O-Red after 16 days of culture. Adipogenic differentiation was confirmed by RT-PCR analysis. The expression of peroxisome proliferation-activated receptor 2 (ppart(2)), a transcription factor that regulates adipogenesis and of lipoprotein lipase was analyzed. Expression of these genes was upregulated in the *ckit*^{pos} cells under adipogenic conditions. *Ckit*^{pos} cells cultured under control conditions and *ckit*^{neg} cells in adipogenic medium did not express either gene at any time point.

Induction of Myogenic Phenotype (Figure 17-2c)

Ckit^{pos} cells were cultured with myogenic medium on Matrigel-coated dishes (Rosenblatt *et al.*, 1995; Ferrari *et al.*, 1998). Induction with 5-azacytidine for 24h promoted the formation of multinucleated cells over a 24–48 h period. After 16 days, the cells grown with myogenic medium formed myofiber-like structures that stained immunocytochemically with desmin and sarcomeric tropomyosin. *Ckit*^{pos} cells grown in control medium and *ckit*^{neg} cells cultured in myogenic medium did not lead to cell fusion or multinucleated cells. Only a few desmin cells were present in the *ckit*^{neg} amniotic cells cultured in myogenic medium at 16 days. Expression of *MyoD*, *Myf5*, *Myf* 6 (MRF4), and *desmin* were analyzed using RT-PCR. *MyoD* and MRF4 were expressed by the *ckit*^{pos} cells in culture at 8 days and suppressed at 16 days. Both these genes were not expressed either at 8 or 16 days in the controls. *Desmin* expression was induced at 8 days and increased by 16 days in the *ckit*^{pos} cells cultured in myogenic medium. In contrast, there was no activation of *desmin* in the control cells at 8 and 16 days. *Myf5* was present at 8 days and increased at 16 days in the *ckit*^{pos} cells. Lower levels of the *Myf5* gene were detected in the cells maintained in culture with the control medium at 16 days.

Induction of Endothelial Phenotype (Figure 17-2d)

Ckit^{pos} cells were cultured with endothelial medium in PBS–gelatin-coated dishes. After 1 week in culture the cells started to change their morphology, and by the second week, were mostly tubular. The cells stained positively for FVIII, KDR, and P1H12. *Ckit*^{neg} cells cultured in the same conditions and *ckit*^{pos} cells cultured in Chang medium for the same period were not able to form tubular structures and did not stain for endothelial specific markers. The cells, once differentiated, were able to grow in culture for more than 1 month.

Induction of Hepatocytes Phenotype (Figure 17-2e)

When cultured in hepatic conditions cells exhibited morphological changes after 7 days showing a change in the morphology from an elongated to a cobblestone appearance (Dunn *et al.*, 1989; Hamazaki *et al.*, 2001). The cells showed positive staining for albumin at day 45 post-differentiation, and were also found to express transcription factor HNF4α, c-met receptor, multidrug resistance gene (MDR) membrane transporter, albumin, and alpha-fetoprotein. RT-PCR analysis further provided evidence of albumin production. The maximum rate of urea production for hepatic differentiation induced cells was 1.21×10^3 ng urea/h/cell as opposed to 5.0×10^1 ng urea/h/cell for control progenitor cell populations.

Induction of Neurogenic Phenotype (Figure 17-2f)

Ckit^{pos} cells cultured in neurogenic conditions changed their morphology within the first 24h (Black and Woodbury *et al.*, 2001; Barberi *et al.*, 2003). Responsive cells progressively assumed neuronal morphological **characteristics**; initially the cytoplasm retracted toward the nucleus, forming contracted multipolar **structures**. Over the subsequent hours, **the cells displayed** primary and secondary branches, **and** cone-like terminal expansions. Induced *ckit*^{pos} cells stained positively for beta-III tubulin and nestin. *Ckit*^{neg} cells cultured in the same conditions and *ckit*^{pos} cells cultured in Chang medium for the same period were not able to form tubular structures and did not stain for endothelial specific markers. The cells, once differentiated, were able to grow in culture for more than 1 month.

Clonal and Proliferative Analyses

Ckit^{pos} cells were able to be expanded clonally. After serial dilution we observed that most of the wells **contained no cells**, and only a few of the 96 wells contained a single cell. Cells from numerous clones showed a similar morphology and growth behavior. Clonal lineages from different patients were tested. All the cells underwent osteogenic, adipogenic, myogenic, neurogenic, and endothelial differentiation. Amniotic stem cells did not show any decrease in their growth ability after more than 100 cell **divisions, and they maintained** their ability to differentiate into different lineages.

FUTURE DIRECTION

Fetal tissue has been used in the past for transplantation and tissue engineering research because of its pluripotency and proliferative ability. Fetal cells maintain a higher capacity to proliferate than adult cells and may preserve their pluripotency longer in culture. However, fetal cell transplants are plagued by problems that are very difficult to overcome. Beyond the ethical concerns regarding the use of cells from aborted fetuses or living fetuses, there are other issues which remain a challenge. Previous studies have shown that it takes almost six fetuses to provide enough material to treat one patient with Parkinson's disease.

In this study we hypothesized that placental and amniotic cells, which have been used for decades for prenatal diagnosis, could represent a viable source of fetal stem cells that could be used therapeutically.

SUMMARY

It is well known that placenta and amniotic fluid contain a large variety of cells. Our aim was to try to identify and isolate cells that still maintained their pluripotential and proliferative abilities. The vast majority of the cells in the placenta and in the amniotic fluid are already differentiated, and, therefore, have a limited proliferative ability. In this study the *ckit*^{pos} cells were induced to different lineages. The ability to induce specific differentiation was initially evident by morphological changes, and was confirmed by immunocytochemical and gene expression analyses.

In conclusion, placenta and amniotic fluid could be an excellent cell source for therapeutic applications. Fetal stem cells have a better potential for expansion than adult stem cells and for this reason they could represent a better source for any **therapeutic** application where large numbers of cells are needed. When compared with ES **cells, fetal** stem cells are easily differentiated **into** specific cell lineages, do not need any feeder layer to grow, **and avoid** the current controversies associated with the use of human ES cells.

REFERENCES

Alvarez-Silva, M., Belo-Diabangouaya, P., Salaun, J. and Dieterlen-Lievre, F. (2003). Mouse placenta is a major hematopoietic organ. *Development 130*: 5437–5444.

Bailo, M., Soncini, M., Vertua, E., Signoroni, P.B., Sanzone, S., Lombardi, G., Arienti, D., Calamani, F., Zatti, D., Paul, P., Albertini, A., Zorzi, F., Cavagnini, A., Candotti, F., Wengler, G.S. and Parolini, O. (2004). Engraftment potential of human amnion and chorion cells derived from term placenta. *Transplantation 78*: 1439–1448.

Barberi, T., *et al.* (2003). Neural subtype specification of fertilization and nuclear transfer embryonic stem cells and application in parkinsonian mice. *Nat. Biotechnol. 21*: 1200–1207.

Black, I.B. and Woodbury, D. (2001). Adult rat and human bone marrow stromal stem cells differentiate into neurons. *Blood Cell Mol. Dis. 27*: 632–636.

Brace, R.A. and Resnik, R. (1999). Dynamics and disorders of amniotic fluid. In: Creasy, R.K. and Renik, R. (eds.), *Maternal Fetal Medicine*. Philadelphia: Saunders, pp. 623–643.

Cousineau, J., Potier, M., Dallaire, L. and Melancon, S.B. (1982). Separation of amniotic fluid cell types in primary culture by Percoll density gradient centrifugation. *Prenat. Diagn. 2*: 241–249.

Crane, J.P. and Cheung, S.W. (1988). An embryogenic model to explain cytogenetic inconsistencies observed in chorionic villus versus fetal tissue. *Prenat. Diagn. 8*: 119–129.

De Coppi, P., *et al.* (2001). Human fetal stem cell isolation from amniotic fluid. *American Academy of Pediatrics National Conference*, San Francisco, pp. 210–211.

De Coppi, P., Filippo, R.D., Soker, S., *et al.* (2002). Human embryonic and fetal stem-cell isolation from amniotic fluid and placenta for tissue reconstruction. *J. Am. Coll. Surg. 195*: S93 (abstract).

De Coppi, P., Bartsch Jr., G., Siddiqui, M.M., Xu, T., Santos, C.C., Perin, L., Mostoslavsky, G., Serre, A.C., Snyder, E.Y., Yoo, J.J., Furth, M., Soker, S. and Atala, A. (2007). Isolation of clonal amniotic stem cell lines with potential for therapy. *Nat. Biotechnol. 25*: 100–106.

Dunn, J.C., Yarmush, M.L., Koebe, H.G. and Tompkins, R.G. (1989). Hepatocyte function and extracellular matrix geometry: long-term culture in a sandwich configuration. *FASEB J. 3*: 174–177.

Elwan, M.A. and Sakuragawa, N. (1997). Evidence for synthesis and release of catecholamines by human amniotic epithelial cells. *Neuroreport 8*: 3435–3438.

Ferrari, G., *et al.* (1998). Muscle regeneration by bone marrow-derived myogenic progenitors. *Science 279*: 1528–1530.

Fuchs, J.R., Kaviani, A., Oh, J.T., LaVan, D., Udagawa, T., Jennings, R.W., Wilson, J.M. and Fauza, D.O. (2004). Diaphragmatic reconstruction with autologous tendon engineered from mesenchymal amniocytes. *J. Pediatr. Surg. 39*: 834–838; discussion 834–838.

Fukuchi, Y., Nakajima, H., Sugiyama, D., Hirose, I., Kitamura, T. and Tsuji, K. (2004). Human placenta-derived cells have mesenchymal stem/progenitor cell potential. *Stem Cell 22*: 649–658.

Gekas, C., Dieterlen-Lievre, F., Orkin, S.H. and Mikkola, H.K. (2005). The placenta is a niche for hematopoietic stem cells. *Dev. Cell 8*: 365–375.

Gosden, C.M. (1983). Amniotic fluid cell types and culture. *Br. Med. Bull. 39*: 348–354.

Haigh, T., Chen, C., Jones, C.J. and Aplin, J.D. (1999). Studies of mesenchymal cells from 1st trimester human placenta: expression of cytokeratin outside the trophoblast lineage. *Placenta 20*: 615–625.

Hamazaki, T., *et al.* (2001). Hepatic maturation in differentiating embryonic stem cells *in vitro*. *FEBS Lett. 497*: 15–19.

Hoehn, H. and Salk, D. (1982). Morphological and biochemical heterogeneity of amniotic fluid cells in culture. *Meth. Cell Biol. 26*: 11–34.

Igura, K., Zhang, X., Takahashi, K., Mitsuru, A., Yamaguchi, S. and Takashi, T.A. (2004). Isolation and characterization of mesenchymal progenitor cells from chorionic villi of human placenta. *Cytotherapy 6*: 543–553.

In't Anker, P.S., Scherjon, S.A., Kleijburg-van der Keur, C., Noort, W.A., Claas, F.H., Willemze, R., Fibbe, W.E. and Kanhai, H.H. (2003). Amniotic fluid as a novel source of mesenchymal stem cells for therapeutic transplantation. *Blood 102*: 1548–1549.

In't Anker, P.S., Scherjon, S.A., Kleijburg-van der Keur, C., de Groot-Swings, G.M., Claas, F.H., Fibbe, W.E. and Kanhai, H.H. (2004). Isolation of mesenchymal stem cells of fetal or maternal origin from human placenta. *Stem Cell 22*: 1338–1345.

Karlmark, K.R., Freilinger, A., Marton, E., Rosner, M., Lubec, G. and Hengstschlager, M. (2005). Activation of ectopic Oct-4 and Rex-1 promoters in human amniotic fluid cells. *Int. J. Mol. Med. 16*: 987–992.

Karsenty, G. (2000). Role of Cbfa1 in osteoblast differentiation and function. *Semin. Cell Dev. Biol. 11*: 343–346.

Kaviani, A., Perry, T.E., Dzakovic, A., *et al.* (2001). The amniotic fluid as a source of cells for fetal tissue engineering. *J. Pediatr. Surg. 36*: 1662–1665.

Kaviani, A., Perry, T.E., Barnes, C.M., *et al.* (2002). The placenta as a cell source in fetal tissue engineering. *J. Pediatr. Surg. 37*: 995–999.

Kaviani, A., Guleserian, K., Perry, T.E., *et al.* (2003). Tissue engineering from amniotic fluid. *J. Am. Coll. Surg. 196*: 592–597.

Kim, J.B., Wright, H.M., Wright, M. and Spiegelman, B.M. (1998). ADD1/SREBP1 activates PPARgamma through the production of endogenous ligand. *Proc. Natl Acad. Sci. USA 95*: 4333–4337.

Kogler, G., Sensken, S., Airey, J.A., Trapp, T., Muschen, M., Feldhahn, N., Liedtke, S., Sorg, R.V., Fischer, J., Rosenbaum, C., Greschat, S., Knipper, A., Bender, J., Degistirici, O., Gao, J., Caplan, A.I., Colletti, E.J., Almeida-Porada, G., Muller, H.W., Zanjani, E. and Wernet, P. (2004). A new human somatic stem cell from placental cord blood with intrinsic pluripotent differentiation potential. *J. Exp. Med. 200*: 123–135.

Li, C.D., Zhang, W.Y., Li, H.L., Jiang, X.X., Zhang, Y., Tang, P. and Mao, N. (2005a). Isolation and identification of a multilineage potential mesenchymal cell from human placenta. *Placenta*.

Li, C.D., Zhang, W.Y., Li, H.L., Jiang, X.X., Zhang, Y., Tang, P.H. and Mao, N. (2005b). Mesenchymal stem cells derived from human placenta suppress allogeneic umbilical cord blood lymphocyte proliferation. *Cell Res. 15*: 539–547.

Lotgering, F.K. and Wallenburg, H.C. (1986). Mechanisms of production and clearance of amniotic fluid. *Semin. Perinatol. 10*: 94–102.

Medina-Gomez, P. and Johnston, T.H. (1982). Cell morphology in long-term cultures of normal and abnormal amniotic fluids. *Hum. Genet. 60*: 310–313.

Miki, T., Lehmann, T., Cai, H., Stolz, D.B. and Strom, S.C. (2005). Stem cell characteristics of amniotic epithelial cells. *Stem Cell 23*: 1549–1559.

Mikkola, H.K., Gekas, C., Orkin, S.H. and Dieterlen-Lievre, F. (2005). Placenta as a site for hematopoietic stem cell development. *Exp. Hematol. 33*: 1048–1054.

Milunsky, A. (1979). Amniotic fluid cell culture. In: Milunsky, A. (ed.), *Genetic Disorder of the Fetus*. New York: Plenum Press, pp. 75–84.

Mosquera, A., Fernandez, J.L., Campos, A., *et al.* (1999). Simultaneous decrease of telomerase length and telomerase activity with ageing of human amniotic fluid cells. *J. Med. Genet. 36*: 494–496.

Okawa, H., Okuda, O., Arai, H., Sakuragawa, N. and Sato, K. (2001). Amniotic epithelial cells transform into neuron-like cells in the ischemic brain. *Neuroreport 12*: 4003–4007.

Olmsted-Davis, E.A., *et al.* (2003). Primitive adult hematopoietic stem cells can function as osteoblast precursors. *Proc. Natl Acad. Sci. USA 100*: 15877–15882.

Ottersbach, K. and Dzierzak, E. (2005). The murine placenta contains hematopoietic stem cells within the vascular labyrinth region. *Dev. Cell 8*: 377–387.

Prusa, A.R., Marton, E., Rosner, M., Bernaschek, G. and Hengstschlager, M. (2003). Oct-4-expressing cells in human amniotic fluid: a new source for stem cell research? *Hum. Reprod. 18*: 1489–1493.

Rosenblatt, J.D., Lunt, A.I., Parry, D.J. and Partridge, T.A. (1995). Culturing satellite cells from living single muscle fiber explants. *In Vitro Cell Dev. Biol. Anim. 31*: 773–779.

Sakuragawa, N., Thangavel, R., Mizuguchi, M., *et al.* (1996). Expression of markers for both neuronal and glial cells in human amniotic epithelial cells. *Neurosci. Lett. 209*: 9–12, 23.

Sakuragawa, N., Enosawa, S., Ishii, T., *et al.* (2000). Human amniotic epithelial cells are promising transgene carriers for allogeneic cell transplantation into liver. *J. Hum. Genet. 45*: 171–176.

Sankar, V. and Muthusamy, R. (2003). Role of human amniotic epithelial cell transplantation in spinal cord injury repair research. *Neuroscience 118*: 11–17.

Sarkar, S., Chang, H.C., Porreco, R.P. and Jones, O.W. (1980). Neural origin of cells in amniotic fluid. *Am. J. Obstet. Gynecol. 136*: 67–72.

Sartore, S., Lenzi, M., Angelini, A., Chiavegato, A., Gasparotto, L., De Coppi, P., Bianco, R. and Gerosa, G. (2005). Amniotic mesenchymal cells autotransplanted in a porcine model of cardiac ischemia do not differentiate to cardiogenic phenotypes. *Eur. J. Cardiothorac. Surg. 28*: 677–684.

Takahashi, N., Enosawa, S., Mitani, T., *et al.* (2002). Transplantation of amniotic epithelial cells into fetal rat liver by *in utero* manipulation. *Cell Transplant. 11*: 443–449.

Takeda, J., Seino, S. and Bell, G.I. (1992). Human Oct-3 gene family: cDNA sequences, alternative splicing, gene organization, chromosomal location, and expression at low levels in adult tissues. *Nucl. Acid Res. 20*: 4613–4620.

Tamagawa, T., Ishiwata, I. and Saito, S. (2004). Establishment and characterization of a pluripotent stem cell line derived from human amniotic membranes and initiation of germ layers *in vitro*. *Hum. Cell 17*: 125–130.

Tsai, M.S., Lee, J.L., Chang, Y.J. and Hwang, S.M. (2004). Isolation of human multipotent mesenchymal stem cells from second-trimester amniotic fluid using a novel two-stage culture protocol. *Hum. Reprod. 19*: 1450–1456.

Tsai, M.S., Hwang, S.M., Tsai, Y.L., Cheng, F.C., Lee, J.L. and Chang, Y.J. (2005). Clonal amniotic fluid-derived stem cells express characteristics of both mesenchymal and neural stem cells. *Biol. Reprod.*

Underwood, M.A., Gilbert, W.M. and Sherman, M.P. (2005). Amniotic fluid: not just fetal urine anymore. *J. Perinatol. 25*: 341–348.

von Koskull, H., Aula, P., Trejdosiewicz, L.K. and Virtanen, I. (1984). Identification of cells from fetal bladder epithelium in human amniotic fluid. *Hum. Genet. 65*: 262–267.

Wulf, G.G., Viereck, V., Hemmerlein, B., Haase, D., Vehmeyer, K., Pukrop, T., Glass, B., Emons, G. and Trumper, L. (2004). Mesengenic progenitor cells derived from human placenta. *Tissue Eng. 10*: 1136–1147.

Yen, B.L., Huang, H.I., Chien, C.C., Jui, H.Y., Ko, B.S., Yao, M., Shun, C.T., Yen, M.L., Lee, M.C. and Chen, Y.C. (2005). Isolation of multipotent cells from human term placenta. *Stem Cell 23*: 3–9.

Zhang, X., Mitsuru, A., Igura, K., Takahashi, K., Ichinose, S., Yamaguchi, S. and Takahashi, T.A. (2005). Mesenchymal progenitor cells derived from chorionic villi of human placenta for cartilage tissue engineering. *Biochem. Biophys. Res. Commun. 340*: 944–952.

Zhao, P., Ise, H., Hongo, M., Ota, M., Konishi, I., Nikaido, T. (2005). Human amniotic mesenchymal cells have some characteristics of cardiomyocytes. *Transplantation 79*: 528–535.

Chapter 18

Nuclear Transfer for Stem Cells

Alan Trounson

INTRODUCTION

The demonstration that pluripotent embryonic stem (ES) cells can be produced from the culture of inner cell mass (ICM) cells of the human blastocyst (Reubinoff *et al.*, 2000; Thomson *et al.*, 1998) has evoked interest in the potential of ES cells for transplantation for cell therapies in human medicine (Trounson, 2001; Trounson and Pera, 2001). Only a small number of human ES cell lines are available at the present time, and although they do not express the major histocompatibility complex (MHC), this antigen appears during ES cell differentiation and would ultimately result in rejection of transplants derived from incompatible ES cell derivatives.

There are a number of strategies for preventing rejection of cell and tissue transplants derived from ES cells. It is possible to generate a very large panel of ES cells with a wide range of human leukocyte antigen (HLA) subtypes. The production of 200,000 or more different ES cell lines of HLA types suitable for tissue matching with patients may be possible by utilizing the large numbers of human embryos disposed of by infertility clinics worldwide. If there were a range of ES cells with HLA types, immune therapies may enable establishment of new stem cell populations, for recovery of function of the tissues of interest. It is also possible that hematopoietic derivatives of ES cells could be used to induce tolerance in patients to allow transplantation of the cell and tissue types that would correct pathologies, diseases, or injuries.

Given the beneficial clinical potential for stem cell therapies, there is a need to consider the option of developing a bank of ES cells with an appropriate range of HLA subtypes, or of customizing ES cells for transplantation compatibility by the technique of nuclear transfer for stem cells (NTSC), using somatic cells derived from the patients. This has been termed "therapeutic cloning," but this term does not adequately describe the manipulations or intentions involved, and there is a strong preference to use the term NTSC.

NUCLEAR TRANSFER USING SOMATIC CELLS TO PRODUCE ES CELLS

Proof of Concept Studies

Embryonic stem cell lines have been established from mouse blastocysts (Table 18-1) generated by nuclear transfer of mouse somatic cell nuclei. The somatic cell types were ovarian cumulus cells (Munsie *et al.*, 2000; Wakayama *et al.*, 2001), tail-tip fibroblasts (Wakayama *et al.*, 2001), and fetal neuronal cells (Kawase *et al.*, 2000).

The ES cells produced in these experiments were shown to differentiate in culture and produced teratomas when injected under the testis capsule of severe combined immune-deficient (SCID) mice. The three primary embryonic lineages were all represented in these teratomas (Kawase *et al.*, 2000; Munsie *et al.*, 2000) and normal chimeric progeny were obtained when the ES cells were injected into host blastocysts (Munsie, 2000; Wakayama *et al.*, 2001). The ES cells contributed to all tissues of the chimeric mice (Figure18-1), including germ line transmission.

The primary difficulty for nuclear transfer in mice is the efficiency of insertion of somatic cell nuclei into oocytes (Munsie, 2000). It is difficult to electrofuse cells that are very different sizes, and despite attempts to vary

TABLE 18-1 Efficiency of NTSC Blastocysts Producing ES Cell Lines

No. of nuclear transfer blastocysts derived from somatic cells	Number of ES cell lines produced	Reference
10	1 (10%)	Munsie *et al.*, 2000
101	5 (5%)	Kawase *et al.*, 2000
398	35 (9%)	Wakayama *et al.*, 2001

(A)
(B)
(C)
(D)
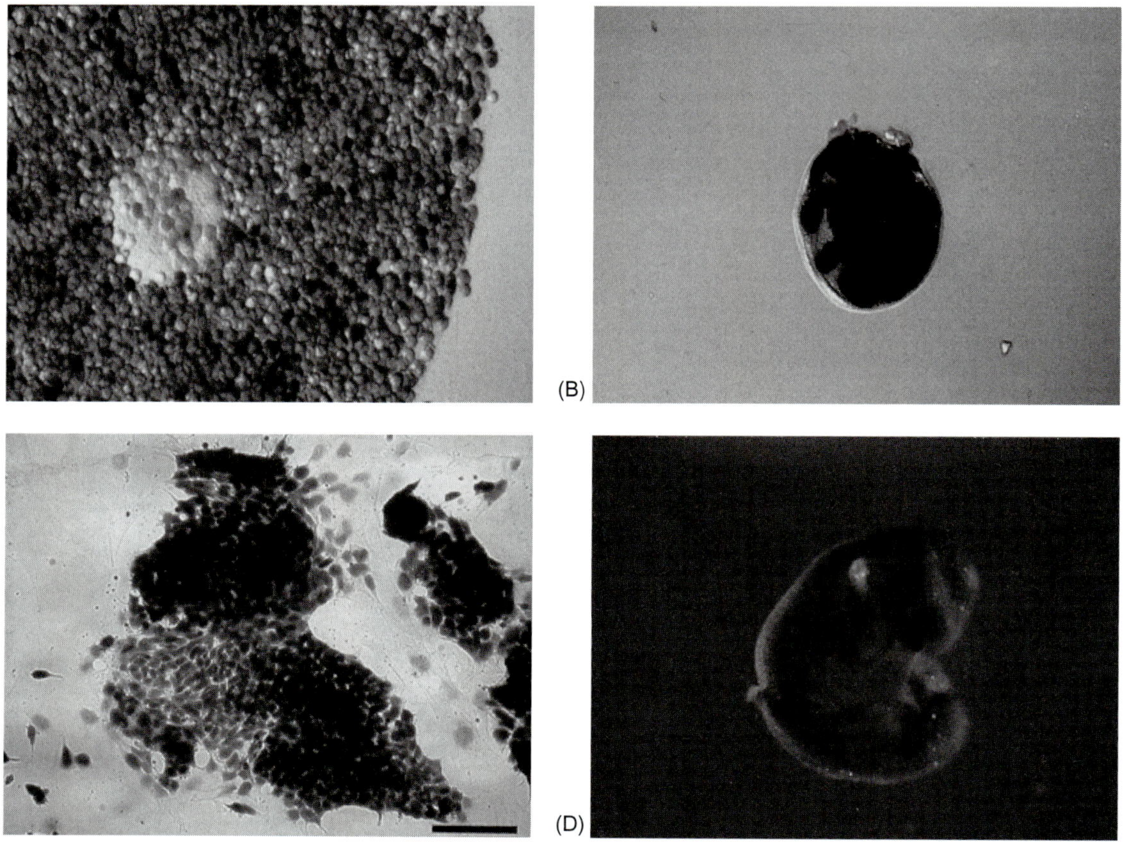

FIGURE 18-1 LacZ expression in ZIN40 transgenic nuclei of (A) cumulus cells recovered from ZIN40 females–somatic cells used for NTSC, (B) a nuclear transfer morula, (C) ES cells produced by NTSC, and (D) chimeric fetus composed of NTSC and wild-type blastocyst cells (Munsie, 2000).

the techniques of electrofusion, reduced size differences, and other parameters, nuclear injection has remained the method of choice (Trounson *et al.*, 1998). The use of piezo electronic impact micromanipulation (Prime Tech Ltd or Burleigh Instruments) enables very improved efficiencies of nuclear transfer to produce viable embryos (Wakayama *et al.*, 1998). This method was used to produce nuclear transfer embryos for ES cell production in all the studies reported to date.

There were no obvious differences between strains of mice used as donor cells nor between cumulus cells or fibroblasts, in the development to blastocysts, establishment of ES cell lines, and number of chimeric offspring produced by combining NTSC cells with wild-type blastocysts (Wakayama *et al.*, 2001). This would suggest that the donor cell origin or tissue type has no effect on the ability to generate ES cells, and this would be an attractive strategy to explore for generation of ES cells for human cell therapies (Figure 18-2). The concern that remains for NTSC in the human is the potential for tissue developmental abnormality that exists for nuclear transfer embryos. These embryos develop to blastocysts *in vitro* at rates equivalent to fertilized embryos, but have a dramatically reduced implantation rate, increased fetal growth, and increased structural abnormalities and losses, birth defects, and postnatal losses (Figure 18-3) (Lewis *et al.*, 2001). Nuclear transfer embryos have been shown to have aberrant epigenetic methylation of CpG-rich islands in their genome, which may result in abnormal gene expression (Kang *et al.*, 2001; Dean *et al.*, 2001). This may explain the aberrant expression of key genes involved in implantation in bovines and the high rate of implantation failure of cloned embryos (Daniels *et al.*, 2000, 2001).

Normality of Nuclear Transfer Embryonic Stem Cells

Embryonic stem cells are frequently used for determining gene function, chimeric mice are used to produce transgenic animals with genetic alteration to the germ line. It is of interest that mouse ES cells produced from fertilized oocytes are extremely unstable in their epigenetic control of some imprinted genes (e.g., *Peg1* and *H19*), with variations in subclones for methylation and expression of these genes (Humpherys *et al.*, 2001). However, this does not impair development of chimeric embryos, fetuses, or offspring created with ES cells and wild-type blastocysts.

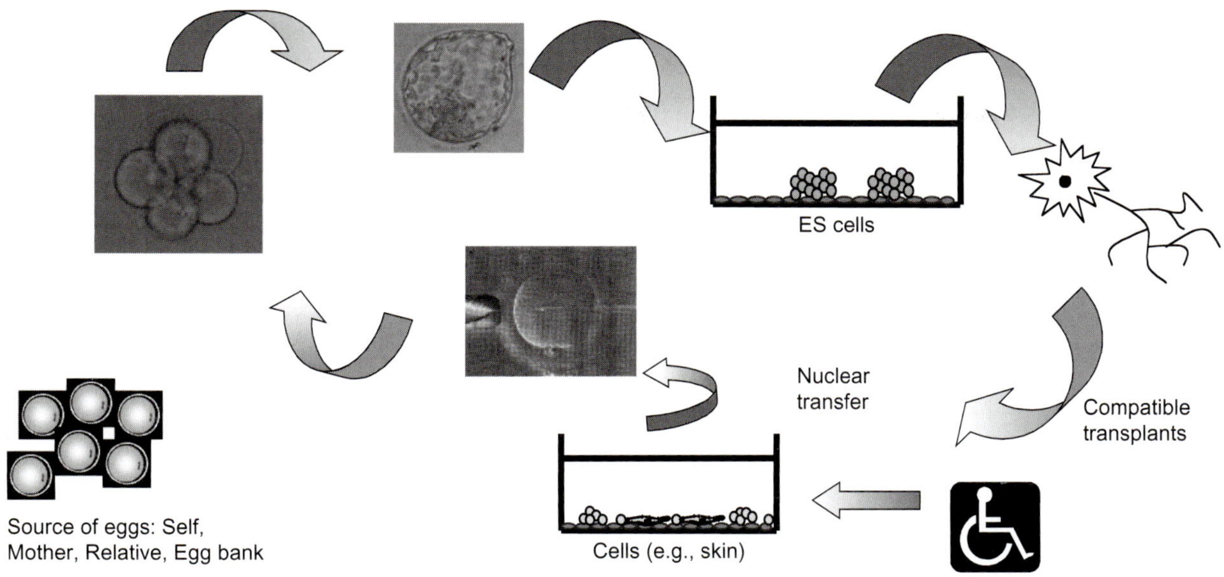

FIGURE 18-2 Nuclear transfer for stem cells. Cells provided by a patient are introduced into enucleated oocytes, and the inner cell mass cells of the nuclear transfer blastocyst produced are used to derive transplant-compatible ES cells. Directed differentiation of the ES cells into the desired lineage (e.g., neuronal cells) (Reubinoff et al., 2000) may be used to correct the existing pathology.

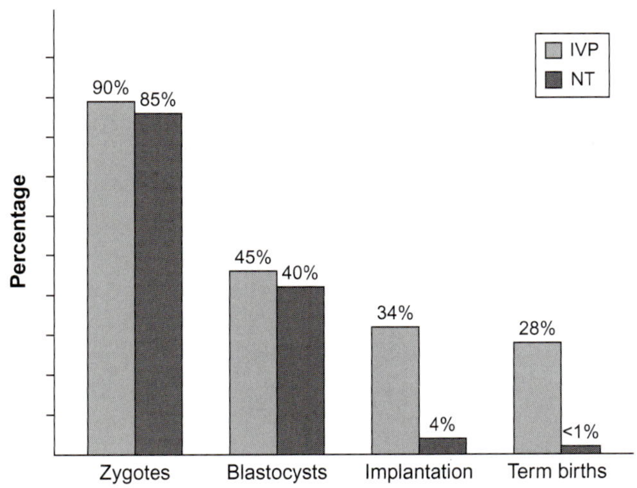

FIGURE 18-3 Survival of *in vitro*-produced (IVP) and nuclear transfer (NT) bovine embryos (data from the Monash IVP and NT programs).

The epigenetic variants among ES cells are transmitted to offspring when these cells are used for nuclear transfer (Humpherys *et al.*, 2001), which explains some of the developmental abnormalities observed in cloned animals.

Given the epigenetic instability in ES cells and the abnormalities in development of cloned embryos, it is of interest that no abnormalities have been reported in nuclear transfer embryonic stem (NTES) cell lines and chimeric fetuses or offspring. The same cells used for cloning showed that only 50% of the live-born young survived to adulthood (Wakayama *et al.*, 2001), which is the usual postnatal loss observed for cloned embryos. This was not substantially altered by creating nuclear transfer embryos with tetraploid cells (Humpherys *et al.*, 2001). This combination of cell types allows developmentally competent transgenic ES cell lines to form the fetal tissues and relegate the tetraploid components to extraembryonic membranes (Nagy *et al.*, 1993).

A detailed analysis of embryo production and normality of development of cloning of mice from inbred and F_1 hybrid ES cell lines was reported by Eggan *et al.* (2001). It was apparent that the rate of blastocyst formation was the same for inbred and F_1 ES cell donor nuclei, but more pups were born from F_1 ES cell nuclei (17%) than from inbred nuclei (8%). None of the inbred pups survived for long after birth, whereas 78% of the F_1 pups survived to adulthood. These data strongly argue for a role for increased heterozygosity for survival of nuclear transfer embryos. There was no benefit for constructing embryos of the inbred strain ES cells with tetraploid cells. The majority (14/15) of pups were unable to sustain respiration after birth. Hence the losses and phenotypes were similar to cloned inbred mouse strains and tetraploid inbred ES cell composites, confirming the developmental incompetence of these cloned ES cell types. The situation for F_1 hybrid ES cell lines is different, and although embryonic losses are high, those that develop to term have a high expectation of development to adulthood. It is again interesting to compare these results with those of creating ES cells with fertilized wild-type blastocysts. Their rates of development to term and survival to adulthood are comparable to those of fertilized embryos.

It is possible that normal developing tissue is capable of reprogramming epigenetically unstable ES cells and even NTES cells that have aberrant epigenetic signatures remaining from their somatic origins. This may be the same mechanism that enables transdifferentiation of some adult stem cells after experimental transplantation (Blau *et al.*, 2001). Further experiments are needed to explore this hypothesis.

Oocyte Availability for NTSC

Presently nuclear transfer requires a mature oocyte for reprogramming a somatic cell nucleus. It is possible that half or less of the oocyte's cytoplasm is necessary for somatic cell reprogramming (Trounson *et al.*, 1998), and it is apparent that the proteins involved in erasure of epigenetic regulatory factors are conserved across species, at least for reprogramming early development (Dominko *et al.*, 1999). It has been suggested that xenotransfer of human somatic cells to enucleated animal oocytes could be utilized for NTSC, but this would be strongly opposed. There would also be concerns about potential developmental problems that occur because of nuclear–cytoplasmic (including mitochondrial) incompatibility, and about epigenetic aberrations that could occur because of this incompatibility.

The availability of human oocytes is a major problem for NTSC because it is likely that only 1 of 10 oocyte fusions will develop to a blastocyst and only 1 of 10 blastocysts will develop to an ES cell line (Table 18-1). As a result, 100 oocytes or more are needed to establish each ES cell line. Even if substantially improved, it is likely that 20–50 oocytes will be needed in the future for each NTSC line. Large numbers of human oocytes will need to be harvested for NTSC, and it is unlikely that such numbers will readily be available for "customizing" ES cell lines for individual patients.

Studies of the reprogramming factors have been initiated. In the frog, demethylation occurs as a consequence of chromatin remodeling due to maternal activity of the nucleosome-dependent ATPase ISWI protein, in cloned somatic cells (Kikyo *et al.*, 2000). Chromatin remodeling occurs after nuclear transfer and cell hybridization, and alterations to the basal transcriptional complex may facilitate loose chromatin and loss of somatic epigenetic memory (Tada *et al.*, 2001).

If the cytoplasmic nuclear reprogramming factors can be identified, these could be synthesized and used to erase epigenetic determinants that characterize somatic cells. Although factors have not as yet been identified in mammals, it is likely that they are conserved across species. These factors may need to be introduced into somatic cells to be effective in producing a pluripotential cell line with desired genomic compatibility for transplantation. The benefit of using reprogramming factors to reconstitute a pluripotential cell line for a patient, rather than the complete oocyte, is the additional homology of mitochondrial DNA (mtDNA). There are concerns that heterogeneous mtDNA produced by nuclear transfer of somatic cells into donor oocytes could produce unpredictable outcomes for mitochondrial function, because of nuclear protein interactions with mitochondria.

STEM CELL HYBRIDIZATION

It has been shown that mouse embryonic germ (EG) cells and mouse ES cells can be used to reprogram somatic cells by cell hybridization (Tada *et al.*, 1997, 2001).

The hybrid cells are able to contribute to the formation of endoderm, mesoderm, and ectoderm when injected into wild-type blastocysts. This pluripotent plasticity is also shared with neuronal stem cells (Clarke *et al.*, 2000). EG cells are able to reprogram somatic methylation patterns, but ES cells are unable to reprogram parental imprints, when used to hybridize with somatic cells. Given the apparent dominance of EG and ES cells in their ability to reprogram the epigenetic status of somatic nuclei, this could be used to induce pluripotency in adult somatic cells of patients, without the need to resort to the use of oocytes for NTSC.

It is known that the quantitative expression of the POU transcription factor Oct4 determines pluripotency in the mouse (Niwa *et al.*, 2000). In mouse ES cells, a two-fold increase in expression causes differentiation into endoderm and mesoderm, whereas repression induces a loss of pluripotency. In the majority of ES cell–somatic cell hybrids, Oct4, coupled with green fluorescent protein (Oct4-GFP), was expressed (Tada *et al.*, 2001) and equates with the pluripotency observed for the hybrid cells in chimeric embryos. The primary methylation imprints for *H19* and *Igf2r* genes were maintained. In this study the full epigenetic profile of developmental genes was not determined, so it is not possible to predict the developmental competence of the stem cell hybrids. It will be necessary to remove the ES cell nucleus to determine developmental normality of ES cell–somatic cell hybrids in due course.

CONCLUSIONS

NTSC, potentially a very important strategy for the production of pluripotent stem cells, can be directed into a wide range of tissue types (Trounson, 2001). The proof-of-concept studies for NTSC have been done in mice, and it is of interest to note that the ES cell lines produced contribute normally to all the tissues of chimeric offspring, including the germ line. The developmental abnormalities observed in cloned embryos, fetuses, and offspring are not observed in the chimeric offspring or in their NTSC genomic progeny. This raises the hypothesis that the normal tissues of wild-type embryos may reprogram epigenetic instability or epigenetic remnants of the original somatic cells. This may be analogous to the epigenetic reprogramming observed in transdifferentiation.

The primary problem for NTSC is the need for large numbers of mature oocytes to serve as recipients of somatic cell nuclei. This may be solved by epigenetic engineering using reprogramming factors isolated from oocytes or synthesized by molecular techniques. It is also likely that hybridization of somatic cells with stem cell cytoplasm will enable pluripotent stem cells to be derived for individual patients in the future. These techniques may be the basis for cell therapies in the future for a wide range of degenerative disorders, serious pathologies, and tissue damage.

REFERENCES

Blau, H. M., Brazelton, T. R., and Weimann, J. M. (2001). The evolving concept of a stem cell: Entity or function? *Cell 105*(7), 829–841.

Clarke, D. L., Johansson, C. B., Wilbertz, J., Veress, B., Nilsson, E., Karlstrom, H., Lendahl, U., and Frisen, J. (2000). Generalized potential of adult neural stem cells. *Science 288*(5471), 1660–1663.

Daniels, R., Hall, V., and Trounson, A. O. (2000). Analysis of gene transcription in bovine nuclear transfer embryos reconstructed with granulosa cell nuclei. *Biol. Reprod. 63*(4), 1034–1040.

Daniels, R., Hall, V. J., French, A. J., Korfiatis, N. A., and Trounson, A. O. (2001). Comparison of gene transcription in cloned bovine embryos produced by different nuclear transfer techniques. *Mol. Reprod. Dev. 60*(3), 281–288.

Dean, W., Santos, F., Stojlcovic, M., Zakhartchenko, V., Walter, J., and Reik, W. (2001). Conservation of methylation reprogramming in mammalian development: Aberrant reprogramming in cloned embryos. *Proc. Natl. Acad. Sci. U.S.A. 98*(24), 13734–13738.

Dominko, T., Mitalipova, M., Haley, B., Beyhan, Z., Memili, E., McKusick, B., and First, N. L. (1999). Bovine oocyte cytoplasm supports development of embryos produced by nuclear transfer of somatic cell nuclei from various mammalian species. *Biol. Reprod. 60*(6), 1496–1502.

Eggan, K., Akutsu, H., Loring, J., Jackson-Grusby, L., Klemm, M., Rideout, W. M., 3rd, Yanagimachi, R., and Jaenisch, R. (2001). Hybrid vigor, fetal overgrowth, and viability of mice derived by nuclear cloning and tetraploid embryo complementation. *Proc. Natl. Acad. Sci. U.S.A. 98*(11), 6209–6214.

Humpherys, D., Eggan, K., Akutsu, H., Hochedlinger, K., Rideout, W. M., 3rd, Biniszkiewicz, D., Yanagimachi, R., and Jaenisch, R. (2001). Epigenetic instability in ES cells and cloned mice. *Science 293*(5527), 95–97.

Kang, Y.-K., Koo, D.-B., Park, J.-S., Choi, Y.-M., Chung, A.-S., Lee, K.-K., and Han, Y.-M. (2001). Aberrant methylation of donor genome in cloned bovine embryos. *Nature Gen. 28*, 173–177.

Kawase, E., Yamazaki, Y., Yagi, T., Yanagimachi, R., and Pedersen, R. A. (2000). Mouse embryonic stem (ES) cell lines established from neuronal cell-derived cloned blastocysts. *Genesis 28*(3–4), 156–163.

Kikyo, N., Wade, P. A., Guschin, D., Ge, H., and Wolffe, A. P. (2000). Active remodeling of somatic nuclei in egg cytoplasm by the nucleosomal ATPase ISWI. *Science 289*(5488), 2360–2362.

Lewis, I. M., Munsie, M. J., French, A. J., Daniels, R., and Trounson, A. O. (2001). The cloning cycle: From amphibia to mammals and back. *Reprod. Med. Rev. 9*, 3–33.

Munsie, M. J. (2000). Nuclear transfer in the mouse: Reprogramming to pluripotency. *In "Institute of Reproduction and Development."* Monash University, Melbourne.

Munsie, M. J., Michalska, A. E., O'Brien, C. M., Trounson, A. O., Pera, M. F., and Mountford, P. S. (2000). Isolation of pluripotent embryonic stem cells from reprogrammed adult mouse somatic cell nuclei. *Curr. Biol. 10*(16), 989–992.

Nagy, A., Rossant, J., Nagy, R., Abramow-Newerly, W., and Roder, J. C. (1993). Derivation of completely cell culture-derived mice from early-passage embryonic stem cells. *Proc. Natl. Acad. Sci. U.S.A. 90*(18), 8424–8428.

Niwa, H., Miyazaki, J., and Smith, A. G. (2000). Quantitative expression of Oct-3/4 defines differentiation, dedifferentiation or self-renewal of ES cells. *Nat. Genet. 24*(4), 372–376.

Reubinoff, B. E., Pera, M. F., Fong, C. Y., Trounson, A., and Bongso, A. (2000). Embryonic stem cell lines from human blastocysts: Somatic differentiation in vitro. *Nat. Biotechnol. 18*(4), 399–404.

Tada, M., Tada, T., Lefebvre, L., Barton, S. C., and Surani, M. A. (1997). Embryonic germ cells induce epigenetic reprogramming of somatic nucleus in hybrid cells. *EMBO J. 16*(21), 6510–6520.

Tada, M., Takahama, Y., Abe, K., Nakatsuji, N., and Tada, T. (2001). Nuclear reprogramming of somatic cells by *in vitro* hybridization with ES cells. *Curr. Biol. 11*(19), 1553–1558.

Thomson, J. A., Itskovitz-Eldor, J., Shapiro, S. S., Waknitz, M. A., Swiergiel, J. J., Marshall, V. S., and Jones, J. M. (1998). Embryonic stem cell lines derived from human blastocysts. *Science 282*(5391), 1145–1147.

Trounson, A. (2001). The derivation and potential use of human embryonic stem cells. *Reprod. Fertil. Dev. 13*, 523–532.

Trounson, A., and Pera, M. (2001). Human embryonic stem cells. *Fertil. Steril. 76*(4), 660–661.

Trounson, A., Lacham-Kaplan, O., Diamente, M., and Gougoulidis, T. (1998). Reprogramming cattle somatic cells by isolated nuclear injection. *Reprod. Fertil. Dev. 10*(7–8), 645–650.

Wakayama, T., Perry, A. C., Zuccotti, M., Johnson, K. R., and Yanagimachi, R. (1998). Full-term development of mice from enucleated oocytes injected with cumulus cell nuclei. *Nature 394*(6691), 369–374.

Wakayama, T., Tabar, V., Rodriguez, I., Perry, A. C., Studer, L., and Mombaerts, P. (2001). Differentiation of embryonic stem cell lines generated from adult somatic cells by nuclear transfer. *Science 292*(5517), 740–743.

Part IV

Application of Stem Cell Biology to Regenerative Medicine

Chapter 19

Cancer Stem Cells

Michael Rothenberg and Michael F. Clarke

The worldwide burden of cancer is immense, yet our ability to cure most cancers sadly remains quite limited. Tumors are clonal, heterogeneous collections of abnormal cells whose development is only partially understood. Recently, the Cancer Stem Cell (CSC) model has re-emerged as a compelling way to reconcile the seemingly contradictory properties of clonality and developmental heterogeneity by postulating that a subset of cells within certain tumors—the cancer stem cells—self-renew, give rise to various differentiated tumor cells, and ultimately drive tumor growth and metastasis. The modern tools of stem cell biology have provided strong evidence to support this model—particularly with the prospective isolation of CSCs from hematologic and solid malignancies in both mice and humans. As molecular pathways underlying self-renewal have been dissected, the potential for cancer stem cells to arise from both normal stem cells as well as from non self-renewing progenitor cells has become apparent. The term "cancer stem cell" is thus an operative definition that does not imply a relationship to a normal tissue stem cell. Ultimately, cancer stem cells will likely have profound effects on our conceptualization and treatment of cancer.

INTRODUCTION

Cancer biologists have long recognized that tumors are complex collections of aberrant cells, each of which ultimately derives from a single ancestor, the cell of origin. The process by which the cell of origin develops into an ever-expanding, heterogeneous tumor that retains many features of its parent tissue remains mysterious. Recently, the cancer stem cell hypothesis has re-emerged as a compelling model to explain both the clonality and some of the heterogeneity of tumors. It postulates that each tumor contains a subset of cells—the cancer stem cells—that are uniquely responsible for tumor growth, heterogeneity, and metastasis. These specialized cells possess two key features that together distinguish them from the remainder of the tumor cells: self-renewal and differentiation potential. Substantial evidence from clinical oncology, as well as from experiments on human and animal tumors, provides strong support for this model. Most importantly, cancer stem cells have now been prospectively identified from leukemias and many different solid tumors in mice and humans. Simultaneously, significant progress in stem cell biology has provided insights into the molecular mechanisms by which normal stem cells self-renew. Interestingly, experiments on non-self-renewing cells have demonstrated that self-renewal is a property that cells can acquire through mutations. This finding sheds light on the possible origins of cancer stem cells, and illustrates how they may originate from either self-renewing or non self-renewing cells. The existence of cancer stem cells has important implications for how we conceptualize and treat cancer. This review provides an overview of the cancer stem cell model, and the data that support it.

CLONALITY AND HETEROGENEITY IN TUMORS

Tumors Have a Single Cell of Origin

The clonality of tumors is a fundamental principle in cancer biology. Each tumor—no matter how large or heterogeneous—ultimately originated from a single transformed cell, called the cell of origin. This principle was first demonstrated in malignancies in women, using the phenomenon of X chromosome inactivation (Fialkow, 1972; Fialkow *et al.*, 1972; Gale and Linch, 1998). During early female development, the process of lyonization causes different somatic cells (and their progeny) to randomly inactivate one of their two X-chromosomes; consequently, an adult woman is a mosaic of two distinct cell populations, expressing either the maternal or the paternal X-chromosome. When tumors from women who are heterozygous for various X-linked loci are characterized, the cancerous cells exhibit only one pattern of X-chromosome inactivation, suggesting a common (clonal) origin. Other analyses using the highly polymorphic X-linked HUMARA (human androgen receptor) locus, as well as distinct stably-inherited patterns of promoter methylation on different X-chromosomes, have been

Stem Cell Anthology

used to further demonstrate this point (Fearon *et al.*, 1987; Vogelstein *et al.*, 1987).

Tumor clonality can also be demonstrated for malignancies in which all the cancerous cells harbor the same chromosomal translocation, such as the Philadelphia chromosome (Yoffe *et al.*, 1987). Similarly, lymphoma cells that all harbor the same genomic rearrangements at the immunoglobulin locus or the T-cell receptor locus must have originated from a common origin. Although a few studies have suggested that tumors may not always be clonal in their earliest stages (Novelli *et al.*, 1996; Thliveris *et al.*, 2005; Halberg and Dove, 2007), the clonality of cancer has emerged as a central principle in tumorigenesis. This idea also has held true for other solid tumors using similar techniques.

Tumors Resemble "Aberrant Organs"

In the context of this clonality, the cancer cells of all but the most highly aggressive tumors appear heterogeneous. This is readily apparent on routine hematoxylin and eosin (H&E) staining, which shows malignant cells in varying stages of differentiation. For example, cancers originating in glandular epithelial tissues (adenocarcinomas) often elaborate rudimentary glands with distinct basolateral and luminal sides, which may even retain the ability to generate tissue-specific secretions: milk proteins in the case of breast cancers; prostatic secretions in the case of prostate cancers; and mucus in the case of colon cancers. Likewise, cancers originating in squamous epithelia (squamous cell carcinomas) form keratinous pearls, suggestive of their tissue of origin. An extreme example of tumor heterogeneity is a teratoma, which contains immature (embryonic carcinoma) cells, as well as differentiated cancerous cells from all three germ layers: endoderm; mesoderm; and ectoderm. On the other side of the spectrum is Burkitt's lymphoma, a highly aggressive, rapidly-dividing B-cell malignancy, characterized by homogeneous sheets of pleiotropic cells.

Not unexpectedly, phenotypic heterogeneity is also evident on analysis of gene expression (for example, by *in situ* hybridization) and immunohistochemistry. Various genes and proteins are specifically and differentially expressed in subsets of tumor cells. Sub-populations within a tumor may also differ from each other in other attributes: their ability to migrate or invade; their proclivity to enter the cell cycle or undergo apoptosis; their responsiveness to drugs or hormones; their ability to exclude vital dyes; their karyotype; and their ability to sustain growth in culture (Scharenberg *et al.*, 2002; Hirschmann-Jax *et al.*, 2004; Singh *et al.*, 2004). This heterogeneity and organization suggests that the development of a tumor from a single cell may recapitulate a developmental program. Thus, tumors can often be seen as caricatures of normal tissue or "aberrant organs", an idea that can be traced back to Rudolf Virchow, the father of modern cellular pathology (Virchow, 1858; Fidler and Hart, 1982; Heppner, 1984; Reya *et al.*, 2001; Lobo *et al.*, 2007).

THE CANCER STEM CELL MODEL

Models to Explain Both Clonality and Heterogeneity

Since tumors are both clonal and heterogeneous, a key question emerges: which cells drive the growth of a tumor? For example, is the property of tumorigenicity passed on from the cell of origin to all progeny, or is tumorigenicity a property possessed by only a subset of tumor cells? This question is relevant not only to understanding the basic biology of cancer, but also to determining the most effective treatment strategy in patients. If all cells in a tumor are potentially tumorigenic, then any curative treatment must eliminate all malignant cells. However, if only a fraction of the cells drive tumorigenesis, then a treatment that eliminates this sub-population while sparing some of the other cancer cells could result in long-term remission in at least some patients (Reya *et al.*, 2001; Al-Hajj *et al.*, 2004).

At least two different models attempt to explain both clonality and tumor heterogeneity: the stochastic model; and the cancer stem cell model. The stochastic model is based on the notion that the transforming events that endow the cell of origin with a malignant phenotype cause all progeny of that cell also to be tumorigenic. In this model, tumor heterogeneity is caused by subclones of tumor cells that result from a combination of different microenvironments and random genetic changes (including point mutations, chromosomal rearrangements, aneuploidy, and gene amplifications) (Nowell, 1976; Lengauer *et al.*, 1998). Ultimately, this model implies that effective treatment for cancer must involve eradication of all cancer cells.

The cancer stem cell (CSC) model postulates that only some tumor cells are tumorigenic—the cancer stem cells. When they divide, these malignant cells self-renew, and also give rise to the non-self-renewing cells that go on to elaborate the heterogeneous cells within a cancer. In this way, cancer stem cells both drive tumor growth, and also initiate the execution of a developmental program (Virchow, 1858; Fialkow, 1972; Mintz and Illmensee, 1975; Illmensee and Mintz, 1976; Fialkow, 1990). Cancer stem cells thus utilize pre-existing developmental hierarchies, and give rise to some of the cellular heterogeneity seen in such tumors.

Cancer Stem Cell, a Functional Definition

Importantly, the term "cancer stem cell" is an operative definition that does not necessarily connote a developmental relationship to normal stem cells; instead, it merely states that a subset of cells within a tumor can self-renew and elaborate tumor heterogeneity (Clarke *et al.*, 2006). Ultimately, the definition depends on the assays of self-renewal and tumorigenicity. Also, the cancer stem cell model does not imply what percentage of cells within a tumor are cancer stem cells. Some very aggressive tumors may have a high percentage of cancer stem cells (Kelly *et al.*, 2007; Kennedy *et al.*, 2007), and chemotherapeutic treatments may increase the frequency of cancer stem cells in a tumor (Dylla *et al.*, 2008); however, neither of these observations affects the operative definition.

The stochastic model and the cancer stem cell model are not mutually exclusive. Since they are malignant cells, cancer stem cells likely exhibit genomic instability, and have a mutator phenotype (Lengauer *et al.*, 1998; Hanahan and Weinberg, 2000). Thus, as they divide and give rise to both tumorigenic and non-tumorigenic progeny, subclones of cancer stem cells may develop. Thus, the cancer stem cell population within a tumor may become heterogeneous (Odoux *et al.*, 2008). Also, different subclones of cancer stem cells may give rise to genomically distinct non-tumorigenic cells.

DEFINING FEATURES OF STEM CELLS IN NORMAL TISSUES

Self-renewal and Differentiation

Tissues are complex, organized collections of cells that conduct specialized functions using multiple distinct terminally-differentiated effector cells. Often, these effector cells have lifespans that are only a fraction of the lifespan of the organism. Thus, in order to maintain tissue homeostasis, and prevent a tissue from degenerating, tissues must have a mechanism for self-renewal; by necessity, this involves cellular proliferation and differentiation. This is accomplished by multipotent tissue stem cells, which are developmentally committed, but relatively undifferentiated, cells that can give rise to one or more differentiated cell types (Figure 19-1).

Stem cells have two features that together distinguish them from the other cells in a tissue: self-renewal; and differentiation potential. Self-renewal is the ability of a cell to divide and give rise to at least one daughter cell that has extensive proliferative capacity and the same developmental potential as its parent (Weissman, 2000; Melton D, 2004). This feature distinguishes stem cells from the other cells in the tissue, including progenitor cells. The rigorous experimental demonstration of long-term self-renewal can be challenging, since the developmental potential of a cell may not become fully evident until several cell divisions have elapsed (Weissman *et al.*, 2001).

Differentiation is the ability of a committed cell to give rise to one or more developmentally-distinct mature cell types. This is a property that is not entirely unique to stem cells. For example, both hematopoietic stem cells and progenitor cells can give rise to all the major hematopoietic lineages (Figure 19-1). However, because only the stem cells self-renew, if the bone marrow of a lethally-irradiated mouse is reconstituted with progenitor cells rather than stem cells, adequate multi-lineage hematopoiesis will only persist for a finite period of time (Spangrude *et al.*, 1988; Morrison *et al.*, 1995). Eventually, the progenitor cells will stop contributing to hematopoiesis, and the mouse will succumb (in approximately eight weeks) as the terminally differentiated blood cells are exhausted.

The size of the stem cell compartment, i.e., the number of stem cells within a tissue at any given time, is under tight genetic control (Muller-Sieburg and Riblet, 1996; Muller-Sieburg *et al.*, 2000). When a stem cell divides symmetrically to give rise to two stem cells, the stem cell compartment is expanded. Alternatively, when the stem cell divides asymmetrically to give rise to a stem cell and a non-stem cell, the stem cell compartment is maintained (Horvitz and Herskowitz, 1992; Morrison and Kimble, 2006). Of course, the stem cell compartment can also shrink if a stem cell divides to give rise to two non-stem cells. Other mechanisms for regulating the size of the stem cell compartment may also be at play. This probably has relevance to cancer biology, for one prediction of the cancer stem cell model would be that tumor suppressor genes might restrict the size of the cancer stem cell compartment. Also, more aggressive tumors may have a more pronounced cancer stem cell compartment (Dylla *et al.*, 2008).

Prospective Isolation of Normal Tissue Stem Cells

Adult tissue self-renewal has been best studied in the hematopoietic system, where hematopoietic stem cells (HSCs) have been unequivocally identified and isolated in both mice and humans (Spangrude *et al.*, 1988; Baum *et al.*, 1992). (A major technique used to achieve this is fluorescence-activated cell sorting (FACS), known as flow cytometry, a process by which different sub-populations of cells are identified and sorted away from a larger population, based on molecules (markers) on the cell surface that can be labeled with fluorescently-labeled antibodies. The sub-population of interest can then be characterized and further studied.) Realization that the human hematopoietic stem could self-renew led to the development of an entire field within medicine, hematopoietic stem cell transplantation. Patients who undergo this procedure develop and maintain a completely reconstituted hematopoietic system, comprised exclusively of the progeny of the donor's hematopoietic stem cells. The hematopoietic stem cells are rare; they constitute only 0.01–0.05% of the mouse bone marrow cells (Spangrude *et al.*, 1988; Baum *et al.*, 1992; Morrison *et al.*, 1995), and less than 1.1% of human bone

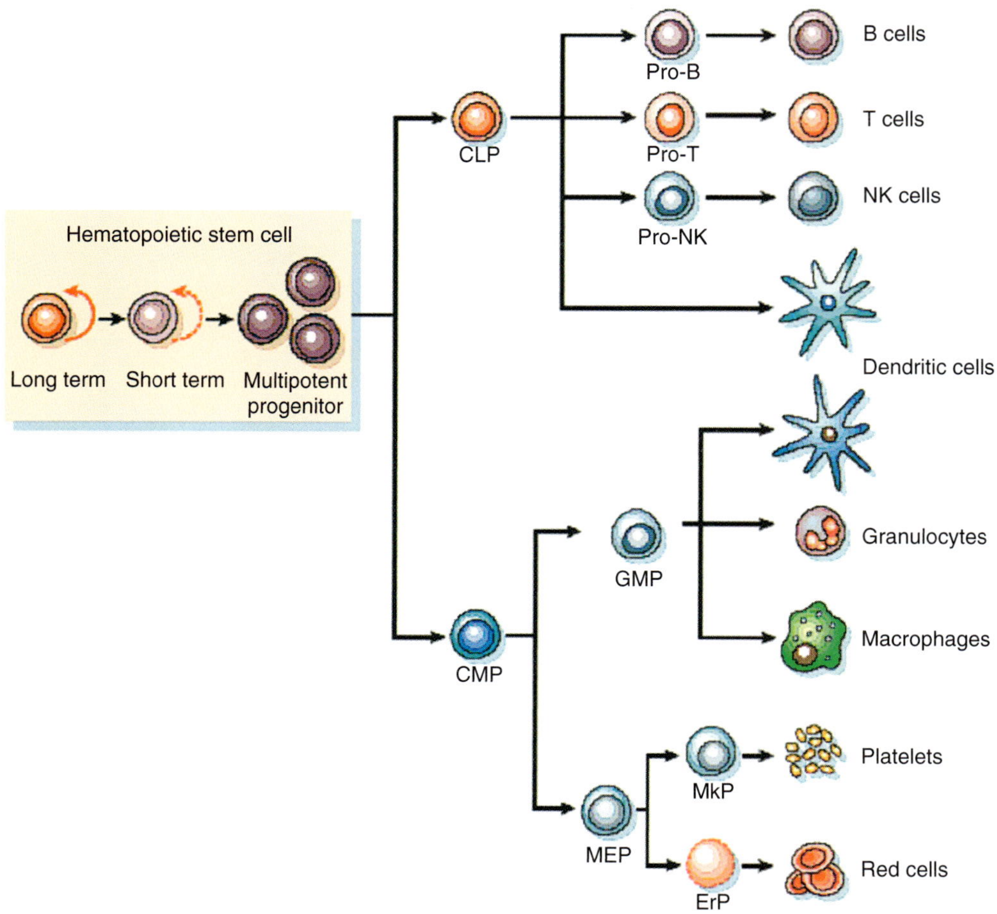

FIGURE 19-1 Self-renewal and differentiation in hematopoiesis. Long-term hematopoietic stem cells (HSCs) can self-renew for the lifetime of an animal (red arrows indicate self-renewal). Short-term HSCs have only limited self-renewal capacity. In contrast, multipotent progenitors do not have self-renewal capacity. However, in reference to differentiation, long-term self-renewing HSCs, short-term self-renewing HSCs, and multipotent progenitors all have the ability to give rise to all lineages of the blood system (shaded box). (CLPs: the precursors of all lymphoid cells; CMPs: the precursors of all myeloid cells; ErP: erythrocyte precursor; MEP: megakaryocyte erythrocyte precursor; GMP: granulocyte/macrophage progenitor; MkP: megakaryocyte precursor; NK: natural killer cell.) Modified from Reya *et al.* (2001). *Nature*, 414, 105–111.

marrow cells (Baum *et al.*, 1992; Fruehauf and Seggewiss, 2003). By contrast, approximately 5% of gut epithelial cells are stem cells, demonstrating that tissue stem cells need not be as rare as blood stem cells in all tissues (Barker *et al.*, 2007).

Aside from the hematopoietic system, stem cells have now been prospectively identified in the adult tissues of both invertebrates and vertebrates (Clarke and Fuller, 2006; Morrison and Kimble, 2006). Relevant to cancer biology, mammalian stem cells have been identified in many of the tissues that commonly give rise to cancers, including the skin, the nervous system, the hematopoietic system, the breast, prostate, and the intestine (Spangrude *et al.*, 1988; Baum *et al.*, 1992; Stemple and Anderson, 1992; Kordon and Smith, 1998; Morrison *et al.*, 1999; Taylor *et al.*, 2000; Blanpain *et al.*, 2004; Stingl *et al.*, 2006; Barker *et al.*, 2007; Lawson *et al.*, 2007; Sangiorgi and Capecchi, 2008). Thus, self-renewal is an ancient and widespread process, and it is likely to involve evolutionarily conserved pathways.

NOT ALL CANCER CELLS ARE TUMORIGENIC: CLINICAL AND EXPERIMENTAL EVIDENCE

Under the CSC model, only a subset of cells with a tumor is able to self-renew and generate at least a part of a tumor's developmental heterogeneity: other malignant cells within the tumor should be non-tumorigenic. A number of independent observations, both in the clinic and in the laboratory, support this prediction. (Of note, these observations taken alone do not prove the CSC model, for the failure of a cancer cell to develop into a tumor could be due to an intrinsic inability of that cell to self-renew, or to the absence of necessary signals in the immediate microenvironment.)

The experiments of Southam and Brunschwig are among the most startling in cancer biology (Southam *et al.*, 1966). At that time, physicians had realized that malignant cells could be found circulating in the blood or lymphatics of cancer patients, but only a small percentage of these seemed to give rise to metastatic tumors. Southam and Brunschwig sought to test this observation directly by harvesting tumor cells from patients with various disseminated or unresectable cancers, dissociating tumor fragments into single cell suspensions, and then subcutaneously autotransplanting varying numbers of cells to roughly determine how many tumor cells were needed to generate a new tumor. In spite of the deep ethical and methodological flaws, this study did suggest that, using their methods, only a subset of cells within a tumor is tumorigenic.

A similar observation has been made by clinicians providing palliative care to cancer patients with malignant ascites. In such patients who have implantation of a peritoneo–venous shunt—a procedure which can help alleviate abdominal distention and discomfort, but also has the unavoidable effect of shunting huge numbers of tumor cells into the systemic circulation—the development of disseminated cancer is an exceedingly rare event (Tarin *et al.*, 1984; Campioni *et al.*, 1986). A related clinical example is the recognition that viable breast cancer cells, the identity of which has been confirmed by immunocytochemical as well as genetic analysis, have been found in the bone marrow of a large percentage of breast cancer patients. Although in many patients the presence of these cells correlates with an increased probability of overt metastasis and a worse overall prognosis, the vast majority of these cells either remain dormant for years or are simply not tumorigenic (Braun *et al.*, 2005; Riethdorf *et al.*, 2008). Another supportive example is the observation that percutaneous biopsy of breast and liver tumors frequently leads to the disruption of the tumor and the deposition of thousands of malignant cells along the biopsy track, yet even with large bore needles, the rate of tumor seeding of the biopsy track is relatively low (Ryd *et al.*, 1983; Eriksson *et al.*, 1984; Smith, 1984; Diaz *et al.*, 1999; Silva *et al.*, 2008). Several possibilities could explain this discrepancy, including the idea that only a fraction of the tumor cells are tumorigenic.

Along these lines, in many different animal models of cancer—both with hematologic malignancies and solid tumors—a large number of tumor cells must be transplanted into a recipient animal order to generate a tumor (Bruce and Van Der Gaag, 1963; Hewitt *et al.*, 1976; Cho *et al.*, 2008). Similarly, only a minority of tumor cells is able to proliferate *in vitro* and generate colonies (Hamburger and Salmon, 1977a; Hamburger and Salmon, 1977b; Hill and Milas, 1989)

SOME CANCERS CAN DIFFERENTIATE INTO NONMALIGNANT CELLS

The CSC model implies that tumor heterogeneity is, at least in part, due to the execution of a developmental program in the progeny of cancer stem cells. As defined, cancer stem cells should be immature, multipotent, and able to give rise to both tumorigenic and non-tumorigenic cells. This has been shown in the case of mouse teratocarcinoma, a malignant germ cell tumor that can give rise to differentiated cell types of all three germ layers. When this tumor develops (it can occur spontaneously or be induced), it will metastasize and kill the host within a matter of weeks. Single cells, when injected subcutaneously into recipient mice, cause tumors to form. However, when single genetically-marked teratocarcinoma cells are isolated and injected into a normal mouse blastocyst, the progeny of the teratocarcinoma cells contribute functioning terminally-differentiated post-mitotic cells to multiple different tissues without inducing more tumors (Mintz and Illmensee, 1975; Illmensee and Mintz, 1976).

Importantly, among cancers, teratocarcinomas are not unique in their ability to execute a program of differentiation. Experiments with a transplantable squamous cell cancer in rats showed that tumor growth is driven mainly by immature, proliferating cells that subsequently develop into differentiated cells that, when transplanted, lose their tumorigenicity (Pierce and Wallace, 1971). Also, supportive data in humans comes from the genetic analysis of leukemias, in which genetic changes found in the leukemic clone can also be seen in mature, well-differentiated cells of various lineages. This suggests that the cancer is driven by immature hematopoietic cells that retain the capability of generating mature, noncancerous progeny (Fialkow, 1972; Fialkow *et al.*, 1977).

EVIDENCE SUPPORTING THE CANCER STEM CELL MODEL: PROSPECTIVE ISOLATION OF CANCER STEM CELLS

Prospective Isolation of Leukemia-initiating Cells

The demonstration that only some cells within many tumors are tumorigenic, and that malignant cells can elaborate developmental programs, provides strong evidence that supports the CSC model. But understanding the molecular controls that permit tumorigenicity requires the prospective isolation and purification of CSC so that they can be characterized and further studied.

Using the tools and strategies used to isolate normal stem hematopoietic stem cells (Spangrude *et al.*, 1988), this was first accomplished in acute myelogenous leukemia (AML) with the isolation of leukemia-initiating cells (L-IC) (Lapidot *et al.*, 1994; Bonnet and Dick, 1997). In AML, the peripheral blood and bone marrow are overpopulated with leukemic blasts, but these cells had demonstrated only a limited proliferative capacity *in vitro*. One possible explanation for this result was the existence of rare leukemia-initiating cells that drive the disease—cancer stem cells distinct from the leukemic blasts. (A potentially confusing point in the literature is that the term "leukemia stem cell" does not necessarily imply an ontogenic relationship to normal hematopoietic stem cells.)

The general experimental approach, the xenograft assay (Figure 19-2), was to use cell surface markers to label primary human leukemia (AML) cells, perform flow cytometry to isolate distinct sub-populations, inject the different sub-populations into immunocompromised NOD–SCID (non-obese diabetic–severe combined immunodeficiency) recipient mice to show tumorigenicity of one sub-population, and to then repeat the process using the reconstituted leukemia to demonstrate that the tumorigenic sub-population was indeed capable of self-renewal. The L-ICs were found in the CD34+CD38− population; the other blasts were not capable of engraftment. Importantly, both the reconstituted and serially-transplanted leukemias that engrafted in the NOD–SCID mice preserved the surface marker heterogeneity of the parent malignancy; thus, this was not simply an assay of proliferation. Rather, it suggested a developmental program may be intrinsic to the L-ICs. And in fact, this was later shown in lineage tracing experiments with individual L-ICs (Hope *et al.*, 2004). Interestingly, CD34+CD38− is the same immunophenotype exhibited by normal hematopoietic stem cells, suggesting an ontogenic relationship between leukemia-initiating cells and normal hematopoietic stem cells. (Actually, more recent evidence suggests that the L-ICs from most patients phenotypically resemble early progenitor cells rather than normal HSCs, since (like progenitors) the L-ICs do not express CD90, i.e., they are CD34+CD38−CD90−, while normal HSCs do express CD90, i.e., they are CD34+CD38−CD90+ (Majeti *et al.*, 2007).) Regardless, because only the L-IC compartment remained small in both the parent leukemia and in the serial transplants into immunocompromised hosts, at least some constraints on self-renewal still remained intact in the leukemia.

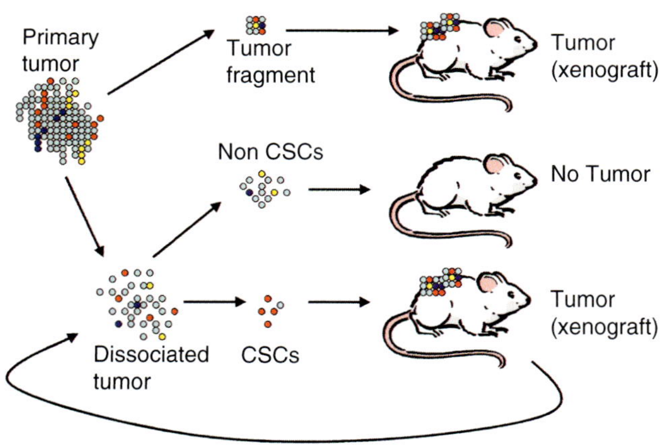

FIGURE 19-2 The xenograft assay (for solid tumors). A heterogeneous primary tumor is removed from a mouse or a human. A piece of the tumor can be directly implanted subcutaneously onto the back of a recipient mouse, and this will reliably lead to tumor growth. Alternatively, the tumor can be mechanically and enzymatically digested into a single-cell suspension, which can then be sorted by flow cytometry into different populations based on surface markers. One population may be enriched for cancer stem cells (CSCs, shown in red) and will give rise to a heterogeneous tumor in a recipient mouse, while the non-CSC population is non-tumorigenic. The xenografted tumor originating from cancer stem cells is then harvested, and the procedure is repeated to demonstrate that the CSCs have generated a heterogeneous tumor, which itself has CSCs and nonCSCs, and thus has the property of self-renewal. The xenograft assay is very similar for leukemias, where CSCs (otherwise known as leukemia initiating cells) are injected into the tail vein of the recipient mouse, rather than injected heterotopically.

Prospective Isolation of Solid Tumor Cancer Stem Cell

Subsequently, cancer stem cells (CSC) were prospectively isolated from solid tumors—first achieved in human breast cancer (Al-Hajj *et al.*, 2003). This was done by mechanically and enzymatically digesting primary tumors into single cell suspensions, and conducting xenograft assays on various sub-populations separated by flow cytometry (Figure 19-2). Such experiments revealed that only a percentage of tumor cells retained tumorigenic capacity. An empiric search for surface markers that would identify tumorigenic or nontumorigenic subpopulations showed that in eight out of nine patients' tumors, the tumorigenic fraction was in the CD44+CD24low/-Lineage-compartment. Cells outside of this compartment were non-tumorigenic. As few as 100 cells with the tumorigenic immunophenotype formed a tumor. Importantly, the xenograft tumors were heterogeneous, and themselves contained both tumorigenic and non-tumorigenic fractions, showing again that this was not merely a proliferation assay. The tumorigenic fraction could be used to passage the tumor serially, suggesting that the tumorigenic cells retained both self-renewal and the ability to execute a developmental program. Furthermore, because serially-passaged tumors continued to resemble their parent tumor, tumor heterogeneity was unlikely to be due exclusively to random mutations in different subclones of tumor cells (Figure 19-3).

Since then, using the xenograft assay, tumorigenic fractions of cells have been found in multiple different human solid tumors, including various nervous system tumors, colon cancer, hepatocellular carcinoma, pancreatic cancer, prostate cancer, and squamous cell cancer of the head and neck (Singh *et al.*, 2004; Collins *et al.*, 2005; Dalerba *et al.*, 2007; Li *et al.*, 2007; O'Brien *et al.*, 2007; Prince *et al.*, 2007; Ricci-Vitiani *et al.*, 2007). Different cell surface markers label the tumorigenic fraction in each system. For example, glioblastoma multiforme is a highly-aggressive human brain tumor containing cells that are both positive and negative for the surface marker CD133. In xenograft assays, the tumorigenic cells are contained within the fraction positive for CD133, comprising less than 30% of the tumor cells (Singh *et al.*, 2004). Similarly, in colon

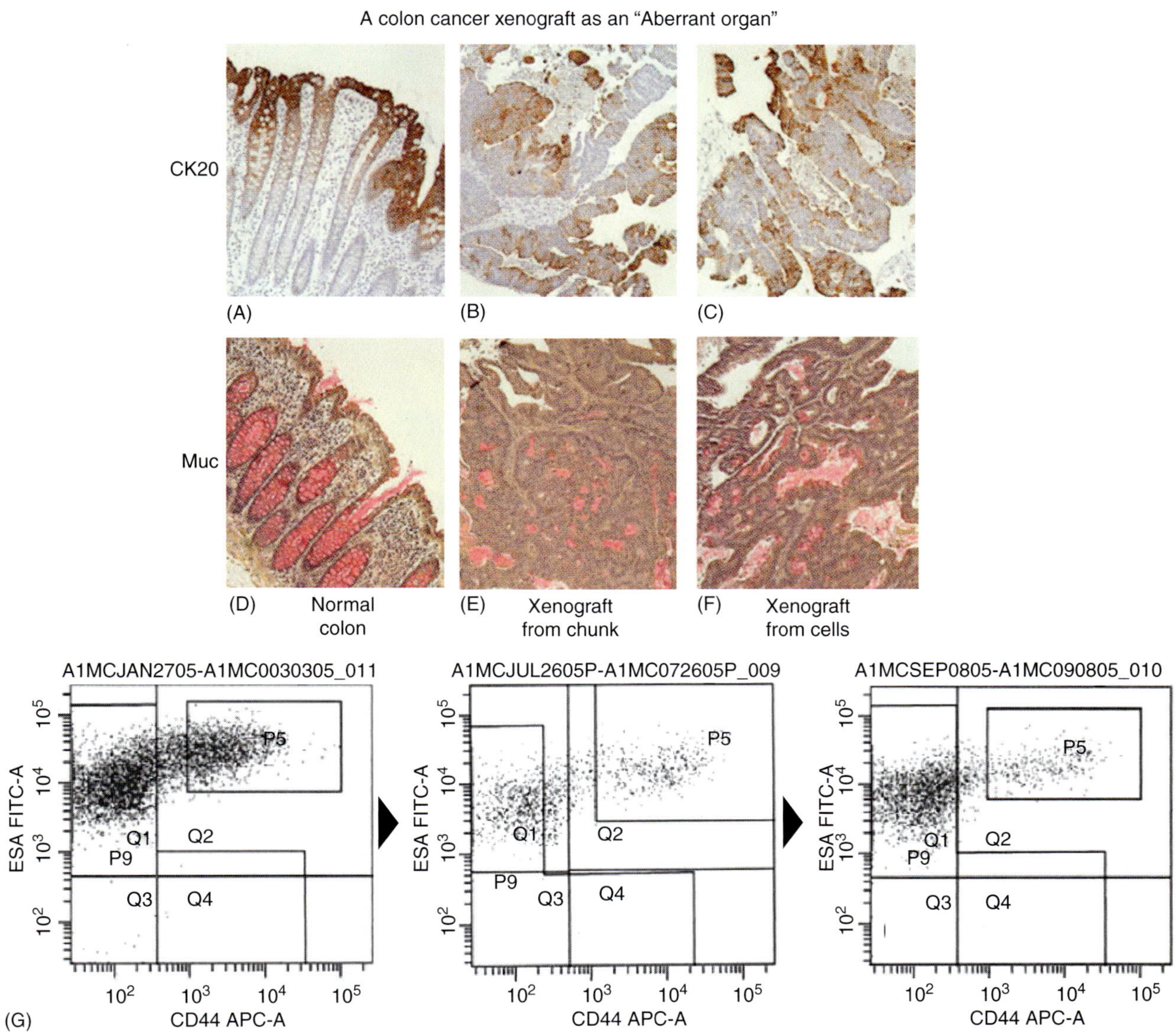

FIGURE 19-3 Shown are stainings of normal and colorectal carcinoma (CRC) samples with cytokeratin 20 (CK20) and neural epithelial mucins (Muc). (A, D) Normal epithelium stained with CK20 and Muc. (B, E) and (C, F) Xenografted human CRCs in NOD–SCID mice. The xenografts are differentiated adenocarcinomas forming gland-like structures reminiscent of normal colon. (B, E) Xenografts resulting from tumor fragments (undissociated pieces of tumor). (C, F) From sorted cancer stem cells. Resultant tumors from cancer stem cells show heterogeneity of cells as shown by staining and morphology. Bottom: (G) FACS plots from serially-passaged xenografts (the same tumor as shown in B, C, E, F), using ESA (y-axis) and CD44 (x-axis), two markers of colon cancer stem cells. Each FACS plot is from a different passage of the same parent tumor, and shows that the overall pattern remains constant. Modified from Dalerba *et al.* (2007). *Proc. Nat Acad. Sci. USA*, 104, 10158–10163

cancer, different groups have found tumorigenic cells within the CD133+ sub-population, while others have observed them in the CD44+CD166+EpCAM+ population (Dalerba *et al.*, 2007; O'Brien *et al.*, 2007; Ricci-Vitiani *et al.*, 2007). In pancreatic cancers, the tumorigenic cells have the immunophenotype CD44+CD24+ESA+ (Li *et al.*, 2007).

It is important to consider that the surface markers merely serve as tools to mark the tumorigenic fraction. They may not actually play a role in self-renewal, tumorigenesis, or differentiation, and they may not mark the tumorigenic cells in all tumors. Furthermore, the markers used for cancer stem cells in most tumors are useful for enriching, but not purifying, the apparent stem cell compartment. Most likely not all of the cells isolated with the present markers are stem cells. Also, the other cellular compartments, although depleted for stem cells, may still contain some stem cells. Importantly, the present markers do not necessarily represent a specific state of differentiation, but rather were preferentially expressed by the stem cells in the tissue from which they were isolated, such as pleural effusions in the case of most breast cancers. Ultimately, cancer stem cells from different tissues may share certain molecular properties; for example, CD44 is expressed in the cancer stem cells of most human epithelial cancers analyzed to date.

Clinical Relevance of Cancer Stem Cells

The tumorigenic cells isolated in the xenograft assay are clinically relevant. Gene expression profiling of the CD44+CD24low/− tumorigenic population in breast tumors has been used to generate an "invasiveness" gene signature (Liu *et al.*, 2007). Interestingly, this profile is associated with overall survival and metastasis-free survival, and predicts clinical outcomes in breast cancer independent of other established clinical and pathological characteristics. A related finding was made in the case of a mouse model of breast cancer, the MMTV-Wnt1 mouse. In spontaneous breast tumors from these mice, the tumorigenic cells exhibit a Thy1+CD24+CD49f+CD45− immunophenotype, and as few as 50 cells can generate a heterogeneous tumor in a syngeneic recipient mouse. Genes differentially expressed in tumorigenic and non-tumorigenic cells were used to generate an expression signature that predicted survival of human breast cancer patients (Cho *et al.*, 2008). Because these experiments dealt with mouse tumor cells being transplanted into syngeneic mouse recipients, they suggest that the identification of tumorigenic sub-populations of human breast cancer cells in xenograft assays are unlikely to be an artifact of transplanting human cells into mice. Furthermore, genes important in cancer stem cells in mouse models are likely to be relevant to human cancer biology.

Surface markers can be used to isolate a tumorigenic population of cells that is likely to be highly-enriched for cancer stem cells, but individual cancer stem cells have not yet been isolated. Undoubtedly, the tumorigenic cell populations are heterogeneous, and future studies will attempt to identify and characterize cancer stem cells more precisely.

ORIGINS OF CANCER STEM CELLS: THE PLASTICITY OF SELF-RENEWAL

Cancer Originating from Stem or Progenitor Cells

As described above, both normal tissue stem cells and cancer stem cells, by definition, are special cells that are able to self-renew and generate a heterogeneous tissue. The finding that leukemic stem cells in AML share the same immunophenotype with normal hematopoietic stem cells, CD34+CD38−, might suggest that cancers result from the oncogenic transformation of normal stem cells. Similarly, observations about the time from carcinogenic exposure to the development of cancer also seem to support the notion that cancers derive from normal tissue stem cells. For example, if a person is exposed to a carcinogen for a finite period of time, many years may elapse between the end of the exposure and the eventual development of cancer. This has been shown for solar ultraviolet light exposure and the later development of malignant melanoma (Brash and Ponten, 1998), as well as cigarette smoke exposure and the later development of lung cancer (Thun *et al.*, 2002). Traditionally, the only cells in a tissue thought to persist long enough to accumulate several oncogenic mutations, and exhibit such a delayed effect, were the tissue stem cells. Because of their rapid turnover, any progenitor cells, i.e., non-stem cells, that acquire mutations would be expected to die off before enough mutations could accumulate to transform them—unless, of course, the very mutations that cause cancer also hinder the mechanisms that normally limit proliferation in progenitor cells and drive their maturation and senescence. This is born out in a mouse model of human intestinal tumors. The initiating mutation of most human colon cancers is loss of the APC tumor suppressor gene (Kinzler and Vogelstein, 1996). Elegant experiments by Clevers and colleagues show that loss of APC in intestinal progenitor cells results in small, benign lesions, whereas loss of APC in colon stem cells gives rise to malignant lesions (Clevers, unpublished data). Although the initiating lesion is likely to arise in the normal stem cells, other recent studies suggest that cancer stem cells are not necessarily descended from normal tissue stem cells. They can also develop from progenitor cells, i.e., non-self-renewing cells that acquire the capacity for self-renewal (Figure 19-4).

Chronic myelogenous leukemia (CML) is a myeloproliferative disorder characterized by the uncontrolled proliferation of granulocytes. It accounts for 15–20% of leukemia in adults. The disease has a relatively indolent course until it enters an accelerated phase, and then a "blast

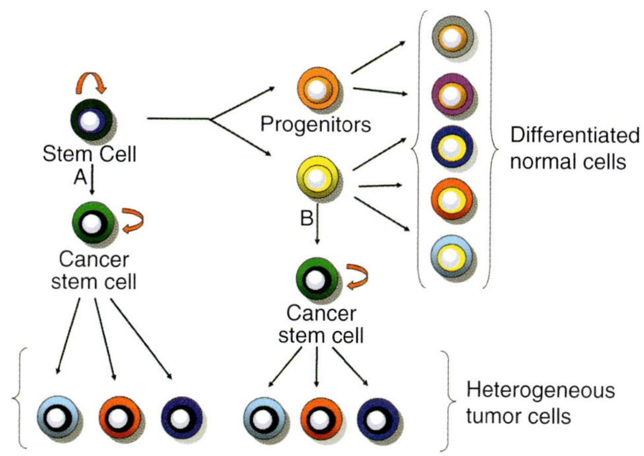

FIGURE 19-4 Origins of cancer stem cells. Cancer stem cells may originate from transformed stem cells (A), or progenitor cells (B). If they originate from progenitor cells, they gain self-renewal properties. In normal tissues, progenitor cells can acquire self-renewal through mutation of several genes.

crisis" phase which resembles acute leukemia. The cell of origin in CML was originally thought to be a hematopoietic stem cell (HSC) since the Philadelphia chromosome, t(9;22), the characteristic chromosomal translocation of CML which leads to the production of the oncogenic fusion protein Bcr-Abl, can be detected in HSCs in CML patients. However, solid evidence for malignant HSCs in CML has been lacking, and a mouse model of CML has shown that blast crisis can develop even when the Bcr-Abl fusion protein is driven in progenitor cells rather than HSCs (Jaiswal *et al.*, 2003). Recently, careful surface marker studies of leukemic cells in blast crisis CML have shown that the self-renewing leukemic cells exhibit a cell surface profile shared by myeloid progenitor cells rather than HSCs (Jamieson *et al.*, 2004).

In AML, further characterization of the immunophenotype of leukemia-initiating cells also supports the hypothesis that L-ICs may originate from progenitors rather than stem cells. Although both HSCs and L-ICs are CD34+CD38−, unlike normal HSCs, L-ICs do not express CD90, and thus more closely resemble progenitors rather than stem cells (Majeti *et al.*, 2007). This principle has also been supported by recent studies in solid tumors. For example, in a mouse model of medulloblastoma driven by Hedgehog signaling, the acquisition of a progenitor cell fate is an essential step in tumorigenesis. This suggests that CNS progenitor cells (i.e., cerebellar granule neuron precursors) rather than neural stem cells (i.e., multipotent ventricular zone cells) are the likely cell of origin in this CNS malignancy (Schuller *et al.*, 2008; Yang *et al.*, 2008).

Mechanisms for Acquiring Self-renewal

If a cancer, which by definition must contain self-renewing cells, can arise from progenitor cells, then self-renewal must be an attribute that non-self-renewing cells can acquire. How can this occur? Recent studies have shown that Bmi1 and its downstream targets may be involved. Bmi1 is a Polycomb group gene that normally functions by epigenetically repressing genes that positively regulate senescence, differentiation, and cell death. When HSCs are mutant for Bmi1, they quickly exhaust after several weeks, rather than continually self-renewing. Thus, Bmi1 is required to maintain self-renewal in HSCs (Lessard and Sauvageau, 2003; Park *et al.*, 2003) (Bmi1 may be a more general self-renewal factor: it is required for maintenance of self-renewal in neural stem cells, marks intestinal stem cells, and is expressed in many tissues that continuously regenerate (Molofsky *et al.*, 2003; Hosen *et al.*, 2007; Sangiorgi and Capecchi, 2008).) Remarkably, when three major downstream targets of Bmi1 are mutated (p53, p16-Ink4a, and p19-ARF), progenitor cells gain the ability to self-renew (Akala *et al.*, 2008). This suggests a mechanism by which progenitor cells might adopt a self-renewing fate to become cancer stem cells (Figure 19-4). Interestingly, mutations of p53, p16-Ink4a, and p19-ARF are commonly found in cancers, hinting that progenitor cells may be the origin of cancer stem cells more broadly.

The loss of function of genes that positively regulate senescence, differentiation, and cell death may not be the only mechanism to enable non stem-cells to obtain a self-renewing phenotype. Several groups have recently demonstrated that the expression of a small number of defined transcription factors (for example, Oct3/4, Sox2, Klf4, and c-Myc) can enable terminally-differentiated somatic cells (fibroblasts) to become pluripotent, adopt an embryonic stem cell-like fate, and thus obtain the property of self-renewal (Takahashi and Yamanaka, 2006; Okita *et al.*, 2007; Wernig *et al.*, 2007). Remarkably, this has been shown with both mouse and human somatic cells (Takahashi *et al.*, 2007; Yu *et al.*, 2007). More recently, this has been achieved in human fibroblasts with only two non-oncogenic transcription factors, Oct3/4 and Sox2, in the presence of a chemical inhibitor of histone deacetylase, valproic acid (Huangfu *et al.*, 2008).

Even though terminally-differentiated cells such as fibroblasts possess the capacity to be reprogrammed and adopt a self-renewing fate, as of now there is no evidence that cancer stem cells actually utilize this mechanism to acquire self-renewal. Indeed they may not, since the *in vivo* reprogramming of somatic cells would likely lead to germ cell tumors (Thomson *et al.*, 1998; Takahashi and Yamanaka, 2006; Maherali *et al.*, 2007; Wernig *et al.*, 2007), but such tumors generally do not arise in differentiated tissues such as breast, colon, prostate, lung, etc. However, it remains possible that the partial reprogramming of a differentiated cell could activate the self-renewal machinery. Several genes that are required for maintaining pluripotency and driving rapid proliferation in embryonic stem cells are frequently up-regulated in cancer, including c-Myc, Beta-catenin, Stat3, Klf4, and E-ras (Niwa *et al.*, 1998; Matsuda *et al.*, 1999; Kielman *et al.*, 2002; Takahashi *et al.*, 2003; Sato *et al.*, 2004; Cartwright *et al.*, 2005; Li *et al.*, 2005).

SIGNALING PATHWAYS IN CANCER STEM CELLS

Evolutionarily divergent metazoans—including the worm *C. elegans*, the fruitfly *Drosophila melanogaster*, the mouse *Mus musculus*, and the human *Homo sapien*—possess stem cells that maintain various adult tissues. This observation has led to the hypothesis that certain highly-conserved genes may contain an evolutionarily-conserved self-renewal machinery. Such reasoning has in turn led stem cell biologists to hypothesize that normal and cancer stem cells share at least some molecular mechanisms of self-renewal. This makes sense, since the maintenance of stem cells—whether normal or cancerous—involves the repression of pathways for differentiation, senescence, and apoptosis (which themselves are highly conserved). Thus, positive regulators or effectors of these pathways should be quiescent in stem cells, while repressors of such pathways should be active or overexpressed. This has been demonstrated in HSCs, where the molecular mechanisms underlying self-renewal have been most extensively characterized (Akala and Clarke, 2006).

Several observations provide indirect support for the idea that mechanisms of self-renewal are shared by CSCs and normal tissue stem cells. For example, in cases where the immunophenotype of CSCs resembles that of the normal tissue stem cells, a parsimonious model of tumorigenesis might suggest that, as the normal stem cell becomes cancerous, the self-renewal machinery simply continues to function. This idea is supported by the observation that both normal hematopoietic stem cells and leukemia stem cells require the same gene, Bmi1, for self-renewal (Lessard and Sauvageau, 2003; Park *et al.*, 2003). Similarly, Wnt signaling has been implicated in self-renewal, in both normal hematopoietic stem cells and leukemia initiating cells (Reya *et al.*, 2003; Jamieson *et al.*, 2004). However, even if cancer stem cells are not always descended directly from normal tissue stem cells (as described above), they may still share mechanisms of self-renewal. Many genes involved in the pathways that regulate self-renewal in normal stem cells (such as Hedgehog, Wnt/Beta-catenin, Notch, and Bmi1), are often implicated in neoplasia.

For example, in the intestine, Wnt signaling has been shown to be an important pathway regulating tissue architecture, possibly (but not definitely) due to effects on stem cells (Gregorieff and Clevers, 2005; Reya and Clevers, 2005). In the intestinal epithelium, reducing Wnt signaling by mutating β-catenin, mutating Tcf-4, or overexpressing Dickkopf-1 all lead to severe tissue disruption and loss of proliferating epithelial cells, as one would predict if Wnt signaling were required for intestinal stem cell renewal (Korinek *et al.*, 1998; Pinto *et al.*, 2003; Ireland *et al.*, 2004; Kuhnert *et al.*, 2004). Recently, the intestinal stem cells were unambiguously identified based on their expression of Lgr5, a Wnt target gene (Barker *et al.*, 2007). As in hematopoietic stem cells, Wnt signaling is likely playing an important role in normal intestinal stem cells, perhaps by regulating self-renewal—although definitive evidence that the pathway regulates self-renewal in the intestine is still lacking. In colon cancer tumorigenesis, Wnt signaling has long been recognized as an important pathway, in large part because patients with Familial Adenomatous Polyposis invariably develop colon adenocarcinoma. These patients are heterozygous for loss of function mutations in APC, a key molecule in canonical Wnt signaling, and their colon cancers exhibit severe loss of APC (by loss of the normal allele) and constitutive Wnt signaling (as indicated by nuclear β-catenin staining) (Gregorieff and Clevers, 2005). Thus, it seems likely that Wnt signaling regulates self-renewal in both normal intestinal stem cells and colon CSCs.

Since cancer stem cells were only recently isolated, very little is definitively known about the molecular pathways involved in their regulation. Many pathways shown to have a role in self-renewal in normal stem cells may or may not have a role in cancer stem cells, and genes involved in self-renewal of normal stem cells may serve a different function in cancer stem cells.

For example, distinct from its role in embryonic patterning, Hedgehog signaling has been implicated in the regulation of self-renewal in adult tissues (Taipale and Beachy, 2001). Sonic Hedgehog (Shh) triggers increased self-renewal and proliferation in HSCs (Bhardwaj *et al.*, 2001), and also in neural stem cells, perhaps via activation of Bmi1 (Leung *et al.*, 2004). (Interestingly, Hedgehog signaling does not appear to be required for self-renewal in normal HSCs, since HSCs mutant for Smoothened, an important cofactor in Hedgehog signaling, are not defective in long-term hematopoietic reconstitution (Dierks *et al.*, 2008).) Hedgehog signaling has also been implicated in the tumorigenesis of several cancers, including basal cell carcinoma, medulloblastoma, and others (Taipale and Beachy, 2001; Liu *et al.*, 2006). In breast cancer stem cells and CML stem cells, Hedgehog signaling is active and involved in regulating proliferation and self-renewal (Liu *et al.*, 2006; Dierks *et al.*, 2008). But cancer stem cells may utilize Hedgehog signaling for other processes besides self-renewal. In pancreatic cancer stem cells, experiments with small molecule inhibitors of Hedgehog signaling suggest that Hedgehog signaling may have a more predominant effect on invasion and metastasis (Feldmann *et al.*, 2007). Many pathways have been implicated in the regulation of self-renewal in normal tissues, including Wnt, Notch, Hedgehog, BMP, Jak-Stat, AKT/PI3K, Polycomb group proteins, and others (Akala and Clarke, 2006; Lobo *et al.*, 2007; Visvader and Lindeman, 2008). Most of these have also been implicated in tumorigenesis. However, because cancer stem cells are not the same as normal tissue stem cells, they may utilize various pathways differently. In the end, it is likely that some mechanisms of self-renewal will be shared, and others will not. Regardless of whether normal and cancer stem cells share mechanisms of self-renewal, the identification of molecular pathways involved in the maintenance of cancer stem cells may present therapeutic opportunities.

CLINICAL IMPLICATIONS OF CANCER STEM CELLS

The cancer stem cell model has several important clinical implications, some of which may be profound. It might not only contribute a novel pathophysiological mechanism to explain tumors' resistance to treatment, but it might also improve clinicians' ability to more effectively diagnose, prognose, and treat cancer.

Since the cancer stem cell model suggests that CSCs are responsible for the sustained and uncontrolled growth of malignancies, treatment failures in the clinic may be due partly to the resistance of cancer stem cells to therapy. If cancer stem cells are intrinsically more resistant to treatment (such as chemotherapy or radiotherapy) than other cells in the tumor, then treatment should increase the percentage of cancer stem cells relative to pretreatment levels. Conversely, if all cells are similarly sensitive to treatment, then the relative percentages of tumorigenic and non-tumorigenic cells should not change as a tumor shrinks.

In support of this prediction, a recent study examining early passage colon cancer xenografts showed that the CD44+ESA+ cells, the population containing the colon cancer stem cells, were significantly enriched after treatment with cyclophosphamide, a chemotherapeutic agent. Tumors that had been exposed to cyclophosphamide had an expanded CD44+ESA+ compartment, and also more tumorigenic cells relative to controls, as shown in a serial dilution study of tumorigenicity. Importantly, even after treatment, equal numbers of CSCs from the treatment and control arms were equally tumorigenic (Dylla *et al.*, 2008). Similarly, in human patients with locally advanced breast cancer, the frequency of CSCs as shown by immunophenotype, as well as the ability to form mammospheres *in vitro* (an assay that correlates with stem cell activity), significantly increased shortly after treatment (Li *et al.*, 2008).

The often-observed initial shrinkage of a tumor soon after treatment might reflect the relative sensitivity of the non-tumorigenic cells—those that comprise the bulk of the tumor—and may not be as important as is currently thought. Thus, the most important clinical variable for assessing treatment response and prognosis may not be absolute tumor size, but rather the absolute number of CSCs remaining after treatment. Similarly, tumors with high proportions of cancer stem cells might be more aggressive and difficult to treat. This is not to say that one should not attempt to eliminate most of the cancer cells in a tumor. The fact that certain mutations can enable normal progenitor cells to self-renew (Akala *et al.*, 2008) suggests that mutations could reactivate self-renewal programs in residual non-tumorigenic cancer cells in some patients. Nonetheless, elimination of the cancer stem cell population in a tumor could lead to dramatically better clinical outcomes.

Aside from their intrinsic resistance to treatment, CSCs are also likely to possess mechanisms for acquiring resistance to treatment, such as genetic instability. The various genetic changes known to occur in tumors—including point mutations, chromosomal rearrangements, aneuploidy, and gene amplifications—may occur within CSCs (Lengauer *et al.*, 1998). In fact, a study on tumorigenic colon cancer cells suggests that they can develop karyotypic changes (Odoux et al., 2008), but whether this occurs *in vivo* and leads to treatment resistance remains unproven.

FUTURE DIRECTIONS

Although the cancer stem cell model has existed for well over 50 years, during the past several years the techniques of modern stem cell biology have facilitated rapid progress in the field. Among the major goals is to further enrich tumorigenic cancer stem cells to a higher level of purity—ideally to a single-cell level—in order to demonstrate formally that a single cell can generate a heterogeneous cancer and self-renew. One current approach to this problem is to discover new markers; however, this is intrinsically difficult because of the time- and labor-intensive nature of the current assays, as well as the lack of simple forward genetic approaches. Finding new markers is empiric and difficult. Ultimately, achieving this goal may depend on the development of entirely new approaches, assays, and techniques—most likely drawing on knowledge from normal stem cell biology and working at the single-cell level.

The purification of CSCs would dramatically change the field. First, cancer stem cell biologists would simply be better able to characterize the cells—to understand more clearly what makes them unique. What genetic pathways do they utilize, and how do they use them? What is their origin? What do they need to maintain self-renewal? How do they go on to generate a developmental hierarchy? Which of the many genes and mutations previously described in cancer are important in the cancer stem cells? How often do they divide, and what regulates their division? Do they divide symmetrically, or asymmetrically, or both—and what regulates the choice? Where in the tumors can they be found, and what do they look like? Can they be imaged in live tumors? Do they migrate, and if so how? How do they interact with their niche? How do they contribute to metastasis?

Ultimately, the goal is to apply these insights to the clinic—to develop better diagnostic tests and more targeted and effective anti-cancer therapies. Developments in the field should lead to a deeper understanding of the mechanisms of resistance to chemotherapy and radiotherapy. And ideally, patients will have more opportunities to experience curative, rather than palliative, therapies.

FURTHER READING

Akala, O. O., & Clarke, M. F. (2006). Hematopoietic stem cell self-renewal. *Curr. Opin. Genet. Dev.*, 16, 496–501.

Akala, O. O., Park, I. K., Qian, D., Pihalja, M., Becker, M. W., & Clarke, M. F. (2008). Long-term haematopoietic reconstitution by Trp53−/−p16Ink4a−/−p19Arf−/− multipotent progenitors. *Nature*, 453, 228–232.

Al-Hajj, M., Becker, M. W., Wicha, M., Weissman, I., & Clarke, M. F. (2004). Therapeutic implications of cancer stem cells. *Curr. Opin. Genet. Dev.*, 14, 43–47.

Al-Hajj, M., Wicha, M. S., Benito-Hernandez, A., Morrison, S. J., & Clarke, M. F. (2003). Prospective identification of tumorigenic breast cancer cells. *Proc. Nat. Acad. Sci. USA*, 100, 3983–3988.

Barker, N., van Es, J. H., Kuipers, J., Kujala, P., van den Born, M., Cozijnsen, M., Haegebarth, A., Korving, J., Begthel, H., Peters, P. J., & Clevers, H. (2007). Identification of stem cells in small intestine and colon by marker gene Lgr5. *Nature*, 449, 1003–1007.

Baum, C. M., Weissman, I. L., Tsukamoto, A. S., Buckle, A. M., & Peault, B. (1992). Isolation of a candidate human hematopoietic stem-cell population. *Proc. Nat. Acad. Sci. USA*, 89, 2804–2808.

Bhardwaj, G., Murdoch, B., Wu, D., Baker, D. P., Williams, K. P., Chadwick, K., Ling, L. E., Karanu, F. N., & Bhatia, M. (2001). Sonic hedgehog induces the proliferation of primitive human hematopoietic cells via BMP regulation. *Nat. Immunol.*, 2, 172–180.

Blanpain, C., Lowry, W. E., Geoghegan, A., Polak, L., & Fuchs, E. (2004). Self-renewal, multipotency, and the existence of two cell populations within an epithelial stem cell niche. *Cell*, *118*, 635–648.

Bonnet, D., & Dick, J. E. (1997). Human acute myeloid leukemia is organized as a hierarchy that originates from a primitive hematopoietic cell. *Nat. Med.*, *3*, 730–737.

Brash, D. E., & Ponten, J. (1998). Skin precancer. *Cancer Surv.*, *32*, 69–113.

Braun, S., Vogl, F. D., Naume, B., Janni, W., Osborne, M. P., Coombes, R. C., Schlimok, G., Diel, I. J., Gerber, B., Gebauer, G., Pierga, J. Y., Marth, C., Oruzio, D., Wiedswang, G., Solomayer, E. F., Kundt, G., Strobl, B., Fehm, T., Wong, G. Y., Bliss, J., Vincent-Salomon, A., & Pantel, K. (2005). A pooled analysis of bone marrow micrometastasis in breast cancer. *N. Engl. J. Med.*, *353*, 793–802.

Bruce, W. R., & Van Der Gaag, H. (1963). A quantitative assay for the number of murine lymphoma cells capable of proliferation *in vivo*. *Nature*, *199*, 79–80.

Campioni, N., Pasquali Lasagni, R., Vitucci, C., Filippetti, M., Spagnoli, A., Pompei, S., Romano, P., Pasqualetti, A., Sorvillo, G., Sega, F. M. *et al.* (1986). Peritoneovenous shunt and neoplastic ascites: a 5-year experience report. *J. Surg. Oncol.*, *33*, 31–35.

Cartwright, P., McLean, C., Sheppard, A., Rivett, D., Jones, K., & Dalton, S. (2005). LIF/STAT3 controls ES cell self-renewal and pluripotency by a Myc-dependent mechanism. *Development*, *132*, 885–896.

Cho, R. W., Wang, X., Diehn, M., Shedden, K., Chen, G. Y., Sherlock, G., Gurney, A., Lewicki, J., & Clarke, M. F. (2008). Isolation and molecular characterization of cancer stem cells in MMTV-Wnt-1 murine breast tumors. *Stem Cells*, *26*, 364–371.

Clarke, M. F., Dick, J. E., Dirks, P. B., Eaves, C. J., Jamieson, C. H., Jones, D. L., Visvader, J., Weissman, I. L., & Wahl, G. M. (2006). Cancer stem cells—perspectives on current status and future directions: AACR Workshop on cancer stem cells. *Cancer Res.*, *66*, 9339–9344.

Clarke, M. F., & Fuller, M. (2006). Stem cells and cancer: two faces of Eve. *Cell*, *124*, 1111–1115.

Collins, A. T., Berry, P. A., Hyde, C., Stower, M. J., & Maitland, N. J. (2005). Prospective identification of tumorigenic prostate cancer stem cells. *Cancer Res.*, *65*, 10946–10951.

Dalerba, P., Dylla, S. J., Park, I. K., Liu, R., Wang, X., Cho, R. W., Hoey, T., Gurney, A., Huang, E. H., Simeone, D. M., Shelton, A. A., Parmiani, G., Castelli, C., & Clarke, M. F. (2007). Phenotypic characterization of human colorectal cancer stem cells. *Proc. Nat. Acad. Sci. USA*, *104*, 10158–10163.

Diaz, L. K., Wiley, E. L., & Venta, L. A. (1999). Are malignant cells displaced by large-gauge needle core biopsy of the breast? *AJR Am. J. Roentgenol.*, *173*, 1303–1313.

Dierks, C., Beigi, R., Guo, G. R., Zirlik, K., Stegert, M. R., Manley, P., Trussell, C., Schmitt-Graeff, A., Landwerlin, K., Veelken, H., & Warmuth, M. (2008). Expansion of Bcr-Abl-positive leukemic stem cells is dependent on Hedgehog pathway activation. *Cancer Cell*, *14*, 238–249.

Dylla, S. J., Beviglia, L., Park, I. K., Chartier, C., Raval, J., Ngan, L., Pickell, K., Aguilar, J., Lazetic, S., Smith-Berdan, S., Clarke, M. F., Hoey, T., Lewicki, J., & Gurney, A. L. (2008). Colorectal cancer stem cells are enriched in xenogeneic tumors following chemotherapy. *PLoS ONE*, *3*, e2428.

Eriksson, O., Hagmar, B., & Ryd, W. (1984). Effects of fine-needle aspiration and other biopsy procedures on tumor dissemination in mice. *Cancer*, *54*, 73–78.

Fearon, E. R., Hamilton, S. R., & Vogelstein, B. (1987). Clonal analysis of human colorectal tumors. *Science*, *238*, 193–197.

Feldmann, G., Dhara, S., Fendrich, V., Bedja, D., Beaty, R., Mullendore, M., Karikari, C., Alvarez, H., Iacobuzio-Donahue, C., Jimeno, A., Gabrielson, K. L., Matsui, W., & Maitra, A. (2007). Blockade of hedgehog signaling inhibits pancreatic cancer invasion and metastases: a new paradigm for combination therapy in solid cancers. *Cancer Res.*, *67*, 2187–2196.

Fialkow, P. J. (1972). Use of genetic markers to study cellular origin and development of tumors in human females. *Adv. Cancer Res.*, *15*, 191–226.

Fialkow, P. J. (1990). Stem cell origin of human myeloid blood cell neoplasms. *Verh. Dtsch. Ges. Pathol.*, *74*, 43–47.

Fialkow, P. J., Jacobson, R. J., & Papayannopoulou, T. (1977). Chronic myelocytic leukemia: clonal origin in a stem cell common to the granulocyte, erythrocyte, platelet and monocyte/macrophage. *Am. J. Med.*, *63*, 125–130.

Fialkow, P. J., Martin, G. M., Klein, G., Clifford, P., & Singh, S. (1972). Evidence for a clonal origin of head and neck tumors. *Int. J. Cancer*, *9*, 133–142.

Fidler, I. J., & Hart, I. R. (1982). Biological diversity in metastatic neoplasms: origins and implications. *Science*, *217*, 998–1003.

Fruehauf, S., & Seggewiss, R. (2003). It's moving day: factors affecting peripheral blood stem cell mobilization and strategies for improvement (corrected). *Br. J. Haematol.*, *122*, 360–375.

Gale, R. E., & Linch, D. C. (1998). Clonality studies in acute myeloid leukemia. *Leukemia*, *12*, 117–120.

Gregorieff, A., & Clevers, H. (2005). Wnt signaling in the intestinal epithelium: from endoderm to cancer. *Genes Dev.*, *19*, 877–890.

Halberg, R. B., & Dove, W. F. (2007). Polyclonal tumors in the mammalian intestine: are interactions among multiple initiated clones necessary for tumor initiation, growth, and progression? *Cell Cycle*, *6*, 44–51.

Hamburger, A., & Salmon, S. E. (1977a). Primary bioassay of human myeloma stem cells. *J. Clin. Invest.*, *60*, 846–854.

Hamburger, A. W., & Salmon, S. E. (1977b). Primary bioassay of human tumor stem cells. *Science*, *197*, 461–463.

Hanahan, D., & Weinberg, R. A. (2000). The hallmarks of cancer. *Cell*, *100*, 57–70.

Heppner, G. H. (1984). Tumor heterogeneity. *Cancer Res.*, *44*, 2259–2265.

Hewitt, H. B., Blake, E. R., & Walder, A. S. (1976). A critique of the evidence for active host defence against cancer, based on personal studies of 27 murine tumours of spontaneous origin. *Br. J. Cancer*, *33*, 241–259.

Hill, R. P., & Milas, L. (1989). The proportion of stem cells in murine tumors. *Int. J. Radiat. Oncol. Biol. Phys.*, *16*, 513–518.

Hirschmann-Jax, C., Foster, A. E., Wulf, G. G., Nuchtern, J. G., Jax, T. W., Gobel, U., Goodell, M. A., & Brenner, M. K. (2004). A distinct "side population" of cells with high drug efflux capacity in human tumor cells. *Proc. Nat. Acad. Sci. USA*, *101*, 14228–14233.

Hope, K. J., Jin, L., & Dick, J. E. (2004). Acute myeloid leukemia originates from a hierarchy of leukemic stem cell classes that differ in self-renewal capacity. *Nat. Immunol.*, *5*, 738–743.

Horvitz, H. R., & Herskowitz, I. (1992). Mechanisms of asymmetric cell division: two Bs or not two Bs, that is the question. *Cell*, *68*, 237–255.

Hosen, N., Yamane, T., Muijtjens, M., Pham, K., Clarke, M. F., & Weissman, I. L. (2007). Bmi-1-green fluorescent protein-knock-in mice reveal the dynamic regulation of bmi-1 expression in normal and leukemic hematopoietic cells. *Stem Cells*, *25*, 1635–1644.

Huangfu, D., Maehr, R., Guo, W., Eijkelenboom, A., Snitow, M., Chen, A. E., & Melton, D. A. (2008). Induction of pluripotent stem cells by defined factors is greatly improved by small-molecule compounds. *Nat. Biotechnol.*, *26*, 795–797.

Illmensee, K., & Mintz, B. (1976). Totipotency and normal differentiation of single teratocarcinoma cells cloned by injection into blastocysts. *Proc. Nat. Acad. Sci. USA*, *73*, 549–553.

Ireland, H., Kemp, R., Houghton, C., Howard, L., Clarke, A. R., Sansom, O. J., & Winton, D. J. (2004). Inducible Cre-mediated control of gene expression in the murine gastrointestinal tract: effect of loss of beta-catenin. *Gastroenterology*, *126*, 1236–1246.

Jaiswal, S., Traver, D., Miyamoto, T., Akashi, K., Lagasse, E., & Weissman, I. L. (2003). Expression of BCR/ABL and BCL-2 in myeloid progenitors leads to myeloid leukemias. *Proc. Nat. Acad. Sci. USA*, *100*, 10002–10007.

Jamieson, C. H., Ailles, L. E., Dylla, S. J., Muijtjens, M., Jones, C., Zehnder, J. L., Gotlib, J., Li, K., Manz, M. G., Keating, A., Sawyers, C. L., & Weissman, I. L. (2004). Granulocyte-macrophage progenitors as candidate leukemic stem cells in blast-crisis CML. *N. Engl. J. Med.*, *351*, 657–667.

Kelly, P. N., Dakic, A., Adams, J. M., Nutt, S. L., & Strasser, A. (2007). Tumor growth need not be driven by rare cancer stem cells. *Science*, *317*, 337.

Kennedy, J. A., Barabe, F., Poeppl, A. G., Wang, J. C., & Dick, J. E. (2007). Comment on "Tumor growth need not be driven by rare cancer stem cells." *Science*, *318*, 1722; author reply 1722.

Kielman, M. F., Rindapaa, M., Gaspar, C., van Poppel, N., Breukel, C., van Leeuwen, S., Taketo, M. M., Roberts, S., Smits, R., & Fodde, R. (2002). Apc modulates embryonic stem-cell differentiation by controlling the dosage of beta-catenin signaling. *Nat. Genet.*, *32*, 594–605.

Kinzler, K. W., & Vogelstein, B. (1996). Lessons from hereditary colorectal cancer. *Cell*, *87*, 159–170.

Kordon, E. C., & Smith, G. H. (1998). An entire functional mammary gland may comprise the progeny from a single cell. *Development*, *125*, 1921–1930.

Korinek, V., Barker, N., Moerer, P., van Donselaar, E., Huls, G., Peters, P. J., & Clevers, H. (1998). Depletion of epithelial stem-cell compartments in the small intestine of mice lacking Tcf-4. *Nat. Genet.*, *19*, 379–383.

Kuhnert, F., Davis, C. R., Wang, H. T., Chu, P., Lee, M., Yuan, J., Nusse, R., & Kuo, C. J. (2004). Essential requirement for Wnt signaling in proliferation of adult small intestine and colon revealed by adenoviral expression of Dickkopf-1. *Proc. Nat. Acad. Sci.USA*, *101*, 266–271.

Lapidot, T., Sirard, C., Vormoor, J., Murdoch, B., Hoang, T., Caceres-Cortes, J., Minden, M., Paterson, B., Caligiuri, M. A., & Dick, J. E. (1994). A cell initiating human acute myeloid leukaemia after transplantation into SCID mice. *Nature*, *367*, 645–648.

Lawson, D. A., Xin, L., Lukacs, R. U., Cheng, D., & Witte, O. N. (2007). Isolation and functional characterization of murine prostate stem cells. *Proc. Nat. Acad. Sci. USA*, *104*, 181–186.

Lengauer, C., Kinzler, K. W., & Vogelstein, B. (1998). Genetic instabilities in human cancers. *Nature*, *396*, 643–649.

Lessard, J., & Sauvageau, G. (2003). Bmi-1 determines the proliferative capacity of normal and leukaemic stem cells. *Nature*, *423*, 255–260.

Leung, C., Lingbeek, M., Shakhova, O., Liu, J., Tanger, E., Saremaslani, P., Van Lohuizen, M., & Marino, S. (2004). Bmi1 is essential for cerebellar development and is overexpressed in human medulloblastomas. *Nature*, *428*, 337–341.

Li, C., Heidt, D. G., Dalerba, P., Burant, C. F., Zhang, L., Adsay, V., Wicha, M., Clarke, M. F., & Simeone, D. M. (2007). Identification of pancreatic cancer stem cells. *Cancer Res.*, *67*, 1030–1037.

Li, X., Lewis, M. T., Huang, J., Gutierrez, C., Osborne, C. K., Wu, M. F., Hilsenbeck, S. G., Pavlick, A., Zhang, X., Chamness, G. C., Wong, H., Rosen, J., & Chang, J. C. (2008). Intrinsic resistance of tumorigenic breast cancer cells to chemotherapy. *J. Nat. Cancer Inst.*, *100*, 672–679.

Li, Y., McClintick, J., Zhong, L., Edenberg, H. J., Yoder, M. C., & Chan, R. J. (2005). Murine embryonic stem cell differentiation is promoted by SOCS-3 and inhibited by the zinc finger transcription factor Klf4. *Blood*, *105*, 635–637.

Liu, R., Wang, X., Chen, G. Y., Dalerba, P., Gurney, A., Hoey, T., Sherlock, G., Lewicki, J., Shedden, K., & Clarke, M. F. (2007). The prognostic role of a gene signature from tumorigenic breast-cancer cells. *N. Engl. J. Med.*, *356*, 217–226.

Liu, S., Dontu, G., Mantle, I. D., Patel, S., Ahn, N. S., Jackson, K. W., Suri, P., & Wicha, M. S. (2006). Hedgehog signaling and Bmi-1 regulate self-renewal of normal and malignant human mammary stem cells. *Cancer Res.*, *66*, 6063–6071.

Lobo, N. A., Shimono, Y., Qian, D., & Clarke, M. F. (2007). The biology of cancer stem cells. *Annu. Rev. Cell Dev. Biol.*, *23*, 675–699.

Maherali, N., Sridharan, R., Xie, W., Utikal, J., Eminli, S., Arnold, K., Stadtfeld, M., Yachechko, R., Tchieu, J., Jaenisch, R., Plath, K., & Hochedlinger, K. (2007). Directly reprogrammed fibroblasts show global epigenetic remodeling and widespread tissue contribution. *Cell Stem Cell*, *1*, 55–70.

Majeti, R., Park, C. Y., & Weissman, I. L. (2007). Identification of a hierarchy of multipotent hematopoietic progenitors in human cord blood. *Cell Stem Cell*, *1*, 635–645.

Matsuda, T., Nakamura, T., Nakao, K., Arai, T., Katsuki, M., Heike, T., & Yokota, T. (1999). STAT3 activation is sufficient to maintain an undifferentiated state of mouse embryonic stem cells. *Embo. J.*, *18*, 4261–4269.

Melton D, C. C. (2004). "Stemness": Definitions, Criteria, & Standards. In *Handbook of Stem Cells*, R. Lanza (Ed.), pp. xxv-xxxi. Oxford: Elsevier.

Mintz, B., & Illmensee, K. (1975). Normal genetically mosaic mice produced from malignant teratocarcinoma cells. *Proc. Nat. Acad. Sci. USA*, *72*, 3585–3589.

Molofsky, A. V., Pardal, R., Iwashita, T., Park, I. K., Clarke, M. F., & Morrison, S. J. (2003). Bmi-1 dependence distinguishes neural stem cell self-renewal from progenitor proliferation. *Nature*, *425*, 962–967.

Morrison, S. J., & Kimble, J. (2006). Asymmetric and symmetric stem-cell divisions in development and cancer. *Nature*, *441*, 1068–1074.

Morrison, S. J., Uchida, N., & Weissman, I. L. (1995). The biology of hematopoietic stem cells. *Annu. Rev. Cell Dev. Biol.*, *11*, 35–71.

Morrison, S. J., White, P. M., Zock, C., & Anderson, D. J. (1999). Prospective identification, isolation by flow cytometry, and *in vivo* self-renewal of multipotent mammalian neural crest stem cells. *Cell*, *96*, 737–749.

Muller-Sieburg, C. E., Cho, R. H., Sieburg, H. B., Kupriyanov, S., & Riblet, R. (2000). Genetic control of hematopoietic stem cell frequency in mice is mostly cell autonomous. *Blood*, *95*, 2446–2448.

Muller-Sieburg, C. E., & Riblet, R. (1996). Genetic control of the frequency of hematopoietic stem cells in mice: mapping of a candidate locus to chromosome 1. *J. Exp. Med.*, *183*, 1141–1150.

Niwa, H., Burdon, T., Chambers, I., & Smith, A. (1998). Self-renewal of pluripotent embryonic stem cells is mediated via activation of STAT3. *Genes Dev.*, *12*, 2048–2060.

Novelli, M. R., Williamson, J. A., Tomlinson, I. P., Elia, G., Hodgson, S. V., Talbot, I. C., Bodmer, W. F., & Wright, N. A. (1996). Polyclonal origin of colonic adenomas in an XO/XY patient with FAP. *Science*, *272*, 1187–1190.

Nowell, P. C. (1976). The clonal evolution of tumor cell populations. *Science*, *194*, 23–28.

O'Brien, C. A., Pollett, A., Gallinger, S., & Dick, J. E. (2007). A human colon cancer cell capable of initiating tumour growth in immunodeficient mice. *Nature*, *445*, 106–110.

Odoux, C., Fohrer, H., Hoppo, T., Guzik, L., Stolz, D. B., Lewis, D. W., Gollin, S. M., Gamblin, T. C., Geller, D. A., & Lagasse, E. (2008). A stochastic model for cancer stem cell origin in metastatic colon cancer. *Cancer Res.*, *68*, 6932–6941.

Okita, K., Ichisaka, T., & Yamanaka, S. (2007). Generation of germline-competent induced pluripotent stem cells. *Nature*, *448*, 313–317.

Park, I. K., Qian, D., Kiel, M., Becker, M. W., Pihalja, M., Weissman, I. L., Morrison, S. J., & Clarke, M. F. (2003). Bmi-1 is required for maintenance of adult self-renewing haematopoietic stem cells. *Nature*, *423*, 302–305.

Pierce, G. B., & Wallace, C. (1971). Differentiation of malignant to benign cells. *Cancer Res.*, *31*, 127–134.

Pinto, D., Gregorieff, A., Begthel, H., & Clevers, H. (2003). Canonical Wnt signals are essential for homeostasis of the intestinal epithelium. *Genes Dev.*, *17*, 1709–1713.

Prince, M. E., Sivanandan, R., Kaczorowski, A., Wolf, G. T., Kaplan, M. J., Dalerba, P., Weissman, I. L., Clarke, M. F., & Ailles, L. E. (2007). Identification of a subpopulation of cells with cancer stem cell properties in head and neck squamous cell carcinoma. *Proc. Nat. Acad. Sci. USA*, *104*, 973–978.

Reya, T., & Clevers, H. (2005). Wnt signalling in stem cells and cancer. *Nature*, *434*, 843–850.

Reya, T., Duncan, A. W., Ailles, L., Domen, J., Scherer, D. C., Willert, K., Hintz, L., Nusse, R., & Weissman, I. L. (2003). A role for Wnt signalling in self-renewal of haematopoietic stem cells. *Nature*, *423*, 409–414.

Reya, T., Morrison, S. J., Clarke, M. F., & Weissman, I. L. (2001). Stem cells, cancer, and cancer stem cells. *Nature*, *414*, 105–111.

Ricci-Vitiani, L., Lombardi, D. G., Pilozzi, E., Biffoni, M., Todaro, M., Peschle, C., & De Maria, R. (2007). Identification and expansion of human colon-cancer-initiating cells. *Nature*, *445*, 111–115.

Riethdorf, S., Wikman, H., & Pantel, K. (2008). Review: Biological relevance of disseminated tumor cells in cancer patients. *Int. J. Cancer.*, *123*, 1991–2006.

Ryd, W., Hagmar, B., & Eriksson, O. (1983). Local tumour cell seeding by fine-needle aspiration biopsy. A semiquantitative study. *Acta. Pathol. Microbiol. Immunol. Scand.*, (A) *91*, 17–21.

Sangiorgi, E., & Capecchi, M. R. (2008). Bmi1 is expressed *in vivo* in intestinal stem cells. *Nat. Genet.*, *40*, 915–920.

Sato, N., Meijer, L., Skaltsounis, L., Greengard, P., & Brivanlou, A. H. (2004). Maintenance of pluripotency in human and mouse embryonic stem cells through activation of Wnt signaling by a pharmacological GSK-3-specific inhibitor. *Nat. Med.*, *10*, 55–63.

Scharenberg, C. W., Harkey, M. A., & Torok-Storb, B. (2002). The ABCG2 transporter is an efficient Hoechst 33342 efflux pump and is preferentially expressed by immature human hematopoietic progenitors. *Blood*, *99*, 507–512.

Schuller, U., Heine, V. M., Mao, J., Kho, A. T., Dillon, A. K., Han, Y. G., Huillard, E., Sun, T., Ligon, A. H., Qian, Y., Ma, Q., Alvarez-Buylla, A., McMahon, A. P., Rowitch, D. H., & Ligon, K. L. (2008). Acquisition of granule neuron precursor identity is a critical determinant of progenitor cell competence to form Shh-induced medulloblastoma. *Cancer Cell*, *14*, 123–134.

Silva, M. A., Hegab, B., Hyde, C. J., Guo, B., Buckels, J. A., & Mirza, D. F. (2008). Needle track seeding following biopsy of liver lesions; a systematic review and meta-analysis. *Gut*, *57*(11), 1592–1596.

Singh, S. K., Hawkins, C., Clarke, I. D., Squire, J. A., Bayani, J., Hide, T., Henkelman, R. M., Cusimano, M. D., & Dirks, P. B. (2004). Identification of human brain tumour initiating cells. *Nature*, *432*, 396–401.

Smith, E. H. (1984). The hazards of fine-needle aspiration biopsy. *Ultrasound Med. Biol.*, *10*, 629–634.

Southam, C. M., Brunschwig, A., Levin, A. G., & Dizon, Q. S. (1966). Effect of leukocytes on transplantability of human cancer. *Cancer*, *19*, 1743–1753.

Spangrude, G. J., Heimfeld, S., & Weissman, I. L. (1988). Purification and characterization of mouse hematopoietic stem cells. *Science*, *241*, 58–62.

Stemple, D. L., & Anderson, D. J. (1992). Isolation of a stem cell for neurons and glia from the mammalian neural crest. *Cell*, *71*, 973–985.

Stingl, J., Eirew, P., Ricketson, I., Shackleton, M., Vaillant, F., Choi, D., Li, H. I., & Eaves, C. J. (2006). Purification and unique properties of mammary epithelial stem cells. *Nature*, *439*, 993–997.

Taipale, J., & Beachy, P. A. (2001). The Hedgehog and Wnt signalling pathways in cancer. *Nature*, *411*, 349–354.

Takahashi, K., Mitsui, K., & Yamanaka, S. (2003). Role of ERas in promoting tumour-like properties in mouse embryonic stem cells. *Nature*, *423*, 541–545.

Takahashi, K., Tanabe, K., Ohnuki, M., Narita, M., Ichisaka, T., Tomoda, K., & Yamanaka, S. (2007). Induction of pluripotent stem cells from adult human fibroblasts by defined factors. *Cell*, *131*, 861–872.

Takahashi, K., & Yamanaka, S. (2006). Induction of pluripotent stem cells from mouse embryonic and adult fibroblast cultures by defined factors. *Cell*, *126*, 663–676.

Tarin, D., Price, J. E., Kettlewell, M. G., Souter, R. G., Vass, A. C., & Crossley, B. (1984). Clinicopathological observations on metastasis in man studied in patients treated with peritoneovenous shunts. *Br. Med. J.* (Clin. Res. Ed.), *288*, 749–751.

Taylor, G., Lehrer, M. S., Jensen, P. J., Sun, T. T., & Lavker, R. M. (2000). Involvement of follicular stem cells in forming not only the follicle but also the epidermis. *Cell*, *102*, 451–461.

Thliveris, A. T., Halberg, R. B., Clipson, L., Dove, W. F., Sullivan, R., Washington, M. K., Stanhope, S., & Newton, M. A. (2005). Polyclonality of familial murine adenomas: analyses of mouse chimeras with low tumor multiplicity suggest short-range interactions. *Proc. Nat. Acad. Sci. USA*, *102*, 6960–6965.

Thomson, J. A., Itskovitz-Eldor, J., Shapiro, S. S., Waknitz, M. A., Swiergiel, J. J., Marshall, V. S., & Jones, J. M. (1998). Embryonic stem cell lines derived from human blastocysts. *Science*, *282*, 1145–1147.

Thun, M. J., Henley, S. J., & Calle, E. E. (2002). Tobacco use and cancer: an epidemiologic perspective for geneticists. *Oncogene*, *21*, 7307–7325.

Virchow, R. (1858). *Die Cellularpathologie in ihrer Begrundung auf Physiologische und Pathologische Gewebelehre*. Berlin: August Hirschwald.

Visvader, J. E., & Lindeman, G. J. (2008). Cancer stem cells in solid tumours: accumulating evidence and unresolved questions. *Nat. Rev. Cancer*, *8*, 755–768.

Vogelstein, B., Fearon, E. R., Hamilton, S. R., Preisinger, A. C., Willard, H. F., Michelson, A. M., Riggs, A. D., & Orkin, S. H. (1987). Clonal analysis using recombinant DNA probes from the X-chromosome. *Cancer Res.*, *47*, 4806–4813.

Weissman, I. L. (2000). Stem cells: units of development, units of regeneration, and units in evolution. *Cell*, *100*, 157–168.

Weissman, I. L., Anderson, D. J., & Gage, F. (2001). Stem and progenitor cells: origins, phenotypes, lineage commitments, and transdifferentiations. *Annu. Rev. Cell Dev. Biol.*, *17*, 387–403.

Wernig, M., Meissner, A., Foreman, R., Brambrink, T., Ku, M., Hochedlinger, K., Bernstein, B. E., & Jaenisch, R. (2007). *In vitro* reprogramming of fibroblasts into a pluripotent ES-cell-like state. *Nature*, *448*, 318–324.

Yang, Z. J., Ellis, T., Markant, S. L., Read, T. A., Kessler, J. D., Bourboulas, M., Schuller, U., Machold, R., Fishell, G., Rowitch, D. H., Wainwright, B. J., & Wechsler-Reya, R. J. (2008). Medulloblastoma can be initiated by deletion of Patched in lineage-restricted progenitors or stem cells. *Cancer Cell*, *14*, 135–145.

Yoffe, G., Chinault, A. C., Talpaz, M., Blick, M. B., Kantarjian, H. M., Taylor, K., & Spitzer, G. (1987). Clonal nature of Philadelphia chromosome-positive and -negative chronic myelogenous leukemia by DNA hybridization analyses. *Exp. Hematol.*, *15*, 725–728.

Yu, J., Vodyanik, M. A., Smuga-Otto, K., Antosiewicz-Bourget, J., Frane, J. L., Tian, S., Nie, J., Jonsdottir, G. A., Ruotti, V., Stewart, R., Slukvin, I. I., & Thomson, J. A. (2007). Induced pluripotent stem cell lines derived from human somatic cells. *Science*, *318*, 1917–1920.

Chapter 20

Neural Stem Cells for Central Nervous System Repair

Rodolfo Gonzalez, Jean Pyo Lee, and Evan Y. Snyder

INTRODUCTION

Despite the presence of endogenous neural stem cells (NSCs) in the mammalian brain, it is recognized that intrinsic "self-repair" activity for the most devastating of injuries is inadequate or ineffective. This poor regenerative ability, particularly in the adult central nervous system (CNS), may be because of the limited number and restricted location of native NSCs, and/or limitations imposed by the surrounding microenvironment, which may not be supportive or instructive for neuronal differentiation. NSCs expanded *ex vivo* in culture, and then implanted into regions needing repair, may overcome those limitations related simply to inadequate numbers of NSCs near the defective region. Whether the environment may also inhibit exogenous NSCs from surviving or differentiating toward replacement cells is a possibility. However, several transplantation experiments have suggested that neurogenic cues are transiently elaborated during degenerative processes (perhaps recapitulating developmental cues), and that exogenous NSCs are able to sense, home in on, and respond appropriately to those. In other words, NSCs appear to respond *in vivo* to neurogenic signals, not only when they occur appropriately during development, but even when induced at later stages by certain neurodegenerative processes.

In this chapter, we review some of the work that has been performed in animal models of CNS diseases where transplanted NSCs have mediated a therapeutic effect. These disorders include rodent models of genetic and acquired (e.g., traumatic and ischemic) neurodegeneration, inheritable metabolic disorders, age-related degeneration, and neoplasms. These conditions often have widespread neural cell loss, dysfunction, or both. The disseminated nature of the pathology in these diseases is not readily treated by conventional transplantation approaches, in which a solid tissue graft or a limited number of non-motile cells are delivered to a restricted area. Similarly, most gene therapy approaches tend to fall short because of their limited "sphere of influence" following injection of the vector into the CNS parenchyma. The use of bone marrow transplantation, even with hematopoietic stem cells, is typically inadequate, because the cells do not efficiently broach the blood–brain barrier. Cells from non-neural organs, if they transdifferentiate into neural cells at all (a controversial prospect), do so far too inefficiently to be therapeutically reliable for cell replacement in these conditions.

NSCs, on the other hand, circumvent many of these obstacles. Because they differentiate robustly into neural cells, integrate seamlessly into neural parenchyma as multiple neural cell types (both neuronal and glial), respond to normal developmental and regeneration cues, and migrate (even long distances) to multiple, disseminated areas of neuropathology, NSCs appear to be ideally suited for the molecular and cellular therapies required by extensive, diffuse (even "global") degenerative processes. Examples of such widespread neurodegenerative conditions include myelin disorders, storage diseases, motor neuron degeneration, dementia conditions such as Alzheimer's disease, and ischemic and traumatic pathologies, such as stroke. Some diseases appear to be restricted in their involvement—Parkinson's disease is localized to the mesostriatum, Huntington's to the caudate, spinal cord contusion to a few spinal segments, and cerebellar degeneration to the hindbrain. However, even these disorders require cell replacement to be distributed evenly over relatively large areas, and that multiple neural cell types be replaced in a specific region, or both. Again, these needs are best accomplished by a migratory, responsive, multipotent neural progenitor, even if transplantation is directed toward a more circumscribed CNS region.

THERAPEUTIC POTENTIAL OF NEURAL STEM CELLS

The clinical potential of the NSC is rooted in its inherent biologic properties. The ability of NSCs and their

Stem Cell Anthology

progeny to develop into integral cytoarchitectural components of many regions throughout the host brain as neurons, oligodendrocytes, astrocytes, or even immature neural progenitors, makes them capable of replacing a range of missing or dysfunctional neural cells. Although the field of neural repair has tended to place emphasis on the replacement of missing neurons in acute and chronic neurodegenerative diseases, or oligodendrocytes in demyelinating disorders, it is becoming increasingly evident that simultaneously replacing a diseased cell's "neighboring cells"—typically astrocytes—may be of equal importance, because of the indispensable trophic, guidance, and detoxification role such "chaperone" cells may play. Fortuitously, part of the biological repertoire of an NSC is to generate the variety of cells that constitute the "fabric" of a given neural region. NSCs in particular differentiation states will spontaneously produce a variety of neurotrophic factors that may serve trophic, protective, or both functions—for example, glial cell line-derived neurotrophic factor (GDNF), brain-derived neurotrophic factor (BDNF), nerve growth factor (NGF), and neurotrophin-3 (NT-3). NSCs also inherently express most of the "housekeeping" enzymes and factors necessary for any cell to maintain normal metabolism.

NSCs may also be readily engineered *ex vivo* to express a variety of molecules that either are not produced by NSCs, or are not produced in therapeutically adequate quantities. The NSCs are amenable to various types of viral vector transduction, as well as to other gene transfer strategies, such as lipofection, electroporation, and calcium-phosphate precipitation. Following transplantation, such engrafted NSCs may then be used as cellular vectors for the *in vivo* expression of exogenous genes of developmental and/or therapeutic relevance. Such gene products can be delivered either to circumscribed regions or, if necessary, to more widely-disseminated areas throughout the host CNS. Because they display significant migratory capacity, as well as an ability to integrate widely throughout the brain when implanted into germinal zones, NSCs may help to reconstitute enzyme and cellular deficiencies in a global fashion (Figure 20-1).

The ability of true NSCs to adjust to their regions of engraftment probably obviates the need for abstracting stem cells from specific CNS regions. Furthermore, NSCs appear to possess a tropism for degenerating CNS regions (Figure 20-2). NSCs may be attracted—across long distances—to regions of neurodegeneration in brains of all ages, including old age. Under some circumstances, an environment in which a particular type of neural cell has degenerated creates a milieu that directs the differentiation of an NSC toward maintaining homeostasis, including replenishment of a specific deficient neural cell type. Through mechanisms yet to be determined, it appears that some neurodegenerative processes (e.g., those associated with apoptosis) elaborate neurogenic signals that recapitulate developmental cues to which NSCs can respond (Figure 20-3).

It would seem that NSC-mediated cell replacement would only be feasible in "cell autonomous" disease

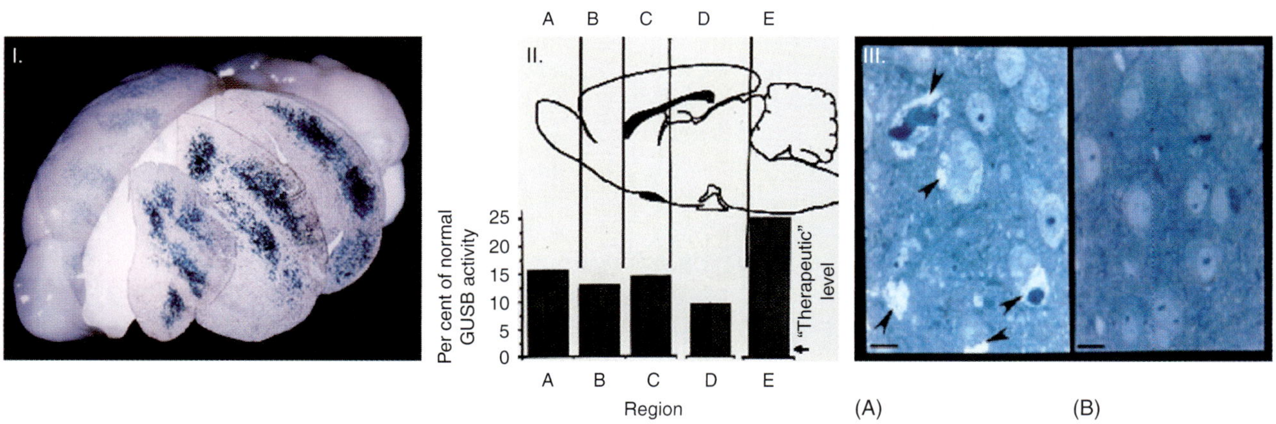

FIGURE 20-1 Widespread engraftment of NSCs expressing GUSB throughout the brain of the MPS VII mouse. (A) Brain of a mature MPS VII mouse after receiving a neonatal intraventricular transplant of murine NSCs expressing GUSB. Donor NSC-derived cells, identified by their X-gal histochemical reaction for expression of the *LacZ* marker gene, have engrafted throughout the recipient mutant brain. Coronal sections—placed at their appropriate level by computer—show these cells to span the rostral–caudal expanse of the brain. (B) Distribution of GUSB enzymatic activity throughout brains of MPS VII NSC transplant recipients. Serial sections were collected from throughout the brains of transplant recipients and assayed for GUSB activity. Sections were pooled to reflect the activity within the demarcated regions. The regions were defined by anatomical landmarks in the anterior-to-posterior plane to permit comparison among animals. The mean levels of GUSB activity for each region ($n = 17$) are presented as the percentage of average normal levels for each region. Untreated MPS VII mice show no GUSB activity, biochemically or histochemically. Enzyme activity of 2% of normal is corrective, based on data from liver and spleen. (C) Decreased lysosomal storage in a treated MPS VII mouse brain at eight months. (1) Extensive vacuolation representing distended lysosomes (arrowheads) in both neurons and glia in the neocortex of an 8-month-old, untransplanted control MPS VII mouse. (2) Decrease in lysosomal storage in the cortex of an MPS VII mouse treated at birth from a region analogous to the untreated control section in panel 1. The other regions of this animal's brain showed a similar decrease in storage compared with untreated, age-matched mutants in regions where GUSB was expressed. Scale bars = 21 μm.

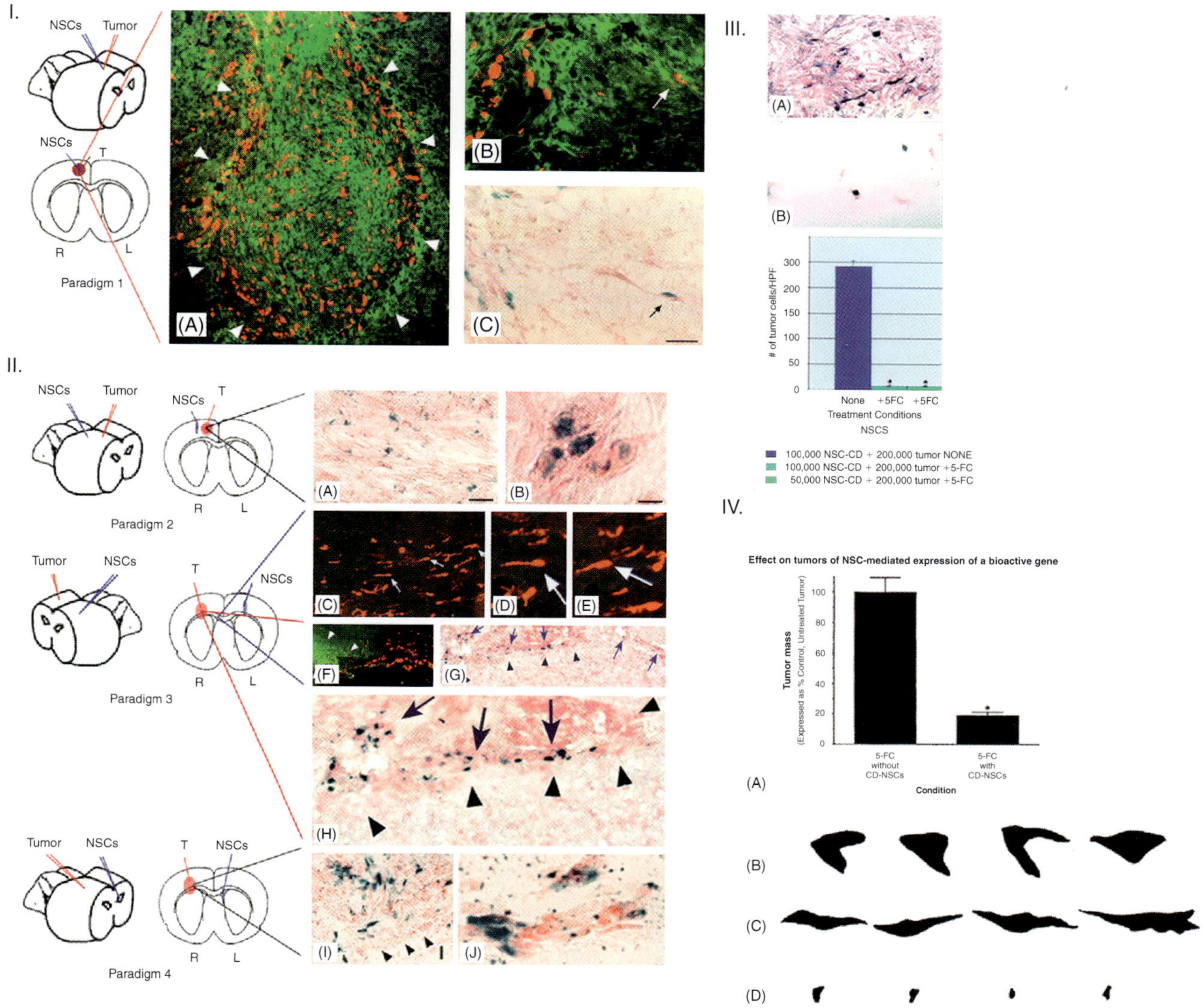

FIGURE 20-2 Neural stem cells display extensive tropism for pathology in the adult brain and can express bioactive genes within such pathological situations: evidence from intracranial gliomas. (A) NSCs migrate extensively throughout a brain tumor mass *in vivo* and "trail" advancing tumor cells. Paradigm (1), in which NSCs are implanted directly into an established experimental intracranial glioblastoma, is illustrated schematically. (1) The virulent and aggressively-invasive CNS-1 glioblastoma cell line, used to create the tumor, has been labeled *ex vivo* by transduction with GFP cDNA. The NSCs stably express *lacZ*, and produce βgal. This panel, processed for double immunofluorescence using an anti-βgal antibody (NSCs) and an anti-GFP antibody (glioblastoma cells), shows a section of brain (under low power) from an adult nude mouse sacrificed 10 days after NSC injection into the glioblastoma; arrowheads mark where the tumor mass interfaces with normal tissue. Donor βgal+ NSCs can be seen extensively distributed throughout the mass, interspersed among the tumor cells. This degree of interspersion by NSCs after injection occurs within 48 hours. Interestingly, although NSCs have extensively migrated and distributed themselves within the mass, they largely stop at the junction between tumor and normal tissue, except where a tumor cell is infiltrating normal tissue; then, NSCs appear to follow the invading tumor cell into surrounding tissue. (1 and 2) Detail of the trailing of individual glioblastoma cells migrating from the main tumor bed. (2) High power view, under fluorescence microscopy, of single migrating infiltrating GFP+ tumor cells in apposition to βgal+ NSCs (white arrow). (3) Costaining with Xgal (the arrow points to the *lacZ*-expressing NSCs) and with neutral red (for the elongated glioblastoma cells). The NSC is in direct juxtaposition to a single migrating, invading the neutral red+, spindle-shaped tumor cell (arrow), with the NSC "piggy-backing" the glioma cell. Scale bar = 60 μm. (B) NSCs implanted at various intracranial sites far from main tumor bed migrate through normal adult tissue toward glioblastoma cells. (1 and 2) Same hemisphere: A section through the tumor from an adult nude mouse six days following NSC implantation caudal to tumor. Panel 1 shows a tumor populated as pictured under low power in Figure 20-1A. Note Xgal+ NSCs interspersed among neutral red+ tumor cells. (2) High-power view of NSCs in juxtaposition to islands of tumor cells. (3–8) Contralateral hemisphere. (2–5) Views through the corpus callosum (c), where βgal+ immunopositive NSCs (arrows) are migrating from their site of implantation on one side of the brain toward tumor on the other. Two NSCs, indicated by arrows in panel 3, are viewed at higher magnification in panels 4 and 5 to show the classic elongated morphology and leading process of a migrating neural progenitor oriented toward its target. (6) βgal+ NSCs are "homing in" on the GFP+ tumor, having migrated from the other hemisphere. In panel 7, and magnified further in panel 8, the Xgal+ NSCs (arrows) have entered the neutral red+ tumor (arrowheads) from the opposite hemisphere. (9 and 10) Intraventricular: A section through the brain tumor of an adult nude mouse six days after NSC injection into the contralateral cerebral ventricle. (9) Xgal+ NSCs are distributed within the neutral red+ main tumor bed (edge delineated by arrowheads). (10) At higher power, the NSCs are in juxtaposition to migrating islands of glioblastoma cells. Fibroblast control cells never migrated from their injection site in any paradigm. All Xgal positivity was corroborated by anti-βgal immunoreactivity. Scale bars = 20 μm (1; also applies to 3) 8 μm (2), 14 μm (4 and 5), 30 μm (6 and 7), 15 μm (8), 20 μm (9), and 15 μm (10). (C) Bioactive transgene (cytosine

FIGURE 20-2 (*Continued*)
deaminase, CD) remains functional (as assayed by *in vitro* oncolysis) when expressed within NSCs. CNS-1 glioblastoma cells were co-cultured with murine CD-NSCs. (1) Co-cultures unexposed to 5-FC grew healthily and confluent, (2) whereas plates exposed to 5-FC showed dramatic loss of tumor cells represented quantitatively by the histograms ($* = p < 0.001$). The oncolytic effect was identical whether 1×10^5 CD-NSCs or half that number were co-cultured with a constant number of tumor cells. (Subconfluent NSCs were still mitotic at the time of 5-FC exposure, and thus subject to self-elimination by the generated 5-FU and its toxic metabolites.) (D) Expression of *CD* delivered by NSCs *in vivo* as assayed by reduction in tumor mass. The size of an intracranial glioblastoma populated with CD-NSCs in an adult nude mouse treated with 5-FC was compared with that of tumor treated with 5-FC, but lacking CD-NSCs. These data, standardized against and expressed as a percentage of a control tumor populated with CD-NSCs receiving no treatment, are in the histograms in panel 1. These measurements were derived from measuring the surface area of tumors; camera lucidas of them are in panels 2–4. Note the large areas of panel 2 a control non-5-FC-treated, tumor-containing CD-NSCs and in panel 3 a control 5-FC-treated, tumor-lacking CD-NSCs as compared to panel 4 the dramatically smaller tumor areas of the 5-FC-treated animal, which also received CD-NSCs (~80% reduction, $* = p < 0.001$), suggesting both activity and specificity of the transgene. (3) The lack of effect of 5-FC on tumor mass when no CD-bearing NSCs were within the tumor was identical to panel 4 the effect of CD-NSCs in the tumor without the gene being employed.

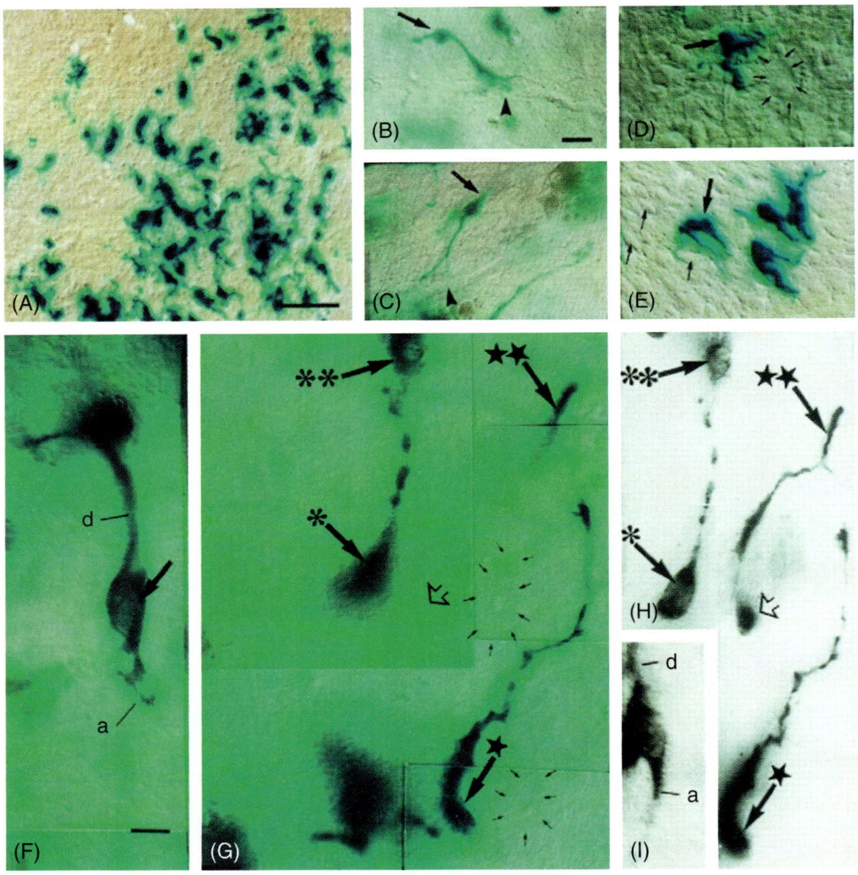

FIGURE 20-3 Multipotent NSCs acquire neuronal morphology in regions of adult neocortex subjected to targeted apoptotic neuronal degeneration. They differentiate into only glia, or remain undifferentiated in intact control cortex. (A) Engrafted X-Gal+ glia at low magnification six weeks following transplantation at 12 weeks; (B–E) higher magnification. (B, C) Donor-derived cells with astroglial features: Small, ovoid cell bodies (arrows) with few, short processes often extending as perivascular end-feet (arrowheads). (D) Small soma of a donor-derived presumptive glial cell (arrow) compared with a much larger (≈30 μm), unlabeled, host pyramidal neuron (small arrows). (E) Donor-derived cells with oligodendroglial features (arrow): Multiprocessed, ensheathing neuronal processes (short arrows). (F–I) A total of 15 ± 7% of engrafted cells in regions of neurodegeneration developed neuronal morphology, resembling pyramidal neurons within layer II/III six weeks following transplantation at 12 weeks. (F, G) Donor-derived cells with neuronal morphology (large arrows): Large somata (20–30 μm diameter) comparable to residual host pyramidal neurons (small arrows in panel G outline two host neurons, visualized under DIC, each with the characteristic large nucleus and prominent nucleolus of a pyramidal neuron), 300–600 μm presumptive apical dendrites positioned between host (small arrows) and donor (large arrows) neurons of similar morphology and size, and presumptive axons. (F) The dark object at the upper end of the presumptive dendrite is another X-Gal+ cell out of the plane of focus. (H) DCM image of panel G, collapsing multiple planes of focus. The cell has the characteristic large nucleus of a pyramidal neuron. All but the terminal dendrite of the soma of the cell is out of the plane of focus in panel G. Cells and neurons are indicated to cross-reference views of the same field in panels G and H. (I) Two overlapping donor-derived pyramidal neurons in different focal planes show a characteristic large nucleus, prominent nucleolus, and axon of the overlying cell and a prominent dendrite of the underlying cell, imaged through multiple focal planes of this thick section under DCM. The identification of the previously-mentioned cells was supported by immunocytochemical and ultrastructural analysis under electron microscopy. In addition, one could determine that donor-derived neurons received synaptic input from the host and were myelinated by host oligodendrocytes, further supporting their incorporation into the host cytoarchitecture (a: axon; d: dendrite; cells: *,★★,** and open arrow; and *: neurons). Bars = 25 μm.

states—that is, diseases in which the pathology is restricted to a particular cell whose lifespan is short-circuited, but within an extracellular milieu otherwise normal. Conversely, a "cell nonautonomous" condition, where normal cells die because of an inhospitable microenvironment, would appear not to be amenable to NSC therapy, given that "replacement" cells would presumably meet a similar fate. Surprisingly, however, NSCs may nevertheless be useful in such cell-extrinsic pathological conditions. It appears that NSCs, particularly in their immature, undifferentiated state, may be more resistant to certain stressors—various toxic metabolites, oxidizing agents—than more mature cells. Furthermore, if the NSCs are not inherently more resistant, they can be readily genetically engineered *ex vivo* to become more resistant.

Despite their extensive plasticity, NSCs never produce cell types inappropriate to the brain (e.g., muscle, bone, or teeth), or yield neoplasms. The use of NSCs as graft material in the CNS may be considered almost analogous to hematopoietic stem cell-mediated reconstitution of the bone marrow. Therefore, the biological repertoire of the NSC, if harnessed, may provide multiple strategies for addressing CNS dysfunction. Some of these approaches have already shown promise experimentally in animal models of neurodegeneration. Some illustrative examples are briefly described in the next sections of this chapter.

GENE THERAPY USING NEURAL STEM CELLS

As stated previously, the ability of NSCs to deliver therapeutic gene products in an immediate, direct, sustained, and perhaps regulated fashion as normal cytoarchitectural components throughout the CNS may overcome some of the limitations of standard viral and cellular vectors (reviewed by Park *et al.*, 2002). The feasibility of this strategy was first demonstrated in a mouse mutant characterized by a single gene defect in all cells, including those in the CNS, leading to their death. The particular mouse modeled the lysosomal storage disease mucopolysaccharidosis type VII (MPS VII), caused by a deletion mutation of the β-glucuronidase (*GUSB*) gene. This incurable inheritable condition is characterized by neurodegeneration in mice, and by progressive mental retardation in humans. Although the particular disease was rare, it served as a model for neurological diseases whose etiology stemmed from a genetically-based loss of function. GUSB-secreting NSCs were implanted into the cerebral ventricles of newborn MPS VII mice, allowing the cells access to the subventricular zone (SVZ), a germinal zone from which the cells were disseminated throughout the brain (creating "chimeric forebrains"). These enzyme-producing cells, now in residence as normal cerebral constituents, not only metabolized lysosomal storage normally for themselves, but also cross-corrected mutant cells throughout the brains of recipient mice (Figure 20-1). Employing a similar strategy, retrovirally-transduced NSCs implanted into the brains of fetal and neonatal mice (particularly into periventricular regions) have successfully mediated brain-wide expression of other enzymes—for example, β-hexosaminidase, a deficiency of which leads to the pathological accumulation of GM_2 ganglioside.

Findings such as these have helped to establish the paradigm of using NSCs for the transfer of other factors of therapeutic or developmental interest into and throughout the CNS. Although NSCs express baseline amounts of particular enzymes and neuroprotective factors, NSCs can be genetically-modified to enhance their production of these molecules, or to produce additional molecules that might enhance their therapeutic potential. For example, NSC have been used to deliver NT-3 within the hemisectioned rat spinal cord, to express NGF and BDNF within the septum and basal ganglia, to provide tyrosine hydroxylase to the parkinsonian striatum, and to express myelin basic protein (MBP) throughout the dysmyelinated cerebrum. In one illustrative set of experiments, NSCs that over-expressed NGF were transplanted into the striatum of a rat lesioned with quinolinic acid, a toxin used to emulate some of the neuronal loss seen in Huntington's disease. Delivery of NGF by the engrafted NSCs appeared to reduce the size of the lesion, and to promote sparing of host striatal neurons. When implanted into the septum of aging rats, NGF-overexpressing NSCs appeared to blunt the typical cognitive decline by preventing age-related atrophy of forebrain cholinergic neurons. Well-integrated into the host tissue, the engrafted NSCs continued to produce NGF until nine months after grafting.

That NSCs appear to have a strong affinity for sites of pathology, and will migrate extensive distances (even from the opposite cerebral hemisphere) to home in on them, makes the NSC unique and valuable as a gene delivery vehicle. For example, a type of pathology particularly elusive to gene therapeutic interventions has been the brain tumor, especially the glioma. Gliomas are so exceptionally migratory and infiltrative that they elude even the most effective surgical, radiation, or gene therapeutic strategy. However, the ability of transgene-expressing NSCs to "track down" and deliver therapeutic gene products directly to these widely-dispersed and invasive neoplastic cells makes them a potentially valuable adjunct in the treatment of these aggressive tumors.

Although this therapeutic approach—exploiting the normal biological behavior of NSCs for transplantation-based gene therapy—is being extended to animal models of many neurological disorders, it is important to recognize that each pathophysiological process, each animal model, and each therapeutic molecule must be assessed and optimized individually.

CELL REPLACEMENT USING NEURAL STEM CELLS

It is postulated that NSCs persist within the CNS long beyond cerebrogenesis to maintain homeostasis following perturbations in the CNS. Presumably, the mechanisms that allow the endogenous NSC to perform this function are preserved when the cells are isolated from the CNS, expanded in culture, and reimplanted into the damaged CNS. Indeed, it appears that when NSCs are so implanted, they respond by shifting their pattern of differentiation toward replenishing the missing cell type.

The study that first demonstrated this phenomenon was performed on a model of experimentally-induced apoptosis of selectively targeted pyramidal neurons in the adult mammalian neocortex. When transplanted into this circumscribed, neuron-depleted region, 15% of the NSCs altered the differentiation path they would otherwise have pursued in the intact adult, non-neurogenic mammalian neocortex (i.e., to become glia); instead, they differentiated specifically into pyramidal neurons, partially replacing that lost neuronal population (Figure 20-3). Of these, a sub-population spontaneously sent axonal projections to their proper targets in the contralateral cortex. Outside the borders of this small region of selective neuronal death, the NSCs yielded only glia. Thus, neurodegeneration appeared to create a milieu that recapitulated embryonic development cues (e.g., for cortical neurogenesis), and the NSCs were sufficiently sensitive to detect and respond to this micromolecular alteration—possibly to therapeutic advantage.

Apoptosis is implicated in a growing number of both neurodegenerative and normal developmental processes. Whether this differentiation shift occurs only in response to signals associated with apoptosis, or in response to other types of cell death as well, remains to be determined. Most neurological diseases are characterized by a mixture of neuropathological processes, making dissection of the key stimuli complex. For example, we have observed that, following an ischemic insult, NSCs will robustly repopulate infarcted regions of the postdevelopmental cortex, and differentiate into cortical neurons in these regions. However, even hypoxic–ischemic cerebral injury—the quintessential example of necrotic and excitotoxic injury—is characterized by an apoptotic phase. An added level of complexity may be the tempo at which apoptotic signals are elaborated—an acute burst may be more instructive than a more languid burst.

Studies are ongoing to identify the key molecules that direct NSC fate during and following various kinds of neural injury. Intriguingly, we are starting to learn that cytokines released during an inflammatory reaction—those emanating from microphages and microglia, as well as from parenchymal damage (e.g., SDF1-α)—play a pivotal role in "beckoning" NSCs. Coincidentally, we are starting to recognize that many processes regarded as neurodegenerative—including amyotrophic lateral sclerosis (ALS), Alzheimer's disease, tumors, and stroke—are characterized by a prominent inflammatory signature. Although each is different in its etiology, region of involvement, and course, inflammation may be their common denominator from the "viewpoint" of the NSC.

"GLOBAL" CELL REPLACEMENT USING NEURAL STEM CELLS

The targeted-apoptosis model described in the previous section exemplifies a circumscribed type of neural cell loss. However, the pathologic lesions of many neurological disorders are often widely—even globally—dispersed throughout the brain. Such conditions include the neurodegenerative disorders of childhood (e.g., inborn errors of metabolism, storage diseases, leukodystrophies, and neuronal ceroid lipofuscinoses) and hypoxic–ischemic encephalopathy, as well as some adult CNS diseases (e.g. multiple sclerosis, Alzheimer's disease, and ALS). Treatment for these disorders requires widespread replacement of genes, cells, or both, as well as the regeneration, protection, or both, of broad networks of neural circuitry. The ability of inherently migratory NSCs to integrate into germinal zones from which they can be "launched," and the inclination of NSCs to travel long distances to home in on pathologic regions, makes these cells ideally—and perhaps uniquely—suited for this task.

Mouse mutants characterized by CNS-wide white matter disease provided the first models for testing the hypothesis that NSCs might be useful against neuropathologies requiring widespread neural cell replacement. The oligodendrocytes—the myelin-producing cells—of the dysmyelinated *shiverer (shi)* mouse are dysfunctional because they lack MBP, a molecule essential for effective myelination. Therapy, therefore, requires widespread replacement with MBP-expressing oligodendrocytes. NSCs transplanted at birth (employing the intracerebroventricular implantation technique described previously for the diffuse engraftment of enzyme-expressing NSCs to treat global metabolic lesions) resulted in engraftment throughout the *shi* brain with repletion of significant amounts of MBP. Of the many donor cells that differentiated into oligodendrocytes, a subgroup myelinated ~40% of host neuronal processes. In some recipient animals, the symptomatic tremor decreased (Figure 20-4). Therefore, "global" cell replacement was shown to be feasible for some pathological conditions if cells with stem-like features are employed. This approach has been extended to other myelin-impaired animal models, such as the experimental allergic encephalomyelitis mouse model of multiple sclerosis, as well as rodent mutant models of Palezeus-Merzbacher, Krabbe, and Canavan leukodystrophies. Recently, it was demonstrated that neural progenitor stem cells of human origin have a similar remyelinating capacity in the *shi* mouse

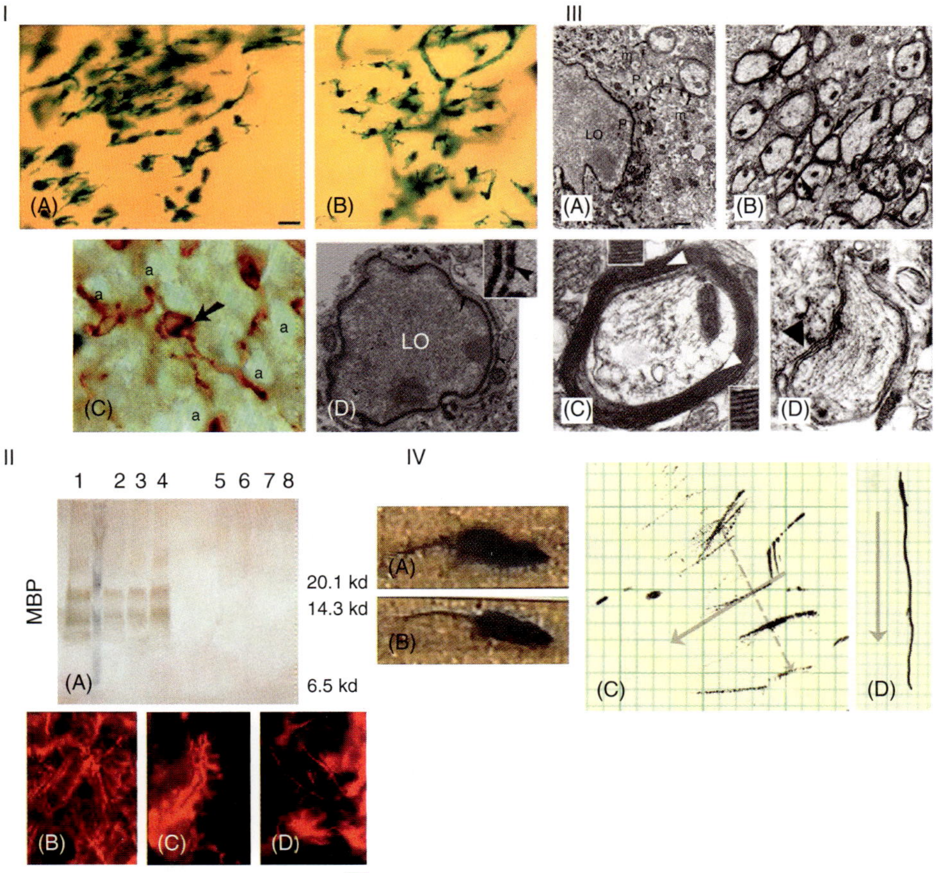

FIGURE 20-4 "Global" cell replacement is feasible using NSC transplantation: evidence from the dysmyelinated *shi* mouse brain. (A) NSCs engraft extensively throughout the *shi* dysmyelinated brain, including within white tracts, and differentiate into oligodendrocytes. *LacZ*-expressing, βgal-producing NSCs were transplanted into newborn *shi* mutants and analyzed systematically at intervals between two and eight weeks following engraftment. Coronal sections through the *shi* brain at adulthood demonstrated widely-disseminated integration of Xgal+ donor-derived cells throughout the neuraxis, similar to the pattern seen in Figure 20-1A in the MPS VII mutant mouse. (1 and 2) Donor-derived Xgal+ cells (higher magnification) in sections through the corpus callosum possessed characteristic oligodendroglial features (small, round, or polygonal cell bodies with multiple fine processes oriented toward the neural fiber tracts). (3) Close-up of a donor-derived anti-β immunoreactive oligodendrocyte (arrow) extending multiple processes toward, and beginning to enwrap, large, adjacent axonal bundles (a) viewed on end in a section through the corpus callosum. That cells such as those in panel A 1–3 and in panel B 2–4 were oligodendroglia was confirmed by the electron micrograph in panel A 4 and in panel C, demonstrating their defining ultrastructural features. A donor-derived Xgal+ oligodendrocyte (LO) can be distinguished by the electron dense Xgal precipitate, typically localized to the nuclear membrane (arrow), ER (arrowhead), and other cytoplasmic organelles. The ER is magnified in the inset to demonstrate the unique crystalline nature of individual precipitate particles. (B) MBP expression in mature transplanted and control brains. (1) Western analysis for MBP in whole brain lysates. The brains of three transplanted *shi* mutants (lanes 2–4) expressed MBP at levels close to those of an age-matched unaffected mouse (lane 1, positive control) and significantly greater than the amounts seen in untransplanted (lanes 7–8, negative control) or unengrafted (lanes 5–6, negative control) age-matched *shi* mutants. (Identical total protein amounts were loaded in each lane.) (2–4) Immunocytochemical analysis for MBP. (2) The brain of a mature unaffected mouse was immunoreactive to an antibody to MBP (revealed with a Texas red-conjugated secondary antibody). (3 and 4) Age-matched engrafted brains from *shi* mice similarly showed immunoreactivity. Because untransplanted *shi* brains lack MBP, MBP immunoreactivity has classically been a marker for normal donor-derived oligodendrocytes in transplant paradigms. (C) NSC-derived "replacement" oligodendrocytes are capable of myelination of *shi* axons. In regions of MBP-expressing NSC engraftment, *shi* neuronal processes became enwrapped by thick, better-compacted myelin. (1) Two weeks after transplant, a donor-derived, labeled oligodendrocyte (LO), recognized by extensive Xgal precipitate (p) in the nuclear membrane, cytoplasmic organelles, and processes, was extending processes (arrowheads) to host neurites, and was beginning to ensheathe them with myelin (m). (2) If engrafted *shi* regions (1) were followed to four weeks, the myelin began to appear healthier, thicker, and better-compacted (arrows) than that in age-matched untransplanted control mutants. (3) By six weeks after transplant, these matured into even thicker wraps; ~40% of host axons were ensheathed by myelin (white arrowheads; MDLs are evident) that was dramatically thicker and better-compacted than that of *shi* myelin (4, black arrowhead) from an unengrafted region of an otherwise successfully engrafted *shi* brain). (D) Functional and behavioral assessment of transplanted *shi* mutants and controls. The *shi* mutation is characterized by the onset of tremor and a "shivering gait" by postnatal week two or three. The degree of motor dysfunction in animals was gauged in two ways: by blindly scoring periods of standardized, videotaped cage behavior of experimental and control animals; and by measuring the amplitude of tail displacement from the body's rostral–caudal axis (an objective, quantifiable index of tremor). Video freeze-frames of (4.4a) unengrafted and (4.4b) successfully engrafted *shi* mice. (4.4a) The whole body tremor and ataxic movement observed in the unengrafted symptomatic animal causes the frame to blur, a contrast with (1) the well-focused frame of the asymptomatic transplanted *shi* mouse. Of transplanted mutants, 60% evinced nearly normal-appearing behavior (2) and attained scores similar to normal controls. (3 and 4) Whole body tremor was mirrored by the amplitude of tail displacement (dotted gray arrow), measured perpendicularly from a line drawn in the direction of the animal's movement. Measurements were made by permitting a mouse, whose tail had been dipped in Indian ink, to move freely in a straight line on a sheet of graph paper. (3) Large degrees of tremor cause the tail to make widely-divergent ink marks from the midline, representing the body's axis (solid gray arrow). (4) Absence of tremor allows the tail to make long, straight, uninterrupted ink lines on the paper congruent with the body's axis. The distance between points of maximal tail displacement from the axis was measured and averaged for transplanted and untransplanted *shi* mutants and for unaffected controls (dotted gray arrow). Panel 3 shows data from a poorly-engrafted mutant that did not improve with respect to tremor, whereas panel 4 reveals lack of tail displacement in a successfully-engrafted asymptomatic mutant. Overall, 64% of transplanted *shi* mice examined displayed at least a 50% decrement in the degree of tremor or "shiver." Several showed no displacement.

brain. Viewed more parochially, the ability of NSCs to remyelinate is of significant importance, because impaired myelination plays such an important role in many genetic (e.g., leukodystrophies and inborn metabolic errors) and acquired (e.g., traumatic, infectious, asphyxial, ischemic, and inflammatory) neurodegenerative processes. However, viewed more broadly, the ability of the NSC to replace this particular type of neural cell throughout the brain bodes well for its ability to replace other classes of neural cell across the broad terrains often demanded by other categories of complex neurodegenerative diseases.

NEURAL STEM CELLS DISPLAY AN INHERENT MECHANISM FOR RESCUING DYSFUNCTIONAL NEURONS

The examples cited previously highlight that an abnormal environment can direct the behavior of a grafted NSC. However, they leave the impression that the exogenous NSC alone fills gaps. The situation is more complex and richer. We are beginning to learn that the NSC and the injured host engage in a dynamic series of ongoing reciprocal interactions, each instructing the other. Under instruction from exogenous NSCs, the injured host nervous system also contributes to its own repair. These important stem cell phenomena were first illustrated in a few examples.

The effect of NSCs in directly rescuing endangered host neurons was first evinced in a series of experiments in aged rodents in which the nigrostriatal system was impaired (Figure 20-5). Parkinson's disease is a degenerative disorder characterized by a loss of midbrain dopamine (DA) neurons, with a subsequent reduction in striatal DA. The disease, in addition to incapacitating thousands of patients, has long served as a model for testing neural cell replacement strategies. Transplantation therapy for this CNS disorder has a long history (for a review, see Isacson *et al.*). It was the neural disease first treated clinically by neural transplantation, using primary tissue from human fetal ventral mesencephalon to replace DA-expressing cells. Indeed, in this disease, the limitations of fetal tissue grafts not only in rodent and primate models of Parkinson's disease, but also in clinical trials, were first recognized. These limitations include, on the one hand, short graft survival and limited integration of the grafts and, on the other hand, the possibility of unregulated DA production in improper regions, leading adversely to dyskinesias. Given the storied role of Parkinson's disease in the development of cellular therapies, it is appropriate that a model of this disease should also have played a pivotal role in revealing a little-suspected but powerful therapeutic action that NSCs may play in preserving degenerating host cells by some heretofore unheralded mechanisms that are nevertheless inherent to stem cell biology.

In the hope that NSCs might spontaneously differentiate into DA neurons when implanted into a DA-depleted region of the CNS, unmanipulated murine NSCs were implanted unilaterally into the substantia nigra of aged mice exposed systemically to high-dose 1-methyl-4-phenyl-1,2,3,6-tetrahydropyridine (MPTP), a neurotoxin that produces a persistent impairment of mesencephalic DA neurons and their striatal projections. The NSCs not only migrated from their point of implantation and integrated extensively within both hemispheres, but also were associated with a dramatic reconstitution of DA function throughout the mesostriatal system. Although there was spontaneous conversion of a sub-population of donor NSCs into dopaminergic neurons in DA-depleted areas, contributing to nigral reconstitution, most (80–90%) dopaminergic neurons in the "reconstituted midbrain" were actually host cells that had been "rescued" by factors produced constitutively by the NSCs with which they were juxtaposed and that had not become neurons. These chaperone cells constitutively produce substantial amounts of neurosupportive agents. One such prominent molecule was GDNF, a factor known to be neuroprotective of ventrally located neurons (including DA neurons and spinal–ventral horn cells). A similar observation is beginning to emerge from the implantation of human NSCs into the MPTP-lesioned, subhuman primate model of Parkinson's disease.

A sense of the extent of the cross talk also became evident when examining rodent models of hypoxia–ischemia, a common cause of neurological disability in adults and children. Hypoxia–ischemia causes much of its damage from extensive loss of cerebral parenchyma, and the cells and connections that reside there. When NSCs are implanted into these regions of extensive degeneration (particularly when transiently supported by biodegradable scaffolds), robust reciprocal interactions ensue spontaneously between the exogenous implant and the injured host brain, substantially reconstituting parenchyma and anatomical connections, as well as reducing parenchymal loss, secondary cell loss, inflammation, and scarring. Similar results are observed in the hemi-resectioned adult rodent spinal cord, in which evidence of an up-regulated host neuronal regenerative response is noted, resulting in significant functional improvement.

Indeed, the ability of engrafted NSCs to exert a protective and regenerative influence on degenerating host neural systems, because of their intrinsic expression of trophic factors, is being observed in an increasing number of conditions. For example, the implantation of murine and human NSCs into the spinal cords of the SOD1 transgenic mouse model of ALS, a disease characterized by virulent motor neuron degeneration, has been pivotal in protecting these ventral horn cells from death, preserving motor and respiratory function, blunting disease progression, and extending life. NSCs can similarly protect other neuron pools, promote motor axonal outgrowth following traumatic spinal cord injury, preserve infarcted regions of

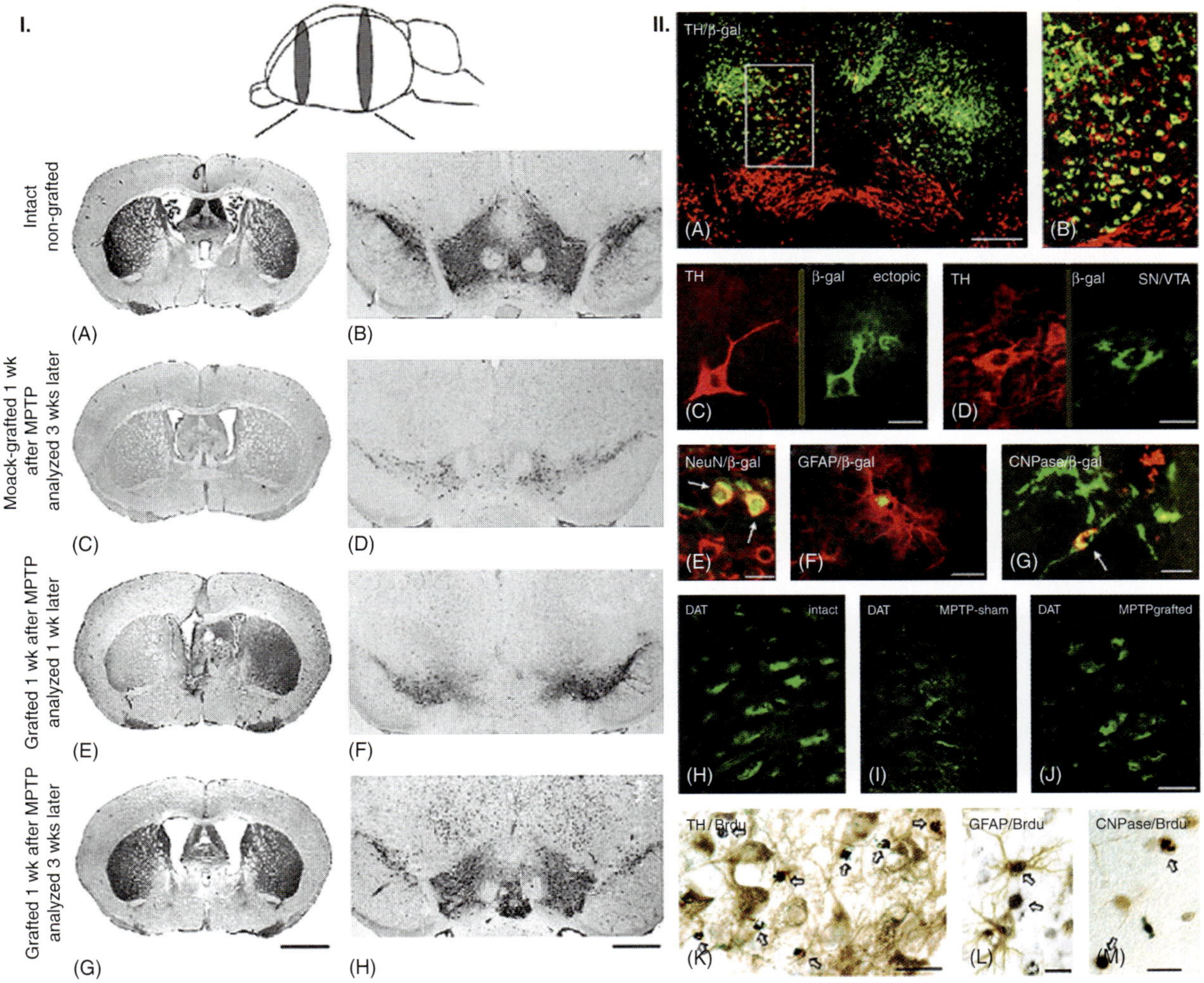

FIGURE 20-5 NSCs possess an inherent mechanism for rescuing dysfunctional neurons: evidence from the effects of NSCs in the restoration of mesencephalic dopaminergic function. (A) TH expression in mesencephalon and striatum of aged mice following MPTP lesioning and unilateral NSC engraftment into the substantia nigra–ventral tegmental area (SN–VTA). A model that emulates the slow dysfunction of aging dopaminergic neurons in SN was generated by giving aged mice repeated high doses of MPTP. Schematic indicates the levels of the analyzed transverse sections along the rostrocaudal axis of the mouse brain. Coronal sections are through the striatum in the left column and through the SN–VTA area in the right column. Immunodetection of TH (black cells) shows the normal distribution of DA-producing TH+ neurons in: (2) coronal sections in the intact SN–VTA; and (1) their projections to the striatum. Within one week, MPTP treatment caused extensive and permanent bilateral loss of TH immunoreactivity in both: (3) the mesostriatal nuclei; and (4) the striatum. Shown in this example, and matching the time point in 7 and 8, is the situation in a mock-grafted animal four weeks after MPTP treatment. Unilateral (right side) stereotactic injection of NSCs into the nigra is associated, within one week after grafting, with substantial recovery of TH synthesis within: (6) the ipsilateral DA nuclei; and (5) their ipsilateral striatal projections. By three weeks after transplant, however, the asymmetric distribution of TH expression disappeared, producing TH immunoreactivity in: (8) the midbrain; and (7) striatum of both hemispheres that approached the immunoreactivity of (1 and 2) the intact controls, and gave the appearance of mesostriatal restoration. Similar observations were made when NSCs were injected four weeks after MPTP treatment (not shown). Bars: 2 mm (left), 1 mm (right). Note the ectopically placed TH+ cells in panel 8. These are analyzed in greater detail with the entire SN in B. (B) Immunohistochemical analyses of TH, DAT, and BrdU+ cells in MPTP-treated and grafted mouse brains. The initial presumption was that the NSCs had replaced the dysfunctional TH neurons. However, examination of the reconstituted SN with dual βgal and TH ICC showed that: (1 and 3) 90% of the TH+ cells in the SN were rescued, host-derived cells; and (4) only 10% were donor-derived cells. Most NSC-derived TH+ cells were just above the SN ectopically (blocked area in panel 1; enlarged in panel 2). These photomicrographs were taken from immunostained brain sections from aged mice exposed to MPTP, transplanted one week later with NSCs and sacrificed after three weeks. The following combinations of markers were evaluated: (1–4) TH with βgal; (5) NeuN with βgal; (6) GFAP with βgal; (7) CNPase with βgal; (8) TH with BrdU; (9) GFAP with BrdU; and (10) CNPase with BrdU. Anti-DAT-stained areas are revealed in the SN of: (11) intact; (12) mock-grafted; and (13) NSC-grafted brains. Fluorescence filters for Alexa Fluor 488 and Texas red and a double-filter for both types of fluorochromes were used to show antibody binding. (3, 4, and 8–10) Single-filter exposures; (1, 2, and 5–7) double-filter exposures. (1) Low-power overview of the SN–VTA of both hemispheres. Most TH+ cells within the nigra are of host origin (~90%), a much smaller proportion (~10%) are donor derived (close-up in panel 4). Although a significant proportion of NSCs differentiated into TH+ neurons, many of these resided ectopically, dorsal to the SN (boxed area in panel 1 enlarged in panel 2; and a high-power view of the donor-derived cell that was also TH+ in panel 3), where the ratio of donor-to-host cells was inverted: ~90% donor derived compared with ~10% host derived. Note the near absence of a βgal-specific signal in the SN–VTA, whereas ectopically, many of the TH+ cells were double labeled and thus NSC derived. (5) NSC-derived non-TH neurons (NeuN+, arrow); (6) astrocytes (GFAP+); and (7) oligodendrocytes (CNPase+, arrow) were also seen, both within the mesencephalic nuclei and dorsal to them. (10) The DAT-specific signal suggests that the reconstituted mesencephalic nuclei in the NSC-grafted mice were functional DA neurons comparable to those seen (8) in intact nuclei, but not (9) in MPTP-lesioned, sham-engrafted controls. This further suggests that the TH+ mesostriatal DA neurons affected by MPTP are functionally impaired. Note that (9) sham-grafted animals contain only punctate residual DAT staining within their dysfunctional fibers, whereas DAT staining (8) in normal and (10) in engrafted animals was normally and robustly distributed, both within processes and throughout their cell bodies. (11–12) Any proliferative BrdU+ cells after MPTP insult, graft, or both were confined to glial cells, whereas (11) the TH+ neurons were BrdU–. This finding suggested that the reappearance of TH+ host cells was not the result of neurogenesis, but rather the recovery of extant host TH+ neurons. Bars = 90 μm (1), 20 μm (3–5), 30 μm (6), 10 μm (7), 20 μm (8–10), 25 μm (11), 10 μm (11), and 20 μm (12).

cerebrum, induce vascularization of reconstituted regions of cortical parenchyma, and inhibit inflammation and scarring following traumatic or ischemic insult.

Together, these observations suggest that exogenous NSCs may not only replenish inadequate pools of endogenous NSCs to compensate for missing neural cells, but also reactivate or enhance endogenous regenerative and protective capacities. It also highlights a heretofore unanticipated mechanism by which NSCs may exert a therapeutic effect—to be added to their more traditional roles in direct cell replacement and gene therapy.

NEURAL STEM CELLS AS THE GLUE THAT HOLDS MULTIPLE THERAPIES TOGETHER

Experimentally, one may eliminate a particular cell type, lesion a particular region of CNS, knockout a particular gene, or choose a mouse strain in which, by chance, certain mutations have spontaneously occurred. However, most human neurodegenerative diseases are not as "clean." They are quite complex. And complex diseases, such as those affecting the nervous system, will require complex and multifaceted solutions—including pharmacological, gene and molecular, cell replacement, tissue engineering, angiogenic, anti-inflammatory, antiapoptotic, pro-regenerative, and pro-neurite outgrowth-promoting therapies. The NSC, as a key player in a set of fundamental developmental mechanisms, may serve as the glue that holds many of these strategies together. How they may be intelligently, effectively, safely, and practically orchestrated in actual patients will require a good deal of careful investigation.

For example, as our sophistication about disease processes grow, we are beginning to learn that more than one neural cell type probably needs to be replaced in a disease. For example, in a disease like ALS, a disorder characterized by progressive motor neuron degeneration, we are beginning to learn that astrocyte replacement may be just as critical as motor neuron replacement. Conversely, in multiple sclerosis, a white matter disease characterized by oligodendrocyte degeneration, replacing neurons and their axonal connections may be critical for the restoration of function. These same caveats may apply to many diseases in which replacement of multiple cell types may be the key to neurological reconstitution and to damaged milieu reconstruction. Because of their ability to develop into multiple integral, cytoarchitectural components of many regions throughout the host brain, NSCs may be able to replace a range of missing or dysfunctional neural cell types within the same region. This is important in the likely situation in which return of function may require the reconstitution of the milieu of a given region (e.g., not just the neurons, but also the chaperone cells—the glia and support cells) to nurture, detoxify, guide, and/or myelinate the neurons. As noted previously, the NSCs, especially in particular differentiation states, express certain genes of intrinsic interest (many neurotrophic factors, lysosomal enzymes, angiogenic factors, anti-inflammatory molecules, antioxidants, etc.), or they can be engineered *ex vivo* to do so. The challenge remains, however, to coordinate these multifaceted therapies so that they work in concert synergistically, and not at cross-purposes.

SUMMARY

The ability of NSCs to migrate and integrate throughout the brain, as well as to disseminate a foreign gene product, is of great significance for the development of new therapies for neurodegenerative diseases in humans. Lethal, hereditary, neurodegenerative diseases of childhood, such as the gangliosidoses, leukodystrophies, neuronal ceroid lipofuscinoses, and other storage diseases, result in lesions throughout the CNS. Diseases of adult onset (e.g., Alzheimer's disease) are also diffuse in their pathology. Even acquired diseases, such as spinal cord injury, head trauma, and stroke, are broader in their involvement than typically assumed. Such abnormalities may benefit from the multifaceted approach NSCs may enable. First, they may replace a range of cell types—not only neurons to reconstitute neural connections, but also oligodendrocytes to elaborate myelin and astrocytes to serve trophic, guidance, protective, and detoxification functions. Second, NSCs may deliver exogenous genes that might restore normal metabolism, complement a factor deficiency, support the survival of a damaged host neuron, neutralize a hostile milieu, counteract a growth-inhibitory environment, promote neurite regrowth, and reform stable and functional contacts. As noted previously, many molecules are produced inherently by NSCs when they are in particular states of differentiation. The production of other molecules might require *ex vivo* genetic engineering. NSCs appear to have their greatest therapeutic effect when used in the early stages of a degenerative process, or in the subacute phase following injury.

A major requirement for the better use of NSCs will be a better understanding of the pathophysiology of the diseases to be targeted, i.e., knowing what aspects require repair, and which cell type or types require replacement or rescue. Another challenge will be determining how to exploit stem cell biology for chronic conditions—in other words, how the acute milieu can be recreated in the chronic environment such that NSCs behave therapeutically there as well.

Among the methodological hurdles will be to devise how, when, where, and with what frequency to deliver NSCs to adults with disseminated diseases in regions not fed by readily accessible germinal zones, e.g., the spinal cord in which the central canal is no longer patent. Linked to this is the need for a better understanding of the methods

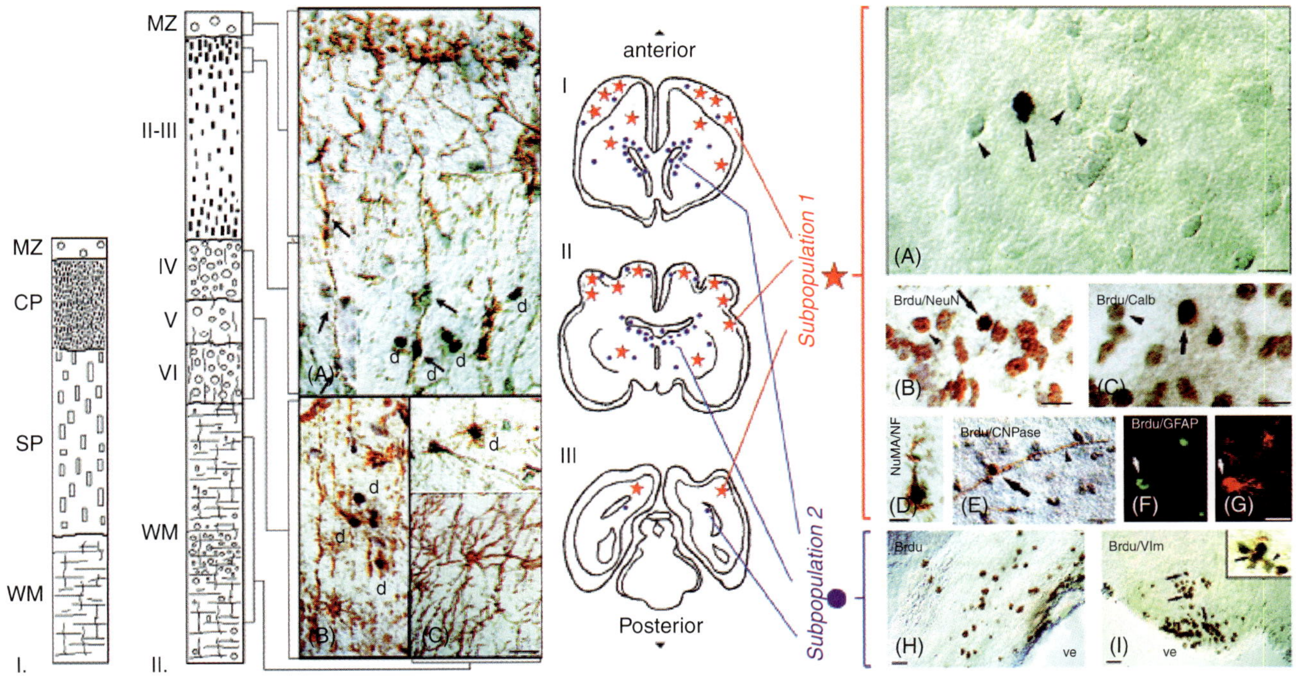

FIGURE 20-6 Clonal human NSCs migrate from the VZ into the developing neocortex. Schematics of the developing monkey neopallium (A) at the time of transplantation. (1) 12–13 wpc and (2) at the time of death (16–17 wpc). (3–5) Photomicrographs from selected locations spanning the neopallium. (Their location relative to the schematic is indicated by brackets.) (3) Injected into the left lateral ventricle and having integrated throughout the VZ, the human NSC-derived cells (d), identified by their BrdU immunoreactivity (black nuclei), migrated along the monkey's radial glial processes through the neopallial wall, to reach their temporally appropriate destination in the nascent superficial layers II and III (3), where they detached from the radial glia and took up residence as neurons. Arrows indicate climbing (donor- and host-derived) cells positioned along the processes of the vimentin-positive host radial glia. Some cells (inset) are still attached to these fibers and in the process of migration. (4 and 5) Immature, donor, human NSC-derived astrocytes intermixed with host-derived astrocytes in deeper cortical lamina, having differentiated as expected for that site and time. (B) Segregation of the fates of human NSCs and their progeny into two sub-populations in the brains of developing Old World monkeys. Schematics (left) and photomicrographs (right) illustrate the distribution and properties of clonal human NSC-derived cells. Human NSCs (labeled with BrdU) dispersed throughout and integrated into the VZ. From there, clonally-related human NSC-derived cells pursued one of two fates. Those donor cells that migrated outward from the VZ along radial glial fibers into the developing neocortex constituted one pool or sub-population. (4–12) The differentiated phenotypes of cells in this sub-population (1) (stars in the schematic), particularly in layers II and III. (4) A human NSC-derived BrdU+ cell (black nucleus, arrow)—likely a neuron according to its size, morphology, large nucleus, and location—is shown (under Nomarski optics) intermingled with the monkey's similar neurons (arrowheads) in neocortical layers II and III. The neuronal identity of such donor-derived cells is confirmed by immunocytochemical analysis. High-power photomicrographs of human donor-derived cells integrated into the monkey cortex double-stained with antibodies against BrdU and cell type-specific markers: (5) NeuN; and (6) calbindin for neurons (arrows, donor-derived cells; arrowheads, host-derived cells). (8) CNPase for oligodendroglia (arrow). (BrdU+ is the black nucleus in the CNPase+ cell; the arrowhead indicates the long process emanating from the soma.) (9, 10) GFAP for astroglia antibody to BrdU revealed through fluorescein; (9) antibody to GFAP revealed through Texas red. The human origin of the cortical neurons is independently confirmed in panel 7, where the human-specific nuclear marker NuMA (black nucleus) is colocalized in the same cell with neurofilament immunoreactivity. Progeny from this same human NSC clone were allocated to a second cellular pool—sub-population 2 (dots in the schematic; arrows in panels 9 and 10)—that remained mainly confined to the SVZ, and stained only for an immature neural marker. (Vimentin colocalized with BrdU is easily seen in inset (arrows); the arrowhead indicates the host vimentin-positive cell.) Some members of sub-population 2 were identified within the developing neocortex (dots) intermixed with differentiated cells. Panels 9 and 10 use immunofluorescence; the other immunostains use a DAB-based color reaction. Scale bars = 30 μm (4–6) and 20 μm (7–12). (d: human NSC-derived cells, MZ: marginal zone, CP: cortical plate, SP: subplate, WM: white matter, and II–VI: cortical layers).

for augmenting, yet controlling, the propagation and the phenotype specification of NSCs *in vitro* and *in vivo*.

Another question has come to dominate the stem cell field: what is the most effective way to obtain NSCs for therapy? Should they be obtained directly from regions of neuroectodermal origin, and if so, should that be from a fetal source? Can an adult source be equally effective? Should stable lines of NSCs be established that can be used for all patients, or should NSCs be abstracted from each patient to be used as an autograft on a case-by-case basis? The degree to which the immune system presents a barrier to stem cell transplantation will likely determine the answers to this question and its subquestions. If ES cells (from the inner cell mass of blastocysts) are directed to become NSCs *in vitro*, will such cells be equally as safe and effective?

The success in isolating stem-like cells from the CNS (the first "solid organ stem cells" discovered), and their therapeutic potential, also gave rise to the search for and successful isolation of stem-like cells from other non-neural organ systems, including bone marrow mesenchyme, muscle, skin, retina, and liver, for the purposes of repairing those tissues. Whether such *non-neural* stem cells can yield *neural* stem cells—through metaplasia or transdifferentiation—remains unresolved and exceptionally controversial. Can NSCs be derived efficiently and effectively from non-neural organs with outcomes as good as from NSCs themselves?

A better understanding of fundamental NSC biology will be required before human NSCs can be transplanted efficaciously in true clinical settings. Nevertheless, progress in this regard is being made. Several NSC lines have been established from the human fetal telencephalon and spinal cord that seem to emulate many of the appealing properties of their rodent counterparts: They differentiate *in vitro* and *in vivo* into neurons, astrocytes, and oligodendrocytes; they follow appropriate developmental programs and migrational pathways similar to endogenous precursors following engraftment into developing rodent and subhuman primate brain (Figure 20-6); they express foreign genes *in vivo* in a widely-disseminated manner; and they can replace missing neural cell types when grafted into various mutant mice or rodent models of injury. The principal differences between human NSCs and mouse NSCs so far seem to be the length of the cell cycle (up to four days in the human), and the predilection of human NSCs to senesce (after ~100 cell divisions), obstacles that are being actively addressed. If human NSCs behave in lesioned subhuman primate brains, with safe and effective engraftment and foreign gene expression, as they seem to do in rodents, then human trials might be warranted for testing their value against certain genuine clinical neurodegenerative diseases. Through a careful and circumspect series of experiments and trials, we may learn whether we, indeed, have found within Nature's own toolbox a powerful and versatile therapeutic tool—one of the goals of experiments that started more than 15 years ago.

KEY WORDS

Acute degeneration Damage is restricted to a circumscribed region (both in space and time), multiple neural cells types must be replaced (for example stroke, CNS injury, and spinal injury).

Chronic degeneration Progressive neurodegenerative diseases requiring the replacement of all neural cell types in large brain regions, in the face of ongoing degeneration and possibly the presence of toxic molecules.

Demyelination disorders Mainly of two classes; disorders in which endogenous myelinating cells are genetically impaired and fail to myelinate during development, and disorders that result from loss/failure of remyelination in the adult (for example, multiple sclerosis, allergic encephalomyelitis, and injury).

Neural stem cells (NSCs) NSCs are the most primordial cells of the nervous system. They generate the array of specialized cells throughout the CNS (and probably the peripheral, autonomic, and enteric nervous systems as well). NSCs are operationally defined by their ability: (1) to differentiate into cells of all neural lineages (i.e., neurons, ideally of multiple subtypes; oligodendroglia, and astroglia) in multiple regional and developmental contexts; (2) to self-renew; and (3) to populate developing and/or degenerating CNS regions.

FURTHER READING

Lindvall, O., Kokaia, Z., & Martinez-Serrano, A. (2004). Stem cell therapy for human neurodegenerative disorders—how to make it work. *Nat. Med.*, *10* Suppl, S42–S50.

Park, K. I., Ourednik, V., Taylor, R. M., Aboody, K. S., Auguste, K. I., Lachyankar, M. B., Redmond, D. E., & Snyder, E. Y. (2002). Global gene and cell replacement strategies via stem cells. *Gene Ther.*, *9*, 613–624.

Snyder, E. Y., Daley, G. Q., & Goodell, M. (2004). Taking stock and planning for the next decade: realistic prospects for stem cell therapies for the nervous system. *J. Neurosci.*, *76*, 257–168.

Chapter 21

Burns and Skin Ulcers

Edward Upjohn, George Varigos, and Pritinder Kaur

INTRODUCTION

Burns and skin ulcers are major causes of morbidity and, in the case of burns, mortality, in both the developed and developing world. Epidermal cells have been used in the therapy of these conditions for decades in the form of autologous skin grafts. For the last 20 years, advanced cell culture techniques have permitted the development of methods to identify and even isolate viable skin stem cells that could potentially be used in the treatment of these two conditions. These advances provide a strong foundation for building more advanced and exciting therapies based on stem cells to treat these challenging conditions.

The epidermis of the skin is a constantly renewing stratified squamous epithelium. It consists mostly of keratinocytes, but also of Langerhans cells, melanocytes, and Merkel cells resting on a supporting dermis that contains the nerve and vascular networks, which nourish the epidermis. The dermis is also the location of epidermal appendages, fibroblasts, mast cells, macrophages, and lymphocytes. Epidermal stem cells are responsible for the ability of the epidermis to replace itself, both in normal circumstances and in traumatic skin loss, such as from burns and skin ulceration.

BURNS AND SKIN ULCERS: THE PROBLEM

As with many medical conditions, it is difficult to quantify the burden and impact of these two conditions in a meaningful and tangible way. A study estimating the lifetime costs for injuries in the United States in 1985 rated fire and burns as the fourth largest cause of lifetime economic loss, at \$3.8 billion. Only motor vehicles, firearms, and poisonings were responsible for greater losses. Burns are also an important cause of injury in the developing world, where more traditional methods of lighting and cooking, such as oil lamps and open fires, are still commonplace. Skin ulcers may be caused by several pathological processes, including infection, trauma, diabetes, and venous ulcer disease. Chronic venous ulceration is a common cause of skin ulcers, with an estimated prevalence of 1–1.3%. Skin ulcers are also difficult and expensive to manage, because of their slow rates of healing and the requirement for expensive and labor-intensive dressing regimes. A cost of care analysis by the Visiting Nurse Association in Boston found the average cost per month in 1992 of an unhealed ulcer to be \$1927.89. Diabetic foot ulcers, like venous ulcers, often have a chronic course. In one study, the cost estimates for foot ulcer care over a two-year period in a population of type 1 and type 2 diabetics who developed an ulcer was \$27 987. These figures, especially when extrapolated to include lost productivity, are almost beyond comprehension. What is easily comprehended, however, is the personal cost and burden borne by patients suffering from burns or skin ulcers.

EPIDERMAL STEM CELLS

The epidermis of the skin is a multilayered, continuously self-renewing tissue, replaced every 30 to 60 days in human skin. *In vivo* cell turnover studies in mice have shown that all proliferative activity is restricted to the basal layer, which generates the mature functional suprabasal keratinocytes. Thus, epidermal stem cells reside within the basal layer, and are characterized by their capacity for self-maintenance and self-renewal. In addition to producing more stem cells, the stem cells generate transient amplifying cells that divide three to five times before producing terminally differentiated keratinocytes *in vivo*. The ability to identify slowly-cycling stem cells *in situ*, visualized as ^{3}H-Tdr or bromodeoxyuridine (BrdU) label-retaining cells, has permitted their localization at specific niches, including the deep rete ridges of the interfollicular epidermis and the bulge region of hair follicles (Figure 21-1). The identification of epidermal stem cells *ex vivo* is a more controversial subject, because there is no unequivocal assay for these cells *in vitro*.

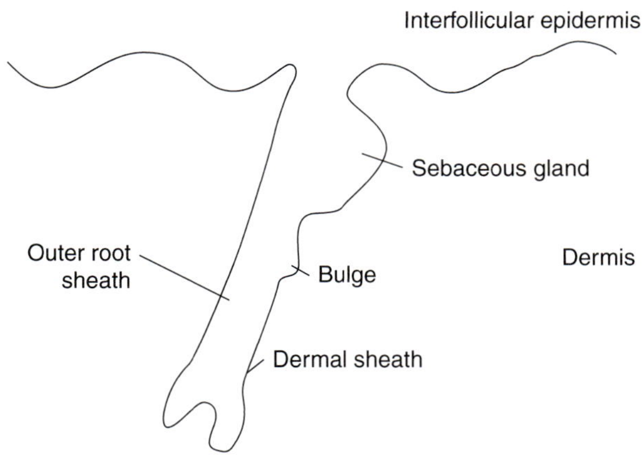

FIGURE 21-1 Epidermis, dermis, and other structures.

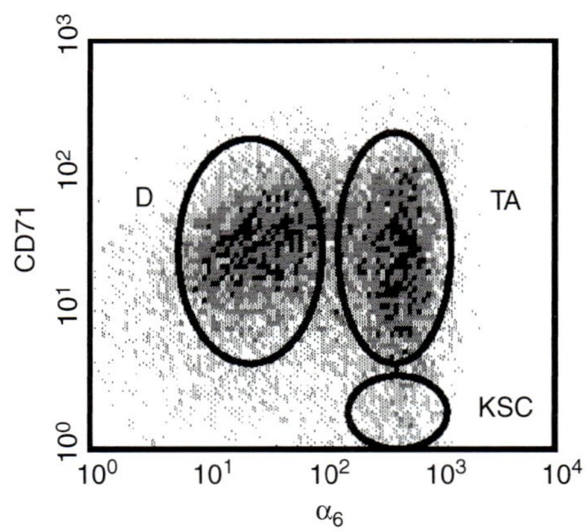

FIGURE 21-2 This FACS plot of primary human, neonatal, foreskin keratinocytes—labeled with antibodies to the cell-surface marker α6 integrin (FITC) and the transferrin receptor (CD71-PE)—reveals phenotypically discrete sub-populations of cells corresponding to stem cells (KSC) with the phenotype α6(bri)CD71(dim), transit-amplifying (TA) cells with α6(bri)CD71(bri), and early differentiating (ED) cells with α6(dim).

Heterogeneity in the growth capacity of keratinocytes was first reported by Barrandon and Green in 1987; they used clonal analysis to identify cells retrospectively capable of generating large colonies of cells exhibiting limited differentiation in culture, termed *holoclones*. These investigators proposed that holoclones were derived from stem cells because of their greater proliferative capacity. This work did not provide a means to prospectively isolate keratinocyte stem cells (KSCs). Subsequent work from Watt and colleagues established that cell-surface markers, specifically β1 integrin, could be used to distinguish basal keratinocytes, with high (β1 bright) or low (β1 dim) proliferative capacity measured in terms of short-term colony-forming efficiency. Despite the initial conclusion that β1 integrin was a marker of KSCs, subsequent work from many laboratories has shown that, although most basal cells are integrin bright, only a small sub-population of these exhibit quiescence, as defined by their ability to retain a ^{3}H-Tdr or BrdU label.

Work from our laboratory has further established that ^{3}H-Tdr label-retaining cells can be purified from integrin-bright keratinocytes on the basis of a second cell-surface marker, i.e., CD71, or the transferrin receptor. Thus, epidermal cells exhibiting the phenotype α6(bri)CD71(dim) represent the stem cell population of both human neonatal and adult murine epidermis. This fraction is enriched for label-retaining cells that are small (~9 μm) with a blast-like morphology, display a high nuclear-to-cytoplasmic ratio, exhibit the greatest long-term proliferative capacity to regenerate keratinocytes *in vitro*, and comprise about 5% of total basal cells (Figure 21-2). Moreover, we have demonstrated that the progeny of KSCs can be distinguished by their cell-surface phenotype: transit-amplifying (TA) cells exhibit high levels of CD71 (α6(bri)CD71(bri)), are enriched for actively-cycling cells, defined as pulse-labeled cells in murine epidermis, and exhibit intermediate keratinocyte cell regeneration capacity; early differentiating (ED) cells are identifiable as α6(dim) cells exhibiting the poorest long-term proliferative capacity and expressing keratin 10 and involucrin—both markers of keratinocyte differentiation (Figure 21-2). This work permits the prospective isolation of KSCs and their immediate progeny by fluorescence-activated cell sorting (FACS), so that the contribution of distinct classes of epidermal progenitors to tissue regeneration during homeostasis or in wound healing can be directly assessed. This work is an important prerequisite to the development of therapeutic strategies using KSCs for skin conditions, including gene therapy. The identification of growth factors that recruit KSCs to proliferate and regenerate tissue will be important to the development of techniques for rapid *ex vivo* expansion that facilitate earlier transplantation for burn victims.

STEM CELLS IN BURNS AND SKIN ULCERS: CURRENT USE

Burns

Autografts

Epidermal cells have been used in the treatment of burns since the introduction of the split skin graft by Karl Thiersh in the late 1800s. Skin grafting to cover defects caused by burns or skin ulceration is limited by the area of skin that may be harvested on any one occasion. Full-thickness grafts (including all of the epidermis and dermis) provide good cosmetic results, but require a primary closure of the donor site, limiting the area that may be grafted. To overcome this problem, the use of split skin grafts was developed, whereby epidermis and underlying dermis is shaved from the donor site to provide a graft. The donor site then re-epithelializes from hair follicles, a process that is easily seen with the unaided eye and that takes two to three weeks, after which the donor site can be reharvested. It is thought that stem cells of the hair follicle are responsible for healing both the donor site and the grafted area. Experimental evidence in favor of this notion was recently provided by elegant experiments performed by Barrandon and colleagues (2001), who demonstrated that microdissected hair follicle bulges (enriched for stem cells) from transgenic mice expressing the β-galactosidase gene could regenerate interfollicular epidermal tissue, as well as entire hair follicles, when transplanted.

The limitations of skin grafting techniques are the area that can be covered by them, and the many weeks it may take to cover a large area of burn with autologous split skin. Burns of 80–90% are survivable in the short-term with resuscitation, but if coverage of the wounds is delayed because of a lack of grafts, then a high morbidity and mortality is the result.

In the mid-1970s, Rheinwald and Green developed a technique for serial cultivation of epidermal cells, producing a 1000- to 10000-fold area of graftable epidermis than the initial biopsy (Figure 21-3 and Figure 21-4). These epidermal sheets can then be grafted onto clean wound beds, but they are sensitive to loss by bacterial infection and blistering. In full-thickness burns where the dermis has been lost, the cultured epidermal autograft may be placed directly onto muscle or fascia (Figure 21-5). These cultured epidermal autografts form a permanent covering, suggesting that the stem cells initially cultured and then transplanted have been maintained as stem cells, and therefore retain their crucial role in epidermal maintenance. Histological examination of cultured epithelium reveals the structural similarity to normal epithelium *in vivo* (Figure 21-6). Culturing epidermis is a time- and labor-intensive exercise, estimated to cost between \$600 and \$13000 per 1% of the body surface area covered (depending on the proportion of the grafts that successfully take).

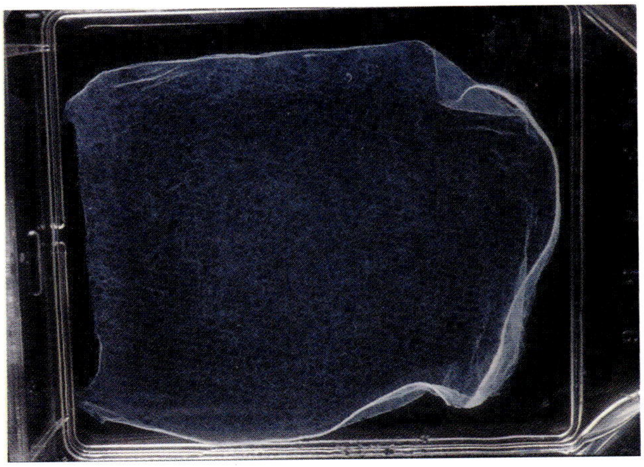

FIGURE 21-3 Detached epithelial graft in a tissue culture flask. Reproduced with the permission of Joanne Paddle-Ledinek.

Allografts

The development of allografts has been driven by the lack of available donor sites for split skin grafting in patients with massive burns, and by the time taken to grow cultured autologous skin from these patients. Burn therapy requires coverage of the burnt areas to prevent secondary sepsis

FIGURE 21-4 Secondary cultures of the epithelium in an incubator. Reproduced with the permission of Joanne Paddle-Ledinek.

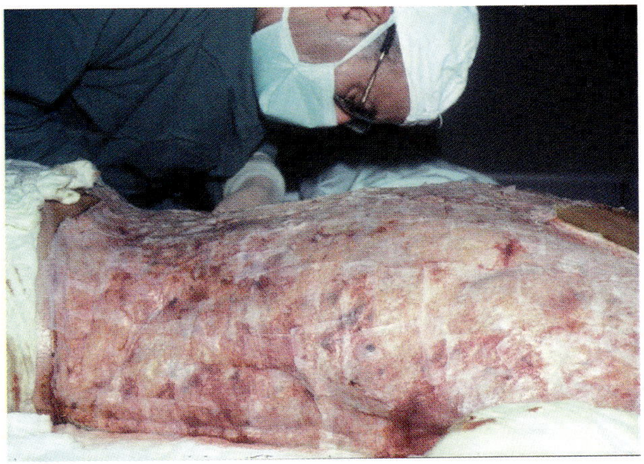

FIGURE 21-5 Application of cultured autograft epidermis to a burns patient. Reproduced with the permission of Joanne Paddle-Ledinek.

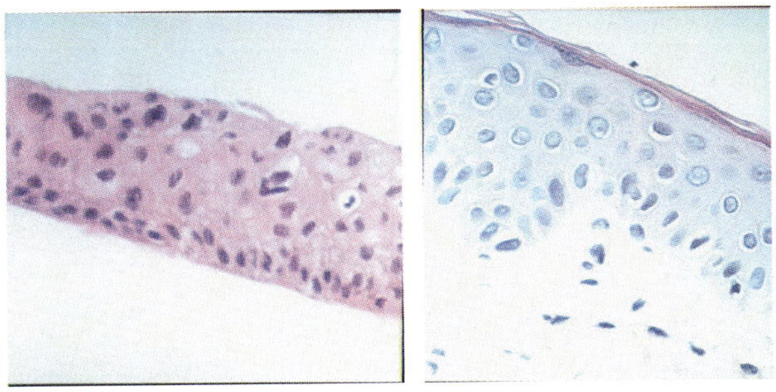

(A) (B)

FIGURE 21-6 (A) Side-by-side histological comparison of cultured epidermis and (B) epidermis *in vivo*. Reproduced with the permission of Joanne Paddle-Ledinek.

and other complications. An alternative to split skin grafts is needed that is immediately available, plentiful, effective, and affordable. Cadaveric skin is such an alternative; it is a true allograft and is always eventually rejected by the recipient. As mentioned earlier, skin that lacks a dermis is less able to resist trauma, and is prone to contraction, resulting in a poor functional and cosmetic outcome. Alloderm (Lifecell, Branchburg, New Jersey) is a processed human dermis from which the epidermal and dermal cells have been removed, leaving only the connective tissue matrix. Alloderm can then be applied to burns, and cultured autograft may be placed on top of it. Integra (Integra LifeSciences, Plainsboro, New Jersey) is another dermal substitute developed through the coprecipitation and lyophilization of bovine collagen, chondroitin-6-sulfate, and an artificial epidermal layer of synthetic, polysiloxane polymer. One to two weeks after application, the artificial epidermis is removed, and an ultrathin epidermal autograft (0.003–0.005 inches thick) is placed on the burn area.

All currently available examples of artificial dermis lack a vascular plexus for the nourishment of the epidermis, and require host vasculogenesis into the dermis graft to supply nourishment to the grafted epidermis. Efforts have therefore been focused on encouraging the process of vasculogenesis by genetic engineering of grafts to produce growth factors and cytokines vital to this process (see later sections of this chapter).

Skin Ulcers

Therapy for skin ulcers is based on treating the precipitating and perpetuating factors. This includes antibiotic use for infective ulcers, rigorous pressure care for decubitus ulcers, and compression stockings for venous ulcers (for examples of skin ulcers, see Figures 21-7 and 21-8). The use of occlusive dressing techniques has greatly expanded in recent decades, and these form the foundation for the treatment of many ulcers. In spite of advances such as occlusive dressings, ulcer healing often takes months, if not years, to achieve. In burns therapy, the impetus behind the use of stem cells or cultured epidermal autografts is the need to cover large areas quickly. In ulcer therapy, the time constraints are not as severe, and for definitive closure, split skin grafts remain the gold standard. Cultured skin has been used in the treatment of skin ulcers—in particular, in the use of cultured allografts as a "living dressing." Cultured, autologous outer root sheath cells used in the treatment of chronic decubitus ulcers have been found to produce an "edging effect"—the contraction of the chronic wounds edges in response to the graft, believed to be caused by a release of growth factors, cytokines, and hormones from the outer root sheath cells. Apligraf (Organogenesis, Canton, Massachusetts) is a cultured, bilayered living skin equivalent derived from neonatal foreskin keratinocytes, fibroblasts, and bovine type I collagen. It is indicated for the treatment of venous ulcers and

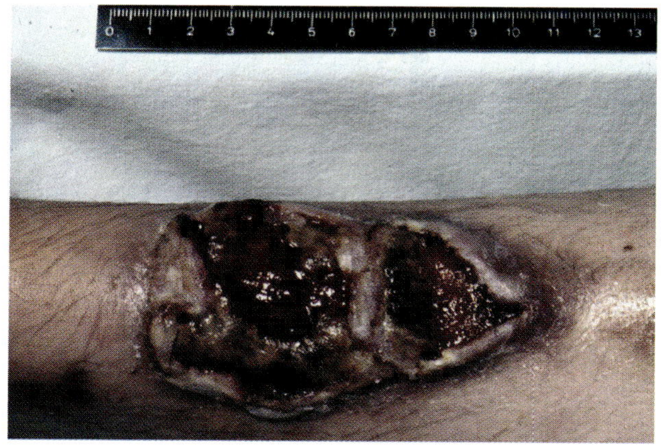

FIGURE 21-7 Leg ulcer. Reproduced with the permission of George Varigos.

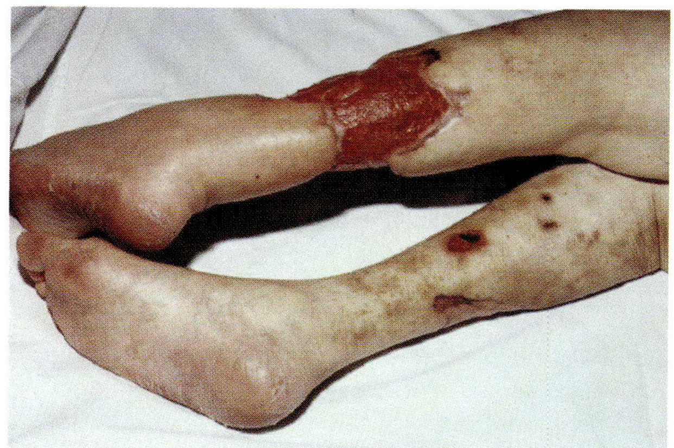

FIGURE 21-8 Circumferential leg ulcer healing by granulation. Reproduced with the permission of George Varigos.

neuropathic diabetic foot ulcers. Chronic wounds (e.g., those with dormant edges) re-epithelialize when exposed to living allograft material. This edge effect, like that seen with outer root sheath cells, is most probably caused by the presence of stimulatory factors. Chronic wounds heal better after repeated application of skin grafts, suggesting that growth factors in the grafts are responsible. Bioengineered tissues, such as Apligraf, probably act as biological systems for delivering growth factors to wounds.

FUTURE DEVELOPMENTS

Gene Therapy Approaches in Wound Healing

Epidermal cells can be genetically modified both *in vivo* and *ex vivo*, by both viral and nonviral methods (e.g., recombinant retro and adenovirus infection, liposomes, plasmid injection, and particle bombardment). Initially, gene therapy of epidermal cells was pursued to correct inherited genetic defects, but it is now used to treat wound healing by using genetically engineered keratinocytes as a source of cytokines and growth factors. Fluid from chronic wounds has been found to be inhibitory to cell proliferation, and to contain degradation products that inhibit keratinocyte migration. The beneficial effect that allografts have on wound healing is thought to be at least partly because of the production of cytokines and growth factors, although the host eventually rejects the allografts. Cultured autografts engineered to produce these cytokines and growth factors in supranormal quantities may expedite wound healing. Epidermal cells can be engineered to express a gene either permanently or transiently. These genes could encode growth factors or cytokines, as mentioned earlier, or they may antagonize some of the inhibitory factors found in chronic wounds.

The vascularization of cultured skin autografts is often delayed when compared to split skin grafting, and may be a contributing factor to the failure of the cultured autograft to take. Cultured skin has been genetically modified *in vitro* to produce vascular endothelial growth factor (VEGF). These modified, cultured skin substitutes have been shown to secrete elevated levels of VEGF, and to have decreased time to vascularization when grafted onto athymic mice.

Tissue Engineering

Tissue engineering of skin is an active area of research and development. Efforts to develop temporary skin substitutes began in the early 1960s. The skin is one of the first organs to have been successfully generated *ex vivo*, and there are several tissue-engineered skin equivalents available today. In bioengineered skin, the epidermal component is either cultured allograft or cultured autograft. The keratinocyte sheets are then combined with the dermal component. The dermal component may be acellular, or it may contain allogeneic or autogenic fibroblasts, or other cells. The optimum skin equivalent would be readily available, could be stored or frozen and ready for use, would be inexpensive to make, would have excellent take, and would give good cosmetic results. Such an ideal skin equivalent does not exist. Split skin grafts fulfill many of these requirements but, unfortunately, they are often not available in a sufficient amount. Currently, the culturing of autografts requires at least three weeks, is labor intensive, and is not available in many developing parts of the world. It would be a great advance if the technique could be made faster and easier to perform. Given our recent ability to use FACS methodology to separate KSC versus TA populations from the epidermis, we have been testing the hypothesis that the stem cell-enriched keratinocytes may provide faster and more reliable skin regeneration in *in vitro* and *in vivo* transplantation model systems. Our recent studies indicate that both epidermal stem cells and TA cells can regenerate epithelial tissue in the short-term fairly rapidly, and that dermal cells, and specific components of the basement membrane (i.e., Laminins), are critical regulators of this tissue-regenerative capacity. Current studies in our laboratory are aimed at understanding how to recruit greater numbers of epidermal stem cells to proliferate *in vitro*, as well as identifying factors capable of promoting stem cell expansion and renewal *ex vivo*. These studies will be important to the development of cellular therapies using epidermal stem cells as vehicles.

SUMMARY

Investigators of epidermal stem cells have the advantage of easy access to their cells of interest. Keratinocytes are also relatively easy to culture and engineer into their normal "organ form" (i.e., sheets of epidermis). As stated earlier, there is an established role for epidermal stem cells in the therapy of human diseases and injuries. The concept of epidermal stem cells, and their application to health care, could form a conceptual framework for the education of the general public about stem cells, and could help to demystify what is commonly perceived to be a complicated science. Epidermal stem cells are excellent targets for gene therapy seeking to correct genetic deficiencies permanently. This could form the basis of new therapies for previously untreatable genodermatoses (such as epidermolysis bullosa), as well as for burns and skin ulcers.

Epidermal appendages such as hair, sweat glands, and sebaceous glands are often destroyed by burns, and their replacement with bioengineered equivalents is still to be achieved. The ability to develop a hair-bearing skin replacement, for example, would address many of the cosmetic problems caused by burns to the face and scalp.

The stem cells of the dermis have yet to be definitively identified and localized within the skin; this can be attributed to the complex cellular heterogeneity of this tissue. Tissue engineering of a replacement dermis to graft onto burns or other defects lacking a dermis would clearly benefit from the incorporation of dermal stem cells. Given that the dermis provides growth factors regulating epidermal growth and morphogenesis, as well as "hair inductive" capacity, further elucidation of its molecular, cellular, and functional components is essential to the development of cellular therapies.

The risk of viral transmission remains a concern with allografts, but they can be frozen, thawed, and used when needed. This is very convenient, and contributes to their lifesaving potential. Tissue engineering with bovine or other sources of collagen must be carefully managed, to reduce the risk of transmission of prion and other diseases to humans. For these reasons, research into cultured autografts will continue, with the goal being a cultured skin equivalent made from epidermal and dermal stem cells, tissue engineered to provide rapid wound coverage with excellent take. The great advances made in the last quarter of a century have saved many lives, in the case of burns, and made lives worth living again, in the case of skin ulcers. Our expanding understanding of stem cells and their manipulation will build on these advances to enable even better treatment of these conditions in the future.

ACKNOWLEDGMENTS

We wish to thank Joanne Paddle-Ledinek and Heather Cleland.

KEY WORDS

Allograft Cultured allogeneic keratinocyte graft grown from a donor's skin cells, which is eventually rejected by the recipient's immune system (may also have a dermal component).

Autograft Cultured autologous keratinocyte graft grown from the recipient's own skin cells, not rejected by the recipient's immune system.

Epidermal stem cells Slow-cycling and self-renewing cells found in the basal layer of the epidermis responsible for its constant repopulation.

Holoclone A colony of cultured epithelial cells thought to be produced by keratinocyte stem cells.

FURTHER READING

Barrandon, Y., & Green, H. (1987). Three clonal types of keratinocyte with different capacities for multiplication. *Proc. Nat. Acad. Sci. USA*, *84*, 2302–2306.

Eaglstein, W., & Falanga, V. (1998). Tissue engineering and the development of Apligraf, a human skin equivalent. *Cutis*, *62*, 1–8.

Eming, S., Davidson, J., & Krieg, T. (2001). Gene transfer strategies in tissue repair. In *The skin and gene therapy*, U. Hengge *et al.* (Eds), pp. 117–137. Berlin: Springer.

Jeschke, M. G., Richter, W., & Ruf, S. G. (2001). Cultured autologous outer root sheath cells: a new therapeutic alternative for chronic decubitus ulcers. *Plastic Reconst. Surg.*, *107*, 1803–1806.

Li, A., Pouliot, N., Redvers, R., & Kaur, P. (2004). Extensive tissue regenerative capacity of neonatal human keratinocyte stem cells and their progeny. *J. Clin. Invest.*, *113*, 390–400.

Li, A., Simmons, P. J., & Kaur, P. (1998). Identification and isolation of candidate human keratinocyte stem cells based on cell surface phenotype. *Proc. Nat. Acad. Sci. USA*, *95*, 3902–3907.

Oshima, H., Rochat, A., Kedzia, C., Kobayashi, K., & Barrandon, Y. (2001). Morphogenesis and renewal of hair follicles from adult multipotent stem cells. *Cell*, *104*, 233–245.

Pellegrini, G., Bondanza, S., Guerra, L., & De Luca, M. (1998). Cultivation of human keratinocyte stem cells: current and future clinical implications. *Med. Biol. Eng. Comp.*, *36*, 778–790.

Rheinwald, J. G., & Green, H. (1975). Serial cultivation of strains of human epidermal keratinocytes: The formation of keratinizing colonies from single cells. *Cell*, *6*, 331–344.

Chapter 22

Cell-Based Repair for Cardiovascular Regeneration and Neovascularization: What, Why, How, and Where Are We Going in the Next 5–10 Years?

Doris A. Taylor and Andrey G. Zenovich

INTRODUCTION

Cardiovascular disease (CVD) has become a major health problem throughout the world, exceeding infection and cancer as the leading cause of death in the Western world (Thom *et al.*, 2006). In the United States, CVD has been the No. 1 killer since 1900, except for 1918, when it momentarily transferred its reign to influenza (Thom *et al.*, 2006). Even though we have experienced a steady decline in mortality in CVD in general, and in acute myocardial infarction (AMI) in particular, since 1980 (Thom *et al.*, 2006), CVD can be viewed as an "impending public health catastrophe" for several reasons.

First, although mortality has declined owing to recent major advances in pharmacological therapy of atherosclerosis, in percutaneous and surgical revascularizations, and in therapies aimed at reducing the degree of hypercholesterolemia, hypertension, diabetes and post-AMI left ventricular (LV) remodeling (Pearson *et al.*, 2002; Smith *et al.*, 2006), CVD still accounts for 1 in every 2.7 deaths in the United States, which cumulatively translates into approximately 2.5 million deaths per year (Thom *et al.*, 2006).

Secondly, changes in incidence have not paralleled the reduction in mortality, largely because of the increased prevalence of the risk factors for CVD, such as hypertension, obesity and type 2 diabetes (Haffner, 2002; Pearson *et al.*, 2002; Appel *et al.*, 2006; Wyatt *et al.*, 2006). Recent data show that the incidence of CVD in the 30–50 year old age group is actually on the rise, particularly owing to growing prevalence of CVD risk factors (Juonala *et al.*, 2006; Yan et al, 2006).

Thirdly, as a result of our getting better at preventing and treating AMI, improved survival post-AMI has been achieved, increasing the prevalence of heart failure (HF) such that nearly 40% of patients manifest HF by year 7 following their first AMI (Miller and Missov, 2001). The root causes of this unsatisfactory dynamic are not only pathophysiological. Pharmacological agents, such as angiotensin converting enzyme (ACE) inhibitors (Jong *et al.*, 2003; Bertrand, 2004), angiotensin receptor blockers (ARB) (Cohn, 2002; Doggrell, 2005; Hernandez *et al.*, 2005), and beta blockers (Thattassery and Gheorghiade, 2004; Jost *et al.*, 2005; Torp-Pedersen *et al.*, 2005) have demonstrated their abilities to decrease LV remodeling and reduce the number of HF-related hospitalizations. But the magnitude of their efficacy is very modest, when evaluated from the public health perspective (Levy *et al.*, 2002). In addition, these therapies are underutilized not only geographically but also across diverse racial and economic groups, therefore the efficacy observed in clinical trials is not necessarily similar when these drugs are applied clinically (Lenzen *et al.*, 2005). Furthermore, as survival of HF patients is also increased by technological advances such as implantable cardioverter defibrillators (ICDs) (Moss *et al.*, 2002), which now allow termination of lethal arrhythmias outside the hospital, an increased number of patients survive with HF fueling a rise of health care expenditures.

Lastly, the number of people over 65 years of age in the United States will double in the next 25 years as a result of the aging of the "baby-boomer" population. It is estimated that nearly 15% of this population will develop HF due to aging, CVD, and type 2 diabetes (Thom *et al.*, 2006).

It is clear that the urgency of this growing public health problem must be solved via a new, more advanced level of understanding of pathophysiology of atherosclerosis and engineering of therapies to be applied throughout the continuum of CVD; that is, to intervene after acute injury

Stem Cell Anthology

to prevent LV remodeling, to treat chronically failing myocardium to stop a progressive loss of cardiac function and worsening of symptomatic status, and also to halt CVD process altogether by restoring vascular health and thereby preventing injury.

All these strategies have led to research efforts directed at cell-based repair to achieve functional regeneration of the vasculature **at large** and of the myocardium. In this chapter, we provide a brief overview of the state of cell-based therapies for vascular repair and provide a perspective on the development of this field in the near future.

THE STATE OF CELL THERAPY

Application of stem cells to achieve vascular repair and regeneration remains novel. Developing any new therapeutic product and translating it from bench experiments to bedside efficacy is a process of multiple interrelated steps. The first step is the idea embodied in basic science. Next, that idea must be tested in clinically relevant animal models of disease. If the data indicate a potential therapeutic benefit, further testing occurs in several consecutive clinical studies. However, if unexpected issues arise or new pieces of the puzzle emerge (e.g. delivery-related issues in the case of cell therapy), the process should move back to the bench and, only when resolved, move again to bedside. The ultimate goal of this iterative process is a "clinical product" with broad applicability to patients, easy administration, and an extremely low frequency of adverse events in follow-up.

Cell therapy at present is in the iterative stage between bedside and bench. The first set of ideas has successfully moved into clinical studies. Early clinical safety and efficacy data are emerging. New insights into the mechanisms of the effects seen (or lack thereof) are being garnered to produce the next generation of preclinical innovations to optimize the "clinical product." The concept of a product itself has been undergoing changes as new cells and therapeutic strategies are being evaluated.

Ten years ago the concept of tissue or organ (in this case, cardiac) repair with exogenous progenitor cells was unheard of. Only a few studies were published **suggesting** that injected cells could actually incorporate into the damaged heart (Marelli *et al.*, 1992; Chiu *et al.*, 1995). There were no studies that demonstrated functional improvement of the myocardium. The first preclinical study was published in 1998 (now 9 years ago) showing functional improvement of injured myocardium after transplantation of skeletal myoblasts (SKMBs) (Taylor *et al.*, 1998). Since then, the field has been growing exponentially. Within the following 2 years, a clinical trial in which SKMBs were delivered as an adjunct to coronary artery bypass grafting (CABG) in HF patients began in Europe (Menasche *et al.*, 2001). Now, multiple clinical trials using 5–6 different muscle-, bone-marrow-, and blood-derived cell types have been reported (Tables 22-2–22-3).

This explosion of trials lies in the simplicity, straightforward nature, and timing of cell therapy. First, cell therapy offers an opportunity for repair of injury, rather than simply an augmentation of the remaining uninjured heart, often to its detriment. In other words, cell therapy provides hope of a permanent solution to a previously unsolvable problem.

Next, the treatment makes sense to clinicians and patients. When cells die after AMI leading to HF (Abbate *et al.*, 2006), it seems logical to prevent them from dying, as is a goal with current bone marrow mononuclear cells (BMNCs) trials early post-AMI, or replace them with new, undamaged cells, a current goal of virtually all HF trials. Because cell therapy primarily involves autologous cells, patients endorse it—since ethical concerns (such as with embryonic stem cell research or live allogeneic donors) are not an issue—and often seek out participation in clinical trials. In addition, as is often the case with anything cutting edge in medicine, it is exciting and prestigious to participate both for investigators and patients.

In addition, cell therapy came just as the hope of angiogenic therapies (vascular endothelial growth factor, fibroblast growth factor, etc.) failed to deliver the new blood vessel growth they so enticingly promised (Henry *et al.*, 2003; Simons, 2005). When therapeutic angiogenesis did not become the mainstream one-shot-fix-all treatment, there was an urgent need for a new frontier: for something that would fulfill the promise and become the "clinical product." Cell-based therapies became "what's next" in cardiovascular medicine. As it did so, the anticipated 5.3 billion dollar market potential in CVD has led scientists, companies, investors, clinicians, and patients into a plethora of first-in-man studies. As a result, within 10 years, we have moved from conception of a field to completed Phase I clinical studies, completed, and ongoing Phase II studies and additional "next generation" bench and pre-clinical studies.

But has cell therapy arrived? The good news is the field is moving rapidly, and the possibility of cell-based therapies joining the clinical armamentarium to repair myocardium has begun to be supported by several Phase II patient studies. The bad news is the field is sometimes moving too fast, without a critical evaluation of the science behind the data or with trials that are outdated even by the time they begin to enroll patients. So it is not at all surprising that we have conflicting clinical outcomes: both negative and positive results with the same cells, in the same patients, and in similar pathophysiological contexts. We do not **yet** understand all the relationships among cells, engraftment, delivery, mechanism of action, and outcomes. We have **yet** to ask all the right questions. The field is just emerging, and with it our knowledge changes fast. Within the next few years, we should witness the completion of both surgical and percutaneous catheter-based Phase II studies with bone marrow, blood, muscle and possibly even fat-derived progenitor cells. We will likely see the initiation of at least one clinical trial utilizing resident cardiac progenitor cells. It is very likely that as this process further unfolds, the applicability and the maximum benefits of the cells types will segregate with a specific disease state; as we gain more molecular/mechanistic insights, we will be enabled to choose the right cell for the right stage of the CVD continuum. The likely major impact will be provided by Phase II (and even Phase III) data regarding the use of BMNCs in the acute and subacute setting of AMI and the use of SKMBs or mesenchymal cells (MSCs) in HF patients. So has cell therapy arrived? The short answer is that it is **arriving**.

In summary, this is a dynamic time in a new and exciting field: treatment strategies are being developed and modified almost every week; new cell types are being reported, novel delivery strategies are emerging, and slowly we are dissecting the mechanistic components of cellular cardiomyoplasty

CELL-BASED REPAIR AS A MEANS OF REGENERATION IN CVD

The goal of regeneration as a therapeutic process is to repopulate diseased area of the tissue with exogenous (by direct application) or endogenous (by stimulation of production and homing) cells that restore function of the organ and halt the disease process that resulted in tissue damage. In AMI and HF, cardiac regeneration—restoration of myocardial function and cessation of LV remodeling—is the ultimate goal. Functional vasculature has to develop and electrical conductance has to occur in underperfused areas and/or fibrous scar to provide optimal contractile force. However, full regeneration is currently not yet achievable, although it may be possible in the future through tissue engineering methods, such as application of cell-based patches to critically injured areas (Liu *et al.*, 2004). Futurism apart, what **can** be claimed as a realistic goal of cell therapy in the next 5 years is **restoring at least some degree of mechanical function and perfusion to the injured, dilated, and decompensated myocardium**—even in the absence of full regeneration.

Endogenous tissue repair is inherent in the human body. We realize now that virtually every organ in the body, including heart, is capable of ongoing maintenance throughout much of our lifespan, and only with ongoing disease, aging, or when overwhelmed by a catastrophic event does this process fail. Yet, repair in the heart and vasculature appears to fail more often than in other organs. As we understand more about the elements of this endogenous cardiac process—what initiates it, what controls it, what allows it to work, and what leads to failure—we will refine our targets for repair. Manipulating those targets first pre-clinically and then clinically will become increasingly important as the field advances. Meanwhile, what can we learn from non-cardiac repair that could apply to heart?

Successful endogenous cell-based repair of many non-cardiac tissues routinely occurs after injury. For example, in bone, endogenous repair occurs **if** three simple premises are fulfilled: reduction of inflammation, fixation of new cells, and perfusion of the tissue (Ott and Taylor, 2006; Taylor *et al.*, 2006). Although bone injuries are not usually ischemic in origin, the cascade of wound healing is somewhat similar to the process in myocardium: inflammation leads to clearing of necrotic (bone) tissue and formation of a fibrous scar. However, unlike in the heart where the scar stays (Kwong *et al.*, 2006), in bone the scar is replaced by regenerating osteoclasts and osteoblasts (Maddi *et al.*, 2006; Wutzl *et al.*, 2006), and in a relatively short period the bone is completely healed (Karladani *et al.*, 2001). So why does endogenous cardiac repair fail? In bone repair, **mechanical stress has to be minimized**. Patients' fractured extremities are routinely immobilized to allow healing. When an injured bone is insufficiently immobilized, an unstable scar forms (and this process frequently takes longer than formation of a stable scar), which could be compared to the process of LV remodeling and progression to HF. **Next, perfusion has to be maintained**. When perfusion is compromised, tissue necrosis leads to sequestration, chronic inflammation, and pain—a process not unlike chronic post-AMI angina (Hansen-Algenstaedt *et al.*, 2006). Furthermore, in the absence of adequate perfusion, a mature bone does not arise (Hansen-Algenstaedt *et al.*, 2005,2006), similar to the need for functional angiogenesis to support new muscle formation in the myocardium (Simons, 2005). Finally, **cells have to home to the site of the lesion**. Bone regeneration is ultimately performed by endogenous progenitor cells (osteoblasts and osteoclasts) that migrate into the lesion (Wright *et al.*, 2005). In the injured myocardium, this process breaks down. Rather than intensive colonization with the desired cardiac progenitor cells, the lesion is colonized primarily by fibroblasts, resulting in collagen deposition and scar formation, instead of nascent myocardium (Lu *et al.*, 2004). This comparison with bone makes endogenous cell-based repair in the heart seem difficult, as the need for immobilization cannot be achieved, we do not understand all the components of homing of cells to the lesion, and adequate perfusion is a product of endothelial health, function, and regulation, which is often missing in CVD.

VASCULAR INTEGRITY IS A BALANCING ACT IN WHICH INFLAMMATION IS KEY: INJURY VERSUS REPAIR

What we do understand, however, is that atherosclerosis represents the failure of **attempts** of endogenous vascular repair in response to repeated injuries to the vessel wall (Goldschmidt-Clermont *et al.*, 2005). After Ross' seminal work introducing the "response to injury" hypothesis whereby the **primum**

movens in atherogenesis was endothelial denudation (Ross, 1993), it became apparent that endothelial cells directly regulate vascular function, transport of solutes, and antithrombotic properties of the blood–tissue interface (Hansson, 2005; Feletou and Vanhoutte, 2006; Schmieder, 2006). More recently, the revised "response to injury" hypothesis focuses on endothelial dysfunction rather than denudation as a trigger for the inflammatory responses and progression of atherosclerosis (Sun *et al.*, 2000; Hansson, 2005). Once regarded simply as a static barrier between tissue and blood, vascular endothelium is now known to play a key regulatory and integrating role in the initiation and progression of atherosclerosis (Feletou and Vanhoutte, 2006; Schmieder, 2006). We now understand that vasodilatory, antiplatelet, and antithrombotic processes are primarily regulated at the endothelial level, and the loss of normal endothelial function may be the most important driver of the balance in favor of inflammation and thrombosis (Feletou and Vanhoutte, 2006; Schmieder, 2006), which, in turn, contributes to the transition between stable and unstable angina (Kostner *et al.*, 2006). Indeed, atherosclerotic plaques have been shown to be more thrombogenic when they express higher levels of tissue factor, a key endothelial regulator in the extrinsic clotting cascade (Marmur *et al.*, 1993; Libby and Theroux, 2005). Therefore, in CVD, the repair of endothelium with exogenously applied cells must be one of the main steps to achieve restoration of perfusion and slowing and possibly halting both LV remodeling and atherosclerotic lesion progression.

We believe that ongoing endogenous repair is a process that reflects the balance between protective and detrimental factors (Goldschmidt-Clermont *et al.*, 2005), and that by providing reparative cells it is possible to tip this balance to allow for repair of the injured tissues (Figure 22-1). Vascular repair is most likely a stepwise process. First, the tissue undergoes injury (e.g. ischemic insult), as shown in Figure 22-1, accompanied by inflammatory response (release of cytokines and chemokines). The inflammatory milieu and injury-specific mediators recruit "detrimental" BMNCs (e.g. $CD45^+$, $CD11^+$, $CD3^+$) and exacerbate injury (Yu *et al.*, 2002; Abbate *et al.*, 2004). At the same time, mobilization of "reparative" BMNCs (e.g. $AC133^+$, $CD34^+$, $CD31^+$, KDR^+) occurs in an endogenous attempt to promote repair (Fadini *et al.*, 2006; Haider, 2006). If sufficient amounts of reparative cells are recruited (or supplied) the balance tips toward repair and inflammations is halted. Three factors play a crucial role: availability of the "reparative" BMNCs, the capability of these cells to home to the site of injury, and their functional capacity to initiate and propagate repair. Obviously, any of these factors could be the weakest link, and that alone would make repair inefficient allowing injury to prevail. The balance between the injury and repair may also reside in the quantitative, functional, and mechanistic/paracrine relationships of **both** "protective" and "detrimental" BMNCs at the site of injury (Figure 22-2). Therefore, from the balance point of view, we believe that these relationships are different in mild, moderate, and severe disease (Figure 22-2). Recent studies have shown that the availability of the bone marrow progenitors is reduced in number in severe CVD (Werner *et al.*, 2005; Kunz *et al.*, 2006). Similar findings as well as a profound reduction of functional capacity of these cells have been observed in patients with HF (Valgimigli *et al.*, 2004). The decline in number and function is most likely not **the** exclusive feature of the advanced stages of atherosclerotic process because similar results have been reported much earlier in the disease process, when endothelial damage is only provoked by hypertension or type 2 diabetes (Hill *et al.*, 2003; Fadini *et al.*, 2005; van Zonneveld, 2006). Inverse associations with aging itself have also been shown (Thorin-Trescases *et al.*, 2005; Shaffer *et al.*, 2006). Our preliminary data in $ApoE^{-/-}$ mice as well as the recently published WISE study (Bairey

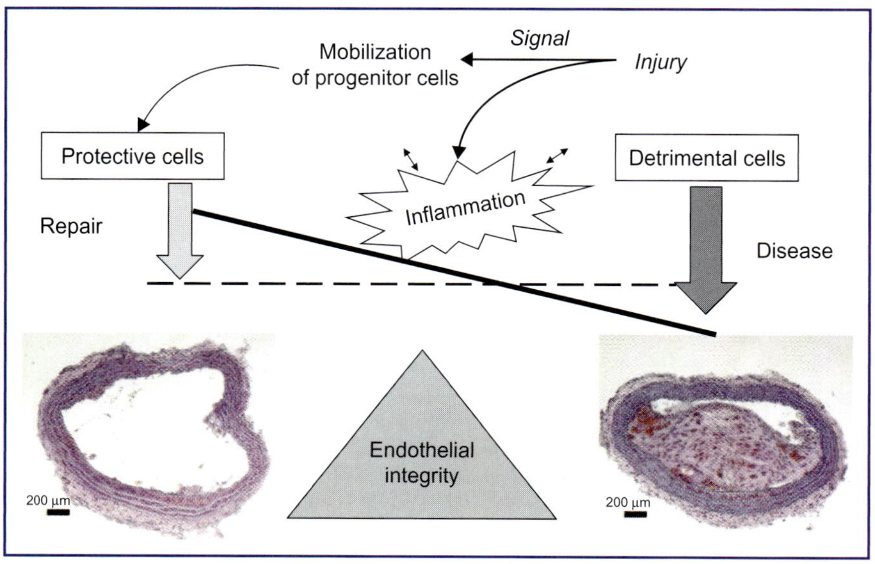

FIGURE 22-1 Schematic representation of the main factors influencing the balance between positive and negative factors (cells, cytokines, chemokines) influencing progression of disease and endogenous repair.

Merz *et al.*, 2006) indicate that males and females display different speeds of progression of disease that correlate with progenitor cell profiles. If these data hold, gender differences in repair will need to be extensively studied as the results may have major implications not only for the field of cell therapy but for the entire cardiovascular medicine. Those insights could not only provide guidance for developing next generation clinical studies, but could greatly advance our understanding of pathophysiological changes in men and women.

Overall, there are several important aspects in cardiac repair. First, the failure of repair occurs earlier in the disease process than previously thought (i.e. before a clinical event) suggesting that owing to age, atherosclerosis, and risk factors, the availability of the progenitor cells and their function decline (Figure 22-2) (Hill *et al.*, 2003). Secondly, repair is most likely evident in the mild–moderate/moderate phase of the disease, where it ultimately fails and injury progresses (Goldschmidt-Clermont *et al.*, 2005). Unfortunately (for detection of abnormalities) many patients in this stage are asymptomatic (Blumenthal *et al.*, 2006). Finally, mobilization of these cells does occur following the acute injury of AMI (Lev *et al.*, 2005; Li *et al.*, 2005; Massa *et al.*, 2005), however, the response is not sufficient for two main reasons: (a) the availability of the cells and their function are greatly reduced (Thum *et al.*, 2006) and (b) a recent study showed that revascularization abolish endogenous mobilization (Muller-Ehmsen *et al.*, 2005).

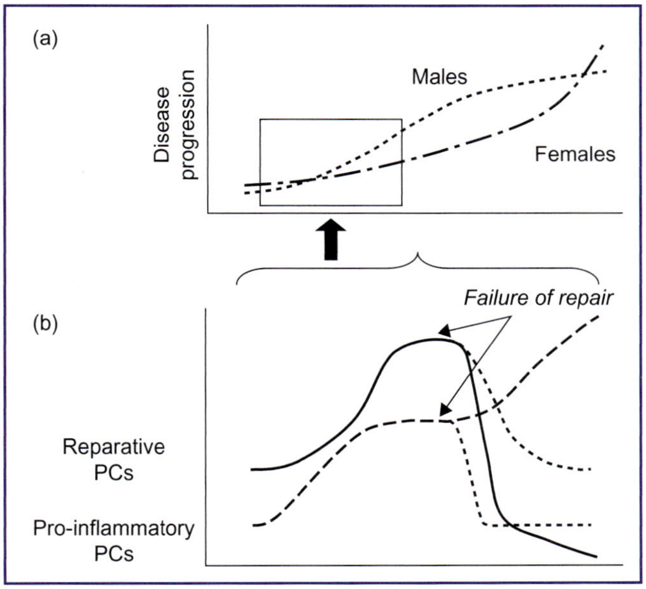

FIGURE 22-2 Schematic representation of relationships between CVD progression (a) and pro-inflammatory (solid line) and reparative progenitor cells (dashed line) number/function (b). As ongoing injury occurs (panel b, left side), the number of pro-inflammatory cells increases (reflecting inflammation), and the reparative progenitor cells compensates in attempt to repair. However, with failure of repair (shown by arrow), reparatory cells falls, and pro-inflammatory cells increase leading to disease progression (panel b, right side). Dotted lines indicate the proposed dynamic of both reparatory and pro-inflammatory cells when repair is successful. Data from animal experiments in our laboratory and data from the Women's Ischemia Syndrome Evaluation (WISE) study (Bairey Merz *et al.*, 2006) were used to reflect the time course of CVD in (a). Relationships of pro-inflammatory and reparative cells were portrayed based on recent publications and the preliminary data of our laboratory (b). *Abbreviations*. PC: progenitor cells. See text for a detailed discussion.

We also now understand that the reduction of inflammation is one of the key steps in successful vascular repair. Recent research has demonstrated that inflammation plays a key role in CVD (Shin *et al.*, 2002; Hansson, 2005; Libby and Theroux, 2005). Indeed, endothelial-dependent relaxation is impaired early in atherogenesis (Feletou and Vanhoutte, 2006). In addition, local administration of proinflammatory cytokines impairs endothelium-dependent dilatation in humans (Bhagat and Vallance, 1997; Shin *et al.*, 2002). Furthermore, clinical measures of coronary endothelial dysfunction are associated with myocardial ischemia in the absence of flow-limiting lesions (Suwaidi *et al.*, 2000) and predict cardiovascular events, including stroke (Targonski *et al.*, 2003). In the early atherosclerotic lesion, immune cells predominate, and their effector molecules significantly contribute to the progression of the plaque (Paoletti *et al.*, 2004; Goldschmidt-Clermont *et al.*, 2005; Libby and Theroux, 2005). Inflammatory response has been shown to be a large part of acute coronary syndromes (ACS)/ AMI (Ikeda, 2003; Paoletti *et al.*, 2004; Libby and Theroux, 2005). The cascade of events leading to clinical atherosclerosis begins with lipid deposition and oxidation, and initiation of an arterial lesion site within the endothelial lining (Libby and Theroux, 2005). These acute processes trigger platelet aggregation and monocyte/macrophage infiltration (Ferns and Avades, 2000; Ikeda, 2003). Smooth muscle cell (SMC) apoptosis is observed within the first few days following the initial assault (Sata *et al.*, 2000). During this time, there is upregulation of pro-inflammatory that recruit various cells including monocytes, macrophages, neutrophils, leukocytes, and smooth muscle precursors to the injury site (Simon *et al.*, 2000; Fujiyama *et al.*, 2003; Ikeda, 2003; Schober *et al.*, 2003; Weber *et al.*, 2004). Studies have demonstrated that by blocking these cytokine signals, a significant reduction in lesion formation is achieved, suggesting that inflammation and/or the cells these signals recruit are a cause of plaque progression (Rogers *et al.*, 1998; Simon *et al.*, 2000; Schober *et al.*, 2003; Weber *et al.*, 2004). A reduction of these signals is associated with re-endothelialization of the artery and the "end to the injury" process (Ferns and Avades, 2000; Schwartz *et al.*, 2004). Extracellular lipids and foam cells form the center of the atherosclerotic plaque, which is surrounded by SMCs and collagen matrix (Paoletti *et al.*, 2004; Libby and Theroux, 2005). However, it is the T-cells, macrophages, and mast cells present in the shoulder region of the plaque that ensure atheroma growth (Frostegard *et al.*, 1999). Many of the immune cells show the signs of activation and release pro-inflammatory cytokines (Shin *et al.*, 2002). Recent analysis of soluble inflammatory mediators has been driven by the expectation that they may be used as indicators of presence and/or progression of CVD (Baldus *et al.*, 2003; Labarrere and Zaloga, 2004; Paoletti *et al.*, 2004). ACS occurs when activation of the atherosclerotic plaque leads to accelerated thrombogenesis and blocks the blood flow to the myocardium downstream from the lesion (Libby and Theroux, 2005). Plaque activation is a complex inflammatory process, in which metalloproteinases along with a number of cytokines directly attack collagen and other components of the tissue matrix, which surrounds the lipid core of the plaque (van der Wal *et al.*, 1994). One of the indicators of inflammation is C-reactive protein, which has been found to correlate with (and reflect) the presence of vascular inflammation and endothelial activation (Labarrere and Zaloga, 2004). CRP is elevated in patients at a high risk and/or experiencing ACS (Tsimikas *et al.*, 2006). Increased levels of CRP indicate upregulation of many inflammatory cytokines (IL-6, IL1β, TNFα) and adhesion molecules (ICAM-1, VCAM-1, E-selectin), facilitating the inflammatory process, and promoting cell recruitment and attachment, ultimately advancing CVD (Shin *et al.*, 2002; Paoletti *et al.*, 2004; Jain and Ridker, 2005; Tsimikas *et al.*, 2006). Other inflammatory markers such as MPO, SDF-1, and MCP-1 have also been associated with CVD (Shin *et al.*, 2002; Baldus *et al.*, 2003; Schober *et al.*, 2003).

However, reduction of inflammation in clinical settings is problematic. There is no single drug therapy which addresses all components of inflammation or results in the "end to the injury" response. Statins have been shown to be beneficial in reduction of hs-CRP, MCP-1 and some inflammatory cytokines (Jain and Ridker, 2005; Chello *et al.*, 2006). However, whether the effect of statins on inflammatory biomarkers is a function of reduction of low-density lipoprotein (LDL) cholesterol or represents a separate pleiotropic effect (including possible mobilization of progenitor cells), represents a matter of considerable debate (Davidson, 2005). Other therapeutics, such as aspirin and ACE inhibitors have also been shown to elicit a beneficial effect on some components of inflammatory process, but their administration does not abolish the entire ravaging cytokine and soluble marker cascade (Peeters *et al.*, 1998; Lauten *et al.*, 2003; Tsikouris *et al.*, 2004). Non-steroidal agents, cyclooxygenase-2 inhibitors, and glucocorticosteroids are potent anti-inflammatory drugs, but their administration alters the course of ACS and combat restenosis yielded more adverse effects than clinical benefits (Lee *et al.*, 1999; Mukherjee *et al.*, 2001; Niederberger *et al.*, 2001; Tamai *et al.*, 2002; Niederberger *et al.*, 2004; Khan and Mehta, 2005; Krotz *et al.*, 2005; Nicolae *et al.*, 2005; Williams *et al.*, 2005).

We have recently shown that repeated intravenous injection of BMNCs from young ApoE$^{-/-}$ mice prevented further progression of atherosclerotic plaque lesions in old ApoE null mice (Rauscher *et al.*, 2003). Injected cells differentiated into endothelial cells and engrafted in atherosclerotic lesions of recipient animals. Comparison of bone marrow progenitor cell (BMC) profiles showed a specific depletion of intermediate vascular progenitor cells (CD31^{+}/CD45^{+}), without parallel changes in more primitive stem cells (sca-1^{+}, c-kit^{+}, or CD34^{+}) or mature vascular cells (VEGFR-2^{+}), most likely accounted for the age-related loss of bone-marrow-derived vascular repair capacity. In addition, the treatment with BMNCs reduced IL-6, and the effect was maintained up until day 15 post-injection, and then the levels would increase again (Rauscher *et al.*, 2003). Significant regression of plaque that occurred in concert with the reduction of IL-6 demonstrated that exogenous BMNCs might mediate regression of atherosclerotic plaque via anti-inflammatory action/positive interference with immune responses that are associated with the atherosclerotic process. However, when BMNCs were given in sex-matched and mismatched fashion only males that received female marrow showed the highest degree of plaque regression (Nelson, unpublished data). Further dissection of the cytokines involved in BMNC-mediated atheroprotection is in its final stages in our laboratory. Preliminary data suggests that plaque regression may be mediated by a distinct group of cytokines, rather than by IL-6 alone (Zenovich, unpublished data). Because atherosclerosis is an inflammatory disease (Paoletti *et al.*, 2004; Hansson, 2005; Libby and Theroux, 2005), the anti-inflammatory properties of BMNC administration and resulting plaque regression offers hope that the treatment with exogenous progenitor cells truly interferes into the pathogenesis of the disease. In addition, recent data on the regression of atherosclerosis with rosuvastatin have demonstrated that this drug is capable of reduction of plaque volume when measured by intravascular ultrasound technique (Nissen *et al.*, 2006). Hopefully one day in the future, a cocktail of a statin drug and bone marrow progenitors will be a therapeutic option in interventional cardiology.

While we only have a conceptual understanding of repair, and many questions are yet unanswered, mediation of this process by progenitor cells makes sense. After all, if cells that originate from the bone marrow did not have the ability to fight inflammation and/or pathogens and participate in healing, the organism would not survive for too long. When the number and/or function of these cells decreases because of a systemic disease (such as type 2 diabetes) (Fadini *et al.*, 2005), aging or toxic substances (such as smoking) (Raupach *et al.*, 2006), healthy and functioning endothelium fails and adverse consequences manifest clinically (as atherosclerosis, thrombosis and its micro- and macrovascular sequelae) (Feletou and Vanhoutte, 2006).

A recent data analysis modeled the potential health effects of bone-marrow-derived progenitor cell therapy using the follow-up data (1950–1996) of the Framingham Study. To model CVD mortality, progenitor cell therapy was applied at age 30, with the effect assumed to be a 10-year delay in atherosclerosis progression. This study suggests that progenitor cell therapy might increase life expectancy in the population by as much as the complete elimination of cancer (in females, an additional 3.67 versus 3.37 years; in males, an additional 5.94 versus 2.86 years, respectively) (Kravchenko *et al.*, 2005). These exciting findings fuel enthusiasm for cell therapy to halt and ultimately prevent CVD.

In summary, in contrast to other tissues in the human body that successfully accomplish the process of repair, cardiac repair becomes inefficient prior to AMI or HF. This failure to endogenously repair/regenerate the vasculature stems from the reduced availability of the circulating bone marrow progenitors and existent inflammatory milieu. Exogenous BMNCs show attractive qualities of plaque regression and reduction of inflammation. Interventions to tip the balance of injury and repair to promote repair may need to be applied earlier in the disease process to ensure efficacy. Further research is needed to optimize that target, as well as to define specific inflammatory mechanisms that play a major role in vascular repair.

REPARATIVE POTENTIAL OF CELL THERAPY

Autologous SKMBs

The idea of using skeletal muscle to repair the heart evolved well before cell therapy emerged as a possible treatment option. In 1987, after being preconditioned by chronic pacing, the latissimus dorsi muscle was surgically wrapped around the failing heart to provide contractile support to the left ventricle—a procedure named "dynamic cardiomyoplasty" (Chachques *et al.*, 1987). Thereafter, cellular cardiomyoplasty was introduced, when cells derived from the C_2C_{12} SKMB transformed cell line were successfully transplanted into **normal** mouse hearts (Koh *et al.*, 1993).

SKMBs, derived from muscle "satellite cells," were first described in 1961 as cells that regenerate damaged skeletal muscle (Mauro, 1961). SKMBs expand and form neofibers after muscle injury. It is not surprising that these cells were the first candidates for cardiac repair. In 1994, Magovern *et al.* reported the first successful transplantation of SKMBs into an **injured** heart (Zibaitis *et al.*, 1994). The critical finding that transplanted SKMBs survived and formed striated muscle grafts within the damaged cardiac tissue was followed by several independent experimental studies investigating the engraftment potential of these cells.

When we showed in 1998, for the first time, that the successful engraftment of SKMBs into injured myocardium improved LV function and attenuated remodeling (Taylor *et al.*, 1998), it was deemed novel by many and unrealistic by some. The mechanisms of how these successfully engrafted SKMBs improved function was unclear. It appeared that the muscle cells could improve contractility of the scarred heart without strict transdifferentiating into cardiomyocytes. SKMBs appeared to yield two populations of cells in injured heart: myogenin-positive transplanted skeletal muscle-like cells in the center of the scar and a second population of myogenin-negative more primitive cardiac muscle-like cells "recruited" around the scar periphery (Atkins *et al.*, 1999a). The transplanted SKMBs appeared to adapt to the surrounding myocardium by forming slow twitch myofibers that were electrically isolated from host cardiomyocytes and yet to improve LV performance (Murry *et al.*, 1996; Atkins *et al.*, 1999b).

These results, stating an improved cardiac function without full integration of transplanted cells into the host myocardium and with recruitment of endogenous cells, raised questions about potential mechanisms that provide functional benefit. Numerous mechanisms have been suggested—from changes in LV wall stress and/or geometry to active mechanically induced contraction of the injected cells—but the exact mechanism(s) underlying the beneficial effect are still the matter of considerable debate (Ott and Taylor, 2006). It is likely that the improvement of function is a result of both a direct effect of the transplanted SKMBs on LV performance, and an indirect "paracrine" effect on endogenous cell recruitment and on LV remodeling (van den Bos and Taylor, 2003). As the mechanisms underlying the positive effects of SKMB transplantation are still not fully understood, there is some discordance of thought as to whether autologous myoblasts improve contraction or just prevent further deterioration of the injured heart—despite much preclinical data showing a direct positive effect of the cells above that seen with other or sham treatments (van den Bos *et al.*, 2004). However, from a clinical point of view, whether the cells do one or the other, may not be the right question to ask, as even attenuation of remodeling is a therapeutic avenue much needed in the HF armamentarium, since the current drug therapies are only capable of a very moderate effect on LV remodeling, only approximately 30%–40% (Reiffel, 2005). Despite these and other remaining questions, there are approximately 15 years of preclinical data available which have shown that autologous SKMBs transplantation can augment both diastolic and systolic myocardial performance in a number of animal models after both acute and chronic injury. These preclinical data opened the field of CVD to this new therapeutic approach.

The advantages of autologous SKMBs for treating patients with CVD/HF extend beyond the evidence of benefit in preclinical studies. By using self-derived SKMBs, it is possible to overcome the major limitations associated with allogeneic cell-based treatments: a critical shortage of donor tissue and the clinical complexities of immunosuppression. By using adult-derived cells, it is possible

TABLE 22-1 Selected Clinically-Relevant Pre-Clinical Models Used for Cell Therapy

Cell Type(s)	Species	Model	Outcomes	Investigator, Country
SKMBs	Rabbit	Cryoinjury	LV systolic function and diastolic relaxation improved in 7/12 rabbits, correlated with engraftment of cells.	Taylor *et al.* (1998) (US)
SKMBs	Pig (targeted cell placement with Da Vinci robotic system)	LAD or LCX embolization	Cells successfully transplanted into apical, anterior and lateral target segments; LVEF, wall thickening, regional wall motion, LV end-diastolic volume improved.	Ott *et al.* (2006) (US)
SKMBs vs. FBs	Rabbit	Cryoinjury	Diastolic performance improved with FBs and SKMBs. FBs reduced but SKMBs increased systolic function.	(Hutcheson *et al.*, 2000) (US)
SKMBs vs. BMNCs	Rabbit	Cryoinjury	Similar improvement of LV systolic function and equal degree of engraftment with SKMBs and BMNCs. Subset of BMNCs differentiated towards a myogenic phenotype.	Thompson *et al.* (2003) (US)
SKMBs vs. CFS vs.CMs	Guinea pig	LAD Ligation	CMs and CFs concentrated in infarction territory; CMs formed gap junctions with native cells, CFs did not; SKMBs proliferated and partially differentiated into cardiac phenotype by 2–3 weeks post-procedure. Gap junctions present.	Horackova *et al.* (2004) (Canada)
Human SKMBs vs. Human AC133$^+$	Rat	LAD Ligation	SKMBs increased LV EF by 15±5%, AC133$^+$ by 7±3% (controls reduced by 8±4%). Engraftment larger with SKMBs.	Agbulut *et al.* (2004) (France)
SKMBs vs. BMNCs	Rat	LAD Ligation	Combination of cell types prevented LV remodeling at 8 weeks post-procedure.	Ott *et al.* (2004) (Austria)
SKMBs vs. VSMCs	Hamster	Cardiomyopathy (Δ-sarcoglycan-deficiency)	Attenuation of LV remodeling was greater with SKMBs vs. VSMCs.	Ohno *et al.* (2003) (Canada)
BMNCs	Mouse	LAD Ligation	Reduction of myocardial infarction size.	Hiasa *et al.* (2004) (Japan)
BMNCs	Pig	Ameroid Occluder, Placement on LCX	BMNCs secreted angiogenic growth factors, augmented perfusion and LV function.	Fuchs *et al.* (2001) (US)

Abbreviations. ***AC133$^+$: immature endothelial progenitor cells; CF: cardiac fibroblast; CM: cardiomyocyte; EF: ejection fraction; FB: fibroblast; LAD: left anterior descending artery; LCX: left coronary circumflex artery; LV: left ventricle; VSMC: vascular smooth muscle cell.***

to avoid the ethical dilemma associated with embryonic stem cells. By using primary cells rather than immortalized or totipotent stem cells, the likelihood of tumor formation after SKMB transplantation is decreased (Tremblay *et al.*, 1991). By using muscle-forming, myogenic cells, the regeneration of contractile muscle in an infarcted cardiac region is more likely. By using relatively ischemia-resistant SKMBs rather than cardiocytes, a higher level of engraftment and survival is likely to occur in infarcted regions—where transplanted cardiocytes perish (Reffelmann *et al.*, 2003). Based on these advantages and the suggestions that regeneration of functional muscle in infarct is possible after autologous SKMBs in pre-clinical models (Taylor *et al.*, 1998; Hutcheson *et al.*, 2000; Fuchs, S. *et al.*, 2001; Ohno *et al.*, 2003; Thompson *et al.*, 2003; Agbulut *et al.*, 2004; Hiasa *et al.*, 2004; Horackova *et al.*, 2004; Ott *et al.*, 2004, 2006) (Table 22-1), clinical studies in both Europe and the United States took off (Field and Reinlib, 2000).

Although the initial clinical data (Herreros *et al.*, 2003; Menasche *et al.*, 2003; Smits *et al.*, 2003; Chachques *et al.*, 2004b; Ince *et al.*, 2004; Siminiak *et al.*, 2004; Dib *et al.*, 2005; Siminiak *et al.*, 2005; Gavira *et al.*, 2006;) (Table 22-2) appear encouraging, myoblast transplantation is not without potential limitations. The first limitation is associated with any autologous cell that has to be expanded in the laboratory (i.e. myoblasts, mesenchymal stem cells, endothelial progenitor cells (EPCs)) that the use of autologous cells necessitates sufficient time between injury and

TABLE 22-2 Cell Therapy Trials with SKMBs

Year	Investigator, Country	Patients No.	Diagnosis	Average SKMB Dose ($\times 10^6$)	Delivery Route	Outcomes
2003	Menasche *et al.* (2003) (France)	10	Post-AMI HF	871	Transepicardial without CABG	NYHA class improved to 1.6 ± 0.1 from 2.7 ± 0.2; LVEF increased to 32 ± 1% from 24 ± 1%.
2003	Herreros *et al.* (2003) (Spain)	11	Ischemia + prior AMI	221	Transepicardial with CABG	LVEF increased to 53.5 ± 5% from 35.5 ± 2.3%, regional wall motion (by E) and viability (glucose update by PET) improved.
2003	Smits *et al.* (2003) (Netherlands)	5	Post-AMI (anterior) HF	196	Transendocardial guided by electromechanical mapping of LV	LVEF increased to 41 ± 9% from 36 ± 11%, regional contractility (by MRI) significantly increased.
2004	Chachques *et al.* (2004b) (France)	20	Post-AMI	300	Transepicardial without CABG	LVEF normalized to 52.0 ± 4.7% (baseline = 28 ± 3%), wall motion score improved, glucose update (by PET) increased.
2004	Siminiak *et al.* (2004) (Poland)	10	>3-month post-AMI	0.4–50 (range)	Transepicardial with CABG	Mean LVEF improved to 42% from 35.2%
2004	Ince *et al.* (2004) (Germany)	6 (+6 controls)	Ischemic HF	210	Transendocardial	LVEF rose to 32.2 ± 10.2% from 24.3 ± 6.7% (in controls decreased to 21.0 ± 4.0%), walking distance and NYHA class significantly improved.
2005	Dib *et al.* (2005) (US)	30	Post-AMI HF	2.2–300	Transepicardial with CABG or LVAD	LVEF increased to 36% at year 2 post-procedure vs. 35% at year 1 vs. 28% at baseline.
2005	Siminiak *et al.* (2005) (Poland)	9	Post-AMI HF	17–106 (range)	Transcoronary	LVEF improved 3–8% in 6 of 9 patients; NYHA class improved in all patients.
2006	Gavira *et al.* (2006) (Spain)	12 (+14 historical controls)	Post-AMI	221	Transepicardial with CABG	LVEF rose to 55.1 ± 8.2% from 35.5 ± 2.3% at baseline (controls 38. 6 ± 11.0%); wall motion score (by E) improved; myocardial viability (by PET) increased.

Studies with ≥5 patients enrolled.
Abbreviations. E: echocardiography; ICM: ischemic cardiomyopathy; LVED: left ventricular end-diastolic diameter; MRI: magnetic resonance imaging; NYHA class: New York Heart Association functional class; PET: positron emission tomography.
Note. (Range) denotes the minimum and the maximum amount of cells given in a trial. Dose-escalation or variation was used in studies where range is provided. For methodological and other details, please refer to original publications.

injection to allow cell expansion *in vitro*. In normal healthy donors and in many HF patients, this ranges from several days to several weeks depending on the cell type, which does not seem problematic if the treatment can wait. In the case of AMI, where an early treatment may be beneficial, either alternative cells can be employed, or if myoblasts are truly superior, allogeneic cells may ultimately offer a solution after the acuity has subsided. Of note, the 2–3-week period after AMI is one of flourishing inflammation (Paoletti *et al.*, 2004; Libby and Theroux, 2005), which in preclinical studies has been associated with increased loss of cells after transplantation (Suzuki *et al.*, 2004). Thus, a delayed treatment with autologous SKMBs could still be beneficial.

A second potential limitation to any cardiac cell therapy is the (**in**)ability of the transplanted cells to electrically integrate with native tissue (Abraham *et al.*, 2005). Myoblasts are the most well studied cell type, yet it is not clear if, and/or how, they integrate into surrounding myocardium. Nor is it understood what impact various degrees of integration may have on either LV function or continuous normal sinus rhythm. Menasche and colleagues did not observe any electrical integration of myoblasts pre-clinically (Scorsin *et al.*, 2000). In support of that, Suzuki *et al.* (2001) reported that in the absence of connexin-43 overexpression, SKMBs did not couple very well with surrounding myocardium. Yet the cells appear to synchronously contract in with surrounding tissue and contribute to overall cardiac performance. Early clinical data (Table 22-2) suggest that SKMBs may be associated with a transient electrical instability in the first weeks after transplantation. How does this occur in the absence of cell coupling? One possibility is suggested by the preliminary modeling data from our group, which shows that the absence of connectivity among SKMBs within the scar provides a tortuous path (through the scar) that can form a nidus for re-entry (Tranquillo and Taylor, unpublished data). Likewise, our data suggest that electrical connection of SKMBs, which have a 10-fold shorter action potential duration than cardiocytes, to the surrounding normal myocardium could similarly provide for re-entry. Which of these hypotheses will ultimately be supported by the majority of clinical data may not be a clear-cut answer, as the location of transplantation (central or peripheral), the homogeneity of the scar itself, the functional properties of the border zone, and the number of cells engrafted (versus administered) are likely to be the major determinants of the outcome. Even in animals, there is data showing increased incidence of arrhythmia in animals who receive SKMBs into the border zone versus center of the infarct (Atkins *et al.*, 1999c), and there is preliminary data showing the exact opposite (McCue, unpublished data).

Finally, and most importantly, the choice of optimal patients may also play a role in electrical outcomes post-SKMBs. The target HF patient population (MADIT-like population) is known to be highly susceptible to arrhythmia (Moss *et al.*, 2002), and it remains to be seen whether the adverse events in clinical trials reflect a cell-associated event (e.g. cell integration, differentiation or even death), related to the electrical status of these seriously ill patients, the HF myocardial environment, or maybe even the location of injections. What appears clear is that the prevalence of arrhythmia may be significantly lowered by administration of low-dose amiodarone (Siminiak *et al.*, 2005), the best clinical agent to normalize inhomogeneous action potential duration. Adverse effects of amiodarone in HF patients are well known to clinicals, so pairing the SKMB therapy with an anti-arrhythmic agent is unlikely to result in a global learning curve. Moreover, a congener of amiodarone—dronedarone—is now in clinical trials, and so far shows a comparable efficacy with a much friendlier safety profile, likely to be associated with removal of two iodine atoms from the molecule (Sablayrolles and Le Grand, 2006). In addition, alterations in the cell culturing) process (e.g. the use of human serum for SKMB culture) may also be beneficial (Chachques *et al.*, 2004a). Administration of these cells in patients with an already implanted ICD has been tested as another strategy. However, the equivocal outcomes observed in the Myogenesis Heart Efficiency and Regeneration Trial so far (although the trial is still ongoing, last follow-up scheduled for middle of 2007) may be associated with the transplantation of SKMBs very late in the disease process, beyond the point where it may be biologically possible to provide engraftment and derive benefit from the functionality of the transplanted cells.

In summary, even though SKMBs transplantation is the most well-defined technique for myocardial repair/regeneration, important questions remain about its long-term safety and efficacy—the questions that will need to be answered for any cell therapy. If cell transplantation becomes clinically a long-term solution to myocardial injury, cells must be able to provide a sustained and functioning revascularization, and mediate a positive contractile effect for years in the heart without eliciting negative sequelae. The clinical outcomes data from 5-year patient follow-up will provide some answers in this regard.

BMNCS

Bone marrow and peripheral blood contain a number of cell populations that have recently been shown capable of differentiation into cells other than blood. They include the hematopoietic stem cells (HSCs), mesenchymal stem cells (MSCs), EPCs, and subsets of each of those, including $CD34^+$ progenitors, multipotent adult progenitor cells and CD14- cells (Saulnier *et al.*, 2005; Verfaillie, 2005). Similar cell populations have been isolated from umbilical cord blood (Zhai *et al.*, 2004). These cells have the potential to become endothelial cells and be the drivers of vascular repair and at the same time spare the researchers, clinicians and patients the ethical and immunological hurdles of embryonic stem cells.

Hematopoietic stem cells: Historically, HSCs have been thought of as those that differentiate only down the erythrocyte and leukocyte lineages (Till *et al.*, 1978). These cells are identified as $CD34^+$ and/or $AC133^+$ for human cells. In mice, these cells were shown to be negative for mature hematopoietic cell lineage markers (lin) and sometimes positive for stem cell antigen-1 ($Sca\text{-}1^+$) and $c\text{-}kit^+$ (also known as CD117). Over the past few years, it has been shown that HSCs can, under appropriate conditions, differentiate into various cell types, including cardiomyocytes

(Yeh *et al.*, 2003). Although HSCs can become cardiomyocytes under strict *in vitro* conditions, there have not been any reports showing differentiation into cardiomyocytes when transplanted into an infarcted myocardium (Murry *et al.*, 2004). Perhaps because of this lack of differentiation in infarct, recent studies suggest that HSCs may not have the potential of some other cell types to improve LV function following transplantation into infarcted myocardium (Deten *et al.*, 2005).

Mesenchymal stem cells: MSCs are rare multipotent progenitor cells, also known as bone marrow stromal cells. In the past, MSCs were shown to differentiate into a number of cell types including, fat, bone, cartilage, and skeletal muscle precursors both *in vitro* and in an infarcted rat myocardium (Jiang *et al.*, 2002). There is also some evidence that after the injection into the myocardium, these MSCs differentiate into cardiomyocyte-like cells (Kawada *et al.*, 2004). However, current studies suggest that this can only happen when MSCs are in contact with native cardiomyocytes and does not happen in the infarct core (Strauer, B.E. *et al.*, 2002). Therefore, the optimal time of therapy using MSCs may be early in the course of the injury, when surviving cardiomyocytes are still present in the infarcted territory. Despite their inability to form cardiomyocytes to a significant degree, transplanted MSCs engraft at high numbers in an infarcted heart (Schuster *et al.*, 2004), and lead to an increased neovascularization and improved regional contractility and the overall LV diastolic function (Schuster *et al.*, 2004). In fact, a recent study from our group suggests that MSCs and SKMBs improve function after ischemia-induced cardiac injury to a similar degree (Thompson *et al.*, 2003). Furthermore, several other studies suggest that MSCs can home to sites of injury following injection into the coronary or even peripheral vasculature (Strauer *et al.*, 2002; Bittira *et al.*, 2003). However, it has also been reported that intra-coronary administration of MSCs can cause microinfarctions and promote damage of otherwise healthy myocardium (Vulliet *et al.*, 2004), which has led to some caution with regards to the design and patient selection in the ongoing clinical trials.

More recently, MSCs have been suggested to be immunoprivileged cells capable of allogeneic administration *in vivo* with very few negative consequences (Jiang *et al.*, 2005), although a certain degree of skepticism about this fact remains in the scientific community. If proven to be true in ongoing trials, this quality could become the most tantalizing aspect in terms of applicability toward cardiac repair. In turn, lack of negative immunological effects and the presence of benefits of functional restoration of the myocardium can lead to a fast development of a commercial cell therapy product for use in many patients. Clinical studies with intravenous administration of allogeneic MSCs in AMI and HF are ongoing.

Endothelial progenitor cells: EPCs are bone-marrow-derived cells that are mobilized into peripheral blood and believed to participate in neoangiogenesis (Kalka *et al.*, 2000). Recent research has shown that the number of EPCs in vascular circulation is increased in patients following AMI (Shintani *et al.*, 2001). EPCs are presumed to be mobilized by the ischemic damage in the heart (and other tissues) and migrate to the damaged areas to induce formation of neovasculature. In support of this, it was recently shown that when EPCs were injected either into the rat's tail vein or LV cavity after an ischemic myocardial injury, a greater than 2-fold increase in the accumulation of infused EPCs was observed when compared to animals undergoing sham surgery (Aicher *et al.*, 2003). LV dimensions, fractional shortening, and regional wall motion improved in rats that received EPCs and were not observed in the control animals injected with culture media (Kawamoto *et al.*, 2001). Although the mechanism of these benefits has not been clearly elucidated, it is likely that improvements seen in this study at least in part depended on improved myocardial perfusion and decreased inflammation. To date, EPCs have not been shown to induce or play a role in neomyogenesis within the injured myocardium, but several paracrine properties have recently been attributed to these cells (Kinnaird *et al.*, 2004).

Human EPCs are typically thought to primarily express CD133 (AC133), CD34, and VEGF-R2. The quantity of the EPCs circulating in humans decreases with age, and mirrors a rapid increase in CVD-related deaths (Hill *et al.*, 2003; Werner *et al.*, 2005). It has been suggested that this correlation is due to the EPCs' contribution to maintaining vascular integrity (Hill *et al.*, 2003). Recent data have shown that the number of circulating EPCs and their ability to migrate is decreased in patients at a high risk for clinical CVD, including AMI (Hill *et al.*, 2003). The reduction in the number and/or functional capacity of EPCs may be a critical factor in the development of major cardiovascular events (Werner *et al.*, 2005). Our group recently published data showing that a reduction in $CD31^+$ $CD45^-$ vascular progenitor cells, thought to be related to EPCs, is associated with aging and disease state in the mouse $ApoE^{-}/^{-}$ model of atherosclerosis (Rauscher *et al.*, 2003). We showed that delivery of functionally viable cells could prevent the progression of atherosclerosis. Completed studies using EPCs and other cell types to treat CVD in humans are shown in Table 22-3.

Umbilical cord blood cells: A relatively new source for progenitor cells is umbilical cord blood, which contains most of the bone-marrow-derived cell types. Cord blood cells are easily obtained, albeit not in large volumes, have the potential to develop into multiple lineages, do not pose a myriad of ethical questions and are less immunogenic than their bone marrow counterparts. As a result, a larger proportion of the population could receive cells from appropriately matched donors. Further, if cord blood cells are isolated and stored at birth, these cells could provide an autologous source of stem cells to treat myocardial damage later

TABLE 22-3 Cell Therapy Trials with BMNCs, MSCs, EPCs, and CPCs

Year	Investigator (country)	Patient's number (type of cell)	Diagnosis	Average SKMB dose($\times 10^6$)	Delivery route	LVEF and other outcomes (maximum follow-up time, month/year)
2001	Hamano *et al.* (2001) (Japan)	5 (BMNCs)	Advanced CAD	300–2,200 (range)	Transepi during CABG	Perfusion improved in three out of five patients (by S) (1 year)
2002	Strauer *et al.* (2002) (Germany)	10 (BMNCs) (+10 controls)	AMI, 5–9 days post	28	IC (after standard Tx)	Infarcted region decreased to 12 ± 7% from 30 ± 13% (by V); LV contractility and EDV improved, perfusion increased (by DE, RV, RC) (3 months)
2002	*TOPCARE-AMI:* Assmus *et al.* (2002); Britten *et al.* (2003) (Germany)	10 (CPCs) + 9 (BMNCs) (+11 controls)	AMI (reperfused)	CPCs: 13; BMNCs: 238	IC, 4-day post-AMI	LVEF improved to 60% from 51.6%; ESV reduced; wall motion in the infarction zone improved (all by V and DE); similar LV functional data by MRI; myocardial viability increased (by PET); CPCs and BMNCs behaved similarly. Migratory capacity predicted LV remodeling in multivariate analysis (4 months)
2003	Fuchs, S. *et al.* (2003) (US)	10 (BMNCs)	Refractory Angina	78	Transendo EMM-guided	Perfusion improved (by SPECT); CCS angina score decreased to 2.0 ± 0.9 from 3.1 ± 0.3 (3 months)
2003	Tse *et al.* (2003) (China)	8 (BMNCs)	Advanced CAD	40 ml BMNC (0.6–8.9% CD34$^+$)	Transendo EMM-guided	Regional wall motion, thickening improved, hypoperfused areas lessened (by MRI); angina reduced to 16.4 from 26.5 episodes per week (3 months)
2003	Stamm *et al.* (2003) (Germany)	6 (AC 133$^+$)	AMI	1.02–1.57 (purified)	Transepi with CABG	LVEF, EDV, EDD improved (by E); perfusion (area at-risk) improved in five out of six patients (3 months)
2003	Perin *et al.* (2003) (study conducted in Brazil)	14 (+7 controls)	Severe Ischemic HF	25.5	Transendo EMM guided	Mean LVEF increased to 35.5% from 30% (by E); perfusion improved (by SPECT); NYHA class decreased to 1.1 ± 0.4 from 2.2 ± 0.9; CCS angina reduced to 1.3 ± 0.6 from 2.6 ± 0.8 class. (2 and 4 months)
2004	*TOPCARE-AMI*: Schachinger *et al.* (2004); Schachinger *et al.* (2006a) (Germany)	30 (CPCs) + 29 (BMNCs)	AMI	CPCs: 13; BMNCs: 238	IC, 4.9 days after AMI	At 1 year, one patient in each cell group died due to cardiogenic shock, no other MACE or malignant arrhythmias; LV functional data similar to prior report for additional patients. MRI at 1 year showed maintenance of LVEF and reducion of infarct size, no reactive LV hypertrophy. Coronary flow normalized in infarct-related arteries (1 year)
2004	Chen *et al.* (2004) (China)	34 (MSCs) (+35 controls)	AMI	8,000–10,000 (range)	IC, 8 ± 4h after AMI	LVEF and regional wall motion increased (by V); per fusion defects decreased (by PET); real-time electromechanical mapping of LV showed improvements in mechanical capabilities, electrical properties and functional indices (6 months)
2004	*BOOST:* Wollert *et al.* (2004); Meyer *et al.* (2006); Schaefer *et al.* (2006) (Germany)	30 (BMNCs) (+ 30controls)	STEMI	2,460	IC 4.8 ± 1.3-days post-PCI	LVEF increased by 6.7% mostly due to improved regional wall motion in the peripheral area (by MRI) LVEDV and infarct size decreased. Diastolic function improved (by E). LV functional benefits did not persist at 1 year (1 year)
2004	Fernandez-Aviles *et al.* (2004) (Spain)	20 (BMNCs)	Extensive AMI (reperfused)	78	IC, 13.5 ± 5.5-day post-AMI	LVEF improved by mean of 6%, contractile reserve increased; ESV decreased in (by MRI, DE) (6 months)
2004	Perin *et al.* (2004) (study conducted in Brazil)	11 (BMNCs) (+9 controls)	End-stage ICM	15 injections, 0.2 ml/each (50 ml aspirated)	Transendo EMM guided	LVEF increased at 2 m, did not change at 6 and 12 m, perfusion improved (by SPECT), NYHA class decreased to a mean of 1.4 from 2.2 and CCS fell to 1.2 from 2.6 class. Exercise capacity improved (by treadmill) (6 and 12 months)

2004	Kuethe *et al.* (2004) (Germany)	5 (BMNCs)	AMI (reperfused, stented)	39	IC, 6.3 ±	0.4 days post-PCI
2004	Silva *et al.* (2004) (US)	5 (BMNCs)	Pre-transplantation HF	15 injections 0.2 ml/each (50 ml aspirated)	Transendo EMM guided	Exercise capacity (by treadmill oxygen consumption) improved in four out of five patients, disqualifying them from listing for transplantation (2 and 6 months)
2005	Bartunek *et al.* (2005) (Belgium)	19(AC133$^+$) (+16 controls)	AMI	12.6	IC, 11.6 ± 1.4 day post-AMI	LVEF increased to 52.1% from 45% similar to controls (by E), perfusion improved (by SPECT); seven patients in cell therapy group developed restenosis (versus four in control group), two patients in cell therapy group had *de novo* lesions (4 months)
2005	Dohmann *et al.* (2005) (Brasil)	14 (BMNCs) (+7 controls)	Severe CAD + HF	25.5	Transendo EMM guided	Perfusion increased (by S), NYHA class, functional capacity, and function improved. (2 and 6 months)
2005	*IACT:* Strauer *et al.* (2005) (Germany)	18 (BMNCs)	Post-AMI (5m–8.5y)	15–22 (range) each infusion, 4–6 total	IC	LVEF increased by 15% (by V), infarct size fell by 30% (by SPECT), myocardial viability of infracted zone increased by 15% (by PET) (3 months)
2005	Blatt *et al.* (2005) (Israel)	6 (BMNCs)	ICM	50 ml of aspirated BMNCs	IC, after induction of ischemia by balloon inflation for 3 min	LVEF improved from mean of 25% to 28% (by E); wall motion (by DE) increased but only in segments with baseline hibernation. NYHA class fell to mean of 2.3 from 3.5; one patient developed post-procedure hypotension and troponin increase (4 months)
2006	*ASTAMI:* Lunde *et al.* (2006) (Norway)	47 (BMNCs) (+50 controls)	AMI treated with PCI	54–130 (range)	IC	No differences in LVEF (by MRI), perfusion (by SPECT), trend toward infarct size reduction (by MRI) (6 months)
2006	*REPAIR-AMI*: Schachinger *et al.* (2006b) (Germany)	101 (BMNCs) (+103 controls)	STEMI (reperfused)	236		IC, 3–6 days after AMI
2006	*TOPCARE-CHD:* Assmus *et al.* (2006) (Germany)	34 (CPCs) + 35 (BMNCs) (+23 controls)	Prior AMI(>3m)	CPCs: *21* BMNCs: 205	IC, with cross-over to the other cell type	LVEF increased significantly in patients that crossed over to BMNCs, absolute increase = 2.9% (by MRI). No changes in LVEF with CPCs. NYHA class improved in BMNC group—reduction of 2.0 ±0.7 from 2.2 ± 0.6, no improvement with CPCs (3 months).
2006	Fuchs *et al.* (2006) (Israel)	27 (BMNCs)	Refractory Angina + Ischemia	28	Transendo EMM Guided	CCS angina class improved to 2.0 ± 0.9 from 3.2 ± 0.5; exercise duration increased to 489 ± 142s from 418 ± 136s; ischemia lessened (by SPECT). At 1 year, five patients had revascularization procedures, functional and symptomatic improvements were maintained in other patients. (3 and 12 months)
2006	Tse *et al.* (2006) (China)	12 (BMNCs)	Advanced CAD	12–16	Transendo EMM guided	No significant changes in LVEF at 3 or 6 months (baseline LVEF 60%). At long-term follow-up, two patients died, one patient received CABG. (3, 6 and 44 ± 10 months)

Studies with ≥ 5 patients enrolled.
Abbreviations. ***CAD: coronary artery disease; CCS: Canadian Cardiovascular Society; DE: dobutamine echocardiography; EDD: end-diastolic (LV) dimension; EDV: end-diastolic (LV) volume; EMM: electromechanical mapping of LV; ; IC: intracoronary; ICM: ischemic cardiomyopathy MACE: major adverse cardiac events; NYHA class: New York Heart Association functional class; PET: positron emission tomography; RC: right heart catheterization; RV: radionuclide ventriculography; S: scintigraphy; SPECT: single photon emission tomography; STEMI: acute ST-elevation myocardial infarction; Transendo: transendocardial; Transepi: transepicardial; Tx: therapy/treatment; V: ventriculography.***
Note: ***(range) denotes the minimum and the maximum amount of cells given in a trial. Dose-escalation/variation was used in studies where range is provided. Controls comprise historical and active randomized participants and those patients who received placebo. For methodological and other details, please refer to original publications.***

in life. Current studies in animal models show that unfractionated cord blood cells injected directly into the infarcted myocardium improve LV ejection fraction, anteroseptal wall thickening, and d*P*/d*t* (max), while decreasing infarct size (Henning *et al.*, 2004). In addition, intravenous injection of cord blood cells in mice following ligation-induced injury resulted in an approximately 20% higher capillary density in the border zones of the infarction—a finding not observed in untreated animals (Ma *et al.*, 2005). Recent data have suggested that human cord blood-derived $CD34^+$ cells may be capable both of preventing injury progression in nude rats and of partially reversing systolic and diastolic dysfunction in the failing heart, if administered shortly after AMI (Leor *et al.*, 2006). No evidence **yet** suggests that cord blood cells injected into the infarcted myocardium are able to produce mature cardiomyocytes. Overall, however, it appears that cord blood cells may appear to be an interesting cell of choice to be used in further studies of treatment of myocardial injury.

Cardiac-Derived Stem Cells

Within the past several years, cardiac-derived stem cells (CSCs) have been identified and are now considered a potential option for cardiac repair. Although the evidence for cardiac repair with these cells is limited, their potential to mature into cardiomyocytes makes them a promising candidate (Laugwitz *et al.*, 2005). These cells have primarily been isolated from neonatal heart (Laugwitz *et al.*, 2005) and, to a very limited extent, from adult myocardium (Anversa and Nadal-Ginard, 2002; Oh *et al.*, 2003). The results of the pre-clinical use are intriguing and suggest that the future of cardiac repair may involve endogenous mobilization or recruitment of these cells—if they can be found in reasonable numbers in the adult myocardium, or can demonstrate adequate trans-differentiation when transplanted into an infarction milieu.

CSCs can be isolated from neonatal rat hearts using LIM-homeodomain transcription factor islet-1 (Laugwitz *et al.*, 2005). It is possible to expand these cells *in vitro* when coupled with a cardiac mesenchymal layer. Further, when these cells are co-cultured with neonatal cardiomyocytes, they are able to electrically integrate with myocardial cells *in vitro* by forming gap junctions (Laugwitz *et al.*, 2005). CSCs isolated from adult hearts, including those from acutely infarcted, failing, and even hearts destined to be replaced by transplantation, have been identified by their expression of c-kit, MRD-1, and Sca-1 and by their lack of expression of hematopoietic lineage markers (Urbanek *et al.*, 2005). These cells have shown the ability to differentiate down myocyte, smooth muscle, and endothelial cell pathways, but their ability to form mature cells of these types (or cardiomyocytes) is not yet known. Endogenous $Sca\text{-}1^+$ CSCs may differentiate into functional cardiomyocytes (Oh *et al.*, 2003), but such potential within an infarct scar has not been elucidated. To date, methods for harvest, expansion, and *in vitro* growth of these precursors are very limited. Therefore, because of these factors, it is difficult to judge the clinical potential of these cells. None-the-less, CSCs biology is interesting enough to make future developments be anticipated with interest and hope. For example, CSCs expanded from endomyocardial biopsies and predifferentiated *in vitro* could become very strong candidates for cardiac repair.

CLINICAL STUDIES

Clinical trials with SKMBs (Table 22-2), BMNCs, MSCs, EPCs and CPCs (Hamano *et al.*, 2001; Assmus *et al.*, 2002; Strauer *et al.*, 2002; Britten *et al.*, 2003; Fuchs *et al.*, 2003; Perin *et al.*, 2003; Stamm *et al.*, 2003; Tse *et al.*, 2003; Chen *et al.*, 2004; Fernandez-Aviles *et al.*, 2004; Kuethe *et al.*, 2004; Perin *et al.*, 2004; Schachinger *et al.*, 2004; Silva *et al.*, 2004; Wollert *et al.*, 2004; Strauer *et al.*, 2005; Bartunek *et al.*, 2005; Blatt *et al.*, 2005; Dohmann *et al.*, 2005; Assmus *et al.*, 2006; Fuchs *et al.*, 2006; Lunde *et al.*, 2006; Meyer *et al.*, 2006; Schachinger *et al.*, 2006a, b; Schaefer *et al.*, 2006; Tse *et al.*, 2006) (Table 22-3) published to date have been summarized. Ongoing trials are listed on the Internet (http://www.clinicaltrials.gov; http://www.thescientist.com). To highlight important points, we chose to comment on several published trials.

SKMBs: The first clinical trial using cell therapy to treat CVD was initiated by Menasche (2003) in 2000. In this trial, an average of 871×10^6 cells (at least 85% were identified as SKMBs by a positive staining for CD56) was injected into non-revascularizable scarred portion of LV as an adjunct to CABG. Over several years following transplantation, significant improvements in LV ejection fraction (EF) and regional wall thickening were observed. Unfortunately, there was no control group. None-the-less, the data are encouraging. However, 4 out of 10 patients experienced ventricular tachycardia requiring ICD implantation. Fortunately, none of the patients experienced intractable/fatal ventricular tachycardia/fibrillation. The data suggests that concurrent administration of amiodarone can minimize untoward electrical events without compromising efficacy of the cells. The data of Menasche *et al.* (2003) provided the impetus to begin a new trial. The Myoblast Autologous Graft in Ischemic Cardiomyopathy trial is a Phase II randomized clinical trial to examine the efficacy and safety of CABG + SKMBs versus CABG alone in approximately 300 patients in North America and Europe. The trial was halted in 2006 in part because the design of the trial was no longer considered state of the art (as the number of CABG cases declined), and as a result recruitment was below projected targets. Although no increase in mortality was reported, the published results of this large study are greatly anticipated.

In a separate US trial (Dib *et al.*, 2005), SKMBs were injected concurrently with CABG (n = 12) or LV assist device as a bridge to transplantation (n = 6), myocardial perfusion improved and left ventricular ejection fraction (LVEF) increased. Upon examination of the explanted hearts (for indicated cardiac transplantation), four of the five specimens showed areas of engrafted myoblasts within the infarcted regions. In another clinical trial (Smits *et al.*, 2003), $196+/-10^5 \times 10^6$ SKMBs were injected directly into the infarcted area (via a NOGA-guided catheter system) as sole therapy in HF patients. These patients showed improved regional wall motion and a trend towards increased LV EF over 3–6 months. Taken together, these data suggest that SKMBs can be delivered in HF patients and survive within the infarcted myocardium to achieve improved LV function.

Early reports of electrical instability in patients after receipt of autologous SKMBs have led to doubts and overt clinical skepticism about the safety of these cells as a treatment option. However, conflicting data exists, and therefore, several considerations should be made. First, patients who received SKMBs in the earliest clinical studies had advanced HF, where electrical events are an inherent part of the pathophysiology of the disease (and therefore these events are expected). In fact, many of the patients met the MADIT-II criteria (Moss *et al.*, 2002) which were presented after those cell therapy trials began and suggested that all patients who meet those criteria be treated with ICDs. As a result, in more recent clinical studies where myoblasts are being used to treat HF, many investigators have only enrolled patients who had already received ICDs or had ongoing treatment with low-dose anti-arrhythmics. This practice may have significantly reduced the incidence of arrhythmogenic events. For example, in the Phase II MAGIC trial, the incidence of electrical instability in patients post-SKMB delivery was approximately 10% (lower than the initial 40% reported by the same group of investigators) (Menasche *et al.*, 2003). Whether this discrepancy occurred because of a better selection of patients in the second study, the co-administration of anti-arrhythmic agents or an improved safety profile of the cells remains to be determined. Furthermore, in clinical studies in the United States, Dib *et al.* (2005) have not reported an increased incidence of electrical instability after SKMB administration, nor have others in pre-clinical studies (Chachques *et al.*, 1987). None-the-less, these data suggest that autologous SKMBs for patients with HF have a potential to be a relatively safe and efficacious product, if such holds true in definitive Phase III trials.

Bone marrow stem cells: In a trial similar to that performed with SKMBs, patients received up to 1.6×10^6 $AC133^+$ BMNCs into the peri-infarct zone concurrent with CABG (Stamm *et al.*, 2003). However, in contrast to SKMB studies, this study examined patients treated shortly after AMI. A total of six patients were treated, and perfusion in treated areas increased and LV dimensions and EF improved. Further, unlike in the SKMB trials, these improvements occurred without electrical abnormalities. Whether this represents a difference in patient population, cell type or even cell dose remains unresolved.

In a more preventive approach, a number of studies have been performed in an attempt to rescue the myocardium and to prevent HF. These studies have primarily focused on percutaneous delivery of bone marrow cells after AMI. In the TOPCARE-AMI studies (Assmus *et al.*, 2002; Britten *et al.*, 2003; Schachinger *et al.*, 2004; Schachinger *et al.*, 2006a), investigators injected $13+/-12 \times 10^6$ circulating progenitor cells (CPCs) or $238+/-79 \times 10^6$ BMCs into the infarct artery of patients 4.9+/−1.5 days (minimum of 4 days) after AMI. At 4 months, LV end-diastolic volume and EF improved in both cell dose groups compared to control patients who underwent standard treatment during the same time but were not randomized into the study. No significant differences between CPC and BM groups were observed. By 1 year, EF remained significantly improved, infarct size was decreased, and no LV remodeling was observed. These data, when combined with the reports by others (Table 22-3) suggest a very favorable response to BMNC therapy following AMI, with improved myocardial performance secondary to improved cardiac perfusion. These encouraging data also provided the impetus for initiation of randomized controlled trials using BMNCs for the treatment of STEMI—REPAIR-AMI, which has brought extremely positive results (Schachinger *et al.*, 2006b).

Although the data for the treatment of AMI with bone marrow cells are encouraging, what remains unclear is the response of the myocardium to these cells, when HF pathophysiology predominates. To begin to address this, the TOPCARE-HF study has been initiated. Given the reduced number and migratory capacity of EPCs shown in preclinical studies and the deficits in EPC quantity seen in patients with advanced CVD (Werner *et al.*, 2005), it will be interesting to see if cells from these patients are capable of improving cardiac function or if the HF milieu only allows ischemia-resistant cells, such as SKMBs, to survive.

In Germany, in a randomized trial entitled BOne marrOw transfer to enhance ST-elevation infarct regeneration (BOOST trial) (Wollert *et al.*, 2004) compared 30 patients under standard care following AMI (percutaneous coronary intervention (PCI) with stent placement) and 30 patients receiving $24.6\times10^8 \pm 9.4\times10^8$ bone marrow cells 4.8 ± 1.8 days after PCI. Six months after therapy, patients receiving cell therapy showed significantly enhanced LVEF when compared to control patients. At 18 months, the speed of LVEF recovery was significant in patients that received cells and PCI. There were no arrhythmic events or increased restenosis in the cell-treated patients. However, in Belgium, a recent clinical controlled trial evaluated the ability of autologous bone marrow cells (mean of 12.6×10^6 AC133$^+$ cells) to improve LV function after AMI (Bartunek *et al.*, 2005). In this study, myocardial perfusion was improved, but no improvement in LV function was seen when compared to controls. Most importantly, seven patients treated with cell therapy developed restenosis (versus four in control group), and two had *de novo* lesions.

Recently, three trials with autologous bone marrow cell administration were published. One of them (Lunde *et al.*, 2006) did not demonstrate a profoundly significant effect of BMNCs on LV function ($p = 0.054$) when injected at a median of 6 days (range: 4–7 days) post-AMI. However, on detailed examination of the data, it is clear that the infarct size measured with MRI was reduced compared to the control group at both 2–3 weeks and at 6 months post-therapy, exhibiting a statistically significant trend ($p = 0.07$), if we take a small sample size into consideration. One important aspect in this trial was that even though the groups were carefully matched at randomization, the patients that received BMNCs were prescribed more diuretics (40% in the cell therapy group versus 26% in the control group), which might have negatively impacted the engraftment of the cells. In the second of the recent three published trials, Dimmeler and Zeiher's group (Schachinger *et al.*, 2006b) achieved a larger sample size and showed a significantly more positive effect in settings and in patient population similar to Lunde *et al.* In the second trial (Schachinger *et al.*, 2006b), intracoronary infusion of BMNCs was associated with reduction in death, recurrence of myocardial infarction and revascularizations. This trial represents a **milestone** achievement for the field of cell therapy. There is no doubt that Phase III trial will take place soon. However, the discrepant results may represent that BMNCs need to be aspirated, expanded, prepared, and infused in strict adherence to small technical details. In the third trial (Assmus *et al.*, 2006) (from the same investigators), the type of cells were similar to the TOPCARE-AMI, but the design was crossover, and the delivery target was somewhat different. Both CPCs and BMNCs were infused into the most dyskinetic area of the healed (at least 3 months since index event) infarcted zone in the LV. Infusion of CPCs was much less successful than administration of BMNCs, and crossover to BMNCs was associated with significantly better outcome in terms of LV EF compared with the crossover to CPCs.

Taken together, these results show a great deal of evidence toward efficacy and preliminary but very encouraging evidence for BMNC therapy in settings of AMI. Although these data do not strictly address the use of these cells to treat HF, they illustrate what could the future be—early intervention to prevent the progression to end-stage HF in addition to optimized pharmacological treatment.

Cell therapy and administration of G-CSF: As the field of cell therapy matures, it is important to step back and evaluate the steps that led to progress. Borrowing from the established practices of bone marrow transplantation, several studies in CVD utilized granulocyte colony-stimulating factor (G-CSF) to stimulate production of bone marrow progenitors, then collected peripheral stem cells and infused them intracoronary. Even though G-CSF did not show major effects on LV function, in recent reports from Kang *et al.* (2004) and Hill *et al.* (2005), G-CSF administered in patients with chronic angina caused two AMIs and one death. The rates of restenosis following G-CSF administration increased, which can be explained by the augmented circulating cytokine milieu. Stimulating the bone marrow to produce progenitor cells (not necessarily exclusively with G-CSF) may be a part of therapeutic armamentarium in the future. Therefore, before we begin employing cytokine to stimulate bone marrow clinically in CVD patients, we need to better understand the entire cytokine cascade during AMI and during exacerbations of HF and cytokine signaling, as these are the likely therapeutic targets where cell therapy is going to be clinically applied. Such investigations may also shine light on the mechanisms behind the effects observed in clinical trials. These efforts are now underway at our institution.

Treating the entire vascular tree with cell therapy: Peripheral arterial disease (PAD) is becoming a specialized focus of cell-based repair. Treatment options in PAD depend on whether the underlying pathology is intermittent claudication or critical limb ischemia. But the current options are limited to exercise, anti-aggregants, thrombolytics, angioplasty, surgical revascularization, and when all fails, limb amputation. Experimental data suggests that the number and/or function of circulating EPCs may reflect progression or stabilization of atherosclerotic lesions and is currently being evaluated as a biomarker for PAD. In earlier stages of the disease, the therapeutic focus clearly lies on repair via decreased claudication and improved vascular endothelium **at large**. As the disease progresses, restoration of perfusion to minimize tissue damage and achieve symptomatic relief becomes of primary importance, whereas regeneration of functional muscle is secondary.

Different cell types and delivery methods are currently evaluated. Preliminary studies applying direct intramuscular injection of BMNCs and MSCs show promising results in increasing microvascular density and tissue perfusion and lead to the move to clinical studies. The Therapeutic Angiogenesis by Cell Transplantation Study investigators (Tateishi-Yuyama *et al.*, 2002) performed a randomized controlled trial in PAD patients and reported a significant increase in transcutaneous oxygen pressure, and pain-free walking time in 22 patients with leg ischemia after intramuscular injection of BMNCs. In a concomitant study, BMNC transplantation improved endothelial dysfunction by increasing endothelium-dependent vasodilation in patients with limb ischemia (Higashi *et al.*, 2004). Currently, a trial evaluating the therapeutic potential of CD34$^+$ cells is underway in patients with intermittent claudication, and results are much anticipated.

WHAT IS REQUIRED FOR SUCCESSFUL REPAIR IN 2007 AND BEYOND?

Today, cardiovascular repair seems to be a reachable goal. With the progress made to date, the field appears very promising. Nonetheless, several obstacles remain before we can declare unfettered success. What have we learned from over 10 years of preclinical and 6 years of clinical research in this area? Moving cell therapy from bench to bedside is complex. As new cell types emerge and old ones find new applications, it is important to design a preclinical path that predicts clinical outcome. It will also be important as the field moves forward to compare cells in head-to-head studies. Beginning to dissect the mechanism by which transplanted cells mediate repair is **crucial**. And finally if we are to ultimately regenerate heart with cell therapy, we must continue to think outside the box and view cells as only one tool in our armamentarium further moving into the 21st century and successfully integrate cell types and delivery routes with new pharmacotherapies. Cells plus genes, small molecules that replace the need for cells, and personalized genomics-based cell therapy are all medicines of the future. All of them may seem too novel and unreachable today, but so did cell therapy in 1998 (just 11 years ago). Considering new options, maximizing our mechanistic understanding of cell effects and standardizing our approaches to cell delivery, conducting clinical trials, and measuring outcomes should provide us the tools to succeed where endogenous repair fails. Below, we have outlined four specific "requirements for success."

Requirement No. 1: Selecting the Appropriate Cell Type for the Appropriate Disease Environment

At present, discrepant clinical trials outcomes exist for different types of cells and numbers of cells administered. For example, SKMBs seem to engraft into the myocardium and result in functional regeneration in HF, and BMNCs show great promise in treating acutely injured myocardium. What is clear from the basic science point of view, is that different environments in the myocardium at the time of injury likely generate different milieus, and therefore the cells that engraft in one environment may not survive in another one. Whether the discrepant clinical results are a result of a rush to clinical trials and applying various cells types in various contexts, or if segregation of the types of cells (at least between SKMBs and BMNCs) for the appropriate types of injury has already happened inadvertently remains to be understood. Because developing a successful therapy, which is based on the biology of the human body and the pathogenesis of disease, requires multiple reiterations between bench and beside, we need to go back to bench research **now** to compare various cell types side by side in various types of ischemic injury **in appropriate animal models**. This may seem to be a simple process, but in reality all available cells, delivery routes, and injury models are taken together would result in approximately 2,400 comparisons to be done. This clearly is a prohibitive number for a promising therapy. Therefore, we need to come to a consensus on the patients to be treated (i.e. concentrate on the most common types of injury, such as reperfused AMI at up to 4h from the onset of symptoms, ischemic HF with a mild-to-moderate ischemic process), and conduct comparisons in those models. As the data becomes available, we can then build hypotheses as to what may or may not work in other types of pathology and models. Such experiments will also bring additional insights into our understanding of how and when repair happens. This suggestion may sound contradictory to reality, considering the number of preclinical studies and clinical studies that have been published in the field (i.e. approximately 1,300 MEDLINE hits on keyword search for "heart" and "cell transplantation"). However, only a few head-to-head comparisons of different cell populations have been performed. We clearly lack direct comparisons of different cell types in clearly defined, clinically relevant models of disease.

In addition, and perhaps **most** importantly, there is an urgent need for a task force to define the nomenclature of progenitor cells to arrive at a consensus of which cells we are going to call "progenitor cells." Similar taxonomy efforts have been recently accomplished by Krumholz *et al.* (2006) for clarification of disease management. A writing group comprising experts in cell biology, taxonomy, cell differentiation, and translational research could very rapidly accomplish this task. Efforts in this direction will advance the field . . . and may help avoid unfortunate outcomes.

Comparing different cell types in various contexts of disease will also help us define how to improve survival of transplanted cells. Currently, one of the largest hurdles of cell therapy is the limited survival of transplanted cells. Most reports suggest that 70–90% of all transplanted cells die within the first few days of transplantation into infarct scar. Studies have shown that a subset of the transplanted cells survive and multiply, but it is unlikely that this multiplication can make up for the massive early necrosis and apoptosis of cells. Preclinical data suggest a dose response for several cell types, indicating that improving the number of surviving cells is critical to maximize functional outcome. Learning from previous fields is important. A confounding inflammatory response secondary to needle punctures during cell delivery is reminiscent of early percutaneous or transmyocardial myocardial revascularization studies where the "injection" *per se* promoted inflammation. Although the inflammatory response to needle stick has been reported as mild in most cases, the possibility that needle-based cell delivery is pro-inflammatory should be explored further; similarly if it is we need to define specific cytokines that might be involved. The problem faced is determining whether inflammation is an initiator of the necrosis of transplanted cells or a response secondary to the apoptosis of the transplanted cells. The most likely hypothesis is that the ischemic environment is the driver of these processes. This hypothesis is strengthened by data showing that survival of neonatal cardiomyocytes more than doubled when injected into 2-week-old cardiac granulation tissue or normal myocardium versus myocardial scar tissue in rats (Zhang *et al.*, 2001). Further, preconditioning of cells before transplantation via heat shock or transfecting cells with prosurvival factors (Akt, heat shock proteins, specific growth factors, or certain signaling molecules to provide protection of cells from hypoxia or glucose deprivation) helps increase their survival rate *in vivo* (Kohin *et al.*, 2001; Zhang *et al.*, 2001). In addition, we have preliminary data indicating that the composition of nutrients in which cells are grown *in vitro* alters survival in an infarct-like milieu (Davis, unpublished data). Ideally, more work will be focused on this area in the future to better define the relationship between the microenvironment of the infarct scar and outcome of the transplanted cells.

In addition to surviving in the ischemic environment at the time of implantation, the ideal cell for myocardial repair will be able to despite ischemia, become a fully functioning cardiomyocyte or an endothelial cell. However, none of the progenitor cells currently used satisfies both of these criteria at the numbers sufficient for maximal repair or recovery of function. Therefore, it is important to keep working toward understanding the differentiation of progenitor cells into a cardiomyocyte phenotype at the bench level. The goal, then, might be to design molecular tools to induce a pathway of directed differentiation prior to implantation, so that achievement of a specific phenotype would occur slowly enough to allow neovascularization to become functional (and not collapse) to support nascent myocardium.

Lastly, injected cells have significantly different electrical properties than cardiomyocytes. These differences have led to ventricular tachycardia observed in some of the clinical trials—especially where SKMBs were delivered in patients with HF. For cardiovascular cell therapy to reach its potential, it will be critical to electrically integrate transplanted cells into the surviving myocardium. This problem may be approached by genetically altering transplanted cells (to promote electrical coupling), by developing new adjunctive safety measures (such as co-administration of anti-arrhythmics), delivery of cells only in patients who meet the MADIT-II criteria and have ICDs, or preferably, by conditioning the transplanted cells to become true cardiomyocytes that can survive in a regenerating milieu.

Only by accomplishing these goals can we better design methods to maximize cell survival and thus increase the benefits versus the risks of cell therapy and achieve quantum leaps of progress in treating and preventing CVD.

Requirement No. 2: Choosing the Best Delivery Route

It is clear that choosing the best delivery route is the second prerequisite for success, after choosing the right cell for the right environment. A major experimental obstacle to the clinical efficacy is the poor engraftment seen when cells are administered by intracoronary, intravenous, and intracardiac routes. This is likely due to multiple factors, out of which technical difficulties of injecting exactly into the center or the periphery of the scar or catheter manipulations in the coronary tree cannot be overemphasized. We have recently published data showing that a direct placement of cells with the Da Vinci robotic system results in very accurately directed cell transplantation and does enhance the outcomes of the procedure in terms of improvement of LV function (Ott *et al.*, 2006). We also have preliminary data highlighting the importance of a very careful, targeted needle injection into the center of the scar versus periphery. Not only does LV remodeling differ with location but the arrhythmogenic potential may highly depend on the accurate placement of the transplanted cells (McCue, unpublished data). It is an established dogma that in the real estate business, location, location, location is the most well-known determinant of the success of a transaction. The same may hold for cell therapy. If so, training of the operators gains a pivotal importance. Recently, concerns of myocardial perforation due to operator error halted the GENASIS trial. As we go forward, creating a specialized network of centers for cell therapy, as currently proposed by the NHLBI, could allow for training of interventional cardiologists by experts in delivery techniques. Alternatively, it may also make sense to restrict the number of centers per region that act as referral centers and deliver cell therapy, at least until the techniques come to solid maturity. We have learned that operator volume and experience was a critical determinant of success in CABG and PCI clinical trials and also in routine clinical practice. As the field of cell therapy goes forward, we cannot ignore importance of appropriately trained specialists.

However, by the same token, we cannot ignore the need for further studies. The data has suggested that intracoronary delivery, at least in the context of AMI, can provide a comparable level of engraftment of cells to surgical delivery. This again points out the need for head-to-head comparisons of various delivery methods in controlled, designed experiments. We may need to consider again the roles that inflammatory factors play, and how the process of endogenous bone marrow mobilization impacts exogenous administration in AMI or in the beginning of HF, and what the covariates of that process might be. Understanding of the biology of these could allow coupling with an optimal situation-specific delivery system to produce several distinctly different products for the field. Achieving such an understanding and creating such a system will take some time.

However, the technological progress in the 20th century has been so fast that it will not be surprising if the next 5 years brings major progress in this regard.

Requirement No. 3: Rigorous Trial Design and Selection of End-Points

Right now clinical trials in cell therapy suffer from several major shortcomings primarily involving design and selection of end-points. Several examples of current limitations can be illustrated. To date, most studies have been accompanied by an additional revascularization procedure, either by PCI or CABG, making any functional improvement due to cellular therapy nearly impossible to distinguish from the current standard of care. The need to establish appropriate controls in a novel area where standard of care is evolving is an active area of debate in the field. We also need to account for the stimulatory effects of drugs, such as statins, PPAR agonists, erythropoietin, estrogen, and possibly others in various disease states. Right now, it is completely unclear which if any of these combinations of drugs alter the number and the function of progenitor cells available for repair. Clearly, if cell therapy is to be adequately evaluated we will need this information for the design of definitive Phase III trials and also to go forward with clinical applications.

Further, we lack data that evaluate time as an additional factor in treatment; time in disease progression as well as time in dynamics of transplanted cells. Overall, there is a lack of standardization in the current pre-clinical approach to cell therapy; for example, cell types, doses, pre-clinical models, and end-points all differ. This may also explain the discrepancy between pre-clinical and clinical results. Attempts to standardize these parameters and to decide on a consensus will move us forward.

What we call an "end-point" in cell therapy matters a great deal. So far, clinical trials have been geared toward measuring functional improvement of the LV by assessing global EF. As we know from the HF trials, improvement of regional contractility may not always translate into better HF numbers because of the differences in loading conditions. In addition, recently, we have begun to appreciate observer dependence of such measurements. Even though cardiac MRI offers the best topographic assessment of the heart, the variability is best minimized by conducting clinical trials with centralized core laboratories where the personnel undergoes regular inter- and intra-observer reproducibility assessments. More attention needs to be paid to regional contractility, peri-infarct zone, and scar volume quantification—all best done in an environment of a core laboratory. In addition, we need to evaluate myocardial perfusion. Over the last 10 years, the field of cardiac MRI has matured to offer quantitative assessment of myocardial perfusion (Jerosch-Herold *et al.*, 2004). Several sensitivity and specificity studies showed that assessment of myocardial flow with MRI may offer an edge of superiority over other techniques. Measuring changes in blood flow was proposed to be used as an end-point (Wilke *et al.*, 2001) and it is now becoming apparent that cell therapy will need a measure of blood flow as well. Concurrently, we need to critically evaluate the end-points that are used at the present time and come to an agreement, most likely through an AHA/ACC-sponsored consensus document, similarly to available data standards for AMI, HF, and atrial fibrillation that would outline the standard sets of data to be captured in the cell therapy trials. As that process goes along, some end-points with high subject variability, such as exercise treadmill time, will be critically evaluated and new, biologically relevant and clinically translatable end-points will be introduced. Such process will also enormously aid acceptance of new end-points by Food and Drug Administration (FDA) and will over time accelerate bringing cell therapies to market.

Requirement No. 4: Establish the Registry for the Results of Trials and the Biorepository for Blood Samples to Fill the Void of Mechanistic Understanding of Cell Therapy

Decades of CVD research have taught us the importance of centralized databases in advancing of our understanding of the disease process. The field of cardiovascular medicine would not have advanced as far as it did in the last 25–30 years if the Framingham Study or the TIMI trials had not been initiated and executed in a centralized matter. Large databases give the power to control for necessary covariates—a step not possible to accomplish even in a study of a few hundreds of patients. The field of cell therapy has arrived to the point when the next great advancement might be employing a large database of all results of clinical trials to serve as a filter for the hypotheses. With the aid of such a tool, ideas will be segregated before hundreds of thousands of dollars are spent only to find out that a specific factor interfered with the outcome. Creation of such a database (Figure 22-3) will pay off a 100-fold over time, as the field of cell therapy is coming out of infancy and maturing into adulthood. We should not underestimate the powers of computer technology and of the Internet available to us to create such a tool. Centralized data collecting efforts in acute HF, such as ADHERE registry (Yancy and Fonarow, 2004), have brought extremely valuable data with regards to the outcomes of clinical management of HF patients. It is time to create a registry for all applications of cell therapy in CVD.

Along with the outcomes, we must conduct population studies to define the role of bone marrow progenitors in vascular repair. As we are learning the importance of gender and race in the pathogenesis of CVD, we also need to

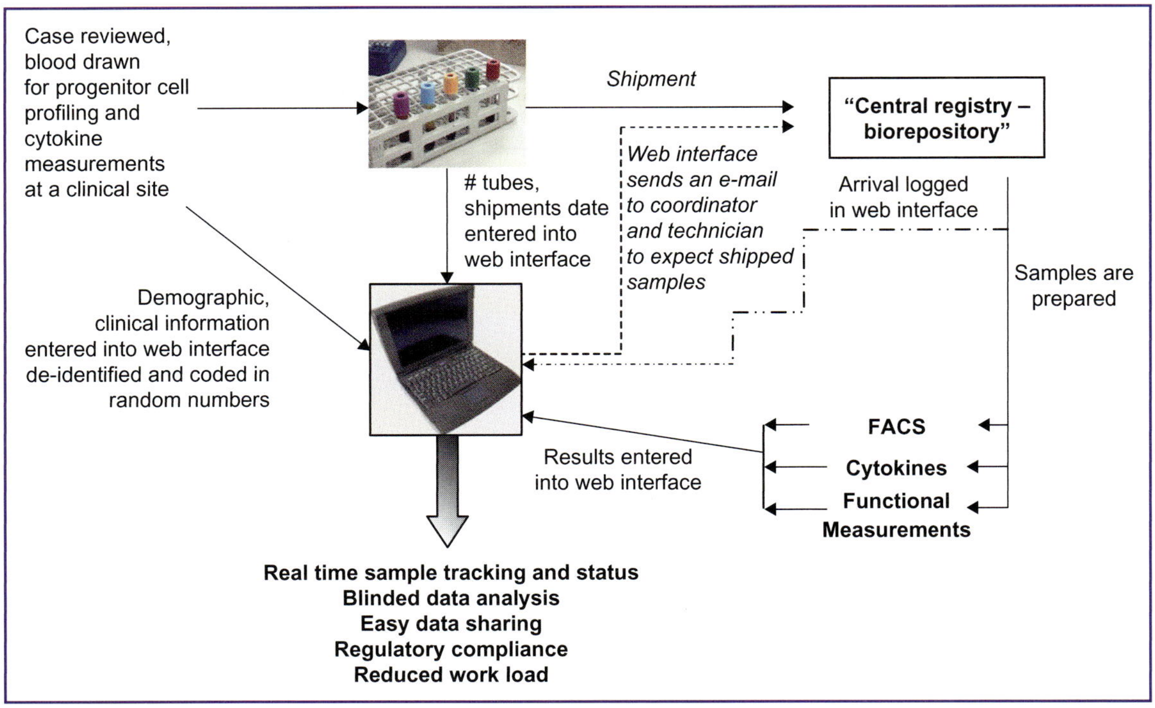

FIGURE 22-3 Schematic representation of the proposed model for central registry/biorepository for clinical trials in cell therapy. *Abbreviations.* FACS: fluorescent-activated cell sorting.

understand the differences in repair across wide age, gender and race groups, and also understand how the process of repair differs in those groups when different degrees of risk factors are superimposed. Conducting studies of this magnitude will help lessen the chances of not capturing significance when it truly exists—a frequent problem of small samples. As clinical studies go forward and the field matures further, we will need to evaluate other cell types involved (not just EPCs) and measure various evolving markers to supplement knowledge in the field of vascular repair. Centralized availability of samples will help reduce cost of repeated clinical trials by several orders of magnitude—something we all care about, especially at times of high national deficit and budget cuts to the NIH. In this regard, a centralized repository of BMCs should be the next national priority (Figure 22-3). Most importantly, measuring various progenitor cell populations, their function, carrying out gene profiling experiments, and performing experiments that require specific cell culture conditions needs to be accomplished with strict adherence to standardized protocols to achieve a desired outcome, as we are learning that even the smallest technical details matter in cell therapy.

Therefore, centralizing sample collection, storage, flow cytometry, and assays makes a great deal of common sense and will help greatly advance the science. Combining the registry for the clinical trials data and the biorepository for the blood and tissue samples seems to be exactly what the field needs to make another decade of major progress and help shape future cell therapy products. The short-term goal of a Repository would be to compare various subsets of circulating bone marrow progenitors in patient populations and to evaluate the impact of age, gender, and race. The long-term goal would be to use the progenitor cell characterization in conjuction with clinical data and examine the dynamics within multiple populations of progenitors in different states of disease when different types of cells are given. Provided that $CD34^+$ cells, for example, only represent approximately 1% of all circulating cells, a centralized Repository will provide in many ways a "win–win" situation by minimizing the effort required and maximizing the benefits yielded.

In summary, we believe that the outlined four requirements represent major issues in the cardiovascular cell therapy field today. As the field develops further and products moves closer to market, resolution of each requirement will increase the likelihood of successful outcome. The ultimate success, however, will be achievement of prevention of atherosclerosis and CVD altogether, reduction of hospitalization and major adverse cardiac events, and in prolonging a healthier life for patients who currently have limited options available to them.

SUMMARY

This is an exciting time in the field of treating CVD. Cell transplantation opened a new frontier, providing physicians with techniques and treatment alternatives for a large patient population that extends beyond revascularization and metabolic control to reverse damage that, in many cases, has already been done and may not truly be controllable. The concept of repairing or regenerating ischemic cardiac tissue is a truly fantastic possibility, and while many question its validity, it has an excellent chance to eventually become a clinical reality, if we address every requirement for its success.

While some more conservative researchers consider large human trials premature at this point, cell therapies, especially the recent trials, have shown clinical benefit. Due to the small study sizes and an inability, at this point, to standardize therapy, we are limited in our power to determine the best cell type, dose, and administration techniques, and to answer many other relevant questions. But the relevance of this therapy is evident to both researchers and clinicians. Both animal studies and clinical trials thus far have evoked the scientific enthusiasm and promising results to warrant large-scale controlled clinical trials to determine the best and safest application of this technology, and to gain a better understanding of its mechanism(s).

To bring this field forward we now have to come together and outline a plan for future studies. The diversity of cell types, application techniques, and disease stages can be a hurdle and an opportunity—only collaborations will allow us to move forward as a field instead of expanding information that cannot be combined or compared. We have the opportunity to create a new era in the treatment of CVD. Doing so will require continued bench to bedside and back to bench evaluations as we learn from early clinical studies, find a consensus on preclinical models and the design of clinical trials to maximize the potential of a 21st century approach to repairing the injured heart.

Finally, even as the field progresses, we have a responsibility to promise patients (and the press) only what we can deliver, that is to tell the truth about cardiac repair. BMNCs, MSCs, SKMBs, or other types of cells hold a great promise to modify pathophysiological process in specific ways. It is crucial to understand for clinicians, patients, and the press that specificity precludes panacea. As we go forward, some applications will succeed, and some will fail. Cells may not be found guilty of failures. On the contrary, the disease contexts may come to be the primary determinants of efficacy. We have already experienced a similar process with angiogenic growth factors in CVD, and we now know that those trials should have more carefully targeted the disease process, as the results uniformly showed that sicker patients had larger therapeutic benefits. As investigators, we need to be realistic of the expectations we place on cell therapy, and ultimately we need to underpromise and overdeliver, based on rigorous science . . . otherwise, the great potential will eventually be destroyed. Cell therapy is, however, a new and very promising alternative that warrants much further exploration, inspiration, and investment of our time and resources.

ACKNOWLEDGMENTS

This work has been supported in part by NHLBI/NIH award to Dr. Taylor (R01-HL-063346), Minnesota Partnership for Biotechnology and Medical Genomics award, and by funding from the Center for Cardiovascular Repair, University of Minnesota.

Authors sincerely thank Harald C. Ott, MD, for his continuous contributions to the ongoing success of the Center for Cardiovascular Repair.

REFERENCES

Abbate, A., Bonanno, E., Mauriello, A., Bussani, R., Biondi-Zoccai, G., Liuzzo, G., Leone, A., Silvestri , F., Dobrina, A., Baldi, F., Pandolfi, F., Biasucci, L., Baldi, A., Spagnoli, L., and Crea, F. (2004). Widespread myocardial inflammation and infarct-related artery patency. *Circulation 110*: 46–50.

Abbate, A., Bussani, R., Amin, M.S., Vetrovec, G.W. and Baldi, A. (2006). Acute myocardial infarction and heart failure: role of apoptosis. *Int. J. Biochem. Cell Biol.* 38: 1834–1840.

Abraham, M., Henrikson, C., Tung, L., Chang, M., Aon, M., Xue, T., Li, R., O' Rourke, B. and Marban, E. (2005). Antiarrhythmic engineering of skeletal myoblasts for cardiac transplantation. *Circ. Res. 97*: 159–167.

Agbulut, O., Vandervelde, S., Al Attar, N., Larghero, J., Ghostine, S., Leobon, B., Robidel, E., Borsani, P., Le Lorc'h, M., Bissery, A., Chomienne, C., Bruneval, P., Marolleau, J.P., Vilquin, J.T., Hagege, A., Samuel, J.L. and Menasche, P. (2004). Comparison of human skeletal myoblasts and bone marrow-derived CD133+ progenitors for the repair of infracted myocardium. *J. Am. Coll. Cardiol. 44*: 458–463.

Aicher, A., Brenner, W., Zuhayra, M., Badorff, C., Massoudi, S., Assmus, B., Eckey, T., Henze, E., Zeiher, A.M. and Dimmeler, S. (2003). Assessment of the tissue distribution of transplanted human endothelial progenitor cells by radioactive labeling. *Circulation 107*: 2134–2139.

Anversa, P. and Nadal-Ginard, B. (2002). Myocyte renewal and ventricular remodelling. *Nature 415*: 240–243.

Appel, L.J., Brands, M.W., Daniels, S.R., Karanja, N., Elmer, P.J. and Sacks, F.M. (2006). Dietary approaches to prevent and treat hypertension: a scientific statement from the American Heart Association. *Hypertension 47*: 296–308.

Assmus, B., Schachinger, V., Teupe, C., Britten, M., Lehmann, R., Dobert, N., Grunwald, F., Aicher, A., Urbich, C., Martin, H., Hoelzer, D., Dimmeler, S. and Zeiher, A.M. (2002). Transplantation of progenitor cells and regeneration enhancement in acute myocardial infarction (TOPCARE-AMI). *Circulation 106*: 3009–3017.

Assmus, B., Honold, J., Schachinger, V., Britten, M., Fischer-Rasokat, U., Lehmann, R., Teupe, C., Pistorius, K., Martin, H., Abolmaali, N., Tonn, T., Dimmeler, S. and Zeiher, A. (2006). Trancoronary transplantation of progenitor cells after myocardial infarction. *N. Engl. J. Med. 355*: 1222–1232.

Atkins, B.Z., Hueman, M.T., Meuchel, J.M., Cottman, M.J., Hutcheson, K.A. and Taylor, D.A. (1999a). Myogenic cell transplantation improves *in vivo* regional performance in infarcted rabbit myocardium. *J. Heart Lung Transplant. 18*: 1173–1180.

Atkins, B.Z., Hutcheson, K.H., Hueman, M.T., Cottman, M.J., Meuchel, J.M., Glower, D.D. and Taylor, D.A. (1999b). Reversing post-MI dysfunction: improved myocardial performance after autologous skeletal myoblast transfer to infarcted rabbit heart. *Circ. Suppl. 100*:I–838.

Atkins, B.Z., Lewis, C.W., Kraus, W.E., Hutcheson, K.A., Glower, D.D., and Taylor, D.A. (1999c). Intracardiac transplantation of skeletal myoblasts yields two populations of striated cells *in situ. Ann. Thor. Surg. 67*, 124–129.

Bairey Merz, C.N., Shaw, L.J., Reis, S.E., Bittner, V., Kelsey, S.F., Olson, M., Johnson, B.D., Pepine, C.J., Mankad, S., Sharaf, B.L., Rogers, W.J., Pohost, G.M., Lerman, A., Quyyumi, A.A. and Sopko, G. (2006). Insights from the NHLBI-Sponsored Women's Ischemia Syndrome Evaluation (WISE) Study: Part II: gender differences in presentation, diagnosis, and outcome with regard to gender-based pathophysiology of atherosclerosis and macrovascular and microvascular coronary disease. *J. Am. Coll. Cardiol. 47*: S21–S29.

Baldus, S., Heeschen, C., Meinertz, T., Zeiher, A.M., Eiserich, J.P., Munzel, T., Simoons, M.L., and Hamm, C.W. (2003). Myeloperoxidase serum levels predict risk in patients with acute coronary syndromes. *Circulation 108*: 1440–1445.

Bartunek, J., Vanderheyden, M., Vandekerckhove, B., Mansour, S., De Bruyne, B., De Bondt, P., Van Haute, I., Lootens, N., Heyndrickx, G. and Wijns, W. (2005). Intracoronary injection of CD133-positive enriched bone marrow progenitor cells promotes cardiac recovery after recent myocardial infarction: feasibility and safety. *Circulation 112*:I178–I183.

Bertrand, M.E. (2004). Provision of cardiovascular protection by ACE inhibitors: a review of recent trials. *Curr. Med. Res. Opin. 20*: 1559–1569.

Bhagat, K. and Vallance, P. (1997). Inflammatory cytokines impair endothelium-dependent dilatation in human veins *in vivo. Circulation 96*: 3042–3047.

Bittira, B., Shum-Tim, D., Al-Khaldi, A. and Chiu, R.C. (2003). Mobilization and homing of bone marrow stromal cells in myocardial infarction. *Eur. J. Cardiothorac. Surg. 24*: 393–398.

Blatt, A., Cotter, G., Leitman, M., Krakover, R., Kaluski, E., Milo-Cotter, O., Resnick, I., Samuel, S., Gozal, D., Vered, Z., Slavin, S. and Shapira, M. (2005). Intracoronary administration of autologous bone marrow mononuclear cells after induction of short ischemia is safe and may improve hibernation and ischemia in patients with ischemic cardiomyopa-thy *Am. Heart J. 150*: 981–987.

Blumenthal, R.S., Becker, D.M., Yanek, L.R., Moy, T.E., Michos, E.D., Fishman, E.K. and Becker, L.C. (2006). Comparison of coronary calcium and stress myocardial perfusion imaging in apparently healthy siblings of individuals with premature coronary artery disease. *Am. J. Cardiol. 97*: 328–333.

Britten, M.B., Abolmaali, N.D., Assmus, B., Lehmann, R., Honold, J., Schmitt, J., Vogl, T.J., Martin, H., Schachinger, V., Dimmeler, S. and Zeiher, A.M. (2003). Infarct remodeling after intracoronary progenitor cell treatment in patients with acute myocardial infarction (TOPCARE-AMI): mechanistic insights from serial contrast-enhanced magnetic resonance imaging. *Circulation 108*: 2212–2218.

Chachques, J.C., Grandjean, P.A., Tommasi, J.J., Perier, P., Chauvaud, S., Bourgeois, I. and Carpentier, A. (1987). Dynamic cardiomyoplasty: a new approach to assist chronic myocardial failure. *Life Support Syst. 5*: 323–327.

Chachques, J.C., Duarte, F., Cattadori, B., Shafy, A., Lila, N., Chatellier, G., Fabiani, J.N. and Carpentier, A.F. (2004a). Angiogenic growth factors and/or cellular therapy for myocardial regeneration: a comparative study. *J. Thorac. Cardiovasc. Surg. 128*: 245–253.

Chachques, J.C., Herreros, J., Trainini, J., Juffe, A., Rendal, E., Prosper, F. and Genovese, J. (2004b). Autologous human serum for cell culture avoids the implantation of cardioverter-defibrillators in cellular cardiomyoplasty. *Int. J. Cardiol. 95*(Suppl 1): S29–S33.

Chello, M., Patti, G., Candura, D., Mastrobuoni, S., Sciascio, G., Agro, F., Carassiti, M. and Covino, E. (2006). Effects of atorvastatin on systemic inflammatory response after coronary bypass surgery. *Crit. Care Med. 108*: 660–667.

Chen, S.L., Fang, W.W., Ye, F., Liu, Y.H., Qian, J., Shan, S.J., Zhang, J.J., Chunhua, R.Z., Liao, L.M., Lin, S. and Sun, J.R. (2004). Effect on left ventricular function of intracoronary transplantation of autologous bone marrow mesenchymal stem cell in patients with acute myocardial infarction. *Am. J. Cardiol. 94*: 92–95.

Chiu, R.C., Zibaitis, A. and Kao, R.L. (1995). Cellular cardiomyoplasty: myocardial regeneration with satellite cell implantation. *Ann. Thorac. Surg. 60*: 12–18.

Cohn, J.N. (2002). Lessons learned from the valsartan-heart failure trial (Val-HeFT): angiotensin receptor blockers in heart failure. *Am. J. Cardiol. 90*: 992–993.

Davidson, M. (2005). Clinical significance of statin pleiotropic effects: hypothesis versus evidence. *Circulation 111*: 2280–2281.

Deten, A., Volz, H.C., Clamors, S., Leiblein, S., Briest, W., Marx, G. and Zimmer, H.G. (2005). Hematopoietic stem cells do not repair the infarcted mouse heart. *Cardiovasc. Res. 65*: 52–63.

Dib, N., Michler, R., Pagani, E., Wright, S., Kereiakes, D., Lengerich, R., Binkley, P., Buchele, D., Anand, I., Swingen, C., Di Carl, M.R., Thomas, J., Jaber, W., Opie, S.R., Campbell, A., McCarthy, P., Yeager, M., Dilsizian, V., Griffith, B., Korn, R., Kreuger, S., Ghazoul, M., MacLellan, W., Fonarow, G., Eisen, H., Dinsmore, J. and Diethrich, E. (2005). Safety and feasibility of autologous myoblast transplantation in patients with ischemic cardiomyopathy: four-year follow-up. *Circulation 112*: 1748–1755.

Doggrell, S.A. (2005). CHARMed—the effects of candesartan in heart failure. *Exp. Opin. Pharmacother. 6*: 513–516.

Dohmann, H., Perin, E., Borojevic, R., Silva, S., Souza, A., Silva, G., Assad, J., Rossi, M.I., Mesquita, C. and Dohmann, H. (2005). (Sustained improvement in symptoms and exercise capacity up to six months after autologous transendocardial transplantation of bone marrow mononuclear cells in patients with severe ischemic heart disease.) [Portuguese]. *Arq. Bras. Cardiol. 84*: 360–366.

Fadini, G.P., Miorin, M., Facco, M., Bonamico, S., Baesso, I., Grego, F., Menegolo, M., de Kreutzenberg, S.V., Tiengo, A., Agostini, C. and Avogaro, A. (2005). Circulating endothelial progenitor cells are reduced in peripheral vascular complications of type 2 diabetes mellitus. *J. Am. Coll. Cardiol. 45*: 1449–1457.

Fadini, G.P., Coracina, A., Baesso, I., Agostini, C., Tiengo, A., Avogaro, A. and de Kreutzenberg, S.V. (2006). Peripheral blood $CD34^+KDR^+$ endothelial progenitor cells are determinants of subclinical atherosclerosis in a middle-aged general population. *Stroke 37*: 2277–2282.

Feletou, M. and Vanhoutte, P. (2006). Endothelial dysfunction: a multifaceted disorder (The Wiggers Award Lecture). *Am. J. Physiol. Heart Circ. Physiol. 291*: H985–H1002.

Fernandez-Aviles, F., San Roman, J.A., Garcia-Frade, J., Fernandez, M.E., Penarrubia, M.J., de la Fuente, L., Gomez-Bueno, M., Cantalapiedra, A., Fernandez, J., Gutierrez, O., Sanchez, P.L., Hernandez, C., Sanz, R., Garcia-Sancho, J. and Sanchez, A. (2004). Experimental and clinical regenerative capability of human bone marrow cells after myocardial infarction. *Circ. Res. 95*: 742–748.

Ferns, G.A. and Avades, T.Y. (2000). The mechanisms of coronary restenosis: insights from experimental models. *Int. J. Exp. Pathol. 81*: 63–88.

Field, L. and Reinlib, L. (2000). Cell transplantation as future therapy for cardiovascular disease?: a workshop of the National Heart, Lung, and Blood Institute. *Circulation 101*, E182–E187.

Frostegard, J., Ulfgren, A.K., Nyberg, P., Hedin, U., Swedenborg, J., Andersson, U. and Hansson, G.K. (1999). Cytokine expression in advanced human atherosclerotic plaques: dominance of pro-inflammatory (Thl) and macrophage-stimulating cytokines. *Atherosclerosis 145*: 33–43.

Fuchs, S., Baffour, R., Zhou, Y., Shou, M., Pierre, A., Tio, F., Weissman, N., Leon, M., Epstein, S. and Kornowski, R. (2001). Transendocardial delivery of autologous bone marrow enhances collateral perfusion and regional function in pigs with chronic experimental myocardial ischemia. *J. Am. Coll. Cardiol. 37*: 1726–1732.

Fuchs, S., Satler, L.F., Kornowski, R., Okubagzi, P., Weisz, G., Baffour, R., Waksman, R., Weissman, N.J., Cerqueira, M., Leon, M.B. and Epstein, S.E. (2003). Catheter-based autologous bone marrow myocardial injection in no-option patients with advanced coronary artery disease: a feasibility study. *J. Am. Coll. Cardiol. 41*: 1721–1724.

Fuchs, S., Kornowski, R., Weisz, G., Satler, L., Smits, P., Okubagzi, P., Baffour, R., Aggarwal, A., Weissman, N., Cerqueira, M., Waksman, R., Serrruys, P., Battler, A., Moses, J., Leon, M. and Epstein, S. (2006). Safety and feasibility of transendocardial autologous bone marrow cell transplantation in patients with advanced heart disease. *Am. J. Cardiol. 97*: 823–829.

Fujiyama, S., Amano, K., Uehira, K., Yoshida, M., Nishiwaki, Y., Nozawa, Y., Jin, D., Takai, S., Miyazaki, M., Egashira, K., Imada, T., Iwasaka, T. and Matsubara, H. (2003). Bone marrow monocyte lineage cells adhere on injured endothelium in a monocyte chemoattractant protein-1-dependent manner and accelerate reendothelialization as endothelial progenitor cells. *Circ. Res. 93*: 980–989.

Gavira, J., Herreros, J., Perez, A., Garcia-Velloso, M., Barba, J., Martin-Herrero, F., Canizo, C., Martin-Arnau, A., Marti-Climent, J., Hernandez, M., Lopez-Holgado, N., Gonzalez-Santos, J., Martin-Luengo, C., Alegria, E. and Prosper, F. (2006). Autologous skeletal myoblast transplantation in patients with nonacute myocardial infarction: 1-year follow-up. *J. Thorac. Cardiovasc. Surg. 131*: 799–804.

Goldschmidt-Clermont, P.J., Creager, M.A., Losordo, D.W., Lam, G.K., Wassef, M. and Dzau, V.J. (2005). Atherosclerosis 2005: recent discoveries and novel hypotheses. *Circulation 112*: 3348–3353.

Haffner, S.M. (2002). Can reducing peaks prevent type 2 diabetes: implication from recent diabetes prevention trials. *Intl J. Clin. Pract. Suppl. 129*: 33–39.

Haider, H. (2006). Bone marrow cells for cardiac regeneration and repair: current status and issues. *Expert Rev. Cardiovasc. Ther. 4*: 557–568.

Hamano, K., Nishida, M., Hirata, K., Mikamo, A., Li, T.S., Harada, M., Miura, T., Matsuzaki, M. and Esato, K. (2001). Local implantation of autologous bone marrow cells for therapeutic angiogenesis in patients with ischemic heart disease: clinical trial and preliminary results. *Jpn Circ. J. 65*: 845–847.

Hansen-Algenstaedt, N., Joscheck, C., Wolfram, L., Schaefer, C., Muller, I., Bottcher, A., Deuretzbacher, G., Wiesner, L., Leunig, M.A., lgenstaedt, P. and Ruther, W. (2006). Sequential changes in vessel formation and micro-vascular function during bone repair. *Acta Orthop. 77*: 429–139.

Hansen-Algenstaedt, N., Schaefer, C., Wolfram, L., Joscheck, C., Schroeder, M., Algenstaedt, P. and Ruther, W. (2005). Femur window—a new approach to microcirculation of living bone *in situ*. *J. Orthop. Res. 23*: 1073–1082.

Hansson, G.K. (2005). Inflammation, atherosclerosis, and coronary artery disease. *N. Engl. J. Med. 352*: 1685–1695.

Henning, R.J., Abu-Ali, H., Balis, J.U., Morgan, M.B., Willing, A.E. and Sanberg, P.R. (2004). Human umbilical cord blood mononuclear cells for the treatment of acute myocardial infarction. *Cell Transplant. 13*: 729–739.

Henry, T.D., Annex, B.H., McKendall, G.R., Azrin, M.A., Lopez, J.J., Giordano, F.J., Shah, P.K., Willerson, J.T., Benza, R.L., Berman, D.S., Gibson, C.M., Bajamonde, A., Rundle, A.C., Fine, J., McCluskey, E.R. and Investigators, V. (2003). The VIVA, trial: Vascular endothelial growth factor in Ischemia for Vascular Angiogenesis. *Circulation 107*: 1359–1365.

Hernandez, A.F., Velazquez, E.J., Solomon, S.D., Kilaru, R., Diaz, R., O'Connor, C.M., Ertl, G., Maggioni, A.P., Rouleau, J.L., van Gilst, W., Pfeffer, M.A., Califf, R.M. and Registry, V (2005). Left ventricular assessment in myocardial infarction: the VALIANT registry. *Arch. Intern. Med. 165*: 2162–2169.

Herreros, J., Prosper, F., Perez, A., Gavira, J.J., Garcia-Velloso, M.J., Barba, J., Sanchez, P.L., Canizo, C., Rabago, G., Marti-Climent, J.M., Hernandez, M., Lopez-Holgado, N., Gonzalez-Santos, J.M., Martin-Luengo, C. and Alegria, E. (2003). Autologous intramyocardial injection of cultured skeletal muscle-derived stem cells in patients with non-acute myocardial infarction. *Eur. Heart. 24*: 2012–2020.

Hiasa, K., Egashira, K., Kitamoto, S., Ishibashi, M., Inoue, S., Ni, W., Zhao, Q., Nagata, S., Katoh, M., Sata, M. and Takeshita, A. (2004). Bone marrow mononuclear cell therapy limits myocardial infarct size through vascular endothelial growth factor. *Basic Res. Cardiol. 99*: 165–172.

Higashi, Y., Kimura, M., Hara, K., Noma, K., Jitsuiki, D., Nakagawa, K., Oshima, T., Chayama, K., Sueda, T., Goto, C., Matsubara, H., Murohara, T. and Yoshizumi, M. (2004). Autologous bone-marrow mononuclear cell implantation improves endothelium-dependent vasodilation in patients with limb ischemia. *Circulation 109*: 1215–1218.

Hill, J.M., Zalos, G., Halcox, J.P., Schenke, W.H., Waclawiw, M.A., Quyyumi, A.A. and Finkel, T. (2003). Circulating endothelial progenitor cells, vascular function, and cardiovascular risk. *N. Engl. J. Med. 348*: 593–600.

Hill, J.M., Syed, M.A., Arai, A.E., Powell, T.M., Paul, J.D., Zalos, G., Read, E.J., Khuu, H.M., Leitman, S.F., Home, M., Csako, G., Dunbar, C.E., Waclawiw, M.A. and Cannon III, R.O. (2005). Outcomes and risks of granulocyte colony-stimulating factor in patients with coronary artery disease. *J. Am. Coll. Cardiol. 46*: 1643–1648.

Horackova, M., Arora, R., Chen, R., Armour, J.A., Cattini, P.A., Livingston, R. and Byczko, Z. (2004). Cell transplantation for treatment of acute myocardial infarction: unique capacity for repair by skeletal muscle satellite cells. *Am. J. Physiol. Heart Circ. Physiol. 287*: H1599–H1608.

Hutcheson, K.A., Atkins, B.Z., Hueman, M.T., Hopkins, M.B., Glower, D.D. and Taylor, D.A. (2000). Comparing the benefits of cellular cardiomyoplasty with skeletal myoblasts or dermal fibroblasts on myocardial performance. *Cell Transplant. 9*: 359–368.

Ikeda, U. (2003). Inflammation and coronary artery disease. *Curr. Vasc. Pharmacol. 1*: 65–70.

Ince, H., Petzsch, M., Rehders, T., Chatterjee, T. and Nienaber, C. (2004). Transcatheter transplantation of autologous skeletal myoblasts in postinfarction patients with severe left ventricular dysfunction. *J. Endovasc. Ther. 11*: 695–704.

Jain, M. and Ridker, P. (2005). Anti-inflammatory effects of statins: clinical evidence and basic mechanisms. *Nat. Rev. Drug Discov. 4*: 977–987.

Jerosch-Herold, M., Seethamraju, R., Swingen, C., Wilke, N. and Stillman, A. (2004). Analysis of myocardial perfusion MRI. *J. Magn. Reson. Imag. 19*: 758–770.

Jiang, X.X., Zhang, Y., Liu, B., Zhang, S.X., Wu, Y., Yu, X.D. and Mao, N. (2005). Human mesenchymal stem cells inhibit differentiation and function of monocyte-derived dendritic cells. *Blood 105*: 4120–4126.

Jiang, Y., Jahagirdar, B.N., Reinhardt, R.L., Schwartz, R.E., Keene, C.D., Ortiz-Gonzalez, X.R., Reyes, M., Lenvik, T., Lund, T., Blackstad, M., Du, J., Aldrich, S., Lisberg, A., Low, W.C., Largaespada, D.A. and Verfaillie, C.M. (2002). Pluripotency of mesenchymal stem cells derived from adult marrow. *Nature 418*: 41–49.

Jong, P., Yusuf, S., Rousseau, M.F., Ahn, S.A. and Bangdiwala, S.I. (2003). Effect of enalapril on 12-year survival and life expectancy in patients with left ventricular systolic dysfunction: a follow-up study. *Lancet 361*: 1843–1848.

Jost, A., Rauch, B., Hochadel, M., Winkler, R., Schneider, S., Jacobs, M., Kilkowski, C., Kilkowski, A., Lorenz, H., Muth, K., Zugck, C., Remppis, A., Haass, M., Senges, J. (2005). Beta-blocker treatment of chronic systolic heart failure improves prognosis even in patients meeting one or more exclusion criteria of the MERIT-HF study. *Eur. Heart J. 26*: 2689–2697.

Juonala, M., Viikari, J.S., Rasanen, L., Helenius, H., Pietikainen, M. and Raitakari, O.T. (2006). Young adults with family history of coronary heart disease have increased arterial vulnerability to metabolic risk factors: the Cardiovascular Risk in Young Finns Study. *Arterioscl. Thromb. Vasc. Biol. 26*: 1376–1382.

Kalka, C., Masuda, H., Takahashi, T., Kalka-Moll, W.M., Silver, M., Kearney, M., Li, T., Isner, J.M. and Asahara, T. (2000). Transplantation of *ex vivo* expanded endothelial progenitor cells for therapeutic neovascularization. *Proc. Natl Acad. Sci. USA 97*: 3422–3427.

Kang, H.J., Kim, H.S., Zhang, S.Y., Park, K.W., Cho, H.J., Koo, B.K., Kim, Y.J., Soo Lee, D., Sohn, D.W., Han, K.S., Oh, B.H., Lee, M.M. and Park, YB. (2004). Effects of intracoronary infusion of peripheral blood stem-cells mobilised with granulocyte-colony stimulating factor on left ventricular systolic function and restenosis after coronary stenting in myocardial infarction: the MAGIC cell randomised clinical trial. *Lancet 363*: 751–756.

Karladani, A., Granhed, H.P., Karrholm, J. and Styf, J. (2001). The influence of fracture etiology and type on fracture healing: a review of 104 consecutive tibial shaft fractures. *Arch. Orthop. Trauma Surg. 121*: 325–328.

Kawada, H., Fujita, J., Kinjo, K., Matsuzaki, Y., Tsuma, M., Miyatake, H., Muguruma, Y., Tsuboi, K., Itabashi, Y., Ikeda, Y., Ogawa, S., Okano, H., Hotta, T., Ando, K. and Fukuda, K. (2004). Nonhematopoietic mesenchymal stem cells can be mobilized and differentiate into cardiomyocytes after myocardial infarction. *Blood 104*: 3581–3587.

Kawamoto, A., Gwon, H.C., Iwaguro, H., Yamaguchi, J.I., Uchida, S., Masuda, H., Silver, M., Ma, H., Kearney, M., Isner, J.M. and Asahara, T. (2001). Therapeutic potential of *ex vivo* expanded endothelial progenitor cells for myocardial ischemia. *Circulation 103*: 634–637.

Khan, Q. and Mehta, J.L. (2005). Relevance of platelet-independent effects of aspirin to its salutary effect in atherosclerosis-related events. *J. Atheroscler. Thromb. 12*: 185–190.

Kinnaird, T., Stabile, E., Burnett, M., Shou, M., Lee, C., Barr, S., Fuchs, S. and Epstein, S. (2004). Local delivery of marrow-derived stromal cells augments collateral perfusion through paracrine mechanisms. *Circulation 109*: 1543–1549.

Koh, G.Y., Klug, M.G., Soonpaa, M.H. and Field, L.J. (1993). Differentiation and long-term survival of C2C12 myoblast grafts in heart. *J. Clin. Invest. 92*: 1548–1554.

Kohin, S., Stary, C.M., Howlett, R.A. and Hogan, M.C. (2001). Preconditioning improves function and recovery of single muscle fibers during severe hypoxia and reoxygenation. *Am. J. Physiol. Cell Physiol. 281*: C142–C146.

Kostner, K., Fahti, R., Case, C., Hobson, P., Tate, J. and Marwick, T. (2006). Inflammation, complement activation and endothelial function in stable and unstable coronary artery disease. *Clin. Chim. Acta 365*: 129–134.

Kravchenko, J., Goldschmidt-Clermont, P.J., Powell, T., Stallard, E., Akushevich, I., Cuffe, M.S. and Manton, K.G. (2005). Endothelial progenitor cell therapy for atherosclerosis: the philosopher's stone for an aging population? *Sci. Aging Knowledge Environ.* 2005: pe18.

Krotz, F., Schiele, T.M., Klauss, V. and Sohn, H.Y. (2005). Selective COX-2 inhibitors and risk of myocardial infarction. *J. Vasc. Res. 42*: 312–324.

Krumholz, H.M., Currie, P.M., Riegel, B., Phillips, C.O., Peterson, E., Smith, R., Yancy, C.W., and Faxon, D. (2006). AHA Scientific Statement. A Taxonomy for Disease Management. A scientific statement from the American Heart Association Disease Management Taxonomy Writing Group. *Circulation 114*:1432–1445.

Kuethe, F., Richartz, B.M., Sayer, H.G., Kasper, C., Werner, G.S., Hoffken, K. and Figulla, H.R. (2004). Lack of regeneration of myocardium by autologous intracoronary mononuclear bone marrow cell transplantation in humans with large anterior myocardial infarctions. *Int. J. Cardiol. 97*: 123–127.

Kunz, G.A., Liang, G., Cuculoski, R., Gregg, D., Vata, K.C., Shaw, L.K., Goldschmidt-Clermont, P.J., Dong, C., Taylor, D.A. and Peterson, E.D. (2006). Circulating endothelial progenitor cells predict coronary artery disease severity. *Am. Heart J. 152*: 190–195.

Kwong, R., Chan, A.K., Brown, K.A., Chan, C.W., Reynolds, H.G., Tsang, S. and Davis, R.B. (2006). Impact of unrecognized myocardial scar detected by cardiac magnetic resonance imaging on event-free survival in patients presenting with signs or symptoms of coronary artery disease. *Circulation 113*: 2733–2743.

Labarrere, C.A. and Zaloga, G.P. (2004). C-reactive protein: from innocent bystander to pivotal mediator of atherosclerosis. *Am. J. Med. 117*: 499–507.

Laugwitz, K.L., Moretti, A., Lam, J., Gruber, P., Chen, Y., Woodard, S., Lin, L.Z., Cai, C.L., Lu, M.M., Reth, M., Platoshyn, O., Yuan, J.X., Evans, S. and Chien, K.R. (2005). Postnatal isl1+ cardioblasts enter fully differentiated cardiomyocyte lineages. *Nature 433*: 647–653.

Lauten, W.B., Khan, Q.A., Rajagopalan, S., Lerakis, S., Rahman, S.T., Parthasarathy, S. and Khan, B.V. (2003). Usefulness of quinapril and irbesartan to improve the anti-inflammatory response of atorvastatin and aspirin in patients with coronary heart disease. *Am. J. Cardiol. 91*: 1116–1119.

Lee, C.W., Chae, J.K., Lim, H.Y., Hong, M.K., Kim, J.J., Park, S.W. and Park, S.J. (1999). Prospective randomized trial of corticosteroids for the prevention of restenosis after intracoronary stent implantation. *Am. Heart J. 138*: 60–63.

Lenzen, M.J., Boersma, E., Reimer, W.J., Balk, A.H., Komajda, M., Swedberg, K., Follath, E., Jimenez-Navarro, M., Simoons, M.L. and Cleland, J.G. (2005). Under-utilization of evidence-based drug treatment in patients with heart failure is only partially explained by dissimilarity to patients enrolled in landmark trials: a report from the Euro Heart Survey on Heart Failure. *Eur. Heart J. 26*: 2706–2713.

Leor, J., Guetta, E., Feinberg, M., Galski, H., Bar, I., Holbova, R., Miller, L., Zarin, P., Castel, D., Barbash, I. and Nagler, A. (2006). Human umbilical cord blood-derived $CD133^+$ cells enhance function and repair of the infarcted myocardium. *Stem Cell 24*: 772–780.

Lev, E., Kleiman, N., Birnbaum, Y., Harris, D., Korbling, M. and Estrov, Z. (2005). Circulating endothelial progenitor cells and coronary collaterals in patients with non-ST segment elevation myocardial infarction. *J. Vasc. Res. 42*: 408–414.

Levy, D., Kenchaiah, S., Larson, M.G., Benjamin, E.J., Kupka, M.J., Ho, K.K., Murabito, J.M. and Vasan, R.S. (2002). Long-term trends in the incidence of and survival with heart failure. *N. Engl. J. Med. 347*: 1397–1402.

Li, M., Nishimura, H., Iwakura, A., Wecker, A., Eaton, E., Asahara, T. and Losordo, D. (2005). Endothelial progenitor cells are rapidly recruited to myocardium and mediate protective effect of ischemic preconditioning via "imported" nitric oxide synthase activity. *Circulation 111*: 1114–1120.

Libby, P. and Theroux, P. (2005). Pathophysiology of coronary artery disease. *Circulation 111*: 3481–3488.

Liu, J., Hu, Q., Wang, Z., Xu, C., Wang, X., Gong, G., Mansoor, A., Lee, J., Hou, M., Zeng, L., Zhang, J.R., Jerosch-Herold, M., Guo, T., Bache, R.J. and Zhang, J. (2004). Autologous stem cell transplantation for myocardial repair. *Am. J. Physiol. Heart Circ. Physiol. 287*: H501–H511.

Lu, L., Zhang, J.Q., Ramires, F.J. and Sun, Y. (2004). Molecular and cellular events at the site of myocardial infarction: from the perspective of rebuilding myocardial tissue. *Biochem. Biophys. Res. Commun. 320*: 907–913.

Lunde, K., Solheim, S., Aakhus, S., Arnesen, H., Abdelnoor, M., Egeland, T., Endresen, K., Ilebekk, A., Mangschau, A., Fjeld, J., Smith, H., Taraldsrud, E., Grogaard, H., Bjornerheim, R., Brekke, M., Muller, C., Hopp, E., Ragnarsson, A., Brinchmann, J. and Forfang, K. (2006). Intracoronary injection of mononuclear bone marrow cells in acute myocardial infarction. *N. Engl. J. Med. 355*: 1199–1209.

Ma, N., Stamm, C., Kaminski, A., Li, W., Kleine, H.D., Muller-Hilke, B., Zhang, L., Ladilov, Y., Egger, D. and Steinhoff, G. (2005). Human cord blood cells induce angiogenesis following myocardial infarction in NOD/scid-mice. *Cardiovasc. Res. 66*: 45–54.

Maddi, A., Hai, H., Ong, S.T., Sharp, L., Harris, M. and Meghji, S. (2006). Long wave ultrasound may enhance bone regeneration by altering OPG/RANKL ratio in human osteoblast-like cells. *Bone* 39: 283–288.

Marelli, D., Desrosiers, C., El-Alfy, M., Kao, R.L. and Chiu, R.C-J. (1992). Cell transplantation for myocardial repair: an experimental approach. *Cell Transpl. 1*: 383–390.

Marmur, J.D., Rossikhina, M., Guha, A., Fyfe, B., Friedrich,V., Mendlowitz, M., Nemerson,Y. and Taubman, M.B. (1993). Tissue factor is rapidly induced in arterial smooth muscle after balloon injury. *J. Clin. Invest. 91*: 2253–2259.

Massa, M., Rosti, V., Ferrario, M., Campanelli, R., Ramajoli, I., Rosso, R., De Ferrari, G. M., Ferlini, M., Goffredo, L., Bertoletti, A., Klersy, C., Pecci, A., Moratti, R. and Tavazzi, L. (2005). Increased circulating hematopoietic and endothelial progenitor cells in the early phase of acute myocardial infarction. *Blood 105*: 199–206.

Mauro, A. (1961). Satellite cell of skeletal muscle fibers. *J. Biophys. Biochem. Cytol. 9*: 493–495.

Menasche, P., Hagege, A.A., Scorsin, M., Pouzet, B., Desnos, M., Duboc, D., Schwartz, K., Vilquin, J.T. and Marolleau, J.R (2001). Myoblast transplantation for heart failure. *Lancet 357*: 279–280.

Menasche, P., Hagege, A.A., Vilquin, J.T., Desnos, M., Abergel, E., Pouzet, B., Bel, A., Sarateanu, S., Scorsin, M., Schwartz, K., Bruneval, P., Benbunan, M., Marolleau, J.P. and Duboc, D. (2003). Autologous skeletal myoblast transplantation for severe postinfarction left ventricular dysfunction. *J. Am. Coll. Cardiol. 41*: 1078–1083.

Meyer, G., Wollert, K., Lotz, J., Steffens, J., Lippolt, P., Fichtner, S., Hecker, H., Schaefer, A., Arseniev, L., Hertenstein, B., Ganser, A. and Drexler, H. (2006). Intracoronary bone marrow cell transfer after myocardial infarction: eighteen months' follow-up data from the randomized, controlled BOOST (BOne marrOw transfer to enhance ST-elevation infarct regeneration) trial. *Circulation 113*: 1287–1294.

Miller, L.W. and Missov, E.D. (2001). Epidemiology of heart failure. *Cardiol. Clin. 19*: 547–555.

Moss, A.J., Zareba, W., Hall, W.J., Klein, H., Wilber, D.J., Cannom, D.S., Daubert, J.P., Higgins, S.L., Brown, M.W. and Andrews, M.L. (2002). Prophylactic implantation of a defibrillator in patients with myocardial infarction and reduced ejection fraction. *N. Engl. J. Med. 346*: 877–883.

Mukherjee, D., Nissen, S.E. and Topol, E.J. (2001). Risk of cardiovascular events associated with selective COX-2 inhibitors. *JAMA 286*: 954–959.

Muller-Ehmsen, J., Scheid, C., Grundmann, F., Hirsch, I., Turan, G., Tossios, P., Mehlhorn, U. and Schwinger, R. (2005). The mobilization of CD34 positive mononuclear cells after myocardial infarction is abolished by revascularization of the culprit vessel. *Int. J. Cardiol. 103*: 7–11.

Murry, C.E., Wiseman, R.W., Schwartz, S.M. and Hauschka, S.D. (1996). Skeletal myoblast transplantation for repair of myocardial necrosis. *J. Clin. Invest. 98*: 2512–2523.

Murry, C.E., Soonpaa, M.H., Reinecke, H., Nakajima, H., Nakajima, H.O., Rubart, M., Pasumarthi, K.B., Virag, J.I., Bartelmez, S.H., Poppa, V., Bradford, G., Dowell, J.D., Williams, D.A. and Field, L.J. (2004). Hematopoietic stem cells do not transdifferentiate into cardiac myocytes in myocardial infarcts. *Nature 428*: 664–668.

Nicolae, M., Tircol, M. and Alexandru, D. (2005). Inhibitory effect of acetylsalicylic acid on matrix metalloproteinase-2 activity in human endothelial cells exposed to high glucose. *J. Cell Mol. Med. 9*: 953–960.

Niederberger, E., Tegeder, I., Vetter, G., Schmidtko, A., Schmidt, H., Euchenhofer, C., Brautigam, L., Grosch, S. and Geisslinger, G. (2001). Celecoxib loses its anti-inflammatory efficacy at high doses through activation of NF-kappaB. *FASEB J. 15*: 1622–1624.

Niederberger, E., Manderscheid, C., Grosch, S., Schmidt, H., Ehnert, C. and Geisslinger, G. (2004). Effects of the selective COX-2 inhibitors celecoxib and rofecoxib on human vascular cells. *Biochem. Pharmacol. 68*: 341–350.

Nissen, S., Nicholls, S., Sipahi, I., Libby, P., Raichlen, J., Ballantyne, C., Davignon, J., Erbel, R., Fruchart, J., Tardif, J., Schoenhagen, P., Crowe, T., Cain, V., Wolski, K., Goormastic, M. and Tuzcu, E.A.I. (2006). Effect of very high-intensity statin therapy on regression of coronary atherosclerosis: the ASTEROID trial. *JAMA 295*: 1556–1565.

Oh, H., Bradfute, S.B., Gallardo, T.D., Nakamura, T., Gaussin, V., Mishina, Y., Pocius, J., Michael, L.H., Behringer, R.R., Garry, D.J., Entman, M.L. and Schneider, M.D. (2003). Cardiac progenitor cells from adult myocardium: homing, differentiation, and fusion after infarction. *Proc. Natl Acad. Sci. USA 100*: 12313–12318.

Ohno, N., Fedak, P., Weisel, R., Mickle, D., Fujii, T. and Li, R. (2003). Transplantation of cryopreserved muscle cells in dilated cardiomyopathy: effects on left ventricular geometry and function. *J. Thorac. Cardiovasc. Surg. 126*: 1537–1548.

Ott, H.C. and Taylor, D.A. (2006). From cardiac repair to cardiac regeneration—ready to translate? *Expert Opin. Biol. Ther. 6*: 867–878.

Ott, H.C., Bonaros, N., Marksteiner, R., Wolf, D., Margreiter, E., Schachner, T., Laufer, G. and Hering, S. (2004). Combined transplantation of skeletal myoblasts and bone marrow stem cells for myocardial repair in rats. *Eur. J. Cardiothor. Surg. 25*: 627–634.

Ott, H.C., Brechtken, J., Swingen, C., Feldberg, T., Matthiesen, T., Nelson, WD. and Taylor, D. (2006). Robotic minimally invasive cell transplantation for heart failure. *J. Thorac. Cardiovasc. Surg. 132*: 170–173.

Paoletti, R., Gotto Jr., A.M. and Hajjar, D.P. (2004). Inflammation in atherosclerosis and implications for therapy. *Circulation 109*: III20–III26.

Pearson, T.A., Blair, S.N., Daniels, S.R., Eckel, R.H., Fair, J.M., Fortmann, S.P., Franklin, B.A., Goldstein, L.B., Greenland, P., Grundy, S.M., Hong, Y., Miller, N.H., Lauer, R.M., Ockene, I.S., Sacco, R.L., Sallis Jr., J.F., Smith Jr., S.C., Stone, N.J. and Taubert, K.A. (2002). AHA guidelines for primary prevention of cardiovascular disease and stroke: 2002 Update: consensus panel guide to comprehensive risk reduction for adult patients without coronary or other atherosclerotic vascular diseases. American Heart Association Science Advisory and Coordinating Committee. *Circulation 106*: 388–391.

Peeters, A., Netea, M., Kullberg, B., Thien, T. and van der Meer, J. (1998). The effect of renin-angiotensin system inhibitors on pro- and anti-inflammatory cytokine production. *Immunology 94*: 376–379.

Perin, E.C., ohmann, H.F., Borojevic, R., Silva, S.A., Sousa, A.L., Mesquita, C.T., Rossi, M.I., Carvalho, A.C., Dutra, H.S., Dohmann, H.J., Silva, G.V., Belem, L., Vivacqua, R., Rangel, F.O., Esporcatte, R., Geng, Y.J., Vaughn, W.K., Assad, J.A., Mesquita, E.T. and Willerson, J.T. (2003). Transendocardial, autologous bone marrow cell transplantation for severe, chronic ischemic heart failure. *Circulation 107*: 2294–2302.

Perin, E.C., Dohmann, H.F., Borojevic, R., Silva, S.A., Sousa, A.L., Silva, G.V., Mesquita, C.T., Belem, L., Vaughn, W.K., Rangel, F.O., Assad, J.A., Carvalho, A.C., Branco, R.V., Rossi, M.I., Dohmann, H.J. and Willerson, J.T. (2004). Improved exercise capacity and ischemia 6 and 12 months after transendocardial injection of autologous bone marrow mono-nuclear cells for ischemic cardiomyopathy. *Circulation 110*: II213–II218.

Raupach, T., Schafer, K., Konstantinides, S. and Andreas, S. (2006). Smoke as an acute threat for the cardiovascular system: a change in paradigm. *Eur. Heart J. 27*: 386–392.

Rauscher, F.M., Goldschmidt-Clermont, P.J., Davis, B.H., Wang, T., Gregg, D., Ramaswami, P., Pippen, A.M., Annex, B.H., Dong, C. and Taylor, D.A. (2003). Aging, progenitor cell exhaustion, and atherosclerosis. *Circulation 108*: 457–463.

Reffelmann, T., Leor, J., Muller-Ehmsen, J., Kedes, L. and Kloner, R.A. (2003). Cardiomyocyte transplantation into the failing heart—new therapeutic approach for heart failure? *Heart Fail. Rev. 8*: 201–211.

Reiffel, J. (2005). Practical algorithms for pharmacologic management of the post myocardial infarction patient. *Clin. Cardiol. 28*:128–137.

Rogers, C., Edelman, E.R. and Simon, D.I. (1998). A mAb to the beta2-leukocyte integrin Mac-1 (CDlib/CD18) reduces intimal thickening after angioplasty or stent implantation in rabbits. *Proc. Natl Acad. Sci. USA 95*: 10134–10139.

Ross, R. (1993). The pathogenesis of atherosclerosis: a perspective for the 1990s. *Nature 362*: 801–809.

Sablayrolles, S. and Le Grand, B. (2006). Drug evaluation: dronedarone, a novel non-iodinated anti-arrhythmic agent. *Curr. Opin. Invest. Drug. 7*: 842–849.

Sata, M., Maejima, Y., Adachi, F., Fukino, K., Saiura, A., Sugiura, S., Aoyagi, T., Imai, Y., Kurihara, H., Kimura, K., Omata, M., Makuuchi, M., Hirata, Y. and Nagai, R. (2000). A mouse model of vascular injury that induces rapid onset of medial cell apoptosis followed by reproducible neointimal hyperplasia. *J. Mol. Cell Cardiol. 32*: 2097–2104.

Saulnier, N., Di Campli, C., Zocco, M., Di Gioacchino, G., Novi, M. and Gasbarrini, A. (2005). From stem cell to solid organ. Bone marrow, peripheral blood or umbilical cord blood as favorable source? *Eur. Rev. Med. Pharmacol. Sci. 9*: 315–324.

Schachinger, V., Assmus, B., Britten, M.B., Honold, J., Lehmann, R., Teupe, C., Abolmaali, N.D., Vogl, T.J., Hofmann, W.K., Martin, H., Dimmeler, S. and Zeiher, A.M. (2004). Transplantation of progenitor cells and regeneration enhancement in acute myocardial infarction: final one-year results of the TOPCARE-AMI Trial. *J. Am. Coll. Cardiol. 44*: 1690–1699.

Schachinger, V., Assmus, B., Honold, J., Lehmann, R., Hofmann, W., Martin, H., Dimmeler, S. and Zeiher, A. (2006a). Normalization of coronary blood flow in the infarct-related artery after intracoronary progenitor cell therapy: intra-coronary Doppler substudy of the TOPCARE-AMI trial. *Clin. Res. Cardiol. 95*: 13–22.

Schachinger, V., Erbs, S., Elsasser, A., Haberbosch, W., Hambrecht, R., Holschermann, H., Yu, J., Corti, R., Mathey, D., Hamm, C., Suselbeck, T., Assmu, s.B., Tonn, T., Dimmeler, S., Zeiher, A. and REPAIR-AMI Investigators. (2006b). Intracoronary bone marrow-derived progenitor cells in acute myocardial infarction. *N. Engl. J. Med. 355*: 1210–1221.

Schaefer, A., Meyer, G., Fuchs, M., Klein, G., Kaplan, M., Wollert, K. and Drexler, H. (2006). Impact of intracoronary bone marrow cell transfer on diastolic function in patients after acute myocardial infarction: results from the BOOST trial. *Eur. Heart J. 27*: 929–935.

Schmieder, R.E. (2006). Endothelial dysfunction: how can one intervene at the beginning of the cardiovascular continuum? *J. Hypertens Suppl. 24*: S31–S35.

Schober, A., Knarren, S., Lietz, M., Lin, E.A. and Weber, C. (2003). Crucial role of stromal cell-derived factor-lalpha in neointima formation after vascular injury in apolipoprotein E-deficient mice. *Circulation 108*: 2491–2497.

Schuster, M.D., Kocher, A.A., Seki, T., Martens, T.P., Xiang, G., Homma, S. and Itescu, S. (2004). Myocardial neovascularization by bone marrow angioblasts results in cardiomyocyte regeneration. *Am. J. Physiol. Heart Circ. Physiol. 287*: H525–H532.

Schwartz, R.S., Chronos, N.A. and Virmani, R. (2004). Preclinical restenosis models and drug-eluting stents still important, still much to learn. *J. Am. Coll. Cardiol. 44*: 1373–1385.

Scorsin, M., Hagege, A., Vilquin, J.T., Fiszman, M., Marotte, E., Samuel, J.L., Rappaport, L., Schwartz, K. and Menasche, P. (2000). Comparison of the effects of fetal cardiomyocyte and skeletal myoblast transplantation on postinfarction left ventricular function. *J. Thorac. Cardiovasc. Surg. 119*: 1169–1175.

Shaffer, R.G., Greene, S., Arshi, A., Supple, G., Bantly, A., Moore, J.S. and Mohler III, E.R. (2006). Flow cytometric measurement of circulating endothelial cells: the effect of age and peripheral arterial disease on baseline levels of mature and progenitor populations. *Cytometr. B Clin. Cytom. 70*: 56–62.

Shin, W., Szuba, A. and Rockson, S. (2002). The role of chemokines in human cardiovascular pathology: enhanced biological insights. *Atherosclerosis 160*: 91–102.

Shintani, S., Murohara, T., Ikeda, H., Ueno, T., Honma, T., Katoh, A., Sasaki, K., Shimada, T., Oike, Y. and Imaizumi, T. (2001). Mobilization of endothelial progenitor cells in patients with acute myocardial infarction. *Circulation 103*: 2776–2779.

Silva, G., Perin, E., Dohmann, H., Borojevic, R., Silva, S., Sousa, A., Assad, J., Vaughn, W., Mesquita, C., Belem, L., Carvalho, A.,

Dohmann, H., Barroso do Amaral, E., Coutinho, J., Branco, R., Oliveira, E. and Willerson, J. (2004). Catheter-based transendocardial delivery of autologous bone-marrow-derived mononuclear cells in patients listed for heart transplantation. *Tex. Heart Inst. J. 31*: 214–219.

Siminiak, T., Kalawski, R., Fiszer, D., Jerzykowska, O., Rzezniczak, J., Rozwadowska, N. and Kurpisz, M. (2004). Autologous skeletal myoblast transplantation for the treatment of postinfarction myocardial injury: phase I clinical study with 12 months of follow-up. *Am. Heart J. 148*: 531–537.

Siminiak, T., Fiszer, D., Jerzykowska, O., Grygielska, B., Rozwadowska, N., Kalmucki, P. and Kurpisz, M. (2005). Percutaneous trans-coronary-venous transplantation of autologous skeletal myoblasts in the treatment of post-infarction myocardial contractility impairment: the POZNAN trial. *Eur. Heart J. 26*: 1188–1195.

Simon, D.I., Dhen, Z., Seifert, P., Edelman, E.R., Ballantyne, C.M. and Rogers, C. (2000). Decreased neointimal formation in Mac-1(−/−) mice reveals a role for inflammation in vascular repair after angioplasty *J. Clin. Invest. 105*: 293–300.

Simons, M. (2005). Angiogenesis: where do we stand now? *Circulation 111*: 1556–1566.

Smith Jr., S.C., Allen, J., Blair, S.N., Bonow, R.O., Brass, L.M., Fonarow, G.C., Grundy, S.M., Hiratzka, L., Jones, D., Krumholz, H.M., Mosca, L., Pasternak, R.C., Pearson, T., Pfeffer, M.A., Taubert, K.A., Aha/Ace, National Heart, L. and Blood, I. (2006). AHA/ACC guidelines for secondary prevention for patients with coronary and other atherosclerotic vascular disease: 2006 update: endorsed by the National Heart, Lung, and Blood Institute.(erratum appears in *Circulation* 2006 Jun 6,113(22): e847). *Circulation 113*: 2363–2372.

Smits, P.C., van Geuns, R.J., Poldermans, D., Bountioukos, M., Onderwater, E.E., Lee, C.H., Maat, A.P. and Serruys, P.W. (2003). Catheter-based intramyocardial injection of autologous skeletal myoblasts as a primary treatment of ischemic heart failure: clinical experience with six-month follow-up. *J. Am. Coll. Cardiol. 42*: 2063–2069.

Stamm, C., Westphal, B., Kleine, H.D., Petzsch, M., Kittner, C., Klinge, H., Schumichen, C., Nienaber, C.A., Freund, M. and Steinhoff, G. (2003). Autologous bone-marrow stem-cell transplantation for myocardial regeneration. *Lancet 361*: 45–46.

Strauer, B., Brehm, M., Zeus, T., Bartsch, T., Schannwell, C., Antke, C., Sorg, R., Kogler, G., Wernet, P., Muller, H. and Kostering, M. (2005). Regeneration of human infarcted heart muscle by intracoronary autologous bone marrow cell transplantation in chronic coronary artery disease: the IACT Study. *J. Am. Coll. Cardiol. 46*: 1651–1658.

Strauer, B.E., Brehm, M., Zeus, T., Kostering, M., Hernandez, A., Sorg, R.V., Kogler, G. and Wernet, P. (2002). Repair of infarcted myocardium by autologous intracoronary mononuclear bone marrow cell transplantation in humans. *Circulation 106*: 1913–1918.

Sun, P., Dwyer, K.M., Merz, C.N., Sun, W., Johnson, C.A., Shircore, A.M. and Dwyer, J.H. (2000). Blood pressure, LDL cholesterol, and intima-media thickness: a test of the "response to injury" hypothesis of atherosclerosis. *Arterioscl. Thromb. Vasc. Biol. 20*: 2005–2010.

Suwaidi, J.A., Hamasaki, S., Higano, S.T., Nishimura, R.A., Holmes Jr., D.R. and Lerman, A. (2000). Long-term follow-up of patients with mild coronary artery disease and endothelial dysfunction. *Circulation 101*: 948–954.

Suzuki, K., Brand, N.J., Allen, S., Khan, M.A., Farrell, A.O., Murtuza, B., Oakley, R.E. and Yacoub, M.H. (2001). Overexpression of connexin 43 in skeletal myoblasts: relevance to cell transplantation to the heart. *J. Thorac. Cardiovasc. Surg. 122*: 759–766.

Suzuki, K., Murtuza, B., Beauchamp, J.R., Smolenski, R.T., Varela-Carver, A., Fukushima, S., Coppen, S.R., Partridge, T.A. and Yacoub, M.H. (2004). Dynamics and mediators of acute graft attrition after myoblast transplantation to the heart. *FASEB 18*: 1153–1155.

Tamai, H., Katoh, K.,Yamaguchi, T., Hayakawa, H., Kanmatsuse, K., Haze, K., Aizawa, T., Nakanishi, S., Suzuki, S., Suzuki, T., Takase, S., Nishikawa, H. and Katoh, O. (2002). The impact of tranilast on restenosis after coronary angioplasty: the second tranilast restenosis following angioplasty trial (TREAT-2). *Am. Heart J. 143*: 506–513.

Targonski, P.V., Bonetti, P.O., Pumper, G.M., Higano, S.T., Holmes Jr., D.R. and Lerman, A. (2003). Coronary endothelial dysfunction is associated with an increased risk of cerebrovascular events. *Circulation 107*: 2805–2809.

Tateishi-Yuyama, E., Matsubara, H., Murohara, T., Ikeda, U., Shintani, S., Masaki, H., Amano, K., Kishimoto, Y., Yoshimoto, K., Akashi, H., Shimada, K., Iwasaka, T. and Imaizumi, T. (2002). Therapeutic angiogenesis for patients with limb ischaemia by autologous transplantation of bone-marrow cells: a pilot study and a randomised controlled trial. *Lancet 360*: 427–435.

Taylor, D.A., Atkins, B.Z., Hungspreugs, P., Jones, T.R., Reedy, M.C., Hutcheson, K.A., Glower, D.D. and Kraus, W.E. (1998). Regenerating functional myocardium: improved performance after skeletal myoblast transplantation. *Nat. Med. 4*: 929–933.

Taylor, D.A., Ott, H.C. and Serruys P.W. (2006). Cell transplantation for cardiovascular repair. In: *Textbook of Interventional Cardiovascular Pharmacology, Vol.*. Taylor & Francis Books LTD. pp. 419–437.

Thattassery, E. and Gheorghiade, M. (2004). Beta blocker therapy after acute myocardial infarction in patients with heart failure and systolic dysfunction. *Heart Fail. Rev. 9*: 107–113.

Thorn, T., Haase, N., Rosamond, W., Howard, V.J., Rumsfeld, J., Manolio, T., Zheng, Z.J., Flegal, K., O'Donnell, C., Kittner, S., Lloyd-Jones, D., Goff Jr., D.C., Hong, Y., Adams, R., Friday, G., Furie, K., Gorelick, P., Kissela, B., Marler, J., Meigs, J., Roger, V., Sidney, S., Sorlie, P., Steinberger, J., Wasserthiel-Smoller, S., Wilson, M., Wolf, P., American Heart Association Statistics, C. and Stroke Statistics, S. (2006). Heart disease and stroke statistics—2006 update: a report from the American Heart Association Statistics Committee and Stroke Statistics Subcommittee. *Circulation 113*: e85–el51.

Thompson, R.B., Emani, S.M., Davis, B.H., van den Bos, E.J., Morimoto, Y., Craig, D., Glower, D. and Taylor, D.A. (2003). Comparison of intracardiac cell transplantation: autologous skeletal myoblasts versus bone marrow cells. *Circulation 108*(Suppl 1): II264–II271.

Thorin-Trescases, N., Voghel, G., Gendron, M., Krummen, S., Farhat, N., Drouin, A., Perrault, L. and Thorin, E. (2005). Pathological aging of the vascular endothelium: are endothelial progenitor cells the sentinels of the cardiovascular system? *Can. J. Cardiol. 21*: 1019–1024.

Thum, T., Fraccarollo, D., Galuppo, P., Tsikas, D., Frantz, S., Ertl, G. and Bauersachs, J. (2006). Bone marrow molecular alterations after myocardial infarction: impact on endothelial progenitor cells. *Cardiovasc. Res. 70*: 50–60.

Till, J., Price, G., Lan, S. and McCulloch, E. (1978). Factors affecting the differentiation of blood cells. *Symp. Soc. Dev. Biol. 35*: 133–147.

Torp-Pedersen, C., Poole-Wilson, P.A., Swedberg, K., Cleland, J.G., Di Lenarda, A., Hanrath, P., Komajda, M., Lutiger, B., Metra, M., Remme, W.J., Scherhag, A. and Skene, A. (2005). Effects of metoprolol and carvedilol on cause-specific mortality and morbidity in patients with chronic heart failure—COMET. *Am. Heart J. 149*: 370–376.

Tremblay, J.P., Roy, B. and Goulet, M. (1991). Human myoblast transplantation: a simple assay for tumorigenicity. *Neuromuscul. Disord. 1*: 341–343.

Tse, H.F., Kwong, Y.L., Chan, J.K., Lo, G., Ho, C.L. and Lau, C.P. (2003). Angiogenesis in ischaemic myocardium by intramyocardial autologous bone marrow mononuclear cell implantation. *Lancet 361*: 47–49.

Tse, H.R., Thambar, S., Kwong, Y.L., Rowlings, P., Bellamy, G., McCrohon, J., Bastian, B., Chan, J.K., Lo, G., Ho, C. L. and Lau, C.P. (2006). Safety of catheter-based intramyocardial autologous bone marrow cells implantation for therapeutic angiogenesis. *Am. J. Cardiol. 98*: 60–62.

Tsikouris, J.P., Suarez, J.A., Simoni, J.S., Ziska, M. and Meyrrose, G.E. (2004). Exploring the effects of ACE inhibitor tissue penetration on vascular inflammation following acute myocardial infarction. *Coron. Artery Dis. 15*: 211–217.

Tsimikas, S., Willerson, J.T. and Ridker, P.M. (2006). C-reactive protein and other emerging blood biomarkers to optimize risk stratification of vulnerable patients. *J. Am. Coll. Cardiol. 47*: C19–C31.

Urbanek, K., Torella, D., Sheikh, R., De Angelis, A., Nurzynska, D., Silvestri, R., Beltrami, C.A., Bussani, R., Beltrami, A.P., Quaini, R., Bolli, R., Leri, A., Kajstura, J. and Anversa, P. (2005). Myocardial regeneration by activation of multipotent cardiac stem cells in ischemic heart failure. *Proc. Natl Acad. Sci. USA. 102*: 8692–8697.

Valgimigli, M., Rigolin, G.M., Fucili, A., Porta, M.D., Soukhomovskaia, O., Malagutti, P., Bugli, A.M., Bragotti, L.Z., Francolini, G., Mauro, E., Castoldi, G. and Ferrari, R. (2004). CD34+ and endothelial progenitor cells in patients with various degrees of congestive heart failure. *Circulation 110*: 1209–1212.

van den Bos, E.J. and Taylor, D.A. (2003). Cardiac transplantation of skeletal myoblasts for heart failure. *Minerva Cardioangiol.* 51: 227–243.

van den Bos, E.J., Davis, B.H. and Taylor, D.A. (2004). Transplantation of skeletal myoblasts for cardiac repair. *J. Heart Lung Transplant. 23*: 1217–1227.

van der Wal, A.C., Becker, A.E., van der Loos, C.M., Tigges, A.J. and Das, P.K. (1994). Fibrous and lipid-rich atherosclerotic plaques are part of interchangeable morphologies related to inflammation: a concept. *Coron. Artery Dis. 5*: 463–469.

van Zonneveld, A.R.T. (2006). Endothelial progenitor cells: biology and therapeutic potential in hypertension. *Curr. Opin. Nephrol. Hypertens. 15*: 167–172.

Verfaillie, C.M. (2005). Multipotent adult progenitor cells: an update. *Novartis Found. Symp. 265*: 55–61.

Vulliet, P.R., Greeley, M., Halloran, S.M., MacDonald, K.A. and Kittleson, M.D. (2004). Intra-coronary arterial injection of mesenchymal stromal cells and microinfarction in dogs. *Lancet 363*: 783–784.

Weber, C., Schober, A. and Zernecke, A. (2004). Chemokines. Key regulators of mononuclear cell recruitment in atherosclerotic vascular disease. *Arterioscler. Thromb. Vasc. Biol. 24*: 1891–1896.

Werner, N., Kosiol, S., Schiegl, T., Ahlers, P., Walenta, K., Link, A., Bohm, M. and Nickenig, G. (2005). Circulating endothelial progenitor cells and cardiovascular outcomes. *N. Engl. J. Med. 353*: 999–1007.

Wilke, N., Zenovich, A., Jerosch-Herold, M. and Henry, T. (2001). Cardiac magnetic resonance imaging for the assessment of myocardial angiogenesis. *Curr. Interv. Cardiol. Rep. 3*: 205–212.

Williams, P.C., Coffey, M.J., Coles, B., Sanchez, S., Morrow, J.D., Cockcroft, J.R., Lewis, M.J. and O'Donnell, V.B. (2005). *In vivo* aspirin supplementation inhibits nitric oxide consumption by human platelets. *Blood 106*: 2737–2743.

Wollert, K.C., Meyer, G.P., Lotz, J., Ringes-Lichtenberg, S., Lippolt, P., Breidenbach, C., Fichtner, S., Korte, T., Hornig, B., Messinger, D., Arseniev, L., Hertenstein, B., Ganser, A. and Drexler, H. (2004). Intracoronary autologous bone-marrow cell transfer after myocardial infarction: the BOOST randomised controlled clinical trial. *Lancet 364*: 141–148.

Wright, L., Maloney, W., Yu, X., Kindle, L., Collin-Osdoby, P. and Osdoby P. (2005). Stromal cell-derived factor-1 binding to its chemokine receptor CXCR4 on precursor cells promotes the chemotactic recruitment, development and survival of human osteoclasts. *Bone. 36*: 840–853.

Wutzl, A., Brozek, W., Lernbass, I., Rauner, M., Hofbauer, G., Schopper, C., Watzinger, F., Peterlik, M. and Pietschmann, P. (2006). Bone morphogenetic proteins 5 and 6 stimulate osteoclast generation. *J. Biomed. Mater. Res. 77*: 75–83.

Wyatt, S.B., Winters, K.R. and Dubbert, P.M. (2006). Overweight and obesity: prevalence, consequences, and causes of a growing public health problem. *Am. J. Med. Sci. 331*: 166–174.

Yan, L.L., Liu, K., Daviglus, M.L., Colangelo, L.A., Kiefe, C.I., Sidney, S., Matthews, K.A. and Greenland, P. (2006). Education, 15-year risk factor progression, and coronary artery calcium in young adulthood and early middle age: the coronary artery risk development in young adults study. *JAMA 295*: 1793–1800.

Yancy, C. and Fonarow, G. (2004). Quality of care and outcomes in acute decompensated heart failure: The ADHERE Registry. *Curr. Heart Fail. Rep. 1*: 121–128.

Yeh, E.T., Zhang, S., Wu, H.D., Korbling, M., Willerson, J.T. and Estrov, Z. (2003). Transdifferentiation of human peripheral blood CD34+-enriched cell population into cardiomyocytes, endothelial cells, and smooth muscle cells *in vivo*. *Circulation 108*: 2070–2073.

Yu, J., Rudic, R.D. and Sessa, W.C. (2002). Nitric oxide-releasing aspirin decreases vascular injury by reducing inflammation and promoting apoptosis. *Lab. Invest. 82*: 825–832.

Zhai, Q., Qiu, L., Li, Q., Meng, H., Han, J., Herzig, R. and Han, Z. (2004). Short-term *ex vivo* expansion sustains the homing-related properties of umbilical cord blood hematopoietic stem and progenitor cells. *Haematologica 89*: 265–273.

Zhang, M., Methot, D., Poppa, V., Fujio, Y., Walsh, K. and Murry, C.E. (2001). Cardiomyocyte grafting for cardiac repair: graft cell death and anti-death strategies. *J. Mol. Cell. Cardiol. 33*: 907-921.

Zibaitis, A., Greentree, D., Ma, F., Marelli, D., Duong, M. and Chiu, R.C. (1994). Myocardial regeneration with satellite cell implantation. *Transplant. Proc. 26*: 3294.

Chapter 23

Pancreatic Stem Cells

Yuval Dor and Douglas A. Melton

INTRODUCTION

From a clinical perspective, the pancreas is an important focus of stem cell research, because it is an attractive target for cell replacement therapy. In type I diabetes, the insulin-producing B-cells that reside in the pancreatic islets of Langerhans are destroyed by autoimmune attack, and it is thought that self-renewing stem cells could provide an unlimited source of B-cells for transplantation. Such therapeutic efforts require the prospective isolation of stem cells, with the potential to produce B-cells and the development of methods to direct their expansion and differentiation.

From a developmental biology perspective, the role of stem cells in the pancreas is a fascinating problem. New cells are produced during adulthood, but their origin is not clear. Much of the field is focused on the identification of B-cell progenitors and the characterization of their molecular requirements, but it is not known what role such cells play during pancreas maintenance and regeneration, or whether the adult pancreas contains a population of true stem cells. Regardless, information about the specification of the B-cell fate from undifferentiated progenitors will be important in directing the differentiation of stem cells *in vitro*. In this chapter, we review evidence for the existence and identity of pancreatic progenitor and stem cells and describe the criteria for experimental demonstration of such cells.

DEFINITION OF STEM CELLS AND OF PROGENITOR CELLS

The most rigorous definition of a stem cell is a cell that, upon proliferation, produces some progeny that have the same developmental potential (a process called self-renewal), as well as other progeny that have a more restricted developmental potential. Such cells may be present transiently during embryonic development, or persistently during the entire life of the organism, but their defining property is self-renewal. A progenitor cell, on the other hand, is any cell that generates another differentiated cell type.

Another important notion in stem cell biology, with potential relevance to the pancreas, is that of the facultative stem cell. These cells are thought to be functional, differentiated cells in an adult organ that can dedifferentiate in response to a specific signal (usually believed to be tissue damage), and then differentiate into another cell type. For example, the liver is thought to contain facultative stem cells that reside in the bile ducts and are capable of producing hepatocytes. Thus, the bile duct cells provide an effective reservoir of liver progenitors in case of tissue damage, although they may not meet the criteria of being stem cells at the single-cell level.

How can we demonstrate that a particular cell is a stem cell? Formally, such a demonstration must be based on clonal analysis: a single cell has to be followed over time in order to show that it can produce more stem cells, as well as differentiated cells. The gold standard for the identification of stem cells was set in the hematopoietic system, where clonal analysis has been carried out *in vitro* (by subcloning individual colonies), as well as *in vivo* (by serial transplantation of single stem cells into lethally irradiated mice). By contrast, the identification of progenitor cells is an easier task. It requires demonstration, using some sort of lineage-tracing experiment, that an undifferentiated cell population generates differentiated cells.

Even when stem cells cannot be identified or isolated in a particular organ, their existence may be inferred from kinetic studies of 5′-bromo-2′-deoxyuridine (BrdU) incorporation. Because stem cells are believed to be slowly-dividing, the presence of label-retaining cells can point to the anatomical location of a stem cell niche. Such an analysis was carried out in self-renewing organ systems, such as skin and hair. In light of these definitions, in the next sections we discuss the evidence for stem cells and progenitors during pancreas development and adult life.

PROGENITOR CELLS DURING EMBRYONIC DEVELOPMENT OF THE PANCREAS

The adult pancreas contains three major cell types: exocrine cells, organized in acini, which secrete digestive enzymes; duct epithelial cells which flush these enzymes to the duodenum; and endocrine cells, organized in the islets of Langerhans, that secrete hormones to the blood. The islets, accounting for ~1% of the cells in the adult pancreas, contain four major cell types that secrete different hormones: A, B, D, and PP-cells secreting glucagon, insulin, somatostatin, and pancreatic polypeptide, respectively. In recent years, a detailed cellular and molecular understanding of pancreas development has emerged (Figure 23-1). Although many questions remain, one recurring theme is the essential role played by progenitor cells in organogenesis. Early patterning of the gut tube generates a sheet of epithelial cells that express the homeobox gene *pdx1* around embryonic day 9 (E9) in the mouse. These epithelial cells bud from the gut tube, proliferate, and branch to form a tubular structure around E12.5, from which cells differentiate and organize into the exocrine and endocrine tissues of the mature pancreas. Numerous experiments demonstrate that this process is based on the proliferation and step-wise differentiation of progenitor cells. As shown by lineage-tracing experiments using tamoxifen-induced Cre recombinase, the early *pdx1*+ cells (which by E10.5 also express the transcription factor p48/ptf1) produce all cell types in the adult pancreas. Later on, progenitors for duct, acinar, and endocrine lineages segregate, proliferate, and differentiate. Endocrine progenitors migrate from the tubular structure of the early pancreas, and coalesce to form the islets of Langerhans just before birth. During that period, they undergo further restrictions in their potential, marked by the transient expression of several genes. Most notably, neurogenin3 expression marks progenitor cells for all endocrine lineages. Down-regulation of progenitor markers and expression of the hormone genes mark the terminal differentiation of endocrine cells. A similar mechanism of gradual commitment, though less well-defined, is thought to act in the developing exocrine pancreas.

How do pancreatic progenitors choose their fates? A long-standing hypothesis is that the axis of mitosis in epithelial cells of the early pancreas correlates with, or even determines, the identity of daughter cells (Figure 23-1). When the plane of division is perpendicular to the lumen, both daughter cells remain epithelial (symmetric division), and may lobulate to form exocrine acini. When the plane of division is parallel to the lumen, one daughter cell detaches from the lumen, and may become an endocrine progenitor cell (asymmetric division). This hypothesis, although untested, supports the popular view that adult pancreatic stem cells reside in the ducts. More recently, genetic evidence has suggested that cell–cell interactions mediated by the Notch pathway affect lateral specification of progenitor cells at multiple developmental junctions. Accordingly, mutations in Notch pathway genes can lead to accelerated, premature endocrine differentiation at the expense of the exocrine pancreas. These two models for specification may be compatible, as the Notch pathway is known to affect the plane of division in neural progenitors.

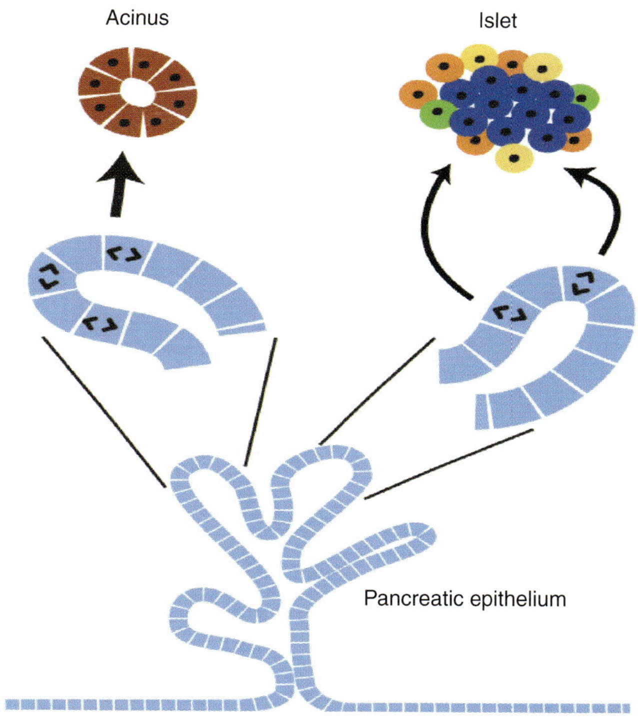

FIGURE 23-1 Embryonic development of the pancreas. Pdx1+ epithelial cells of the early pancreas (pale blue); acinar cells (red); and A-, B-, D-, and PP-cells (indicated by orange, blue, yellow, and green in the islet). The proposed plane of division of endocrine and exocrine progenitors is indicated.

What is the relative contribution of progenitors and differentiated cells to proliferation of the embryonic pancreas? Although this question has not been addressed directly, the fate of hormone-expressing cells that appear early in pancreas development (E9.5–10.5) was recently examined. These cells, many of which coexpress multiple endocrine hormones, were initially thought to replicate and generate the mature islets. However, lineage-tracing and ablation experiments suggest that the multiple-hormone-expressing cells do not contribute to the adult endocrine pancreas. Rather, mature endocrine cells appear to be derived from midgestation progenitors.

Following this progenitor-based formative stage, rapid growth of the pancreas in late gestation and early postnatal life is thought to involve gradually less differentiation of progenitor cells, and more replication of fully-differentiated cells (see later sections of this chapter).

Although the identity and importance of progenitors during pancreas development is clear, there is no indication for true self-renewal of cells during this period. The recent demonstration of heterogeneity among the Pdx1+ cells of the early pancreas suggests that lineage segregation occurs very early during pancreas organogenesis. Furthermore, no clonal analysis of the embryonic pancreas has been carried out *in vivo*. *In vitro* clonal analysis demonstrated the common origin of the exocrine and endocrine pancreas, but not self-renewal. Therefore, it is fair to conclude that progenitor cells play a major role in pancreas formation. However, there is no evidence for self-renewing pancreatic stem cells during embryonic development. We now turn to the postnatal growth and maintenance of the pancreas, with special attention to B-cells. Do progenitor cells of the embryonic type persist in adult life in an active or latent form? Is there evidence for stem cells capable of generating new pancreatic cells in the adult?

PROGENITOR CELLS IN THE ADULT PANCREAS

When describing stem/progenitor cells in the adult pancreas, it is important to deal separately with two questions: first, what is the turnover rate of the different components of the pancreas during postnatal growth, throughout adult maintenance, and in response to injury? Second, are the new cells derived from stem/progenitor cells (neogenesis), or from replicating differentiated cells? It must be remembered that an impressive capacity for homeostatic maintenance, or even regeneration, does not indicate neogenesis. For example, dramatic regeneration of the injured liver can occur solely by proliferation of differentiated hepatocytes, without a requirement for stem cells.

Evidence from Cell Dynamics

After birth, the growth of the pancreas continues, but slows significantly around one month of age in the mouse and rat. However, even in old animals there is a measurable rate of cell birth in all pancreatic compartments. In the B-cell compartment, where most studies have been done, the replication rate falls from ~5% in four-week-old animals to ~0.1% in mice older than three months. Embryonic-type progenitors (based on expression patterns) are not seen in the normal adult animal (with the possible exception of rare islet cells generated from neurogenin3+ progenitors).

In spite of its low basal turnover rate, significant hyperplasia is seen in the adult pancreas under certain physiological and pathological conditions. For example, during pregnancy the B-cell mass doubles, a response attributed to a combination of cell hypertrophy and cell proliferation. More dramatically, several reports have documented an ability of the B-cell compartment to recover from genetically-programmed, autoimmune, surgical, or chemical damage.

Can the new cells, in normal homeostasis or in a regeneration setting, be fully accounted for by the replication of differentiated cells? Empirically, for the B-cell compartment, can the number of BrdU+ pulse-labeled B-cells explain the accumulation of BrdU+ B-cells following continuous administration of BrdU? If not, the presence of undifferentiated progenitors must be invoked. This type of analysis has, however, proven difficult because it requires reliable values for several elusive parameters: what is the duration of the S phase and the total cell cycle in a specific compartment? What is the death rate, and how long does it take for a dying cell to be cleared? Furthermore, cell number is not easily inferred from total cell area as obtained by immunostaining, because of cellular hypertrophy. Thus, a study of B-cell dynamics must include B-cell counting by fluorescence-activated cell sorting or by careful histological analysis across the whole pancreas, a criterion that is not always met.

The most comprehensive effort in this direction was carried out by Finegood *et al.* (1995) who studied B-cell dynamics throughout the lifespan of the rat. Their results imply a significant contribution of progenitor cells to the B-cell mass in the first weeks after birth, and then a shift to tissue maintenance by slow replication of B-cells. In addition, significant B-cell neogenesis was deduced in a similar study of chronic hyperglycemia in rats. These kinetic studies suggest that adult pancreatic progenitors exist. However, they do not help determine the molecular and anatomical origins of these cells.

Another argument for the existence of B-cell progenitors is the identification of single B-cells embedded in the adult exocrine tissue. These isolated cells are reportedly more frequent following insults, which led to the notion that they are newly-generated from progenitors that reside in ducts or acini (see later sections of this chapter). The new B-cells are then believed to coalesce into islets in a mechanism resembling developmental islet morphogenesis. However, careful analysis is required to distinguish this interpretation from other potential explanations (e.g., disintegration of existing islets). In fact, recent genetic lineage tracing experiments suggest that the small clusters of B-cells are derived from B-cells, rather than from stem/progenitor cells.

A combination of BrdU pulse-chase experiments and genetic labeling of differentiated cells (for example, using an inducible Crerecombinase) may allow a direct comparison between the contribution of progenitors and the contribution of differentiated cells to pancreas growth and maintenance.

Many experiments neglect the kinetic aspects of neogenesis, and focus on histological identification of progenitors, based on expression markers in certain anatomical locations. However, without lineage tracing, these studies cannot demonstrate the fate of the putative progenitors, or determine their importance. In the next sections, we describe the most notable proposals regarding the identity of adult pancreatic stem-progenitor cells.

Ducts

It is widely believed that adult pancreatic stem/progenitor cells are located in the duct epithelium. Indeed, cells expressing B-cell markers (insulin, glut-2, pax6, isl1, and

HNF3b) can often be found embedded in or adjacent to adult ducts following an insult to the pancreas.

Moreover, duct cell replication (as assessed by BrdU incorporation) is increased after such insults. In addition, adult duct preparations are claimed to be capable of endocrine differentiation *in vitro* (see later sections of this chapter). Conceptually, the appearance of endocrine cells in or near ducts is interpreted as recapitulation of embryonic pancreas morphogenesis, where endocrine progenitors bud from the epithelium. However, adult ducts are not necessarily identical in their genetic program to the embryonic tubes that should be referred to as "duct-like structures." Indeed, recent lineage experiments indicate that the definitive ductal and endocrine lineages are separated as early as E12.5. With regard to endocrine cells seen budding from ducts, caution is required when suggesting dynamic interpretations to static histological snapshots (Figures 23-2 and 23-3).

The preceding description shows the importance of specific duct markers for the assessment of the ductal origins of endocrine cells. Although several duct markers have been found (such as carbonic anhydrase, cystic fibrosis transmembrane-conductance regulator, and cytokeratin-19), so far none has been translated to a useful lineage marker in transgenic mice.

In summary, it seems that ducts can elicit a proliferative response to tissue damage, and that endocrine markers are occasionally expressed in cells of the ductal epithelium. However, the fate of these cells and their relative contribution to the endocrine or exocrine pancreas has yet to be demonstrated.

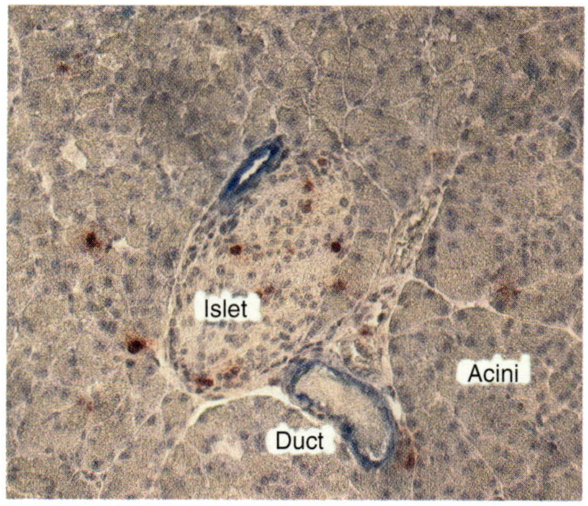

FIGURE 23-2 Histology of adult pancreas. Exocrine tissue (acini and ducts) and endocrine tissue (islet of Langerhans) are shown. Section from a four-week-old mouse pancreas, pulse labeled with BrdU (red). Ducts are stained blue. Note the close association between the ducts and the islet. Original magnification = 200×.

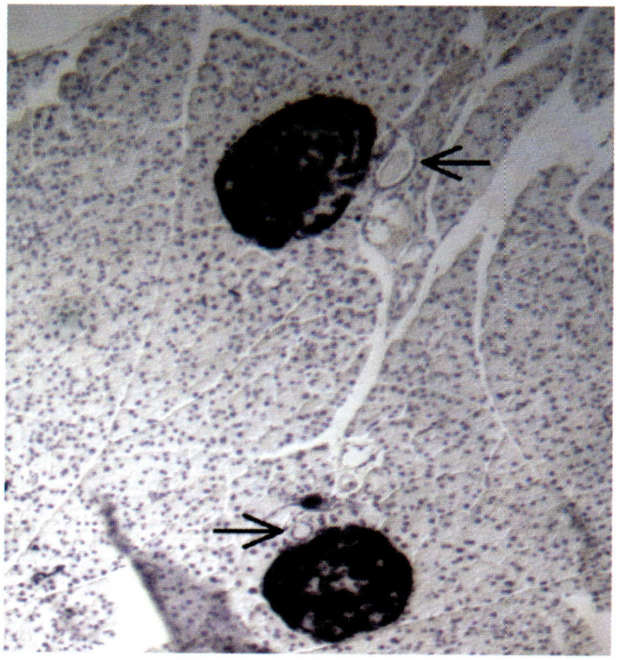

FIGURE 23-3 Lineage labeling of the endocrine pancreas. Section from the pancreas of a four-week-old ngn3-cre; Z/AP double-transgenic mouse, stained for alkaline phosphatase activity (dark blue). In this mouse, cre-mediated recombination in *ngn3+* embryonic endocrine progenitor cells leads to heritable expression of the human placental alkaline phosphatase gene. Only islet cells are labeled, suggesting that ducts (arrows) and acini never expressed the *ngn3* gene. Original magnification = 100×.

Acini

Starting from the observation of isolated B-cells embedded in exocrine tissue, other investigators have proposed that acinar cells can transdifferentiate, with or without replication, into endocrine cells. This possibility is supported by *in vitro* studies showing that an acinar cell line can adopt endocrine features under certain conditions. However, as in the case of duct-embedded progenitors, the lineage of these cells was not followed *in vivo*; therefore, their origin and fate could not be determined. In this case, however, good lineage markers exist, so the possible acinar origin of endocrine cells can be directly tested. Preliminary experiments with an Elastase promoter-driven Crerecombinase have suggested that, under normal conditions, Elastase-expressing cells do not produce endocrine or duct cells. It remains to be tested, however, whether acinar cells contain "facultative" endocrine progenitors activated upon injury.

Intraislet Progenitors

Several groups have suggested the existence of intraislet progenitor cells capable of proliferation and differentiation into B-cells. These reports are based on the expression of putative stem-progenitor cell markers in islets. One such proposal is that the expression of Nestin, a marker of neuronal progenitors, labels intraislet endocrine progenitors with a potential for *in vitro* differentiation into several fates. However, a recent lineage analysis using Crerecombinase driven by the Nestin promoter showed that pancreatic endocrine cells do not form by the differentiation of Nestin+ cells. In addition, Nestin was recently shown to be expressed in the mesenchymal, but not in epithelial, cells of the embryonic pancreas, further questioning its relevance for pancreatic lineages.

Others have documented the coexpression of pancreatic hormones in islets following a diabetogenic insult. For example, It was proposed that following streptozotocin treatment, somatostatin+ cells and glucagon+ cells generate B-cells by proliferation and differentiation through somatostatin+Pdxl+, and glucagon+glut-2+ intermediate cell types. These proposals have yet to be confirmed by lineage analysis.

Finally, clinical transplantation of islets may provide important clues about islet cell dynamics. Islet grafts into the portal vein (as done with diabetic patients), or under the kidney capsule (as done routinely with rodents), survive for many months and provide glycemic control. Although it is clear that cells in the grafts die and proliferate, it is not known if new endocrine cells in the graft are derived only from replicating differentiated cells or also from progenitors. Since purified islets with minimal exocrine tissue are used for transplantations, any indication for the existence of progenitors in the graft would point to an intraislet source.

Bone Marrow

A recent study found a significant contribution of bone marrow-derived cells to the B-cell compartment in the absence of tissue damage other than irradiation. To facilitate the detection of marrow-derived B-cells, marrow was taken from transgenic mice expressing Crerecombinase under the insulin promoter, as well as enhanced green fluorescent protein (eGFP), preceded by a Cre-removable transcriptional stop signal. Thus, bone marrow cells of the donor mice should express eGFP only if they turn on the insulin promoter. By sorting eGFP+ cells in pancreatic islets of recipient mice, up to 3% of all B-cells were identified from a donor origin two months after marrow transplantation. The elegant design of the experiment was such that the fusion of marrow cells with recipient B-cells could be ruled out. These surprising results still await independent confirmation.

Other Evidence for Adult Pancreatic Progenitor Cells

It has been reported that cells in the adult pancreas can reconstitute a degenerating liver deficient in the enzyme fumarylacetoacetate hydrolase. This observation suggested the existence of progenitors in the pancreas capable of liver differentiation. However, more recent results indicate that liver reconstitution by bone marrow cells represents complementation of defective hepatocytes by cell fusion, rather than transdifferentiation of marrow cells to liver. Thus, it is likely that liver reconstitution by pancreatic cells represents the "fusibility" rather than the plasticity of these cells.

In summary, indirect evidence suggests that endocrine progenitor cells exist, or can appear, in the adult pancreas following injury. Careful lineage analysis has yet to demonstrate the existence, origin, fate, and importance of these cells to pancreas homeostasis. Similar to the situation in the embryo, there is no evidence for the existence of self-renewing stem cells in the adult pancreas.

Very recently, we have used a genetic lineage analysis system to study the dynamics of adult pancreatic B-cells. This analysis shows that adult B-cells are maintained largely by self-duplication, rather than stem cell differentiation during normal life in the mouse, as well as during regeneration from pancreatectomy. The results cast doubt on the idea that stem cells contribute significantly to the formation of new B-cells during adult life. We cannot rule out the existence of adult pancreatic stem cells; however, it seems that the contribution of such putative cells to new B-cells is minimal. In addition, facultative pancreatic stem cells might exist and be recruited following other types of injury. The same system can now be used for testing claims about stem cell differentiation into B-cells, *in vitro* and *in vitro*.

FORCING OTHER TISSUES TO ADOPT A PANCREATIC PHENOTYPE

From a therapeutic perspective, artificial induction of a B-cell fate in other cell types is as useful as finding and expanding endogenous pancreatic progenitors. Moreover, nonpancreatic cells, if indeed competent to adopt a B-cell program, may be more accessible for expansion and *in vitro* manipulation. Efforts in this direction have focused on tissues thought to be capable of pancreatic differentiation because of a common lineage origin or a similar genetic program. Here, we explain the pancreatic connection of the most popular starting tissues for such experiments.

Liver to Pancreas

Several observations led to the notion that the adult liver may be a close relative of the pancreas in terms of the steps required for switching fates. First, both organs are derived from the endoderm of the embryonic gut tube. Moreover, studies by Zaret and colleagues suggest that a bipotential liver–pancreas progenitor cell population exists by E8 in the ventral foregut endoderm. Using embryo tissue explantation experiments, they have shown that fibroblast growth factor signaling from the cardiac mesoderm can divert these progenitors from a default pancreatic fate to a liver fate. However, lineage or clonal analysis was not performed to formally demonstrate that bipotential cells exist.

Second, transdifferentiation of pancreas to liver, in which clusters of hepatocytes emerge in the pancreas, occurs spontaneously in humans and can be induced experimentally in rats. In cultured pancreatic cells, liver transdifferentiation was demonstrated by activation of a single transcription factor (C/EBP-b) without a requirement for cell division. These observations suggest that adult liver and pancreas cells may retain a degree of plasticity that allows transdifferentiation to occur. Alternatively, a rare stem-progenitor cell population in these organs, yet to be identified, may be responsible for the metaplastic phenomenon.

Notably, no evidence has been found for deep similarities in the genetic program of the adult pancreas and liver. Based on these observations, several groups have tried experimentally to convert liver to pancreas by enforced expression of key pancreatic genes. One candidate for imposing a pancreatic fate is Pdxl, the most upstream component of the pancreatic program. Indeed, adenoviral-mediated transfer of Pdxl to adult mice was reported to induce low-level expression of insulin or exocrine pancreas markers in the liver.

A conceptual problem with simple overexpression of Pdxl is the need for specific transcriptional cofactors to be present. For example, Pdxl is known to require the Pbx and Meis proteins for proper function. With the hope of bypassing the need for Pdxl cofactors absent from the liver, a Pdxl-VP16 fusion gene was recently constructed. Transgenic expression of Pdxl-VP16 (but not unmodified pdxl) in livers of *Xenopus* tadpoles led to significant expression of pancreatic endocrine and exocrine genes in the liver, concomitant with down-regulation of liver genes. The use of a liver-specific promoter (from the transthyretin gene), and the finding that the same construct induces a pancreatic phenotype in a hepatocyte cell line, suggests that the responding cells are differentiated hepatocytes rather than uncommitted progenitors or stem cells.

In summary, it seems that ectopic expression of pancreatic transcription factors in the embryonic and perhaps the adult liver can lead to activation of a pancreatic program. The available data suggest that the mechanism is direct hepatocyte transdifferentiation, rather than activation of a common liver–pancreas progenitor. In addition to the therapeutic potential of the approach, analysis of the molecular details may provide significant insights into the problems of competence and potential.

Neurons to Pancreas

Although arising from different lineages during embryogenesis, the endocrine pancreas and the nervous system share a significant portion of their developmental programs. For example, most transcription factors participating in the step-wise differentiation of the endocrine pancreas are also expressed in the developing brain. Some examples are isl1, Hb9, neurogenin3, NeuroD, Nkx2.2, Nkx6.1, pax6, and brn4. Notable exceptions are the absence of Pdxl, as well as insulin, expression in the brain. However, the insulin promoter does contain neuronal regulatory elements that are repressed in the context of the endogenous insulin locus, but give rise to neural expression when driving transgenes. The significance of this intriguing similarity is not fully understood. One possibility is that it reflects a common ancestral lineage for brain and endocrine pancreas, which diverged during evolution as the endocrine program came under Pdxl control. This idea is supported by the finding that insulin and glucagon are expressed in specific neurons in the fly (which lacks a pancreas). Thus, neuronal progenitors are the closest known relatives of pancreatic progenitors in terms of their genetic program. This suggests that neuronal progenitors (easily derived from ES cells) may be only a few steps from adopting a pancreatic fate. However, attempts to induce ES-derived neural progenitors to a pancreatic fate have so far failed.

Gut to Pancreas

Several recent studies explored the potential of the nonpancreatic segments of the digestive tract to adopt a pancreatic fate. Ectopic expression of Pdxl in the embryonic gut outside the normal expression domain of Pdxl led to a partial adoption of a pancreatic fate, judged by gene expression and cell budding out of the epithelium. Similarly, ectopic expression of neurogenin3 in the embryonic chick or mouse led to the appearance of endocrine cells expressing glucagon and somatostatin (but not insulin). Inactivation of ptfla-p48, a transcription factor normally expressed in pancreatic progenitors, led to a cell-autonomous switch in the reverse direction, namely, the conversion of pancreas to duodenal epithelium. These results suggest that, even after the onset of segment-specific genetic programs, the embryonic gut epithelium remains plastic. It is not known if such plasticity is retained in the adult intestine.

IN VITRO STUDIES

Numerous attempts have been made to generate B-cells *in vitro* from dissociated pancreatic tissue using the available information on the identity of putative pancreatic stem-progenitors. Accordingly, pancreatic duct preparations were cultured, and the appearance of differentiated cell types was monitored. Some groups have reported on the generation of endocrine cells from duct cultures, and some even claimed to correct hyperglycemia in animals by grafting the *in vitro*-generated islet-like clusters. Others have used islet preparations as a starting material. These studies, however, share several methodological problems that complicate their interpretation. First, the number of endocrine cells at the beginning and at the end of the culture period was not always measured carefully. Second, all studies were carried out on enriched rather than pure cell populations, thus not ruling out the presence of contaminating cells from unintended compartments. Third, clonal analysis was not performed. Fourth, many of these studies used insulin immunostaining as a marker for B-cell differentiation, and supplied large amounts of exogenous insulin in the culture medium. Therefore, they may have scored for insulin uptake rather than *de novo* synthesis.

Perhaps a more promising approach to the problem is the use of embryonic pancreas as a starting material, since it is clear that it contains progenitor cells that can proliferate and differentiate. This, however, has also proved a difficult task. Much of the difficulty may result from the lack of proper culture conditions that mimic the embryonic environment.

SUMMARY

The search for pancreatic stem or progenitor cells may provide insights into the basic mechanisms of organ homeostasis, as well as a promising therapeutic approach for diabetes. It is clear that pancreatic progenitor cells exist and play a major role during embryonic development, although much has to be learned about the mechanisms and molecules controlling them. There is no evidence, however, for pancreatic stem cells capable of self-renewal.

As for the adult pancreas, no conclusive evidence shows the importance or mere existence of pancreatic progenitor or stem cells in the adult organism. On the contrary, it appears that, at least for pancreatic beta cells, the major mechanism for tissue maintenance is self-duplication rather than stem cell differentiation. It is possible, however, that under certain stresses progenitor cells appear and may contribute to the maintenance of the pancreas. Most evidence points to the pancreatic ducts as the likely niche of such "facultative" progenitors. Development of culture conditions that support the growth and differentiation of dissociated pancreatic tissue, as well as *in vivo* lineage analysis systems, are the tools most needed for tackling this problem.

Other cell types may retain a degree of plasticity that allows for pancreatic differentiation under artificial conditions. Embryonic stem cells are the only cells available *in vitro* that can definitely differentiate into pancreatic cell types; however, this potential has yet to be demonstrated *in vitro*. Theoretical considerations and some experimental evidence suggest that neuronal progenitors, liver cells, and intestinal epithelial cells may also be forced to adopt a pancreatic fate.

KEY WORDS

Genetic lineage tracing The use of permanent labeling, usually DNA recombination, to mark a defined cellular population and trace its progeny.

Neurogenin3 A transcription factor expressed in endocrine progenitor cells of the embryonic pancreas.

Pdx1 Pancreatic and duodenal homeobox gene-1; a transcription factor essential for pancreas development. During embryogenesis, Pdx1 is expressed in stem/progenitor cells, giving rise to all cell types in the adult pancreas. During postnatal life, its expression becomes restricted to beta cells.

FURTHER READING

Bonner-Weir, S., & Sharma, A. (2002). Pancreatic stem cells. *J. Pathol.*, *197*, 519–526.

Bouwens, L. (1998). Transdifferentiation versus stem cell hypothesis for the regeneration of islet beta-cells in the pancreas. *Microsc. Res. Tech.*, *43*(4), 332–336.

Deutsch, G., Jung, J., Zheng, M., Lora, J., & Zaret, K. S. (2001). A bipotential precursor population for pancreas and liver within the embryonic endoderm. *Development*, *128*(6), 871–881.

Dor, Y., Brown, J., Martinez, O. I., & Melton, D. A. (2004). Adult pancreatic b cells are formed by self-duplication rather than stem cell differentiation. *Nature*, *429*, 41–46.

Finegood, D. T., Scaglia, L., & Bonner-Weir, S. (1995). Dynamics of beta-cell mass in the growing rat pancreas. Estimation with a simple mathematical model. *Diabetes*, *44*(3), 249–256.

Gu, G., Dubauskaite, J., & Melton, D. A. (2002). Direct evidence for the pancreatic lineage: NGN3+ cells are islet progenitors and are distinct from duct progenitors. *Development*, *129*(10), 2447–2457.

Kodama, S., Kuhtreiber, W., Fujimura, S., Dale, E. A., & Faustman, D. L. (2003). Islet regeneration during the reversal of autoimmune diabetes in NOD mice. *Science*, *302*, 1223–1227.

Murtaugh, L. C., & Melton, D. A. (2003). Genes, signals, and lineages in pancreas development. *Annu. Rev. Cell. Dev. Biol.*, *19*, 71–89.

Pelengaris, S., Khan, M., & Evan, G. I. (2002). Suppression of Myc-induced apoptosis in beta cells exposes multiple oncogenic properties of Myc and triggers carcinogenic progression. *Cell*, *109*(3), 321–334.

Shapiro, A. M., Lakey, J. R., Ryan, E. A., Korbutt, G. S., Toth, E., Warnock, G. N., Kneteman, N. M., & Rajotte, R. V. (2000). Islet transplantation in seven patients with type 1 diabetes mellitus using a glucocorticoid-free immunosuppressive regimen. *N. Engl. J. Med.*, *343*(4), 230–238.

Chapter 24

Use of Embryonic Stem Cells to Treat Heart Disease

Michael Rubart and Loren J. Field

INTRODUCTION

It is now well-established that cardiomyocytes can be stably transplanted into normal or injured adult hearts. Moreover, transplanted donor cells can form a functional syncytium with the host myocardium. It is also well-established that embryonic stem (ES) cells can differentiate into functional cardiomyocytes *in vitro*, and that these ES-derived cardiomyocytes form stable grafts when transplanted into the myocardium. ES cells might thus represent a suitable source of donor cardiomyocytes for cell transplantation therapies aimed at restoring lost myocardial mass in diseased hearts. In this chapter, an overview of the use of cell transplantation as a paradigm for the treatment of diseased hearts is provided. This is followed by a discussion of ES-derived cardiomyocytes, and factors that enhance cardiomyogenic differentiation. Transplantation of ES-derived cardiomyocytes in preclinical studies, and challenges for their clinical implementation, are then considered.

CARDIOMYOCYTE TRANSPLANTATION AS A PARADIGM FOR THE TREATMENT OF DISEASED HEARTS

Cell transplantation has emerged as a potential therapeutic intervention to enhance angiogenesis, provide structural support, and perhaps even to restore lost myocardial mass in diseased hearts. To date, a number of donor cell types have been tested in preclinical studies for their ability to attain these goals, including skeletal myoblasts, fetal cardiomyocytes, angioblasts, vascular precursors, bone marrow stem cells, mesenchymal stem cells, ES-derived cardiomyocytes, and a variety of cardiac-resident stem cells with alleged cardiomyogenic activity. In many cases, cell transplantation was associated with improvement in cardiac function secondary to beneficial post-injury myocardial remodeling, and perhaps, to contractile activity of the donor cells. Clinical studies using skeletal myoblasts or crude mononuclear hematopoietic stem cell preparations from either bone marrow or peripheral blood have also been initiated. These initial clinical studies collectively established that cell transplantation in injured hearts is relatively safe, although the presence of life-threatening tachyarrhythmias in patients receiving skeletal myoblast transplants has necessitated defibrillator implantation. However, the impact of cell transplantation on cardiac function in patients has been mixed, with some studies demonstrating improvement in cardiac function and clinical outcome, and other studies demonstrating no effect.

Cardiomyocyte transplantation as a potential intervention to restore cardiac mass in diseased hearts was initially explored in the early 1990s. In these studies, single cell suspensions generated by enzymatic digestion of fetal hearts were transplanted directly into the ventricle of histocompatibility-matched or immune-suppressed recipient animals. The transplanted cardiomyocytes were identified by virtue of reporter transgene activity, or alternatively by the expression of an endogenous gene product that was absent in the recipient myocardium. The donor cells stably engrafted into the recipient hearts, underwent terminal differentiation, and expressed many of the molecular and morphological attributes typical of normal adult cardiomyocytes. Transmission electron microscopic analyses indicated that the transplanted donor cardiomyocytes formed mature intercalated discs with host cardiomyocytes comprised of fascia adherens, desmosomes, and gap junctions.

Subsequent studies demonstrated that transplanted cardiomyocytes formed a functional syncytium with the host myocardium. In these experiments, fetal cardiomyocytes from transgenic mice expressing enhanced green fluorescent protein (EGFP) were transplanted into the ventricle of histocompatibility-matched non-transgenic recipient animals. An image of a typical graft is shown in Figure 24-1A. To demonstrate the formation of a functional syncytium, transplanted hearts were harvested, and perfused on a Langendorff apparatus with rhod-2 (a calcium indicator dye that exhibits increased fluorescence with increased

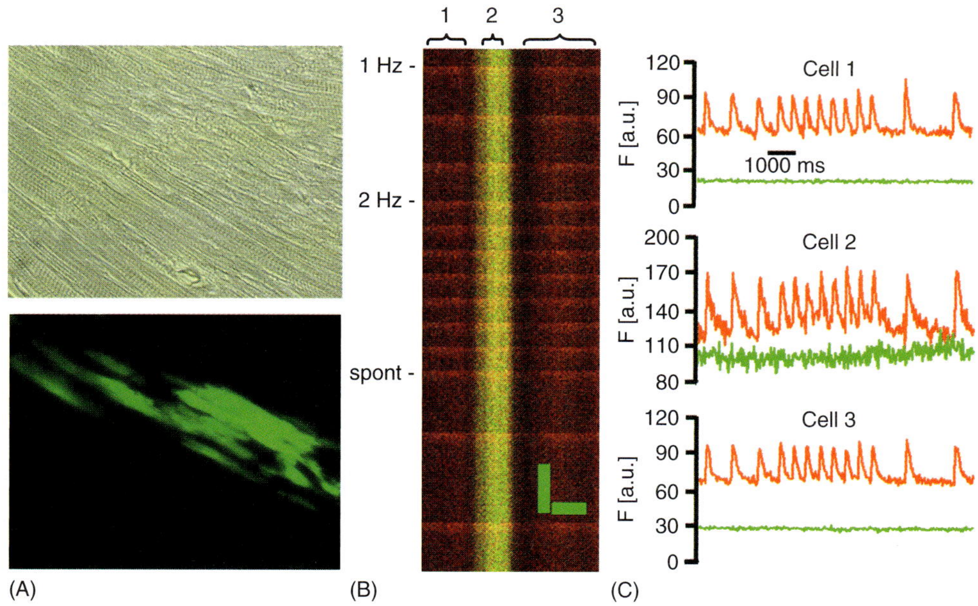

FIGURE 24-1 Functional coupling of donor cardiomyocytes with the host myocardium. (A) Brightfield (top) and epi-fluorescence (bottom) image of an area of a mouse left ventricle containing EGFP expressing donor cardiomyocytes and non-expressing host cardiomyocytes. The heart was harvested 27 days post-engraftment. (B) Simultaneous imaging of rhod-2 (red) and EGFP (green) fluorescence in a non-transgenic heart carrying an MHC-EGFP fetal cardiomyocyte graft. The image shows vertically-stacked line scans that traverse three juxtaposed cardiomyocytes. Cells 1 and 3 are host cardiomyocytes, whereas cell 1 is an EGFP-expressing donor cardiomyocyte. The heart was paced via point stimulation at a remote site at the rates indicated. (spont indicates spontaneous $[Ca^{2+}]_i$ transient.) Donor cells appear yellow, due to superimposition of red rhod-2 fluorescence and the green EGFP signal. Scale bars = 20 microns horizontally, 1000 microns vertically. Time runs from top to the bottom. (C) Spatially integrated traces of the changes in rhod-2 (red) and EGFP (green) fluorescence for the three cardiomyocytes in panel B. The signal across the entire cell was averaged, and plotted as a function of time. Note that both stimulated and spontaneous $[Ca^{2+}]_i$ transients occur simultaneously in donor and host cardiomyocytes, and exhibit similar kinetics.

intracellular calcium content) and cytochalasin D (an excitation-contraction uncoupler). The perfused hearts were then subjected to two-photon laser scanning microscopy under conditions which permitted simultaneous imaging of EGFP status (to distinguish the donor and host cardiomyocytes), and rhod-2 fluorescence (to monitor intracellular calcium transients, and, by inference, action potential propagation). Examination of intracellular calcium transients in neighboring donor and host cardiomyocytes revealed that they occurred simultaneously with, and were indistinguishable from, one another (Figure 24-1B and 24-1C). Thus, donor fetal cardiomyocytes were able to participate in a functional syncytium with the host heart.

EMBRYONIC STEM CELL-DERIVED CARDIOMYOCYTES, AND FACTORS WHICH ENHANCE CARDIOMYOGENIC DIFFERENTIATION

The studies described above indicated that cardiomyocyte transplantation might be a useful mechanism to restore lost function in diseased hearts. Since utilization of fetal donor cells is not practical, a surrogate source of donor cardiomyocytes would be required for clinical application. Embryonic stem (ES) cells are multipotent cells derived from pre-implantation embryos. ES cells have been amplified and maintained indefinitely in an undifferentiated state when grown in monolayer culture under appropriate conditions. When grown in suspension culture, ES cells formed multi-cellular aggregates known as embryoid bodies (EBs). Stochastic cell–cell interactions during EB formation mimicked normal developmental induction cues, and resulted in the random differentiation of ecto-, endo-, and meso-dermal cell lineages. Cardiomyogenic differentiation was frequently observed in differentiating EBs, and was easily recognized due to the presence of spontaneous contractile activity. EB formation has been successfully utilized to generate ES-derived cardiomyocytes from a large variety of species.

Cardiomyocytes derived from ES cells have been fairly well-characterized *in vitro*. ES-derived cardiomyocytes with molecular and electrophysiological characteristics of atrial, ventricular, or conduction system cells were readily identified. The temporal pattern of terminal differentiation of ES-derived cardiomyocytes *in vitro* mirrored that during embryonic development. Moreover, gene expression for the most part qualitatively paralleled that seen during *bona fide* cardiac development. When maintained under typical culture conditions, ES-derived cardiomyocytes did not acquire the gross morphologic attributes, or the quantitative levels of gene expression, characteristic of neonatal or adult cardiomyocytes. However, since the introduction of ES cells into tetraploid blastocysts resulted in the generation of entirely stem cell-derived mice with normal hearts, murine ES cells clearly undergo *bona fide* cardiomyogenic differentiation when placed in the appropriate developmental environment. Thus, the differentiated status of ES-derived cardiomyocytes observed *in vitro* likely reflected limitations in the cell culture conditions. It remains to be seen if utilization of tissue bioengineering approaches that enhance fetal cardiomyocyte differentiation *in vitro* will have a similar effect on ES-derived cardiomyocytes.

The ability to form multiple tissue lineages following transplantation is a hallmark of pluripotent ES cells. Indeed, transplantation of undifferentiated ES cells into healthy or damaged hearts resulted in teratoma formation. Accordingly, the use of ES cells to treat heart disease would require *in vitro* differentiation and purification of donor cardiomyocytes (or alternatively, committed cardiomyogenic precursors) prior to transplantation. Unfortunately, cardiomyocytes typically comprise only 0.5% to circa 5% of the cells present in randomly differentiating EBs. Considerable effort has thus been invested to identify methodologies to enhance cardiomyogenic yield during *in vitro* differentiation.

Early efforts relied on genetic modification of the progenitor cells, such that the resulting cardiomyocytes could be separated from non-myocytes (Figure 24-2). For example, undifferentiated ES cells carrying a transgene comprising the cardiac-restricted alpha-myosin heavy chain promoter driving expression of a cDNA encoding aminoglycoside phosphotransferase were allowed to differentiate *in vitro*. After cardiomyogenic differentiation was apparent (as evidenced by the appearance of spontaneous contractile activity), the cultures were treated with G418. Since the myosin heavy chain promoter was only active in cardiomyocytes, only those cells expressed aminoglycoside phosphotransferase and survived G418 treatment. Consequently, virtually all of the non-myocytes were eliminated from the culture. A similar approach was used to isolate cardiomyocytes by virtue of targeted expression of EGFP, followed by fluorescence activated cell sorting (FACS). Both approaches have been used to isolate human ES-derived cardiomyocytes.

Additional strategies to enhance cardiomyogenic yield have been developed. Of note, Murry and colleagues demonstrated that monolayer cultures of human ES undergo high levels of cardiomyogenic induction when treated with activin A and bone morphogenetic protein-4. Following density centrifugation, an average cardiomyocyte content of 82% was obtained. Using a different approach, Keller and colleagues noted that cardiomyogenic progenitors transiently expressed vascular endothelial growth factor receptor 2. These progenitors could be isolated by sequential rounds of FACS. Remarkably, cardiomyocytes, vascular smooth muscle cells, and endothelial cells were obtained following *in vitro* or *in vivo* differentiation, suggesting the presence of a common progenitor cell for these three cardiovascular cell lineages. Similar progenitors have been obtained based on other markers, and also from human ES cells.

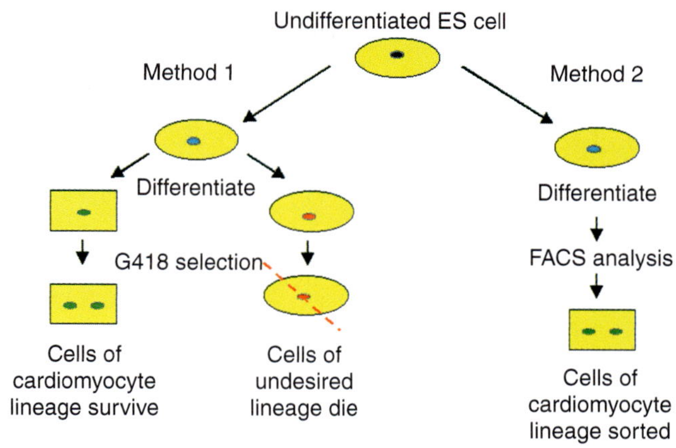

FIGURE 24-2 Selection scheme to produce pure ES-derived cardiomyocyte cultures. The cartoon illustrates two methods to separate ES-derived cardiomyocytes from non-myocytes, by virtue of cardiac-specific expression of a transgene. In Method 1, undifferentiated ES cells carrying a transgene comprising the cardiac-restricted alpha-myosin heavy chain promoter driving expression of aminoglycoside phosphotransferase were allowed to differentiate *in vitro*. Cultures were treated with G418 following cardiomyogenic differentiation to select for cardiomyocytes, since the transgene promoter was only active in cardiomyocytes. In Method 2, undifferentiated ES cells were transfected with a transgene comprising the ventricular-specific myosin light chain-2v (MLC-2v) promoter driving expression of enhanced green fluorescent protein (EGFP), and allowed to differentiate *in vitro*. Following cardiomyogenic differentiation, FACS was used to isolate a cardiomyocyte pure population.

TRANSPLANTATION OF EMBRYONIC STEM CELL-DERIVED CARDIOMYOCYTES IN PRECLINICAL STUDIES, AND CHALLENGES FOR THEIR CLINICAL IMPLEMENTATION

To date, numerous studies have been reported wherein ES cells or ES-derived cardiomyocytes were transplanted into the heart of experimental animals, on some occasions with contradictory results. The initial study transplanted mouse ES-derived cardiomyocytes (generated using the G418 selection scheme described above) into the hearts of adult mice. The donor cardiomyocytes formed stable grafts in the recipient heart, and were well aligned with host cardiomyocytes. Importantly, teratomas were not observed in this study, indicating that the purification scheme efficiently removed any residual multipotent cells from the differentiating ES cultures. Numerous studies have also transplanted human ES-derived cardiomyocytes into recipient hearts. Figure 24-3 shows the results from a study in which human ES-derived cardiomyocytes expressing an EGFP reporter transgene were transplanted into rat hearts following ischemia/reperfusion injury with minimal infarction. The hearts were analyzed four weeks post-transplantation. The

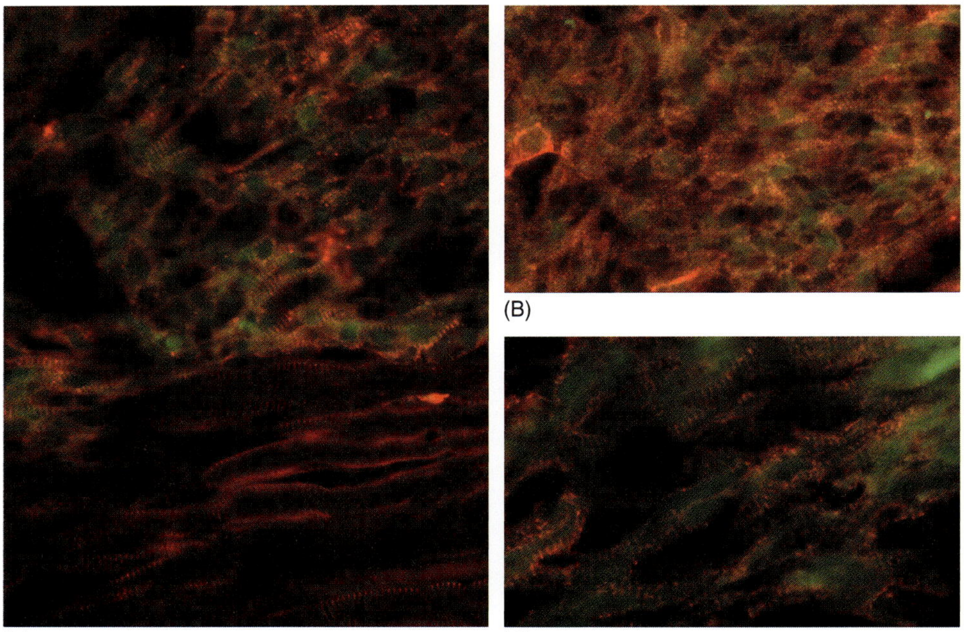

FIGURE 24-3 Stable engraftment of EGFP-expressing human ES-derived cardiomyocytes in rat hearts. Merged images of EGFP (green; epi-fluorescent signal) and anti-cardiac alpha-actinin immune reactivity (red; rhodamine-conjugated secondary antibody) are shown. Note the presence of organized myofibers within the human donor cells.

sections were stained with an anti-cardiac alpha-actinin antibody (rhodamine-conjugated secondary antibody, red signal). Panel A shows the border between the donor human ES-derived myocardium (upper region, donor cells exhibit green epi-fluorescence), and the host rat myocardium (lower region, no green signal). Panels B and C show regions within the donor cell graft. Myofibers in the human ES-derived cardiomyocytes were readily detected (in some cases, the myofibers were highly organized, as in Panel C).

Given their cardiomyogenic potential, it is not surprising that a large number of studies have tested undifferentiated ES cells, ES-derived cardiomyogenic progenitors, or ES-derived cardiomyocytes for their ability to improve cardiac function following transplantation into injured hearts. Typically, animals were subjected to some form of experimental cardiac injury. Control groups were injected with the appropriate vehicle, or alternatively non-cardiomyogenic cell types, while experimental groups were injected with the cell population being tested. Cardiac function and structure were then monitored at subsequent time points. The control groups exhibited a progressive deterioration in cardiac function and structure, commensurate with the degree of myocardial injury. This progressive deterioration was often lessened in the experimental groups receiving cardiomyogenic donor cells.

While quite promising, these studies also highlight the current limitations encountered with cell transplantation-based interventions. Most studies failed to ascertain if the donor cells were functionally coupled to the host myocardium following transplantation. This is an important caveat, particularly when transplanting into injured hearts, as the donor cardiomyocytes were often surrounded by fibrous tissue (and thus were less likely to be electrically coupled to the remote myocardium). Moreover, the impact on cardiac function was often disproportionate to the size of donor cell grafts (which were typically quite small). Given this, indirect mechanisms were likely to have contributed to the improved cardiac function in these studies. For example, cell transplantation could have rescued at-risk myocardium, enhanced healing of the infarct, and/or altered adverse post-injury remodeling, via paracrine-based mechanisms similar to those proposed for hearts transplanted with non-cardiomyogenic stem cells. Presently it is not clear if donor cardiomyocytes can repopulate the heart at sufficient density to directly impact cardiac function. This limitation, as well as a number of additional obstacles, must be overcome for clinical implementation of ES-derived cardiomyocyte transplantation. These are considered below.

Overcoming Limitations in Graft Size

The relatively low levels of donor cell seeding in the host myocardium, particularly following engraftment into injured hearts, is a major limitation of current cardiomyocyte transplantation protocols. Using fetal cardiomyocytes as donor cells, several groups demonstrated that the preponderance of donor cardiomyocytes die following transplantation into the heart, with typically less than 5% of the cells successfully seeding the myocardium. Interventions aimed at enhancing donor cell survival, seeding, and/or post-transplantation proliferation are required to achieve transmural myocardial replacement in injured hearts. Towards that end, Murry and colleagues recently developed a combinatorial approach, wherein donor cells were pre-treated with reagents aimed at blocking ischemia-, anoikis-, apoptosis-, and inflammation-mediated cell death. Pre-treatment of human

ES-derived cardiomyocytes with these reagents markedly enhanced donor cell survival following transplantation into rat hearts with ischemia/reperfusion injury.

Even with this pro-survival pre-treatment regimen, cell transplantation only resulted in remuscularization of, on average, 4% of the infarct (corresponding to only 0.4% of the left ventricle mass). Additional interventions will thus be required to achieve a more extensive (i.e., transmural) remuscularization. Towards that end, numerous cardiomyocyte pro-survival pathways have been identified. Moreover, pathways that impact on cardiomyocyte proliferation during development, and in some cases result in cell cycle re-entry in adult cardiomyocytes, have been identified. Genetic modification of the ES cells, or alternatively development of small molecule modulators of the genetic pathways in question, could be exploited to enhance seeding efficiencies and/or graft size. Use of committed progenitors might also enhance graft size, provided that they exhibit greater post-transplantation proliferation compared with more mature donor cells. Delivery of donor cells in the context of *in vitro* engineered tissue constructs constitutes another viable alternative to enhance graft size. Relatively large volumes of engraftment in infarcted hearts were obtained by surgical attachment of experimental heart tissue which was generated *in vitro* by seeding collagen-based casts with neonatal rat cardiomyocytes. Interestingly, collagen-based casts with complex replacement structures were easily-generated, and appear to undergo rapid vascularization following surgical attachment. These approaches, alone or in combination, may help achieve transmural remuscularization of infarcted hearts with ES-derived cardiomyocytes.

Overcoming Limitations in Donor Cell Production

The ability to generate clinically relevant cell numbers constitutes another significant hurdle for implementation of clinical applications with ES-derived cardiomyocytes. In this context it is interesting to make some projections of cell number requirements. It has been estimated that the left ventricle of the human heart contains about 6×10^9 cardiomyocytes. Consequently, cell transplantation therapies aimed at replacing tissues lost to large myocardial infarction could require the successful incorporation of as many as 3×10^9 donor cardiomyocytes. Generation of this number of donor cells, with concomitant elimination of contaminating cells with teratoma potential, is a formidable task. While the identification of cardiomyogenic growth factors described above will greatly facilitate this process, large-scale production of donor cells will almost certainly require transition from two-dimensional tissue culture-based approaches, to three-dimensional bioreactor-based approaches.

The main challenge to achieve large-scale generation of ES-derived cells is to establish conditions that would initially support the amplification and initial differentiation of ES cells (as for example via EBs formation), and which on subsequent culture would support the further growth and enrichment of the desired cell type. This is particularly challenging, due to the fact that characteristics of the ES cells (and their derivatives) change continuously during the differentiation process. A single-step culture protocol to generate ES-derived cardiomyocytes in an actively-aerated, fully-controlled tank reactor can accomplish these tasks. It utilized a common two-liter bioreactor tank equipped with a blade impeller creating axial flow, coupled with the G418 genetic selection scheme described above. The bioreactor was initially seeded with undifferentiated ES cells. After nine days of culture under conditions suitable for cell expansion, EB formation and differentiation, G418 was added to eliminate the non-cardiomyocytes. After nine days of culture in the presence of G418, more than 109 cardiomyocytes (>99% pure) were obtained from a two-liter reaction vessel. Although quite preliminary, this type of approach provided cell numbers nearing those which are likely to be required for clinical applications.

Overcoming Donor Cell Immune Rejection

Issues associated with the use of allogenic donor cells constitute an additional hurdle for the widespread application of ES-derived cardiomyocyte transplantation. Towards this end, the development of comparably mild immune suppression protocols, used for type-1 diabetic patients following cadaveric beta-cell transplantation, are quite encouraging. Hope, in this regard, is bolstered by the observation that ES-derived cardiomyocytes did not appear to undergo acute rejection when transplanted into rat hearts with experimental infarctions. Moreover, a novel system based on ablation of IL-15 responsive T-cells might ultimately be useful to induce tolerance to allogenic ES-derived donor cells.

The generation of autologous donor cells would bypass immune suppression/rejection concerns. Although cumbersome, nuclear transfer has been successfully used to generate autologous ES-like cells, which could be used to generate ES-derived cardiomyocytes that would escape immune rejection. More recently, Yamanaka and colleagues demonstrated that transient expression of four gene products was sufficient to reprogram transduced fibroblasts into induced pluripotent stem (iPS) cells. Differentiation of iPS cells *in vitro* resulted in the formation of *bona fide* cardiomyocytes. Further refinement of iPS technology (in particular, the ability to generate iPS cells via cytokine treatment, rather than via transgenesis) would constitute an ideal source of donor cells. Indeed, some cells are able to undergo reprogramming into ES-like cells. For example, Hasenfuss and colleagues demonstrated that spermatogonial stem cells spontaneously reprogrammed into ES-like cells during *in vitro* culture. When induced to differentiate, these ES-like cells gave rise to *bona fide* cardiomyocytes.

Thus, there appear to be multiple approaches with which to generate autologous ES-like cells suitable for generation of donor cardiomyocytes. The additional time requirements for the reprogramming event and validation of ES-like characteristics constitute an additional challenge for the clinical implementation of such tailor-made autologous donor cells.

SUMMARY

The potential use of cellular transplantation for the treatment of heart disease is gaining greater acceptance. The notion of using stem cell-derived, and in particular ES-derived, donor cardiomyocytes for partial restoration of lost cardiac mass in diseased hearts has been validated in experimental animals. Additional research is required to determine if it is possible to seed sufficient numbers of donor cells to directly impact on cardiac function, and also to develop viable strategies to prevent donor cell rejection. If these latter hurdles are overcome, cellular transplantation with ES-derived cardiomyocytes will emerge as a promising intervention for the treatment of heart disease.

ACKNOWLEDGMENTS

We thank the NHLBI for support (MR & LJF).

KEY WORDS

Cardiomyocytes Muscle cells in the heart which are interconnected via gap junctions and which participate in a functional syncytium.

Embryoid bodies Three-dimensional structures obtained when ES cells are allowed to differentiate while grown in suspension culture.

Embryonic stem (ES) cells Multipotent cells derived from the inner cell mass of blastocysts and able to differentiate under appropriate conditions into cells of ecto-, endo-, and mesodermal origin.

FURTHER READING

Dai, W., Field, L. J., Rubart, M., Reuter, S., Hale, S. L., Zweigerdt, R., Graichen, R. E., Kay, G. L., Jyrala, A. J., Colman, A., Davidson, B. P., Pera, M., & Kloner, R. A. (2007). Survival and maturation of human embryonic stem cell-derived cardiomyocytes in rat hearts. *J. Mol. Cell Cardiol.*, *43*, 504–516.

Guan, K., Nayernia, K., Maier, L. S., Wagner, S., Dressel, R., Lee, J. H., Nolte, J., Wolf, F., Li, M., Engel, W., & Hasenfuss, G. (2006). Pluripotency of spermatogonial stem cells from adult mouse testis. *Nature*, *440*, 1199–1203.

Klug, M. G., Soonpaa, M. H., Koh, G. Y., & Field, L. J. (1996). Genetically selected cardiomyocytes from differentiating embronic stem cells form stable intracardiac grafts. *J. Clin. Invest.*, *98*, 216–224.

Laflamme, M. A., Chen, K. Y., Naumova, A. V., Muskheli, V., Fugate, J. A., Dupras, S. K., Reinecke, H., Xu, C., Hassanipour, M., Police, S., O'Sullivan, C., Collins, L., Chen, Y., Minami, E., Gill, E. A., Ueno, S., Yuan, C., Gold, J., & Murry, C. E. (2007). Cardiomyocytes derived from human embryonic stem cells in pro-survival factors enhance function of infarcted rat hearts. *Nat. Biotechnol.*, *25*, 1015–1024.

Nakagawa, M., Koyanagi, M., Tanabe, K., Takahashi, K., Ichisaka, T., Aoi, T., Okita, K., Mochiduki, Y., Takizawa, N., & Yamanaka, S. (2008). Generation of induced pluripotent stem cells without Myc from mouse and human fibroblasts. *Nat. Biotechnol.*, *26*, 101–106.

Rubart, M., Pasumarthi, K. B., Nakajima, H., Soonpaa, M. H., Nakajima, H. O., & Field, L. J. (2003). Physiological coupling of donor and host cardiomyocytes after cellular transplantation. *Circ. Res.*, *92*, 1217–1224.

Soonpaa, M. H., Koh, G. Y., Klug, M. G., & Field, L. J. (1994). Formation of nascent intercalated disks between grafted fetal cardiomyocytes and host myocardium. *Science*, *264*, 98–101.

Xu, X.Q., Zweigerdt, R., Soo, S. Y., Ngoh, Z. X., Tham, S. C., Wang, S. T., Graichen, R., Davidson, B., Colman, A., & Sun, W. (2008). Highly enriched cardiomyocytes from human embryonic stem cells. *Cytotherapy*, *10*, 376–389.

Yang, L., Soonpaa, M. H., Adler, E. D., Roepke, T. K., Kattman, S. J., Kennedy, M., Henckaerts, E., Bonham, K., Abbott, G. W., Linden, R. M., Field, L. J., & Keller, G. M. (2008). Human cardiovascular progenitor cells develop from a KDR+ embryonic-stem-cell-derived population. *Nature*, *453*, 524–528.

Zimmermann, W. H., Melnychenko, I., Wasmeier, G., Didie, M., Naito, H., Nixdorff, U., Hess, A., Budinsky, L., Brune, K., Michaelis, B., Dhein, S., Schwoerer, A., Ehmke, H., & Eschenhagen, T. (2006). Engineered heart tissue grafts improve systolic and diastolic function in infarcted rat hearts. *Nat. Med.*, *12*, 452–458.

Chapter 25

Regeneration of Epidermis from Adult Keratinocyte Stem Cells

Ariane Rochat and Yann Barrandon

Keratinocyte stem cells, along with hematopoietic stem cells, have the longest record in cell therapy. Following the seminal work by Green and colleagues in the early 1980s, hundreds of burned patients worldwide had their lives saved by the transplantation of cultured autologous keratinocyte stem cells. Reconstitution of a functional epidermal barrier, self-renewal of the epidermis, and the unexpected regeneration of a papillary dermis are major accomplishments. Gene therapy of hereditary, disabling skin diseases using genetically modified keratinocyte stem cells is down the road. Nevertheless, many more challenges lie ahead. The functionality, the mechanical properties and the aesthetic of the regenerated skin must improve. Sweat glands and hair follicles must be reconstructed, and the pigmentation of the skin better-controlled. It is then necessary to thoroughly comprehend the cellular and molecular mechanisms involved in skin morphogenesis, epidermal renewal, stem cell interactions, dermal remodeling, and fetal wound healing. Only then will stem cell therapy become a major therapeutic option in plastic and reconstructive surgery. Nevertheless, several recent developments are extremely promising, such as *ex vivo* keratinocyte stem cell gene therapy, the capacity to derive adult-like keratinocyte from embryonic stem cells, and the capacity to derive induced pluripotent stem cells from adult keratinocytes.

INTRODUCTION

The skin of vertebrates is covered with scales, feathers, or hairs, whose main function is to protect the body from environmental hazards; the skin also contributes to protection, tactile sensation, thermoregulation, and camouflage. Furthermore, it participates in body homeostasis, for example, through water balance in the frog, or vitamin D synthesis in the human. Most importantly, the appearance of the skin shapes emotional and social attitudes, which is best illustrated in the human by the incessant advertising campaigns promoting a young and smooth skin, and by the distress of patients with disabling skin diseases or scars.

The human skin represents about 16% of the body weight and an area of 1.7 m^2 in the adult. It is composed of three layers that are, from the outer to the inner surface, the epidermis, the dermis, and the subcutis. The epidermis is a stratified, keratinized epithelium in continuity with the epithelial part of the skin appendages (i.e., the hair follicles, the sebaceous glands, and the sweat glands (Figure 25-1). The epidermis contains several specialized cell types that contribute to body protection. The keratinocytes, which represent 90% of the epidermal cells, are of ectodermal origin and are responsible for epidermal renewal, cohesion, and barrier function. The melanocytes are pigment-forming cells of neural crest origin, and are responsible for the color of the skin and UV protection, whereas the Langerhans and epidermal T-cells are of marrow origin, and are involved in immunological defense. The nonvascularized epidermis and epidermal appendages are separated from the highly vascularized dermis by a complex basement membrane, which the nutrients originating from the dermal vessels cross. The dermis is a connective tissue that contains fibroblasts producing proteoglycans, collagen, and elastic fibers, and it is responsible for the mechanical properties of the skin. The subcutis, that mainly contains adipocytes (fat cells), is involved in thermoregulation and lipid storage. The skin is also highly innervated, with numerous free nerve endings in the epidermis and specialized nerves endings in the dermis. The function of the neuroepithelial Merkel cells, which rest in the basal layer of the epidermis, is largely unknown. Altogether, the skin contains more than 25 specialized cell types, originating from at least six different stem cells. Most importantly, the skin is constantly renewed and remodeled. In the human, the epithelial part of the epidermis is renewed once every three weeks, and hair follicles undergo regular cycles of growth, regression, and rest about 25 times in a lifetime. Similarly, the dermis is constantly remodeled with new collagen fibers being produced, as old ones are degraded.

Stem Cell Anthology

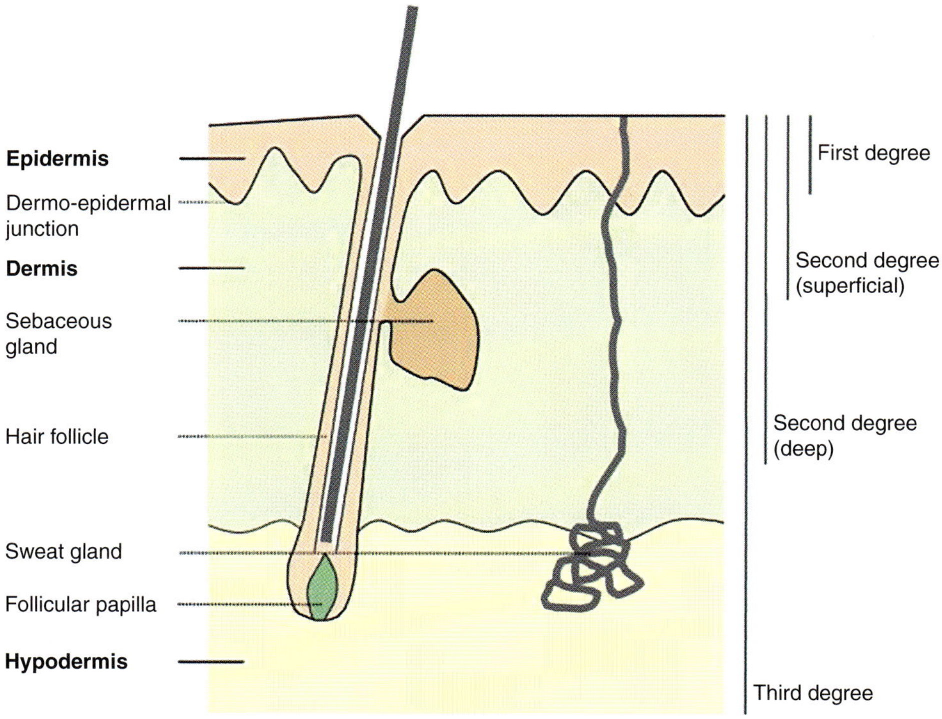

FIGURE 25-1 Depth of burns. First-degree burns heal spontaneously without a scar, second-degree burns heal with a scar. Third-degree burns do not heal, and may necessitate the transplantation of autologous split-thickness skin grafts and autologous cultured keratinocyte stem cells, depending on the extent of the burns.

The Keratinocyte Stem Cell

The keratinocyte stem cell is one of the adult stem cells that inhabits the skin and contributes to skin function and renewal. Adult stem cells are best defined by their capacity to self-renew, and to maintain tissue function for a long period. Keratinocyte stem cells, like hematopoietic stem cells, match this functional and rigorous definition. Moreover, unipotent (single-lineage) and multipotent (multiple-lineages) keratinocyte stem cells coexist in adult skin. Compiling evidence demonstrates that the basal cell layer of the epidermis contains unipotent keratinocyte stem cells that are each responsible for the renewal of a tiny portion of the epidermis. Multipotent keratinocyte stem cells are located in the upper region of the hair follicle below the sebaceous glands, but sweat glands and sweat ducts may also contain multipotent keratinocytes (our unpublished results). Unipotent and multipotent stem cells are clearly related to each other. Indeed, keratinocytes can migrate out of a hair follicle or a sweat gland, to adopt an epidermal fate and to regenerate a wounded epidermis. Most importantly, keratinocyte stem cells, which can divide every 16 to 18 hours, have a remarkable growth capacity in culture, far better than that of dermal fibroblasts. This remarkable growth capacity is at the origin of cutaneous cell therapy.

Keratinocyte stem cells have other remarkable, although circumstantial, properties. They divide infrequently *in vivo*, in contrast to their transient amplifying progeny, which multiply rapidly, and they express high levels of $\beta1$ and $\alpha6$ integrin, and low levels of CD71 (transferrin receptor). Theoretically, a keratinocyte stem cell can generate a progeny with variable fate depending on circumstances, the most probable fates being those of a stem cell (symmetrical divisions), of a transient amplifying cell, or both (asymmetrical division). However, more research is needed to fully-comprehend the mechanisms involved in stem cell decision-making.

Clonal Analysis

The growth potential of individual multiplying keratinocytes is best evaluated by clonal analysis, which can be performed either with cells freshly isolated from the skin or with cultured cells. A clonal analysis experiment typically implies the isolation and the cultivation of 100 to 200 individual cells. Briefly, each single cell is isolated with an elongated Pasteur pipette, under direct vision using an inverted microscope, before it is carefully deposited at the centre of a 35 mm Petri dish containing lethally irradiated 3T3 cells. After seven days of cultivation, each dish is carefully searched for the presence of a clone under an inverted microscope. Each clone is then dissociated with trypsin to obtain a single cell suspension, which is inoculated into several indicator dishes (usually 2 × 100 mm Petri dishes) containing lethally irradiated 3T3 cells. Cultures are fixed after 12 days of cultivation, stained with Rhodamine B, and examined under a dissecting microscope (Figure 25-2). Three types of colonies are easily distinguished: (1) growing

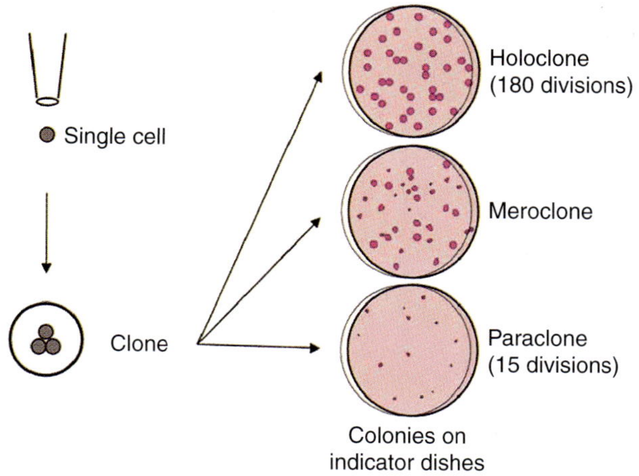

FIGURE 25-2 Clonal types. A single cell is isolated with a Pasteur pipette and cultivated onto a feeder layer of irradiated 3T3 cells. If a clone forms, it is passaged while still growing exponentially. The number and shape of colonies on the indicator dishes determine the clonal type of the founding cell. The progeny of a holoclone generates large, progressively-growing colonies, the progeny of a meroclone generates a mixture of large, progressively-growing colonies, and aborted colonies, and that of a paraclone generates only aborted colonies or no colony. The proportion of those clonal types varies with donor age. Fetal skin mostly contains holoclones, young skin mostly contains holoclones and meroclones, and old skin mainly contains meroclones and paraclones.

colonies that are large and round, have a smooth perimeter, and mostly contain small cells. These colonies can be passaged many times, and are formed by keratinocytes with significant growth potential. (2) Aborted colonies are small, have an irregular shape, and contain large, flattened differentiated squame-like cells. These colonies cannot be passaged, and are formed by cells with restricted growth potential. (3) Mixed colonies usually have a jagged perimeter, and contain intermingled areas of small and large differentiated cells. These colonies can only be passaged a limited number of times, and are formed by cells with limited growth potential.

Clonal Types

The growth potential of a clonogenic keratinocyte is best evaluated by the frequency of aborted colonies formed by its progeny in indicator dishes (Figure 25-2). A cell whose progeny forms almost exclusively growing colonies is termed "holoclone." A cell whose progeny forms growing, mixed colonies, and aborted colonies, is termed "meroclone," and a cell whose progeny forms exclusively aborted colonies or no colony is termed "paraclone." Serial cultivation experiments have demonstrated the tremendous growth capacity of holoclones (up to 180 divisions), whereas transplantation experiments have demonstrated the ability of holoclones to form an epidermis. Indeed, a single holoclone can generate a large enough progeny to entirely restore the epidermis of an adult human. Moreover, clonal analysis experiments have demonstrated the self-renewal capacity of holoclones, in culture and *in vivo* (our unpublished results). Hence, holoclones fulfill the fundamental definition of adult stem cells. The epithelial cells of stratified squamous epithelia are all termed keratinocytes, even if the latter differ in their differentiation capacities. Therefore, all stratified epithelia contain holoclones that theoretically should be referred to as epidermal holoclones, vaginal holoclones, or limbal holoclones. Most importantly, the term holoclone only refers to the growth capacity of a keratinocyte, and not to the lineage capability of the cell. Therefore, some holoclones are multipotent (i.e., follicular holoclones), whereas others are unipotent (i.e., epidermal holoclones). Meroclones, whose growth potential is always lower than that of holoclones, are best characterized by an increased capacity to generate paraclones, possibly by asymmetrical division. Most importantly, the growth potential of meroclones directly correlates with the frequency of paraclone formation. Meroclones with a significant growth potential generate few paraclones, and can form an epidermis when transplanted. Meroclones that generate a significant number of paraclones have a limited growth potential, and are unlikely to contribute to long-term renewal of the epidermis. Thus, meroclones are not strictly self-renewing, and can be viewed as the skin equivalent of short-term hematopoietic stem cells. The growth potential of paraclones is restricted to 1–15 divisions, meaning that the size of the progeny of a paraclone ranges between two cells and 32 000 cells. Consequently, paraclones are transient amplifying cells. However, paraclones are unlikely to be the transient amplifying cells identified *in vivo* by kinetic experiments. Indeed, the number of paraclones in a suspension of freshly-isolated keratinocytes, whatever the age of the donor, is far below the expected number of transient amplifying cells (our unpublished results), which supposedly includes most multiplying cells in the epidermis. Paraclones may then be a population of multiplying cells, generated in response to special circumstances such as wound healing and cultivation.

Clonal Conversion

A culture of human keratinocytes is thus composed of cells with differing capacity for growth, which are termed paraclone, meroclone, and holoclone, that are best identified by clonal analysis. Clonal analysis experiments demonstrate that holoclones self-renew for a few passages in culture, but they stop self-renewing for unknown reasons at some point and convert entirely to meroclones, which in turn generate paraclones (our unpublished results). Ultimately, the culture ends without multiplying cells. This phenomenon is termed "clonal conversion" (Figure 25-3). The conversion of holoclone to meroclone, and to paraclone, is irreversible under normal conditions, and is greatly influenced by the conditions of cultivation. For instance,

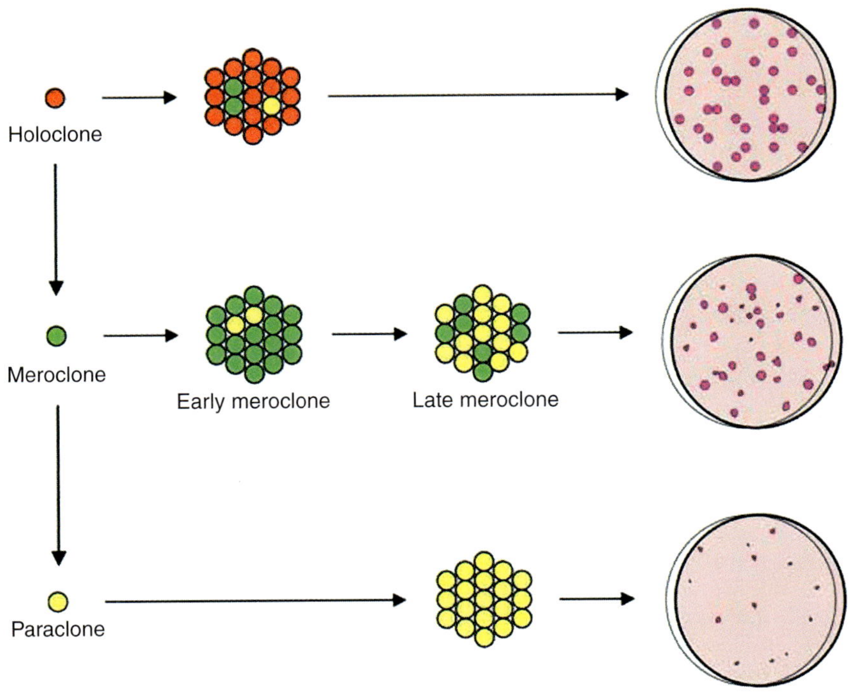

FIGURE 25-3 Clonal conversion. A holoclone generates holoclones, few meroclones, and almost no paraclones. A meroclone generates more paraclones with time. A paraclone generates only aborted colonies. Clonal conversion increases with passages, and is drastically enhanced by suboptimal culture conditions.

it takes more than 23 weeks to exhaust the growth potential of a holoclone cultivated onto a feeder layer of irradiated 3T3 cells in presence of fetal calf serum, an otherwise impossible achievement with suboptimal culture conditions or in defined medium. It is why culture conditions of keratinocytes must be constantly monitored to maintain the highest level of cultivation, especially when the cells are to be transplanted. Clonal conversion is also accelerated by many parameters, including age. In our hands, the number of paraclones in a suspension of freshly-isolated keratinocytes is independent of donor age, but becomes age-related after the primary keratinocytes are passaged (our unpublished data). Indeed, secondary cultures initiated from aged donors contain many more paraclones than cultures initiated from young donors, and they cannot be passaged as many times. Hence, the formation of paraclones reveals the effect on cell growth of age-related genetic modification, or that of an adverse epigenetic phenomenon. This is of paramount importance when cells are cultivated for therapy, as clonal conversion can adversely affect the engraftment (take) of cultured epithelium grafts, and the long-term renewal of the regenerated epidermis.

Keratinocyte Cultivation

Generalities

In 1975, Rheinwald and Green reported the cultivation of human keratinocytes using a feeder layer of irradiated 3T3 cells in a seminal paper, and envisioned its potential application for the treatment of extensive skin defects. It took Green and his colleagues eight more years of intensive work before the first transplant of cultured autologous keratinocytes onto a human was accomplished. This period was paved with several landmarks, including the use of the enzyme dispase to detach the cultured epithelium as a transplantable homogenous sheet, and the demonstration that cultured human keratinocytes can generate an epidermis once transplanted onto athymic mice. The technology has been successfully used worldwide to treat hundreds of extensively burned patients. Green's method to cultivate transplantable keratinocytes is extremely efficient, and we have always been able to initiate cultures, even from tiny biopsies. The cultivation of epidermal stem cells for therapy requires the use of a feeder layer of lethally irradiated mouse 3T3 cells and serum. To our knowledge, no clinical success has been obtained with cells grown in defined medium or in other conditions. The use of cells of animal origin (xenotransplantation) can theoretically induce adverse reactions and transmit pathogens. However, the risk is significantly minimized by using 3T3 cells originating from a master bank that tested negative for a battery of viruses, according to guidelines of regulatory agencies. Animal serum, widely used in the biotechnology industry, is also a potential life hazard, both through the induction of adverse reactions to foreign proteins, and the transmission of pathogens. Again, the risk is minimal if lots of serum with traceable origin, and for which the manufacturer maintains a master file in accordance with the guidelines of regulatory agencies, are purchased.

Most importantly, no adverse reaction to 3T3 cells and serum has been reported in the hundreds of patients treated with epidermal stem cells cultivated in the presence of serum and 3T3 cells. The monitoring of culture conditions include the testing of serum lots, media, and supplements, and is best accomplished by using a reference strain of keratinocytes, the growth characteristics of which are well-known. Indeed, any modification in the method of cultivation (e.g., a new lot of serum or supplements) should be systematically tested in a colony-forming assay using a reference strain, to determine the extent to which the modification affects the ratio of growing colonies (stem cells and close progeny) to aborted colonies (paraclones). It also necessitates strict adherence to instructions and protocols. For instance, 3T3 cells are best grown in the presence of bovine serum, and switching to fetal bovine serum is a common mistake. In our experience, fiddling with protocols in an uncontrolled manner rarely results in better culture performances, and often has deleterious consequences, such as enhanced clonal conversion resulting in rapid exhaustion of growth capacity of the stem cells and, consequently, in failure of engraftment.

A cutaneous wound ruptures skin integrity, and can have serious consequences, such as infection, loss of proteins, and even death. Moreover, chronic wounds favor improper scarring and the development of skin cancers. Burn wounds are a perfect illustration (Figure 25-1). A first-degree burn wound which only affects the epidermis heals spontaneously, and without a scar. Reconstitution of the epidermal integrity then results from the migration of keratinocytes originating from the nearby epidermis or appendages. A second-degree burn wound, defined as superficial or deep on the depth of the dermal burn, destroys the epidermis and part of the dermis. It heals spontaneously by the migration of cells originating from the epidermal appendages. However, a permanent scar will form. A third-degree burn destroys the entire thickness of the skin (full-thickness burn). Although a tiny third-degree burn wound can possibly heal with a scar by migration of cells from the surrounding unwounded skin, a large one will not. The transplantation of split-thickness autografts obtained from unburned donor areas is then the unique therapeutic option, with the added risk of improper healing of donor sites, including infection, hypertrophic scars, and cheloids. Moreover, the recurrent cropping of donor sites can become a serious problem, especially for patients burned over 60% of the total body surface area (TBSA). Progress in medical resuscitation and intensive care allows patients whose skin has almost completely burned, to remain alive. The transplantation of autologous epidermal stem cells then becomes life-saving for those patients who do not have enough donor skin for conventional split thickness autografts.

Methodology

The methodology has been reported in great detail. Briefly, a full thickness skin biopsy is obtained aseptically from normal skin, to isolate stem cells from both the epidermis and the epidermal appendages. The biopsy is then stored in culture medium and transferred to the laboratory, where it is minced with scissors and treated with trypsin to obtain a single-cell suspension. The cells are seeded at a density of 2.4×10^4 cells/cm^2 onto a feeder layer of irradiated 3T3 cells in medium containing fetal bovine serum, and supplemented with cholera toxin, hydrocortisone, insulin, and triodothyronine, according to published protocols. Cultures are supplemented with epidermal growth factor (10 ng/ml) at the first medium change. Under these conditions, clonogenic keratinocytes divide every 16 to 18 hours, and colonies quickly expand and push away the irradiated 3T3 cells, which then detach from the culture vessel. If the culture lasts long enough, the colonies of keratinocytes will fuse and form a stratified epithelium. The method of cultivation selects for clonogenic keratinocytes, but some melanocytes and dermal fibroblasts can also adhere and multiply, although at a much slower rate than the keratinocytes. Primary cultures are usually passaged while still in exponential growth—i.e., before confluence, or after nine days of cultivation, whichever comes first. If primary cultures yield enough cells, cultures for transplantation are then initiated at a seeding density of 1.2×10^4 cells/cm^2, and the leftover cells are frozen for future use. If the primary cultures do not yield enough cells, it is then necessary to further expand the cells by setting up secondary cultures at a seeding density of 1.2×10^3 cells/cm^2. Cells are then cultivated for no more than a week before they are used to set-up cultures for transplantation. Under optimal conditions, a 10 000-fold expansion is obtained within two to three weeks. Confluent cultured epithelia then need to be detached enzymatically from the bottom of the culture vessels, using dispase or thermolysine. However, the enzymatic treatment significantly shrinks the cultured epithelium. Once rinsed, each epithelial graft is mounted onto petroleum gauze backing, with special care being taken to maintain the correct orientation of the epithelium, (i.e., the basal layer facing away from the backing). This step of the procedure is cumbersome, tedious, and manpower-demanding, especially when hundreds of cultured epidermal autografts (CEAs) need to be prepared in a reasonable amount of time. This is why many groups have explored alternative methods to cultivating keratinocytes.

Transplantable Matrices

Clonogenic human keratinocytes can be cultivated with variable success on various matrices, including human fibrin, bovine or human collagen; hyaluronic acid; shark proteoglycans; and polymers. Therefore, it is crucial to thoroughly evaluate the effects of each of these matrices on the growth of the keratinocyte stem cells, since suboptimal culture conditions may result in enhanced clonal conversion, a reduced keratinocyte lifespan, and an irreversible degradation of the quality of the graft. This is best accomplished by clonal analysis and by transplantation of the cultured cells onto athymic or severe-combined immunodeficiency mice, using reference strains of keratinocytes, the performances of which are well-known. Using this approach, we have demonstrated that fibrin matrices alleviate the need for the detachment of the grafts, eliminate shrinkage, and hence have considerable advantages over plastic for the culture of epidermal stem cells for grafting. Use of a fibrin matrix considerably shortens the duration of cultivation required to generate many CEAs, making it possible to transplant the entire body area of an adult human with CEAs as soon as 15 days after injury. Most importantly, the long-term regeneration of human epidermis on third degree burns transplanted with autologous cultured epithelium grown on a fibrin matrix has been demonstrated. Fibrin matrices are also extremely useful for transplantation of autologous limbal stem cells to restore a corneal epithelium.

The transplantation of small colonies of keratinocytes (preconfluent cultures) grown on fibrin matrices, or of individual keratinocytes suspended in fibrin gels has been proposed. This may be useful to quickly transplant small full-thickness burns, or to enhance the healing of superficial burn wounds. However, it remains to be demonstrated that individual cells or small colonies generate an epidermal barrier faster than confluent and organized cultures when directly exposed to the hostile environment of the grafting bed in large full-thickness burns. Composite grafts combining autologous keratinocytes and fibroblasts embedded in a collagen gel, a fibrin matrix, or polymers may also be useful to generate a closer-to-normal skin by enhancing the formation of dermis. However, with today's technology, the population of fibroblasts in a small skin biopsy cannot be sufficiently expanded in a timeframe compatible with the transplantation of acute burn wounds. This is why composite grafts, associating autologous fibroblasts and keratinocytes, are reserved for chronic wounds. Many groups have proposed using allogeneic fibroblasts, large stocks of which can be prepared in advance. However, allogeneic fibroblasts do not permanently engraft, despite contradictory claims in the literature. To our knowledge, composite grafts may be efficient to treat chronic skin defects such as leg ulcers, but they have not proven useful for the successful treatment of extensive burn wounds. Hopefully, progress in the cultivation of human fibroblasts will be made; then, composite grafting will have a major impact.

Transplantation of Keratinocyte Stem Cells

Various surgical procedures have been developed over the years to transplant cultured keratinocytes. CEA were initially transplanted directly on the granulation tissue that spontaneously forms after full-thickness burns were excised to fascia. Success was then variable, ranging from excellent to poor, and surgeons quickly stressed that the quality of the wound bed was crucial. In that regard, one of the most appreciated procedures is known as Cuono's, in which full-thickness burns are excised to fascia, and temporarily covered with meshed, expanded, fresh or cryopreserved split-thickness human allografts obtained from organ donors. At grafting time, which usually occurs two to three weeks after the initial biopsy was taken for expansion of stem cells, allografts are dermabraded or tangentially excised, and a careful hemostasis is performed. Cultured epithelium grafts, with the basal cells oriented toward the grafting bed, are then gently applied and covered with petroleum gauze dressings to prevent desiccation (Figure 25-4). Sliding and displacement of the graft should be strictly avoided. CEA are then overlaid with bridal veil gauze, sterile dry compresses, and elastic bandage. Skin substitutes like Integra® or Biobrane® are useful alternatives to cadaver skin, when the latter is not available or cannot be used for religious reasons.

Cuono's procedure has greatly increased the rate of CEA engraftment, which is best-appreciated clinically. At the time of the first dressing ("take-down"), the engrafted epithelium appears shiny and translucid. With time, the regenerated epithelium matures to resemble a normal epidermis (Figures 25-4 and 25-5). CEAs perform best in areas in which shear forces or frictions are minimal, and in non-weight-bearing surfaces. Successful engraftment now averages 70% of the transplanted area, but it is not unusual to obtain a 100% take. In a recent study, younger age was significantly associated with better CEA take, possibly reflecting clonal type differences between young and old people. Most importantly, the transplantation of CEA necessitates highly-trained surgical and nursing staff. For instance, four to six nurses require from 60 to 90 minutes to perform each dressing change following CEA application at the Percy Burn Center in France, which has one of the largest and most successful records in CEA transplantation. Infection of the grafting bed is the enemy of CEAs, and provokes a quick disappearance of the grafts. The fight of infection relies on systemic antibiotics and local antiseptics, which are clinically efficient, and usually not harmful to the CEAs, despite an alleged toxicity *in vitro*. Fragility and blistering of CEAs are also commonly observed during the weeks following transplantation, most likely as a consequence of a delayed maturation of the anchoring fibrils. In some cases, lysis of CEAs, which had otherwise nicely engrafted, occurs after a few weeks with no rational explanation. Bleeding of the grafting bed can also result in loss of CEA, further emphasizing the importance of thorough hemostasis at grafting time. Another complication of CEAs is epidermal hyperkeratosis, which may need repeated treatment with moisturizing and keratolytic ointments. Wound contracture is also a problem, as in all patients with large burns, but there seems to be less hypertrophic scarring with CEAs than with conventional split-thickness autografts.

Regeneration of Epidermis

Cultured epithelia prepared according to the methods of Green and colleagues are stratified, but lack the uppermost-differentiated epidermal layers necessary for barrier function.

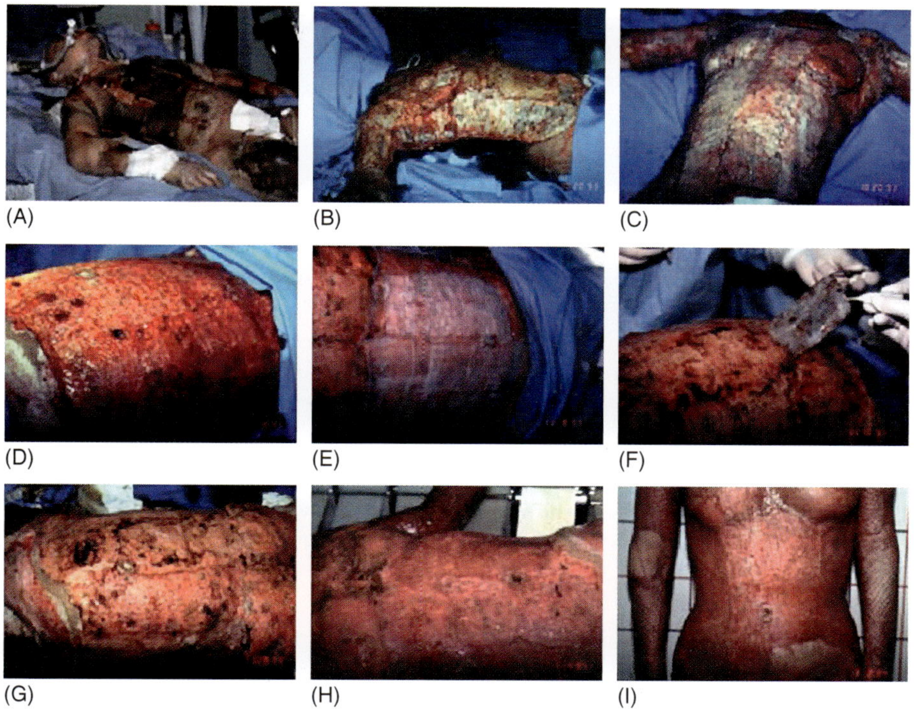

FIGURE 25-4 Transplantation of human cultured epithelium autografts. A 35-year-old woman was burned by flames; over 50% of her body has third degree burns. (A) Admission; (B) Excision to fascia of full-thickness thoracic and abdominal burn wounds the day after injury. (C) Temporary coverage of the excised area with split-thickness meshed human cadaver allografts. (D) Appearance of the dermabraded grafting bed. (E) Transplantation of control cultured epithelia (top) and cultured epithelia grown on fibrin matrices (bottom). (F) Appearance of the transplanted area eight days later (take-down). (G) Appearance of the same area 21 days later. Note the excellent take of both control (bottom) and experimental (top) cultured epithelia (the photograph was taken from the opposite side to that shown in panel (E)). (H, I) Appearance of the same transplanted area 57 days and four months later. Reproduced with permission from Ronfard *et al.*, (2000).

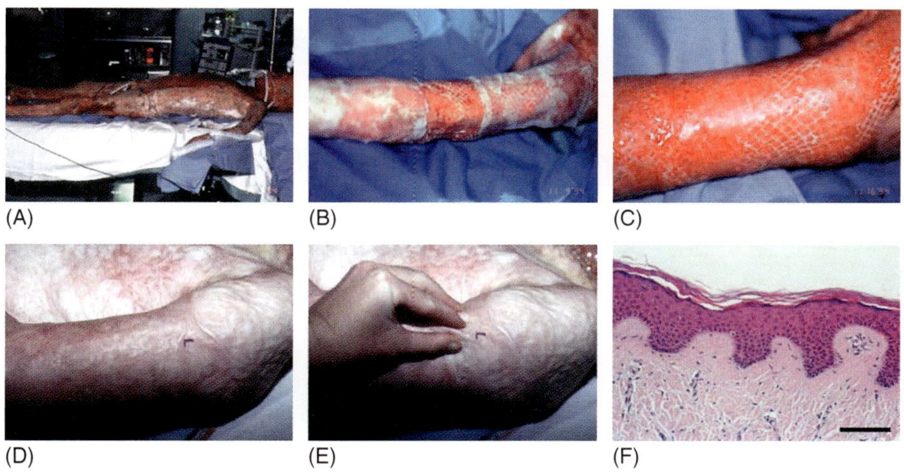

FIGURE 25-5 Long term follow-up of cultured epithelia transplanted on a fibrin matrix. A nine-year-old boy was burned by flames; over 95% of his body has third degree burns. (A) Admission at Percy Burn Centre a month after injury. (B) Transplantation of cultured epithelia grown on a fibrin matrix on the left arm. (C) Appearance of the transplanted area at takedown. (D, E) Clinical appearance of the skin 3.5 years after the transplantation. The skin is elastic when pinched, and has a smoother appearance than the neighbouring split-thickness skin autografts. (F) Histological appearance of the skin 3.5 years after transplantation. The epidermis is histologically normal. Note the presence of rete ridges, and a superficial neodermis with vascular arcades. Similar results were obtained with cultured epithelia grown in absence of fibrin. Bar: 100 μm. Reproduced with permission from Ronfard *et al.*, (2000).

However, the horny layers mature rapidly once the CEA are transplanted, indicating that the cells adequately express their program of differentiation. This epidermis always contains suprabasal Langerhans cells, and may also contain melanocytes, although at a variable density in the basal layer. The presence of Merkel cells has been reported. The most surprising observation is the presence of rete ridges, vascular arcades, and elastic fibers, which are normally not observed during normal wound healing. However, the formation of a neodermis that closely resembles the papillary dermis is a slow process that takes years (see Figure 25-5). This extremely intriguing observation suggests that mesenchymal stem cells may have been transplanted along with the epidermal stem cells, and that epidermo–mesenchymal interactions reminiscent of fetal wound healing have occurred. This awaits confirmation. Most importantly, hair follicles and sweat glands are not regenerated. There are several explanations for this: multipotent epithelial stem cells are not present in CEAs, possibly because current culture conditions favor the epidermal lineage, or because multipotent stem cells are absent in adult skin, and the inductive signals necessary for appendage morphogenesis are missing. However, recent results from our laboratory indicate that the skin of adult mammals contains multipotent epithelial cells with the capacity to generate epidermis, sebaceous glands, and all follicular lineages. Other groups have demonstrated that dermal papilla cells obtained from adult hair follicles have inductive capacities. These are important findings that open the path to the reconstruction of hair follicles and other epidermal appendages.

The long-term self-renewal of epidermis generated from CEA is a concern, in view of recent findings demonstrating a shortening of telomeres in keratinocytes obtained from transplanted patients. This further illustrates that cells should be cultivated according to state-of-the-art technology, and that cell culture, as good as it can be, is a stress to cells. It is impossible as yet to distinguish between the consequences of clonal conversion, which can occur very quickly in relation to poor cultivation and independently of the mitotic clock, and the shortening of telomeres, a difference expressed in a recent review as mitotic clock or culture shock. The expansion of the epidermal stem cells in culture represents at the most 22 cell doublings, a small number when holoclones can undergo at least 180 doublings, meaning that holoclones and early meroclones have plenty of growth potential left to renew their share of epidermis for a long time. The number of stem cells (holoclones) or cells with significant growth potential (meroclones) present in the CEAs, that can engraft and self-renew the epidermis becomes important. Surprisingly, there are no data in the literature on this subject, even though hundreds of patients have been transplanted. Nevertheless, clinical follow-up indicates that epidermis generated from CEAs self-renews for years, suggesting that holoclones and meroclones are present in sufficient numbers.

Indications of Cultured Epidermal Autografts

Cell Therapy

Clinical indications of CEAs have been extensively reviewed. The main indications for CEA are full thickness burns over 60% of TBSA, in which CEAs are highly beneficial and lifesaving. CEAs can also be useful in reconstructive surgery to enhance the healing of donor sites, to prevent recurrence of cheloids, and to cover wounds made by the excision of giant congenital nevi, and of tattoos. CEAs may also be useful in urology and oral surgery. But one of the most impressive uses of CEAs is in ophthalmology. Pellegrini and colleagues have beautifully demonstrated that limbal stem cells (limbal holoclones) can be cultivated on a fibrin matrix and transplanted to restore the integrity of the corneal epithelium. This approach, coupled with elective keratoplasty, has restored vision to patients whose sight was impaired because of corneal burn wounds.

Niche market, customized manufacturing, and high costs have made it difficult for CEAs to become a blockbuster product. As an example, Epicel®, a commercial CEA manufactured according to good manufacturing practices is considered a humanitarian device by the Food and Drug Administration (www.genzyme.com); it costs about \$14 US per cm^2 (including biopsy kit, culture, and shipping), which prices the transplantation of a square meter of CEA at \$140 000 US. However, this cost should be compared to that of other transplant procedures. None-theless, progress has been made. For instance, use of a fibrin matrix that diminishes the fragility of the CEA, and is user-friendly, should reduce costs as the number of CEAs necessary to cover an area is two-fold lower, the duration of cultivation is significantly shortened, and there are no large personnel requirements to prepare the grafts. Composite grafts, including autologous mesenchymal and keratinocyte stem cells, are also foreseen, but they are more complex to manufacture, and their cost will certainly be higher than that of currently available CEAs.

Ex Vivo *Gene Therapy*

Gene therapy of hereditary disabling diseases of the skin or of the cornea will certainly benefit from the experience gained with the treatment of large burns. Indeed, epidermal stem cells or limbal stem cells can be successfully transduced by defective retroviral or lentiviral vectors, and engineered to express a protein of medical interest. For instance, transduction of a cDNA-encoding laminin 5 or collagen VII in epidermal stem cells of patients with junctional epidermolysis bullosa or dystrophic epidermolysis bullosa, respectively, should permit the treatment of elective areas to enhance the quality of life of those patients. Mavilio and De Luca's groups have recently demonstrated the feasibility of *ex vivo* gene therapy. In this breakthrough, the authors used a straightforward retroviral strategy to stably transduce a *LAMB3* cDNA encoding laminin 5 β3-chain into keratinocyte stem cells of a patient with a rare form of junctional epidermolysis bullosa (JEB), a devastating skin disease characterized by skin fragility, chronic blistering, painful wounds, and the emergence of carcinomas. Genetically corrected *LAMB3* CEA were then produced and transplanted to restore epidermal integrity. Importantly, the transplanted areas have remained healed since (De Luca, personal communication), unambiguously demonstrating stem cell engraftment. Mastering the cultivation and the transplantation of keratinocyte stem cells was assuredly the key to success, as the authors have thorough experience in transplanting cultivated keratinocyte stem cells to treat burns or corneal defects. The choice to transplant chronic non-healing wounds depleted in resident stem cells was also crucial, since it likely removed competition between those stem cells present in skin appendages (i.e., in hair follicles and sweat glands) and the transplanted stem cells.

Despite the fact that the transplantation of autologous cultivated keratinocytes has contributed to save the lives of thousands of patients worldwide, some have questioned the origin of the healing cells, suggesting that stem cells other than epidermal (e.g., hematopoietic stem cells) were responsible for epidermal reconstitution in large burns. Mavilio's data, together with our own data demonstrating that the progeny of a single EGFP keratinocyte multipotent stem cell of the rat can reconstitute hair follicles and sebaceous glands for years in serial transplantation, should definitely close the argument on whether keratinocyte stemness is maintained in culture. Another important finding of Mavilio's is that few stem cells are necessary to reconstruct and renew a recombinant epidermis. This opens the door to the transplantation of a genetically-homogenous CEA produced from a single transduced holoclone, in which one can demonstrate safe integration of the recombinant provirus; this together with the use of a self-inactivating shuttle vector should further contribute to safe epidermal gene therapy.

FUTURE DEVELOPMENT

A promising development is the derivation of keratinocytes from embryonic stem cells (ESCs) and, interestingly, the transcription factor p63 appears important for epidermal commitment. Several groups have demonstrated the feasibility of the approach in the mouse and the human, but the ESC-derived keratinocytes, at least in the human, do not yet seem to fully recapitulate the properties expected from adult keratinocyte stem cells. Another exciting development is the recent derivation of induced pluripotent stem cells (iPS) from adult human keratinocytes, which, in combination with the isolation of keratinocytes from a small skin biopsy or even a plug hair, is a powerful means to obtain customized iPS cell lines. Nevertheless, the derivation of keratinocytes from ES cells or iPS does not preclude mastering the cultivation and transplantation of adult keratinocyte stem cells, and it surely does not alleviate the problem of engraftment.

SUMMARY

A challenge in *ex vivo* skin cell therapy is to obtain a massive expansion of the stem cells from a small skin biopsy in the shortest time possible, while preserving stemness. The method of cultivation described by Rheinwald and Green in the seventies achieves this goal, and it has not been matched. Cell therapy using autologous keratinocyte stem cells, cultivated according to the preceding method, has saved the lives of hundreds of burned patients, has restored sight to vision-impaired patients, and has opened the door to *ex vivo* gene therapy. However, the technology relies on a feeder layer of irradiated mouse cells, and is therefore considered a xenotransplantation by regulatory agencies, which adds a significant burden to an already cumbersome approval process. Hence, one has to improve the cultivation of keratinocyte stem cells (and also that of skin mesenchymal stem cells), to improve engraftment, to reconstruct hair follicles and sweat glands, and to improve the mechanical properties and the esthetic of the regenerated skin. The answer to these challenges resides in a better understanding of the cellular and molecular events involved in skin renewal and morphogenesis; in a better understanding of fetal and adult wound healing; in that respect, embryonic stem cell lines might be helpful. It is also of paramount importance to study stem cell engraftment in animal models, e.g., the pig, the skin biology of which is closer to the human than that of rodent. In conclusion, future progress in skin cell therapy necessitates: (1) continuous scientific knowledge to improve stem cell engraftment and transplant functionality; (2) the development of a cultivation technology that meets regulatory requirements without affecting stemness; and (3) a thorough evaluation of the economic constraint, as it can affect the commercial development of cell therapy.

ACKNOWLEDGMENTS

This work was supported by funds from the Swiss Federal Institute Lausanne (EPFL), the Lausanne University Hospital (CHUV) and from the EEC (Therapeuskin, EuroSyStem, and Optistem).

KEY WORDS

Clonal conversion Conversion of a holoclone to a meroclone or to a paraclone irreversible under normal conditions; affected by culture conditions and age.

Holoclone (Gk *holos* whole + Gk *klôn* growing) Colony-forming cell with tremendous growth potential (up to 180 divisions), whose progeny forms almost exclusively growing colonies.

Meroclone (Gk *mero* part + Gk *klôn* growing) Colony-forming cell whose growth potential is lower than the growth potential of a holoclone but superior to 15 divisions; forms growing and aborted colonies.

Paraclone (Gk para *akin* + Gk *klôn* growing) Colony-forming cell committed to a maximum of 15 divisions; forms only aborted colonies.

FURTHER READING

Aasen, T., Raya, A., Barrero, M. J., Garreta, E., Consiglio, A., Gonzalez, F., Vassena, R., Bilić, J., Pekarik, V., Tiscornia, G., Edel, M., Boué, S., & Belmonte, J. C. (2008). Efficient and rapid generation of induced pluripotent stem cells from human keratinocytes. *Nat. Biotechnol.*, *26*, 1276–1284.

Barrandon, Y. (2007). Genetic manipulation of skin stem cells: success, hope, and challenges ahead. *Mol. Ther.*, *15*, 443–444.

Carsin, H., Ainaud, P., Le Bever, H., Rives, J., Lakhel, A., Stephanazzi, J., Lambert, F., & Perrot, J. (2000). Cultured epithelial autografts in extensive burn coverage of severely traumatized patients: a five year single-center experience with 30 patients. *Burns*, *26*, 379–387.

Gallico, G. G., 3rd, O'Connor, N. E., Compton, C. C., Kehinde, O., & Green, H. (1984). Permanent coverage of large burn wounds with autologous cultured human epithelium. *N. Engl. J. Med.*, *311*, 448–451.

Green, H. (2008). The birth of therapy with cultured cells. *Bioessays*, *30*, 897–903.

Mavilio, F., Pellegrini, G., Ferrari, S., Di Nunzio, F., Di Iorio, E., Recchia, A., Maruggi, G., Ferrari, G., Provasi, E., Bonini, C., Capurro, S., Conti, A., Magnoni, C., Giannetti, A., & De Luca, M. (2006). Correction of junctional epidermolysis bullosa by transplantation of genetically modified epidermal stem cells. *Nat. Med.*, *12*, 1397–1402.

Medawar, A., Virolle, T., Rostagno, P., de la Forest-Divonne, S., Gambaro, K., Rouleau, M., Aberdam, D. (2008). DeltaNp63 is essential for epidermal commitment of embryonic stem cells. *PLoS ONE*, *3*, e3441.

Pellegrini, G., Traverso, C. E., Franzi, A. T., Zingirian, M., Cancedda, R., & De Luca, M. (1997). Long-term restoration of damaged corneal surfaces with autologous cultivated corneal epithelium. *Lancet*, *349*, 990–993.

Pellegrini, G., Ranno, R., Stracuzzi, G., Bondanza, S., Guerra, L., Zambruno, G., Micali, G., & De Luca, M. (1999). The control of epidermal stem cells (holoclones) in the treatment of massive full-thickness burns with autologous keratinocytes cultured on fibrin. *Transplantation*, *68*, 868–879.

Ronfard, V., Rives, J. M., Neveux, Y., Carsin, H., & Barrandon, Y. (2000). Long-term regeneration of human epidermis on third degree burns transplanted with autologous cultured epithelium grown on a fibrin matrix. *Transplantation*, *70*, 1588–1598.

Chapter 26

Implantation of Myogenic Cells in Skeletal Muscles

Daniel Skuk and Jacques P. Tremblay

INTRODUCTION

The intramuscular implantation of myogenic cells is an approach to develop a therapeutic tool for myopathies, mainly those of genetic recessive etiology. The point of departure of this approach can be traced to 1978, when Partridge, Grounds, and Sloper proposed that "in subjects suffering from inherited recessive myopathies, muscle function might be restored if normal myoblasts could be made to fuse with defective muscle fibres" (Partridge *et al.*, 1978). Among these myopathies, Duchenne muscular dystrophy (DMD) is the main target of this potential therapeutic tool. This is due to the combination of DMD's relative frequency (a prevalence of 50 cases per million in the male population) and severity: progressive generalized skeletal muscle degeneration during the childhood and adolescence, leading to paralysis and death.

The history of myogenic-cell transplantation in the field of myology is an example of the importance of an appropriate pre-clinical basis to design clinical applications. It was only after few animal experiments in the 1980s (some of which were probably not appropriately conducted) that several groups undertook clinical trials in the early 1990s, most of which on DMD patients (for a review see Skuk, 2004). Lacking appropriate pre-clinical support to plan the strategies of cell implantation and control of acute rejection, these clinical trials reported scarce and very modest results at the molecular level. The error was to expect too much from the grafted cells: these trials were conducted on the hope that few myogenic cells injected in few sites of a skeletal muscle would be able to diffuse throughout the muscle and spontaneously fuse with most myofiber regions. The subsequent research demonstrated that this hope was unrealistic. An important lesson from this experience is that researchers need to know the actual behavior of the cells, in appropriate animal models, in order to use them for clinical applications.

This chapter wishes to introduce the actual characteristics of myogenic-cell transplantation for clinical purposes. Priority will be given to observations obtained in humans and non-human primates. Observations in rodents will be considered when they complemented or supported the observations in humans and monkeys. This is because there is a motley literature concerning myogenic-cell and "stem-cell" implantation in rodents that, either has no clinical relevance, or did not prove so far to be reproducible in appropriate large animal models. The chapter will be organized through three main challenges of cell transplantation: (1) to obtain appropriate cells for implantation, (2) to properly deliver them to the target tissues, and (3) to insure their survival into the recipient.

CELLS TO GRAFT

In cell transplantation, the implantable elements could be either differentiated cells or precursor cells with the ability to differentiate into the former. In the skeletal muscle, the differentiated cells of the parenchyma (i.e. the myofibers) are useless for transplantation: they are very long syncytia that cannot be properly manipulated nor implanted. However, this possibility is offered by the mononucleated stem cell that is specific to the skeletal muscle (i.e. the satellite cell). One of the main functions of satellite cells in the mature skeletal muscle is to repair the partial or total damage of the myofibers. They are normally quiescent until an injury produces focal or total necrosis of a nearby myofiber. This necrosis triggers a process of regeneration, which involves the removal of the myofiber debris by phagocytic cells and the activation of the satellite cells. Activated satellite cells enter mitosis and give rise to mononucleated muscle precursor cells generally referred as **myoblasts** (Betz *et al.*, 1966), the adjective "adult" being sometimes added (Yablonka-Reuveni and Nameroff, 1990) to differentiate them from the embryonic or fetal myoblasts that give rise to skeletal muscles during his-togenesis. Adult myoblasts proliferate and fuse among themselves to form syncytial myotubes that would give rise to myofibers. Satellite cells can be isolated from skeletal muscle biopsies by standard cell culture techniques, and can be expanded as myoblasts *in vitro*, maintaining their capacity to fuse into myotubes that will differentiate into myofibers (Konigsberg, 1960). It was this easiness of satellite cells to be isolated from muscle biopsies, and to be proliferated *in vitro* in order to obtain large quantities of adult myoblasts able to fuse in myofibers, which opened the possibility to use them for strategies of myogenic-cell transplantation.

Useful Properties of the Implanted Cells

One of the pioneer studies of myogenic-cell transplantation was published by Lipton and Schultz (1979). They reported the two main properties of exogenous myogenic cells implanted into skeletal muscles of mice, that is, (1) they fuse with the myofibers of the recipient and (2) they can form new small myofibers. The first property allows the phenomenon of "gene complementation," (i.e. myofibers) in which exogenous myogenic cells fused will express at the same time proteins coded by the exogenous and the host nuclei (Watt *et al.*, 1982). Through this property, the implanted myogenic cells can act as vehicles of therapeutic genes (e.g. by introducing a normal genome in the genetically abnormal myofiber of a muscular dystrophy patient). The second property opens the door to the possibility of regenerating skeletal muscle parenchyma when it is lost. A third property, more recently described, is the possibility of giving rise to new satellite cells.

Gene Complementation

Myotubes or myofibers containing nuclei with different genomic backgrounds are referred to as "mosaic" or "hybrid" (Kikuchi *et al.*, 1980). This is the case of a myofiber in which exogenous myogenic cells have fused, since it will contain a mixture of nuclei of recipient's and donor's origin, and will express proteins from both origins (Watt *et al.*, 1982). By the way of gene complementation, a protein whose deficiency in the recipient's genome causes a myopathy can be expressed by the normal donor's genome. The first experimental demonstration of this principle was reported by Partridge *et al.* (1989). Following transplantation of normal mouse myoblasts in *mdx* mice (which lack the protein called **dystrophin**, as DMD patients) they observed later that several myofibers expressed dystrophin in its normal subsarcolemmal position. The same observation was repeated by other researchers (Kinoshita *et al.*, 1994; Vilquin *et al.*, 1995) and is presently a routine in myogenic-cell transplantation research. Other proteins, which were restored by normal myoblast transplantation in mouse models of muscular dystrophies, were **merosin** in *dy/dy* mice (a model of congenital muscle dystrophy with merosin deficiency) (Vilquin *et al.*, 1996) and **dysferlin** in SJL mice (a model of limb-girdle muscle dystrophy with dysferlin deficiency) (Leriche-Guerin *et al*, 2002). In humans, occasional observations of improved dystrophin expression following normal myoblast implantation in DMD patients were reported during the clinical trials conducted in the 1990s (Huard *et al.*, 1992; Karpati *et al.*, 1993; Tremblay *et al.*, 1993; Mendell *et al.*, 1995), although these results were not conclusive and most of the patients gave negative results at that time. A recent clinical trial, designed with the basis of pre-clinical experiments in nonhuman primates, showed that donor-derived dystrophin can be observed systematically in the muscles of DMD patients implanted with normal myoblasts (Figure 26-1) (Skuk *et al.*, 2004; Skuk *et al.*, in press).

An important factor that conditions the technique of myogenic-cell implantation is that the intracellular proteins coded by a single nucleus into a myofiber do not diffuse throughout the syncytium. On the contrary they remain localized in a region close to its nucleus of origin, this region being known as a "nuclear domain" (Pavlath *et al.*, 1989). This restriction is produced by the limited diffusion of the mRNA, which was reported to be of only 100 μm from the nucleus (Ralston and Hall, 1992), and of the proteins (Hall and Ralston, 1989). Thus, proteins from donor origin will be expressed only in the segments of myofibers where fusion of donor's myogenic cells was produced. The more-or-less wide expression of the exogenous protein will depend on its capacity to diffuse or to remain anchored to stationary cellular components (Hall and Ralston, 1989). As an example, single injections of (β-galactosidase-labeled normal myoblasts into *mdx* mice produced dystrophin expression throughout segments of roughly 500 μm in the myofibers in contrast with 1500 μm for β-galactosidase (Kinoshita *et al.*, 1998). The wider expression of β-galactosidase may be attributed to the solubility of this enzyme, thus being able to diffuse more than dystrophin, which remains attached to stationary cellular components.

Formation of New Myofibers

In DMD and other myopathies, increasing muscle weakness is produced by a progressive and irreversible loss of myofibers. An ideal treatment for patients in advanced stages of these diseases may include not only molecular correction but also restoration of muscle mass. The potential of myoblast transplantation to restore functional skeletal muscle mass in mice was reported following acute severe muscle damage, principally when damage was irreversible. Myoblast implantation significantly restored the muscle mass and force, lost after skeletal muscle destruction (Alameddine *et al.*, 1994; Wernig *et al.*, 1995, 2000; Irintchev *et al.*, 1997). However, these models were not strictly similar to the conditions of a progressive muscular dystrophy, and the manner in which myogenic-cell

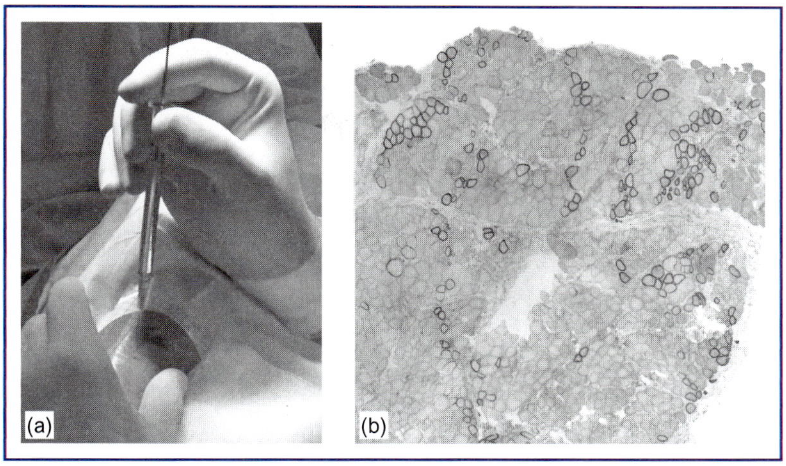

FIGURE 26-1 Allotransplantation of normal adult myoblasts in DMD patients, as done in our recent clinical trial. Several parallel intramuscular cell injections were done in the tibialis anterior using a 100 μl Hamilton syringe and 25–27-gauge needles (a). The cells were delivered homogeneously during the needle withdrawal, and the density of cell injections was controlled by placing on the skin a sterile transparent dressing with a grid. A cross-section of a biopsy done 1 month later in a cell-grafted site in one of these patients is shown (b). Fluorescent immunodetection of dystrophin was done, but the negative of the original image is shown for clarity. Most of the myofiber profiles are dystrophin-negative and can be seen due to the presence of some background, while others show an immunolabeling in their periphery, corresponding to the normal location of dystrophin. The distribution of these dystrophin-positive myofibers, created by the cell graft, follows roughly the original injection trajectories, aligned from the top to the bottom of the image.

transplantation could form new functional tissue in muscles that degenerated to fibrosis and/or fat substitution (as is the case in DMD) remains insufficiently studied. Some studies in mice suggest that it could be possible to create myotubes within the adipose tissue (Satoh *et al.*, 1992). Neo-muscles can be formed ectopically after subcutaneous myoblast implantation in mice, in spite of the absence of a previous endomysial support (Irintchev *et al.*, 1998). In *mdx* mice, formation of new dystrophin-positive myofibers through the fusion of the implanted myoblasts among themselves was observed following irradiation of the recipient muscle (Kinoshita *et al.*, 1996b). Progressing from those observations to a clinically functional procedure remains a challenge, among other factors because these results were obtained in mice, which have intrinsically a greater muscle regeneration capacity than primates (Borisov, 1999). However, a recent clinical observation encouraging this research was the presence of neo-formed dystrophin-positive small myofibers in DMD patients transplanted with normal myoblasts (Skuk *et al.*, in press).

Formation of Donor-Derived Satellite Cells

Mouse studies showed that some of the myoblasts injected into skeletal muscles remain as mononuclear myogenic cells, able to participate later in muscle regeneration (Yao and Kurachi, 1993; Gross and Morgan, 1999) and that they give rise specifically to new satellite cells (Heslop *et al.*, 2001). Some observations suggest that this phenomenon could be produced also in humans. Donor-derived mononuclear cells were detected in the muscles of DMD patients, which received myoblast transplantations (Gussoni *et al.*, 1997; Skuk *et al.*, in press), and some of these nuclei were observed in locations susceptible to correspond to satellite cells (Skuk *et al.*, in press). Similar observations were made following human myoblast transplantation in immunodeficient mice (Brimah *et al.*, 2004). This means that the potential therapeutic effect of myogenic-cell implantation is not limited to the early fusion of the implanted cells, and should also ensure a permanent source of normal satellite cells able to participate later in muscle hypertrophy and regeneration.

Undesirable Properties to be Avoided

The risk of implanting cells capable of developing a neoplasm should be considered in cultured cell transplantation strategies. Rhabdomyosarcomas were observed following myogenic-cell transplantation in mice, but only when permanent cell lines were used for implantation (Wernig *et al.*, 1991,1995). Although cell lines are often tumorigenic, the propagation of primary cultures into cell lines is difficult (Freshney, 1987), thus the risk of tumorigenicity following primary cultured cells can be considered very low. Indeed, neoplasia was never observed in muscles of monkeys transplanted with primary-cultured myoblasts (Kinoshita *et al.*, 1995, 1996a; Skuk *et al.*, 1999b, 2000, 2002). However, to take precautions in clinical trials of myoblast transplantation, we tried to exclude a tumorigenic potential in the cells to be implanted by using two tests (Skuk *et al.*, 2004). One of them was an *in vitro* assay comparing the growth of the donor's cells in soft-agar medium with that of a rhabdomyosarcoma cell line (Tremblay *et al.*, 1991). In this assay, normal cells survive without proliferation, while rhabdomyosarcoma cells proliferate in clusters. An *in vivo* test was also used, which consisted of grafting the donor's cells in muscles of immunodeficient mice (Huard *et al.*, 1994; Skuk *et al.*, 1999a). This confirmed that the donor's cells were able to fuse *in vivo*, and eliminated a tumorigenic potential.

CELL IMPLANTATION

Once appropriate cells for transplantation are isolated and possibly proliferated, the following challenge is to deliver them appropriately to the target tissue. Intramuscular injection remains so far the only method of myogenic-cell delivery that uniformly and reproductively gives rise to hybrid myofibers in human and non-human primates.

Density of Cell Injections

The main factor conditioning the strategy of intramuscular cell injections is that the injected cells fuse mainly with the myofibers damaged by the injection trajectories. This is clearly observed in non-human primates, where each single myoblast injection leads to a narrow track of hybrid myofibers (Figures 26-2 and 26-3) (Skuk *et al.*, 2000, 2002). An almost similar pattern, although less clear, was observed following injections of normal myoblasts in DMD patients (Figure 26-1) (Skuk *et al.*, 2004; Skuk *et al.*, in press). Since the expression of donor-derived proteins is limited to nuclear domains, cell injections must be very close to each other and must reach the whole muscle to obtain an important and homogeneous expression of donor-derived proteins throughout a skeletal muscle. Experiments in monkeys showed that the volume of muscle expressing a donor's protein is proportional to the density of cell injections (Skuk *et al.*, 2002) (Figure 26-3). The percentage of myofiber profiles expressing (β-galactosidase 1 month after the intramuscular injection of (β-galactosidase-labeled myoblasts was of 6–15% when the density of cell injections was 25/cm^2, and 23–67% when it was 100/cm^2 (Skuk *et al.*, 2002). The highest percentages (up to 26%) of dystrophin-positive myofibers observed following normal-myoblast allotransplantation in DMD patients were obtained also when 100 cell injections/cm^2 were done (Skuk *et al.*, in press). Such a protocol of cell implantation was denominated "high-density injections" (Skuk, 2004) in order to establish a difference from those used in the former unsuccessful clinical trials, which performed few and distant injections throughout large skeletal muscles.

Risks of the Procedure

A protocol of high-density injections involves risks that need to be determined in order to be avoided. These risks could be local and systemic and, according to the experience in non-human primates, should be limited to the first days post-implantation.

Locally, a monkey's biceps brachium becomes distended the day post-transplantation but reaches its pre-transplantation diameter after 5 days (Skuk *et al.*, 2000). This results in a risk of developing a compartment syndrome in muscles enclosed in a rigid osteofascial space. The biceps brachium of monkeys tolerates well this treatment, but muscles such as the tibialis anterior may probably need to be injected in different sessions.

Systemically, an extensive muscle damage (rhabdomyolysis) releases intracellular metabolites such as myoglobin and potassium. This implies risks of acute cardiac arrhythmia in the case of severe hyperkalemia and acute renal failure if myoglobinuria is produced. Both phenomena were not observed following high-density cell injections in the biceps brachii of monkeys (Skuk *et al.*, 2000). This problem, thus, may be controlled by maintaining the muscle damage for a single session of cell transplantation under the limits of potentially dangerous. As an example, using high-density injections of myoblasts throughout two biceps brachii of a monkey produced an increase of 2000U/l in serum creatine kinase levels (Skuk *et al.*, 2000), while the risk of developing an acute renal failure is considered to be produced at creatine kinase levels of 16,000 U/l (Ward, 1988).

Trying to Improve the Efficiency of Cell Injections

Lower densities of cell injections are desirable but, to reach this objective, the volume of muscle expressing the therapeutic protein (e.g. dystrophin) following a single cell injection must be increased. This could be obtained by: (1) developing methods allowing the implanted cells to fuse with myofibers other than those reached by the injection and/or (2) increasing the nuclear domain of the therapeutic protein. The last possibility was rarely investigated: only one study in *mdx* mice reported a three-fold increase in the nuclear domain of dystrophin after transplantation of myoblasts overexpressing dystrophin fifty-folds (Kinoshita *et al.*, 1998).

Two factors explain why the implanted cells fuse mainly with the myofibers reached by the injection: they lack the capacity to move through the tissue and/or there is absence of myofiber damage out of the injection sites; this damage triggering a regeneration process permitting the fusion of the transplanted cells with the damaged fibers. Some mouse studies aimed to promote the diffusion of the grafted myoblasts throughout the tissue, generally by inducing the secretion of enzymes degrading the extracellular matrix (Ito *et al.*, 1998; Caron *et al.*, 1999; El Fahime *et al.*, 2002). Other experiments have been done to increase the number of regenerating myofibers in order to favor the uptake of

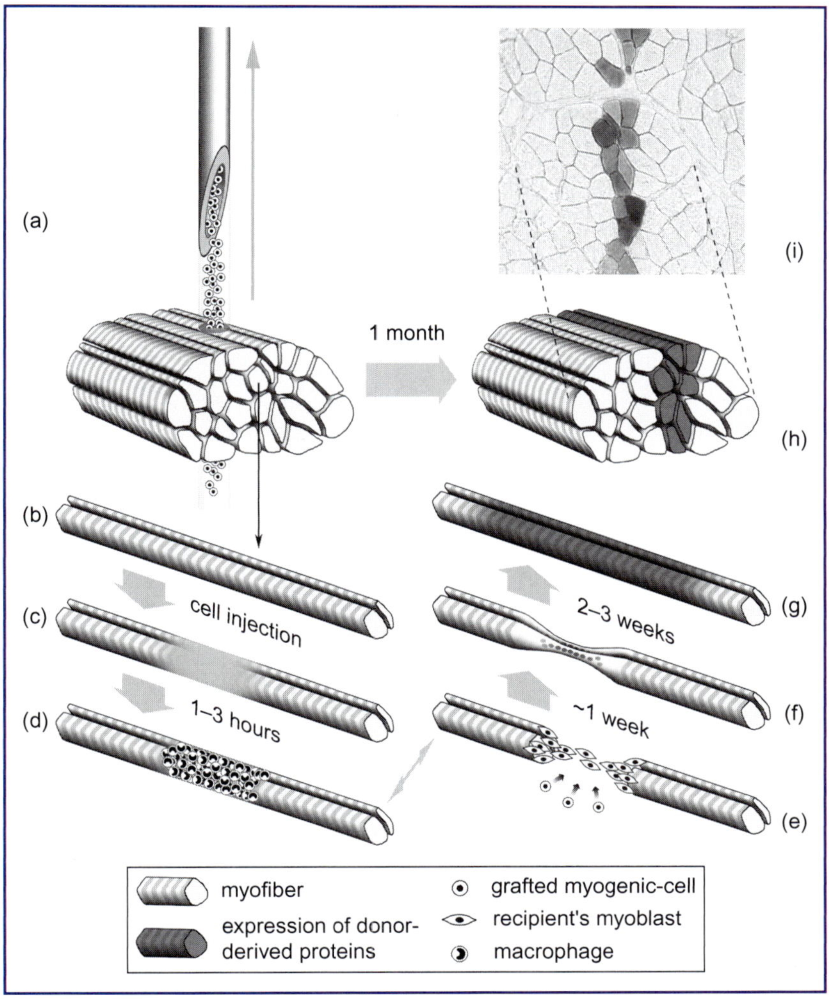

FIGURE 26-2 A representation of the mechanism allowing the incorporation of the grafted myogenic cells into the recipient's myofibers in monkeys is shown. In monkeys, the donor's cells are labeled by introducing a gene coding for β-galactosidase, and the result of the cell transplantation 1 month later (i) is observed through the expression of β-galactosidase into the recipient myofibers (dark staining within the myofibers). A single cell injection, traversing a skeletal muscle fascicle and delivering the cells homogeneously during the needle withdrawal is represented (a). The process of grafted cell uptake into the myofibers is represented in a couple of myofibers isolated from this fascicle (b-g). These two myofibers (b) are physically damaged by the needle and suffer a segmental necrosis (c). This region of necrosis is invaded by circulating monocytes (d), which became macrophages with two main functions: to phagocyte the debris of the necrosed segment and to release factors that help the regenerative process in the myofiber. This regeneration is done by the activation of the recipient satellite cells, which proliferate as myoblasts, migrating to the center of the region being "cleaned" by the macrophages in order to fuse together (e). The grafted myogenic cells placed in the proximity will be recruited in this process (e). The nuclei of the grafted cells recruited into this fusing process will be integrated in the new myotubes that fill the gap lead by the necrosed segment (f), and will later allow the expression of donor-derived proteins throughout a restricted length of the myofiber (j). This process leads basically to restricted regions of donor-protein expression in the fascicle (h), which will be expressed as "tracks" of β-galactosidase expression in the non-human primate muscles (i).

the grafted cells. For example, local injection of myotoxic substances, such as phospholipases derived from snake venoms (Kinoshita *et al.*, 1994; Vilquin *et al.*, 1995) and local anesthetics (Cantini *et al.*, 1994; Pin and Merrifield, 1997), were used efficiently in mice. Inhibiting the capacity of the recipient's satellite cells to proliferate could favor the participation of the grafted myoblasts to the regeneration of the myofibers. This was obtained in mice by submitting the recipient muscle to high doses of ionizing radiation prior to cell transplantation (Morgan *et al.*, 1990; Alameddine *et al.*, 1994; Kinoshita *et al.*, 1994; Vilquin *et al.*, 1995; Wernig *et al.*, 2000). Cryoinjury of the recipient muscle necroses myofibers and satellite cells, and was also used in mice as a pre-treatment to favor the implanted cells (Wernig *et al.*, 1995; Irintchev *et al.*, 1997; Brimah *et al.*, 2004). With the exception of local anesthetics, it seems difficult to expect that the other procedures would be accepted for human use, and even in this case increasing muscle damage would reduce the volume of muscle to be treated in a single session, considering the risks of threatening rhabdomyolysis.

So far, only the co-injection of myoblasts and of myotoxic phospholipases improved the success of myoblast

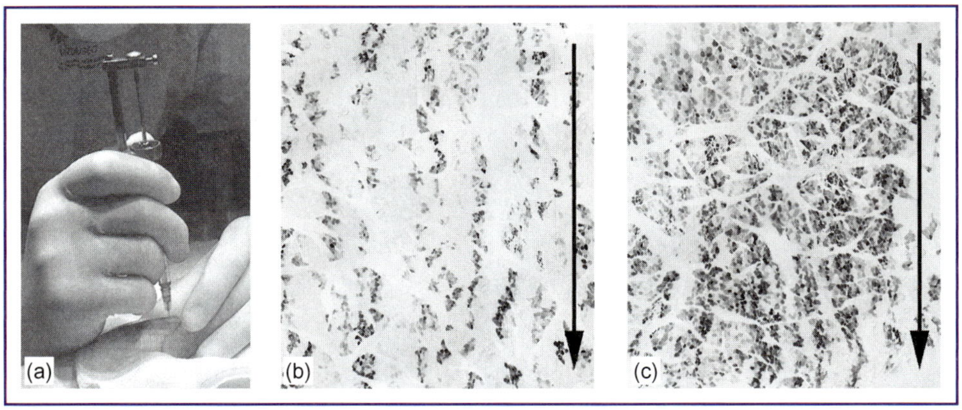

FIGURE 26-3 Transplantation of adult β-galactosidase-labeled myoblasts in non-human primates. The cells are delivered by parallel close intramuscular injections using Hamilton syringes and 27-gauge needles (a). In the figure, the syringe is attached to a repetitive dispenser to accelerate the procedure without loss of precision. As in humans, the density of cell injections is controlled by placing on the skin a sterile transparent dressing with a grid. One month later, the fusion of the grafted cells with the recipient's myofibers is analyzed by histochemical detection of β-galactosidase on histological cross-sections at the cell-grafted sites (b and c). The distribution of the β-galactosidase-positive myofibers follows the pattern of the original cell-injection trajectories (indicated by the arrows). The density of β-galactosidase-positive myofibers is higher in "c" than in "b," because the density of cell injections was higher: 25/cm^2 in "b" against 100/cm^2 in "c".

transplantation in non-human primates. However, this improvement was only observed when cells and the myotoxin were highly concentrated in a small volume of muscle (Skuk *et al.*, 1999b, 2000).

CELL SURVIVAL IN THE RECIPIENT

Once a good delivery of the cells is obtained, their survival in the recipient must be ensured. The post-transplantation survival of myogenic cells should be analyzed at two periods: early and long term.

Early Survival

The prevailing evidences in the field of myogenic-cell transplantation are that most myogenic cells die quite rapidly (during the first 2–3 days) after their intramuscular implantation, independently of the specific immune response. This phenomenon does not prevent the success of myoblast transplantation, because not all cells die (Beauchamp *et al.*, 1999; Skuk *et al.*, 2003) and the proliferation of the surviving cells compensates totally (Skuk *et al.*, 2003) or partly (Beauchamp *et al.*, 1999) the cell death. The process of death and proliferation of the grafted cells during the first days post-transplantation is not well understood, and the studies approaching the subject show contradictions, probably caused by methodological differences (Skuk *et al.*, 2003). Some observations in mice implicated the acute inflammatory reaction in killing the implanted cells (Guerette *et al.*, 1997), however others challenged this hypothesis (Sammels *et al.*, 2004). It was also postulated that the survival of the whole population of grafted myoblasts could be due to a special small subpopulation of cells that specifically avoid the early cell death and proliferate a great deal (Beauchamp *et al.*, 1999; Cousins *et al.*, 2004), but the existence of this subpopulation was not demonstrated, and the hypothesis does not identify the mechanisms responsible for the early cell death.

Long-Term Survival

The principal challenge of the long-term survival of myogenic-cell grafts in primates is acute rejection, obvious in inadequately immunosuppressed allotransplantations (Kinoshita *et al.*, 1996a; Skuk *et al.*, 2000). Acute rejection in the context of myogenic-cell transplantation was extensively studied in mice, since the first description of lymphocyte infiltration and disappearance of the grafted myoblasts soon after allogeneic transplantation (Jones, 1979). Subsequent studies identified CD8+ and CD4+ lymphocytes in these infiltrates (Guerette *et al.*, 1995a; Irintchev *et al.*, 1995; Wernig and Irintchev, 1995) and expression of IL-2 receptors, Th-1 cytokine, and granzyme B (Guerette *et al.*, 1995b, 1996). Non-immunosuppressed or insufficiently immunosuppressed monkeys also exhibited CD4+ and CD8+ infiltration following myoblast allotransplantation, with lymphocyte invasion of myofibers expressing donor proteins (Kinoshita *et al.*, 1996a; Skuk *et al.*, 1999b, 2002).

Ensuring Cell Survival in the Recipient

The phenomenon of the early death among grafted cells is presently misunderstood and so far cannot be significantly prevented. However, since this mechanism does not devastate the population of implanted cells (which seems well restored by the proliferation of the surviving cells) the only potential benefit of inhibiting it would be, in theory, a reduction in the number of cells to be injected.

Acute rejection, on the other hand, precludes the success of myogenic-cell transplantation in allogeneic conditions. Acute rejection is controlled in humans by pharmacological immunosuppression, but a careful selection of the immunosuppressive drug is required for myogenic-cell transplantation, because some of them kill and/or inhibit the differentiation of the grafted cells (for a review, see Skuk and Tremblay, 2003; Skuk, 2004). The best results of myoblast allotransplantation in mice were reported using tacrolimus (Kinoshita *et al.*, 1994), and for this reason this drug became the immunosuppressant of choice for myoblast allotransplantation in monkeys (Kinoshita *et al.*, 1995, 1996a; Skuk *et al.*, 1999b, 2000, 2002) and humans (Skuk *et al.*, 2004; Skuk *et al.*, in press).

Since pharmacological immunosuppression has severe secondary effects, one of the main objectives in clinical transplantation is to develop long-term specific unresponsiveness to grafts with preservation of immune reactions against other foreign antigens (immune tolerance). In the context of myogenic-cell transplantation, immune tolerance developed in some mouse strains after a transient immunosuppression (Pavlath *et al.*, 1994), while in others acute rejection was delayed for months (Pavlath *et al.*, 1994; Wernig *et al.*, 1995). The relevance of these observations is relative, because immune tolerance is more easily obtained in mice than in monkeys or humans. In fact, withdrawal of immunosuppression in monkeys caused acute rejection of the myoblast graft (Skuk *et al.*, 2000). Specific protocols to develop immune tolerance in the context of myoblast transplantation are under investigation (Camirand *et al.*, 2002, 2004). Otherwise, another approach proposed to avoid immunosuppression is the autotransplantation of myoblasts genetically corrected *ex vivo* (e.g. by introducing the dystrophin gene in myoblasts of DMD patients). Mouse experiments support a future development of this approach (Floyd *et al.*, 1998; Moisset *et al.*, 1998).

CONCLUSIONS

To reach a therapeutic objective, cell transplantation needs basically three conditions: the use of an appropriate cell for implantation, a good method to deliver it to the target tissue, and a method to prevent the specific immune response thus insuring their long-term survival in the recipient in allogenic conditions. Non-human primates were used to define these three conditions in a model appropriate for human extrapolation. Concerning the first item, myoblasts derived from the satellite cells constitute so far the only myogenic cells that are easily isolated and proliferated *in vitro*, and successfully implanted in the skeletal muscles. Concerning the second item, intramuscular implantation through high-density injections is the only method that proved so far to give rise to high percentages of hybrid myofibers in the skeletal muscles. Finally, an appropriate immunosuppression (tacrolimus-based) is the only method successfully tested in monkeys to ensure the survival of myoblast allografts. These three parameters have permitted to consistently restore the normal expression of dystrophin in many myofibers of patients suffering an inherited myopathy (DMD).

Some important challenges remain, and the potential treatment requires further improvements. A main challenge is to reduce the density of cell injections needed for an efficient distribution of the grafted cells throughout a skeletal muscle. Another challenge overpasses the specific field of myogenic-cell transplantation (it is one of the main problems to solve in the global field of transplantation) and is to reduce as much as possible the toxicity of the methods needed to control acute rejection. The identification of the factors that condition the early survival (death and proliferation) of the grafted cells still challenges the researchers in the field. Finally, the capacity to restore the functional parenchyma in skeletal muscles that degenerated to fibrosis and/or fat substitution (the only possibility to give hope of recovering muscle force in advanced myopathic patients) remains unsolved and barely studied.

REFERENCES

Alameddine, H.S., Louboutin, J.P., Dehaupas, M., Sebille, A. and Fardeau, M. (1994). Functional recovery induced by satellite cell grafts in irreversibly injured muscles. *Cell Transplant. 3*: 3–14.

Beauchamp, J.R., Morgan, J.E., Pagel, C.N. and Partridge, T.A. (1999). Dynamics of myoblast transplantation reveal a discrete minority of precursors with stem cell-like properties as the myogenic source. *J. Cell Biol. 144*: 1113–1122.

Betz, E.H., Firket, H. and Reznik, M. (1966). Some aspects of muscle regeneration. *Int. Rev. Cytol. 19*: 203–227.

Borisov, A.B. (1999). Regeneration of skeletal and cardiac muscle in mammals: do nonprimate models resemble human pathology? *Wound Repair. Regen. 7*: 26–35.

Brimah, K., Ehrhardt, J., Mouly, V., Butler-Browne, G.S., Partridge, T.A. and Morgan, J.E. (2004). Human muscle precursor cell regeneration in the mouse host is enhanced by growth factors. *Hum. Gene Ther. 15*: 1109–1124.

Camirand, G., Caron, N.J., Turgeon, N.A., Rossini, A.A. and Tremblay, J.P. (2002). Treatment with anti-CD154 antibody and donor-specific transfusion prevents acute rejection of myoblast transplantation. *Transplantation 73*: 453–461.

Camirand, G., Rousseau, J., Ducharme, M.E., Rothstein, D.M. and Tremblay, J.P. (2004). Novel Duchenne muscular dystrophy treatment through myoblast transplantation tolerance with anti-CD45RB, anti-CD154 and mixed chimerism. *Am. J. Transplant. 4*: 1255–1265.

Cantini, M., Massimino, M.L., Catani, C., Rizzuto, R., Brini, M. and Carraro, U. (1994). Gene transfer into satellite cell from regenerating muscle: bupivacaine allows beta-Gal transfection and expression *in vitro* and *in vivo*. *In Vitro Cell Dev. Biol. Anim.* 30A: 131–133.

Caron, N.J., Asselin, I., Morel, G. and Tremblay, J.P. (1999). Increased myogenic potential and fusion of matrilysin-expressing myoblasts transplanted in mice. *Cell Transplant. 8*: 465–476.

Cousins, J.C., Woodward, K.J., Gross, J.G., Partridge, T.A. and Morgan, J.E. (2004). Regeneration of skeletal muscle from transplanted immortalised myoblasts is oligoclonal. *J. Cell Sci. 117*: 3259–3269.

El Fahime, E., Mills, P., Lafreniere, J.F., Torrente, Y. and Tremblay, J.P. (2002). The urokinase plasminogen activator: an interesting way to improve myoblast migration following their transplantation. *Exp. Cell Res. 280*: 169–178.

Floyd Jr., S.S., Clemens, P.R., Ontell, M.R., Kochanek, S., Day, C.S., Yang, J., Hauschka, S.D., Balkir, L., Morgan, J., Moreland, M.S., *et al.* (1998). *Ex vivo* gene transfer using adenovirus-mediated full-length dystrophin delivery to dystrophic muscles. *Gene Ther. 5*: 19–30.

Gross, J.G. and Morgan, J.E. (1999). Muscle precursor cells injected into irradiated mdx mouse muscle persist after serial injury. *Muscle Nerve 22*: 174–185.

Guerette, B., Asselin, I., Vilquin, J.T., Roy, R. and Tremblay, J.P. (1995a). Lymphocyte infiltration following allo- and xeno-myoblast transplantation in mdx mice. *Muscle Nerve 18*: 39–51.

Guerette, B., Roy, R., Tremblay, M., Asselin, I., Kinoshita, I., Puymirat, J. and Tremblay, J.P. (1995b). Increased granzyme B mRNA after alloincompatible myoblast transplantation. *Transplantation 60*: 1011–1016.

Guerette, B., Tremblay, G., Vilquin, J.T., Asselin, I., Gingras, M., Roy, R. and Tremblay, J.P. (1996). Increased interferon-gamma mRNA expression following alloincompatible myoblast transplantation is inhibited by FK506. *Muscle Nerve 19*: 829–835.

Guerette, B., Skuk, D., Celestin, F., Huard, C., Tardif, F., Asselin, I., Roy, B., Goulet, M., Roy, R., Entman, M., *et al.* (1997). Prevention by anti-LFA-1 of acute myoblast death following transplantation. *J. Immunol. 159*: 2522–2531.

Gussoni, E., Blau, H.M. and Kunkel, L.M. (1997). The fate of individual myoblasts after transplantation into muscles of DMD patients. *Nat. Med. 3*: 970–977.

Hall, Z.W. and Ralston, E. (1989). Nuclear domains in muscle cells. *Cell 59*: 771–772.

Heslop, L., Beauchamp, J.R., Tajbakhsh, S., Buckingham, M.E., Partridge, T.A. and Zammit, P.S. (2001). Transplanted primary neonatal myoblasts can give rise to functional satellite cells as identified using the Myf5(nlacZl+) mouse. *Gene Ther. 8*: 778–783.

Huard, J., Bouchard, J.P., Roy, R., Malouin, F., Dansereau, G., Labrecque, C., Albert, N., Richards, C.L., Lemieux, B. and Tremblay, J.P. (1992). Human myoblast transplantation: preliminary results of 4 cases. *Muscle Nerve 15*: 550–560.

Huard, J., Verreault, S., Roy, R., Tremblay, M. and Tremblay, J.P. (1994). High efficiency of muscle regeneration after human myoblast clone transplantation in SCID mice. *J. Clin. Invest. 93*: 586–599.

Irintchev, A., Zweyer, M. and Wernig, A. (1995). Cellular and molecular reactions in mouse muscles after myoblast implantation. *J. Neurocytol. 24*: 319–331.

Irintchev, A., Langer, M., Zweyer, M., Theisen, R. and Wernig, A. (1997). Functional improvement of damaged adult mouse muscle by implantation of primary myoblasts. *J. Physiol. (Lond.) 500*: 775–785.

Irintchev, A., Rosenblatt, J.D., Cullen, M.J., Zweyer, M. and Wernig, A. (1998). Ectopic skeletal muscles derived from myoblasts implanted under the skin. *J. Cell Sci. 111*: 3287–3297.

Ito, H., Hallauer, P.L., Hastings, K.E. and Tremblay, J.P. (1998). Prior culture with concanavalin A increases intramuscular migration of transplanted myoblast. *Muscle Nerve 21*: 291–297.

Jones, P.H. (1979). Implantation of cultured regenerate muscle cells into adult rat muscle. *Exp. Neurol. 66*: 602–610.

Karpati, G., Ajdukovic, D., Arnold, D., Gledhill, R.B., Guttman, R., Holland, P., Koch, P.A., Shoubridge, E., Spence, D., Vanasse, M., *et al.* (1993). Myoblast transfer in Duchenne muscular dystrophy. *Ann. Neurol. 34*: 8–17.

Kikuchi, T, Doerr, L. and Ashmore, CR. (1980). A possible mechanism of phenotypic expression of normal and dystrophic genomes on succinic dehydrogenase activity and fiber size within a single myofiber of muscle transplants. *J. Neurol. Sci. 45*: 273–286.

Kinoshita, I., Vilquin, J.T., Guerette, B., Asselin, I., Roy, R. and Tremblay, J.P. (1994). Very efficient myoblast allotransplantation in mice under FK506 immunosuppression. *Muscle Nerve 17*: 1407–1415.

Kinoshita, I., Vilquin, J.T., Gravel, C., Roy, R. and Tremblay, J.P. (1995). Myoblast allotransplantation in primates. *Muscle Nerve 18*: 1217–1218.

Kinoshita, I., Roy, R., Dugre, F.J., Gravel, C., Roy, B., Goulet, M., Asselin, I. and Tremblay, J.P. (1996a). Myoblast transplantation in monkeys: control of immune response by FK506. *J. Neuropathol. Exp. Neurol. 55*: 687–697.

Kinoshita, I., Vilquin, J.T. and Tremblay, J.P. (1996b). Mechanism of increasing dystrophin-positive myofibers by myoblast transplantation: study using mdx/beta-galactosidase transgenic mice. *Acta Neuropathol. 91*: 489–493.

Kinoshita, I., Vilquin, J.T., Asselin, I., Chamberlain, J. and Tremblay, J.P. (1998). Transplantation of myoblasts from a transgenic mouse overexpressing dystrophin produced only a relatively small increase of dystrophin-positive membrane. *Muscle Nerve 21*: 91–103.

Konigsberg, I.R. (1960). The differentiation of cross-striated myofibrils in short term cell culture. *Exp. Cell Res. 21*: 414–420.

Leriche-Guerin, K., Anderson, L.V., Wrogemann, K., Roy, B., Goulet, M. and Tremblay, J.P. (2002). Dysferlin expression after normal myoblast transplantation in SCID and in SJL mice. *Neuromuscul. Disord. 12*: 167–173.

Lipton, B.H. and Schultz, E. (1979). Developmental fate of skeletal muscle satellite cells. *Science 205*: 1292–1294.

Mendell, J.R., Kissel, J.T., Amato, A.A., King, W., Signore, L., Prior, T.W., Sahenk, Z., Benson, S., McAndrew, P.E., Rice, R., *et al.* (1995). Myoblast transfer in the treatment of Duchenne's muscular dystrophy. *N. Engl. J. Med. 333*: 832–838.

Moisset, P.A., Skuk, D., Asselin, I., Goulet, M., Roy, B., Karpati, G. and Tremblay, J.P. (1998). Successful transplantation of genetically corrected DMD myoblasts following *ex vivo* transduction with the dystrophin minigene. *Biochem. Biophys. Res. Commun. 247*: 94–99.

Morgan, J.E., Hoffman, E.P. and Partridge, T.A. (1990). Normal myogenic cells from newborn mice restore normal histology to degenerating muscles of the mdx mouse. *J. Cell Biol. 111*: 2437–2449.

Partridge, T.A., Grounds, M. and Sloper, J.C. (1978). Evidence of fusion between host and donor myoblasts in skeletal muscle grafts. *Nature 273*: 306–308.

Partridge, T.A., Morgan, J.E., Coulton, G.R., Hoffman, E.P. and Kunkel, L.M. (1989). Conversion of mdx myofibres from dystrophin-negative to -positive by injection of normal myoblasts. *Nature 337*: 176–179.

Pavlath, G.K., Rich, K., Webster, S.G. and Blau, H.M. (1989). Localization of muscle gene products in nuclear domains. *Nature 337*: 570–573.

Pavlath, G.K., Rando, T.A. and Blau, H.M. (1994). Transient immunosuppressive treatment leads to long-term retention of allogeneic myoblasts in hybrid myofibers. *J. Cell Biol. 127*: 1923–1932.

Pin, C.L. and Merrifield, P.A. (1997). Developmental potential of rat L6 myoblasts *in vivo* following injection into regenerating muscles. *Dev. Biol. 188*: 147–166.

Ralston, E. and Hall, Z.W. (1992). Restricted distribution of mRNA produced from a single nucleus in hybrid myotubes. *J. Cell Biol. 119*: 1063–1068.

Sammels, L.M., Bosio, E., Fragall, C.T., Grounds, M.D., van Rooijen, N. and Beilharz, M.W. (2004). Innate inflammatory cells are not responsible for early death of donor myoblasts after myoblast transfer therapy. *Transplantation 77*:1790–1797.

Satoh, A., Labrecque, C. and Tremblay, J.P. (1992). Myotubes can be formed within implanted adipose tissue. *Transplant. Proc. 24*: 3017–3019.

Skuk, D. (2004). Myoblast transplantation for inherited myopathies: a clinical approach. *Expert Opin. Biol. Ther. 4*: 1871–1885.

Skuk, D. and Tremblay, J.P. (2003). Myoblast transplantation: the current status of a potential therapeutic tool for myopathies. *J. Muscle Res. Cell Motil. 24*: 285–300.

Skuk, D., Furling, D., Bouchard, J.P., Goulet, M., Roy, B., Lacroix, Y., Vilquin, J.T., Tremblay, J.P. and Puymirat, J. (1999a). Transplantation of human myoblasts in SCID mice as a potential muscular model for myotonic dystrophy. *J. Neuropathol. Exp. Neurol. 58*: 921–931.

Skuk, D., Roy, B., Goulet, M. and Tremblay, J.P. (1999b). Successful myoblast transplantation in primates depends on appropriate cell delivery and induction of regeneration in the host muscle. *Exp. Neurol. 155*: 22–30.

Skuk, D., Goulet, M., Roy, B. and Tremblay, J.P. (2000). Myoblast transplantation in whole muscle of nonhuman primates. *J. Neuropathol. Exp. Neurol. 59*: 197–206.

Skuk, D., Goulet, M., Roy, B. and Tremblay, J.P. (2002). Efficacy of myoblast transplantation in nonhuman primates following simple intramuscular cell injections: toward defining strategies applicable to humans. *Exp. Neurol. 175*: 112–126.

Skuk, D., Caron, N.J., Goulet, M., Roy, B. and Tremblay, J.P. (2003). Resetting the problem of cell death following muscle-derived cell transplantation: detection, dynamics and mechanisms. *J. Neuropathol. Exp. Neurol. 62*: 951–967.

Skuk, D., Roy, B., Goulet, M., Chapdelaine, P., Bouchard, J.P., Roy, R., Dugré, F.J., Lachance, J.G., Deschênes, L., Senay, H., *et al.* (2004). Dystrophin expression in myofibers of Duchenne muscular dystrophy patients following intramuscular injections of normal myogenic cells. *Mol. Ther. 9*: 475–482.

Skuk, D., Goulet, M., Roy, B., Chapdelaine, P., Bouchard, J.P., Roy, R., Dugré, F.J., Lachance, J.G., Deschênes, L., Senay, H., *et al.* Dystrophin expression in muscles of Duchenne muscular dystrophy patients following high-density injections of normal myogenic cells. *J. Neuropathol. Exp. Neurol.* (In press)

Tremblay, J.P., Roy, B. and Goulet, M. (1991). Human myoblast transplantation: a simple assay for tumorigenicity. *Neuromuscul. Disord. 1*: 341–343.

Tremblay, J.P., Malouin, F., Roy, R., Huard, J., Bouchard, J.P., Satoh, A. and Richards, C.L. (1993). Results of a triple blind clinical study of myoblast transplantations without immunosuppressive treatment in young boys with Duchenne muscular dystrophy. *Cell Transplant. 2*: 99–112.

Vilquin, J.T., Asselin, I., Guerette, B., Kinoshita, I., Roy, R. and Tremblay, J.P. (1995). Successful myoblast allotransplantation in mdx mice using rapamycin. *Transplantation 59*: 422–426.

Vilquin, J.T., Kinoshita, I., Roy, B., Goulet, M., Engvall, E., Tome, F., Fardeau, M. and Tremblay, J.P. (1996). Partial laminin alpha2 chain restoration in alpha2 chain-deficient dy/dy mouse by primary muscle cell culture transplantation. *J. Cell Biol. 133*: 185–197.

Ward, M.M. (1988). Factors predictive of acute renal failure in rhabdomyolysis. *Arch. Intern. Med. 148*: 1553–1557.

Watt, D.J., Lambert, K., Morgan, J.E., Partridge, T.A. and Sloper, J.C. (1982). Incorporation of donor muscle precursor cells into an area of muscle regeneration in the host mouse. *J. Neurol. Sci. 57*: 319–331.

Wernig, A. and Irintchev, A. (1995). "Bystander" damage of host muscle caused by implantation of MHC-compatible myogenic cells. *J. Neurol. Sci. 130*: 190–196.

Wernig, A., Irintchev, A., Hartling, A., Stephan, G., Zimmermann, K. and Starzinski-Powitz, A. (1991). Formation of new muscle fibres and tumours after injection of cultured myogenic cells. *J. Neurocytol. 20*: 982–997.

Wernig, A., Irintchev, A. and Lange, G. (1995). Functional effects of myoblast implantation into histoincompatible mice with or without immunosuppression. *J. Physiol. (Lond.) 484*: 493–504.

Wernig, A., Zweyer, M. and Irintchev, A. (2000). Function of skeletal muscle tissue formed after myoblast transplantation into irradiated mouse muscles. *J. Physiol. (Lond.) 522*: 333–345.

Yablonka-Reuveni, Z. and Nameroff, M. (1990). Temporal differences in desmin expression between myoblasts from embryonic and adult chicken skeletal muscle. *Differentiation 45*: 21–28.

Yao, S.N. and Kurachi, K. (1993). Implanted myoblasts not only fuse with myofibers but also survive as muscle precursor cells. *J. Cell Sci. 105*: 957–963.

Chapter 27

Cell-Based Therapies for Musculoskeletal Repair

Wan-Ju Li, Kiran Gollapudi, David P. Patterson, George T.-J. Huang, and Rocky S. Tuan

INTRODUCTION

Common musculoskeletal disorders include osteoarthritis (OA), rheumatoid arthritis (RA), intervertebral disk (IVD) degeneration, anterior cruciate ligament (ACL) injuries, and muscular dystrophy. These diseased conditions can result from trauma, work-related injuries, immunological malfunction, aging, or genetics. Arthritis is highly prevalent among adults. As the population ages, arthritis is expected to affect an estimated 67 million adults in the United States by 2030, according to the report by Centers for Disease Control and Prevention (CDC, 2006). Approximately 13–16 million people are diagnosed with OA per year in the US or Europe, and joints are replaced due to OA at the rate of one every 1–2 min. A total of ~300,000 and ~500,000 joint replacements are performed per year in Europe and the US, respectively. Although OA is not normally life threatening, it is progressive, disabling, and can greatly impact an individual's quality of life.

Current approaches to musculoskeletal care emphasize prevention, medical treatment, and surgical intervention as a final resort. Progress has been made in the past decades with biological therapeutic approaches to reverse or slow disease progression by specifically targeting molecules involved in the disease process. Despite advances in these approaches, once tissue damage reaches a certain stage, self-repair does not take place. Particularly with cartilage, the repair mechanism is almost non-existent. Currently, there is no available remedy to repair the eroded cartilage in the arthritic joints and the degenerated IVD. While joint replacements provide significant functional restoration and symptomatic improvement, they are compromised by the possibility of prosthetic failure and associated complications (e.g. peri-implant osteolysis).

The emerging disciplines of cell-based therapy and tissue engineering have suggested the prospect of regenerative medicine as a promising approach to the treatment of damaged and diseased musculoskeletal tissues. This chapter will outline mesenchymal cell biology in the context of musculoskeletal tissues, including osteogenec, chondrogenec, myogenec, tenogenec, and ligament cell types. Particular emphasis will be put on the definition of mesenchymal stem cells (MSCs), their niches, isolation and *in vitro* characterization, lineage differentiation and regulation, and immunomodulatory properties. Cell-based applications using these cells to produce specific tissues will be reviewed, covering clinical disorders, gene therapy approaches, *in vivo* studies, and current applications, as well as potential pitfalls and future improvement.

BIOLOGY OF CELLS IN MUSCULOSKELETAL TISSUES

Osteoblasts and Osteocytes

Bone provides structural support for the body, facilitates movement, protects internal organs, and acts as a mineral reserve. Derived from embryonic mesoderm, bone contains a matrix consisting of hydroxyapatite mineral and macromolecular extracellular matrix (ECM) components. Osteoblasts (OB) are the bone forming cells found on all surfaces of bone. They produce osteoid consisting of collagen type I, fibronectin, proteoglycans, and other specialized proteins, which mineralizes by the deposition of hydroxyapatite crystals $[Ca_{10}(OH)_2(PO_4)_6]$. Osteocytes are former OB that have become encased by the bone matrix and are the most abundant cells in mature bone. Their presence highlights the fact that bone is a living, dynamic tissue. Osteocytes cease to produce osteoid but communicate with other cells (osteocytes and OB) through small pores called canaliculi, and are believed to play a role in mechanotransduction activities of bone with feedback to the remodeling process.

Osteoprogenitors, located in the bone marrow and periosteum, can be induced to become OB via growth factors, such as the transforming growth factor-β (TGF-β) superfamily and in particular the bone morphogenetic proteins (BMPs). BMPs regulate chemotaxis, mitosis and differentiation, and are critical in initiating fracture healing. The transcription factors Cbfa1/Runx2 and Osterix are essential for OB differentiation in both intramembranous and endochondral ossification. Runx2 is involved in activating gene expression of collagen type I and other bone proteins, such as osteopontin and osteocalcin. Osterix is believed to be downstream of Runx2 as it is not expressed in Runx2 null mice. Other transcription factors shown to be involved in OB proliferation and differentiation include Msx and Dlx proteins. However, mice containing null alleles for these genes do produce bone, unlike Runx2 and Osterix null mice (Eames *et al.*, 2003).

Chondrocytes

There are three main types of cartilage: hyaline cartilage, elastic cartilage, and fibrocartilage. The external ear is an example of elastic cartilage, and the meniscus and annulus fibrosus of the IVD, which will be discussed later, are examples of fibrocartilaginous tissues. Articular cartilage is an example of hyaline cartilage, the most common type. The major function of articular cartilage is to provide a smooth surface for reduced friction and to support large loads during movement. Articular cartilage consists of a fluid phase composed of water and electrolytes as well as a solid phase consisting of ECM and chondrocytes. Chondrocytes, making up less than 10% by volume of articular cartilage, are the only cell type in the tissue and are responsible for the maintenance of the ECM. Collagen type II is the predominant collagen in articular cartilage and is responsible for its tensile strength; however, other minor collagens, such as collagen types V, VI, IX, and X, are also present. The most abundant component of articular cartilage is water (60–85% by volume), which is held in place by the highly charged proteoglycans, and allows for its compressive behavior. Thus, collagens and proteoglycans provide cartilage with its mechanical properties, and their content is a key component in assessing the functional quality of engineered articular cartilage.

During developmental chondrogenesis, mesenchymal cells are recruited and migrate to areas of chondrogenesis, and subsequent mesenchymal–epithelial cell and cell–cell matrix interactions promote cellular condensation. Mesenchymal cells then differentiate into chondrocytes, heralding the deposition of ECM proteins and the activation of transcription factors such as Sox-9. Sox-9 is required for the expression of cartilage-specific collagen type II in normal skeletal development (Bi *et al.*, 2001). L-Sox-5 and Sox-6, which are other members of the Sox family, cooperate with Sox-9 to turn on the collagen type II gene and are also essential for cartilage formation (Lefebvre and Smits, 2005).

Muscle Cells

Muscle is a specialized contractile tissue, derived from embryonic mesoderm and allows for force exertion and locomotion. There are three types of muscle: skeletal (voluntary) muscle found in the musculoskeletal system, smooth muscle (involuntary) found within walls of organs and structures, and cardiac muscle, a specialized tissue found only in the heart. Skeletal muscle cell differentiation from embryonic mesoderm is driven by transcription factors MyoD, Myf5, myogenin, Mrf4, and MEF2 in a highly coordinated fashion (Sartorelli and Caretti, 2005). Skeletal muscle is composed of bundles of dense muscle fibers that are highly oriented and able to generate longitudinal contraction. These bundles (muscle cells) are multinucleated and derived from myoblasts (muscle cell precursors). Satellite cells are specialized myoblast populations capable of muscle regeneration. They can be obtained and cultured from muscle biopsy. Upon initiation by local growth factors, these normally quiescent and undifferentiated cells become mitotic, differentiate, and eventually self-assemble into muscle fibers themselves. M-Cadherin, Pax7, and neural cell adhesion molecule (NCAM) are known satellite cell markers that can be used to localize and follow satellite cells *in vivo*. Pax7 null mice lacked satellite cells but retained a unique population of interstitial stem cells in muscle that express the stem cell markers, CD 34, and Sca-I (Tamaki *et al.*, 2002). These stem cells, tracked by expression of green fluorescent protein (GFP), were found to originate in the bone marrow (Dreyfus *et al.*, 2004).

Tendon and Ligament Cells

Tendons and ligaments are dense fibrous structures that connect muscle to bone and bone to bone, respectively. Both are composed mostly of collagen type I, produced by specialized elongated fibroblasts (known as tenocytes in tendons) that lie between the collagen fibers. These tissues are very hypocellular compared to other connective tissues, presenting problems for injury repair. Both arise from mesodermal compartments distinct from those that give rise to myogeneic cells. Not many cellular markers have been identified, but the transcription factor Scleraxis appears promising as a specific and early marker for tendons and ligaments (Tozer and Duprez, 2005).

EMBRYONIC AND ADULT STEM CELLS

Embryonic Stem Cells

Embryonic stem (ES) cells are pluripotent cells, derived experimentally from the inner cell mass of the embryonic blastocyst. Human ES (hES) cells are typically obtained from 4- to 5-day-old blastocysts of embryos after *in vitro* fertilization. These cells are potentially immortal *in vitro* without loss of differentiation potential, and when reimplanted into a host embryo, they give rise to pluripotent daughter cells that differentiate into all tissue types. The use of hES cells for research and clinical applications is complicated by controversies surrounding the legal and ethical status of human embryos and is currently restricted by regulations on federal funding. Despite these challenges, both mouse and hES cells have been examined for applications in musculoskeletal regeneration, albeit limited.

Human ES cell differentiation and organization can be influenced by a supportive three-dimensional (3D) environment such as poly(lactic-co-glycolic acid) (PLGA) polymer scaffolds and directed by growth factors such as retinoic acid, TGF-β, activin-A, or insulin-like growth factor (IGF). These growth factors induce differentiation into 3D structures with characteristics of developing neural tissues, cartilage, or liver, respectively (Levenberg *et al.*, 2003). hES cells have been injected into the joint space of immunocompromised rats to promote cartilage repair (Wakitani *et al.*, 2004). Osteogeneic potential of hES cells in the presence of chemical stimuli *in vitro* has also been demonstrated (Karp *et al.*, 2006). In a mouse model, chondrogene differentiation was observed using mES cells via embryoid bodies (EBs) modulated by members of the TGF-β family (TGF-β1, BMP-2 and -4) (Kramer *et al.*, 2000). mES cells differentiate into chondrocytes, which progressively develop into hypertrophic and calcifying cells. At a terminal differentiation stage, cells expressing an OB-like phenotype appear either by transdifferentiation from hypertrophic chondrocytes or directly from OB precursor cells. Under the influence of ascorbic acid, β-glycerophosphate, and 1,25-dihydroxy vitamin-D_3, mES cells are induced to differentiate *in vitro* into OB that produce a mineralized matrix. In 3D scaffold systems such as hydrogels, both hES and mES cells were shown to have chondrogene capacity (Elisseeff *et al.*, 2006).

Mesenchymal Stem Cells

While ES cells hold promise as a cell source for tissue regeneration, ethical issues and the possibility of teratoma formation need to be resolved before they can be used in clinical applications. Notably, our body has other stem cell populations that possess the capabilities of self-renewal and multidifferentiation to repair damaged tissues or maintain tissue homeostasis when the repair is needed. With our improved knowledge of stem cells combined with advances in culture techniques, it is possible that we can harness their potential for the treatment of degenerative musculoskeletal diseases. Unlike pluripotent ES cells which have the ability to form tissues from all three germ layers, adult stem cell populations are believed to be limited in their differentiation capacity. In general, stem cells derived from a particular tissue are programmed to differentiate into various progenies that belong to the same developmental germ layer origin. Although recent studies suggest that adult stem cells can differentiate across germ line boundaries, it remains debatable if differentiation plasticity between the cells of different germ layers actually exists or is simply an artifact resulting from contamination of heterogeneous cell populations or cell fusion. More evidence is needed to assess the possibility of differentiation across the three germ layers. However, plasticity within a germ layer is more strongly supported by evidence from several research groups. The ethically acceptable nature of adult stem cells, combined with proven differentiation abilities, as compared to ES cells and tissue-committed cells, makes them an attractive option for use in cell-based therapy.

For cell-based musculoskeletal tissue regeneration, adult MSCs are an attractive candidate progenitor cell type since they may be isolated from various adult tissues, and can differentiate into different mesenchymal lineage cells, such as bone, cartilage, fat, muscle, ligament, tendon, and stroma (reviewed by Tuan *et al.*, 2003) (Figure 27-1). Reflecting their origin and cell functions, these cells have also been named or described as marrow stromal cells, or mesenchymal stromal cells, which are all abbreviated as MSCs. MSCs were first discovered by a German pathologist in the 1860s and he described the cell morphology as "fibroblast-like." More than a century later, in 1976, Friedenstein further identified MSCs as colony-forming unit-fibroblasts, which were able to commit to osteogene differentiation. The multidifferentiation potential of MSCs

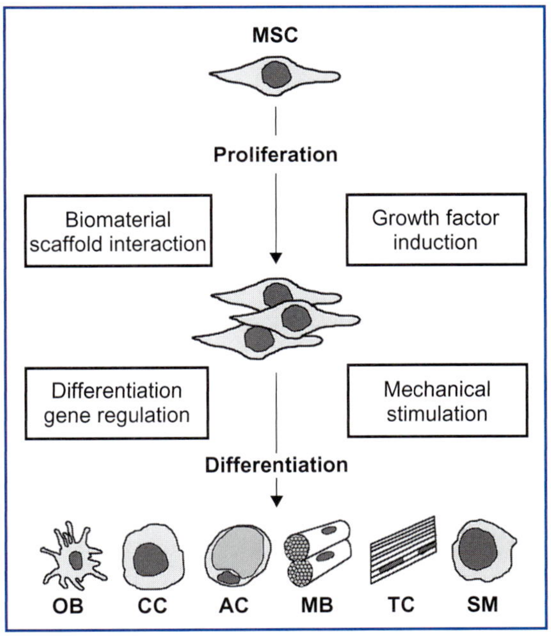

FIGURE 27-1 Multidifferentiation potential of MSCs and factors regulating their biological activities *in vitro*. CC, chondrocyte; AC, adipocyte; MB, myoblast; TC, tenocyte; SM, stroma.

was demonstrated *in vivo* as early as in 1980, in which MSCs were induced to become bone and cartilage. Caplan (1991) and Pittenger *et al.* (1999) showed that MSCs underwent osteogene, chondrogene, and adipogene differentiation in response to different biochemical signals. Numerous subsequent studies have produced a significant body of evidence demonstrating the phenotypic and functional characteristics of MSCs. According to the statistic data retrieved from Medline, research publications containing the key word "MSC" have increased 15-fold in the past 5 years, highlighting the increasing interest in MSC studies. While many studies are focused on characterization and differentiation potential of MSCs, few are concentrated on the molecular regulation of MSCs. Future MSC research should focus on the intrinsic and extrinsic mechanisms mediating the molecular switch between undifferentiated and differentiated MSCs.

MSC Isolation

MSCs can be isolated from a variety of mesenchymal tissues such as bone marrow, fat, trabecular bone, cartilage, muscle, peripheral blood, and umbilical cord blood (reviewed by Tuan *et al.*, 2003). Dependent on the species and tissue types, different isolation protocols and culture methods have been developed. Among these MSC sources, bone marrow is the best studied tissue. The isolation process for bone marrow-derived MSCs includes several steps aimed at reducing contamination by other cell types. Erythrocytes can be removed by density gradient centrifugation using Percoll or Ficoll after bone marrow aspirate is obtained from the iliac crest, tibia, or femur. Fluorescence-activated cell sorting (FACS) and magnetic-activated cell sorting (MACS) are sometimes used to select a more defined cell population based on cell surface markers. With their ability to adhere to tissue culture plastic, MSCs are further discriminated from non-adherent hematopoietic cells after several medium changes. The low frequency of 1 MSC per 10,000–100,000 bone marrow cells *in vitro* indicates that the MSC is a rare cell population (Pittenger *et al.*, 1999). Previous studies have shown that MSC yield is affected by age and health of a donor. The trend is that MSC yield decreases with donor age. Patients with degenerative diseases, such as osteoporosis and OA, tend to have lower MSC yield. Unfortunately, it is this group of people who would benefit most from MSC-based treatment. Therefore, an alternative could be the use of allogeneic MSCs. Although immune reaction is a concern associated with using allogeneic cells, the finding that MSCs have low immunogeneic potential as well as immunosuppressive properties suggests that this concern may not be significant. Immunoregulation of MSCs will be discussed in detail later.

Another MSC source gaining recent attention is adipose tissue, since fat is abundant, easy to access, and considered surgical waste during a cosmetic surgery operation. The procedure of marrow harvesting is painful, and over-harvesting bone marrow may be a risk to the patient's health. In contrast, liposuction is less painful and considered a relatively safe procedure. The method to isolate MSCs is similar to a typical primary cell isolation, in which MSCs are released from collagenase-digested fat and purified after being plated in plastic culture to remove unattached hematopoietic cells. A study comparing MSC proliferation between bone marrow and fat tissue suggests that fat tissue is a more effective source than bone marrow for MSC yield (Lee *et al.*, 2004).

As mentioned above, bone marrow-derived MSC number, proliferation, and differentiation potential may decrease with the donor age. A possible solution is to isolate MSCs from tissue of a younger donor. Fetal tissues, such as umbilical cord blood (Mareschi *et al.*, 2001), cord vein (Romanov *et al.*, 2003), placenta (Fukuchi *et al.*, 2004), and amniotic fluid (In't Anker *et al.*, 2003), have been processed to isolate MSCs. A comprehensive study by Kern *et al.* (2006) in which they compared the morphology and functions of MSCs isolated from bone marrow, adipose tissue, and umbilical cord blood showed that MSCs isolated from these three tissue sources are morphologically and immunophenotypically similar. However, umbilical cord blood-derived MSCs form the fewest colonies and show the highest proliferation capacity, whereas adipose tissue-derived MSCs form the greatest number of colonies and bone marrow-derived MSCs show the lowest proliferation capacity. The findings suggest both adipose tissue and umbilical cord blood can be used as alternatives to bone marrow for MSC isolation.

Recently, various tissue-specific adult stem cells have been successfully isolated and characterized from non-mesenchymal tissues. These cells from ectoderm (epidermis) or endoderm origin (pancreas) morphologically and functionally behave similarly to bone marrow-derived MSCs. Although the lack of specific markers makes verification of their identity as MSCs difficult, and their phenotype could be a result of *in vitro* culture conditions, the possibility of MSCs residing in different tissues throughout the body raises a number of interesting questions. For example, how and where do these cells reside in the different tissues? Is there a common tissue pool, such as bone marrow, that houses the MSCs which are able to integrate to different tissues in response to specific biological needs or activities?

In Vitro MSC Behavior

After being plated in culture, MSCs adhere to the substratum and start to divide, forming colonies. The typical growth pattern is that MSCs experience a few days of a lag phase before undergoing a log phase of growth and then reach a stationary phase. Generally, these cells can undergo 50 population doublings in 10 weeks without losing their multidifferentiation capability. Colter *et al.* (2000) reported that the initial cell seeding density affects the expansion capacity and doubling time of mouse MSCs. When plated at low initial plating density (1.5–3 cells/cm^2), MSCs have 2,000-fold expansion in 10 days, whereas when plated at high density (12 cells/cm^2), the cell number only increases 60-fold in the same period of time. In addition, the doubling time increases from 12 to 24 h with high cell seeding density. MSCs are a heterogeneous population, demonstrated by the finding that there are two MSC morphologies in a colony; small spindle-shaped and large flat fibroblast-like cells. Interestingly, a third type of extremely small, rapidly self-renewing (RS) cells was recently identified in MSC colonies. During the three phases of growth, RS cells give rise to large flat cells in the log phase. It is believed that the large flat MSCs are more mature cells, replicating slowly, whereas small RS cells proliferate rapidly.

MSC Identification

Despite a significant number of studies, to date, there is no specific marker available for the identification of MSCs. There have been some cell surface markers available for MSC identification, but none are exclusive to MSCs. Because MSCs can not be positively identified, their biological activities *in vivo* cannot be verified as with hematopoietic stem cells (HSCs). Our knowledge of MSCs is thus primarily from *in vitro* experimental results. The *in vitro* results of MSC identification are likely dependent on culture environment and purity of cell population, which could explain the differences between the findings reported in the literature. Nevertheless, the use of multiple markers, such as cell surface cluster of differentiation (CD) markers, ECM proteins, cell adhesion molecules, integrins, and cytokines as well as genetic or proteomic fingerprinting can help one identify MSCs.

CD cell surface molecules are the most commonly used markers to identify MSCs. Both positive and negative CD markers have been used to identify MSCs and exclude endothelial cells, HSCs, and hematopoietic lineage cells. Positive MSC markers include Stro-1, SH2 (CD105), SH3 (CD73), SH4 (CD73), CD 29, CD 44, CD 54, CD 90, CD 105, CD 133, CD 166, and p75LNGFR, whereas negative markers are CD 11, CD 14, CD 19, CD 31, CD 34, CD 45, CD 79, and HLA-DR (Deans and Moseley, 2000). To shorten the list, the International Society for Cellular Therapy (ISCT) has provided minimum criteria for defining MSCs (Dominici *et al.*, 2006). Acceptable MSCs meet the minimum requirements of CD 73, CD 90, and CD 105 positive and CD 14, CD 34, CD 45, and HLA-DR negative expression.

MSC Niche

The term MSC niche refers to both the physical and chemical environments that MSCs reside in, which includes other cell types, ECM molecules directly or indirectly contacting with MSCs, and soluble factors regulating MSC activities. The cellular and non-cellular components interact with each other in this highly complex 3D environment, responsible for the maintenance of MSC stemness properties as well as the regulation of symmetric and asymmetric cell division. The concept of the stem cell niche was first introduced in the 1970s (Schofield, 1978) and has been elucidated by *in vitro* co-culture experiments. Ball *et al.* (2004) demonstrated that, by co-culturing MSCs and endothelial cells, MSCs were induced to a phenotype similar to smooth muscle cells, whereas MSCs became myofibroblast-like cells when co-cultured with dermal fibroblasts. Their study suggests that the interaction between MSCs and their neighboring cells may regulate the fate of MSCs and that the type of progeny of MSCs may be determined by the interacting cells in the niche.

The bone marrow microenvironment is a principal MSC niche in the body. It is not only a complex 3D structure, but also allows for many interactions between cellular and non-cellular components, such as HSCs, MSCs, stroma, hematopoietic cells, mesenchymal origin cells, ECM components, growth factors, and cytokines. Tellingly, the success rate of HSC engraftment improves when co-transplanting with MSCs *in vivo*. The *ex vivo* expansion of HSCs increases dramatically when MSCs are co-cultured with HSCs, suggesting that HSCs and MSCs maintain a close biological interaction in the naïve marrow niche. The signals needed for MSC proliferation and differentiation come from soluble factors as well as both cell–cell and cell–matrix interactions. OB have been known to play an important role in the regulation of hematopoiesis but their role in MSC osteogenesis still remains inconclusively defined. A previously published report shows that OB have synergistic interactions on MSC proliferation and alkaline phosphatase activity but not calcium deposition (Kim *et al.*, 2003). In comparison, our preliminary results show that MSC co-culturing with OB enhances their osteogene differentiation (unpublished observation).

MSC–ECM interactions, both physical and chemical, are likely to be critical in the regulation of MSC physiology in the niche. Matsubara *et al.* (2004) cultured and maintained MSCs on basement membrane-like ECM and observed profound effects on MSC proliferation and differentiation. With the support of basement membrane-like ECM, MSCs better maintain their multidifferentiation potential after many cell divisions, which suggests that the interactions between MSCs and basement membrane-like ECM may recapitulate some of the MSC–ECM interactions in bone marrow. Cell–ECM interactions are also the center of the study in biomaterial-based cell therapy and tissue engineering. Artificial ECM, a biomaterial scaffold used to replace damaged or malfunctioning ECM, is designed to fully function as native ECM and interact with MSCs for successful tissue regeneration. The interactions of MSC and a biomaterial scaffold will be discussed later.

MSC Regulation

Soluble factors such as growth factors and cytokines play a significant role in physiological regulation of MSCs in the niche. These biochemical signals guide MSCs either to stay as undifferentiated cells or to differentiate into tissue-specific progenitor cells by activating specific signal pathways. A recent important finding is the involvement of the Wnt signaling pathway. Gregory *et al.* (2003) demonstrate that MSCs enter the cell cycle and inhibit osteogeneic differentiation after Dickkopf-1 (Dkk-1) deactivates the Wnt pathway. Boland *et al.* (2004) further identify that Wnt 3a working through the canonical pathway promotes MSC proliferation but discourages osteogenesis, whereas Wnt 5a via the non-canonical pathway promotes osteogenesis. Chondrogenesis of MSCs also involves the activation of the Wnt signaling pathway (Tuli *et al.*, 2003). During chondrogene differentiation of MSCs, TGF-β1 activates the mitogen activated protein (MAP) kinase pathway which is demonstrated to be involved in cross-talk signaling with the Wnt signaling pathway. The regulation of the Wnt signaling pathway likely induces chondrogenesis by enhancing cell–cell interaction through N-cadherin expression.

The most important and valuable characteristic of MSCs is their multipotential differentiation capacity. Previous studies have shown that MSC differentiation can be induced and regulated by soluble signal factors, both protein- and non-protein-based molecules. TGF-βs and BMPs induce MSCs to undergo chondrogenesis in a serum-free medium. Each member of the TGF-β family has a different level of induction efficiency. TGF-β3 has a higher efficiency in inducing chondrogenesis compared to TGF-β1 but they both contribute similarly to chondrogenesis in long-term culture. To induce osteogenesis, β-glycero-2-phosphate, ascorbic acid, dexamethasone, and 1,25-dihydroxy vitamin-D_3 are required to enhance alkaline phosphatase activity and matrix mineralization. For adipogenesis, MSCs are treated with isobutyl-1-methylxanthine and insulin, resulting in adipocytes with the presence of lipid droplets in the cytoplasm (Pittenger *et al.*, 1999).

In addition to biochemical factors, physical factors, such as mechanical loading as well as matrix geometry and elasticity, have also been found to play a role in MSC biology. Cells generally receive and transduce physical cues from the surrounding environment to the cell nucleus through cytoskeletal changes or signaling pathways. McBeath *et al.* (2004) demonstrated that cell shape regulates commitment of MSCs by using a micropatterned substrate to control cell shape and size of cultured MSCs. They found that cell shape is a key regulator in MSC differentiation with the shape-dependent control of lineage commitment mediated by ROCK-mediated cytoskeletal tension. MSCs forced to spread and flatten in the large substrate pattern differentiate into OB, whereas those forced to unspread and become round in the small pattern differentiate into adipocytes. Recently, Engler *et al.* (2006) demonstrated the effect of matrix stiffness on differentiation of MSCs, mediated by non-muscle myosin II. MSCs on a soft matrix with stiffness similar to brain stiffness differentiate into neurons, on a matrix with intermediate stiffness close to muscle stiffness become myoblasts, and on a stiff matrix comparable to bone commit to OB. In addition, they also show that typical soluble factors for lineage commitment do not alter the differentiation lineage previously activated by matrix stiffness. This study suggests matrix stiffness appears to be important in MSC lineage commitment.

MSC Immunoregulation

Immunoregulation by and of MSCs can be viewed from two perspectives: (1) immunosuppressive effects of allogeneic MSCs, and (2) inflammatory cytokine effect on MSC activity and differentiation. Due to the interest in using allogeneic or xenogeneic MSCs to compensate for the paucity and time constraints associated with expanding autologous MSCs, there has been considerable progress in the understanding of the MSC immunoregulatory effect. While xenogeneic MSCs are rejected by the host after transplantation, allogeneic MSCs are well tolerated by the recipient hosts. Many *in vivo* studies have confirmed the immunosuppressive effects of MSCs (Chen XI *et al.*, 2006). The potential mechanisms underlying this immunosuppression can be explained by downregulation of T, dendritic, natural killer (NK), and B cells. This immunosuppressive characteristic suggests that MSCs can be potentially used *in vivo* for enhancing the engraftment of other tissues (e.g. HSCs), or for prophylactic prevention and even possibly as a treatment of graft-versus-host-disease or autoimmune diseases such as RA (Jorgensen *et al.*, 2003).

Limited information is available on the effects of pro-inflammatory cytokines on MSCs. Liu and Hwang (2005) demonstrated that human cord blood-derived MSCs secrete cytokines and growth factors. Most importantly, continuous supplementation of IL-1β in the cord blood-derived-MSC culture facilitates adipogenec maturation of cord blood-MSCs. A preliminary study using porcine MSCs showed that interferon-α-2b may act to differentiate MSCs into OB (Abukawa *et al.*, 2006). Our recent study suggests that MSCs are relatively resilient to pro-inflammatory cytokines in terms of apoptotic response (Okafor *et al.*, 2006). In the context of autologous implantation for chondrogenesis, a study has shown that inflammatory reactions against scaffold materials and serum components lead to the production of cytokines such as IL-1α that may inhibit cartilage tissue formation (Rotter *et al.*, 2005). These few studies suggest the importance of understanding the effect of tissue inflammation, either caused by diseases or in response to scaffold materials after implantation, on the differentiation and cell behavior of MSCs.

CELL-BASED THERAPIES AND TISSUE ENGINEERING

Tissue Engineering

Three general strategies have been adopted for the engineering of new tissues: (1) delivery of cells, (2) local or systemic delivery of tissue inducing agents, and (3) delivery of biomaterial scaffolds containing both cells and inductive agents. The use of a biomaterial scaffold-based carrier facilitates the delivery of therapeutic cells or agents to the target site and subsequent growth and regeneration of new tissue. A general strategy is that target cells (differentiated/undifferentiated) expanded *in vitro* are cultured in 3D, biomaterial scaffolds (natural/synthetic) under conditions that favor the desired phenotype. After the appropriate culture period, the cell-seeded composite exposed to biological and physical stimuli develops into a natural tissue-like cellular construct (Figure 27-2). The biomaterial scaffolds are believed to play a critical role in the tissue engineering process, by providing a 3D structure for cellular functions such as attachment, migration, proliferation, and differentiation. The ultimate success of this process is determined by the biological and functional similarity of the engineered tissue to native tissue.

Despite the promising prospects of tissue engineering, regenerating tissues that serve a predominantly biomechanical function, such as bone, articular cartilage, and tendon, presents significant challenges. The concept of "functional tissue engineering" has emerged, specifically prescribing the production of a tissue that meets functional and in our case especially, mechanical requirements of the repair tissue. This requires a clear delineation of the structure and functions of living tissues, design of biomaterial scaffolds, and mechanical optimization of regenerative tissues. It is necessary to incorporate each of these steps to regenerate a mechanically sound tissue for safer and more efficacious repairs and replacements for the patient.

Gene Therapy

Gene therapy involves the gene-based modification of cells for the correction of defective genes or the regulation of gene expression. It has a great potential in cell therapy-based treatments due to the fact that a number of musculoskeletal diseases are caused by genetically malfunctioning cells; thus successful tissue engineering strategies must include the repair of the "defective" cells in any engineered construct.

The self-renewal capacity and multipotentiality of MSCs suggest their suitability for cell-based and gene therapy applications in regenerative medicine. For gene therapy, viral transduction has a high efficiency of delivering genes into MSCs. Adeno-associated viral-mediated gene transfer has been tested to be effective in delivering genes into MSCs and to repair bone disorders such as osteogenesis imperfecta (Chamberlain *et al.*, 2004). Lentiviral vectors have also been shown to be effective in delivering genes into MSCs (Lu *et al.*, 2005). Non-viral methods such as transfection using Nucleofection™ has been demonstrated to be promising in delivering functional genes into MSCs (Haleem-Smith *et al.*, 2005).

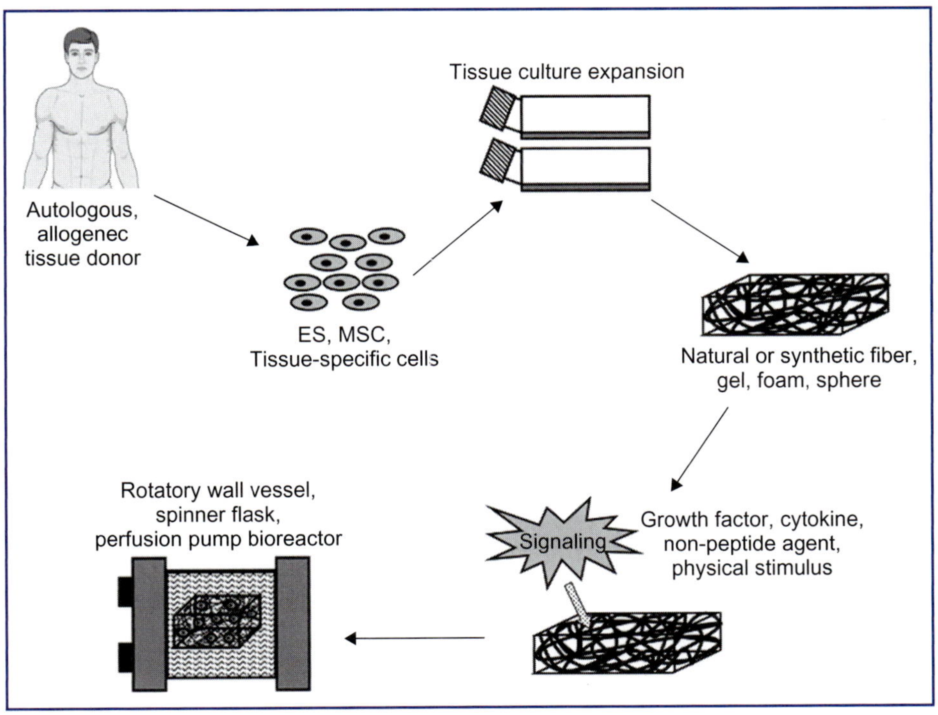

FIGURE 27-2 General strategy of cell-based therapy and tissue engineering.

We will now discuss the application of these cell-based strategies towards regenerating musculoskeletal tissues. A brief list of representative studies in the field of musculoskeletal regeneration is provided in Table 27-1.

Bone

A wide variety of patients with significant bone defects that necessitated amputation in the past now benefit from various orthopedic strategies. Congenital defects of bone, growth plate fractures and defects, fractures resulting in malunion or non-union, the genetic disorder osteogenesis imperfecta (brittle bone disease), and bone loss from tumor resection (primary bone tumors or tumors metastatic to bone) are just a handful of musculoskeletal problems that could be addressed by regenerative medicine. Currently, using orthopedic prosthetics is a severe but highly functional option. Distraction osteogenesis, a surgical procedure for bone reconstruction and lengthening, was developed in the 1950s by Ilizarov and is still used today. Bone autografting is a therapeutic option developed in the 19th century and considered to be the current gold standard, but has limitations, particularly in the size of the defect to be grafted. Autografts contain the patient's own OB and osteocytes, but require a second surgical site for the bone harvesting, most often the iliac crest of the pelvis. This increases patient morbidity such as post-operative pain and risk of infection. Allograft bone from bone banks or cadavers avoids the pitfalls of autograft bone but does not possess the strength or cellularity of auto-graft bone. As the number of surgeries requiring bone grafting continues to rise, the development of functional tissue-engineered bone grafts becomes increasingly significant.

Four critical factors to successful bone tissue engineering are osteoconduction, osteoproduction, osteoinduction, and mechanical stimulation. Osteoconduction refers to the integration of the scaffold or graft material into the site and its eventual remodeling and replacement. Osteoproduction is the production of bone material by cells, and osteoinduction is the use of growth factors that draw additional osteogene cells to the site. For both *in vivo* and *ex vivo* bone tissue, mechanical stimulation appears to be a critical factor in the development of biologically and mechanically optimal bone tissue. Tissue engineering of bone must take into account the tremendous mechanical strength

TABLE 27-1 Representative Reports of *In Vivo* Cell-Based Musculoskeletal Repair

Tissue type	Model	Cells	Strategy and observation	Reference
Bone	Rat	MSCs	Cells transduced with BMP-2 improved healing of a critical defect	Lieberman *et al.* (1999)
	Human	MSCs	Porous ceramic seeded with MSCs repaired large bone defects	Quarto *et al.* (2001)
	Human	MSCs	Direct grafting of cells to non-union defects achieved union	Hernigou *et al.* (2005)
	Human	MSCs	Cells with platelet-rich plasma were polymerized and used successfully for alveolar graft osteoplasty	Hibi *et al.* (2006)
Growth plate	Rabbit	MSCs	Direct loading of cells into growth plate defects reduced growth arrest in the tibia	Chen *et al.* (2003)
Cartilage	Human	Chondrocytes	Autologous cells injected into deep cartilage defects produced good to excellent results in 14 of 16 patients	Brittberg *et al.* (1994)
	Pig	Chondrocytes	Cells seeded in gelatin microbeads mixed with type I collagen gel repaired full-thickness cartilage defects	Chiang *et al.* (2005)
	Human	MSCs	Collagen gel containing cells implanted into cartilage defects improved arthroscopic and histologic scores	Wakitani *et al.* (2002)
Muscle	Rat	MSCs	Myogenic-induced cells were directly injected to damaged muscle, contributing to muscle repair	Dezawa *et al.* (2005)
	Dog	Mesangioblasts	Vascular delivery of wild-type mesangioblasts led to significant clinical amelioration of muscular dystrophy	Sampaolesi *et al.* (2006)
Tendon	Rabbit	MSCs	Collagen gels seeded with MSCs implanted into patellar tendon defects showed improved biomechanics	Awad *et al.* (2003)
Meniscus	Rat	MSCs	MSCs embedded in fibrin glue contributed to the healing of meniscal defects	Izuta *et al.* (2005)
Intervertebral disk	Human	Chondrocytes	Autologous cells delivered to the nucleus pulposus improved pain and disability scores	Meisel *et al.* (2006)

and elasticity of bone. For load-bearing long bones such as the femur, mechanical stability of the construct is crucial, whereas for finer tissues such as fingers or craniofacial applications, plasticity takes an increased significance. Tissue-engineered bone used for clinical applications should meet both biological and mechanical requirements.

Various scaffold strategies have been used for MSC-based bone tissue engineering. Matching the strength of bone is a leading concern, and many strategies have been employed. Fully or partially demineralized bone matrix (DBM) from processed allograft bone contains collagen, growth factors, and other proteins and has been seeded with MSCs to create promising engineered constructs. DBM shares many structural and functional similarities to autologous bone and, as expected, supported osteogene function of MSCs (Mauney *et al.*, 2004). Coral, composed mostly of calcium carbonate and with a similar structure to bone, has been seeded with periosteum as a therapeutic strategy (Vacanti *et al.*, 2001). Porous ceramics, such as those composed of tricalcium phosphate and hydroxyapatite, have been used in conjunction with MSCs to produce bone replacement tissues successfully in patients who failed traditional therapies (Quarto *et al.*, 2001). Optimization of the scaffold strategy will require understanding the mechanism of its action. DBM is capable of withstanding shear forces and does not impair elasticity in the implant, and partially mimics the autologous environment in bone, although allogeneic antigens and pathogens may not have been fully removed. Ceramics provide good osteoconductivity and good integration into the defect site by bonding to tissues without rejection or inflammatory reactions, but unfortunately lack tensile strength, limiting applications involving torsion, shear stress, or bending. Natural coral has been investigated for decades as a bone graft substitute, and is biocompatible, osteoconductive and biodegradable. Improved *ex vivo* construct manufacturing requires combining biomaterial strategies with bioreactors that can produce shear and compressive forces to provide a dynamic culture system. Dynamic culture of cell-seeded scaffolds, for example, using spinner flasks, has been shown to result in more even cell distribution and a 121% increase in cell density (Mauney *et al.*, 2004). Direct cell therapy has also been tested for musculoskeletal applications. Percutaneous autologous bone-marrow grafting, the re-introduction of aspirated bone marrow directly to the site of a non-union in the tibia, has been described in human patients with good results (Hernigou *et al.*, 2005). Growth plate (physis) injuries in children can result in shortening or angular deformity with the formation of bony bridges across the growth plate between the epiphysis and metaphysis. Direct implantation of MSCs into growth plate defects resulted in a significant reduction of growth arrest in rabbit tibia (Chen *et al.*, 2003).

Gene therapy also holds promise for bone tissue engineering. A number of strategies have been tested for bone repair. Proof of concept was established with improved healing of a critical defect in a rat femur with delivery of rat MSCs transduced with the gene for BMP-2 to the site of the defect (Lieberman *et al.*, 1999). In mice, it was demonstrated that systemically injected MSCs transduced with IGF-1 established themselves in bone marrow. The MSCs demonstrated chemotactic ability by responding to the local fracture environment and locating preferentially to the fracture site, where they also accelerated the healing process (Shen *et al.*, 2002). Gene therapy with an MSC-based vehicle is also being harnessed for the treatment of a genetic disease. Engineered adeno-associated viral vectors were successfully used to disrupt the expression of mutated collagen type I gene in MSCs derived from individuals with osteogenesis imperfecta. Subcutaneous implantation of transduced MSCs into immunodeficient mice produced improved bone matrix (Chamberlain *et al.*, 2004). If host MSCs could be augmented or replaced, future OB could then produce osteoid of higher quality. A gene therapy approach has also been developed with muscle-derived mesenchymal progenitor cells. These cells act as vehicles producing osteoinductive proteins and have been demonstrated to heal critically sized bone defects (Young *et al.*, 2002). Mouse primary myoblasts over-expressing Runx2 via aretroviral system were implanted in conjunction with collagen scaffolds in the hind legs of mice and resulted in trabecular bone growth (Gersbach *et al.*, 2006).

Future improvement of bone tissue engineering depends critically on understanding the biological signals necessary for bone induction and optimizing the pharmacokinetics of their delivery. Optimized vascularization is essential, as cell labeling experiments show a considerable loss of OB in the first week following transplantation in porous cancellous bone matrices, presumably due to suboptimal initial vascularization (Kneser *et al.*, 2006). Scaffolds that incorporate growth factors such as vascular endothelial growth factor (VEGF) as well as endothelial cells have been shown to increase vascular formation in constructs *in vivo*, but integration with the host vascular system remains a challenge (Rouwkema *et al.*, 2006).

Articular Cartilage

The demand for engineered articular cartilage arises from the prevalence of traumatic injuries and degenerative diseases of articular cartilage such as OA. Other than total joint arthroplasty, which is associated with surgical risks as well as a finite life span, current therapeutic modalities for OA patients include pharmacologic intervention, lavage, shaving, laser abrasion, drilling or microfracture of subchondral bone to stimulate healing, autologous periosteal and perichondrial grafting, autogeneic or allogeneic osteochondral transplantation, and autologous chondrocyte implantation. These procedures, although variably effective, often cannot repair cartilage to a disease-free state. With an increasing elderly population and the predicted rise in the incidence of OA, novel cell-based therapies are a promising avenue to meet the therapeutic needs of these patients.

Ultimate success of a tissue-engineered cartilage construct requires the presence of cells which can produce the proper cartilaginous ECM. In principle, fully differentiated chondrocytes are the ideal cell candidate, since they are programmed for the cartilage phenotype. Indeed, Brittberg *et al.* (1994) showed that full thickness chondral defects could be repaired using autologous chondrocytes derived from a minor load-bearing area and injected under a periosteal flap. This procedure, commonly referred to as autologous chondrocyte implantation/transplantation, is marketed as Carticel™ by Genzyme Biosurgery and is the only Food and Drug Administration (FDA) approved cell-based therapy for cartilage repair. This therapy has had promising results; however, the cost effectiveness as well as the superiority of this procedure over other available procedures is debatable. Furthermore, the use of chondrocytes is hampered by their limited availability, the potential donor site morbidity, and the propensity for chondrocytes to dedifferentiate when in monolayer culture. Alternatively, stem cells, including ES cells and MSCs, may be induced to differentiate into the chondrogene lineage. EB-derived ES cells cultured as pellets with TGF-β3 showed increased gene expression of cartilage-specific ECM markers as well as a significant increase in collagen and proteoglycan production after 14 days compared to untreated controls. These changes toward a chondrogenec phenotype were accompanied by a downregulation of hematopoietic and neural markers. MSCs derived from various sources present another stem cell source for cartilage repair. Wakitani *et al.* (2002) showed that in OA patients undergoing high tibial osteotomy, autologous bone marrow MSCs seeded in a collagen gel had histological and arthroscopic improvement compared to controls; however, there was no statistically significant clinical improvement.

A variety of both natural and synthetic scaffolds have been tested to serve as carriers for the aforementioned cells (reviewed by Kuo *et al.*, 2006). The advantage of using a carrier scaffold is demonstrated by a study by Chiang *et al.* (2005) in which chondrocytes seeded in gelatin microbeads mixed with a collagen type I gel improved repair in a porcine cartilage defect model and maintained a chondrocyte phenotype better than cells delivered alone. Examples of scaffolds that have been studied for cartilage tissue engineering include fibrous scaffolds of biodegradable polymers such as poly-glycolic acid (PGA), poly-lactic acid (PLA), and their copolymer PLGA, as well as fibrin, agarose, collagen, alginate, gelatin, poly-ethylene glycol (PEG), and hyaluronan-based gels. Combinations of these materials have also been used. However, the currently available scaffolds do not fully meet the necessary requirements and can have variable effects on cell behavior and function. Novel techniques are being developed to engineer scaffolds with biomimetic properties to optimize biomaterial framework for cartilage regeneration. For example, bovine chondrocytes cultured in a PEG-based hydrogel cross-linked with a matrix metalloproteinase (MMP) sensitive peptide, which allows for remodeling by the seeded cells, had increased expression of collagen type II and aggrecan, compared to hydrogels lacking the MMP sensitive peptide (Park *et al.*, 2004). In addition, electrospun nanofibers, structurally similar to natural ECM, have recently been developed as a novel scaffold for tissue engineering. MSCs seeded into a poly(ε-caprolactone) (PCL) nanofibrous scaffold and treated with TGF-β1 were able to differentiate into a chondrocytic phenotype (Li *et al.*, 2005b). More recently it has been shown that changing the fiber diameter in these scaffolds can influence chondrocyte proliferation and ECM matrix production, demonstrating that even slight dimensional changes in scaffolds of the same material can affect seeded cells (Li *et al.*, 2006).

Supplementation of growth medium with specific signaling molecules and growth factors is commonly used to promote cell proliferation, differentiation, and ECM production. These molecules include members of the TGF-β superfamily (TGF-βs, BMPs, and growth differentiation factors (GDFs), IGFs, platelet-derived growth factors (PDGFs), and Wnts. These factors activate intracellular signaling pathways that are presumably similar to those involved in developmental morphogenesis (reviewed by Chen FH *et al.*, 2006). Isoforms of TGF-β as well as BMPs are some of the most potent positive modulators used in tissue engineering. Although TGF-β1 has been widely used as an anabolic factor for both chondrocytes and MSCs, TGF-β2 and TGF-β3 have been shown to be superior inducers of chondrogenesis in MSCs (Barry *et al.*, 2001). Among the BMPs, BMP-2 has been shown to be more effective than other BMPs (-12 and -13, and -4 and -6) for chondrocyte and MSC-based engineered constructs, respectively (Sekiya *et al.*, 2005). Similar to scaffold design, the current trend in growth factor supplementation is to use a combination approach; however, not all combinations are favorable. For example, FGF-2 and IGF-1 showed no synergism, and actually decreased protein synthetic rate by canine articular chondrocytes seeded in a collagen type II–glycosaminoglycan scaffold (Veilleux and Spector, 2005). The challenge of applying soluble factors for cartilage regeneration is that the effects of these factors depend not only on their optimal combination but also on their amount, the timing of administration, and the target cell type.

Since cartilage is a tissue whose function and form are related to physical stimuli, recent attempts to optimize the functional properties of engineered cartilage have focused on using bioreactors to incorporate mechanical loading environments *in vitro* to mimic *in vivo* environments. Mechanical loading environments tested include dynamic deformation, hydrostatic pressure, fluid flow, and shear stress. Beneficial effects of mechanical loading on cartilage constructs have been widely reported. For example, dynamic deformational loading of cell-seeded agarose disks improved the mechanical properties, as well as the sulfated glycosamino-glycan (sGAG) and hydroxyproline contents, compared to controls (Mauck *et al.*, 2000). Recent evidence also supports a positive role for mechanical loading on MSCs. Human MSCs grown in pellet culture with TGF-β3 showed increased expression of cartilage markers in the presence of cyclic hydrostatic pressure. These positive effects were dependent on both the dose and length of time cells were exposed to loads (Miyanishi *et al.*, 2006). This study underscores the need for further studies to determine the optimal type, timing, and amount of mechanical loading, and to establish the appropriate cells, scaffolds, and soluble factors with which it should be combined in order to regenerate truly functional cartilage.

Gene therapy is gaining recognition as a tool for cartilage regeneration. Lapine articular chondrocytes transfected with plasmid vectors expressing IGF-1 were combined with alginate and implanted into a rabbit osteochondral defect model (Madry *et al.*, 2005). IGF-1 expressing implants improved articular cartilage and subchondral bone repair compared to controls. Although gene therapy strategies are promising, long-term studies are needed to evaluate repaired cartilage. It is likely that in the future, optimal cartilage regeneration will result from the combination of genetically modified cells with tissue engineering.

Meniscus

The menisci are two semilunar fibrocartilaginous structures located between the tibia and femur in the knee joint. The meniscus functions as a shock absorber, joint stabilizer, and joint lubricator. Furthermore, it has an outer vascular area which possesses the ability to repair itself and an inner avascular zone where the majority of tears occur. The cells in meniscus have both fibroblast- and chondrocyte-like properties and are termed fibrochondrocytes. Also, the cell phenotype as well as the ECM components vary between the different zones. In addition to having a different cell type, fibrocartilage differs from hyaline cartilage in that the cells predominantly secrete collagen type I and it has a relatively lower sGAG content giving it different mechanical properties. Despite these differences, many attempts at meniscal regeneration have incorporated some of the same scaffolds and growth factors used in articular cartilage repair and have been tested in a variety of animal models. Recent evidence has pointed to the potential for the use of MSCs in meniscal regeneration. GFP-labeled rat MSCs were seeded in fibrin glue and used to treat meniscal defects (Izuta *et al.*, 2005). GFP-positive cells were detected up to 8 weeks post implantation and promoted meniscal repair. Successful cell-based meniscal repair would be of substantial therapeutic value, but further optimization of regenerative conditions and factors is still needed. Another challenge for meniscus tissue engineering is to regenerate a meniscus exhibiting significant anisotropic mechanical properties reflective of a highly oriented underlying ECM. We recently applied electrospinning technology to fabricate a biomaterial scaffold mimicking meniscal ECM fiber alignment, which directs fibrochondrocyte orientation and has controllable, anisotropic properties. These findings suggest the potential application of aligned-nanofiber-based scaffolds for meniscal tissue engineering (Li *et al.*, 2007).

Osteochondral Tissue

Defects of both articular cartilage and the underlying subchondral bone are often associated with pain and joint instability, risk factors for the development of OA. Using a tissue-engineered osteochondral graft is a promising alternative to the current use of autologous osteochondral grafts which are limited by tissue availability, lack of appropriate geometric configuration, and donor site morbidity. Engineering an osteochondral graft has been challenging due to the technical difficulties involved in generating a single unit consisting of two tissues with different properties, which naturally require different conditions for development and optimal functionality. General approaches to osteochondral tissue engineering have involved choosing a scaffold for the bone layer alone, two separate scaffolds for the bone and cartilage layer, or a single scaffold for both layers and combining these scaffolds with one or two cell sources. These cell sources may have either chondrogene, osteogene, or bipotential differentiation capacity. The disadvantage to using two separate scaffolds or cells is that there may be impaired integration between the osteo and chondral components. Various approaches have been employed to circumvent this problem. For instance, a pellet of trabecular bone-derived mesenchymal progenitor cells previously induced to undergo chondrogenesis was press-coated onto a PLA scaffold (Tuli *et al.*, 2004). Progenitor cells from the same patient which were induced to undergo osteogenesis were then seeded onto the other end of the scaffold, followed by culturing under conditions that supported both osteogenesis and chondrogenesis, resulting in an interface resembling the native osteochondral junction. Ideally, a single cell source would be able to differentiate into both lineages on a single scaffold. Recent evidence that electrospun PCL nanofibrous scaffolds allow MSCs to undergo multilineage differentiation suggests that this scaffold is a promising candidate for tissue engineering osteochondral grafts (Li *et al.*, 2005a).

Skeletal Muscle

While muscle-based disorders are not as prevalent as OA or bone defects, there is a serious clinical need for muscle tissue engineering. These needs include muscle atrophy and muscular dystrophy, as well as muscle loss from trauma or surgical resection. Several disease states, such as cancer, infectious disease, heart failure and AIDS can produce a body wasting syndrome known as cachexia, in which muscle atrophy is severe. Muscular dystrophy is a group of mostly inherited neuromuscular disorders that cause muscle wasting, and can lead to death in patients with severe mutations in the dystrophin gene. For the treatment of traumatic muscle loss, free tissue transfer is an option, but autologous muscle transfer causes not only donor site morbidity, but can produce loss of function at the donor site.

Requirements of functional engineering of muscle would be the recapitulation of functional motion and integration with host connective tissues. On the cellular level, parallel alignment of fibers and integration of functional neuromuscular junctions are also important to achieve. As with other musculoskeletal tissues, mechanical stimulation is essential during myogenesis, and influences metabolic activity and gene expression, as well as fiber alignment. Also, myoblasts, similar to many types of cells, become difficult to differentiate as they are culture-expanded. Myoblast cell lines, once immortalized, cannot approximate myogenesis as well as primary myoblasts. Therefore, understanding the biology of precursor cell types, such as satellite cells and other muscle-derived stem cells, is crucial. While satellite cells act as the local regenerative cells in muscle, their limited expansion potential *in vitro* limits their current usefulness relative to MSCs. The therapeutic use of MSCs in muscle disorders was demonstrated when GFP-labeled human bone marrow-derived MSCs were induced to undergo myogenesis and then transplanted by local injection into muscles of immunosuppressed rats. Histological section 4 weeks later showed mature muscle characteristics in most GFP-positive myofibers. In addition, some of the cells appeared to become Pax7 expressing satellite cells, which could respond to local damage and contribute to muscle repair (Dezawa *et al.*, 2005). For genetic disorders such as muscular dystrophy, there have been many attempts to reintroduce a wild type dystrophin gene into muscle via cell-based *ex vivo* gene therapy. Mesangioblasts, a type of vessel-associated stem cell, from wild type canine were delivered intra-arterially in a muscular dystrophy canine model with significant recovery of dystrophin gene expression, muscle morphology, and muscle function. Autologous genetically corrected mesangioblasts were not as effective (Sampaolesi *et al.*, 2006).

As with all tissue engineering, uncovering the optimal growth factors and stimulation environment to effectively produce muscle tissue is essential. As would be expected for the tissue responsible for motor functions, mechanical stimulation is crucial in the development of *in vitro* functional muscle tissue. Mechanical forces applied *in vitro* yield significant differences in morphological and functional appearance of muscle tissues. Mean myofiber diameter and elasticity both improve, as well as the ratio of muscle fiber to ECM (Bach *et al.*, 2004). Electrical stimulation mimicking nerve stimulation during myogenesis and regeneration of injured skeletal muscle also drives differentiation. Chronic electrical stimulation can change gene expression of muscle-specific genes, with increased VEGF expression and blood flow also shown after stimulation. Due to the unique contractile requirement of skeletal muscle, optimal scaffold strategies may differ from those of bone or cartilage production. PGA meshes, collagens, and alginates have been used, as in other tissues. Acellular muscle has also been exploited as a potential scaffold, and 3 weeks after seeding with myoblasts, isometric contractile force testing revealed longitudinal contractile forces upon stimulation (Borschel *et al.*, 2004). Additionally, co-culture of fibroblasts with myoblasts has been explored as a way to produce a primary matrix.

Tendon and Ligament

Injuries to ligaments and tendons heal by forming tissues of inferior quality, due to hypocellularity of the tissue as well as scar formation that is weaker than normal tissue. Current strategies to replace tendons and ligaments consist of tissue autografts and allografts. Both have the traditional problems associated with graft technology, but additionally exhibit slow or poor functional integration to the surgical site, requiring at minimum several months for recovery. In addition, tendons and ligaments are prime examples of how in the musculoskeletal system, mechanical and structural properties are crucial to the function of the tissue. There are significant challenges in developing tissue engineering strategies, due to the fact that using autologous tenocytes can cause a tendon defect, limiting options for cell-based therapy.

MSCs have been used to produce engineered tendon and ligament tissues. Collagen gels seeded with autologous MSCs were implanted into full thickness, full length, central defects in patellar tendons of rabbits. This approach significantly improved the biomechanical properties of tendon repair tissues as compared to natural repair controls (Awad *et al.*, 2003). In one approach to overcoming poor functional integration of graft tissues, MSCs were used at the tendon–bone junction in tendon grafting for ACL repair in the rabbit knee. This resulted in a zone of fibrocartilage closely resembling that of the normal ACL

(Lim *et al.*, 2004). Dermal fibroblasts have also been used to engineer tendon. Autologous dermal fibroblasts, seeded on PGA fibers, were implanted in a porcine model and shared similar tensile strength with constructs developed with autologous tenocytes (Liu *et al.*, 2006). Gene therapy using MSCs as a vehicle was demonstrated with transduction of BMP-12 into a mouse MSC cell line, leading to tendon and cartilage-like tissue formation after injection in the thigh muscle of nude mice (Lou *et al.*, 1999).

Tensile strength and stretch loading are believed to be vital in producing the proper alignment of ligament and tendon tissues in *ex vivo* engineering. It is known that collagen is arranged along the axis of the loading force, implying its requirement for structurally competent tissues (Liu *et al.*, 2006). Tensional and torsional stimulation of MSC-based constructs, mimicking cues the cells receive *in vivo*, enhanced ECM fiber deposition. Our recent findings reveal that both mechanical and biological stimuli act to regulate differentiation and matrix synthesis and remodeling during MSC tenogenesis (Kuo and Tuan, submitted for publication). Designing bioreactors that replicate the most essential of these conditions is critical to allow for *ex vivo* production of ready-to-implant tendons and ligaments with increased load to failure over endogenous tissues.

Intervertebral Disk

Degenerative disk disease is a leading cause of back pain and disability. The IVD has a complex structure comprised of a proteoglycan-rich nucleus pulposus and an outer annulus fibrosus. Disk degeneration is characterized by proteoglycan loss in the nucleus pulposus with concurrent degradation of the annulus fibrosus. As cell-based regenerative efforts for the IVD are still in its infancy, appropriate cell, scaffold, and growth factor combinations are under investigation and have been tested in various animal models. In Europe, based on results from a canine model and pilot studies in humans, a prospective, randomized trial for comparing autologous disk chondrocyte transplantation (ADCT) plus discectomy to discectomy alone is being performed. Nucleus pulposus cells derived from therapeutic discectomy are cultured and delivered 12 weeks after discectomy. An interim analysis after 2 years shows promising results with sustained pain relief and improvement of disability scores in the ADCT patients (Meisel *et al.*, 2006). Autologous disk cells are, however, in limited supply. MSCs have been recently shown to differentiate into cells similar to IVD cells, and present an alternative cell source for regenerative purposes (Steck *et al.*, 2005). In addition, gene therapy strategies can be used to repair IVD, as demonstrated by a study in which adenovirus expressing Sox-9 was used to infect human IVD cells *in vivo*, and was also injected directly into rabbit IVD following a stab injury (Paul *et al.*, 2003). *In vitro*, infected disk cells exhibited increased production of collagen type II, and *in vivo*, IVDs injected with these cells showed decreased scarring compared to those that were untreated or injected with mock adenovirus.

Craniofacial Tissues

The temporomandibular joints (TMJ) in the craniofacial system, connecting the mandible to the cranium, play a vital role in coordinating our eating and speech activities. Osteoarthritic TMJ is correlated with not only aging but also dental function. Craniofacial defects can result from trauma, neoplasm, and most commonly from infection. Additionally, teeth penetrating through the mucosal epithelial barrier are exposed to the moist harsh environment of the mouth and subject to infection via dental caries and periodontal pathogens. Loss of teeth resulting from severe periodontal diseases or root canal infection is a common cause of losing the jaw bone. Recent clinical success of the dental implant renders the need for augmenting lost mandible before implant even more important, because the success of the dental implant is dependent upon sufficient quality bone.

Cell-based therapy to regenerate periodontal defects using autologous MSCs has been recently demonstrated in a dog model (Hasegawa *et al.*, 2006), in which 4 weeks after MSC transplantation, the periodontal defects were almost regenerated with periodontal tissue. Cementoblasts, OB, and osteocytes are found in the regenerated periodontal tissue, suggesting that transplanted MSCs could survive and differentiate into periodontal tissue cells and repair the tissue. Similar accomplishments have also been observed in humans except the assessment is only performed clinically with radiographs and periodontal probing (Yamada *et al.*, 2006). More extensive mandibular repair using cell therapy has been recently demonstrated in both animal models and humans. Mandibular distraction osteogenesis is enhanced by direct delivery of MSCs into the defect in rats (Qi *et al.*, 2006). Autologous MSCs from patient's iliac crest plus platelet-rich plasma (PRP) was applied to reconstruct an alveolar cleft defect or to augment alveolar bone with success (Hibi *et al.*, 2006). In addition, the mixture of MSCs and PRP facilitates osteointegration in dental implants, and may replace the use of autologous particulate cancellous bone and marrow for the same purpose (Yamada *et al.*, 2004).

Ex vivo gene therapy for *de novo* jaw bone regeneration has been demonstrated in a swine model using adenovirus BMP-2-mediated gene transfer to expanded autologous MSCs. Functional bone capable of sustaining axial compressive loads is formed in the maxillary bone defect filled with the BMP-2-transduced cells cast in collagen gel (Chang *et al.*, 2003). Jaw bone regeneration utilizing recombinant protein therapy is considered more straightforward and simpler. The recombinant human BMP-2/collagen sponge implant converts undifferentiated MSCs into OB and promotes an intense local neovascular response. The type of bone synthesis depends upon the mesenchymal substrate and the local cellular environment. Using this simple technique, bone defects can be resynthesized with good outcomes and a significant reduction in donor site morbidity. A drawback of recombinant protein therapy is that a single dose of exogenous protein is not as robust for osteoinduction when compared to the results from preclinical animal studies (Nussenbaum and Krebsbach, 2006).

A proof of concept to regenerate cartilage of TMJ has been demonstrated using rat MSCs to engineer human-shaped mandibular condyle in immunocompromised mice (Alhadlaq and Mao, 2003). Challenges for TMJ tissue engineering include promotion of ECM synthesis and tissue maturation by stem cell-derived chondrogene and osteogene cells that have been encapsulated in biocompatible and bioactive scaffolds. Enhancement of the mechanical properties of a tissue-engineered mandibular condyle for ultimate *in situ* implantation into the human TMJ is another challenge for tissue engineers. Tooth regeneration with cells from tooth buds, and other dental tissue regeneration including dentin and periodontal ligament using stem cells from the respective tissues, has also been proposed and tested in animal models (Ohazama *et al.*, 2004).

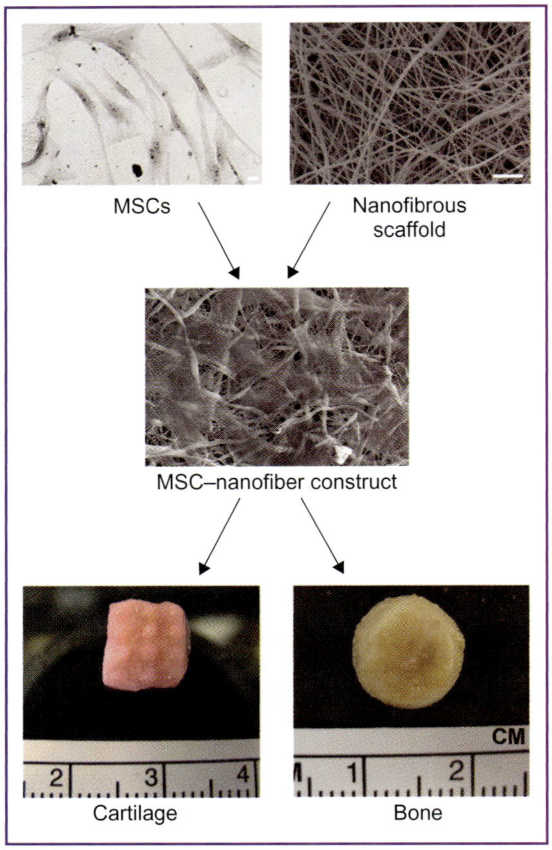

FIGURE 27-3 Production of tissue-engineered cartilage and bone based on MSC-seeded nanofibrous scaffolds. MSCs were cultured in chondrogene and osteogene media for 3 and 6 weeks, respectively. Bar = 25 mm.

CONCLUSIONS AND PROSPECTS

It appears promising that regenerative medicine via cell-based therapy and tissue engineering will become a widespread therapeutic modality. The essential procedures consist of utilizing and manipulating *ex vivo* expanded multipotent stem cells, and delivering them into hosts under a designed condition or package to grow new musculoskeletal tissues (Figure 27-3). However, there are many challenges before reaching this goal. Specifically, we need to understand: (1) the native environment of stem cells, that is, the niche and stem cell markers, to allow stem cell isolation and culture expansion in a more specific and predictable manner; (2) the molecular regulation at various stages of the stem cell differentiation program such that a better control of cell lineage commitment can be achieved; (3) which delivery method or scaffold system is optimal for regeneration of a specific tissue; and (4) the interactions of cells with their carrier or scaffold system and the newly produced ECM, so that optimal tissue regeneration can be accomplished. There is also a need for testing the long-term compatibility of allogeneic MSCs in the host before cell banks can be established to provide an adequate cell source for cell-based applications and tissue engineering. There has been significant experience accumulated in the regeneration of bone, whereas other musculoskeletal tissue regeneration, such as cartilage, ligament, and tendon, is still at its infancy. Another aspect of tissue engineering that requires further development is the design of bioreactors that can replicate conditions to allow *ex vivo* production of ready-to-implant musculoskeletal tissues, which is critical to achieve functional tissue engineering.

ACKNOWLEDGEMENTS

Supported by NIH NIAMS Intramural Research Program (AR Z01411131) and Howard Hughes Medical Institute-NIH Research Scholar Program.

REFERENCES

Abukawa, H., Kaban, L.B., Williams, W.B., Terada, S., Vacanti, J.P. and Troulis, M.J. (2006). Effect of interferon-alpha-2b on porcine mesenchymal stem cells. *J. Oral Maxillofac. Surg. 64*: 1214–1220.

Alhadlaq, A. and Mao, J.J. (2003). Tissue-engineered neogenesis of human-shaped mandibular condyle from rat mesenchymal stemcells. *J. Dent. Res. 82*: 951–956.

Awad, H.A., Boivin, G.P., Dressler, M.R., Smith, F.N., Young, R.G. and Butler, D.L. (2003). Repair of patellar tendon injuries using a cell–collagen composite. *J. Orthop. Res. 21*: 420–431.

Bach, A.D., Beier, J.P., Stern-Staeter, J. and Horch, R.E. (2004). Skeletal muscle tissue engineering. *J. Cell. Mol. Med. 8*: 413–422.

Ball, S.G., Shuttleworth, A.C. and Kielty, C.M. (2004). Direct cell contact influences bone marrow mesenchymal stem cell fate. *Int. J. Biochem. Cell Biol. 36*: 714–727.

Barry, F., Boynton, R.E., Liu, B. and Murphy, J.M. (2001). Chondrogenic differentiation of mesenchymal stem cells from bone marrow: differentiation-dependent gene expression of matrix components. *Exp. Cell Res. 268*: 189–200.

Bi, W., Huang, W., Whitworth, D.J., Deng, J.M., Zhang, Z., Behringer, R.R. and de Crombrugghe, B. (2001). Haploinsufficiency of Sox9 results in defective cartilage primordia and premature skeletal mineralization. *Proc. Natl Acad. Sci. USA 98*: 6698–6703.

Boland, G.M., Perkins, G., Hall, D.J. and Tuan, R.S. (2004). Wnt 3a promotes proliferation and suppresses osteogenic differentiation of adult human mesenchymal stem cells. *J. Cell. Biochem. 93*: 1210–1230.

Borschel, G.H., Dennis, R.G. and Kuzon Jr., W.M. (2004). Contractile skeletal muscle tissue-engineered on a cellular scaffold. *Plast. Reconstr. Surg. 113*: 595–602; discussion 603–604.

Brittberg, M., Lindahl, A., Nilsson, A., Ohlsson, C., Isaksson, O. and Peterson, L. (1994). Treatment of deep cartilage defects in the knee with autologous chondrocyte transplantation. *N. Engl. J. Med. 331*: 889–895.

Caplan, A.I. (1991). Mesenchymal stem cells. *J. Orthop. Res. 9*: 641–650.

CDC (2006). Prevalence of doctor-diagnosed arthritis and arthritis-attributable activity limitation – United States, 2003–2005. *MMWR Morb. Mortal. Wkly Rep. 55*: 1089–1092.

Chamberlain, J.R., Schwarze, U., Wang, P.R., Hirata, R.K., Hankenson, K.D., Pace, J.M., Underwood, R.A., Song, K.M., Sussman, M., Byers, P.H., *et al.* (2004). Gene targeting in stem cells from individuals with osteogenesis imperfecta. *Science 303*: 1198–1201.

Chang, S.C., Chuang, H.L., Chen, Y.R., Chen, J.K., Chung, H.Y., Lu, Y.L., Lin, H.Y., Tai, C.L. and Lou, J. (2003). *Ex vivo* gene therapy in autologous bone marrow stromal stem cells for tissue-engineered maxillofacial bone regeneration. *Gene Ther. 10*: 2013–2019.

Chen, F., Hui, J.H., Chan, W.K. and Lee, E.H. (2003). Cultured mesenchymal stem cell transfers in the treatment of partial growth arrest. *J. Pediatr. Orthop. 23*: 425–429.

Chen, F.H., Rousche, K.T and Tuan, R.S. (2006). Technology insight: adult stem cells in cartilage regeneration and tissue engineering. *Nat. Clin. Pract. Rheumatol. 2*: 373–382.

Chen, X.I., Armstrong, M.A. and Li, G. (2006). Mesenchymal stem cells in immunoregulation. *Immunol. Cell Biol. 84*: 413–421.

Chiang, H., Kuo, T.F., Tsai, C.C., Lin, M.C., She, B.R., Huang, Y.Y., Lee, H.S., Shieh, C.S., Chen, M.H., Ramshaw, J.A., *et al.* (2005). Repair of porcine articular cartilage defect with autologous chondrocyte transplantation. *J. Orthop. Res. 23*: 584–593.

Colter, D.C., Class, R., DiGirolamo, C.M. and Prockop, D.J. (2000). Rapid expansion of recycling stem cells in cultures of plastic-adherent cells from human bone marrow. *Proc. Natl Acad. Sci. USA 97*: 3213–3218.

Deans, R.J. and Moseley, A.B. (2000). Mesenchymal stem cells: biology and potential clinical uses. *Exp. Hematol. 28*: 875–884.

Dezawa, M., Ishikawa, H., Itokazu, Y., Yoshihara, T., Hoshino, M., Takeda, S., Ide, C. and Nabeshima, Y. (2005). Bone marrow stromal cells generate muscle cells and repair muscle degeneration. *Science 309*: 314–317.

Dominici, M., Le Blanc, K., Mueller, I., Slaper-Cortenbach, I., Marini, F., Krause, D., Deans, R., Keating, A., Prockop, D. and Horwitz, E. (2006). Minimal criteria for defining multipotent mesenchymal stromal cells. The International Society for Cellular Therapy position statement. *Cytotherapy 8*: 315–317.

Dreyfus, P.A., Chretien, F., Chazaud, B., Kirova, Y., Caramelle, P., Garcia, L., Butler-Browne, G. and Gherardi, R.K. (2004). Adult bone marrow-derived stem cells in muscle connective tissue and satellite cell niches. *Am. J. Pathol. 164*: 773–779.

Eames, B.F., de la Fuente, L. and Helms, J.A. (2003). Molecular ontogeny of the skeleton. *Birth Defects Res. C Embryo Today 69*: 93–101.

Elisseeff, J., Ferran, A., Hwang, S., Varghese, S. and Zhang, Z. (2006). The role of biomaterials in stem cell differentiation: applications in the musculoskeletal system. *Stem Cells Dev. 15*: 295–303.

Engler, A.J., Sen, S., Sweeney, H.L. and Discher, D.E. (2006). Matrix elasticity directs stem cell lineage specification. *Cell 126*: 677–689.

Fukuchi, Y., Nakajima, H., Sugiyama, D., Hirose, I., Kitamura, T. and Tsuji, K. (2004). Human placenta-derived cells have mesenchymal stem/progenitor cell potential. *Stem Cells 22*: 649–658.

Gersbach, C.A., Le Doux, J.M., Guldberg, R.E. and Garcia, A.J. (2006). Inducible regulation of Runx2-stimulated osteogenesis. *Gene Ther. 13*: 873–882.

Gregory, C.A., Singh, H., Perry, A.S. and Prockop, D.J. (2003). The Wnt signaling inhibitor dickkopf-1 is required for reentry into the cell cycle of human adult stem cells from bone marrow. *J. Biol. Chem. 278*: 28067–28078.

Haleem-Smith, H., Derfoul, A., Okafor, C., Tuli, R., Olsen, D., Hall, D.J. and Tuan, R.S. (2005). Optimization of high-efficiency transfection of adult human mesenchymal stem cells *in vitro*. *Mol. Biotechnol. 30*: 9–20.

Hasegawa, N., Kawaguchi, H., Hirachi, A., Takeda, K., Mizuno, N., Nishimura, M., Koike, C., Tsuji, K., Iba, H., Kato, Y., *et al.* (2006). Behavior of transplanted bone marrow-derived mesenchymal stem cells in periodontal defects. *J. Periodontol. 77*: 1003–1007.

Hernigou, P., Poignard, A., Beaujean, F. and Rouard, H. (2005). Percutaneous autologous bone-marrow grafting for nonunions. Influence of the number and concentration of progenitor cells. *J. Bone Joint Surg. Am. 87*: 1430–1437.

Hibi, H., Yamada, Y., Ueda, M. and Endo, Y. (2006). Alveolar cleft osteoplasty using tissue-engineered osteogenic material. *Int. J. Oral Maxillofac. Surg. 35*: 551–555.

In 't Anker, P.S., Scherjon, S.A., Kleijburg-van der Keur, C., Noort, W.A., Claas, F.H., Willemze, R., Fibbe, W.E. and Kanhai, H.H. (2003). Amniotic fluid as a novel source of mesenchymal stem cells for therapeutic transplantation. *Blood 102*: 1548–1549.

Izuta, Y., Ochi, M., Adachi, N., Deie, M., Yamasaki, T. and Shinomiya, R. (2005). Meniscal repair using bone marrow-derived mesenchymal stem cells: experimental study using green fluorescent protein transgenic rats. *Knee 12*: 217–223.

Jorgensen, C., Djouad, F., Fritz, V., Apparailly, F., Plence, P. and Noel, D. (2003). Mesenchymal stem cells and rheumatoid arthritis. *Joint Bone Spine 70*: 483–485.

Karp, J.M., Ferreira, L.S., Khademhosseini, A., Kwon, A.H., Yeh, J. and Langer, R.S. (2006). Cultivation of human embryonic stem cells without the embryoid body step enhances osteogenesis *in vitro*. *Stem Cells 24*: 835–843.

Kern, S., Eichler, H., Stoeve, J., Kluter, H. and Bieback, K. (2006). Comparative analysis of mesenchymal stem cells from bone marrow, umbilical cord blood, or adipose tissue. *Stem Cells 24*: 1294–1301.

Kim, H., Lee, J.H. and Suh, H. (2003). Interaction of mesenchymal stem cells and osteoblasts for *in vitro* osteogenesis. *Yonsei Med. J. 44*: 187–197.

Kneser, U., Stangenberg, L., Ohnolz, J., Buettner, O., Stern-Straeter, J., Mobest, D., Horch, R.E., Stark, G.B. and Schaefer, D.J. (2006). Evaluation of processed bovine cancellous bone matrix seeded with syngenic osteoblasts in a critical size calvarial defect rat model. *J. Cell. Mol. Med. 10*: 695–707.

Kramer, J., Hegert, C., Guan, K., Wobus, A.M., Muller, P.K. and Rohwedel, J. (2000). Embryonic stem cell-derived chondrogenic differentiation *in vitro*: activation by BMP-2 and BMP-4. *Mech. Dev. 92*: 193–205.

Kuo, C.K., Li, W.J., Mauck, R.L. and Tuan, R.S. (2006). Cartilage tissue engineering: its potential and uses. *Curr. Opin. Rheumatol. 18*: 64–73.

Lee, R.H., Kim, B., Choi, I., Kim, H., Choi, H.S., Suh, K., Bae, Y.C. and Jung, J.S. (2004). Characterization and expression analysis of mesenchymal stem cells from human bone marrow and adipose tissue. *Cell. Physiol. Biochem. 14*: 311–324.

Lefebvre, V. and Smits, P. (2005). Transcriptional control of chondrocyte fate and differentiation. *Birth Defects Res. C Embryo Today 75*: 200–212.

Levenberg, S., Huang, N.F., Lavik, E., Rogers, A.B., Itskovitz-Eldor, J. and Langer, R. (2003). Differentiation of human embryonic stem cells on three-dimensional polymer scaffolds. *Proc. Natl Acad. Sci. USA 100*: 12741–12746.

Li, W.J., Tuli, R., Huang, X., Laquerriere, P. and Tuan, R.S. (2005a). Multilineage differentiation of human mesenchymal stem cells in a three-dimensional nanofibrous scaffold. *Biomaterials 26*: 5158–5166.

Li, W.J., Tuli, R., Okafor, C., Derfoul, A., Danielson, K.G., Hall, D.J. and Tuan, R.S. (2005b). A three-dimensional nanofibrous scaffold for cartilage tissue engineering using human mesenchymal stem cells. *Biomaterials 26*: 599–609.

Li, W.J., Jiang, Y.J. and Tuan, R.S. (2006). Chondrocyte phenotype in engineered fibrous matrix is regulated by fiber size. *Tissue Eng. 12*: 1775–1785.

Li, W.J., Mauck, R.L., Cooper, J.A., Yuan, X. and Tuan, R.S. (2007). Engineering controllable anisotropy in electrospun biodegradable nanofibrous scaffolds for musculoskeletal tissue engineering. *J. Biomech. 40*: 1686–1693.

Lieberman, J.R., Daluiski, A., Stevenson, S., Wu, L., McAllister, P., Lee, Y.P., Kabo, J.M., Finerman, G.A., Berk, A.J. and Witte, O.N. (1999). The effect of regional gene therapy with bone morphogenetic protein-2-producing bone-marrow cells on the repair of segmental femoral defects in rats. *J. Bone Joint Surg. Am. 81*: 905–917.

Lim, J.K., Hui, J., Li, L., Thambyah, A., Goh, J. and Lee, E.H. (2004). Enhancement of tendon graft osteointegration using mesenchymal stem cells in a rabbit model of anterior cruciate ligament reconstruction. *Arthroscopy 20*: 899–910.

Liu, C.-H. and Hwang, S.-M. (2005). Cytokine interactions in mesenchymal stem cells from cord blood. *Cytokine 32*: 270–279.

Liu, W., Chen, B., Deng, D., Xu, F., Cui, L. and Cao, Y. (2006). Repair of tendon defect with dermal fibroblast engineered tendon in a porcine model. *Tissue Eng. 12*: 775–788.

Lou, J., Tu, Y., Ludwig, F.J., Zhang, J. and Manske, P.R. (1999). Effect of bone morphogenetic protein-12 gene transfer on mesenchymal progenitor cells. *Clin. Orthop. Relat. Res.* 333–339.

Lu, F.-Z., Fujino, M., Kitazawa, Y., Uyama, T., Hara, Y., Funeshima, N., Jiang, J.-Y., Umezawa, A. and Li, X.-K. (2005). Characterization and gene transfer in mesenchymal stem cells derived from human umbilical-cord blood. *J. Lab. Clin. Med. 146*: 271–278.

Madry, H., Kaul, G., Cucchiarini, M., Stein, U., Zurakowski, D., Remberger, K., Menger, M.D., Kohn, D. and Trippel, S.B. (2005). Enhanced repair of articular cartilage defects *in vivo* by transplanted chondrocytes overexpressing insulin-like growth factor I (IGF-I). *Gene Ther. 12*: 1171–1179.

Mareschi, K., Biasin, E., Piacibello, W., Aglietta, M., Madon, E. and Fagioli, F. (2001). Isolation of human mesenchymal stem cells: bone marrow versus umbilical cord blood. *Haematologica 86*: 1099–1100.

Matsubara, T., Tsutsumi, S., Pan, H., Hiraoka, H., Oda, R., Nishimura, M., Kawaguchi, H., Nakamura, K. and Kato, Y. (2004). A new technique to expand human mesenchymal stem cells using basement membrane extracellular matrix. *Biochem. Biophys. Res. Commun. 313*: 503–508.

Mauck, R.L., Soltz, M.A., Wang, C.C., Wong, D.D., Chao, P.H., Valhmu, W.B., Hung, C.T. and Ateshian, G.A. (2000). Functional tissue engineering of articular cartilage through dynamic loading of chondrocyte-seeded agarose gels. *J. Biomech. Eng. 122*: 252–260.

Mauney, J.R., Blumberg, J., Pirun, M., Volloch, V., Vunjak-Novakovic, G. and Kaplan, D.L. (2004). Osteogenic differentiation of human bone marrow stromal cells on partially demineralized bone scaffolds *in vitro*. *Tissue Eng. 10*: 81–92.

McBeath, R., Pirone, D.M., Nelson, C.M., Bhadriraju, K. and Chen, C.S. (2004). Cell shape, cytoskeletal tension, and RhoA regulate stem cell lineage commitment. *Dev. Cell 6*: 483–195.

Meisel, H.J., Ganey, T., Hutton, W.C., Libera, J., Minkus, Y. and Alasevic, O. (2006). Clinical experience in cell-based therapeutics: intervention and outcome. *Eur. Spine J. 15*(Suppl 15): 397–405.

Miyanishi, K., Trindade, M.C., Lindsey, D.P., Beaupre, G.S., Carter, D.R., Goodman, S.B., Schurman, D.J. and Smith, R.L. (2006). Dose- and time-dependent effects of cyclic hydrostatic pressure on transforming growth factor-beta3-induced chondrogenesis by adult human mesenchymal stem cells *in vitro*. *Tissue Eng. 12*: 2253–2262.

Nussenbaum, B. and Krebsbach, P.H. (2006). The role of gene therapy for craniofacial and dental tissue engineering. *Adv. Drug Deliv. Rev. 58*: 577–591.

Ohazama, A., Modino, S.A., Miletich, I. and Sharpe, P.T. (2004). Stem-cell-based tissue engineering of murine teeth. *J. Dent. Res. 83*: 518–522.

Okafor, C.C., Haleem-Smith, H., Laqueriere, P., Manner, P.A. and Tuan, R.S. (2006). Particulate endocytosis mediates biological responses of human mesenchymal stem cells to titanium wear debris. *J. Orthop. Res. 24*: 461–473.

Park, Y., Lutolf, M.P., Hubbell, J.A., Hunziker, E.B. and Wong, M. (2004). Bovine primary chondrocyte culture in synthetic matrix metalloproteinase-sensitive poly(ethylene glycol)-based hydrogels as a scaffold for cartilage repair. *Tissue Eng. 10*: 515–522.

Paul, R., Haydon, R.C., Cheng, H., Ishikawa, A., Nenadovich, N., Jiang, W., Zhou, L., Breyer, B., Feng, T., Gupta, P., *et al.* (2003). Potential use of Sox9 gene therapy for intervertebral degenerative disc disease. *Spine 28*: 755–763.

Pittenger, M.F., Mackay, A.M., Beck, S.C., Jaiswal, R.K., Douglas, R., Mosca, J.D., Moorman, M.A., Simonetti, D.W., Craig, S. and Marshak, D.R. (1999). Multilineage potential of adult human mesenchymal stem cells. *Science 284*: 143–147.

Qi, M., Hu, J., Zou, S., Zhou, H. and Han, L. (2006). Mandibular distraction osteogenesis enhanced by bone marrow mesenchymal stem cells in rats. *J. Craniomaxillofac. Surg. 34*: 283–289.

Quarto, R., Mastrogiacomo, M., Cancedda, R., Kutepov, S.M., Mukhachev, V., Lavroukov, A., Kon, E. and Marcacci, M. (2001). Repair of large bone defects with the use of autologous bone marrow stromal cells. *N. Engl. J. Med. 344*: 385–386.

Romanov, Y.A., Svintsitskaya, V.A. and Smirnov, V.N. (2003). Searching for alternative sources of postnatal human mesenchymal stem cells: candidate MSC-like cells from umbilical cord. *Stem Cells 21*: 105–110.

Rotter, N., Ung, F., Roy, A.K., Vacanti, M., Eavey, R.D.,Vacanti, C.A. and Bonassar, L.J. (2005). Role for interleukin 1alpha in the inhibition of chondrogenesis in autologous implants using polyglycolic acid–polylactic acid scaffolds. *Tissue Eng. 11*: 192–200.

Rouwkema, J., de Boer, J. and Van Blitterswijk, C.A. (2006). Endothelial cells assemble into a 3-dimensional prevascular network in a bone tissue engineering construct. *Tissue Eng. 12*: 2685–2693.

Sampaolesi, M., Blot, S., D'Antona, G., Granger, N., Tonlorenzi, R., Innocenzi, A., Mognol, P., Thibaud, J.L., Galvez, B.G., Barthelemy, I., *et al.* (2006). Mesangioblast stem cells ameliorate muscle function in dystrophic dogs. *Nature 444*: 574–579.

Sartorelli, V. and Caretti, G. (2005). Mechanisms underlying the transcriptional regulation of skeletal myogenesis. *Curr. Opin. Genet. Dev. 15*: 528–535.

Schofield, R. (1978). The relationship between the spleen colony-forming cell and the haemopoietic stem cell. *Blood Cells 4*: 7–25.

Sekiya, I., Larson, B.L., Vuoristo, J.T., Reger, R.L. and Prockop, D.J. (2005). Comparison of effect of BMP-2, -4, and -6 on *in vitro* cartilage formation of human adult stem cells from bone marrow stroma. *Cell Tissue Res. 320*: 269–276.

Shen, F.H., Visger, J.M., Balian, G., Hurwitz, S.R. and Diduch, D.R. (2002). Systemically administered mesenchymal stromal cells transduced with insulin-like growth factor-I localize to a fracture site and potentiate healing. *J. Orthop. Trauma 16*: 651–659.

Steck, E., Bertram, H., Abel, R., Chen, B., Winter, A. and Richter, W. (2005). Induction of intervertebral disc-like cells from adult mesenchymal stem cells. *Stem Cells 23*: 403–411.

Tamaki, T., Akatsuka, A., Ando, K., Nakamura, Y., Matsuzawa, H., Hotta, T., Roy, R.R. and Edgerton, V.R. (2002). Identification of myogenic-endothelial progenitor cells in the interstitial spaces of skeletal muscle. *J. Cell Biol. 157*: 571–577.

Tozer, S. and Duprez, D. (2005). Tendon and ligament: development, repair and disease. *Birth Defects Res. C Embryo Today 75*: 226–236.

Tuan, R.S., Boland, G. and Tuli, R. (2003). Adult mesenchymal stem cells and cell-based tissue engineering. *Arthr. Res. Ther. 5*: 32–45.

Tuli, R., Tuli, S., Nandi, S., Huang, X., Manner, P.A., Hozack, W.J., Danielson, K.G., Hall, D.J. and Tuan, R.S. (2003). Transforming growth factor-beta-mediated chondrogenesis of human mesenchymal progenitor cells involves N-cadherin and mitogen-activated protein kinase and Wnt signaling cross-talk. *J. Biol. Chem. 278*: 41227–41236.

Tuli, R., Nandi, S., Li, W.J., Tuli, S., Huang, X., Manner, P.A., Laquerriere, P., Noth, U., Hall, D.J. and Tuan, R.S. (2004). Human mesenchymal progenitor cell-based tissue engineering of a single-unit osteochondral construct. *Tissue Eng. 10*: 1169–1179.

Vacanti, C.A., Bonassar, L.J., Vacanti, M.P. and Shufflebarger, J. (2001). Replacement of an avulsed phalanx with tissue-engineered bone. *N. Engl. J. Med. 344*: 1511–1514.

Veilleux, N. and Spector, M. (2005). Effects of FGF-2 and IGF-1 on adult canine articular chondrocytes in type II collagen–glycosaminoglycan scaffolds *in vitro*. *Osteoarthr. Cartilage 13*: 278–286.

Wakitani, S., Imoto, K., Yamamoto, T., Saito, M., Murata, N. and Yoneda, M. (2002). Human autologous culture expanded bone marrow mesenchymal cell transplantation for repair of cartilage defects in osteoarthritic knees. *Osteoarthr. Cartilage 10*: 199–206.

Wakitani, S., Aoki, H., Harada, Y., Sonobe, M., Morita, Y., Mu, Y., Tomita, N., Nakamura, Y., Takeda, S., Watanabe, T.K., *et al.* (2004). Embryonic stem cells form articular cartilage, not teratomas, in osteochondral defects of rat joints. *Cell Transplant. 13*: 331–336.

Yamada, Y., Ueda, M., Naiki, T. and Nagasaka, T. (2004). Tissue-engineered injectable bone regeneration for osteointegrated dental implants. *Clin. Oral Implants Res. 15*: 589–597.

Yamada, Y., Ueda, M., Hibi, H. and Baba, S. (2006). A novel approach to periodontal tissue regeneration with mesenchymal stem cells and platelet-rich plasma using tissue engineering technology: a clinical case report. *Int. J. Periodont. Rest. Dent. 26*: 363–369.

Young, B.H., Peng, H. and Huard, J. (2002). Muscle-based gene therapy and tissue engineering to improve bone healing. *Clin. Orthop. Relat. Res.* S243–S251.

Chapter 28

Retinal Pigment Epithelium Derived from Embryonic Stem Cells

Irina Klimanskaya

Blindness is one of the most devastating ailments in humans, and in the United States alone, over 9 million people suffer loss of vision from retinal hereditary and degenerative diseases which lead to the loss of photoreceptors. Among these, are ailments such as retinal dystrophies, macular degeneration, and **retinitis pigmentosa**, which are commonly associated with dysfunctions of the retinal pigment epithelium (RPE) that plays a key role in the support of the photoreceptor which does not have its own blood supply. RPE is a highly specialized tissue located between the choroids and the neural retina, and its functions include absorption of stray light that allows a better resolution of images; ion and metabolite transport between the neurosensory retina and the choroids; storage, metabolism, and delivery of vitamin A to the photoreceptor; phagocytosis of shed photoreceptor fragments and serving as a barrier between the neurosensory retina and the choroids (Besharse and Defoe, 1998; Thompson and Gal, 2003).

DEVELOPMENT OF RPE

In vertebrate development RPE shares the same progenitor, neuroectoderm of the optic vesicle, with the neurosensory retina. The cells competent to give rise to RPE or neural retina are morphologically very similar and express Otx2, Pax6, Rx1, and Six3, transcription factors necessary for eye development (Carpenter *et al.*, 2001; Reubinoff *et al.*, 2001; Ying *et al.*, 2003; Ben-Hur *et al.*, 2004; Meyer et al,. 2005; Takagi *et al.*, 2005, reviewed by Ben-Hur, 2006). Signals coming from the tissues surrounding the optic vesicle seem to be instructing different populations of these cells to selectively differentiate into RPE or neural retina. The cells of the distal part of the optic vesicle next to ectoderm which produces fibroblast growth factor (FGF)1 and FGF2 are thought to be giving rise to neural retina, as FGF signaling was shown to support neural retina formation (Pittack *et al.*, 1997; Nguyen and Arnheiter, 2000) and convert RPE into neural retinal cells (Pittack *et al.*, 1997; Vogel-Hopker *et al.*, 2000; Zhao *et al.*, 2001). The dorsal part of the optic vesicle adjacent to mesoderm receives RPE-inductive signals, such as activin A expressed by extraocular mesenchyme (Feijen *et al.*, 1994; Fuhrmann *et al.*, 2000; Chow and Lang, 2001).

There are several transcription factors involved in the signaling cascade, essential for RPE specification. In the following model suggested by Martinez-Morales *et al.* (2004) the RPE induction is triggered by Pax6 and Otx1/Otx2 combined activity, when Pax6 and Wnt signaling induce expression of microftalmia-associated transcription factor (Mitf) which plays an important role in development of melanin-containing cells, melanocytes, and RPE, and can support the further formation of RPE together with Otx proteins, and according to Baumer and co-authors, expression of Mitf is controlled by the redundant activities of Pax6 and Pax2 (Baumer *et al.*, 2003). The paired and homeodomain transcription factor Pax6 is known to be a key regulator of eye formation in multiple species (Gehring and Ikeo, 1999; Ashery-Padan and Gruss, 2001; Kumar, 2001) and is expressed throughout the optic cup (Walther and Gruss, 1991; Grindley *et al.*, 1995). Its overexpression is sufficient to induce ectopic eyes formation in fly and frog embryos (Halder *et al.*, 1995; Chow *et al.*, 1999) and no functional eye structures develop in its absence, as shown in experiments in mouse, human, rat, frog, and fly (reviewed by Gehring and Ikeo, 1999). Its activity seems to be sufficient to induce RPE formation (Baumer *et al.*, 2003), and in mature RPE it is downregulated, remaining in the lens, corneal, and conjunctive epithelia, iris, and amacrine and ganglion cells of the neural retina (Walther and Gruss, 1991; Hitchcock *et al.*, 1996).

CHARACTERISTICS OF RPE

Mature RPE is characterized by the polygonal morphology of the cells forming a "cobblestone" monolayer, granules of melanin in the cytoplasm, and several specific molecular markers. A water-soluble 36 kD cellular retinalde-hyde-binding protein

(CRALBP) is found in apical microvilli of RPE and in Muller glia (Bunt-Milam and Saari, 1983). It is a product of RLBP1 gene, and its mutations were shown to be associated with rod–cone dystrophy (Eichers *et al.*, 2002), **retinistis pigmentosa** (Maw *et al.*, 1997; Morimura *et al.*, 1999), and impaired dark adaptation in a mouse model (Saari *et al.*, 2001). Another important RPE hallmark is RPE65, a 65 kD cytoplasmic protein involved in retinoid metabolism (Hamel *et al.*, 1993; Redmond *et al.*, 1998; Ma *et al.*, 2001). Mutations in RPE65 were shown to be associated with Leber's congenital amaurosis and retinitis pigmentosa as well as with similar dystrophies in animal models of childhood blindness (Marlhens *et al.*, 1997; Morimura *et al.*, 1999; Acland *et al.*, 2001; Van Hooser *et al.*, 2000; reviewed by Perrault *et al.*, 1999) and in a canine RPE65–/– model. Bestrophin, another specific RPE marker localized in the basolateral plasma membrane, is a 68 kD product of the best vitelliform macular dystrophy gene (VMD2) (Petrukhin *et al.*, 1998; Marmorstein *et al.*, 2000) and is associated with macular dystrophies (Marquardt *et al.*, 1998; Allikmets *et al.*, 1999; Kramer *et al.*, 2000). Mer tyrosine kinase protooncogene (MERTK) is associated with retinal dystrophies and phagocytosis pathways (D'Cruz *et al.*, 2000; Gal *et al.*, 2000; Feng *et al.*, 2002). One more molecular marker of RPE is pigment epithelium-derived factor (PEDF) a 48 kD secreted protein with angiostatic properties (Steele *et al.*, 1993; Jablonski *et al.*, 2000; Karakousis *et al.*, 2001). Bestrophin and RPE65 are subject to translational control, and RPE65 protein was shown to be absent from cultured RPE, while the gene expression product can be detected by reverse transcriptase polymerase chain reaction (RT-PCR) (Hamel *et al.*, 1993; Liu and Redmond, 1998; Bakall *et al.*, 2003).

TRANSDIFFERENTIATION OF RPE IN CULTURE

A specific feature of the RPE is its apparent plasticity. The common origin of RPE and neurosensory retina may be the reason why under certain conditions both *in vivo* and *in vitro* RPE can transdifferentiate into neuronal progenitors (Opas and Dziak, 1994), neurons (Vinores *et al.*, 1995; Chen *et al.*, 2003), and lens epithelium (Eguchi, 1986). Amphibian RPE is able to regenerate the retina via dedifferentiation into neural progenitors followed by their differentiation into retinal neurons (Moshiri *et al.*, 2004). So far there have been no indications of mammalian RPE being capable of regenerating retina, although *in vitro* it can undergo a similar transdifferentiation process. One of the factors which can stimulate the change of RPE into neurons is basic fibroblast growth factor (bFGF) (Opas and Dziak, 1994), and this process is associated with the expression of transcriptional activators normally required for the eye development, including rx/rax, chx10/vsx-2/a/x, ots-1, otx-2, six3/optx, six6/optx2, Mitf, and Pax6 (Fischer and Reh, 2001). These authors also showed the presence of proliferative pigmented cells in the margin of the post-natal chicken retina expressing proliferating cell nuclear antigen and incorporating bromdeoxyuridine; these cells also express Pax6 and Mitf and can lose pigmentation and turn into neuronal cells in response to simultaneously added FGF and insulin (Fischer and Reh, 2001). *In vitro,* depending on the combination of growth factors and substratum, RPE can be maintained as an epithelium, or rapidly dedifferentiate and become proliferative (Opas and Dziak, 1994; Zhao, 1997); the epithelial phenotype can then be reestablished in long-term quiescent cultures (Grierson *et al.*, 1994).

TRANSPLANTATION OF RPE

While there is no current technology allowing to restore degenerated photoreceptors which would establish connections with the optic nerve, transplantation of RPE has the potential to prevent the loss of the remaining photoreceptors before a degenerative disease would result in blindness, and it has been extensively studied for over 30 years in various animal models, including rabbits (Lopez *et al.*, 1987; Brittis *et al.*, 1987; El Dirini *et al.*, 1992; Crafoord *et al.*, 1999), monkey (Gouras *et al.*, 1985; Berglin *et al.*, 1997), dogs (Veske *et al.*, 1999; Verdugo *et al.*, 2001), and Royal College of Surgeons (RCS) rats, the latter being an excellent working model of retinal degeneration which is likely to be associated with the impaired phagocytosis function of RPE and its detachment from Bruch's membrane (Lopez *et al.*, 1989; Sauve *et al.*, 2002; Wang *et al.*, 2005a; Philips *et al.*, 2003; Girmann *et al.*, 2003, 2005).

Several studies have been conducted in human subjects (Benson *et al.*, 1998; Algevre *et al.*, 1999; Binder *et al.*, 2004). The potential sources of RPE for transplantation to treat retinal degenerative diseases and prevent the loss of the photoreceptor are autologous (Binder *et al.*, 2004; van Meurs *et al.*, 2004) and fetal RPE (Weisz *et al.*, 1999; Radtke *et al.*, 2002, 2004). Each source is not free of potential problems: autologous RPE may already have impaired function due to the patient's own age and disease and fetal cells can vary from batch to batch and need to be characterized for safety before transplantation, which means that they need to be propagated in culture for a certain time and may lose some of their important functions. Ethical issues related to all fetal material should also be considered. Sourcing suitable donor cells is critical to any transplantation, and such cells need to be non-tumorigenic, free of pathogens and of contaminating other cell types, and every batch of donor tissue meant for clinical applications needs to be assessed for all these parameters. Donor tissue should be readily available with minimal batch-to-batch variation, and it is desirable that each batch be tested first in an animal model (Lund *et al.*, 2001a; Wang *et al.*, 2005b). Cultured cell lines could be an attractive cell source if they meet all of the above criteria.

Lund and co-authors have shown that RPE cell lines can be very effective in supporting the photoreceptor function in RCS rats (Lund *et al.*, 2001b). Two RPE cell lines established from adult donors, spontaneously derived ARPE-19 and SV40 large T (tumor) antigen-transformed h1RPE7, were used in the study. Both lines showed attenuation of the photoreceptor and visual function loss up to 5 months after transplantation into subretinal space. This demonstrates the potential of these or similar cell lines for treatment of retinal degenerative diseases associated with RPE dysfunction. However, their application for human therapy depends on further evaluation of their performance at extended passages, stability of karyotype, preservation of RPE phenotype, and function in long-term cultures through multiple passages.

One potential hurdle for therapeutic applications of RPE, in contrast to the success of transplantation experiments in laboratory animals, is that there are age-related changes in Bruch's membrane in humans which may lead to poor RPE survival and differentiation, as has been shown in organ culture experiments (Gullapalli *et al.*, 2004, 2005). Bruch's membrane is a multi-layered structure underlying RPE, which consists of the RPE basement membrane, inner and outer collagenous layers and elastin layer between them, and choricapillaris basement membrane (Zarbin, 2003). The RPE basement membrane appears to be the most favorable matrix for RPE attachment, compared to other layers of Bruch's membrane (Tezel and Del Priore, 1999; Tezel *et al.*, 1999). Age- and disease-related changes in Bruch's membrane include increased thickness, exposure of underlying collagenous or even elastic layers, deposition of extracellular matrix (ECM) proteins and lipids which impair the attachment, and survival of cultured RPE because the graft may be interacting with less favorable substrate (Zarbin, 2003; Gullapalli *et al.*, 2004). These authors have studied resurfacing of the Bruch's membrane by fetal RPE, using eyes from old donors and donors with age-related macular degeneration (AMD). The explants were treated to expose different layers of the Bruch's membrane to imitate possible age and disease-related changes, and cultured fetal RPE cells between passages 2 and 4 were plated on the explants and incubated in organ culture up to 7 days, then examined histologically by light and scanning electron microscopy. It appeared that on aged submacular Bruch's membrane there were multiple defects in RPE surface coverage and morphology, notable cell death, and the poorest resurfacing was noted in the eye of a donor with AMD. Other experiments performed by the same group compared integrin profiles in cultured versus uncultured fetal and aged RPE and have shown that there is a downregulation of integrin subunits forming adhesion receptors for laminin, collagens, and fibronectin in uncultured aged RPE (Zarbin, 2003).

These studies suggest that certain "adjustment" of the cells' phenotype may be required prior to grafting in order to provide a better "match" between the RPE adhesion receptors and the ECM composition of the recipient's Bruch's membrane. Culturing RPE can allow to modulate integrin expression and thus make the cells attach to Bruch's membrane more efficiently and improve their survival and differentiation. However, multiple passages and prolonged time in culture can amend many RPE features and lead to cell senescence; for instance, loss of $\alpha 5$ integrin was reported in "post-confluent" quiescent RPE cultures (Proulx *et al.*, 2003, 2004); therefore, multiple passages in culture can produce phenotypically or functionally impaired cells. These factors may impede production of fetal or adult RPE cells of high quality and in high enough quantities for characterization, safety assessment, and transplantation.

An attractive solution to the shortage of reproducibly safe and functional cells for therapeutic applications could be the use of human embryonic stem cells (hESC), which can serve as a nearly unlimited cell source for regenerative medicine. After the first work reporting derivation of stable hESC lines was published (Thomson *et al.*, 1998), this field has been almost exponentially growing, many new hESC lines having been derived around the world (Andrews *et al.*, 2005; Hoffman and Carpenter, 2005), and various differentiation derivatives of these lines generated and characterized *in vitro* (Conley *et al.*, 2004; Gerecht-Nir and Itskovitz-Eldor, 2004; Liew *et al.*, 2005; Sathananthan and Trounson, 2005).

GENERATION OF RPE FROM ES CELL

ES cells are progeny of the inner cell mass (ICM) of a blastocyst, the same totipotent entity that gives rise to a whole new organism, and can remain pluripotent virtually indefinitely. Pluripotent hESC express a set of molecular markers, such as octamer-binding protein 4 (Oct-4), stage-specific embryonic antigens (SSEA)-3 and SSEA-4, tumor rejection antigens (TRA)-1-60, TRA-1-80, alkaline phosphatase, Nanog (Carpenter *et al.*, 2003; Hoffman and Carpenter, 2005) and when growing in culture on a feeder layer or feeder-free in defined conditions maintain a very specific morphology (Figure 28-1a). They form flat colonies comprised of small, tightly packed cells with a high ratio of nuclei to cytoplasm, with clear boundaries between the cells and usually sharp, refractile colony borders. Similar to the cells of the ICM that differentiate into predetermined lineages, ES cells in culture easily differentiate, and the first detectable signs of such commitment are the loss of the unique ES cell morphology (Figure 28-1c and d) and downregulation of molecular markers of pluripotency.

Differentiation *in vitro* can be spontaneous when formation of various lineages happens rather unpredictably in

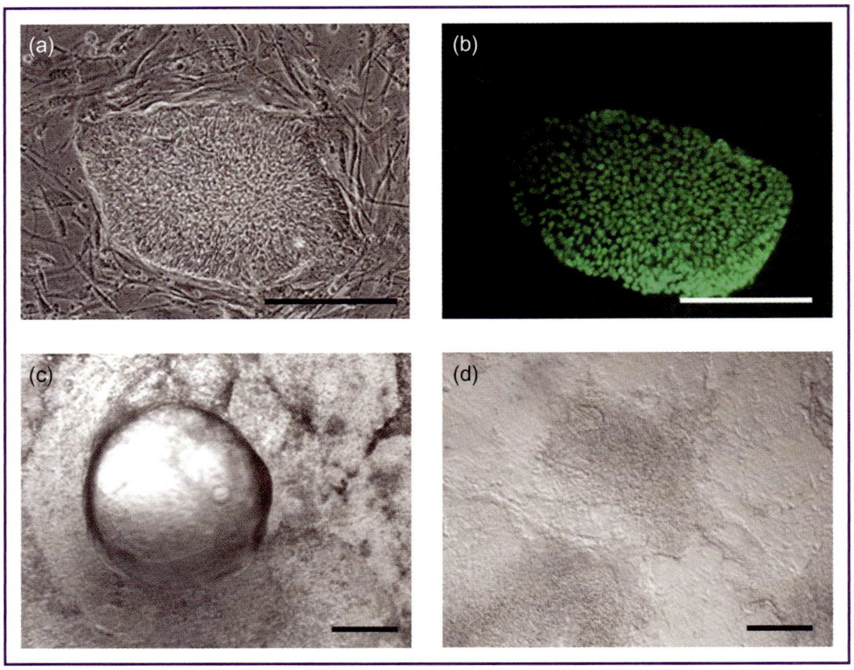

FIGURE 28-1 Growth and differentiation of hESC. (a, b) undifferentiated colonies of hESC. (a) phase contrast, (b) Oct-4 staining, (c, d) spontaneous differentiation of hESC, Scale bar: (a, b) 200 μm, (c, d) 500 μm.

different cultures, or it can be directed toward a variety of derivatives by the use of various agents that can selectively activate different pathways. To date, the differentiation of human and mouse ES cells into numerous cell types has been reported (reviewed by Smith, 2001). However, such directed differentiation does not usually produce a pure population of a single type of derivatives because its orchestration usually requires a cross-talk of signals from different co-differentiating cell types which are close to each other in culture, similar to patterning *in vivo*. Thus most differentiation models currently published have only partial efficacy, and isolation of the desired differentiation derivative is required for obtaining a pure cell population.

There are currently several reports on derivation of RPE-like cells from ES cells of various any species. In 2002 Kawasaki *et al.* (2002) published a study describing derivation of dopamine neurons and pigmented epithelia from primate ES cells in the presence of stromal cell-derived inducing activity (SDIA) coming from a feeder layer of mouse skull bone marrow PA6 cells. The authors found patches of pigmented epithelial cells in their cultures, which stained positively for Pax6 and showed cortical actin distribution; such cells could be mechanically isolated under the microscope and passaged to confluency on PA6 cells or on collagen. This work was continued, and 2 years later another report was published describing *in vitro* and *in vivo* characterization of such pigmented epithelial cells (Haruta *et al.*, 2004). The cells expressed mRNA for RPE65, CRALBP, and MERTK, which are all characteristic markers for RPE cells, and were able to perform phagocytosis of latex beads and support the function of the photoreceptor when transplanted in RCS rats. In another study, SDIA promoted the formation of various structures, pigmented epithelial cells among them, in the cultures of mouse ES cells differentiating in the presence of PA6 feeder cells (Hirano *et al.*, 2003).

Our group has generated RPE from hESC and found that hESC can reliably differentiate into RPE without any such aid, presumably as the "default" neural lineage commitment taken another step further (Klimanskaya *et al.*, 2004). Such pigmented cells usually appear within 6–8 weeks after passaging ES cells, independently of the presence of mouse feeder cells in the starting culture. We were able to isolate putative RPE cells from every hESC line we used for these experiments (the total is currently over 20), establish primary cultures of up to 8 passages, and evaluate the molecular profiles of such cultured cells by immunostaining, RT-PCR, real-time PCR, Western blot, and gene expression analysis.

In our system hESC are routinely cultured in serum-free medium, containing serum replacement (Invitrogen) and Plasmanate (Bayer) on mouse embryonic fibroblast feeder layers in the presence of 10–20 nm/ml human leukemia inhibitory factor (LIF) and 8–16 ng/ml of human bFGF; for passaging we use trypsin or mechanical dispersion of the colonies (Klimanskaya and McMahon, 2005). After 5–7 days in culture, we usually see signs of differentiation, when the typical ES cell morphology is lost and various differentiated cell types appear. Most colonies usually show signs of neural lineage commitment, including cells which stain positively for tubulin βIII, Pax6 (Figures 28-1b and 23-2), and glial fibrillary acidic protein (GFAP). These observations are in agreement with numerous observations

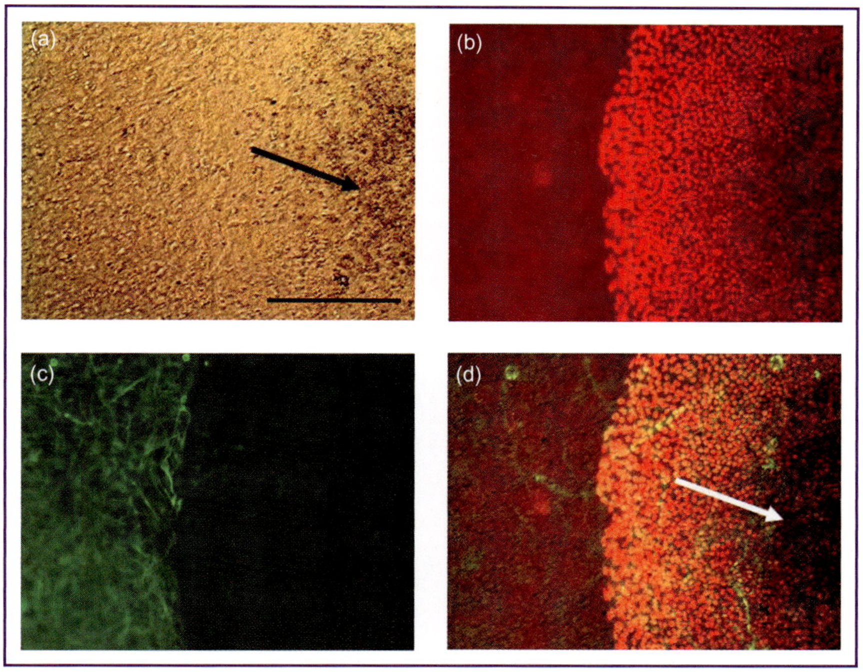

FIGURE 28-2 Formation of RPE in spontaneously differentiating hESC cultures. (a) phase contrast, (b) Pax6 staining, (c) tubulin βIII staining, (d) merged, (a–c, e) petri dish with pigmented culters, (f) embryoid body culture with pigmented cells on the surface. The arrows show the position of pigmented cells. Scale bar: (a–d) 100 μm, (e) 3 mm, (f), 200 μm. The parts of the figure are reproduced from Klimanskaya *et al.* (2004). with the permission from the publisher Mary Ann Liebert, Inc.

in the literature that ES cells in culture select the neuronal pathway of differentiation most readily, which could be chosen by default (Tropepe *et al.*, 2001; Smukler *et al.*, 2006) or in response to autocrine activity of FGF (Ying *et al.*, 2003; Bouhon *et al.*, 2005) or as a result of elimination of other inductive signals (Ying *et al.*, 2003). After 7–10 days of culture, when the majority of the cells have lost their ES cell morphology and molecular markers of pluripotency, the medium is changed to "differentiation medium" which has no LIF, no FGF, no Plasmanate. It is possible that one of the reasons for massive neural induction is the presence of bFGF in the ES culture medium when the initial commitment is being made.

The plates are then cultured until the clusters of pigmented epithelial cells begin to appear, which usually happens in 6–8 weeks. Such clusters keep slowly increasing in size, while new clusters continue to emerge. The same process can be initiated in conventional embryoid body culture (EB), in which case pigmented epithelial cells would appear on the surface of EBs, and then this transition of non-pigmented cells to pigmented epithelium would slowly take over the whole EB.

Of note: in such differentiating systems we found clusters of cells positively stained for neural lineage markers Pax6 and/or tubulin βIII, often in close conjunction with pigmented epithelium; moreover, we frequently saw a pattern that looks like a transition from neural progenitors to RPE within the same cluster: as Pax6 and tubulin βIII staining becomes weaker and disappears toward the center in such clusters (Figure 28-2), pigmentation and epithelial morphology become more prominent (Figure 28-2). Such Pax6-positive cells could be similar to the multipotent cells of the optic vesicle, and the loss of Pax6 from more mature cells may be analogous to the same process during RPE maturation in eye development. The cells stained positively for tubulin βIII may be precursors of neural retina cells, and our histological examination of the same specimens revealed clusters that resemble disorganized rods and cones. RT-PCR performed on the same samples showed the presence of these cell types (data not shown). In addition, we usually saw cells of various types, still unidentified, in the same differentiating cultures of hESC, surrounding the clusters of RPE and their presumptive progenitors. It was possible that the cells producing signals promoting RPE specification in clusters of Pax6-positive progenitors, similar to the signaling of ocular mesoderm in patterning ocular tissues, could be found among such differentiated cells next to Pax6-positive clusters.

These weeks-old cultures were comprised of several layers of cells with a lot of ECM deposition, which made it difficult to disperse them into a single cell suspension to select the desired cell type using fluorescence-activated cell sorter (FACS) or magnetic beads. Instead, we used an approach when the multi-layer of cells was loosened with trypsin or collagenase and the pigmented cells were picked under the dissecting microscope using a glass capillary. Collected cells were plated on laminin or gelatin in RPE culture medium containing serum replacement and fetal bovine serum (FBS) with optional bFGF, and in 48h we found clusters of attached, spread cells which were beginning to proliferate (Figure 28-3b). Proliferating cells lost pigment and acquired a fibroblastic phenotype (Figure 28-2c), strongly resembling the transdifferentiated RPE which dedifferentiated as they proliferated and returned to typical RPE morphology after they established a monolayer (Figure 28-3d), which usually took 2–3 weeks (Reh *et al.*, 1987; Vinores *et al.*, 1995; Sakaguchi *et al.*, 1997; Chen *et al.*, 2003). Such RPE transdifferentiation has been shown to result in formation of neuronal, amacrine, and photoreceptor cells (Zhao *et al.*, 1995), glia (Sakaguchi *et al.*, 1997), neural retina (Galy *et al.*, 2002), and neuronal progenitors (Opas and Dziak, 1994). bFGF accelerated transdifferentiation and RPE proliferation, thus allowing the cells to reach confluence and begin to revert to the RPE phenotype much sooner. ES-derived RPE (ES–RPE) in the transdifferentiated state expressed the neural markers Pax6 and tubulin βIII, strongly resembling immature neural cells, and our comparative gene expression profiling showed their similarity to neural stem cells (Klimanskaya *et al.*, 2004).

CULTURE AND PROPERTIES OF HES-DERIVED RPE

For any regenerative medicine application it is important that the cells can be produced in quantities large enough for thorough characterization and safety testing, and because the life span of differentiated cells is limited, these quantities need to be obtained before the cells get close to senescence and lose any of their RPE functions. RPE are relatively "slow" cells: even in the presence of bFGF that accelerates their transdifferentiation and proliferation, it may take up to 2–3 weeks at each passage at a 1:3 ratio before they "mature" and fully re-gain the RPE phenotype. We tested whether it is possible to passage the ES–RPE while they are still transdifferentiated, so the desired cell quantity can be achieved in less time, and after that allow them to differentiate, rather than letting the cells to go through a full cycle of dedifferentiation–proliferation–differentiation at every passage. We found that the cells that were "rushed"—passaged immediately after they established confluency—lost their ability to re-gain the RPE phenotype several passages earlier than the RPE which went through the whole cycle prior to each passaging.

Such cell behavior requiring slow propagation may seem to impede scaling up the cells for characterization and therapy, especially because the yield of RPE from any given differentiating dish is quite low: from several clusters of RPE cells usually found in one 35 mm plate of differentiating ES cells (each cluster usually has several hundred cells; some large older ones may have several thousand) two or three confluent wells of a four-well plate can be produced in 3–4 weeks. After that the cells are usually subcultured at 1:3–1:6 ratio at 2–3 week intervals, so as to generate several million of cells at P2–P3, which would barely be enough for thorough assessment and one

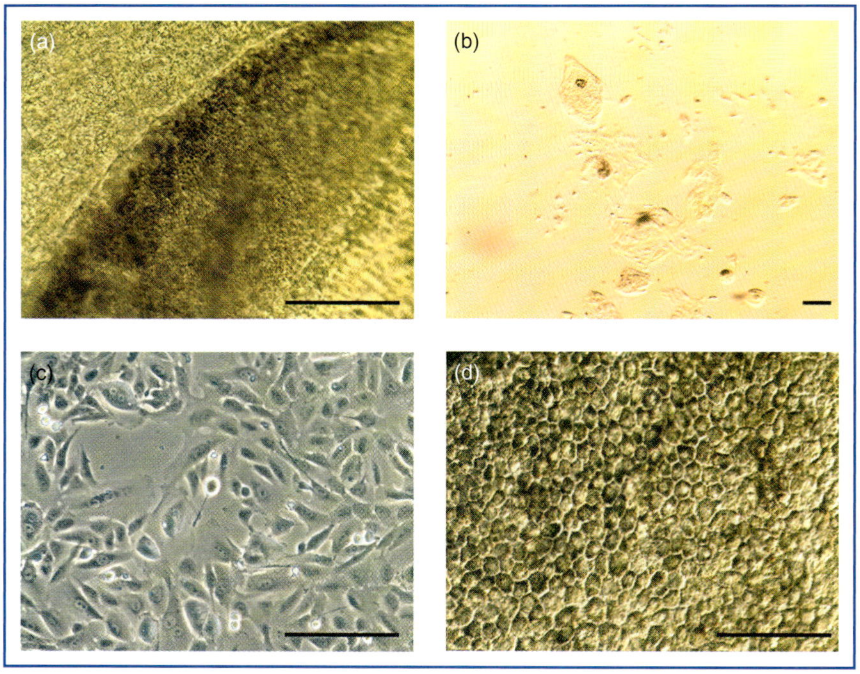

FIGURE 28-3 Morphology of hES-derived RPE in culture. (a) Differentiated cluster of RPE cells surrounded by other cell types, (b) isolated RPE cells after 5 days in culture, (c) transdifferentiated passaged hES–RPE, (d) morphology of hES–RPE monolayer in culture. Scale bar: (a, b, c) 200 μm, (d) 100 μm. The parts of the figure are reproduced from Klimanskaya *et al.* (2004). with the permission from the publisher Mary Ann Liebert, Inc.

transplantation application, takes several weeks. However, with ES cells as starting material there is a simple solution: setting up large scale differentiating ES cell cultures, high numbers of RPE can be obtained at the earliest passages.

Furthermore, better understanding of the signaling mechanisms guiding the differentiation of ES cells in culture would allow us to devise a step-wise procedure employing an efficient combination of growth factors and ECM to manipulate the differentiation process and increase the yield of RPE cells in less time. Since animal co-culture, as in studies with PA6 mouse stromal cells, is undesirable for potential therapeutic applications, some of the approaches might, for instance, focus on producing Pax6-positive retinal progenitors and exposing them to activin A or co-culturing ES cell-derived neuroectoderm with "autologous" (derived from the same ES cell line) mesoderm.

After the culture of differentiation derivatives of ES cells is established, the next important step is to characterize the cells at the molecular and functional level. In our studies we are using bestrophin, CRALBP, RPE65, and PEDF as RPE molecular markers. Pax6, although seen by some authors as a molecular marker of RPE (Kawasaki *et al.*, 2002), is normally downregulated in mature RPE, so it could rather indicate the presence of immature cells. Our ES–RPE cells showed a remarkable resemblance to cultured or intact human RPE: CRALBP, PEDF, and bestrophin were detected by Western blot and by immunofluorescence (bestrophin and CRALBP) and enzyme-linked immunosorbent assay (ELISA) (PEDF). Translationally controlled RPE65 was not found at the protein level, although the real-time RT-PCR has detected high levels of RPE65 mRNA. Interestingly, the level of its expression correlated with the differentiation: in more mature cultures its expression was several times higher than in recently passaged cells (Klimanskaya *et al.*, 2004).

Functional tests for characterization of potential transplantation candidates could include RPE-specific phagocytosis using an assay with labeled rod fragments (Finnemann *et al.*, 1997; Finnemann, 2003) and vitamin A metabolism assay. Another useful "quality assurance" assay could be cell adhesion assessment: literature and our own data indicate that there may be certain variations in integrin expression, and it is yet to be determined whether this is genetic variation or a result of culture condition and in this case could be modulated by plating cells on various ECMs or by using bioactive substances. For instance, loss of integrin alpha 5 in post-confluent culture was described by Proulx *et al.* (2003) and this may impair adhesion of the cells used for transplantation. In our ES–RPE cells we detected a certain variation in expression of α_5 integrin which could be the outcome of such post-confluent cultures. Recently $\alpha_V\beta_5$ was identified as a receptor involved in retinal adhesion (Nandrot *et al.*, 2006) in addition to its role in phagocytosis (reviewed by Finnemann, 2003), and it is unclear yet how much variation in expression of various integrins may be tolerated by the RPE cells without significant loss of their adhesion qualities and functionality.

However, all the molecular and functional *in vitro* characterization data need to be assessed with understanding of tissue culture limitations which may be reversible. The same cells that fail one criteria in culture may turn out to be excellent performers in the natural eye environment. Currently there is not enough data in the literature correlating the molecular profile and *in vitro* assayed functions of cultured and freshly isolated RPE with its ability to support the function of the photoreceptor in animal models. This is applicable to any stem cell derivative: the molecular profiles, morphology, and function could appear quite similar to their *in vitro* counterparts, but even a seemingly subtle difference may turn out to be crucial for performance of the transplanted derivatives *in vivo*—or it may be

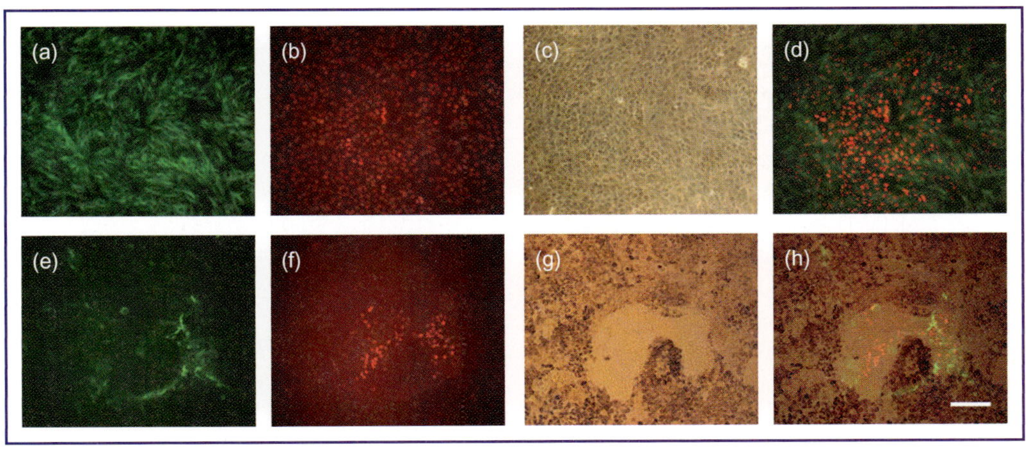

FIGURE 28-4 Transdifferentiation in hES-RPE cultures: immunolocalization of tubulin βIII and Pax6 in recently passaged dedifferentiated ((a–d) 3 days after passaging) and "mature" ((e–h) 3 weeks after passaging) cultures of RPE-like cells. (a, e) tubulin βIII, (b, f) Pax6, (c, g) corresponding phase contrast microscopy field, (d) merged, (a–c), (h) merged (e–g). Scale bar, 100 μm. The figure is reproduced from Klimanskaya *et al.* (2004). with the permission from the publisher Mary Ann Liebert, Inc.

reversible when the shortcomings of the culture system are replaced with natural environment.

Gene expression profiling performed on hEC-derived RPE versus fetal human RPE tissue showed their remarkable similarity (Klimanskaya *et al.*, 2004). The data were also compared with previously published data (Rogojina *et al.*, 2003) on human RPE cell ARPE-19 (spontaneously derived and shown to attenuate the loss of visual function in RCS rats, Lund *et al.*, 2001) and D407 (transformed), and it appeared that hES–RPE showed much more resemblance to fetal RPE than any of these *in vivo* originated RPE cell lines, the latter two lacking expression of many RPE-specific genes. For instance, neither of these two lines expressed bestrophin, lecitin retinol acyltransferase, retinal G-protein coupled receptor, or RPE65. Our preliminary data show that different hES–RPE lines used for transplantation in RCS rats had different efficacy lines in photoreceptor support, and we are currently correlating these differences with the phenotypes of the lines, number of passages after differentiation, molecular markers, and adhesion receptor expression. Some of the possible reasons for different transplant performance could be genetic differences between the lines, or differences in culture conditions and number of passages, which in turn would determine the differentiation potential of the cells. We found that at early passages after derivation hES–RPE transdifferentiate and differentiate to RPE phenotype quite easily, producing a homogeneous monolayer of pigmented polygonal cells. With more divisions and higher passage numbers we start seeing more elongated non-pigmented cells, and after the cells undergo 6–8 passages, they lose the ability to revert to RPE phenotype after transdifferentiation.

For any regenerative medicine applications it is of utmost importance to be able to predict the therapeutic value of any particular culture of stem cell derivatives, and for RPE, assessment of morphology and expression of molecular markers, even at the gene chip scale, may not be sufficient due to high plasticity of these cells. For instance, fully differentiated cells may have a disadvantage attaching to the Bruch's membrane and proliferating, while highly transdifferentiated cells may act unpredictably in the subretinal space environment. On the other hand, the *in vivo* microenvironment may be able to instruct the cells much better than can be done in culture conditions. If the transplanted cells attach and survive, they still have to prove functionality by restoring the function of the photoreceptor, and this may depend on their intrinsic abilities, which may not be necessarily directly dependent on their molecular profiles. It is important to perform a series of experiments correlating the performance of the cells in animal models with their *in vitro* assessment, and find parameters that would ensure their grafting, survival, and function. A possible solution would be to use gene expression profiling approach in combination with *in vitro* functional assessment of differently isolated and cultured derivatives, and/or on progenies of several different ES cell lines and to compare the behavior of the same cells in animal models in order to identify the crucial molecular markers for the survival and function of the candidate cells.

Still the paramount test of the cells' quality is their performance in clinical and preclinical trials. RPE considered for clinical trials need to be proven safe and efficient in appropriate animal models. Several animal studies with other hES-derived cells (neural progenitors) showed that they formed tumors, even though they were Oct-4 negative (Arnhold *et al.*, 2004), and RPE would need to successfully pass safety tests in animals in addition to pathogen clearance. Controlled efficacy studies are also a part of the preclinical trials that can be performed in an adequate animal model. Our own studies have shown that in RCS rat model hES-derived RPE are able to extensively resuce the photoreceptor in this model of inherited retinal degeneration (Lund *et al.*, 2006) reaching 100% improvement over the untreated control. If hES-derived RPE can successfully pass preclinical safety and efficacy tests, it can become one of the first hES-derived cell products for regenerative medicine applications.

ACKNOWLEDGMENTS

I thank Sandy Becker for critical reading of the manuscript and helpful comments and apologize to all those whose work was not mentioned due to space limitations.

REFERENCES

Acland, G.M., Aguirre, G.D., Ray, J., Zhang, Q., Aleman, T.S., Cideciyan, A.V., Pearce-Kelling, S.E., Anand, V., Zeng, Y., Maguire, A.M., Jacobson, S.G., Hauswirth, W.W. and Bennett, J. (2001). Gene therapy restores vision in a canine model of childhood blindness. *Nat. Genet.* May, *28*(1): 92–95.

Algvere, P.V., Gouras, P. and Dafgard Kopp, E. (1999). Long-term outcome of RPE allografts in non-immunosuppressed patients with AMD. *Eur J Ophthalmol.* July–September, *9*(3): 217–230.

Allikmets, R., Seddon, J.M., Bernstein, P.S., Hutchinson, A., Atkinson, A., Sharma, S., Gerrard, B., Li, W., Metzker, M.L., Wadelius, C., Caskey, C.T., Dean, M. and Petrukhin, K. (1999). Evaluation of the best disease gene in patients with age-related macular degeneration and other maculopathies. *Hum. Genet.* June, *104*(6): 449–453.

Andrews, P.W., Benvenisty, N., McKay, R., Pera, M.F., Rossant, J., Semb, H. and Stacey, G.N. (2005). Steering committee of the international stem cell initiative. The international stem cell initiative: toward benchmarks for human embryonic stem cell research. *Nat. Biotechnol.* July, *23*(7): 795–797.

Arnhold, S., Klein, H., Semkova, I., Addicks, K. and Schraermeyer, U. (2004). Neurally selected embryonic stem cells induce tumor formation after long-term survival following engraftment into the subretinal space. *Invest Ophthalmol Vis Sci.* December, *45*(12): 4251–4255.

Ashery-Padan, R. and Gruss, P. (2001). Pax6 lights-up the way for eye development. *Curr. Opin. Cell Biol.* December, *13*(6): 706–714.

Bakall, B., Marmorstein, L.Y., Hoppe, G., Peachey, N.S., Wadelius, C. and Marmorstein, A.D. (2003). Expression and localization of bestrophin during normal mouse development. *Invest. Ophthalmol. Vis. Sci.* August, *44*(8): 3622–3628.

Baumer, N., Marquardt, T., Stoykova, A., Spieler, D., Treichel, D., Ashery-Padan, R. and Gruss, P. (2003). Retinal pigmented epithelium determination requires the redundant activities of Pax2 and Pax6. *Development* July, *130*(13): 2903–2915.

Berglin, L., Gouras, P., Sheng, Y., Lavid, J., Lin, P.K., Cao, H. and Kjeldbye H. (1997). Tolerance of human fetal retinal pigment epithelium xenografts in monkey retina. *Graefes Arch. Clin. Exp. Ophthalmol.* February, *235*(2): 103–110.

Besharse, J.C. and Defoe, D.M. (1998). In: Marmor, M.F. and Wolfensberger, T.J. (eds.), *Role of the Retinal Pigment Epithelium in Photoreceptor Membrane Turnover* New York, Oxford: Oxford University Press, p. 152–172.

Binder, S., Stanzel, B.V., Krebs, I. and Gillenberg, C. (2007). Transplantation of the RPE in AMD. *Prog Retin Eye Res.* September, *9*(3): 217–230.

Binder, S., Krebs, I., Hilgers, R.D., Abri, A., Stolba, U., Assadoulina, A., Kellner, L., Stanzel, B.V., Jahn, C. and Feichtinger, H. (2004). Outcome of transplantation of autologous retinal pigment epithelium in age-related macular degeneration: a prospective trial. *Invest. Ophthalmol. Vis. Sci.* November, *45*(11): 4151–4160.

Bouhon, I.A., Kato, H., Chandran, S. and Allen, N.D. (2005). Neural differentiation of mouse embryonic stem cells in chemically defined medium. *Brain Res. Bull.* December 15, *68*(1–2): 62–75.

Bunt-Milam, A.H. and Saari, J.C. (1983). Immunocytochemical localization of two retinoid-binding proteins in vertebrate retina. *J. Cell Biol.* September, *97*(3): 703–712.

Carpenter, M.K., Rosier, E. and Rao, M.S. (2003). Characterization and differentiation of human embryonic stem cells. *Cloning Stem Cells* *5*(1): 79–88.

Chen, S., Samuel, W., Fariss, R.N., Duncan, T., Kutty, R.K. and Wiggert, B. (2003). Differentiation of human retinal pigment epithelial cells into neuronal phenotype by *N*-(4-hydroxyphenyl)retinamide. *J. Neurochem.* March, *84*(5): 972–981.

Chow, R.L. and Lang, R.A. (2001). Early eye development in vertebrates. *Annu. Rev. Cell Dev. Biol.* *17*: 255–296.

Chow, R.L., Altmann, C.R., Lang, R.A. and Hemmati-Brivanlou, A. (1999). Pax6 induces ectopic eyes in a vertebrate. *Development* October, *126*(19): 4213–4222.

Conley, B.J., Young, J.C., Trounson, A.O. and Mollard, R. (2004). Derivation, propagation and differentiation of human embryonic stem cells. *Int. J. Biochem. Cell Biol.* April, *36*(4): 555–567.

Cowan, C.A., Klimanskaya, I., McMahon, J., Atienza, J., Witmyer. J., Zucker, J.P., Wang, S., Morton, C.C., McMahon, A.P., Powers, D. and Melton, D.A. (2004). Derivation of embryonic stem-cell lines from human blastocysts. *N. Engl. J. Med.* March 25, *350*(13): 1353–1356.

Crafoord, S., Algvere, P.V., Seregard, S. and Kopp, E.D. (1999). Long-term outcome of RPE allografts to the subretinal space of rabbits. *Acta Ophthalmol. Scand.* June, *77*(3): 247–254.

D'Cruz, P.M., Yasumura, D., Weir, J., Matthes, M.T., Abderrahim, H., LaVail, M.M. and Vollrath, D. (2000). Mutation of the receptor tyrosine kinase gene Mertk in the retinal dystrophic RCS rat. *Hum. Mol. Genet.* March 1, *9*(4): 645–651.

Eguchi, G. (1986). Instability in cell commitment of vertebrate pigmented epithelial cells and their transdifferentiation into lens cells. *Curr. Top. Dev. Biol.* *20*: 21–37.

Eichers, E.R., Green, J.S., Stockton, D.W., Jackman, C.S., Whelan, J., McNamara, J.A., Johnson, G.J., Lupski, J.R. and Katsanis, N. (2002). Newfoundland rod-cone dystrophy, an early-onset retinal dystrophy, is caused by splice-junction mutations in RLBP1. *Am. J. Hum. Genet.* April, *70*(4): 955–964.

El Dirini, A.A., Wang, H.M., Ogden, T.E. and Ryan, S.J. (1992). Retinal pigment epithelium implantation in the rabbit: technique and morphology. *Graefes Arch. Clin. Exp. Ophthalmol.* *230*(3): 292–300.

Feijen, A., Goumans, M.J. and van den Eijnden-van Raaij, A.J. (1994). Expression of activin subunits, activin receptors and follistatin in postimplantation mouse embryos suggests specific developmental functions for different activins. *Development* December, *120*(12): 3621–3637.

Feng, W., Yasumura, D., Matthes, M.T., LaVail, M.M. and Vollrath, D. (2002). Mertk triggers uptake of photoreceptor outer segments during phagocytosis by cultured retinal pigment epithelial cells. *J. Biol. Chem.* May 10, *277*(19): 17016–17022.

Finnemann, S.C. (2003). Role of alphavbeta5 integrin in regulating phagocytosis by the retinal pigment epithelium. *Adv. Exp. Med. Biol.* *533*: 337–342.

Finnemann, S.C., Bonilha, V.L., Marmorstein, A.D. and Rodriguez-Boulan, E. (1997). Phagocytosis of rod outer segments by retinal pigment epithelial cells requires alpha(v)beta5 integrin for binding but not for internalization. *Proc. Natl Acad. Sci. USA* November 25, *94*(24): 12932–12937.

Fischer, A.J. and Reh, T.A. (2001). Transdifferentiation of pigmented epithelial cells: a source of retinal stem cells? *Dev. Neurosci.* *23*(4–5): 268–276.

Fuhrmann, S., Levine, E.M. and Reh, T.A. (2000). Extraocular mesenchyme patterns the optic vesicle during early eye development in the embryonic chick. *Development* November, *127*(21): 4599–4609.

Galy, A., Li, Y., Thompson, D.A., Weir, J., Orth, U., Jacobson, S.G., Apfelstedt-Sylla, E. and Vollrath, D. (2000). Mutations in MERTK, the human orthologue of the RCS rat retinal dystrophy gene, cause retinitis pigmentosa. *Nat. Genet.* November, *26*(3): 270–271.

Galy, A., Neron, B., Planque, N, Saule, S. and Eychene, A. (2002). Activated MAPK/ERK kinase (MEK-1) induces transdifferentiation of pigmented epithelium into neural retina. *Dev. Biol.* August 15, *248*(2): 251–264.

Gehring, W.J. and Ikeo, K. (1999). Pax 6: mastering eye morphogenesis and eye evolution. *Trends Genet.* September, *15*(9): 371–377.

Gerecht-Nir, S. and Itskovitz-Eldor, J. (2004). The promise of human embryonic stem cells. *Best Pract. Res. Clin. Obstet. Gynaecol.* December, *18*(6): 843–852.

Girman, S.V., Wang, S. and Lund, R.D. (2003). Cortical visual functions can be preserved by subretinal RPE cell grafting in RCS rats. *Vision Res.* August, *43*(17): 1817–1827.

Girman, S.V., Wang, S. and Lund, R.D. (2005). Time course of deterioration of rod and cone function in RCS rat and the effects of subretinal cell grafting: a light- and dark-adaptation study. *Vision Res.* February, *45*(3): 343–354.

Gouras, P., Flood, M.T., Kjeldbye, H., Bilek, M.K. and Eggers, H. (1985). Transplantation of cultured human retinal epithelium to Bruch's membrane of the owl monkey's eye. *Curr. Eye Res.* March, *4*(3): 253–265.

Grierson, I., Hiscott, P., Hogg, P., Robey, H., Mazure, A. and Larkin, G. (1994). Development, repair and regeneration of the retinal pigment epithelium. *Eye* *8*(Pt 2): 255–262.

Grindley, J.C., Davidson, D.R. and Hill, R.E. (1995). The role of Pax-6 in eye and nasal development. *Development* May, *121*(5): 1433–1442.

Gullapalli, V.K., Sugino, I.K., Van Patten, Y., Shah, S. and Zarbin, M.A. (2004). Retinal pigment epithelium resurfacing of aged submacular human Bruch's membrane. *Trans. Am. Ophthalmol. Soc. 102*: 123–37; discussion 137–138.

Gullapalli, V.K., Sugino, I.K., Van Patten, Y., Shah, S. and Zarbin, M.A. (2005). Impaired RPE survival on aged submacular human Bruch's membrane. *Exp. Eye Res.* February, *80*(2): 235–248.

Halder, G., Callaerts, P. and Gehring, W.J. (1995). New perspectives on eye evolution. *Curr. Opin. Genet. Dev.* October, *5*(5): 602–609.

Hamel, C.P., Tsilou, E., Pfeffer, B.A., Hooks, J.J., Detrick, B. and Redmond, T.M. (1993). Molecular cloning and expression of RPE65, a novel retinal pigment epithelium-specific microsomal protein that is post-transcriptionally regulated *in vitro*. *J. Biol. Chem.* July 25, *268*(21): 15751–15757.

Haruta, M., Sasai, Y., Kawasaki, H., Amemiya, K., Ooto, S., Kitada, M., Suemori, H., Nakatsuji, N., Ide, C., Honda, Y., Takahashi, M. (2004). *In vitro* and *in vivo* characterization of pigment epithelial cells differentiated from primate embryonic stem cells. *Ophthalmol Vis Sci.* March *45*(3): 1020–1025.

Hirano, M., Yamamoto, A., Yoshimura, N., Tokunaga, T., Motohashi, T., Ishizaki, K., Yoshida, H., Okazaki, K., Yamazaki, H., Hayashi, S. and Kunisada, T. (2003). Generation of structures formed by lens and retinal cells differentiating from embryonic stem cells. *Dev. Dynam.* December, *228*(4): 664–671.

Hitchcock, P.F., Macdonald, R.E., VanDeRyt, J.T. and Wilson, S.W. (1996). Antibodies against Pax6 immunostain amacrine and ganglion cells and neuronal progenitors, but not rod precursors, in the normal and regenerating retina of the goldfish. *J. Neurobiol.* March, *29*(3): 399–413.

Hoffman, L.M. and Carpenter, M.K. (2005). Characterization and culture of human embryonic stem cells. *Nat. Biotechnol.* June, *23*(6): 699–708.

Jablonski, M.M., Tombran-Tink, J., Mrazek, D.A. and Iannaccone, A. (2000). Pigment epithelium-derived factor supports normal development of photoreceptor neurons and opsin expression after retinal pigment epithelium removal. *J. Neurosci.* October 1, *20*(19): 7149–7157.

Karakousis, P.C., John, S.K., Behling, K.C., Surace, E.M., Smith, J.E., Hendrickson, A., Tang, W.X., Bennett, J. and Milam, A.H. (2001). Localization of pigment epithelium derived factor (PEDF) in developing and adult human ocular tissues. *Mol. Vis.* June 30, *7*:154–163.

Kawasaki, H., Suemori, H., Mizuseki, K., Watanabe, K., Urano, F., Ichinose, H., Haruta, M., Takahashi, M., Yoshikawa, K., Nishikawa, S., Nakatsuji, N. and Sasai, Y. (2002). Generation of dopaminergic neurons and pigmented epithelia from primate ES cells by stromal cell-derived inducing activity. *Proc. Natl Acad. Sci. USA* February 5, *99*(3): 1580–1585. Epub January 29.

Klimanskaya, I. and McMahon, J. (2004). Approaches for derivation and maintenance of human ES cells: detailed procedures and alternatives. In: Lanza, R., Gearhart, J., Hogan, B., Melton, D., Pedersen, R., Thomson, J., *et al* ., editors. Handbook of Stem Cells. Vol 1: *Embryonic Stem Cells,* San Diego: Elsevier/Academic Press, 437–449.

Klimanskaya, I., Hipp, J., Rezai, K.A., West, M., Atala, A. Lanza, R. (2004). Derivation and comparative assessment of retinal pigment epithelium from human embryonic stem cells using transcriptomics. *Cloning Stem Cells 6*(3): 217–245.

Kramer, F., White, K., Pauleikhoff, D., Gehrig, A., Passmore, L., Rivera, A., Rudolph, G., Kellner, U., Andrassi, M., Lorenz, B., Rohrschneider, K., Blankenagel, A., Jurklies, B., Schilling, H., Schutt, F., Holz, F.G. and Weber, B.H. (2000). Mutations in the VMD2 gene are associated with juvenile-onset vitelliform macular dystrophy (Best disease) and adult vitelliform macular dystrophy but not age-related macular degeneration. *Eur. J. Hum. Genet.* April, *8*(4): 286–292.

Kumar, J.P. (2001). Signalling pathways in Drosophila and vertebrate retinal development. *Nat. Rev. Genet.* November, *2*(11): 846–857.

Liew, C.G., Moore, H., Ruban, L., Shah, N, Cosgrove, K., Dunne, M. and Andrews, P. (2005). Human embryonic stem cells: possibilities for human cell transplantation. *Ann. Med. 37*(7): 521–532.

Liu, S.Y. and Redmond, T.M. (1998). Role of the 3′-untranslated region of RPE65 mRNA in the translational regulation of the RPE65 gene: identification of a specific translation inhibitory element. *Arch. Biochem. Biophys.* September 1, *357*(1): 37–44.

Lopez, R., Gouras, P., Brittis, M. and Kjeldbye, H. (1987). Transplantation of cultured rabbit retinal epithelium to rabbit retina using a closed-eye method. *Invest. Ophthalmol. Vis. Sci.* July, *28*(7): 1131–1137.

Lopez, R., Gouras, P., Kjeldbye, H., Sullivan, B., Reppucci, V., Brittis, M., Wapner, F. and Goluboff, E. (1989). Transplanted retinal pigment epithelium modifies the retinal degeneration in the RCS rat. *Invest. Ophthalmol. Vis. Sci.* March, *30*(3): 586–588.

Lund, R.D., Adamson, P., Sauve, Y., Keegan, D.J., Girman, S.V., Wang, S., Winton, H., Kanuga, N., Kwan, A.S., Beauchene, L., Zerbib, A., Hetherington, L., Couraud, P.O., Coffey, P. and Greenwood, J. (2001a). Subretinal transplantation of genetically modified human cell lines attenuates loss of visual function in dystrophic rats. *Proc. Natl Acad. Sci. USA* August 14, *98*(17): 9942–9947.

Lund, R.D., Kwan, A.S., Keegan, D.J., Sauve, Y., Coffey, P.J. and Lawrence, J.M. (2001b). Cell transplantation as a treatment for retinal disease. *Prog. Retin. Eye Res.* July, *20*(4): 415–449.

Lund R.D., Wang S., Klimanskaya I., Holmes T., Ramos-Kelsey R., Li B., Girman S., Bischoff N., Sauve Y. and Lanza R. (2006). Human embryonic stem cell-derived cells rescue visual function in dystrophic RCS rats. *Cloning and Stem Cells, 8*(3).

Ma, J., Zhang, J., Othersen, K.L., Moiseyev, G., Ablonczy, Z., Redmond, T.M., Chen, Y. and Crouch, R.K. (2001). Expression, purification, and MALDI analysis of RPE65. *Invest. Ophthalmol. Vis. Sci.* June, *42*(7): 1429–1435.

Marlhens, F., Bareil, C., Griffoin, J.M., Zrenner, E., Amalric, P., Eliaou, C., Liu, S.Y., Harris, E., Redmond, T.M., Arnaud, B., Claustres, M. and Hamel, C.P. (1997). Mutations in RPE65 cause Leber's congenital amaurosis. *Nat. Genet.* October, *17*(2): 139–141.

Marmorstein, A.D., Marmorstein, L.Y, Rayborn, M., Wang, X., Hollyfield, J.G. and Petrukhin, K. (2000). Bestrophin, the product of the Best vitelliform macular dystrophy gene (VMD2), localizes to the basolateral plasma membrane of the retinal pigment epithelium. *Proc. Natl Acad. Sci. USA* November 7, *97*(23): 12758–12763.

Marquardt, A., Stohr, H., Passmore, L.A., Kramer, F., Rivera, A. and Weber, B.H. (1998). Mutations in a novel gene, VMD2, encoding a protein of unknown properties cause juvenile-onset vitelliform macular dystrophy (Best's disease). *Hum. Mol. Genet.* September, *7*(9): 1517–1525.

Martinez-Morales, J.R., Rodrigo, I. and Bovolenta, P. (2004). Eye development: a view from the retina pigmented epithelium. *Bioessays* July, *26*(7): 766–777.

Maw, M.A., Kennedy, B., Knight, A., Bridges, R., Roth, K.E., Mani, E.J., Mukkadan, J.K., Nancarrow, D., Crabb, J.W. and Denton, M.J. (1997). Mutation of the gene encoding cellular retinaldehyde-binding protein in autosomal recessive retinitis pigmentosa. *Nat. Genet.* October, *17*(2): 198–200.

Morimura, H., Berson, E.L. and Dryja, T.P. (1999). Recessive mutations in the RLBP1 gene encoding cellular retinaldehyde-binding protein in a form of retinitis punctata albescens. *Invest. Ophthalmol. Vis. Sci.* April, *40*(5): 1000–1004.

Moshiri, A., Close, J. and Reh, T.A. (2004). Retinal stem cells and regeneration. *Int. J. Dev. Biol. 48*(8–9): 1003–1014.

Nandrot, E.E, Anand, M., Sircar, M. and Finnemann, S.C. (2006). Novel role for {alpha}vbeta5-integrin in retinal adhesion and its diurnal peak. *Am. J. Physiol. Cell Physiol.* April, *290*(4): C1256-C1262.

Nguyen, M. and Arnheiter, H. (2000). Signaling and transcriptional regulation in early mammalian eye development: a link between FGF and MITF. *Development* August, *127*(16): 3581–3591.

Opas, M. and Dziak, E. (1994). bFGF-induced transdifferentiation of RPE to neuronal progenitors is regulated by the mechanical properties of the substratum. *Dev. Biol.* February, *161*(2): 440–454.

Perrault, I., Rozet, J.M., Gerber, S., Ghazi, I., Leowski, C., Ducroq, D., Souied, E., Dufier, J.L., Munnich, A. and Kaplan, J. (1999). Leber congenital amaurosis. *Mol. Genet. Metab.* October, *68*(2): 200–208.

Petrukhin, K., Koisti, M.J., Bakall, B., Li, W., Xie, G., Marknell, T., Sandgren, O., Forsman, K., Holmgren, G., Andreasson, S., Vujic, M., Bergen, A.A., McGarty-Dugan, V., Figueroa, D., Austin, C.P., Metzker, M.L., Caskey, C.T. and Wadelius, C. (1998). Identification of the gene responsible for Best macular dystrophy. *Nat. Genet.* July, *19*(3): 241–247.

Phillips, S.J., Sadda, S.R., Tso, M.O., Humayan, M.S., de Juan Jr., E. and Binder, S. (2003). Autologous transplantation of retinal pigment epithelium after mechanical debridement of Bruch's membrane. *Curr. Eye Res.* February, *26*(2): 81–88.

Pittack, C., Grunwald, G.B. and Reh, T.A. (1997). Fibroblast growth factors are necessary for neural retina but not pigmented epithelium differentiation in chick embryos. *Development* February, *124*(4): 805–816.

Proulx, S., Guerin, S.L. and Salesse, C. (2003). Effect of quiescence on integrin alpha5beta1 expression in human retinal pigment epithelium. *Mol. Vis.* October 3, *9*: 473–481.

Proulx, S., Landreville, S., Guerin, S.L. and Salesse, C. (2004). Integrin alpha5 expression by the ARPE-19 cell line: comparison with primary RPE cultures and effect of growth medium on the alpha5 gene promoter strength. *Exp. Eye Res.* August, *79*(2): 157–165.

Radtke, N.D., Seiler, M.J., Aramant, R.B., Petry, H.M. and Pidwell, D.J. (2002). Transplantation of intact sheets of fetal neural retina with its retinal pigment epithelium in retinitis pigmentosa patients. *Am. J. Ophthalmol.* April, *133*(4): 544–550.

Radtke, N.D., Aramant, R.B., Seiler, M.J., Petry, H.M. and Pidwell, D. (2004). Vision change after sheet transplant of fetal retina with retinal pigment epithelium to a patient with retinitis pigmentosa. *Arch. Ophthalmol.* August, *122*(8): 1159–1165.

Redmond, T.M., Yu, S., Lee, E., Bok, D., Hamasaki, D., Chen. N., Goletz, P., Ma, J.X., Crouch, R.K. and Pfeifer, K. (1998). Rpe65 is necessary for production of 11-cis-vitamin A in the retinal visual cycle. *Nat. Genet.* December, *20*(4): 344–351.

Reh, T.A., Nagy, T. and Gretton, H. (1987). Retinal pigmented epithelial cells induced to transdifferentiate to neurons by laminin. *Nature* November 5–11, *330*(6143): 68–71.

Rogojina, A.T., Orr, W.E., Song, B.K. and Geisert Jr., E.E. (2003). Comparing the use of Affymetrix to spotted oligonucleotide microarrays using two retinal pigment epithelium cell lines. *Mol. Vis.* October 6, *9*: 482–196.

Saari, J.C., Nawrot, M., Kennedy, B.N., Garwin, G.G., Hurley, J.B., Huang, J., Possin, D.E. and Crabb, J.W. (2001). Visual cycle impairment in cellular retinaldehyde binding protein (CRALBP) knockout mice results in delayed dark adaptation. *Neuron* March, *29*(3): 739–748.

Sakaguchi, D.S., Janick, L.M. and Reh, T.A. (1997). Basic fibroblast growth factor (FGF-2) induced transdifferentiation of retinal pigment epithelium: generation of retinal neurons and glia. *Dev. Dynam.* August, *209*(4): 387–398.

Sathananthan, A.H. and Trounson, A. (2005). Human embryonic stem cells and their spontaneous differentiation. *Ital. J. Anat. Embryol. 110*(2 Suppl 1): 151–157.

Sauve, Y., Girman, S.V., Wang, S., Keegan, D.J. and Lund, R.D. (2002). Preservation of visual responsiveness in the superior colliculus of RCS rats after retinal pigment epithelium cell transplantation. *Neuroscience 114*(2): 389–401.

Smith, A.G. (2001). Embryo-derived stem cells: of mice and men. *Annu. Rev. Cell Dev. Biol. 17*: 435–162.

Smukler, S.R., Runciman, S.B., Xu, S. and van der Kooy, D. (2006). Embryonic stem cells assume a primitive neural stem cell fate in the absence of extrinsic influences. *J. Cell. Biol.* January 2, *172*(1): 79–90.

Steele, F.R., Chader, G.J., Johnson, L.V. and Tombran-Tink, J. (1993). Pigment epithelium-derived factor: neurotrophic activity and identification as a member of the serine protease inhibitor gene family. *Proc. Natl Acad. Sci. USA* February 15, *90*(4): 1526–1530.

Tezel, T.H. and Del Priore, L.V. (1999). Repopulation of different layers of host human Bruch's membrane by retinal pigment epithelial cell grafts. *Invest. Ophthalmol. Vis. Sci.* March, *40*(3): 767–774.

Tezel, T.H., Kaplan, H.J. and Del Priore, L.V. (1999). Fate of human retinal pigment epithelial cells seeded onto layers of human Bruch's membrane. *Invest Ophthalmol Vis Sci.* February, *40*(2): 467–476.

Thompson, D.A. and Gal, A. (2003). Vitamin A metabolism in the retinal pigment epithelium: genes, mutations, and diseases. *Prog. Retin. Eye Res.* September, *22*(5): 683–703.

Thomson, J.A., Itskovitz-Eldor, J., Shapiro, S.S., Waknitz, M.A., Swiergiel, J.J., Marshall, V.S. and Jones, J.M. (1998). Embryonic stem cell lines derived from human blastocysts. *Science* November 6, *282*(5391): 1145–1147. Erratum in: *Science* December 4, *282*(5395): 1827.

Tropepe, V., Hitoshi, S., Sirard, C., Mak, T.W., Rossant, J. and van der Kooy, D. (2001). Direct neural fate specification from embryonic stem cells: a primitive mammalian neural stem cell stage acquired through a default mechanism. *Neuron* April, *30*(1): 65–78.

Van Hooser, J.P., Aleman, T.S., He, Y.G., Cideciyan, A.V., Kuksa, V., Pittler, S.J., Stone, E.M., Jacobson, S.G. and Palczewski, K. (2000). Rapid restoration of visual pigment and function with oral retinoid in a mouse model of childhood blindness. *Proc. Natl Acad. Sci. USA* July 18, *97*(15): 8623–8628.

van Meurs, J.C., ter Averst, E., Hofland, L.J., van Hagen, P.M., Mooy, C.M., Baarsma, G.S., Kuijpers, R.W., Boks, T. and Stalmans, P. (2004). Autologous peripheral retinal pigment epithelium translocation in patients with subfoveal neovascular membranes. *Br. J. Ophthalmol.* January, *88*(1): 110–113.

Verdugo, M.E., Ailing, J., Lazar, E.S., del Cerro, M., Ray, J. and Aguirre, G. (2001). Posterior segment approach for subretinal transplantation or injection in the canine model. *Cell Transplant. 10*(3): 317–327.

Veske, A., Nilsson, S.E., Narfstrom, K. and Gal, A. (1999). Retinal dystrophy of Swedish briard/briard-beagle dogs is due to a 4-bp deletion in RPE65. *Genomics* April 1, *57*(1): 57–61.

Vinores, S.A., Derevjanik, N.L., Mahlow, J., Hackett, S.F., Haller, J.A., deJuan, E., Frankfurter, A. and Campochiaro, P.A. (1995). Class III beta-tubulin in human retinal pigment epithelial cells in culture and in epiretinal membranes. *Exp Eye Res.* 1995 Apr; *60*(4): 385–400.

Vogel-Hopker, A., Momose, T., Rohrer, H., Yasuda, K., Ishihara, L. and Rapaport, D.H. (2000). Multiple functions of fibroblast growth factor-8 (FGF-8) in chick eye development. *Mech. Dev.* June, *94*(1–2): 25–36.

Walther, C. and Gruss, P. (1991). Pax-6, a murine paired box gene, is expressed in the developing CNS. *Development* December, *113*(4): 1435–1449.

Wang, S., Lu, B. and Lund, R.D. (2005a). Morphological changes in the Royal College of Surgeons rat retina during photoreceptor degeneration and after cell-based therapy. *J. Comp. Neurol.* October 31, *491* (4): 400–417.

Wang, S., Lu, B., Wood, P. and Lund, R.D. (2005b). Grafting of ARPE-19 and schwann cells to the subretinal space in RCS rats. *Invest. Ophthalmol. Vis. Sci.* July, *46*(7): 2552–2560.

Weisz, J.M., Humayun, M.S., De Juan, Jr., E., Del Cerro, M., Sunness, J.S., Dagnelie, G., Soylu, M., Rizzo, L. and Nussenblatt, R.B. (1999). Allogenic fetal retinal pigment epithelial cell transplant in a patient with geographic atrophy. *Retina 19*(6): 540–545.

Ying, Q.L., Stavridis, M., Griffiths, D., Li, M. and Smith, A. (2003). Conversion of embryonic stem cells into neuroectodermal precursors in adherent monoculture. *Nat. Biotechnol.* February, *21*(2): 183–186.

Zarbin, M.A. (2003). Analysis of retinal pigment epithelium integrin expression and adhesion to aged submacular human Bruch's membrane. *Trans. Am. Ophthalmol. Soc. 101*: 499–520.

Zhao, S., Thornquist, S.C. and Barnstable, C.J. (1995). *In vitro* transdifferentiation of embryonic rat retinal pigment epithelium to neural retina. *Brain Res.* April 24, *677*(2): 300–310.

Zhao, S., Rizzolo, L.J. and Barnstable, C.J. (1997). Differentiation and transdifferentiation of the retinal pigment epithelium. *Int. Rev. Cytol. 171*: 225–266.

Zhao, S., Hung, F.C, Colvin, J.S., White, A., Dai, W, Lovicu, F.J., Ornitz, D.M. and Overbeek, P.A. (2001). Patterning the optic neuroepithelium by FGF signaling and Ras activation. *Development* December, *128*(24): 5051–5060.

Part V

Ethical and Regulatory Considerations

Chapter 29

Ethical Considerations

Ronald M. Green

INTRODUCTION

Human embryonic stem (hES) cell research is ethically controversial, because it entails the destruction of human embryos. Here, I consider and discuss a series of key ethical questions associated with this research. These include whether it is morally permissible to destroy a human embryo, and whether we should clone human embryos for therapeutic cloning research. I conclude by stating my own position, that hES cell and therapeutic cloning research are both acceptable—and I offer some guidelines that have been developed by several authoritative bodies for the actual conduct of this research.

There is consensus in the scientific community that hES cell research holds great promise for developing new treatments for a variety of serious and currently untreatable disease conditions (National Research Council, 2001; Office of Science Policy, 2001). However, because hES cell research requires the manipulation and destruction of human embryos, it has also been a focus of ethical controversy and opposition. In the course of these debates, several challenging ethical questions have been raised. Scientists, clinicians, or patients involved in hES cell research or therapies must formulate their answers to these questions. Society, too, must address them to determine the extent to which hES cell research may require oversight and regulation. Here, I present these questions, and examine some of the leading answers that have been proposed to them.

IS IT MORALLY PERMISSIBLE TO DESTROY A HUMAN EMBRYO?

Human embryonic stem cell lines are currently made by chemically and physically disaggregating an early, blastocyst-stage embryo, and removing its inner cell mass. At this stage, the embryo is composed of approximately 200 cells, including an outer layer of differentiated placental material, and the undifferentiated (pluripotent) cells of the inner cell mass. The embryo dies as a result of this procedure. Hence the question: can we intentionally kill a developing human being at this stage to expand scientific knowledge, and potentially provide medical benefits?

It is true that a method involving single-cell blastomere biopsy has been developed that could obviate the need to destroy the embryo in order to develop an hES cell line (Klimanskaya *et al.*, 2006). New technologies for producing induced pluripotent stem (iPS) cells (see below) hold out the promise of replacing the use of embryos altogether. Nevertheless, many existing hES cell lines were created by destroying embryos. There are also hundreds of thousands of frozen embryos that could still be used to create new lines (Weiss, 2003). These realities make the question of the embryo's moral status, and the ethics of its destruction, still relevant.

At one end of the spectrum are those who believe that, in moral terms, human life begins at conception. For those holding this view, the early embryo is morally no different from a child or an adult human being. It cannot be used in research that is not to its benefit, and it cannot be used without its consent (The Linacre Centre for Healthcare Ethics, 2001; Pope John Paul II, 2003). Furthermore, proxy consent by parents in such cases is inadmissible, since it is an accepted rule of pediatric research that parents may not volunteer a child for studies that are not potentially beneficial, and that risk the child's life. Many Roman Catholics, evangelical Protestants, and some Orthodox Jews take the position that life (morally) begins at conception, and they oppose hES cell research. At the other end of the spectrum are those who believe that the embryo is not yet fully a human being in a moral sense. They hold a "developmental" or "gradualist" view of life's beginning (Ford, 1988; McConnick, 1991; Warren, 1997; Green, 2001; Shannon, 2001). They do not deny that the early embryo is alive, and has the biological potential to become a person. Nevertheless, they believe that other features are needed for the full and equal protection we normally accord children and adults, and that these features only develop gradually across the full term of gestation. These features include such things as bodily form, and the ability to feel or think.

The early embryo, they maintain, does not have these features or abilities. They note also that the very early embryo lacks human individuality, since it can still undergo twinning at this early stage, and two separate embryos with distinct genomes can fuse to become a single individual (Strain *et al.*, 1998). The very high mortality rate of such embryos (most never implant) also reduces the force of the argument from potentiality (Hardy *et al.*, 2001; Norwitz *et al.*, 2001). Those who hold this view do not agree on the classes of research that warrant the destruction of embryos, but most support some form of hES cell research. Their reasoning is that, although the early embryo may merit some respect as a nascent form of human life (Lebacqz, 2001), the lives and health of children and adults outweigh whatever claim it possesses.

Each individual faced with involvement in hES cell research must arrive at his or her own answer to this first question. Legislators and others must also wrestle with these issues. Because American law (and the laws of most other nations) does not regard the early embryo as a person meriting the legal protections afforded to children and adults, it is hard to see how one can justify legal or regulatory prohibitions on privately-financed hES cell research or clinical applications. Such prohibitions would interfere with individual liberty, on grounds inconsistent with the reduced status of the embryo shown elsewhere in the law. However, because publicly-funded research rests on more narrowly political considerations, including the way that a majority chooses to spend public funds, it may be expected that public support for hES cell research will depend on how a majority of citizens answer this first question.

SHOULD WE POSTPONE HUMAN EMBRYONIC STEM CELL RESEARCH?

Some who oppose the destruction of human embryos maintain that hES cell research should be deferred, at least until science provides a better view of the likely benefits of adult stem cell research. They maintain that this research direction is as promising as hES cell research, and they argue that the moral acceptability of the alternative justifies any delay in the availability of therapies (Doerflinger, 1999). Others have argued that the scientific uncertainty, and ethically controversial nature of hES cell research, warrants a moratorium on it in both the private and the public sectors (President's Council on Bioethics, 2002). Scientific studies using adult stem cells continue to be equivocal. They have sometimes supported, and at other times undermined, claims to the plasticity of adult stem cells, their ability to immortalize, and their usefulness (Pearson, 2002). This raises the question of whether it is justified to delay the development of therapies and cures for children and adults to protect embryos and respect the sensitivities of those opposed to embryo research. Many people feel that such a delay is not warranted, and that it is scientifically and morally preferable to keep multiple pathways to stem cell therapies open. In the words of the National Research Council, "The application of stem cell research to therapies for human disease will require much more knowledge about the biological properties of all types of stem cells" (National Research Council, 2001).

CAN WE BENEFIT FROM OTHERS' DESTRUCTION OF EMBRYOS?

It might seem that a negative answer to the first question ends discussion. If it is morally wrong to destroy a human embryo, what could justify the use of cells derived from its deliberate destruction for other people's benefit? However, research or therapy has many steps, not all of which involve the destruction of embryos. This raises the question of whether downstream researchers, clinicians, or patients may use the stem cell lines that others have derived. Ethically, this is a question of whether we can ever benefit from deeds with which we disagree morally, or regard as morally wrong (Kaveny, 2000; Green, 2002).

This question arises partly because most embryos used to produce stem cell lines are left over from infertility procedures. Couples using in vitro fertilization (IVF) routinely create more embryos than can safely be implanted. There are hundreds of thousands of these embryos in cryogenic freezers around the country, and around the world (Weiss, 2003). Since few frozen embryos are made available for adoption (Stolberg, 2001), most of these supernumerary embryos will be destroyed. In 1996, British law mandated the destruction of 3600 such embryos (Ibrahim, 1996). This destruction continues, regardless of whether some embryos are diverted to hES cell research.

Why is it morally wrong to benefit from others' wrongdoing? One answer is that by doing so, we encourage similar deeds in the future. This explains why we are morally and legally prohibited from receiving stolen goods, or why it may be wrong to benefit from research produced by scientists who choose to ignore human-subject constraints (Green, 2002). However, it seems less objectionable to benefit from others' wrongdoing when their deeds are independently undertaken, when our choices are not connected to theirs, and when the choices do not encourage wrongful deeds. For example, few people would object to using the organs from a teenage victim of a gang killing to save the life of another dying child. The use of such organs benefits one young person, and does not encourage teen violence. Similar logic might apply to stem cell research, using spare embryos remaining from infertility procedures. A downstream researcher, clinician, or patient may abhor the deeds that led to the existence of an hES cell line, including the creation and destruction of excess human embryos in infertility medicine. But nothing that a recipient of an hES cell line chooses to do is likely to alter, prevent, or discourage the continuing creation or destruction of human embryos, or make the existing lines go away. Those who use such embryos also know that if they refuse to use an hES cell line, they could forego great therapeutic benefit. A researcher will fail to develop a lifesaving or health-restoring therapy. A clinician's decision may risk a patient's life. People in this position will struggle with the question of whether it is worthwhile to uphold a moral ideal, when doing so has no practical effect, and when it threatens injury to others.

Religious views on the question of whether one may ever benefit from others' wrongdoing are diverse. The Roman Catholic moral tradition, despite its staunch opposition to complicity with wrongdoing, presents different answers, including some that permit one to derive benefit in particular cases (Miller, 1967; Ashley and O'Rourke, 1989; Green, 2002). This suggests that some researchers, clinicians, or patients who morally oppose the destruction of human embryos might ethically use hES cell lines derived from embryos otherwise slated for destruction.

It is noteworthy that in his August 2001 address to the nation, President George W. Bush adopted a conservative version of the position that allows one to benefit from acts one morally opposes. Stating his belief that it is morally wrong to kill a human embryo for others' benefit, the president nevertheless permitted the use of existing stem cell lines on the grounds that the deaths of the embryos had already occurred (Bush, 2001). Presumably, the president believed that it would not encourage further destruction of embryos to permit this use. The president did not go so far as to permit the use of lines derived from embryos certainly slated for destruction, because their progenitors had chosen never to use them. However, in April 2009, the National Institutes of Health, did just this. Implementing an Executive order issued in March by President Barack Obama, the NIH authorized funding for the use, but not the derivation of new hES cell lines from embryos remaining from infertility procedures (National Institutes of Health, 2009).

CAN WE CREATE AN EMBRYO TO DESTROY IT?

A fourth question takes us into even more controversial territory. Is it ever morally permissible to create an embryo deliberately to produce a stem cell line? This was done in the summer of 2002 at the Jones Institute in Norfolk, VA (Kolata, 2001). Those in favor of this research defend it on at least two grounds. First, they say that in the future, if we seek to develop stem cell lines with special properties, and perhaps closer genetic matches to tissue recipients, it will be necessary to produce stem cell lines to order using donor sperm and eggs. Second, they argue that it is ethically better to use an hES cell line created from embryos that have been produced just for this purpose, with the full and informed consent of their donor progenitors, than to use cell lines from embryos originally created for a different, reproductive purpose.

Those who believe the early embryo is our moral equal oppose the deliberate creation of embryos for research or clinical use. They are joined by some that do not share this view of the embryo's status, but who believe that it is morally repellant to deliberately create a potential human being only to destroy it (Annas *et al.*, 1996). Such people argue that this research opens the way to the "instrumentalization" of all human life, and the use of children or adult human beings as commodities. Some ask whether such research does not violate the Kantian principle that we should never use others as "a means only" (Green, 2001).

On the other side of this debate are those who believe that the reduced moral status of the early embryo permits its creation and destruction for lifesaving research and therapies (National Institutes of Health, 1994; Davis, 1995). The proponents of this research direction ask why it is morally permissible to create supernumerary embryos in IVF procedures to help couples have children, but morally wrong to do the same thing to save a child's life. They are not persuaded by the reply that the status of the embryo is affected by its progenitors' intent, and that it is therefore permissible to create excess embryos for a "good" (reproductive) purpose, but not for a "bad" (research) purpose. They point out that the embryo is the same entity, and its status should not depend on its progenitors' intentions. Thus, we do not ordinarily believe that a child's rights are dependent on its parents' intent or degree of concern (Parens, 2001). They conclude that it is not parental intent that warrants the creation of excess of embryos in such cases, but the embryos' lesser moral status and the likelihood of significant human benefit from their use. These same considerations, they believe, justify deliberately creating embryos for stem cell research.

SHOULD WE CLONE HUMAN EMBRYOS?

A fifth question is whether we are willing to support human cloning for stem cell research. This question arises in connection with a specific stem cell technology known as "human therapeutic cloning." It involves the deliberate creation of an embryo by somatic cell nuclear transfer technology (cloning) to produce an immunologically compatible (isogenic) hES cell line (Lanza *et al.*, 2000).

Immune rejection could occur if the embryo used to prepare a line of hES cells for transplant does not share the same genome as the recipient. This would be the case whether the cell line was created from a spare embryo, or from one made to order. Therapeutic cloning offers a way around this problem. In the case of a diabetic child, the mother could donate an egg whose nucleus would then be removed. A cell would be taken from the child's body, and its nucleus inserted into the egg cytoplasm. With stimulation, the reconstructed cell would divide, just like a fertilized egg. If the resulting embryo were transferred back to a womb, it could go on to birth and become a new individual—a clone of the child. But in therapeutic cloning, the blastocyst would be dissected and an hES cell line prepared. Growth factors could be administered to induce the cells to become replacement pancreatic cells for the child. Because these cells contain the child's own DNA, and even the same maternal mitochondrial DNA, they would not be subject to rejection. Research has also shown that the small amount of alien RNA from the mitochondria of an egg other than the mother's would probably not provoke an immune rejection (Lanza *et al.*, 2002).

Although this is a very promising technology, it raises a host of novel ethical questions. One is whether the embryonic organism produced in this way should be regarded as a "human embryo" in the accepted sense of that term (Nature, 2001). Those who believe that "life begins at conception" tend to answer this question affirmatively, even though cloned "embryos" are not the result of sexual fertilization (Doerflinger, 2003). They base their view on the biological similarities between cloned and sexually produced embryos, and on the argument that both have the potential to become a human being. Nevertheless, the very high mortality rate of cloned embryos suggests significant biological differences from sexually produced embryos. Furthermore, if the embryos' status rests on their potential, this potential is more than significantly reduced; in an era of cloning, some degree of potentiality attaches to all bodily cells.

The promise of this technology rests on the ability to make stem cell lines "to order" for a specific patient. If you hold the view that the cloned organism is morally equivalent to a human embryo, therapeutic cloning research and therapies again raise the question of whether it is morally permissible to create an embryo deliberately to destroy it.

Another question often raised in this context is whether therapeutic cloning will create an enormous demand for human eggs. If it does, some believe this may create substantial problems of social justice, since collecting these eggs may involve hundreds or thousands of women, many of whom are likely to be poor women of color attracted by the financial rewards of egg donation (Dresser, 2001; Holland, 2001). Those who discount this problem offer several arguments. They point out that even if many eggs are needed for therapeutic cloning procedures, they are likely to be provided by relatives of patients. This reduces the magnitude of a market in oocytes. Others observe that therapeutic cloning is likely to be "transitional research" (Chief Medical Officer's Expert Group, 2000). As such, it may eventually be replaced by direct somatic cell reprogramming, eliminating the need for eggs altogether.

Finally, there is a moral question specific to cloning itself. The more scientists are able to perfect therapeutic cloning, the more likely it is that they will sharpen the skills needed to accomplish reproductive cloning, which aims at the birth of a cloned child. There is a broad consensus in the scientific and bioethics communities that, for the foreseeable future, the state of cloning technology poses serious health risks to any child born as a result of it (Jaenisch and Wilmut, 2001). There are also serious, unresolved questions about the psychological welfare of such children (National Bioethics Advisory Commission, 1997). Finally, there is the possibility that embryos created for therapeutic cloning research might be diverted to reproductive cloning attempts. All these concerns raise the question: do we really want to develop cloning technology for the production of isogenic stem cells, if doing so hastens the advent of reproductive cloning (Weiss, 1999)?

In 2001, and again in 2003, the US House of Representatives answered "no" to this question, and passed a bill introduced by Rep. James Weldon that banned both reproductive and therapeutic cloning. Similar bills have been introduced in the Senate (Stolberg, 2001, 2003). Although the Senate initiatives have been stalled for some time, this may change (New York Times, 2003). If a Senate bill passes, therapeutic cloning research and therapies will be outlawed in the United States. Similar prohibitions are either in effect, or being considered for passage in continental Europe (BBC News, 2006). This would leave only a relatively small number of countries, including Great Britain, Israel, and China, in which such research would be allowed (Israel Academy of Sciences and Humanities, 2001).

Those who oppose these prohibitions believe that therapeutic and reproductive cloning research can be decoupled. They observe that strict regulations and governmental oversight, of the sort provided in Great Britain by the Human Fertilisation and Embryology Authority, make it unlikely that embryos produced for therapeutic cloning will be diverted to reproductive purposes. They also point out that several researchers or groups with minimal qualifications in cloning research have announced their intent to clone a child, or have even tried to do so. Such attempts are likely to continue, regardless of whether therapeutic cloning research is banned. As a result, a ban on therapeutic cloning will not protect children, and will only have the negative effect of interrupting beneficial stem cell research.

DOES INDUCED PLURIPOTENT STEM CELL RESEARCH ELIMINATE THE ETHICAL QUESTIONS SURROUNDING HUMAN EMBRYONIC STEM CELL RESEARCH?

In late 2007, teams led by Yamanaka in Japan and Thomson in the US announced success in the use of gene transfer technology to produce pluripotent stem cell lines (Takahashi *et al.*, 2007; Yu *et al.*, 2007). Some believe that this development has ended debates about the appropriateness of using hES cell lines (Krauthammer, 2007). If we can directly manipulate somatic cells, why should we engage in the ethically controversial activity of destroying embryos? However, there are scientific and ethical issues that challenge this view.

On the scientific side, there is a question of whether iPS cell lines will prove suitable for use in human transplant and cell regeneration therapies. Current iPS cell lines exhibit high rates of tumorigenicity in mice, a result of the use of retroviral vectors to carry pluripotency-inducing transcription factors, including the cancer-related factor c-Myc (Hyun *et al.*, 2007). Direct up-regulation of the relevant genes and replacement of c-Myc may eventually overcome this problem, but until this is accomplished, iPS cell lines cannot ethically be used for human transplant purposes. They may be of value, however, for the creation of model lines of disease-related cells.

Also questionable is the widespread assumption that inducing pluripotency in somatic cells raises no ethical questions. It is true that protecting every body cell from gene manipulation would seem absurd. But is this so if the genetic manipulation of a cell renders it capable of eventual development into a human being? The work of Nagy and others shows that ES cells, when inserted into tetraploid embryos, are able to develop the placental material needed for further development (Nagy *et al.*, 1990, 1993; Magill and Neaves, 2009). This possibility raises complex ethical questions. Is potentiality morally relevant if it is accompanied by such intensive technical interventions? IVF or cloned embryos also require intensive technical interventions. Those who favor iPS cell research, but oppose the use of IVF or cloned embryos as sources of pluripotent stem cells will have to examine the consistency of their views. If it is acknowledged that the potentiality for full human development, however assisted, confers moral status on an entity, then iPS research may not be the panacea that hES cell research opponents believe it to be.

WHAT ETHICAL GUIDELINES SHOULD GOVERN HUMAN EMBRYONIC STEM CELL, THERAPEUTIC CLONING, AND INDUCED PLURIPOTENT STEM CELL RESEARCH?

Mention of the need to prevent the diversion of cloned embryos to reproductive purposes raises the larger question of what guidelines should apply to the conduct of hES cell, therapeutic cloning, and iPS research. In August 2000, the US National Institutes of Health released a series of guidelines for hES cell research that never went into effect, because they were largely preempted by President Bush's decision to limit hES cell research to existing cell lines (National Institutes of Health, 2000). Guidelines have also been developed by the Chief Medical Officer's Expert Group in Great Britain (Holland, 2001), by private ethics boards at the Geron Corporation (Geron Advisory Board, 1999) and Advanced Cell Technology (Green *et al.*, 2002) in the US, by committees of the National Research Council (Committee on Guidelines for Human Embryonic Stem Cell Research, National Research Council 2005; National Research Council and Institute of Medicine of the National Academies, 2007, 2008), and by the International Society for Stem Cell Research (ISSCR, 2006). These various recommendations share several features.

Escro Review

First, there is agreement that, in addition to the usual scientific and review of research proposals by an Institutional Review Board, it is appropriate that there be another layer of review provided by an Embryonic Stem Cell Research Oversight (ESCRO) committee. ESCROs are needed to review protocols involving special issues, such as the transplanting of hES cell materials into human subjects, and research that generates chimeric animals using human cells. This research is particularly sensitive, because of the unknown effects of differentiation and integration of the human cells into nonhuman animal tissues, especially cerebral cortex or germline. There are special cautions about breeding chimeric animals in which human pluripotent cells have been introduced into germline tissues (Krauthammer, 2007).

Donor and Procurement Issues

Second, because hES cell, therapeutic cloning, and iPS cell research require a supply of human gametes and embryos, any of which raise sensitive issues of moral status, steps must be taken to elicit the informed consent of donors, to protect their privacy, and to minimize any risks to which they might be subject. Because non-embryo-derived iPS cells are derived from human material, their derivation is covered by existing IRB regulations concerning review and informed consent, and no ESCRO committee review may be necessary. Nevertheless, the possibility of the use of such cells in chimeric animals and special transplantation issues probably recommend ESCRO review. To aid researchers, both the NRC and ISSCR have offered the outlines of model consent documents for these purposes (Hyun *et al.*, 2007; Krauthammer, 2007).

In general, informed consent requires that donors fully understand the nature of the research being undertaken (for example that it may require the destruction of an embryo), and that they explicitly consent to that research. Whenever possible, they should be given the option of agreeing to some forms of hES cell research, but not others (e.g. the creation of stem cells, but not cloned embryos). It is morally impermissible to elicit somatic cells, sperm, eggs, or embryos for the production of stem cell lines without informing donors that an immortalized, pluripotent cell line might result that could be widely used in research or therapeutic applications. If there are likely to be commercial benefits flowing from the research, donors must also be informed of this, and their rights (if any) in such benefits should be clearly specified. If the research involves therapeutic cloning, both egg and somatic cell donors must be informed that a cloned embryo and a cell line with the egg donor's mitochondrial genetic material and a somatic cell donor's nuclear DNA will result. IPS cell somatic cell donors should also be informed about the developmental and commercial possibilities of their cellular materials. Special questions also arise about the use of somatic cells, embryos, or sperm whose provenance is uncertain, including embryos created with sperm whose donor is unknown (Sugarman and Siegel, 2008).

In conducting research, efforts should be made to preserve donor privacy, by removing identifying information from gametes, embryos, and cell lines, and keeping this information apart in a secure location. In view of the controversy surrounding much of this research, donors can be subjected to harassment or embarrassment if their association with the research is revealed without their consent.

Therapeutic cloning research raises the special issue of ovulation induction. This is an invasive medical procedure, with both known and undetermined risks (Paulson, 1996; Rossing *et al.*, 1996). Not only must egg donors be informed of these risks, but steps also must be taken to preserve the voluntary nature of their consent. This includes avoiding undue financial incentives. It also includes preventing them from being pressured into producing excess eggs or embryos for research in return for discounts on infertility services (Cohen, 2001). Although it is unreasonable to expect even altruistically-motivated donors to undergo the inconvenience and risks of these procedures without some form of compensation, payment should not be so high as to lead a donor to ignore the risks involved. For Advanced Cell Technology's therapeutic cloning egg donor program, its Ethics Advisory Board set a payment level similar to that established for reproductive egg donors in the New England area. Payments were also prorated for the degree of participation in the program, to allow donors to drop out at any time. Levels of stimulatory medications were maintained on the low side of current regimens, and payments were never attached to the number of eggs harvested. Additional protections ensured that donors had the educational level and background needed to appreciate the risks involved. A study monitor was employed to ensure donors' consent was free and informed.

Research Conduct

Guidelines also apply to the overall conduct of research. These include the requirement that no embryo used in hES cell or therapeutic cloning research be allowed to develop beyond fourteen days. This limit is based on the substantial changes that occur at gastrulation, which marks the beginning of individualization and organogenesis (O'Rahilly and Mülller, 1992). Supervision and accountability of all staff and scientists involved in this research to prevent any diversion of gametes or embryos to reproductive purposes is also required.

Transplantation Research

If and when hES cell lines become available for research in transplant therapies, researchers and institutional review boards will have to address ethical questions raised by their use. For example, if such lines were cultured on mouse or other feeder layers, as almost all of the federally-approved lines were, there will be a safety question regarding the possible introduction of exogenous retroviruses or other pathogens into the human population. Basing their judgments on adequate preliminary animal studies, researchers and oversight bodies will have to assess the risks of rejection, including graft-versus-host disease. Issues surrounding the tumorigenicity of hES and iPS cells will also have to be resolved. Experience with fetal cell transplants for Parkinson's disease shows that cells can behave in unexpected ways when transplanted into the human body (Kolata, 2001). None of these problems are insurmountable, but their existence constitutes an argument for continued research in this area under careful ethical oversight.

SUMMARY

Fully answering all of the questions I have raised would require an ethical treatise. However, in my view, the moral claims of the very early embryo do not outweigh those of children and adults that can be helped by hES cell and therapeutic cloning technologies. While iPS cell research is promising, reliance on it should not, in my view, replace the use of hESC lines produced through proven techniques, and that are currently available. I also do not believe that therapeutic cloning research will lead to reproductive cloning, which we should take firm steps to forbid. I recognize that others may disagree with these conclusions. As our debates move forward, continuing dialogue about these questions, and clearer scientific research results will bring us closer to a national consensus on these issues.

KEY WORDS

Bioethics The study of the norms of conduct that should govern research and clinical applications of biomedical technologies.

Stem cells Self-renewing (immortalized), unspecialized cells that can give rise to multiple types of all specialized cells.

Therapeutic cloning The deliberate creation of a human organism by nuclear transfer (cloning) technology in order to produce an embryo that can be used to create a stem cell line that will not provoke an immune response and rejection.

FURTHER READING

Annas, G., Caplan, A., & Elias S. (1996). The politics of human-embryo research—avoiding ethical gridlock. *New Engl. J. Med.*, *334*(20), 1329–1332.

Ashley, B. M., & O'Rourke, K. D. (1989). *Health Care Ethics: A Theological Analysis*, 3rd edn. St. Louis, MO: Catholic Health Care Association of the United States.

BBC News. (2006). MEPs vote against stem cell research, Thursday, 10 April. Available at: http://news.bbc.co.uk/1/hi/health/2932421.stm.

Bush, G. W. (2001). President's statement on funding stem cell research. *NY Times*, 10 Aug. (late edition-final), A16.

Chief Medical Officer's Expert Group. (2000). "Stem Cell Research: Medical Progress with Responsibility." Available at: http://www.dh.gov.uk/en/Publicationsandstatistics/Publications/Publications PolicyAndGuidance/DH_4065084.

Cohen, C. B. (2001). Leaps and boundaries: Expanding oversight of human stem cell research. In S. Holland *et al.* (Eds), *The Human Embryonic Stem Cell Debate*, pp. 209–222. Cambridge, MA: MIT Press.

Committee on Guidelines for Human Embryonic Stem Cell Research, National Research Council. (2005). *Guidelines for Human Embryonic Stem Cell Research*. Washington, DC.: National Academies Press. Available at: http://www.nap.edu/catalog/11278.html.

Davis, D. S. (1995). Embryos created for research purposes. *Kennedy Inst. Ethics J.*, *5*(4), 343–354.

Doerflinger, R. (1999). The ethics of funding embryonic stem cell research: a Catholic viewpoint. *Kennedy Inst. Ethics J.*, *9*, 137–150.

Doerflinger, R. M. (2003). Testimony of Richard M. Doerflinger on behalf of the Committee for Pro-Life Activities. National Conference of Catholic Bishops, Testimony of Richard Doerflinger, January 14, 2003 before the Health and Human Development Committee of the Delaware House of Representatives concerning Senate Bill No. 55. "Cloning Prohibition and Research Protection Act." Available at: http://www.cloninginformation.org/congressional_testimony/doerflinger_ de.htm.

Dresser, R. (2001). Letter to the editor. *JAMA*, *285*(11), 1439.

Ford, N. M. (1988). *When Did I Begin?* Cambridge: Cambridge University Press.

Geron Advisory Board. (1999). Research with human embryonic stem cells: Ethical considerations. *Hastings Cent. Rep.*, *29*(2), 31–36.

Green, R. M. (2001). *The Human Embryo Research Debates*. New York: Oxford University Press.

Green, R. M. (2001). What does it mean to use someone as "a means only": Rereading Kant. *Kennedy Inst. Ethics J.*, *11*(3), 249–263.

Green, R. M. (2002). Benefiting from "evil": an incipient moral problem in human stem cell research. *Bioethics*, *16*(6), 544–556.

Green, R. M., Olsen DeVries, K., Bernstein, J., Goodman. K. W., Kaufmann, R. W., Kiessling, A. A., Levin. S. R., Moss, S. L., & Tauer, C. A. (2002). Overseeing research on therapeutic cloning: A private ethics board responds to its critics. *Hastings Cent. Rep.*, *32*(3), 27–33.

Hardy, K., Spanos, S., Becker, D., Iannelli, P., Winston. R. M. L., & Stark, J. (2001). From cell death to embryo arrest: mathematical models of human preimplantation embryo development. *Proc. Nat. Acad. Sci. USA*, *98*(4), 1655–1660.

Holland, S. (2001). Beyond the embryo: A feminist appraisal of human embryonic stem cell research. In S. Holland *et al.* (Eds), *The Human Embryonic Stem Cell Debate*, pp. 73–86. Cambridge, MA: MIT Press.

Hyun, I., Hochedlinger, K., Jaenisch, R., & Yamanaka, S. (2007). New advances in iPS cell research do not obviate the need for human embryonic stem cells. *Cell Stem Cell*, *1*, 367–368.

Ibrahim, Y. M. (1996). Ethical furor erupts in Britain: should embryos be destroyed? *NY Times*, 1 Aug. (late edition), A1.

Israel Academy of Sciences and Humanities. (2001) *Report of the Bioethics Advisory Committee of the Israel Academy of Sciences and Humanities on the Use of Embryonic Stem Cells for Therapeutic Research*. Israel Academy of Sciences and Humanities.

ISSCR (International Society for Stem Cell Research). (2006). *Guidelines for the Conduct of Human Embryonic Stem Cell Research*. Version 1: 21 December, 2006. Available at: http://www.isscr.org/guidelines/ISSCRhESCguidelines2006.pdf.

Jaenisch. R., & Wilmut, I. (2001). Don't clone humans! *Science*, *291* (5513), 2552.

Kaveny, M. C. (2000). Appropriation of evil: Cooperation's minor image. *Theol. Stud.*, *61*, 280–313.

Klimanskaya, I., Chung, Y., Becker, S., Lu, S. J., & Lanza, R. (2006). Human embryonic stem cell lines derived from single blastomeres. *Nature*, *444*(7118), 481–485. Epub 23 Aug. 2006.

Kolata, G. (2001). Parkinson's research is set back by failure of fetal cell implants. *NY Times*, 8 March, (late edition-final), A1.

Kolata, G. (2001). Researchers say embryos in labs are not available. *NY Times*, 26 Aug. (late edition-final), A1.

Krauthammer, C. (2007) Stem Cell Vindication. *Wash. Post*, 30 Nov., A23.

Lanza, R. P., Caplan, A. L., Silver, L. M., Cibelli, J. B. West, M. D., & Green, R. M. (2000).The ethical validity of using nuclear transfer in human transplantation. *JAMA*, *284*(24), 3175–3179.

Lanza, R. P., Chung, H. Y., Yoo, J. J., Wettstein, P. J., Blackwell, C., Borson, N., Hofmeister, E., Schuch, G., Soker, S., Moraes, C. T., West, M. D., & Atala, A. (2002). Generation of histocompatible tissues using nuclear transplantation. *Nat. Biotechnol.*, *20*, 689–696.

Lebacqz, K. (2001). On the elusive nature of respect. In S. Holland *et al.* (Eds), *The Human Embryonic Stem Cell Debate*, pp. 149–162. Cambridge, MA: MIT Press.

Magill, G., & Neaves, W. B. (2009). Ontological and ethical implications of direct nuclear reprogramming. *Kennedy Inst. Ethics J.*, In press.

McConnick, R. A. (1991). Who or what is the preembryo? *Kennedy Institute of Ethics Journal*, *1*, 1–15.

Miller, L. G. (1967). Scandal. In *New Catholic Encyclopedia*, Vol. XII, pp. 1112–1113. New York: McGraw Hill.

Nagy, A., *et al.* (1990). Embryonic stem cells alone are able to support fetal development in the mouse. *Development*, *110*, 815–821.

Nagy, A., *et al.* (1993). Derivation of completely cell culture-derived mice from early-passage embryonic stem cells. *Proc. Nat. Acad. Sci.* USA, 90, 8424–8428.

National Bioethics Advisory Commission (1997). *Cloning Human Beings*. Rockville, MD: National Bioethics Advisory Commission.

National Institutes of Health. (1994) *Report of the Human Embryo Research Panel*. Bethesda, MD: National Institutes of Health.

National Institutes of Health. (2000). *Guidelines for Research Using Human Pluripotent Stem Cells* (Effective August 25 2000. 65 FR 5 1976) (Corrected November 21, 2000. 65 FR 69951). Bethesda, MD: National Institutes of Health.

National Institutes of Health. (2009). *Draft National Institutes of Health Guidelines for Human Stem Cell Research*. Available at: http://stemcells.nih.gov/policy/2009draft.

National Research Council and Institute of Medicine of the National Academies. (2007). *2007 Amendments to the National Academies' Guidelines for Human Embryonic Stem Cell Research*. Washington, DC.: National Academies Press. Available at: http://books.nap.edu/openbook.php?record_id=11871&page=R1.

National Research Council and Institute of Medicine of the National Academies. (2008). *2008 Amendments to the National Academies' Guidelines for Human Embryonic Stem Cell Research*. Washington, DC.: National Academies Press. Available at: http://www.nap.edu/catalog/12260.html.

National Research Council. (2001). *Stem Cells and the Future of Regenerative Medicine*. Washington, DC: National Academy Press.

Nature (2001). The meaning of life (editorial). *Nature*, *412*(6844), 255.

New York Times. (2003). Cloning countdown (editorial), *NY Times*, 1 Mar. (late edition-final), A18.

Norwitz, E. R., Shust, D. J., & Fisher, S. J. (2001). Implantation and the survival of early pregnancy. *New Engl. J. Med.*, *345*, 1400–1408.

O'Rahilly, R., & Mülller, F. (1992). *Human Embryology and Teratology*. New York: Wiley-Liss.

Office of Science Policy. (2001). *Stem Cells: Scientific Progress and Future Research Directions*. Bethesda, MD: National Institutes of Health.

Parens, E. (2001). On the ethics and politics of embryonic stem cell research. In S. Holland *et al.* (Eds), *The Human Embryonic Stem Cell Debate*, pp. 37–50. Cambridge, MA: MIT Press.

Paulson, R. J. (1996). Fertility drugs and ovarian epithelial cancer is there a link? *J. Assist. Reprod. Genet.*, *13*(10), 751–756.

Pearson, H. (2002). Stem cells: articles of faith adulterated. *Nature, 420*, 734–735.

Pope John Paul II. (2003). Address of John Paul II to the Members of the Pontifical Academy of Sciences, Monday, 10 November 2003. Available at: http://www.Vatican.va/holy_father/john_paul_ii/ speeches/ 2003/November/documents/hf_pii_spe~20011110–academy-sciences_en.html.

President's Council on Bioethics. (2002). *Human Cloning and Human Dignity: An Ethical Inquiry*. President's Council on Bioethics, Washington, DC. Available at: http://www.bioethics.gov.

Rossing, M. A., Daling, J. R., Weiss, N. S., Moore, D. E., & Self, S. E. (1996). Ovarian tumors in a cohort of infertile women. *New Engl. J. Med., 331*(12), 771–776.

Shannon, T. (2001). From the micro to the macro. In S. Holland *et al.* (Eds), *The Human Embryonic Stem Cell Debate*, pp. 177–184. Cambridge, MA: MIT Press.

Stolberg, S. G. (2001). House backs ban on human cloning for any objective. *NY Times*, 1 Aug. (late edition-final), A1.

Stolberg, S. G. (2001). Some see new route to adoption in clinics full of frozen embryos. *NY Times*, 25 Feb., A1.

Stolberg, S. G. (2003). House votes to ban all human cloning. *NY Times*, 28 Feb. (late edition-final), A22.

Strain, L., Dean, J. C. S., Hamilton, M. P. R., & Bonthron, D. T. (1998). A true hermaphrodite chimera resulting from embryo amalgamation after *in vitro* fertilization. *New Engl. J.* Med., *338*, 166–169.

Sugarman, J., & Siegel, A. (2008). How to determine whether existing human embryonic stem cell lines can be used ethically. *Cell Stem Cell, 3*, 238–239.

Takahashi, K., Tanabe, K., Ohnuki, M., Narita, M., Ichisaka, T., Tomoda, K., & Yamanaka, S. (2007). Induction of pluripotent stem cells from adult human fibroblasts by defined factors. *Cell, 131,* 861–872.

The Linacre Centre for Healthcare Ethics. (2001). A theologian's brief on the place of the human embryo within the Christian tradition. Available at: http://www.lifeissues.net/writers/mis/mis_02christiantradition1.html#b1.

Warren, M. A. (1997). *Moral Status: Obligations to Persons and Other Living Things*. New York: Oxford University Press.

Weiss, R. (1999). Stem cell discovery grows into a debate. Wash. Post, 9 Oct., A1, A8–A9.

Weiss, R. (2003). 400,000 human embryos frozen in U.S. Number at fertility clinics is far greater than previous estimates, survey finds. Wash. Post, Thursday, 8 May, 2003, A10.

Yu, J., Vodyanik, M. A., Smuga-Otto, K., Antosiewicz-Bourget, J., Frane, J. L., Tian, S., Nie, J., Jonsdottir, G. A., Ruotti, V., Stewart, R., Slukvin, I. I., & Thomson, J. A. (2007). Induced pluripotent stem cell lines derived from human somatic cells. Science, 318(5858), 1917–1920. Epub Nov. 2007.

Chapter 30

Stem Cell Research: Religious Considerations

Margaret A. Farley

INTRODUCTION

Religious traditions have endured in large part because they help people to make sense of their lives. Major religious traditions become major because they offer some response to the large human questions of suffering and death, hope and transcendence, history and community, as well as the everyday issues of how we are to live together with some modicum of harmony and peace. Adherents of traditions of faith generally experience their shared beliefs and practices, not as irrational, but as part of the effort of reason to understand the actual and the possible in human life, and all that is around it. Although religious sources of insight may reach beyond empirical data and logical reasoning, they need not ultimately do violence to either. When we ask profound questions such as how humans ought to reproduce themselves, or how extreme suffering can be alleviated, or whether the meaning of the human body changes when we exchange its parts, we can be helped by looking to the source and the substance of the attitudes, beliefs, and practices which animate people's lives.

Of course, when we look for guidance in the faith of believers, we do not find univocal voices—not from believers in general, and not from believers within particular traditions. Roman Catholics will often not agree with spokespersons from the United Church of Christ; many Jews will not agree with many Catholics about abortion; and so on. But the division within faith traditions today is often wider than between or among them. So, there is more than one view regarding stem cell research among Roman Catholics; more than one among the different strands of Judaism; more than one among Muslims. Such disagreements do not make religious voices useless for societal and religious discernment, no more than disagreements among scientists render scientific research and the voices of scientists useless.

This chapter aims to review religious perspectives on stem cell research. It is not possible here to survey every religious or theological argument for and against stem cell research. Something of an overview is presented, and then particular issues of concern articulated by representatives of some religious traditions are examined.

MAPPING THE TERRAIN

Sources for Stem Cells

The major disputed questions surrounding stem cell research are focused, not on stem cells as such, but on the sources from which stem cells are derived. Hence, debates rage regarding the moral status of human embryos and the permissibility of taking stem cells from already dead fetuses. Religious thinkers who consider the human embryo sacrosanct from its inception will consistently oppose the extraction of embryonic stem cells as a form of killing (of embryos). Those who do not appraise the value of an embryo as being on a par with that of a human person are much more willing to favor embryo research in general, including the taking of stem cells for research. Still others who emphasize the ambiguity of the moral status of embryos are likely to weigh the advantages of embryo cell research over the possible violations of the embryo as an entity in itself. All religious traditions recognize an aborted fetus as a cadaver; hence, the derivation of stem cells from the gonadal ridge of dead fetuses may be as permissible as the harvesting of tissues or organs from any human cadavers. Nonetheless, representatives of traditions that prohibit abortion worry about complicity with, and support of, the moral evil of abortion when stem cells are derived from electively aborted fetuses.

Alternative sources for human stem cells are championed by those who are opposed to embryo stem cell and fetal stem cell research. The least objectionable sources often cited are umbilical cord blood, placentas, amniotic fluid, and adult somatic stem cells (extracted, for example, from bone marrow). Other sources considered in 2005 by the President's Council on Bioethics include embryos that are already dead but have some remaining normal cells; blastomeres biopsied from living embryos without destroying the embryos; deliberately engineered biological artifacts that are not embryos, but

Stem Cell Anthology

function like embryos minus any potential to develop into a person. A compromise source (although still controversial) is lines of cells derived from stem cells that have already been harvested (that is, the life of the original embryo has already been taken, and no new moral agency is involved in the killing of more embryos). Despite significant progress on some of these alternative sources—especially adult tissues and cells—there remain serious scientific, religious, and ethical problems.

Scientists note that research on adult sources is still less advanced than research on embryonic cells, and in any case, many scientists argue that it is better to proceed on all fronts. The use of embryo cadavers raises new ethical questions about *in vitro* fertilization and the cryopreservation of embryos (about half of which die when thawed). Removing cells from living embryos is not without risk to the embryos. The creation of biological artifacts can involve cloning (which, even when it is for research, not reproduction, raises another set of ethical issues). Moreover, the artifacts may be "functionally equivalent" to embryos in ways that fail to settle all questions of their moral status—that is, maybe they should be respected in the same way as embryos are respected, since their incapacity to become human persons may be overcome at some point in the further development of reproductive technologies. In addition, a move to "create" these artifacts crosses another moral boundary—manufacturing entities precisely for research, as opposed to simply using "spare" ones—that is troubling for many people. Finally, the proposed use of already existing embryonic cell lines encounters the same objections of complicity with moral evil that is involved in the use of cells taken from aborted fetuses. For those who take seriously all of these objections to otherwise promising alternative sources for human stem cells, the alternatives do not appear as yet to resolve ongoing deep-seated scientific, ethical, or religious concerns.

Other issues complicate moral assessments of the sources of stem cells, and the aims of stem cell research. For example, almost all religious thinkers argue that some form of respect is due to embryos, even if it is not the same form of respect required for human persons. Yet there is lack of clarity about what respect can mean when it is aligned with the killing of the respected object. It is not difficult to get minimal consensus on prohibitions against buying and selling human embryos or on safeguards of informed consent from the donors of embryos for research. For many religious believers, it makes a difference that these embryos are "spare" embryos, left over from *in vitro* fertilization procedures, and destined to be discarded if they are not used in research. From other religious perspectives, however, this fact in itself does not lift the prohibition against killing them.

Questions of Justice

Quite another kind of ethical issue is raised by people from almost all religious traditions. This is the issue of distributive justice. Who will be expected to be the primary donors of embryos or aborted fetuses? Will gender, race, and class discrimination characterize the whole process of research on stem cells? And will the primary goals of research be skewed toward profit rather than toward healing? Who will gain from the predicted marvelous therapeutic advances achieved through stem cell research—the wealthy but not the poor? The powerful, but not the marginalized? What will be the overall results for respect for human persons if human embryos and stem cells are commodified, wholly instrumentalized, and in some ways trivialized?

All of these questions are questions for secular philosophical, scientific, and medical ethics, as well as for religious ethics. When they are identified by religious and theological thinkers, however, they are always lodged in the larger questions with which religious traditions are concerned, and they almost always appeal to sacred texts and faith community traditions, teachings, and practices, as well as to secular sources and forms of moral reasoning.

Religious Beliefs and Ethical Questions

Anchoring religious concerns for the moral status of the human embryo, the nature of moral evil and immoral complicity, and questions of distributive justice, are beliefs about the nature and destiny of human life, the meaning of bodily existence, the interaction between divine and human in the progress and healing of creation, the importance of consequences in moral reasoning along with the importance of religious laws and ontologically based norms, the basic equality of all human people before God, and so forth. Even when there is no direct line between the most profound beliefs of a person and the answer to a particular moral question, such a belief makes a difference for moral discernment and conviction. There are no world religions today that oppose all human intervention in "nature," yet all religions recognize some limitations on what humans may do—either to themselves or to other created beings. All consider humility before concrete reality an important antidote to arrogance or pride (and in the wake of ecological disasters, all now take account of the risks of some interventions made only for the sake of the flourishing of present, but not future, human generations). No religions favor illness over health, or death over life, yet each has a perspective in which health may not be an absolute value, and death may sometimes even be welcomed. In the present context of stem cell research, there is no religious tradition that does not take the beneficial results promised by such research seriously.

Hence, positions of religious believers regarding stem cell research tend to be highly complex. Each tradition needs to be understood in its complexity and its diversity. This chapter cannot track all of this, but it can try to show how some beliefs are coherent, whether or not others can agree on them in the same tradition, or across faiths and cultures.

PARTICULAR TRADITIONS: AN OVERVIEW

A basic understanding of positions taken by religious scholars and spokespersons in mainline traditions is, fortunately, available to some extent in the testimony and writings that have appeared in relation to the public debate on stem cell research. We have here, of course, just the proverbial tip of the iceberg. None-the-less, these documents and essays point us to both the concrete moral positions being debated, and the larger rationales behind them. In brief summaries, then, I survey some of these positions and rationales. Following this, I focus on specific lines of argument taken by representatives of the Roman Catholic tradition. I do this not to privilege this tradition, but because its articulation of arguments is more prolific than most; it claims to be trying to be persuasive in a secular public sphere; its central positions are representative of strands of other religious, as well as philosophical traditions; and it offers an ongoing critical conversation on both sides of the stem cell debate.

Hinduism and Buddhism

Representatives of the Hindu and Buddhist traditions have not been major players in the contemporary debate about stem cell research, at least not in the United States. There are historical reasons for this no doubt, including the fact that neither is as yet a majority religion in the West. Yet there is much in both of these traditions that may, for their adherents, be relevant to issues of stem cell research. Given the complexity of the traditions, it remains for their scholars and representatives to render accessible to outsiders the foundational beliefs and moral concerns that may shed light on such issues. Briefly, however, some elements in these faith traditions can be noted.

Both Hinduism and Buddhism incorporate beliefs in reincarnation. This complicates issues of what can be done with human embryos. According to some Hindu traditions, human life begins prior to conception; the "soul" may be present even in sperm, or it may be in some other life form before and after its human existence. Human reincarnation, however, offers a unique opportunity to influence the future of an individual. Great caution must be taken whenever actions are considered that may destroy this opportunity. On the other hand, for Hindus, a belief in reincarnation is also tied to motivation to be compassionate towards others. Good actions that will change the course of suffering for oneself and others are exhorted, not only so that one can advance morally (and thereby improve one's karma), but because other people are worthy of beneficent deeds. Although the forms of compassion include nonviolence, there is in Hinduism also a broad tradition of sacrifice, wherein one human life can be taken for the sake of a higher cause. It is not impossible that these convictions provide a rationale for embryonic stem cell research. As one example of this, in 2001 the Hindu Endowments Board of Singapore expressed willingness to accept the extraction of stem cells from human embryos in their earliest stages, as long as this did not go beyond 14 days of embryonic development.

Buddhists are divided in their opinions about human embryonic stem cell research. For some, the precept against harming any living thing is determinative, ruling out the extraction of stem cells from, and the concomitant destruction of, embryos. Yet other Buddhists appeal to another element in Buddhism that offers a possible rationale for embryonic stem cell research. Buddhism's goal of self-transcendence through a process of self-forgetting undergirds a requirement for compassion, one of the Four-fold Holy Truths of Buddhism. In Mahayana Buddhism, the ideal is the Bodhisattva, one who achieves self-emptying, but then returns to help those in need. Again, here is a possible rationale for embryonic stem cell research. Moreover, some Buddhists believe that a soul has some choice as to where it will be incarnated. It has even been suggested that souls will not elect to embed in "spare" embryos that will only be destroyed. Hence, early embryos may not ever have a potential to become a human person, and can therefore be used for some other purpose.

Whether such considerations are of overall significance in decisions to pursue embryonic or fetal stem cell research or not, remains to be seen. At the very least, it can be said that they represent large concerns for the value of human life—whether at its inception, or in response to later injury and illness.

Islam

Like other world religions, there are many schools of thought among Muslims. There are also multiple sources for moral guidance, including the Qur'an and its commentaries, the hadiths (a second source of moral indicators supplementary to the Qur'an), Muslim philosophies that range from a form of ontological and moral realism to a version of divine command theory, Sharia (formulations of Islamic law), and centuries of juridical literature. Testifying before the US National Bioethics Advisory Commission in 1999, the Islamic scholar Abdulaziz Sachedina attempted to provide some general insights from the tradition, taking into account the diverse interpretations of major Sunni and Shi'i schools of legal thought. His analysis of various texts led him to infer guidelines for stem cell research from the rulings of the Sharia on fetal viability and the moral status of the embryo.

Islamic traditions have given serious attention to the moral status of the fetus, and its development to a particular point when it achieves human personhood with full moral and legal status. The early embryo has been variously valued in different eras, depending on the information available from science. Throughout the history of legal rulings in this regard, however, a developmental view of the fetus—according to a divine plan—has been sustained. This founds the view that the moral status of personhood is not achieved at the earliest stages of embryonic life, but only after sufficient biological development has taken place—the kind of development that includes a recognizable human anatomy and the possibility of "voluntary" movement. Most Sunni and some Shi'ite scholars therefore distinguish two stages in pregnancy. The first stage, "pre-ensoulment," is human biological life, but not yet human personal life. It is only after the fourth month (120 days, or the time of quickening) that the "biological human" becomes a "moral person." This is to say that, at this stage of development, the fetus achieves the status of a moral and legal person. In the first stage, the biological entity is to be respected, yet abortion is allowed for grave reasons. In the second stage, killing the fetus is homicide. The conclusion to be drawn from this is that there is room for early embryonic stem cell extraction, without violating divinely given laws or the embryo itself. In Islam, therefore, there are strong elements of the tradition that affirm embryonic stem cell research.

Judaism

In the twenty-first century, Orthodox, Conservative, and Reform Jews take different positions on many issues of applied ethics. These views are not easily specified when it comes to stem cell research. What is possible here is to present the opinions of some Jewish scholars, with the caution that not all views are represented. In the context of debates about stem cells, Reform Judaism may be the clearest supporter of this research, even when it entails the derivation of stem cells from human embryos, or from aborted fetuses. Jewish law does not give legal status to the fertilized ovum outside a mother's womb. When the embryo is inside a uterus, it has legal status but not that of a human person. In general, however, this strand of Judaism favors the use only of so-called spare embryos, those that will otherwise be discarded.

Rabbi Elliott Dorff has identified theological assumptions useful for understanding a Jewish response to stem cell research. These include the following: moral discernment (of what God wants of God's people) must be based on both Jewish theology and Jewish law; all human beings are created in the image of God and are to be valued as such; human bodies belong to God and are only on loan to the individuals who have them; human agency is important in responses to human illness, so that both "natural" and "artificial" means are acceptable in overcoming disease. Indeed, there is a duty to develop and use any therapies that can help in the care of the human body. Yet, because humans are not, like God, omniscient, humility and caution are essential, especially when human science and technology press to the edges of human knowledge.

There are grounds in Jewish law and theology for permitting the derivation of stem cells, both from aborted fetuses and from human embryos. Just as abortion (although generally forbidden) can be justified for serious reasons, so also can the use of fetuses for important research. A fetus *in utero* is considered not as a human person but as the "thigh of its mother." One is not allowed to amputate a part of one's body (in this case, one's thigh) except for good reasons (for example, to remove a gangrenous limb, save the life of a mother, or remove other serious risk, and perhaps remove a fetus that is genetically malformed or diseased). When an abortion is thus justified, the abortus may be used as a source of stem cells.

Embryos can also be legitimate sources for stem cells extracted for purposes of research. During the first 40 days of gestation, *in utero* embryos (and fetuses) are only like "water." They are "non-souled," with only a liminal status. Although abortion during this time is permitted only for good reasons, the situation is different when embryos are not in the womb. Here there is no potential for embryos to develop into human persons; hence, they may be discarded, frozen, or used for promising research. Both Rabbi Dorff and Jewish scholar Laurie Zoloth argue that the duty to care for human bodies, the duty to heal, may undergird not just permission to pursue research on stem cells, but an ethical duty to do so.

The Jewish tradition places all of these considerations in a wider context of responsibilities to community, fairness in distribution of benefits and burdens, and caution about the connections of stem cell research to invidious programs of eugenics. Ethical norms that safeguard the common good, as well as the good of individuals, provide boundaries for the development and future uses of stem cell research; they focus it, perhaps obligate it, but do not forbid it.

Christianity

The diversity and pluralism to be expected in Christian responses to stem cell research mirrors to some extent the historical institutional divisions within Christianity. From the early schisms between East and West to the great Reformations in the West and the subsequent proliferations of Protestant denominations, to the Anglican and Roman Catholic Church traditions, to the rise of new forms of Pentecostalism in the world's South, Christians have diversified in their beliefs, and in their moral convictions and practices. There remains some important commonality in basic doctrines, largely expressed in creeds, and these are not irrelevant to issues of stem cell research. Examples include belief in divine revelation, particularly through sacred scripture; reconciliation made possible between humans and God through Jesus Christ; affirmation of the goodness of creation despite the damage sustained as the result of moral evil or sin; acceptance of the importance of human agency (in practice if not always in theology); a call to unconditional love of God, and love and service of neighbor; a basic view of human equality; and the obligation to promote justice. Theological anthropologies vary importantly in terms of understandings of freedom and grace, virtue and sin.

Strands of Christian moral theologies diverge when it comes to the basis and force of moral norms—depending primarily either on God's command or God's will manifest in creation itself, and holding to absolute moral norms or only *prima facie* norms relativized according to circumstances, consequences, or priorities among aims. Law and gospel are important to each Christian tradition, although the emphasis on one or the other may vary.

Articulating an Eastern Orthodox view of embryonic stem cell research, Demetrios Demopulos begins with concern for the alleviation of suffering. "Medical arts" are to be encouraged, with the proviso that they be practiced as gifts from God ordered to divine purposes. Created in the image and likeness of God, people are destined for participation in the life of God—hence the telos of the human person is referred to as theosis or "deification." People are authentic insofar as they struggle to grow into this life; they remain "potential" persons until this is achieved. But even zygotes are potential persons in this sense; that is, they, too, are in a developmental process that ends in deification. Hence, human life is sacred from the beginning, and at every step along the way, entitled to protection and the opportunity to seek its destiny. Even though it is not yet a person, an embryo should not be used for, or sacrificed in, research. Correlatively, practices of *in vitro* fertilization that yield "surplus" embryos which will be discarded cannot be condoned.

Although the development of alternative techniques and sources is preferable, Demopulos allows the use of already-harvested embryonic stem cells for research. Wishing that something had not been done will not undo it, he maintains. The compromise position, then, of continuing research on available cell lines is accepted, so long as the research does not violate other norms of justice (by maximizing profits rather than the health of people, fostering trivial medical procedures for cosmetic purposes alone, or contributing to a questionable agenda for eugenics). The use of cadaver fetal cells is also acceptable if the abortion that provides them is spontaneous, not elected. But human life—even potential human personal life—from zygote to a life beyond this world, remains sacred; what ultimately ends in God ought to be inviolable among humans.

Among the Protestant denominations in the United States, some oppose and some support embryonic stem cell research. Many churches do not yet have an official policy on this issue, although their representatives indicate a kind of majority view. Southern Baptists are, as a group, generally opposed to this research, although there is some openness to the compromise position of using cell lines developed from cells harvested in the past. Some leaders among United Methodists, without an official church position, none-the-less asked President George W. Bush to oppose federal funding for embryonic stem cell research. Presbyterians in the General Assembly articulated their support of research on cells extracted from embryos that would otherwise be discarded, and affirmed the use of federal funds to make this possible. Acknowledging the significance of concepts like "potential personhood," and the need to "respect" embryos, they none-the-less determined that, whatever form this respect takes, it ought not to have priority over alleviating the suffering of actual people. Members of the United Church of Christ tend to approve embryonic stem cell research, as long as it is motivated by a clear and attainable healing benefit.

Ethicists from a variety of Protestant backgrounds (although seldom speaking for their denominations) also vary in their approaches to stem cell research. Ronald Cole-Turner agrees with others in the United Church of Christ, by approving the extraction of stem cells from embryos, but sets conditions for research in terms of justice issues. Ted Peters of the Pacific Lutheran School of Theology comes down on the side of supporting embryonic stem cell research, because he believes that, on balance, concern for the human dignity not of embryos but of future real persons is more compelling. Gene Outka and Gilbert Meilaender, both Lutherans, share certain theological and philosophical convictions about the irreducible value of the early embryo. Both want to affirm the continuity of human life between an early embryo and an actualized full person. This means that both will consider the taking of the life of an embryo as a violation of the prohibition against killing. Both also want to distinguish the question of embryo stem cell extraction from the question of abortion (in a variety of ways, but primarily because in the former there is no direct conflict between a woman and a fetus).

Outka, however, finally approves of the use of some embryos for the derivation of stem cells. He does so by introducing two conditions or principles that can exempt one from the absolute force of the prohibition. The first and most distinctive is the "nothing is lost" condition. That is, it is possible to relativize the prohibition against killing when those who are to be killed will die anyway. This is precisely the condition that characterizes the situation of spare (or excess) embryos (from *in vitro* fertilization procedures) destined to be discarded. The second condition is that "other lives will be saved" through this act of killing. Given the ultimate therapeutic aims of stem cell research, this condition is also present. It is the nothing is lost condition that, in the end, allows Outka both to continue to affirm the status of the embryo as an end, and not a means only, and yet to allow the taking of its life for purposes of research. It becomes, then, the basis of ethical acceptance of the derivation of stem cells from some human embryos, and therefore a carefully-circumscribed, but none-the-less positive approval of embryonic stem cell research.

Meilaender, on the other hand, remains firm (though cautious and reluctant) in the conviction that embryonic stem cell research ought not to go forward, or better, that it is not morally justifiable to use human embryos as the source for stem cells. Referencing three other Protestant theologians (Karl Barth, John Howard Yoder, and Stanley

Hauerwas), Meilaender poses three arguments: (1) people should be considered persons not on the basis of potentiality or actuality, not on the basis of capacities, but simply as members of the human community. To think inclusively about the human species should lead us to honor the dignity of even the weakest of living human beings—the embryo—and thus be able to appreciate the mystery of humanity and the mystery of our own individuality. (2) To opt immediately for embryonic stem cell research is to prevent us from finding better solutions to the medical problems this research is designed to address. Given the ethical compromises involved in following this "handy" route, it would be better to "deny ourselves" this remedy, and look to ones that do not denigrate the dignity of people however and wherever they live. (3) The church should bear witness to its beliefs by refusing to make sophistic distinctions, such as those between funding embryonic stem cell research and funding the "procurement" of these cells (the actual extraction of cells from embryos). Meilaender is skeptical that only excess embryos or available cell lines will be used.

Along with evangelical Protestants, Roman Catholic voices have been prominent in the debates surrounding embryonic stem cell research. The position most frequently articulated in the public forum is that of opposition to the derivation of stem cells from embryos. This is the position of the Catholic hierarchy, promulgated through both Vatican and national episcopal channels. It is a position ably supported theologically by spokespersons such as Richard Doerflinger. Yet it is not the only position espoused by Roman Catholic ethicists and moral theologians, or by all Roman Catholics. An alternate position held by many Catholic theologians is basically the "14 day" position, or the theory that an embryo in the first 14 days, prior to implantation in the uterus, is not yet even a "potential" person, and hence need not be protected in the same way as human persons (or even the same as fetuses, considered potential persons after implantation). Neither of these positions goes the route of declaring absolutely that the early embryo is a person, or the route of approving the exploitation of fetuses *in utero* in later stages of development. Since the debate between these two positions reflects similar philosophical and theological debates in other religious traditions and in the public forum, it is worth looking at it in closer detail.

ROMAN CATHOLIC CONTRIBUTIONS AND THE 14 DAY THEORY

Before turning to the technicalities of the 14 day theory, it should be noted that Roman Catholics (including the official leaders of the church) tend to worry, like people in other traditions, about issues of justice, ecology, and the well-being of the whole earth. Apart from the moral status of the embryo, Catholic concerns are focused on questions of equity in the shared lives of persons across the world. For all the pressures to go forward with embryonic and fetal tissue stem cell research, there is little assurance anywhere that the benefits of this research will be shared in the human community among the poor as well as the wealthy, across racial and gender lines, and in countries of the world's south as well as north. Moreover, as the research goes forward (for it surely will), some safeguards will need to be in place to keep the goals of the research focused on the healing of human persons, rather than on the commodification of human bodies, their tissues and cells. And the specter of human genetic engineering for enhancement purposes (and not only the treatment of diseases and injuries) is never far from the horizon. The other side of tremendous positive advances of modern technology is not to be ignored, not even if our present anxieties lead us desperately to seek means to lift the burdens of one generation or one group of people.

It is important to take preliminary note, also, of the shared community of discourse among Roman Catholic scholars, church leaders, and people. No matter how divided they may be among themselves—regarding the moral status of the early embryo or other particular moral questions—common moral convictions are none-the-less expressed in common language. The Catholic tradition is undivided in its affirmation of the goodness of creation, the role of human persons as agents in cooperation with ongoing divine creative activity, the importance of not only the individual, but also the community, the responsibilities of human persons to promote the health and wellbeing of one another. At the same time, Catholics do disagree profoundly; a key example of disagreement is the debate about the status of the embryo, and the moral evil or goodness of extracting stem cells from it in a way that leads to its demise.

The debate within the Catholic community regarding stem cell research incorporates opposition over something like the 14 day theory, because the Roman Catholic moral tradition has consistently embraced a form of moral realism. At the heart of this tradition is the conviction that creation is itself revelatory. That is, embedded in an ultimately intelligible (though only partially so to humans) created reality is the possibility of perceiving and discerning moral claims of respect in response to every entity, especially human persons. This is what is at stake in the Catholic tradition of natural law. Although there are historical inconsistencies, and even contradictions, in this long tradition, for the most part, natural law has not meant that morality can simply and fully be "read off" of nature, not even with the help of the special revelation of the Bible (although this is a help). What natural law theory does is tell persons where to look—that is, to the concrete reality of the world around us, to the basic needs and possibilities of human persons, and to the world as a whole. "Looking" involves discerning, deliberating, structuring insights, interpretating meanings, taking account of not only what is similar among entities, but also what is particular in their histories, contexts, and relationships.

Roman Catholic thinkers engaged in discernment about human embryonic stem cell research tend, therefore, to "look" to the reality of stem cells and their sources, i.e., human embryos and aborted fetuses, as well as adult tissues. All answers are not in the Bible, official church teachings, or individual experiences of the sacred. Discernment incorporates all of these sources of insight, but it also requires the knowledge available from the sciences, philosophy, and whatever other secular disciplines provide some access to the reality that is being studied. The ongoing intensity of the debate about the status of embryos bears witness to the notion that not all inquiries can be settled by one discipline, or one interpreter, or within one epistemological perspective.

The argument against procuring stem cells from human embryos is primarily that it entails the death of the embryos. No one disputes this fact, but disagreement rages as to whether this death can be justified. Those who oppose its justification argue that for each human person there is a biological and ontological continuum from the single-cell stage to birth, to whatever one has of childhood and adult life, and then to death. Since a new and complete genetic code is present after fertilization, there is already a unique individual human person, potential in an important sense but concretely and really (inherently) already directed to full actualization. The zygote itself, as a living organism, is self-organizing. The moral status of the early embryo is, therefore, that of a human person; it does not achieve this status by degrees or at some arbitrary point in development; nor is this status simply bestowed upon it by social recognition or convention.

But, if the embryo has the status of a human person, killing it even in order to treat the illnesses of other people cannot—from this Catholic point of view—be justified. No human person, not even an inchoate one like an embryo, can be reduced to a pure means in relation to other ends. Creating or using embryos for research purposes is wrong, because it treats this distinct human being as a disposable instrument for the benefit of someone else. Those who oppose the extraction of stem cells from embryos on the grounds that it involves the destruction of a potential human person are not unsympathetic to the suffering that may be alleviated ultimately through research on stem cells. Part of the argument against the use of embryonic sources for stem cells rests on the identification of alternative sources (particularly adult cells).

On the other side of this debate are Catholic moral theologians, who do not consider the human embryo in its earliest stages (prior to the primitive streak or to implantation) to be a human person, potential or actual. They hold the same meaning for "potency" as do their opponents—that is, these theologians hold, too, an Aristotelian and Thomistic meaning, signifying not an extrinsic or "sheer possibility" potential, but rather an intrinsic principle, already as such actual within a being, directing it to individualized species-specific development (even though the being may not yet be developmentally fully actualized). These moral theologians argue, then, precisely against the view that the early embryo has this inherent potency to become a person. Their argument harks back to a centuries-old Catholic position that a certain degree of biological development is necessary in order for the human spirit (or soul) to be embodied. The conceptus, in other words, must have a baseline organization before it can embody a human person; it must embody the potentiality as well as actuality that will lead to the formation of a human person. Although previous theories of embryological development were grounded in minimal (and to a great extent, erroneous) scientific evidence, proponents of embryonic stem cell research find in contemporary embryology insights that tend to support this developmental theory, rather than defeat it.

Australian Catholic moral theologian Norman M. Ford has presented perhaps the most detailed argument in support of the view that an early embryo is not a potential human person. The plausibility of "delayed personhood" is based on scientific evidence that suggests that the embryo prior to implantation is not yet a self-organizing organism. Fertilization itself does not take place in a "moment," as is sometimes rhetorically claimed; it comes about through a process that takes approximately 24 hours, finally issuing in the one-celled zygote. When the zygote then divides, there comes to be not a definitive, singular organism, but a collection of cells, each with a complete genetic code. In other words, the first two cells, even if they interact, appear to be distinct cells, not a two-cell ontological individual. The same is true as division continues to four-, eight-, and sixteen-cells, each with its own life cycle and nutrients for sustaining its life. Cells multiply and differentiate as they are gradually specified in their potential. Some cells form membranes that will finally enclose an organized individual entity, approximately 14 days after fertilization. It is only then that a unified being can be said to be "self-organizing." Until this time, cells are "totipotential," in that they can become any part of what may ultimately be a human being, or even a human being as a whole.

The genotype of the zygote does not have a built-in blueprint for development, as if it were a miniature person. Only with the formation of a single primitive streak (after implantation) does the totipotency of the embryo actually become a potency to develop into one human person. As Ford puts it, fertilization is not the beginning of the development of the human individual but the beginning of the formative process and development into one (or more human individuals). Many who find this argument plausible tend to express it more simply: since at its early stages an embryo can twin and recombine, there is not here an individualized human entity with the inherent settled potential to become a human person. Critics of the position respond that if an embryo does not twin or recombine, then it may be supposed that an individual already exists. Ford's analysis goes further than this, however, suggesting that the very undifferentiation of cells prior to implantation disallows the identification of individuality, whether or not there is twinning and recombining.

Since this articulation of the 14 day theory is dependent on the current state of mammalian embryological studies, Catholic moral theologians on both sides of the debate take any new scientific information seriously regarding just what is happening in the first stages of embryological development. For example, studies of the role of first cleavage (of cells) in the development of mouse embryos suggest that an organizational process is already underway. Those who argue that the identity of the embryo is determined right after fertilization use this information to demonstrate a unitary process of self-organization. Those who support the 14 day theory do not deny that something is happening in the early stages of cell cleavage; they maintain, however, that the level of complexity and of organization is as yet too low to settle the individuality of a developing potential person. The embryo in its first stages must develop, in order to be able to implant; only when it reaches this stage is its singularity and individuality settled.

In addition to the argument from lack of individuation, some theologians have noted that the high rate of spontaneous early embryo loss undermines the credibility of the claim that human individuals begin at fertilization. In unassisted human reproduction, the development and loss of early embryos is estimated to be as high as 50–80%. From a Catholic theological point of view, it seems unbelievable that more than half of the individual human persons created by God should populate heaven without ever having seen the light of day in this world. This is a speculative observation, however, and it is generally not used today by those who favor embryonic stem cell research.

There are, then, two opposing cases articulated within the Roman Catholic tradition. This need not leave Catholics (or others) with a kind of "draw." Rather, it opens and sustains a significant conversation that takes account of scientific discoveries, as well as of a larger set of theological and ethical insights. Moral theologians and ethicists attempt to provide reasons for their views that will be open to the scrutiny of all. Both sides claim a certain amount of epistemic humility, since there are only degrees of certainty available regarding the ontological status of the early embryo. Indeed, even official church documents acknowledge that there is no definitive answer to the question of when human individual life begins. In 1974, the "Declaration on Procured Abortion" stated in a footnote that the question of the moment when the spiritual soul is infused is an open question. It acknowledged that there is not a unanimous tradition on this point, and authors are as yet in disagreement. In 1987, an instruction on "The Gift of Life" admitted that its authors were "aware of" current debates concerning the beginning of human life, the individuality of the human being, and the identity of the human person. Pope John Paul II made no decision on the question of delayed ensoulment (or "hominization") in his 1995 encyclical letter, "*Evangelium Vitae*." And the US Catholic bishops, even in statements made in 2008, have not attempted to settle this question.

Uncertainty is dealt with in different ways by those who oppose and those who support embryonic stem cell research. Those who oppose it argue that even if the presence of the "personal soul" in an embryo is only "probable," this suffices to prohibit risking the killing of a human person. In other words, probability (if not certitude) warrants the safer course. Hence, the embryo from fertilization on is to be treated as a human person, with the kind of unconditional respect due to all members of the human community. This argument is reinforced by a concern for a "slippery slope" along which human persons will cease to be respected, no matter what their actuality or their needs.

For those who conclude that embryonic stem cell research can be permitted, uncertainty works the other way. First, the probability of an early embryo's actually representing an individualized human being is much lower than is argued by opponents of embryonic stem cell research. Although there is no absolute certitude to be gained at this point in time from scientific evidence, the weight of the evidence against there being an individualized human being (incarnated in an embryo prior to implantation) is greater and more persuasive than the evidence for it. Moreover, the low level of uncertainty (for those who hold this position), when placed in relation to the prospect of great benefits for human healing (and some assurance that these benefits are not promised unrealistically) makes it possible—without yielding to a full-blown utilitarianism—to justify using early embryos as sources for stem cells for research. To prohibit all such research on the basis of a low probability of fact, or on a theological stipulation of moral status in the face of questionable accuracy of appraisal, seems itself not ethically justifiable.

Neither side in this debate wants to sacrifice the tradition's commitments to respect the dignity inherent in human life, promote human well-being, and honor the sacred in created realities. When a move forward is advocated for embryonic stem cell research, it need not soften the tradition's concerns to oppose the commercialization of human life and to promote distributive justice in the provision of medical care. The ongoing Roman Catholic conversation on all of these matters can be of assistance to others in a pluralistic society, as long as it remains open to wider dialogue, respectful of all dialogue partners, while retaining its own integrity. This is probably a lesson for all religious and secular traditions.

KEY WORDS

Moral realism Theory of knowledge that affirms the possibility of at least partial corresponence between what is in reality and what is in the mind.

Ontological Having to do with the structure of being.

Primitive streak The initial band of cells from which the individuated embryo begins to develop; it is present at approximately 14 days after fertilization.

FURTHER READING

Ford, Norman M. (2002). The Prenatal Person: Ethics from Conception to Birth. Oxford: Blackwell Publishing.

Holland, S., Lebacqz, K., & Zoloth, L. (Eds). (2001). The Human Embryonic Stem Cell: Science, Ethics, and Public Policy. Cambridge, MA: MIT Press.

National Bioethics Advisory Commission. (2000). Ethical Issues in Stem Cell Research. Vol III. Rockville, MD: National Bioethics Advisory Commission.

Outka, Gene. (2002). The ethics of human stem cell research. Kennedy Institute of Ethics Journal, *12*, 175–213.

President's Council on Bioethics. (2005). Alternative Sources of Human Pluripotent Stem Cells. Washington, DC.

Shannon, T., & Walter, J. (1990). Reflections on the moral status of the pre-embryo. Theological Studies, *51*, 603–626.

Walters, L. (2004). Human embryonic stem cell research: An intercultural perspective. Kennedy Institute of Ethics Journal, *14*, 3–38.

Warnock, M. (1984). A Question of Life: The Warnock Report on Human Fertilization and Embryology. Oxford: Basil Blackwell.

Waters, B., & Cole-Turner, R. (Eds). (2003). God and the Embryo: Religious Voices on Stem Cells and Cloning. Washington, DC: Georgetown University Press.

Chapter 31

Current Issues in US Patent Law

Patrea L. Pabst

Intellectual property rights provide a means for the owners of technology to recover their investment in the technology and, in some cases, to make a profit. More importantly, intellectual property rights provide a means for financing the incredibly expensive research and development and testing required for commercialization of new products and processes in the medical and biotechnology field. When the intellectual property rights have been lost, many times it is not possible to obtain the money required to see a product or process reach the clinic and benefit those for whom it is intended. It is only by protecting the technology that it can be used to help those who need it the most.

WHAT ARE INTELLECTUAL PROPERTY RIGHTS?

Intellectual property is intangible. It is embodied in patents, trademarks, service marks or trade names, copyrights, and trade secrets. Companies may have rights in such intellectual property by virtue of various licensing agreements, joint venture agreements, or by virtue of employment agreements.

The Constitution of the United States gives Congress the power to enact laws relating to patents, in Article I, Section 8, which reads "Congress shall have power . . . to promote the progress of science and useful arts, by securing for limited times to authors and inventors the exclusive right to their respective writings and discoveries." Trademarks and trade secrets are also typically protected under state law. Further protection can also be obtained using employment non-compete agreements, confidentiality agreements, and material transfer agreements. Extensions of patent rights and exclusive rights may also be obtained for delays in obtaining regulatory approval by the Food and Drug Administration (FDA), for Orphan Drug Act and under the Waxman-Hatch act. These provisions are not dealt with in detail in this chapter.

Patents

Patents are grants by a government entity that gives the patent owner the right to exclude competitors from making or using that which is defined by the claims of the patent. Patents have basically the same requirements throughout the world. In the United States, the requirements for obtaining and asserting a patent are defined by Chapter 35 of the United States Code ("USC"). Patents are governed exclusively by federal law. US Patents are granted exclusively by the US Patent and Trademark Office (PTO). Patents are granted on applications for new, useful, and non-obvious inventions. A patent is issued by the US PTO after it has examined the application for the patent and determined that it meets various criteria. Most other countries have similar laws governing granting of patents. In some cases, patents are obtained for a "region" such as the European Patent Convention countries, through a centralized patent office. The granted patents are then registered in the individual member countries in which they are enforceable.

A patent for an invention is the grant of a property right to the inventor, issued by the PTO. Currently the term of a new patent is 20 years from the date on which the utility application for the patent was filed in the United States or, in special cases, from the date an earlier related application was filed, subject to the payment of maintenance fees and disclaimers to earlier expiring patents. Effective May 29, 2000, the patent term for subsequently filed utility applications, was extended to compensate for delays in prosecution due to the Patent Office. Design patents are granted for new, ornamental, and non-obvious designs, and have a maximum term of 14 years from the date of grant. US patent grants are effective only within the US, US territories, and US possessions. The right conferred by the patent grant is, in the language of the statute and of the grant itself, the right to exclude others from making, using, offering for sale, or selling the invention in the United States

Stem Cell Anthology

or importing the invention into the United States.[1] What is granted is not the right to make, use, offer for sale, sell, or import, but the right to exclude others from making, using, offering for sale, selling, or importing the invention.

In general, technology relating to tissue engineering is subject to the same rules as are other compositions and methods of use and manufacture thereof. As long as the "thing itself" is new, non-obvious, and subject to written description, it is patentable.

In the language of the statute, any person who "invents or discovers any new and useful process, machine, manufacture, or composition of matter, or any new and useful improvement thereof, may obtain a patent,"[2] subject to the conditions and requirements of the law. The word "process" is defined by law as a process, act or method, and primarily includes industrial or technical processes. The term "machine" used in the statute needs no explanation. The term "manufacture" refers to articles which are made, and includes all manufactured articles. The term "composition of matter" relates to chemical compositions and may include mixtures of ingredients as well as new chemical compounds. These classes of subject matter taken together include practically everything which is made by man and the processes for making the products. The patent law specifies that the subject matter must be "useful." The term "useful" in this connection refers to the condition that the subject matter has a useful purpose and also includes operativeness, that is, a machine which will not operate to perform the intended purpose would not be called useful, and therefore would not be granted a patent.

In general, patentable subject matter includes compositions, methods of manufacture, and methods of use.[2] Compositions may include, for example, biodegradable polymeric matrices seeded with cells, or isolated cells for implantation in the body. Methods of manufacture may be directed to, for example, a process for creating a unique prosthetic device. Methods of use may entail methods for administration of therapeutic compositions, or surgical implantation of, for example, a synthetic tissue matrix containing implanted isolated cells that secrete insulin.

The most significant problems we observe in patenting in pharmacology and biotechnology are with the issue of written description and enablement. The PTO has provided guidelines for the written description requirement.[3] The fundamental factual inquiry is "whether the specification conveys with reasonable clarity to those skilled in the art that, as of the filing date sought, applicant was in possession of the invention as now claimed".[3] Answering this question is not a single, simple determination, but rather is a factual determination reached by considering a number of factors. The PTO guidelines list the following factors which should be considered by an examiner in his written description analysis: (1) the level of skill and knowledge in the art; (2) partial structure, physical, and/or chemical properties disclosed in the specification; (3) functional characteristics disclosed in the specification, alone or coupled with a known or disclosed correlation between structure and function; and (4) the method of making the claimed invention disclosed in the application.

The PTO has also provided guidelines for the enablement requirement under 35 USC §112 and its relationship to the utility requirement under 35 USC §101.[4] The PTO guidelines focus on a three pronged test for determining whether an invention is "useful" within the meaning of the law: Does the invention have a utility that is specific, substantial, and credible? A specific utility is one that is particular to the subject matter claimed. A substantial utility is one that defines a "real world" use. Utilities that require or constitute carrying out further research to identify or reasonably confirm a "real world" context of use are not substantial utilities. A utility is credible unless the logic underlying the assertion is seriously flawed, or the facts upon which the assertion is based are inconsistent with the logic underlying the assertion.

Inventions are defined by claims, which are supported by a detailed disclosure in a specification that tells one how to make and use that which is defined by the claims. For product claims that do not recite any utilities, disclosure or assertion in the specification of one specific, substantial, and credible utility meets the criteria of 35 USC §101. If no credible, specific, and substantial utility is asserted in the specification and none is well established, then the claims are rejected under 35 USC §101. Utilities that constitute curing or preventing a condition are sometimes not credible to one of the skill in the art and thus may raise a question under 35 USC §101. However, any rejection based on lack of credible utility must be supported by documentary evidence or sound technical reasoning by the examiner. Since most diseases or conditions can be treated (although not necessarily "cured"), rejections under 35 USC §101 for treatment claims rarely should be made. Each case is decided based on the specific facts. For example, since vaccines are regularly prepared to combat various viruses and organisms, vaccines would have a credible utility to one of the skill in the art. Thus, vaccines should not raise a question under 35 USC §101. Materials to be used for research, or methods of using those materials for research, raise issues of whether the utilities require or constitute carrying out further research to identify or reasonably confirm a "real world" context of use.

One of the areas people frequently have questions on is when does something have utility when it is still in a

[1] 35 USC §271 (a) (2005).

[2] 35 USC §101 (2005).

[3] Manual of Patent Examining Procedure (MPEP) §2163 (8th Edition, August 2001) (revised August 2005).

[4] MPEP §2107.02 (8th Edition, August 2001) (revised August 2005).

research stage. For example, if one makes a matrix out of a particular experimental polymer, seeds it with cells, and shows the cells attach and proliferate in a petri dish under laboratory conditions, is that sufficient to make a claim to a matrix with cells seeded thereon to form an organ equivalent for transplantation? The key question is whether those skilled in the art would have a reasonable expectation of success—if they thought it more likely than not that the laboratory tests were predictive of success in a human or animal, then one has met the criteria for utility. What about when one clones a portion of a gene, the so-called "expressed sequence tags" or "ESTs." Since one does not know what is encoded, even using the various analytical programs, one can only claim the isolated sequence, not the protein that may be encoded by the full length gene, nor its use. The ESTs themselves can be claimed for use as probes but this is a very limited use. Another frequent question is with regard to isolated cells, especially stem cells or fetal cells. These are clearly patentable subject matter, if they meet the other requirements of being novel and non-obvious, and one can either make a deposit with an approved depository such as the American Type Culture Collection, or provide an adequate written description of how to obtain the cells.

Although the law provides for patenting of compositions, methods of manufacture, and methods of use, biotechnology can present a problem under US patent law when the subject matter moves away from the realm of the artificial or "things engineered by the hand of man" to a blend or chimera of "artificial" and "natural".[5] An example is when one blends cells and a matrix to form a cell matrix structure that is then implanted in a patient. Then, the matrix degrades to leave only implanted cells and/or the patient's own tissue grows into an implanted matrix structure which then degrades. At what point do these materials become patient and not patentable subject matter? Ethical issues may arise due to overlap between patient material and traditional subject matter, particularly in those cases involving dissociated isolated cells, biodegradable matrices for implantation, polymeric materials for altering cell/cell interaction (such as adhesion or restenosis), as well as materials for implantation that are designed to remain in the body, such as stainless steel hip replacements or cryopreserved pig valves.

Interpretations of the statute by the courts have defined the limits of the field of subject matter which can be patented. Examples of materials that cannot be patented include the laws of nature, physical phenomena, and abstract ideas.[6] Naturally occurring materials cannot be patented **unless altered "by the hand of man". Id**. Thus, isolated cells are patentable; while the naturally occurring source—harvested tissue—is not patentable subject matter. A patent cannot be obtained upon a mere idea or suggestion. A complete description of the actual machine or other subject matter for which a patent is sought is required.

Outside of the United States and Australia, methods of treatment of humans or other animals are generally not patentable subject matter. For example, although surgical instruments, drugs, or devices used in surgery are patentable, surgical treatments are not patentable subject matter. Therefore, one cannot obtain a patent on a method for surgically treating a patient. Typically, while this subject matter is not patentable, the compositions and methods of manufacture for use in treating patients are patentable subject matter. Claims may be obtained to the composition *per se*, which is to be implanted. In Europe, claims can be obtained to a first, or even a second, use of the material when the material itself is known. However, the patentability is quite limited in individual countries and in the European Patent Office, for policy and ethical reasons. Generally, Patent Offices in Asian countries are far less flexible than the European Patent Office in this matter. As a result, patent attorneys have adopted a number of strategic approaches to obtain protection equivalent to that which is available in the United States. For example, one may draft claims directed to methods of manufacture of such materials, as well as to methods of use that are defined by the composition rather than the method of use steps.

An exception to the statute of what constitutes patent infringement in the United States was created by the legislature to prevent enforcement of patents for methods of surgery not involving a patentable drug or device. The patent cannot be enforced against the doctor performing the surgery or against the patient. The legislation arose as a result of a controversy over enforcement of a patent on a method for cataract surgery, which solely involved the way a surgical incision was made and available lenses inserted into the eye. The patent would have been enforceable if it utilized a patent device for cutting or inserting the lens, or a drug for treating the eye.

Novelty

Another requirement for patentability is novelty. In order for an invention to be patentable, it must be new, that is, it cannot be patented if: (a) "the invention was known or used by others in this country, or patented or described in a printed publication in this or a foreign country, before the invention thereof by the applicant for patent" or (b) "the invention was patented or described in a printed publication in this or a foreign country or in public use or on sale in this country more than 1 year prior to the application for patent in the United States . . .".[7]

[5] *Diamond v. Chakrabarty*, 447 US 303, 206 U.S.P.Q. 193 (1980).

[6] Diamond, 447 US at 309.

[7] 35 USC §102 (2005).

Novelty means that no one, including the applicant for the patent, has publicly used or described that which is being claimed, prior to filing an application for patent. In the United States, there is an exception when the publication is made less than 1 year prior to filing of the patent application or when an application by a third party filed before the application of interest is published or granted after the application of interest is filed.[8] The publication can be "removed" as prior art if the applicants are able to demonstrate that, prior to the publication, they conceived and diligently reduced to practice what they are claiming. In the United States, one can also get into a proceeding between multiple parties who have filed patent applications on the same invention, called an interference, to determine who was first to invent, regardless of who was first to file for a patent, which in turn determines who is entitled to the patent. Outside of the United States, rights are awarded to the first to file, and patent rights are lost immediately upon publication of the invention if made prior to filing for the patent.

What constitutes a publication? Generally, a publication is any oral, written, or physical description that conveys to the public that which an applicant would like to claim. It may be a talk at the proceedings of a society (including any slides presented), an article in a scientific journal, a grant application that is awarded (effective as prior art at the time of award, not of application), a thesis (effective at the time of cataloging), or even an offer for sale or a press release. A critical requirement is that the publication must be enabling, that is, it must convey to one of the ordinary skill in the art how to make and use that which is being claimed. Public use means more than using the composition or method in one's laboratory. However, it can include even a one patient study that is reported during clinical rounds or at a presentation at which a drug company or surgical supply representative is present. The courts, in many cases, have had to interpret what it means to be publicly available. A frequent question is when is a student's thesis available as prior art. Courts have now held that once the thesis is cataloged, it is publicly available, because it has been entered into a computer database that one searching the database will be able to access.[9] Accordingly, the publication date of a thesis is the date on which the thesis is cataloged, not the date on which it is defended or signed by the thesis committee. Slides that are not distributed, but that are shown at an oral presentation, are considered to be publications, particularly if the meeting is attended by those skilled in the art who would be able to understand and use the information in the slides. Additionally, a poster presented at a scientific meeting constitutes a "printed publication" under 35 USC §102.[10]

Disclosures to another party under the terms of a confidentiality agreement are not publications. Uses that are strictly experimental may not be public disclosures, if, among other aspects, they are designed to determine if that which is to be claimed will work, and if any other parties who are involved are clearly informed that the studies are experimental in nature. If an announcement is made publicly, which does not enable one of the ordinary skill in the art to use or make that which is later claimed, then the announcement is not a publication. For example, an announcement could be a statement made to the press that researchers X and Y have discovered a cure for cancer. Since the announcement does not tell one of the ordinary skill in the art how to cure cancer, it is not enabling. However, enablement can be difficult to prove and standards may change over time. A 1995 case involved the question of whether a publication was enabling for the development of a transdermal patch for delivery of nicotine.[11] The court found that a prior publication referring to transdermal patches for drug delivery mentioned that the drug in the transdermal patch for treatment of heart disease could be replaced with nicotine for assisting patients in quitting smoking. The court held that the article disclosed or made obvious the transdermal patch for delivery of nicotine claimed by the applicant, because the applicant merely took the transdermal patch described in the article, put nicotine in it, and demonstrate that the nicotine was delivered and would work exactly as predicted based on the delivery of the drug for treatment of heart diseases. Even though there was no information relating how the drug was to be incorporated into the transdermal patch, the publication was enabling because those of ordinary skill in the art would have been able to determine how to put the nicotine in the transdermal patch without undue experimentation.

If the invention has been described in a printed publication anywhere in the world, or if it has been in public use or on sale in this country before the date that the applicant made his/her invention or more than 1 year before the date on which an application for patent is filed in this country, a patent cannot be obtained.[7] In this connection, it is immaterial when the invention was made, or whether the printed publication or public use was by the inventor himself/herself or by someone else. If the inventor describes the invention in a printed publication or uses the invention publicly, or places it on sale, he/she must apply for a patent before 1 year has gone by, otherwise any right to a patent will be lost.

[8] 35 USC §102 (a) and (e) (2005).

[9] *Philips Elec. & Pharmaceutical Indus. Corp. v. Thermal & Elec. Indus., Inc.*, 450 F.2d 1164, 1169–72,171 U.S.P.Q. 641 (3d Cir. 1971); Gulliksen v. Halberg, 75 U.S.P.Q. 252 (Pat. Off. Bd. Int'f. 1937).

[10] In re Klopfenstein, 380 F.3d 1345, 72 U.S.P.Q.2d 1117 (Fed. Cir. 2004).

[11] *Ciba-Geigy Corp. v. Alza Corp.*, 1995 US App. LEXIS 28214, 37 U.S.P.Q.2d 1337 (Fed. Cir. 1995).

Obviousness

Even if the subject matter sought to be patented is not exactly shown by the prior art, and involves one or more differences over the most nearly similar thing already known, a patent may still be refused if the differences would be obvious. The subject matter sought to be patented must be sufficiently different from what has been used or described before that it may be said to be non-obvious to a person having ordinary skill in the area of technology related to the invention. The claimed method or composition must be non-obvious to those of ordinary skill in the art from what is publicly known.[12] This is usefully referred to outside of the United States as a requirement for an "inventive step." In the 1960s, the United States Supreme Court carefully analyzed non-obviousness and those factors that are to be considered in determining whether that which is claimed is obvious from the prior art.[13] This analysis is a fact-based determination, involving not only the elements which are claimed, but also the level of skill in the art and the expectation or predictability that the claimed method or composition would perform as predicted, actual success in the marketplace, long felt need, and whether there are unexpected results. If one has no better than a 50–50 chance that a particular method may work, and the method works, it is arguably not obvious, although it may be obvious to try. If one tries something and the results are vastly different from what was expected, then the results are not obvious. For example, if one administered two drugs each in the dosage known to yield a particular effect and the combination results in a substantially greater effect than the sum of the individual effects of the two drugs, resulting in the ability to use a much lower dosage of each drug than expected, then one would have unexpected results or "synergy." If the prior art teaches away from what the applicant has done, this result would support a finding of non-obviousness. For example, if the prior art states that one cannot proliferate implanted hepatocytes, then it may be non-obvious if the applicant for a patent finds that he can administer particular growth factors with his hepatocytes and see proliferation. Many other considerations factor into whether a claimed composition or method is obvious in view of the prior art.

Enablement, Written Description, and Best Mode

Additional requirements for patentability are defined by 35 USC §112, first and second paragraphs. These include enabling one of the ordinary skill in the art to make and use that which is claimed; a written description that clearly describes that which is claimed; and, in the United States only, disclosure of the best mode for practicing the invention.[14]

The applicant must describe that which is claimed in sufficient detail, and with appropriate methods and sources of reagents or other materials or equipment, to enable one of the ordinary skill in the art to make and use that which is claimed. This sounds far simpler than it actually is in practice. In many cases, particularly when coming out of a university study or a start-up company, the invention that applicants would like to claim is that which the applicant intends to develop over the next several years, based on a limited amount of data available at the time of filing. Particularly in the case of universities, where the applicant must publish or has submitted grant applications (which in and of themselves constitute prior art once they are awarded), the difficulty is in describing that which has not yet been done. The application must not only describe a specific limited example, but must describe the various ways in which one intends to practice that which is claimed.

"Invention" usually consists of two steps: "conception" and "reduction to practice." There are two kinds of reduction to practice: actual and constructive. "Constructive reduction to practice" means that the applicant has described in the application for patent **how** to make and use that which is claimed, but has not actually made and used what is claimed. This may be as simple as stating that although a biodegradable polymer such as polylactic acid-co-glycolic acid is preferred for making a matrix for culturing cells, other biodegradable polymers such as polyorthoesters or polyanhydrides could also be used. It may be less obvious that other cell types or shapes or methods of processing may be used when only one example showing reduction to practice is available. The rule of thumb in this case is the level of predictability. Therefore, in stating what kind of drugs one could deliver using the claimed technology, one might list a wide variety of drugs based on the data available with one type of drug. However, one may not be able to include in a list of drugs to be delivered peptides or very hydrophobic compounds, which usually are viewed as difficult to deliver, based on data obtained with a drug that is "easy" to deliver, such as a sugar or small molecular weight dye. Being too predictive (i.e. engaging in extensive constructive reduction to practice), which includes "non-enabling" or non-enabled technology, may in some cases be a detriment during prosecution of subsequently filed applications, because the examiner may cite the earlier work as making obvious the applicant's subsequent work. Patent attorneys frequently must play a balancing game in determining how far to go with constructive reduction to practice in order to exclude competitors while not eliminating the applicant's own ability to obtain additional, subsequent patent protection.

[12] 35 USC §103 (2005).

[13] *Graham v. John Deere Co.*, 383 US 1, 148 U.S.P.Q. 459 (1966).

[14] 35 USC §112 (2005).

In the United States, there is a requirement to disclose the best mode for practicing that which is claimed at the time of filing the application. No similar requirement exists outside of the United States. Because most applicants file the same application in the United States as outside of the United States, US applicants frequently disclose their best mode in foreign-filed applications.

Patent Rights

The purpose of a patent is to exclude the competition from making and using that which is claimed, not to "protect" a product—a frequent misconception of patents. In order to exclude competition, one must describe and claim not only that which one intends to practice, but that which another party could practice in competition with the patentee. What does this mean in real terms? It means that the applicant for a patent must describe his preferred method, which is known as of the date of filing, the preferred embodiments that he or his company intends to market, as well as any embodiments that a competitor could make and use in competition with the applicant's product.

Patent Term and Patent Term Extension

Under the revised US patent law that was enacted as a result of implementation of the General Agreement for Trade and Tariffs ("GATT"), the term of a patent is 20 years from the original date of filing or the filing date of the earliest utility application to which priority is claimed. Applicants therefore have more incentive to prosecute all claims in a single application in order to minimize costs for prosecuting and maintaining the patent. Under the law in effect prior to June 8, 1995, the patent term was 17 years from the date of issue in the United States. Divisional applications were a commonly used method to extend patent protection to encompass different aspects of the technology over a period of time much greater than 17 years. For example, an application would be filed in 1990, and a single inventive concept (e.g. the composition) would be prosecuted in the first application. Three years later, when those claims were allowable and a patent was to issue, a divisional application would be filed with another set of the claims that had been restricted out of the original application. This divisional application would be prosecuted for another 2 to 3 years, the claims would be determined to be allowable, the second patent would issue with a 17-year term, and a third divisional application would be filed. The result is that patents on related technology would issue sequentially over several years, increasing the effective term of patent protection beyond 20 years. Under the new law, this mechanism to extend patent protection is not possible.

The GATT was signed into law in the United States on December 7, 1994, and the initial provisions affecting US patent practice were implemented on June 8, 1995. The most significant changes arising from enactment of that agreement, now called Uruguay Round Act, were changes in the patent term in the United States, the implementation of provisional patent applications, and the broadening of what constitutes infringement in the United States. The change in patent term has been discussed above. For those applications filed before June 8, 1995, the term of any issuing patent is 17 years from the date of issue or 20 years from the filing date, whichever is longer. The term of any patent issued on an application filed June 8, 1995 or later is 20 years from the earliest claimed non-provisional priority date (i.e. the filing date of the first utility application to which priority is claimed or the filing date of a Patent Cooperative Treaty (PCT) application designating the United States). Extensions of terms are available upon delays in issuance arising from appeals or interferences. Additional extensions of terms are available for delays in obtaining regulatory approval by the Food and Drug Administration (FDA) for a device or a drug. Recent legislation has provided for an increased patent term due to unreasonable delays by the US Patent Office. Under changes implemented in late 2000, the US Patent Office recognized that losses in patent term were resulting from delays in prosecution of applications through no fault of the applicants. Diversion of user fees to other government agencies, in particular, as well as the initial "up time" associated with contracting out the publication and other services at the US Patent Office resulted in some cases in delays of months to even years. Excessive delays may now be the basis for extensions of patent term.

Another result of the GATT was the implementation of provisional patent applications. These provide a means for applicants to make a preliminary filing, with limited claims or even no claims, and less stringent requirements as to form (but exactly the same requirements as to enablement, best mode, and written description), to preserve their initial filing date, while collecting additional data or financing or evaluating the technology. As long as a utility application is filed within 1 year claiming priority to the provisional application, one can benefit from the provisional application priority date. Provisional applications are not themselves examined and expire exactly 1 year after filing. They constitute a convenient way to preserve priority dates and provide a particular advantage in the medical and biotechnology fields, since patent term runs from the date of the utility application, not the provisional application, and therefore the effective patent term of an application claiming priority to a provisional application can be 21 years from the earliest priority date, not 20 years.

Failure in the provisional application to completely disclose and enable that which is subsequently claimed in an utility application can result in a loss of the claim to priority to the provisional application, if that which is claimed is not enabled. Merely filing as a provisional application an article that will be published or presented in order to avoid loss of foreign rights usually will not comply with the enablement requirements, and therefore will not serve as an adequate basis for priority. It is essential that applicants who file provisional applications based on an article amplify the description to encompass other embodiments and to provide the basis by which one of the ordinary skills in the art can practice that which is claimed. Application sections that are not required for enablement, which are typically included in a utility application, include the background of the invention, the problems that the claimed invention addresses, and the claims. These sections can be omitted from the provisional application, thus saving time and money in preparing the application. In many cases, fairly standard language can be used to expand or broaden the description in an article in order to meet the enablement requirements, providing a means for those with limited amounts of time or money to protect that which they are disclosing with minimum risk and expenditure.

The World Intellectual Property Organization ("WIPO"), which implements the provisions of the "PCT" and the European Patent Office, has confirmed that US provisional applications serve as an adequate basis for a claim to priority in corresponding foreign file applications. However, under the Patent Convention, all foreign applications that claim priority from an earlier filed application must still be filed within 1 year of the US filing date or the filing date of the country in which the first application is originally filed.

Because biotechnology is a complex field, especially in the areas of patentable subject matter and enablement, many of the general observations relating to patents may not be as directly applicable to more "conventional" patentable subject matter. For example, the US Patent Office consistently maintained that the change from a 17-year term from the date of issue to a 20-year term from the earliest priority date would not result in a significant loss of patent term. However, because the Patent Office applies such a stringent examination proceeding under §112 (written description and enablement) in the biotechnology area, issue time in these cases typically has been considerably longer, not uncommonly taking as many as 5–7 years from the original priority date. The result is that these complex biotechnology patents have a substantially shortened term as compared to many other types of patents. This is the case even if one obtains some extension due to delays by the US Patent Office. Because a patent extension can still be obtained for delays due to regulatory issues involving the FDA, as well as for appeals to the Board of Patent Appeals and Interferences, those in the United States who believe that their patent rights will be limited in term due to delays in prosecution should avail themselves of the Patent Extension Act, if at all possible. One must bear in mind, however, that an extension for regulatory delays can only be obtained on **one** patent for any particular product or process; thus, the inventor or licensee with multiple, related patents clearly should choose the most important patent or the patent subject to the greatest increase in patent term, when facing such a situation. The patent that is to be extended must be brought to the attention of the FDA. Following FDA approval of the claimed product or process, the extension must be applied for in a timely fashion.

Trademarks, Service Marks, and Trade Names

A trademark is a word, design, shape, number or slogan, or a combination of those elements, that identifies and distinguishes one company's products from others. It is a symbol of goodwill (e.g. Coca-Cola®, PC Paintbrush® and Budweiser®). A service mark is to services what a trademark is to products (e.g. SAKS® (retail department store services), Holiday Inn® (hotel services), and Orkin® (pest control services)). A trade name is a name used to identify a business, (i.e. a "commercial name"). A trade name can also be used as a trademark to identify products and as a service mark to identify the services (e.g. IBM®).

Trademarks and service marks can be registered on the Federal Register with the US PTO if they are used in interstate commerce, or they can be registered in a state if they are only used within that state. Trade names cannot be registered federally.

A trademark is created by use, not by filing of an application. Lack of registration is not a bar to enforcing the mark, but when a trademark owner has obtained a registration, it obtains significant procedural and substantive advantages. For example, federal registration provides constructive nationwide notice to all users of a trademark. This eliminates a second user's defense of "innocent adoption" which might otherwise permit such a second user to continue to use the mark. Federal registration can also protect an internet domain name or email address from a disruptive challenge.

"Intent to use" trademark applications allow a trademark owner to obtain "pre-approval" prior to investing money in the use of the mark. The date of filing of the intent to use application is the priority date for the mark and actual use must be shown within a period of not greater than 36 months following the approval of the mark by the US PTO.

Copyrights

A copyright protects original expression. It is totally a creature of statute, 17 USC, the US Copyright Act. The author of a copyrighted work has the exclusive right to use, distribute, modify, display, reproduce, and perform the work. The copyright protects its owner from actual copying or unauthorized use of the protected expression in the copyrighted work. It does not prohibit the independent creation of a similar work by a third party if that third party does not have access to, or knowledge of the work.

A copyright also does not grant the copyright owner a monopoly on the ideas contained within the work. Copyright protection does not extend to any "idea, procedure, process, system, method of operation, concept, principal, or discovery" embodied in the work.[15] Such concepts may be protected only through patents or trade secrets.

Trade Secrets

"Trade secret" means information including, but not limited to, technical or non-technical data, a formula, a pattern, a compilation, a program, a device, a method, a technique, a drawing, a process, financial data, financial plans, product plans, or a list of actual or potential customers or suppliers which derives economic value, actual or potential, from not being generally known to, and not being readily ascertainable by proper means by, other persons who can obtain economic value from its disclosure or use; and is the subject of efforts that are reasonable under the circumstances to maintain its secrecy.

This definition suggests three main characteristics of a trade secret. First, a trade secret generally relates to some sort of data, formula, device, method, or other similar type of information. Second, a trade secret must be valuable. This quality is generally self-evident. If a trade secret were not valuable, then a company would not consider it to be worth keeping secret.

Third, and most importantly, the information must be the subject of efforts that are reasonable under the circumstances to maintain its secrecy. An owner need not take every steps possible to maintain a secret but must take steps that are reasonable under the circumstances to maintain the secrecy. This typically means that employees will have employee agreements with non-compete provisions that the trade secrets are kept in a restricted area or under conditions wherein the documents cannot be downloaded by others and that parties accessing the information are aware that it is both valuable and confidential.

Trade secrets are becoming increasingly valuable in the biotechnology and pharmaceutical areas, with the advent of databases containing information that allows one to accumulate and collate information on a regular basis. An example of a trade secret may be a database including information regarding drug design or what amino acids can be substituted or deleted, to achieve a particular activity. It may also be what kinds of genes are in a microarray that is being marketed or what kind of controls are required to generate useful data.

Trade secrets are undervalued assets by many investors who do not always understand that the trade off for a patent is that one must disclose who to make and use the claimed invention. To enforce rights in a patent, one must be prepared to enter into costly and time-consuming litigation. In contrast, trade secrets do not need to be enforced since the competitor never learns how to practice the trade secret. Of course, this only works if the information that is being maintained in secrecy is not subject to "reverse engineering," that is, one cannot figure out the secret from the product that is being sold. In the case of a database, it is the product that is designed using the information that is being sold, not the information *per se*. Trade secrets can last much longer than a patent, and do not cost hundreds of thousands of dollars to obtain and maintain.

WHY ARE INTELLECTUAL PROPERTY RIGHTS IMPORTANT?

Advancements in the medical and biotechnology fields often are achieved only through substantial investment of industrial, academic, and governmental resources. Patenting of these technological advancements frequently is employed to recoup that investment, to create profits which are used in part to develop new or improved products, and to enhance a competitive commercial edge. Other forms of intellectual property protection, such as trade secrets, copyrights, and trademarks, may also be used to further protect and exploit drug delivery processes, products, and services. In the rapidly expanding fields of microarray screening and genomics, where databases are revised on a daily basis and much of the information does not need to be distributed, protection of only the results obtained using the data (i.e. trade secret) is far more effective than obtaining patent protection.

One of the most frequently asked questions is why do we need to go to the trouble and expense of patenting a composition or method. The most common reason is that protecting a new composition or method of manufacture or use provides a means for obtaining the revenue required to develop a new drug or medical treatment. With the cost of developing and obtaining regulatory approval for a new drug approaching $200 million dollars in the United States, patent rights are essential to recovering expenses. For small companies that spend more time raising money than selling products, patents and patent applications represent the company's only tangible assets which it can show to potential investors. For universities and other non-profit research institutions, patents and associated know-how and, in some limited cases, trade secrets can be used to obtain royalties from license agreements, sponsored research funding in many cases, and equity in new companies started for the purpose of exploiting the technology.

[15] 17 USC §102(b) (2005).

Patents and other intellectual properties are valued in many different ways. For example, a process for manufacture typically would be licensed for 2–3% of the gross selling price of a product of the process. This price would be decreased if multiple licenses had to be obtained to use the process. Patents claiming compositions tend to have a greater market value, for example, between 5% and 10% of the gross selling price, due to the perception that these patents are easier to enforce than process patents. These numbers of course are affected by the stage in development, the market cap, and the number of competing technologies in the same field.

Enforcement is a risky business. A good patent strategy is to obtain patents that claim a product, methods of manufacture, and methods of use, broadly and specifically, so that a patentee is able to assert multiple patents against an alleged infringer. Patents with broad claims generally will be easier to invalidate than more specific patents. Faced with the prospect of fighting several patents, most parties will opt for settlement. The alternative, litigation, is extraordinarily expensive for both parties and can result in the patents being invalidated, or the infringer being liable not only for damages for infringement, but also for attorney's fees and punitive damages.

Patents give the patent owner the right to exclude competition. This is accomplished by asserting the patent against third parties who are marketing a product or service which falls within the scope of the claims. Referred to as "infringement," the criteria are totally different from the criteria for obtaining a patent, referred to as "patentability." In simple terms, a patent claim consists of "elements" in a defined relationship. Certain phrases expand or limit the scope of the claim. For example, the term "comprising" can be translated as "including at least," while "consisting" means "including only." If a claim reads:
Composition comprising:

- A
- B
- C

then the claim would cover any composition including A, B, C, and any other component. Use of the term "consisting" in place of "comprising" would restrict the claim to a composition including **only** A, B, and C.

In determining infringement, one must look to the claims of the patent. Claims may be clear on their face, or require reference to the specification, or description, of the patent. Claims also may be limited by agreements made during prosecution, a doctrine referred to as "file wrapper estoppel." For example, if the prosecuting attorney argues that the claims distinguish over the prior art on the basis that the prior art does not disclose a particular feature that the attorney argues is essential to the claims in the patent, then the claims will be construed to require that limitation, even if not explicitly recited in the claims as issued.

Asserting a patent allows the alleged infringer to file an action for declaratory judgment in a federal district court, asking the judge to declare the patent invalid or non-infringed. Litigation is very expensive and especially detrimental to small companies, thus providing a great deal of incentive to license the technology on terms favorable to both parties.

Since patents are now published as applications, 18 months after the original filing date, the public gets the benefit of this knowledge prior to the period of exclusivity, since at least in the United States, patent rights are not enforceable until the patent is granted, and then only to the extent that the alleged infringing method or composition falls within the issued claims, although it is now possible even in the United States to backdate damages if the claims as published are essentially the same as the issued claims. A major limitation on the ability of a patent owner to enforce his patents lies in his ability to pay for the cost of litigation. In my experience, the cost of patent infringement litigation in the United States has gone from approximately $1 million in the mid-1980s, to $3 million by 1990, to five times or more that amount by 2000. This effectively prevents small companies and individuals from enforcing their patent rights, even with legislation allowing recovery of treble damages and attorneys' fees in the event of willful infringement.

Further, as discussed briefly above, in some cases, the rights to enforce a patent have been limited by legislative or other political action. In the United States, these rights are defined by 35 USC §271. 35 USC §287, which defines the limitations on damages and other remedies, was amended on September 30, 1996, in an apparent effort to prevent issuance of patents directed to new medical treatments. The amendment, which applies only to patents issued after September 29, 1996, deprives patentees of remedies for direct infringement and induced infringement of patents to surgical or medical procedures that do not involve patented drugs or devices.[16] Although the exact scope of the exclusion is subject to interpretation of numerous terms in the amendment, it appears that the exclusion applies only to the performance by a medical practitioner (defined as a person licensed to provide medical activity or a person under the direction of such a person) of a medical or surgical procedure not involving the use of a patented machine, manufacture, or composition of matter (e.g. medical devices, implants, and drugs) or the patented use of a composition of matter. Thus, it appears that the use of patented medical devices and patented drugs will still be subjected to infringement remedies.

The amendment does not apply to activities of persons engaged in the commercial development, manufacture,

[16] 35 USC §287 (2005).

sale, importation, or distribution of a machine, or composition of matter or the provision of pharmacy or clinical laboratory services where such activities are (1) directly related to the commercial development, manufacture, sale, importation, or distribution of a machine, manufacture, or composition of matter or the provision of pharmacy or clinical laboratory services, and (2) regulated under the Federal Food, Drug, and Cosmetic Act, the Public Health Service Act, or the Clinical Laboratories Improvement Act. Although it is not clear how this exclusion will be interpreted, and as of 2005, this clause has not been discussed in any published legal opinion, it appears that infringement remedies will not be limited for performance of a medical or surgical procedure that is related to the commercial development or exploitation of a product or service where the medical or surgical procedure is subject to Federal regulation.

From a practical perspective, patents are usually only enforced against manufacturers or distributors, not doctors, since it would be unduly burdensome, and typically not cost effective to pursue actions against many individuals rather than a single large entity. It is therefore doubtful that the change in law will have any detrimental impact on the health care industry in the United States.

The absence of literal infringement does not necessarily mean that a process or device does not infringe a patent. According to the judicially created "doctrine of equivalents," even though the language of a claim cannot be read literally upon a process or device, a claim can be infringed if the process or device "performs substantially the same function in substantially the same way to obtain the same result."[17] In 2002, the Supreme Court rendered its decision in *Festo Corp. v. Shoketsu Kinzoku Koygo Kabushiki Co.*, addressing the relationship between prosecution history estoppel and the doctrine of equivalents.[18] In an unanimous opinion, the Court held that prosecution history estoppel may apply to any claim amendment made to satisfy the Patent Act's requirements (e.g. novelty, obviousness, enablement, written description, etc.); it is not limited to amendments made to avoid the prior art. The Court also held that prosecution history estoppel creates a flexible bar to the doctrine of equivalents. The Court explained that amending a claim to satisfy the Patent Act's requirements need not bar suit against every equivalent to the amended claim element. By amending a claim, the inventor is deemed to concede that the patent does not extend as far as the original claim. A rebuttable presumption is made that the amendment to the claim limitation bars the patentee from asserting infringement against an equivalent to the narrowed limitation. The patentee can rebut this presumption and show that the amendment does not surrender the equivalent in question. For example, the patentee could establish that the equivalent was unforeseeable at the time the claim was drafted, the amendment did not surrender the particular equivalent in question or there was some reason why the patentee could not have recited the equivalent in the claim.

In some cases it is possible to pursue alternative remedies, such as third party reexamination proceedings in the US patent office to have a party's patent invalidated over art not considered during prosecution, or in Europe or Japan, by third party opposition proceedings. The latter are considerably less expensive than litigation in the United States and decisions are reached typically in less than 2 years, which is also quite quick compared to litigation in the United States. The results of litigation are also quite unpredictable but very risky in the medical and biotechnology fields. Especially in the biotechnology area, the late 1990s were known for the swift and drastic decisions by the Court of Appeals for the Federal Circuit, invalidating patents on the grounds the claims were not enabled by the specifications.

HOW DO YOU VALUE INTELLECTUAL PROPERTY?

A common misconception is that one must have a patent in order to license the technology or for investors to be willing to invest in the technology. This is just not the case. In most cases where the technology arises in a university and many times with individuals, licenses are entered into with only a single application having been filed, and no indication of whether or not a patent will actually issue. Numerous start-up companies have been formed based on technology which is not the subject of US patents, and may in fact be better protected solely by trade secret and collaborative agreements.

What kind of products is most attractive to investors? Certainly, in the medical area, drugs that are for known indications are attractive because the issues are simpler; one does not need PhD investors in order for the investor to understand the technology or the market. The regulatory issues are straightforward, unlike with technologies such as gene therapy and stem cells where the political and ethical issues are in a constant state of flux. The development costs can be calculated with a higher degree of certainty.

A single product usually is not enough to start a company with venture or angle funding. A single technology may be sufficient if it is basic, broad, and exclusive. There is a tremendous difference. A single product may be sufficient to start a company if the product is ready to go to market in a relatively short time frame and does not require huge amounts of funding. Certainly there are numerous

[17] *Graver Tank & Mfg. Co., Inc. v. Linde Air Products Co.*, 339 US 605,608, 85 U.S.P.Q. 328, 330 (1950).

[18] *Festo Corp. v. Shoketsu Kinzoku Kogyo Kabushiki Co.*, 535 US 722, 122 S. Ct. 1831 (US 2002).

Chapter 32

Stem Cell-Based Therapies: Food and Drug Administration Product and Pre-Clinical Regulatory Considerations

Donald W. Fink and Steven R. Bauer

ABSTRACT

The expectation that cellular biologic therapies either consisting of or derived from embryonic, fetal or adult stem cells (stem cell-based products) will provide effective treatments for a variety of current unmet medical problems requiring replacement, restoration, repair, or regeneration of damaged or diseased tissues and organ systems continues to drive expanding research efforts. Previously limited primarily to the domain of stem cells isolated from bone marrow or peripheral blood used for reconstitution of the hematopoietic and lymphopoietic systems following myeloablative and immunosuppressive treatments, the phenotypes, anatomic sources and potential clinical indications for stem-cell based products has expanded considerably in recent years. Beyond the promise of direct therapeutic efficacy achieved through seeding and repopulating or promulgation of bystander effects that provide trophic support of endogenous cells in areas ravaged by trauma, stem cells are projected to serve as biologic vehicles that may be engineered to deliver functional genes and gene products to discrete target sites where degenerative damage and disease is the consequence of genetic anomalies. Moreover, there is growing recognition that tissue-specific *in vitro* differentiation of stem cells could result in generation of important new model systems suited for investigating the underlying biological basis of disease as well as provide critical testing paradigms to screen potential lead candidate pharmaceuticals for toxicological safety and putative therapeutic efficacy. The scientific underpinnings necessary to support initiation of clinical safety trials in humans involving stem cell-based products have been buttressed by important research findings generated over the past several years. Sustaining ongoing science-driven efforts to identify and critically assess issues pertaining to stem cell safety is key to achieving the ultimate goal of producing novel cellular products that are both safe and effective. The challenge of drawing upon the spectrum of scientific evidence available to assess and guage the safety of biologics produced from stem cells is the responsibility of the Center for Biologics Evaluation and Research (CBER) within the Food and Drug Administration (FDA). Detailed information pertaining to the safety and efficacy of investigational stem cell-based products is reviewed by staff within the Office of Cellular, Tissue, and Gene Therapies (OCTGT), CBER. The stated mission of CBER is to ensure the safety, purity, identity, potency and efficacy of new biological products for the prevention, diagnosis, and treatment of human diseases, conditions, or injury. Mission objectives are achieved, in significant measure, using a regulatory review process that is founded squarely on scientific principles thus making available to the public innovative new treatments in as timely a manner as is feasible. This chapter describes the approach used by CBER to evaluate investigational stem cell-based products.

INTRODUCTION

Stem cell-based products are cellular constructs consisting of, or derived from, populations of stem cell progenitors. They represent examples of complex and dynamic biological entities. Stem cells are characterized by their capacity for expansion, through self-renewing proliferation, and their capacity for differentiation into a variety of cell types, limited primarily by constraints imposed by the phenotype lineage-forming potential of the stem cell progenitors. Once transplanted into patients, stem cell-based preparations are expected to interact intimately with, and be influenced by, the physiological milieu of the recipient. Elucidation of the molecular mechanisms that control stem cell self-renewal and commitment to any of a variety of cell fates pose vexing challenges to the field of developmental biology, and it is the responsibility of the Food and Drug Administration (FDA) to confront these complexities, as the agency executes regulatory oversight of investigational stem cell-based products. Numerous fundamental questions pertaining to regulation of gene expression, cell–cell interaction, and control of cellular differentiation remain under active investigation within the scientific community. The science-based regulatory review approach implemented by Center for Biologics Evaluation and Research, Office of Cellular, Tissue and Gene Therapies (CBER/OCTGT) for evaluation

Stem Cell Anthology

of novel stem cell-based products is consistent with practices in evidence throughout the FDA. This regulatory approach is designed to facilitate progress in the development of new biological treatments, ensuring to the greatest extent possible the safety of patients enrolled in investigational clinical studies conducted to support potential future licensure of these innovative biological products.

RESOURCES TO DEVELOP RECOMMENDATIONS FOR THE MANUFACTURE AND CHARACTERIZATION OF STEM CELL-BASED PRODUCTS

Irrespective of their source, constructs derived from stem cells are categorized as somatic cellular products. The legal authority for FDA to regulate stem cell-based biological products is derived from two Congressional Acts, the Federal Food, Drug and Cosmetic Act (FD&C Act), and the Public Health Service Act (PHS Act). Stem cell-based somatic cellular products are regulated as Biologics, under Section 351 of the Public Health Service Act. To initiate first-in-human clinical studies, designed to generate information pertaining to the safety and possible effectiveness of a previously-untested stem cell-based product intended for the treatment, mitigation, or cure of disease, an Investigational New Drug application (IND) must be compiled and submitted to CBER/FDA for review and assessment. An IND is a request for permission to be granted by the FDA to proceed with patient administration of an investigational stem cell-based product. Among the essential items included in the IND submission document is detailed information that describes the manufacture and characterization of the investigational stem cell-based therapy, along with a compilation of results obtained from preclinical studies conducted in animals, and a well-designed, prospectively written, adequately controlled clinical protocol. Due to the inherent complexity of stem cell-based products, coupled with continuous accumulation of research-generated information about their characteristic biological properties, individual investigators and corporate sponsors, alike, are encouraged to request a pre-meeting with CBER/FDA before submitting a completed IND application for evaluation. Within the context of a pre-IND meeting, sponsors are able to receive input from CBER/FDA staff that represents current thinking with respect to regulatory expectations that will be addressed by the information provided in an IND application. FDA recommendations and requirements regarding characterization of stem cell-based products are derived from a variety of sources, including product-specific regulations stipulated in the Code of Federal Regulations (CFR), relevant guidance documents, input received from select FDA advisory committees, and direct interactions with the scientific community.

Code of Federal Regulations (CFRs)

The CFRs represent the structural backbone of the FDA's regulatory requirements. Regulations encode details of the laws enacted by Congress, and serve as the blueprint for their implementation. Regulations applicable to the clinical investigation of unproven stem cell-based products include those outlining requirements that govern submission of an investigational new drug application (21 CFR 312), and compliance with current good manufacturing practices (21 CFR 210 & 211). These regulations were drafted by the Agency, and approved by the Executive Branch. During early clinical development, the primary objective of this specific set of regulations is to ensure, to the greatest extent possible, that patients enrolled in clinical trials receive investigational stem cell-based products with an acceptable safety profile, taking into account the balance between identified risk signals, clinical indication, and possible patient benefit. Ultimately, the goal of performing clinical trials is to collect sufficient safety and efficacy data on the investigational product, in accordance with regulatory requirements, to permit license of the product, which allows for widespread interstate commercial distribution to patients.

Relevant Guidance Documents

Regulations outlined in the CFRs that are germane to the submission of an IND may be applied broadly to a diversity of therapeutic products. In contrast, guidance documents are often developed for the purpose of reflecting the FDA's current thinking and expectations concerning the development of specific product classes. Guidance documents written and approved in accordance with intra-organization procedures represent the Agency's interpretation of published regulations. In addition to its own guidance development efforts, the FDA interacts with outside regulatory organizations participating in wider global initiatives, such as the International Conference on Harmonization (ICH), resulting in development of instructive documents that pertain to the manufacture, preclinical and clinical testing of many types of biological products. The principles outlined in a number of these guidance documents have been developed collaboratively, and are applicable to the evaluation of stem cell-based products.

Scientific Interactions

To facilitate the development of guidance documents and regulatory review approaches that are pertinent to the development of stem cell-based biological products, CBER/OCTGT review staff participate in regularly-scheduled internal meetings, to discuss relevant scientific issues pertaining to specific biological products, and to promote consistency in their regulation. In addition, the OCTGT staff is involved in various outreach activities, including regulatory and scientific society meetings where interested parties are afforded the opportunity to interact directly with individuals involved in the regulation of stem cell-based products. FDA staff and members of the regulated community benefit mutually from interactions that occur outside of the formal regulatory process. These exchanges allow for an enhanced understanding of, and greater appreciation for, each other's perspectives. A number of CBER/OCTGT review scientists also maintain active laboratory-based research programs in cellular or developmental biology, as well as in other areas relevant to the regulation of stem cell-based products. This strategic staffing approach, designated the research-reviewer model, is critical to the FDA's effort to sustain state-of-the-art scientific expertise and credibility, as it conducts the complex task of regulating stem cells and developing regulatory policy that affects the stem cell therapy field.

The practical experiences of academic clinical institutions and corporate entities involved in the development of stem cell products play a key role in shaping FDA recommendations pertaining to the regulation of stem cell-based products. The communication of these broader experiences through meetings with CBER/OCTGT staff, and presentations made at scientific society meetings, or in front of CBER's Cellular, Tissue and Gene Therapy Advisory Committee (CTGTAC) is vital to informing the regulatory process. Within the context of a CTGTAC meeting, the FDA is able to obtain advice in a public forum on scientific issues that may affect its approach to the regulation of stem cell-based products. The contributions of the CTGTAC to execution of the CBER's regulatory oversight of stem cell-based products are not to be underestimated or overlooked. Two notable meetings of the CTGTAC, one conducted in July of 2000 convened to consider use of stem cell-derived products for treatment of neurological disorders, and the second held in April of 2008 focused on safety assessment of products derived from human embryonic stem cells, represent examples of the CTGTAC's contribution to the CBER's regulation of stem cell-derived products. Transcripts of these meetings are available through the CBER's public website (http://www.fda.gov/cber/advisory/advisory.html).

Another important example of scientific resource leveraging, that facilitates development of recommendations pertaining to the regulation of stem cell-based products, is collegial interagency interaction between the FDA and the National Institutes of Health (NIH), under the umbrella of formalized Memoranda of Understanding agreements (MOUs) that allow for information exchange on the subject of translational research. The FDA benefits from the scientific expertise provided by the NIH extramural program managers and intramural research scientists, while the NIH derives benefit from the FDA's advice related to development of important NIH grant funding initiatives designed to expedite translation of scientific research from bench to bedside. These MOU agreements serve the FDA-regulated community, the NIH-funded scientific research community, and the public by focusing critical resources on the development of novel technologies, such as stem cell-based products, to address challenges posed to the public health in the form of significant unmet medical need.

In summary, the FDA receives input from a variety of sources that facilitates formulation of recommendations germane to the evaluation of stem cell-based products for evidence of their safety and effectiveness. Guidance documents are developed that contain recommendations encompassing the Agency's current thinking with respect to the regulation of cellular products, including those composed of stem cells. These documents are written to accommodate updating, as appropriate, in order to reflect continuing scientific advances, and the accumulation of regulatory experience. When making specific decisions pertaining to investigational new drug applications for stem cell-based products, the FDA takes into account evidence of risks posed versus any possible benefit that may be realized for each stem cell-based product being evaluated within the context of an adequately designed clinical trial.

CRITICAL ELEMENTS FOR DEVELOPING A SAFE STEM CELL-BASED PRODUCT

The current, comprehensive approach the CBER uses to evaluate investigational stem cell-based products is depicted schematically in Figure 32-1. Highlighted are essential elements sponsors need to address in order for CBER/OCTGT review staff to perform thorough safety assessment of a stem cell-based product, prior to authorizing permission for initiation of an early phase clinical safety trial. The multifaceted review process encompasses the primary objective of assuring the safety of stem cell-based cellular products. Beginning with "Source Controls," each individual step involved in production of an investigational stem cell-based cellular product intended for clinical testing is subject to careful scrutiny. At the core, evaluating the safety of a stem cell-based cellular product is about establishing and maintaining control over every facet of the process developed to produce the product designated for clinical study. It is only in this manner that the degree of safety is maximized, and consistency from one cell preparation to the next assured.

The individual elements of this comprehensive review approach are not to be viewed as isolated one from the other, but rather as representative of interconnected aspects that constitute the review framework. Acquisition of the starting biological materials obtained from eligible consented donors; derivation, expansion, manipulation, and characterization of established stem cell lines; generation of differentiated cellular preparations; and preclinical proof-of-concept/toxicological testing conducted in appropriate animal models are all items considered in an interwoven review process. An example of the nature of this review approach is demonstrated in the principle of traceability, or linkage. In those situations when an unanticipated

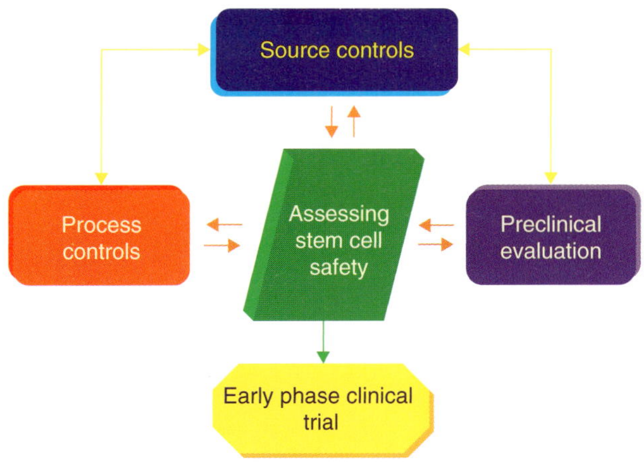

FIGURE 32-1 Representation of elements deemed necessary for assessing the safety of novel human embryonic stem cell-derived cellular products.

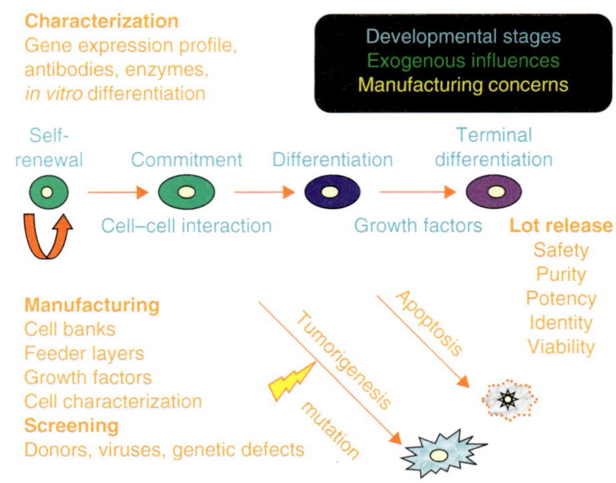

FIGURE 32-2 A hypothetical cellular differentiation pathway is used to illustrate the complexity of regulatory oversight of stem cell products. Mechanisms that control biological processes, such as self-renewal or lineage commitment, include changes in gene and protein expression that lead to differentiation (steps in development, blue text), and interactions with the host through microenvironment cell–cell interactions or growth factors (influences on development, green text). Regulatory concerns associated with these processes during *in vitro* stem cell manufacture are shown (yellow text).

adverse event occurs during the course of an investigational clinical trial, traceability allows investigators and study reviewers alike the opportunity to link back from the patient through the stem cell preparation process, all the way to the initial acquisition of biological materials used to generate the initial starting material. Traceability permits evaluation of an unexpected adverse event in the context of clinical, manufacturing, and preclinical testing issues, allowing informed decisions to be made that assist in ensuring the status of patient safety. Finally, to increase the degree of safety assurance attainable through product characterization, it is essential to develop analytical methods that are sufficiently sensitive for the detection of cells within a preparation that exhibits undesirable characteristics. To achieve this end, it is necessary to identify characteristic cellular markers that function as reliable indices of product safety. Accordingly, during the development phase that precedes initiation of a first-in-human clinical study, efforts are undertaken to identify product characteristics associated with positive outcomes, and that could be capable of predicting clinical effectiveness, as well as establishing those product features that anticipate the potential for adverse events to occur.

Manufacturing and Characterization Issues Pertaining to Stem Cell Products

Ensuring the safety of patients receiving an investigational therapeutic product not tested previously in humans is a primary objective of the FDA's oversight of stem cell-based products. Attention focused on the details of product manufacturing and characterization is essential to safety assurance. The information provided in this section describes current CBER/OCTGT reviewer recommendations intended to ensure that investigational stem cell products administered to patients during the course of a clinical trial meet agency expectations in terms of safety, purity, identity, and potency. Figure 32-2 provides illustration of the complexities inherent in stem cell biology, and depicts challenges associated with development of stem cell-based products.

Control of Source Materials is the First Step in Safety Assurance

Providing demonstration of an acceptable safety profile for an investigational, stem cell-based product prior to initiating an investigational clinical study begins at the stage of determining tissue donor eligibility. For cellular products derived from human embryonic stem (hES) cells, this encompasses screening and testing of gamete donors providing the eggs and sperm used to generate embryos from which embryonic stem cells are isolated. A description of the specific sources from which stem cells are acquired includes information about the tissue of origin and cell type, such as hematopoietic, neuronal, or embryonic stem cells, as well as details indicating whether donor cells are obtained following induced mobilization from an anatomical location that is difficult to access.

Determining Donor Eligibility

One of the FDA's principal public health obligations is to prevent the introduction, transmission, and spread of communicable diseases. Biological therapies comprised of stem cells pose a risk for transmission of communicable disease from donor to recipient. In the FDA's final rule on determining the eligibility of human cells, tissues, and cellular and tissue-based products donors, acceptance of tissues obtained from allogeneic (non-self) donors as a suitable source material for producing stem cell-based products depends on the screening and appropriate testing of donors for a panel of relevant communicable diseases. This includes testing for human immunodeficiency virus types 1 and 2, hepatitis B virus (surface and core antigen), hepatitis C virus, and *Treponema pallidum* (syphilis). In addition, for donors of leukocyte-rich cells of tissues testing for human T-lymphotropic virus types I and II, and cytomegalovirus (CMV) is required. Testing is performed by qualified laboratories, using FDA-licensed, approved, or cleared test kits. Within the autologous setting, where patients serve as their own source of stem cells, testing for infectious disease or adventitious agents is relevant when *ex vivo* culture methods are used during processing which could promote propagation of an adventitious agent within a given stem cell product, or increase the risk for cross-contamination of other biological products manufactured within the same facility. An additional level of safety is provided by required labeling of cellular and tissue products. In clinical settings when the products are intended for autologous use, they should be labeled "FOR AUTOLOGOUS USE ONLY." Circumstances may arise when it is not possible to either conduct or complete screening and testing necessary for determining donor eligibility. In this case, cellular and tissue products should bear the label "NOT EVALUATED FOR INFECTIOUS SUBSTANCES."

In addition to testing for evidence of infectious diseases, IND applications to the Agency should include a description of other serologic, diagnostic, and clinical history information obtained for the donor. It may be important to conduct further donor characterizations, such as typing for genetic polymorphisms, or assessment of major histocompatibility complex (MHC) loci that may play a role in determining whether transplanted stem cell products will engraft successfully, given that engraftment failure may occur due to donor–host immunological incompatibility.

Donor eligibility determination is required for non-sexually intimate gamete donors providing unfertilized eggs or sperm to be used specifically for generation of human embryonic stem cells (hES) isolated from the inner cell mass of fresh 5–7-day-old blastocysts. When the collection of gametes is not for the expressed purpose of reproductive use, volunteer gamete donors are to be screened for high-risk behaviors, and tested for evidence of infectious diseases, using FDA-licensed, approved, or cleared test kits. On the occasions when extra, unused, cryopreserved fertilized embryos obtained from a fertility clinic serve as source material for derivation of the human embryonic stem cell lines, donor eligibility determination is not required for embryos created and stored prior to 25 May 2005. The requirement for gamete donor screening and testing is complicated by the fact that sexually intimate partners providing egg and sperm initially intended for personal reproductive use are exempted from donor eligibility testing requirements. In these situations, gamete donors may not be available for relevant communicable disease screening and testing. Under these circumstances, once cultures of derivative undifferentiated hES cells have expanded to numbers sufficient to support adequate sampling, the cells may be assessed directly for evidence of infectious diseases using PCR-based tests qualified and validated for this purpose. Direct testing of derivative cell lines for relevant communicable diseases is also performed for human embryonic stem cell lines isolated from embryos created and cryopreserved prior to 25 May 2005. Archiving of gamete donor-derived blood samples permits screening for additional disease markers as new information and diagnostic techniques become available, providing an added measure of safety assurance.

In addition to screening for infectious diseases, a role for molecular genetic testing has been contemplated as a component of an overall assessment program to establish the suitability of particular hES cell preparations for use in developing a stem cell-based product. Information obtained from this type of evaluation could serve to determine whether a preparation of hES cells represents the best-qualified source material for generating a cellular product, in the context of a particular disease setting. As a hypothetical example, genetic testing could detect, in theory, mutations in specific target genes, such as α-synuclein, that have been linked to autosomal dominant familial Parkinson's disease. Detection of such genetic anomalies in a population of hES cells could negate their use in generating neuronal progenitor cells for treating a number of neurodegenerative conditions, including Parkinson's disease. Presently, a relatively small number of genes are known to be directly responsible for causing disease in a monogenetic fashion, or to operate as suspected contributors to aberrant physiological function. Undoubtedly, this number will expand, due to advances in techniques for identifying, isolating, and analyzing genes, coupled with an increasing abundance of information certain to become available as one outcome of the human genome sequencing projects. In addition, it is expected that a great deal more will be learned as to how multiple gene products, each contributing incrementally to an overall outcome, will predispose individuals to development of certain diseases. Barring significant improvements to existing technology, it is impractical to expect screening of every donor or hES cell preparation for the entire panoply of disease-associated genes. In this context, molecular genetic testing would be conducted only for targeted genes that are relevant to a specific clinical indication. It is important to note that, at present, molecular genetic testing is not required by the FDA as part of the approach used to evaluate the safety of an investigational stem cell-based therapy intended for use in a clinical trial.

Manufacturing Components and Procedures

The approach adopted by CBER/OCTGT for regulation of biological products, including stem cell-based products, involves scrutiny of the manufacturing process used to produce the stem cell-based product, and analysis of final product testing procedures. Often the production of a stem cell-based product requires a multi-step process that must be performed using aseptic techniques, as terminal sterilization of the final product is not feasible. An error in the execution of virtually any manufacturing process step could affect the safety, purity, and potency of the stem cell product, along with allowing for the introduction of undesired adventitious agents. CBER/OCTGT review staff evaluate all procedures used during the production, purification, and harvest of a cellular therapy product. Details describing methods used for cell collection, processing, and culturing conditions, including mechanical or enzymatic dispersion, cell selection methods, and a description of the optimized cell culture system, are important. In addition, CBER/OCTGT review staff examine details pertaining to the quality of all reagents used in the manufacturing process. Examples include reagents that are required for promoting cellular growth, eliciting differentiation, and selection of target cell populations, specifically, bovine serum, trypsin, peptide growth factors, cytokines, monoclonal antibodies, antibiotics, cell selection devices, and specified culture media. Use of FDA-approved reagents, when available, is encouraged, however, when this is not possible, additional testing of a reagent may be necessary to ensure its safety and purity. Add-on testing could include assessment of sterility, endotoxin levels, mycoplasma contamination, and evaluation for the presence of adventitious agents. Analysis of reagent purity, as well as additional analytical tests designed to demonstrate the absence of potentially harmful substances, might be indicated.

One issue that receives considerable attention is the inclusion of bovine-sourced serum or other critical components as media supplements. Of prominent concern is the potential risk for contamination by prions, the causative agent of bovine spongiform encephalopathy (BSE). Several options are available to safeguard against inadvertent transmission of BSE to patients through transplantation of contaminated stem cell-based products. First, cell cultures may be maintained in medium supplemented with bovine serum, once it is demonstrated that the serum is sourced from donor cows reared for the entirety of their lives in herds maintained in countries certified as BSE-free. A second alternative is to replace bovine serum with clinical grade serum sourced from humans. In this case, testing of the serum donors or final pooled serum product for infectious human diseases is essential. The third option is to develop a chemically defined, serum-free medium, thus obviating altogether risks associated with serum supplementation.

Derivation of Stem Cell Lines

General Principles

An IND application should contain information about the origin of any stem cells used to produce an investigational therapy, and include details about the source tissue, manner of collection, and methods for cell expansion. Demonstrating rigorous control over standardized practices and procedures used during isolation, derivation, and maintenance of stem cell lines ensures the integrity, uniformity, and reliability of candidate stem cell-based products. When a stem cell therapy is derived from a non-human source, it meets the definition of a xenotransplantation product. As defined in the PHS guideline on this topic, xenotransplantation is any procedure that involves the transplantation, implantation, or infusion into human recipients of either: (1) live cells, tissues, or organs from a nonhuman animal source; or (2) human body fluids, cells, tissues, or organs that have had *ex vivo* contact with live nonhuman animal cells, tissues, or body organs. Guidance documents provided by the agency describe additional testing requirements necessary to address potential risks posed to humans resulting from the use of xenotransplantation products. Details that pertain to animal husbandry practices, herd health surveillance, adventitious agent testing of derivative stem cell products, and archiving of animal and patient specimens are all taken into consideration by the agency in association with the use of xenotransplantation products.

Finally, once a stem cell line is established, a description of prospective testing used to characterize that stem cell line should be provided in an IND application submitted to the FDA/CBER. A representative panel of tests that includes, but is not limited to, the following parameters is recommended: (1) for hES cell lines, demonstration of pluripotency; (2) karyotypic and chromosomal analysis; (3) assessment of proliferative potential and growth characteristics; (4) evaluation for the expression of markers indicative of undifferentiated stem cells versus differentiated derivatives; (5) analysis of the stability of cell lines over time in tissue culture and following prolonged periods of cryopreservation storage. These recommendations also apply to cell lines comprised of inducible pluripotent stem cells (iPS). Inducible pluripotent stem cells are generated by nuclear reprogramming of adult, functionally-differentiated somatic cells, such as skin-derived fibroblasts, and appear to replicate the biological properties of embryonic stem cells. For iPS cell lines created by vector-mediated delivery of transcriptional regulating factors, regulatory requirements governing gene transfer products become a component of the review process.

Human Embryonic Stem Cells

The derivation of an hES cell line requires additional considerations be taken into account. Because hES cells represent a complex, dynamic biological entity characterized by robust proliferative and differentiation potential, failure to standardize procedures for expansion and maintenance of cultured hES cells could result in unintended alterations of their intrinsic properties. For example, the initial cell seeding density, the frequency with which culture medium is replenished, the quality of the reagents used to supplement the culture media, and the degree of confluency cell cultures are permitted to achieve prior to subdividing could all impact cultured hES cell characteristics. Standardization of procedures used to acquire egg and sperm from eligible volunteer donors, perform *in vitro* fertilization, and isolate cleavage-stage embryos is important to the generation of quality, undifferentiated hES cell cultures. When cryopreserved embryos serve as the starting source material for deriving self-renewing hES cell cultures, it is essential that critical information is documented and records maintained regarding the reagents and procedures used to perform embryo cryopreservation. This is the case particularly when more than one infertility clinic is providing cryopreserved embryos as source material for hES cell derivations. Differences in the concentration and type of cryoprotectant used, rate of initial cell freeze down, and routine length of cryopreservation storage may result in variability that precludes consistency from preparation to preparation. Perhaps the most significant issue with respect to use of frozen fertilized embryos is specification of prospective criteria used to ascertain the acceptability of a thawed, cryopreserved embryo as source material for generating hES cell cultures.

Historically, undifferentiated hES cells teased from the inner cell mass of cleavage-stage embryos by mechanical or enzymatic means have been propagated on layers of irradiated, non-replicating feeder cells, frequently comprised of fibroblasts obtained from fetal mice. This is due to the fact that simply adding growth factors and cytokines to the culture media appears insufficient to maintain hES cells in an undifferentiated state, thus the presence of the underlying feeder cells is required. Accordingly, as indicated above, stem cell therapies derived from hES cells cultured on layers of nonhuman cells meet the definition of xenotransplantation products, and are subject to requirements specified in various guidance documents pertaining to the regulation of xenotransplantation products. This circumstance persists even if hES founder cell lines are migrated from nonhuman feeder layers to feeder-free conditions during the process of developing a stem cell-based product. Regardless of the species of origin for feeder layers used to support hES cell cultures, detailed information should be supplied that indicates whether or not feeder cells are derived from primary cell cultures, or established cell lines, describes maintenance of the feeder cells, provides an assessment of feeder line stability (if derived from established cell lines), and includes details regarding the establishment of feeder cell banks (passage number for cells stored as a cell bank, as well as the passage number for cells used during production of the stem cell product).

Histocompatibility constitutes an additional safety issue meriting consideration with respect to the characterization of cell lines established following differentiation of hES cells. Initial reports suggest undifferentiated hES cells derived from the inner cell mass of blastocyst stage embryos exhibit very low levels of major histocompatability complex class I proteins (MHC-I), thus raising the possibility that hES cells might be immune-privileged. This has engendered the hope that products derived from hES cells will be immunologically silent, and thus remain undetected by the recipient's immune surveillance system following transplantation. Moderate increases in MHC-I are noted in conjunction with *in vitro* or *in vivo* differentiation of undifferentiated hES cells. Inflammatory cytokines, generated as part of the inflammatory response, up-regulate MHC expression considerably. In the context of cellular transplantation, induction of an inflammatory response and subsequent up-regulation of MHC-I could enhance the risk for *in vivo* rejection. At this point it is uncertain whether the level of MHC class I expressed in undifferentiated hES cells is sufficient to elicit a vigorous, let alone even a tepid immunologic rejection response, however, it does appear that characterization of MHC profiles for stem cell products derived from hES cells is warranted. Generation of autologous iPS cells is intended to mitigate potential risk posed by development of immunohistoincompatibility associated with a stem cell-based product resulting from the differentiation of allogeneic, undifferentiated hES cells.

Testing of Stem Cell Products: Safety, Identity, Purity, and Potency

A challenge to the evaluation of stem cell-based products is accurate assessment of their safety, identity, purity, and potency. Testing of process intermediates and final stem cell products intended for patient administration is comprehensive and includes, but is not limited to, microbiological testing (sterility, mycoplasma, and adventitious viral agents), as well as assessments of various product parameters, such as unique product identity attributes, purity (freedom from contaminants introduced during the manufacturing process), viability, and potency (biological activity). Appropriate analytical tests selected from this battery of representative tests should be performed at various points throughout the manufacturing process, providing a check on the process itself in order to ensure the quality and reproducible consistency of stem cell products produced over time.

Identity and purity testing may involve biological, biochemical, and/or functional characterization. Tests to demonstrate stem cell product identity and purity should be designed to determine the degree to which characteristics of a manufactured stem cell-based product conform to expected and desired properties. Potency, as defined in the Code of Federal Regulations (21 CFR §600.3[s]), is the "specific ability or capacity of the product, as indicated by appropriate laboratory tests or by adequately controlled clinical data obtained through the administration of the product in the manner intended, to effect a given result." In many circumstances, the mechanism of action for a stem cell-based product is unknown, however, it is expected that the outcome of testing for product potency will correlate with the likelihood to elicit a desired clinical effect. Due to the fact that multipotent stem cells are expected to undergo additional differentiation following patient administration, prior to mediating a measurable clinical effect, the problem of deciding what to evaluate as an index of potency, and how to correlate particular measurements with a stem cell product's final outcome, is a notable challenge. Since direct measurement of the potency of a stem cell-based product may prove daunting, because of a lack of appropriate *in vitro* or *in vivo* assays, alternative approaches that assess surrogate biomarkers or activities providing an estimate of potency are considered adequate for initiation of early phase clinical trials. In order to comply with regulations for FDA licensure, development of a valid potency assay that is used for release of the final formulated stem cell-based product is required. It is expected that a suitable potency assay be in place prior to commencing a pivotal phase 3 clinical trial to confirm safety and determine efficacy of the investigational stem cell-based product.

GENERAL EXPECTATIONS FOR CONTROL OF THE MANUFACTURING PROCESS

Control of the Manufacturing Process

A principle that applies broadly to the entire range of complex biologics, including stem cell-based products, is demonstration of control over the manufacturing process used to produce a biological product. In terms of significance, this principle is equivalent to the objectives of thorough product characterization and adequate final product testing for ensuring the safety, identity, purity, and potency of stem cell products. Consequently, strict adherence to standard operating procedures (SOPs) and quality control (QC) of raw materials, reagents, and manufactured product intermediates, in addition to comprehensive final product testing, is essential to ensure that the final formulated stem cell-based product will be acceptable for administration to human subjects. Exerting control over the manufacturing process is crucial for the consistent production of stem cell-based products.

Current Good Manufacturing Practices (cGMPs)

Current Good Manufacturing Practices (cGMPs), as set forth in 21 CFR 210 and 211 of the Code of Federal Regulations, convey principles that apply to stem cell products encompassing both the physical attributes of the facility where products are manufactured, and the processes and procedures specific to the cellular products prepared within a particular manufacturing facility. The following are representative examples of expectations set forth in the cGMPs. Evidence of appropriate written protocols applicable to each stage of product manufacturing and characterization is expected. Standard operating procedures documenting all critical information pertaining to stem cell-based therapy production should be established and adhered to. Quality oversight exercised by the sponsor is important at each stage of product development and involves two elements, QC and quality assurance (QA). Through these mechanisms, individuals assure that manufacturing and characterization testing have been properly performed, that finished product meets specified release criteria, and they maintain independence from and are not directly subordinate to individuals responsible for conducting the specific tests and filing detailed reports. During the final stage of biological product development, namely execution of a pivotal phase 3 clinical study, validation testing of the processes and assays used during product manufacturing and characterization should be performed. The FDA recently issued a final rule that exempts compliance with cGMPs for the manufacture of investigational drugs, including stem cell-based products, used in early stage pilot and phase 1 clinical trials. A guidance for industry on GMPs for phase 1 clinical trials, issued in conjunction with the rule discussed above, describes how manufacturers may implement manufacturing controls to comply with the statutory requirement (Section 501 (a)(2)(B) of the PHS Act) for complying with current good manufacturing practice. It is expected the manufacture of investigational stem cell-based products for early phase clinical testing will be performed in accordance with well-defined written procedures, using adequately-controlled equipment within a suitable manufacturing environment, as documented by accurately and consistently recorded data generated during manufacture and in-process/final product testing.

PRECLINICAL EVALUATION SUPPORTS SAFETY ASSESSMENT: PROOF-OF-CONCEPT AND TOXICITY TESTING IN ANIMAL MODELS

In addition to careful management of the process used to manufacture a stem cell-based product along with intentionally-designed analytical testing used to perform characterization and assure its quality, preclinical evaluation of cellular products consisting of, or derived from, stem cells in appropriate animal models is a key element in the overall safety assessment process. Preclinical testing focuses on two primary objectives: supportive scientific justification for the use of a particular cellular product for a specified disease indication (i.e., demonstrating "proof-of-concept") and assessment of the potential toxicity of the stem cell-based product. Given

ES-LIKE CELLS IN OTHER SPECIES

As shown in Table 1-1, cell lines that can be maintained for variable periods *in vitro* in a morphologically undifferentiated state have been obtained from morulae or blastocysts of a variety of species of mammals in addition to the mouse. They have also been obtained from the stage X blastoderms in the chick, and from blastulae in several different species of teleost fish. The criteria that have been employed to support claims that such lines are counterparts of murine ES cells are quite varied and, not infrequently, far from unequivocal. They range from maintenance of an undifferentiated morphology during propagation or expression of at least some ES cell markers, through differentiation into a variety of cell types *in vitro*, to production of histologically diverse teratomas or chimerism *in vivo*.

TABLE 1-1 Vertebrates Other Than Mouse From Which ES-like Cells Have Been Obtained

Species	Basis of Validation*
Rat	M&M & CP
Golden hamster	IVD
Rabbit	M&M, IVD & CP
Mink	M&M, IVD & T (with wide range of cell types in one study)
Dog	IVD
Cat	IVD
Pig	M&M, IVD & CP
Sheep	M&M & **
Cow	M&M, IVD & CP
Horse	IVD
Marmoset	IVD
Rhesus monkey	T
Human	T
Chicken	IVD & CP (including germline with low passage cells)
Medaka	IVD & CP
Zebra fish	IVD (limited) & CP (with short-term cultured cells)
Gilthead sea bream	IVD & (CP with short-term cultured cells)

** M & M = morphology and ES cell markers; IVD = differentiation in* vitro*; T = teratoma production in* vivo*; CP = chimera production by morula aggregation or blastocyst injection.*
*** Exhibited an ES-like morphology initially but rapidly acquired a more epithelial one thereafter*

What such ES-like (ESL) cells lines have in common with murine ES cells, in addition to a morphologically undifferentiated appearance and expression of various genes associated with pluripotency, is a high nuclear/cytoplasmic ratio. Among the complications in assessing cell lines in different species is variability in morphology of the growing colonies. While colonies of ESL cells in the hamster and rabbit are very similar to those of murine ES cells, those of most other mammals are not. This is particularly true in the human whose undifferentiated ESL cell colonies closely resemble those formed by human EC cells of testicular origin, as also are those of ESL cells from other primates. In the marmoset, rhesus monkey, and human, ESL cells not only form relatively flattened colonies, but also exhibit a number of differences from mouse ES cells in the markers they express. Since they closely resemble human EC cells in all these respects, the differences were assumed until recently to relate to species rather than cell type.

In two studies in the sheep, colonies are reported to look like those formed by murine ES cells initially, but then to adopt a more epithelial-like appearance rapidly thereafter. This change in morphology bears an intriguing similarity to the transition in conditioned medium of murine ES to so-called epiblast-like (EPL) cells which is accompanied by loss of their ability to colonize the blastocyst. Given that this transition is said to be completely reversible, whether a comparable one is occurring spontaneously in sheep clearly warrants further investigation.

In no species has production of chimeras with ESL cells rivalled that obtained with murine ES cells. Where it has been attempted, both the rates and levels of chimerism are typically much lower than those found with murine ES cells. An apparent exception is one report for the pig, in which 72% of offspring were judged to be chimeric. However, this figure is presented in an overview of work that remains unpublished, and no details are provided regarding the number of times the donor cells had been passaged before being injected into blastocysts. In a subsequent study in this species using ES-like cells that had been through 11 passages, one chimera was recorded among 34 offspring. However, as the authors of this latter study point out, rates of chimerism of only 10–12% have been obtained following direct transfer of ICM cells in the pig. Hence, technical limitations may have contributed to the low success with ES-like cells in this species.

The only species listed in Table 1-1 in which colonization of the germline has been demonstrated are the chicken and the zebra fish, but in both cases this was with cells that had been passaged only 1–3 times before being injected into host embryos. Stem cells from early chick embryos that have been passaged for longer can give strong somatic chimerism, but have not yet been shown to be able to yield gametes. Consequently, in conformity with the terminology discussed earlier, morphologically undifferentiated cell lines in all species listed in Table 1-1 except the mouse should be assigned the status of ES-like (ESL) cells rather than ES cells.

Generally, the strategy for attempting to derive ES cell lines in other species has been initially to follow more or less closely the conditions that proved successful in the mouse, namely the use of enriched medium in conjunction with growth-inactivated feeder cells and either leukemia inhibitory factor (LIF) or a related cytokine. Various modifications that have been introduced subsequently include same-species rather than murine feeder cells and, in a number of species including the human, dispensing with LIF. Optimal conditions for deriving cell lines may differ from those for maintaining them. Thus, in one study in the pig use of same-species feeder cells was found to be necessary to obtain cell lines, although murine STO cells were adequate for securing their propagation thereafter. Feeder-free conditions were found to work best in the case of both the medaka and the gilthead sea bream. Moreover, the cloning efficiency of human ESL lines was improved in serum-free culture conditions.

Unexpectedly, despite being closely related to the mouse, the rat has proved particularly refractory to derivation of ES cell lines (see Table 1-1). So far, the only cell lines that have proved to be sustainable in longer term in this species seem to lack all properties of mouse ES cells apart from colony morphology. Indeed, except for the 129 strain of mouse, establishing cells lines that can be propagated *in vitro* in a morphologically undifferentiated state seems almost more difficult in rodents than in most of the other vertebrates in which it has been attempted.

Overall, one is struck by species variability in the growth factors, status of conceptus or embryo, and other requirements for obtaining pluripotential cell lines in species other than the mouse. So far, one can discern no clear recipe for success. Of course, obtaining cells that retain the capacity to colonize the germline following long-term culture is essential only for the purpose of genetically-modifying animals in a controlled manner. Having cells that fall short of this but are nevertheless able to differentiate into a range of distinct types of cells *in vitro* may suffice for many other purposes, including regenerative medicine.

Recent Findings on Mouse Epiblast Cells

Recent findings in the mouse which may help to explain the differing experiences in other species emerged from attempts to derive ES cells from the epiblast of early post-implantation conceptuses. Stem cells exhibiting pluripotency could be obtained thus, but these clearly differed from true ES cells from pre-implantation conceptuses in conditions for their derivation and maintenance, their colony morphology, and also in how their differentiation was induced. Most interestingly, they not only resembled human ESL cells in these respects, but also almost entirely lacked the ability to yield chimerism following introduction into pre-implantation conceptuses. This raises the intriguing possibility that the mouse is peculiar in being permissive for derivation stem cells at an earlier stage in the epiblast lineage than other species. These novel pluripotential mouse cell lines have been termed "epiblast stem cells" or EpiSc.

EMBRYONIC GERM CELLS

The pre-implantation conceptus is not the only source of pluripotential stem cells in the mouse. Sustainable cultures of undifferentiated cells that resemble ES cells strikingly in their colony morphology have also been obtained from primordial germ cells and very early gonocytes in this species. These cells, termed embryonic germ (EG) cells, have also been shown to be capable of yielding high rates of both somatic and germline chimerism following injection into blastocysts.

The above findings have prompted those struggling to derive ES cell lines in other species to explore primordial germ cells as an alternative for achieving controlled genetic modification of the germ line. As shown in Table 1-2, EG-like (EGL) cells have been obtained in several mammals as well as the chick, but as with ES-like cells, their ability to participate in chimera-formation has, with one exception, only been demonstrated at low passage. Moreover, while donor cells have been detected in the gonad of a chimera obtained from low passage EG-like cells in the pig, no case of germline colonization has been reported except with cells from chick genital ridges that were cultured for only five days. Even here, the proportion of offspring of donor type was very low.

TABLE 1-2 Vertebrates From Which Embryonic Germ (EG) Cells Have Been Obtained

Species	Basis of validation *	Reference
Mouse	M&M	Resnick *et al.*
	CP	Matsui *et al.*
	CP (including germline)	Laborski *et al.*
	CP (including germline)	Stewart *et al.*
	IVD	Rohwedel *et al.*
Pig	CP	Shim *et al.*
	CP (with transfected cells)	Piedrahita *et al.*
Cow	IVD (& short-term CP)	Cherny *et al.*
Human	IVD	Shamblott *et al.*
Chicken	CP (including germline, but cells cultured for only five days)	Chang *et al.*
	CP	Park and Han

** Abbreviations as listed in the footnote to Table 1-1*

examples in the e-commerce and software fields. In the medical areas, this is more difficult.

An example of a company that is typical of the models that are being used successfully uses the combination of a very broad technology, at least one product in or close to clinical trials within a few months of funding, funding, and experienced investors who evaluate the technology, the scientists and the market, and who then put together the deal with the university or other source of the technology, form the scientific advisory board, and hire the business people to run the company. Absent this essential element, experienced business people who can put together the market plan, the people, and the money, even the best technology will fail.

In summary, there are three things one must have to start a successful company: a business plan which describes a technology base, specific products, cost of development, marketing plans, and means of generating income; the individuals who can develop the technology; and the business people who can run the company and obtain funding. Failure to have anyone of the three will cause the company to fail.

A frequent question is what are typical licensing terms. Generally speaking, I suggest the following guidelines.

For a new drug, which is patented or patentable as a compound *per se*, percentages range from 5% to 15%, with 8% not being unreasonable. Variables include whether or not there is animal testing data, clinical data, formulation issues, regulatory issues, and a crowded market. For example, a new Hepatitis C drug commands a high price—there are few if any good drugs on the market and the market is huge. A new HIV drug that is another variation of known drugs commands less of a price—market is crowded and it does not represent a radically different solution to the problem.

For a diagnostic assay, antibody with desirable specificity to difficult antigen, medical device, and screening assay with novel reagents, percentages typically range from 3% to 6% with 4–5% being common.

For a method of manufacture or screening, the ranges are typically from 1/2% up to about 3–4%. Major variables limiting price include the need for multiple licenses to practice method, particularly if they have to be obtained from multiple third parties.

Up front fees and milestones are highly variable. In the pharmaceutical and medical areas, minimum royalties are primarily designed to insure the technology comes back to the owner if it is not developed, but are low enough not to divert the needed resources from development. Highest milestones are on completion of regulatory steps, such as filing of a new drug application or completion of Phase I clinical studies. Initial fees are usually set at a number to help offset or recover patent expenses. These fees are usually the most contentious; if the product is a success, then both licensor and licensee win. In the case of advances, the licensor has put up money for patenting expenses and development which it wants to recover; the licensee wants to defer as much as possible while it evaluates the technology, so it loses as little as possible if the technology does not pan out.

Advantages of licensing versus forming a company include:

- Less risk to the technology owner.
- Greater likelihood of having experienced partner developing, testing, and obtaining regulatory approval. More likely to have sales and distribution team in place.
- No need to provide separate manufacturing and packaging facilities.
- Partner responsible for insurance and indemnification.
- Partner assumes liability and cost of defending technology against competitors.
- Partner assumes responsibility for obtaining licenses to practice the technology, if any.
- Partner assumes cost of patenting, which can be in the tens to hundreds of thousands of dollars, especially if foreign rights are pursued.

Advantages of forming a company include:

- The party developing the technology retains control over the development.
- Some technology is too early stage to license. The company may be able to fund sufficient development of the product to license or partner the technology and both profit from the technology and see technology developed that would not otherwise be developed.

Potential for Higher Profit

There are many successful business models. Some of those that I have seen failed because they did not have sufficient initial investment, did not immediately identify products, or focused on products that took too long to get to market. These were not cases where the technology was not good; rather, it was a case of where the business people failed.

In contrast, several years after my experiences with "boom and bust" startups, I was privileged to watch a different model developed by Terry McQuire of Polaris Ventures in Cambridge, Massachusetts. Terry helped form a company called AIR (Advanced Inhalation Research), licensing in some very basic technology developed by Bob Langer of MIT and David Edwards of Penn State University. They had realized that pulmonary drug delivery was the next big target for drug delivery—potentially a means of delivering drugs currently deliverable only by injection, such as insulin. A number of other companies were working on other means for delivery such as transdermal drug delivery or ultrasound mediated drug delivery, but technical problems were delaying progress. The company was funded with very little investment money, and Terry, Bob, and David immediately set out to establish corporate deals as a source of revenue. In an extremely short time, they had deals with half a dozen companies to take their drugs and turn them into pulmonary formulations. The attraction of the technology was that it was quick to market—reformulating already approved drugs, using simple technology based on a single discovery—that "aerodynamic diameter" not "actual diameter" was determinative of whether or not the drug particles were delivered by inhalation into the "deep lung." The funding for all of the development came from the corporate partners, the time to clinicals was less than 1 year, several products were being developed simultaneously, extremely broad, dominating patents were obtained, and the company was positioned for either sale or an IPO in a little over a year. Approximately 1.5 years after formation of the company, AIR was sold to Alkermes for an excellent return on their investment.

A different model that I have seen implemented by David Scheer of Connecticut, over the last few years, has also been extremely successful. David identified markets with good growth potential and high market caps, companies that had done extensive work in developing products for that market but had made a business decision to move out of the market, then acquired the technology, intellectual property, and people to form new companies. The two companies I have had the most experience with are Esperion and Orapharma, both of which went public less than 2 years after being funded, and were then acquired shortly thereafter by Pfizer and Johnson & Johnson, respectively.

Esperion initially acquired technology based on Apolipoprotein A1 Milano from Pharmacia Upjohn, including an extensive patent portfolio, and the people involved in manufacturing, regulation, and development. Esperion then expanded to obtain rights to some small molecule technology for treatment of atherosclerosis, then rights to liposomal formulations for treatment of atherosclerosis. Phase I data had been obtained by Pharmacia so that at least one product was positioned for Phase II clinical trials within a year of funding.

Orapharma was set up similarly, with an initial product including extensive patent portfolio and people experienced with the science and regulatory sides of the technology, for treatment of periodontal disease using polymer-microsphere-based technology for administration of minocycline into the periodontal pockets. Phase I clinical trials had already been completed by American Cyanamid when it decided not to pursue the technology. Orapharma acquired all rights to the technology and used this as a basis for its IPO about 1.5 years after formation. The money from the IPO allowed the company to enter into phase II/III trials almost immediately, with regulatory approval and product on sale about 1 year after completion of its IPO. Simultaneously, it acquired rights to university technology which was much earlier stage for recruitment and regrowth of the bone in the periodontal pockets.

No matter how the property is protected, or the technology valued, the important issue is that it be developed. The amount of innovation in the medical field has been truly extraordinary, even as the cost of developing it has escalated exponentially. Indeed, we now see companies that appear to conduct their entire research and development based on identifying products such as Eli Lilly's Prozac with huge market shares and profits, reviewing the relevant patents to identify weaknesses, then spending millions of dollars to invalidate the patents, thereby not only achieving the right to market the product generically, but having a period of 180 days exclusivity under the Waxman-Hatch Act in which to recover the litigation costs. One can only hope that the current spate of litigation by generic companies to knock out patents will not discourage further innovation and development of new technologies.

the core biological attributes of stem cells, namely a propensity for self-renewal and differentiation potential that could result in outcomes of teratoma and ectopic tissue formation (Figure 32-3), experimental animals should be thoroughly and carefully examined following transplantation of stem cell products, for evidence of unregulated growth, genesis of inappropriate cell types, and formation of ectopic tissues.

Demonstrating Proof-of-Concept

Investigations performed using animal transplant models of human disease constitute an essential component of the assessment paradigm used for judging the safety and rationale for administration of stem cell-based products in humans. Results from these proof-of-concept animal model studies serve to substantiate a rationale for conducting a proposed clinical trial with an investigational stem cell product. Although the ideal animal model should reflect the anatomical features and pathophysiology of the targeted human disease as closely as possible, it is recognized that, unless the disease occurs spontaneously in an animal species, such models will not replicate the clinical situation in an identical manner. Thus, chemical, surgical, and immunological methods are routinely employed to create these models, resulting in damage to neural tissue; induction of diabetes; simulation of myocardial infarction, stroke and hypertension; or compromised organ function. In selected situations when focal monogenetic lesions are known to cause disease, it is possible to create animal models in which the culpable gene is either eliminated, or overexpressed, resulting in the generation of models that recapitulate features of human disease-specific pathologies.

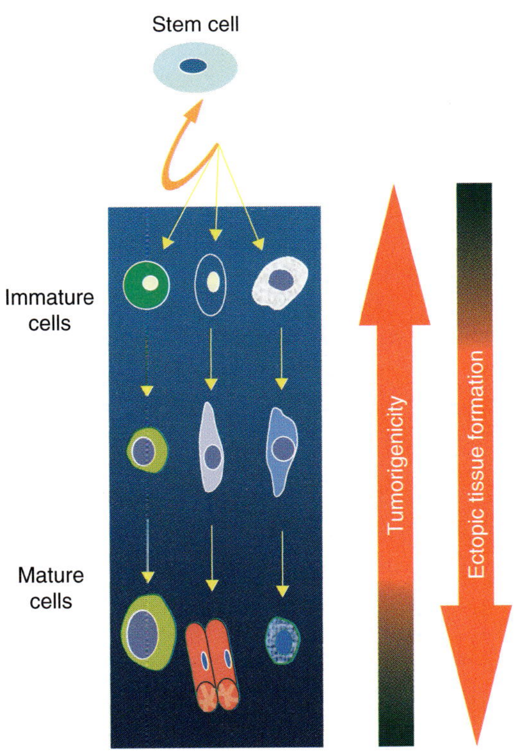

FIGURE 32-3 Schematic representation of undesired outcomes that could occur subsequent to administration of a stem cell-based product, due to the intrinsic proliferative and differentiation properties of stem cells.

Proof-of-concept demonstration for a stem cell-based product in an animal model of human disease is encouraged for the following reasons:

1. to establish a rationale that supports initiation of clinical trials;
2. to permit concurrent measurement of bioactivity and safety endpoints;
3. to explore the dose–response relationship between a stem cell-based product and an activity/safety outcome; and
4. to facilitate optimization of the route of administration.

Demonstrating that a restorative effect occurs following implantation of an investigational stem cell-based product into an animal model of human disease provides circumstantial evidence for anatomical/functional integration within the host physiology. If corroborated by histopathological data, this proof-of-concept information serves to provide support for hypotheses underlying the mechanism of action of a stem cell-based product. It is important that the clinically relevant route of administration, and intended clinical delivery system are used in the animal models to the extent feasible. Immune tolerance to the implanted investigational human stem cell product, and comparability between cell phenotypes are important considerations when determining whether the use of an analogous stem cell-based product derived from animals is necessary and/or beneficial.

Toxicological Assessment

Appropriate evaluation of the safety of an implanted stem cell-based product is an important aspect of the overall development program. Toxicity endpoints are selected that focus on the potential for both local and systemic effects to occur. Depending on the nature of the stem cell-based product being tested, studies conducted in normal animals or animal models of disease should take into consideration assessment of the following, in response to implantation of the stem cell-based product:

1. implant site reaction;
2. any inflammatory response observed in target or non-target tissues;
3. differentiation/ phenotype expression of the stem cell-based product post-implant;
4. the capacity for the cells to survive, migrate and integrate into the host physiology post-implant;
5. observed behavioral changes;
6. morphological alterations in either target or non-target tissues; and
7. evidence of tumorigenicity or ectopic tissue formation.

Completing a comprehensive microscopic evaluation following implantation of a stem cell-based product into animals provides a fundamental assessment of the overall safety of the product intended for administration into humans. The morphological data obtained from such an analysis provides insight into other, in-life, non-terminal evaluations of the animals (i.e., behavioral, imaging modalities neurologic, cardiac, and clinical laboratory changes). Histopathological evaluation can encompass assessment of acute and long-term cell survival, the extent and pattern of cell migration, evidence for differentiation and plasticity, indication of hyperplastic growth or tumorigenicity, and anatomical/functional integration. The extent of interpretable safety information extracted from investigations conducted in animals will be a direct function of the experimental study design, as well as technical limitations associated with the specific test methodologies. For example, for methodologies used to identify and track implanted populations of stem cells, investigators recently reported success related to monitoring stem cell migration *in vivo* following implantation using high resolution *in vivo* magnetic resonance imaging or sensitive methods for detecting visible light emitted from marked stem cells. These types of non-invasive monitoring tools will enable determinations to be made as to whether migration of a stem cell-based product occurs, and if so, whether it is associated with a deleterious outcome.

Tumorigenicity or hyperplastic, unregulated cellular proliferation, and ectopic tissue formation are important safety issues for stem cell-based products, particularly those that are derived from hES cells. Concern as to the ramifications of dysregulated cellular replication following implantation arises from one of the fundamental characteristics of true stem cells, namely their capacity to effect expansion through self-renewing proliferation. For hES cells, concern is heightened, due to their pluripotential character and proclivity for generating malignant teratomas when injected in sufficient numbers into immunodeficient strains of mice. This property, alone, provides a persuasive argument against the administration into humans of stem cell-based products consisting exclusively of unmodified, undifferentiated hES cells. The critical issue is to determine at what point during stem cell differentiation do risks attributable to tumorigenic potential and hyperplasia become insignificant, if ever. Successful identification of the precise stage in the differentiation process of these hES cells, when the risk for tumor formation is no longer a safety concern, will depend on whether or not cellular differentiation proceeds in a unidirectional manner, or is reversible. Due to inherent inefficiencies in the biological process of differentiation, it is unlikely that phenotypic maturation induced in cultures of undifferentiated stem cells will be complete. Accordingly, it is essential that analytical techniques are developed that permit careful evaluation of cell preparations derived from stem cells, in order to determine the number of residual undifferentiated cells, and partially differentiated intermediates, that are present. Judicious preclinical studies of appropriate duration that involve implantation of undifferentiated or partially-differentiated stem cells into immunocompromised animals should be carried out prior to initiating clinical trials, in order to address the issue of unregulated growth potential of a candidate stem cell-based therapy and its relationship to the process of cell differentiation. The propensity for more pluripotent, less differentiated stem cells to form histologically "benign" structures in anatomically sensitive locations, such as the enclosed space of the spinal cord or the neuropil of the central nervous system, presents a significant safety concern that must be evaluated.

As previously stated, proof-of-concept studies are important for characterizing the dose–response relationship of the intended stem cell-based therapy as it pertains to both activity and safety outcomes. Unlike small molecule pharmaceuticals, or recombinant therapeutic proteins, standard kinetic profiles for product metabolism and excretion are not applicable to stem cell-based products. Moreover, once delivered the stem cell-based product is not readily retrievable and thus, dose-dependent adverse effects may be difficult to control. It is important to bear in mind that the fate of cells postimplant, i.e., survival, migration, structural and functional integration, and phenotypic plasticity could potentially impact the safety and/or activity profile for a stem cell-based product. Thus, employing a range of doses of the stem cell-based product in all animal studies is highly recommended, to establish both a safety and activity profile. Similarly, developing strategies for effectively controlling cell dose levels after implantation of stem cell-based products could serve to improve the safety profile for this category of products. Potential solutions include ongoing efforts to introduce genetic switches into preparations of stem cells capable of regulating cellular proliferation that may be turned on or off using noninvasive, extrinsic means after the cells have been implanted.

SUMMARY

Stem cells are among the most complex biological therapeutic entities proposed for clinical use to date. In this chapter, a regulatory review approach used by the FDA to assess the safety of novel stem cell-based products has been described. Primary responsibility for evaluating the safety of investigational new treatments comprised of stem cells and their derivatives is assigned to the FDA's Center for Biologics Evaluation and Research, Office of Cellular, Tissue and Gene Therapies. Outfitted with a comprehensive approach developed for the regulation of tissue and cellular products, staffed by scientific reviewers with expertise in cell and developmental biology, and endowed with a wealth of experience in the evaluation of submissions involving administration of stem cell preparations to patients, the FDA is well-positioned to successfully meet its obligation for assuring that investigational human clinical trials are conducted in as safe a manner as is reasonably achievable. With respect to experimental treatments involving stem cell-based products, safety assurance is fundamentally about demonstrating control over every facet of the cell therapy preparation process, beginning with the acquisition of source donor materials, comprehensive, detailed characterization of the derivative stem cell product, and ultimately implementation of adequate and appropriate preclinical testing. In counseling academic clinical investigators and biotechnology sponsors about particulars concerning individual stem cell development programs, the FDA recognizes the need to maintain flexibility in its recommendations, taking into consideration a variety of impact factors including the intended target patient population, the severity of the specific disease under study, along with the potential risks and benefits attributable to the investigational stem cell-based product. The agency regularly updates and re-evaluates recommendations applicable to production and testing of stem cell products, based on the accumulation of scientific and clinical experience along with feedback obtained from a variety of valued sources. In order to be credible and effective, any approach taken for assessing the safety of stem cell-based products must consider the dynamic complexity of their biology. Operating within the jurisdiction of its mandated legal authority, the FDA strives to meet its obligation of ensuring the safety of nascent biological products consisting of or derived from stem cells, by relying on an approach that encourages iterative and collaborative interactions with the innovators, and which is updated based on the best available science.

ACKNOWLEDGMENTS

The authors wish to acknowledge the thoughtful contributions of numerous colleagues in the Office of Cellular, Tissue, and Gene Therapies. We are especially grateful to Drs Kimberly Benton, Keith Wonnacott, Richard McFarland, and to Mercedes Serabian, MS, for their critical review of the manuscript during its preparation.

KEY WORDS

cGMPs (current Good Manufacturing Practices) The most up-to-date scientifically sound methods, practices or principles that are implemented and documented during product development and production to ensure consistent manufacture of safe, pure and potent products.

Identity The use of physical or chemical characteristics of the product, inspection by macroscopic or microscopic methods, specific culture tests, of *in vivo* or *in vitro* tests sufficient to identify a product and distinguish it from any other product being processed in the same laboratory [21 CFR 610.14].

Manufacture All steps involved in the propagation or processing and preparation of therapeutic products including, but not limited to filling, testing, labeling, packaging, and storage as performed by the manufacturer [21 CFR 600.3(t)].

Potency The specific ability or capacity of a therapeutic product to elicit an expected result, as indicated by appropriate laboratory tests or by adequately controlled clinical data obtained through the administration of the product in the manner intended [21 CFR 600.3(s)].

Purity Relative freedom from extraneous matter in the finished product introduced during the manufacturing process, whether or not harmful to the recipient or deleterious to the product [21 CFR 600.3(r)].

Stability Ability of a product to retain its sterility, identity, purity, and potency over time when stored and shipped in designated packaging and under specified conditions.

FURTHER READING

Moos, M. Jr. (2008). Stem-cell-derived products: an FDA update. Trends in Pharmacological Sciences, Epub ahead of print, 18 October 2008.

Pal, R., Mandal, A., Rao, H. S., Rao, M. S., & Khanna, A. (2007) A panel of tests to standardize the characterization of human embryonic stem cells. *Regenerative Medicine, 2*(2), 179–192.

Halme, D. G., & Kessler, D. A. (2006). FDA Regulation of Stem Cell-Based Therapies. *N Eng J Med, 355*(16), 1730–1735.

Drukker, M., Katz, G., Mandelboim, O., & Benvenisty, N. (2004). Immunogenicity of human embryonic stem cells. In R. Lanza, J. Gearhart, B. Hogan, D. Melton, R. Pedersen, J. Thomson, M. West (Eds), Handbook of Stem Cells—Volume 1, Embryonic Stem Cells, pp. 663–674. US: Elsevier Academic Press.

Gardner, R. L. (2004). Pluripotential stem cells from vertebrate embryos: present perspective and future challenges. In R. Lanza, J. Gearhart, B. Hogan, D. Melton, R. Pedersen, J. Thomson, M. West (Eds), Handbook of Stem Cells—Volume 1, Embryonic Stem Cells, pp. 15–26. US: Elsevier Academic Press.

Weber, D. J. (2004). Biosafety considerations for cell-based therapies. Available at http://www.biopharminternational.com/biopharm/article/articleDetail.jsp?id=104336. Accessed 31 October, 2008.

Guidance for Industry: CGMP for Phase 1 Investigational Drugs. July 2008. Available at http://www.fda.gov/cber/gdlns/indcgmp.pdf. Accessed 19 December, 2008.

Guidance for FDA Reviewers and Sponsors: Content and Review of Chemistry, Manufacturing, and Control (CMC) information for Human Somatic Cell Therapy Investigational New Drug Applications. April 9, 2008. Available at http://www.fda.gov/cber/gdlns/cmcsomcell.pdf. Accessed 31 October, 2008.

NARA. (2004). Eligibility determination for donors of human cells, tissues, and cellular and tissue-based products—Final rule. Federal Register,69, 101, 29786–29833.

Guidance for Industry: Eligibility Determination for Donors of Human Cells, Tissues and Cellular and Tissue-Based Products (HCT/Ps). August 2007. Available at http://www.fda.gov/cber/gdlns/tissdonor.pdf. Accessed 31 October, 2008.

Guidance for Industry: Guidance for Human Somatic Cell Therapy and Gene Therapy. March 1998. Available at http://www.fda.gov/cber/gdlns/somgene.pdf. Accessed 31 October, 2008.

PHS Guidelines on Infectious Disease Issues in Xenotransplantation. January 19, 2001. Available at http://www.fda.gov/cber/gdlns/xenophs0101.pdf. Accessed 31 October, 2008.

Guidance for Industry: Source Animal, Product, Preclinical, and Clinical Issues Concerning the Use of Xenotransplantation Products in Humans. April 2003. Available at http://www.fda.gov/cber/gdlns/clinxeno.pdf. Accessed 19 December, 2008.

Points to Consider in the Characterization of Cell Lines Used to Produce Biologicals. July 12, 1993. Available at http://www.fda.gov/cber/gdlns/ptccell.pdf. Accessed 31 October, 2008.

Index

Note: Page numbers followed by *f* indicate figure; page numbers followed by *t* indicate table.

D

E

F

G

H

I

J

K

N

O

T

U

V

W

X

Y

Z